INTRACELLULAR
PATHOGENS I

Chlamydiales

INTRACELLULAR PATHOGENS I

Chlamydiales

EDITED BY

Ming Tan
Departments of Microbiology & Molecular Genetics, and Medicine, University of California, Irvine, CA

Patrik M. Bavoil
Departments of Microbial Pathogenesis, and Microbiology & Immunology, University of Maryland, Baltimore, MD

***Lead Editor,* Ming Tan**

ASM PRESS

Washington, DC

Cover image: 3D model of *Chlamydia*-infected cell. EM reconstruction of a HeLa cell infected with *C. trachomatis* serovar L2 (L2/434/Bu), based on 3View serial block face SEM with 390 sections (each 60 nm thick, for a total thickness of ~23 μm). A representative EM section is shown below the 3D model. The inclusion membrane is shown in green, the nucleus is light blue, and the plasma membrane of the infected cell is pink. Elementary bodies are blue, and reticulate bodies are yellow. Courtesy of Jennifer Lee, Christine Suetterlin, and Ming Tan (University of California, Irvine, CA) and Masako Terada, Eric Bushong, Andrea Thor, Mark Ellisman, and Daniela Boassa (National Center for Microscopy and Imaging Research, Center for Research in Biological Systems, University of California San Diego, La Jolla, CA).

Library of Congress Cataloging-in-Publication Data

Intracellular Pathogens I: Chlamydiales / edited by Ming Tan, Patrik M. Bavoil; lead editor, Ming Tan.
p. ; cm.
Chlamydiales
Includes bibliographical references and index.
ISBN-13: 978-1-55581-674-2 (alk. paper)
I. Tan, Ming, M.D. II. Bavoil, Patrik M. III. Title: Chlamydiales.
[DNLM: 1. Chlamydia Infections. 2. Chlamydiales—pathogenicity.
WC 600]

LC
614.5'735—dc23

2011043432

eISBN: 978-1-55581-732-9

10 9 8 7 6 5 4 3 2

Printed in the United States of America

Address editorial correspondence to ASM Press, 1752 N St., N.W., Washington, DC 20036-2904, USA

Send orders to ASM Press, P.O. Box 605, Herndon, VA 20172, USA
Phone: 800-546-2416; 703-661-1593
Fax: 703-661-1501
E-mail: books@asmusa.org
Online: http://estore.asm.org

We dedicate this book to our families, who had to share us with the book in the summer of 2011.

To Ru-ching, Julien, and Lei-Lei

To Christine, Toby, and Lucas

CONTENTS

CONTRIBUTORS

Bernard P. Arulanandam
South Texas Center for Emerging Infectious Diseases,
Department of Biology, University of Texas at San Antonio,
San Antonio, TX 78249

Byron E. Batteiger
Department of Medicine, Division of Infectious Diseases,
Indiana University School of Medicine, Indianapolis, IN 46202

Patrik M. Bavoil
Department of Microbial Pathogenesis, University of Maryland,
Baltimore, MD 21201

Wandy L. Beatty
Department of Molecular Microbiology, Washington University at
St. Louis, St. Louis, MO 63110

Gerald I. Byrne
Department of Microbiology, Immunology, and Biochemistry,
University of Tennessee Health Science Center, Memphis, TN 38163

Jonathan Crabtree
Institute for Genome Sciences, University of Maryland School of
Medicine, Baltimore, MD 21201

Andrew P. Craig
The Kirby Institute, University of New South Wales, Sydney,
NSW 2010, Australia

Toni Darville
Children's Hospital of Pittsburgh of UPMC, Rangos Research Center,
Pittsburgh, PA 15224

Kenneth A. Fields
Department of Microbiology and Immunology,
University of Miami Miller School of Medicine, Miami, FL 33136

Ted Hackstadt
Host-Parasite Interactions Section, Laboratory of Intracellular Parasites, Rocky Mountain Laboratories, NIAID, NIH, Hamilton, MT 59840

Johannes H. Hegemann
Institut für Funktionelle Genomforschung der Mikroorganismen, Heinrich-Heine-Universität Düsseldorf, 40225 Düsseldorf, Germany

Matthias Horn
Department of Microbial Ecology, University of Vienna, 1090 Vienna, Austria

Heather Huot Creasy
Institute for Genome Sciences, University of Maryland School of Medicine, Baltimore, MD 21201

Brendan M. Jeffrey
Department of Biomedical Sciences and Molecular and Cellular Biology Program, Oregon State University, Corvallis, OR 97331-4804

Marcela Kokes
Department of Molecular Genetics and Microbiology and Center for Microbial Pathogenesis, Duke University, Durham, NC 27710

Anthony T. Maurelli
Department of Microbiology and Immunology, F. Edward Hébert School of Medicine, Uniformed Services University, Bethesda, MD 20814-4799

Katja Moelleken
Institut für Funktionelle Genomforschung der Mikroorganismen, Heinrich-Heine-Universität Düsseldorf, 40225 Düsseldorf, Germany

Ashlesh K. Murthy
Department of Pathology, Midwestern University, Downers Grove, IL 60515

Garry S. A. Myers
Institute for Genome Sciences, University of Maryland School of Medicine, Baltimore, MD 21201

Uma M. Nagarajan
Children's Hospital of Pittsburgh of UPMC, Rangos Research Center, Pittsburgh, PA 15224

David E. Nelson
Department of Biology, Indiana University, Bloomington, IN 47405

Catherine M. O'Connell
Children's Hospital of Pittsburgh of UPMC, Rangos Research Center, Pittsburgh, PA 15224

Roger G. Rank
Department of Microbiology and Immunology, University of Arkansas for Medical Sciences and Arkansas Children's Hospital Research Institute, Little Rock, AR 72202

Daniel D. Rockey
Department of Biomedical Sciences and Molecular and Cellular Biology Program, Oregon State University, Corvallis, OR 97331-4804

Alexander Siegl
Department of Microbial Ecology, University of Vienna, 1090 Vienna, Austria

Ming Tan
Departments of Microbiology and Molecular Genetics, and Medicine, University of California, Irvine, CA 92697-4025

Raphael H. Valdivia
Department of Molecular Genetics and Microbiology and Center for Microbial Pathogenesis, Duke University, Durham, NC 27710

David P. Wilson
The Kirby Institute, University of New South Wales, Sydney, NSW 2010, Australia

Guangming Zhong
Department of Microbiology and Immunology, University of Texas Health Science Center, San Antonio, TX 78229

PREFACE

More cases of *Chlamydia* infection are reported to the CDC each year than all other infectious diseases combined. This dubious distinction is due to a steady increase in the number of chlamydial infections while other major infectious diseases have become less common because of successful diagnosis, treatment, and prevention. Reported chlamydial infections almost doubled over 10 years to 1.2 million cases for 2009, which is the latest year for which statistics are available (CDC, 2011). In contrast, rates of gonorrhea have declined about fourfold since the mid-1970s. As a result of these opposite trends, *Neisseria gonorrhoeae* is no longer the most common bacterial cause of sexually transmitted infection and has ceded that "honor" to *Chlamydia trachomatis* since the mid-1990s (CDC, 2010).

The burden of chlamydial infections is even higher because the CDC numbers are almost exclusively for genital infections caused by *Chlamydia trachomatis* and do not include other infections caused by *Chlamydia* spp. Tens of millions in underdeveloped parts of the world suffer from trachoma, which is an infectious form of blindness that is also caused by *C. trachomatis*. In addition, the majority of individuals will have a *Chlamydia pneumoniae* respiratory infection at some point in their lifetime even though it may not be formally diagnosed. *Chlamydia* spp. are also a significant cause of disease in animals, and new evidence suggests that human chlamydial isolates have been acquired in our evolutionary past from animal hosts. To set the stage for this book, Byron Batteiger discusses the range of chlamydial infections in chapter 1 ("*Chlamydia* infection and epidemiology"), with an emphasis on relating clinical knowledge to the fundamental biology of *Chlamydia*.

A fascinating aspect of chlamydial biology is how these organisms have evolved to become such successful intracellular parasites while having one of the smallest bacterial genomes. In chapter 2 ("Deep and wide: comparative genomics of *Chlamydia*"), Garry Myers offers a glimpse of the enormous impact genomic analysis has had and continues to have on our understanding of chlamydial biology and evolution. At last count, 33 chlamydial genome sequences were publicly available. Most of the sequenced genomes have come from reference strains, but many more clinical isolates will be sequenced in the coming years. The high level of sequence coverage with modern whole-genome sequencing methods ("deep sequencing") suggests that individual chlamydial isolates are not homogenous but rather consist of a "metapopulation" of genomic variants. Comparative genome sequencing of different chlamydial species and isolates ("wide sequencing") has demonstrated strain-specific differences within the overall context of genus-wide conservation and has provided a powerful means to learn about chlamydial biology in the absence of an experimental genetic system.

In chapter 3 ("Lessons from environmental chlamydiae"), Alexander Siegl and Matthias Horn discuss *Chlamydia*-like organisms within the order *Chlamydiales*. This is an expanding group of intracellular bacteria, such as the *Parachlamydiaceae*, with several new families identified in just the last few years. As descendants of an ancestral bacterium that learned to survive and replicate in eukaryotic cells, *Chlamydia* and the environmental chlamydiae are cousins, and much can be learned by comparing the biology of these two groups. For example, the genomes of environmental chlamydiae are two to three times larger than those of *Chlamydia*; many metabolic pathways that are truncated in *Chlamydia* are more completely represented in the environmental chlamydiae, supporting the notion that *Chlamydia* spp. have undergone reductive evolution of their genomes.

The first step in the intracellular chlamydial infection is adherence of elementary bodies (EBs) to epithelial cells at specific mucosal surfaces in the body. Binding between ligands on the chlamydial envelope and receptors on the

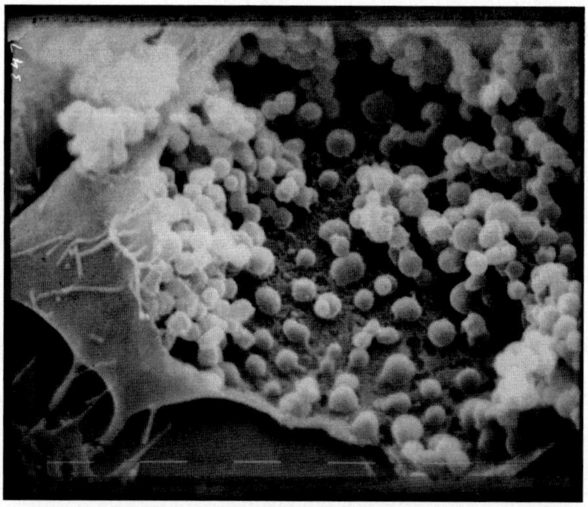

FIGURE 1 Chlamydial inclusion. Scanning electron micrograph of a McCoy cell infected with *C. trachomatis* serovar B (TW-5/0T) at 36 h postinfection. The image shows a ruptured host cell with multiple elementary bodies in the cytoplasmic inclusion. Magnification, ×5,000. Courtesy of Luis de la Maza, University of California, Irvine.

surface of epithelial cells facilitates the internalization of chlamydiae. Chapter 4 ("The chlamydial cell envelope") by David Nelson and chapter 5 ("Chlamydial adhesion and adhesins") by Johannes Hegemann and Katja Moelleken give the most up-to-date reviews of the chlamydial envelope and describe how specific envelope components mediate these surface interactions between chlamydiae and susceptible cells. The proposed two-step binding process represents the culmination of 4 decades of painstaking research by many researchers. This model is elegant in its simplicity and has clarified what was once a confusing aspect of chlamydial pathogenesis.

Once inside a eukaryotic cell, *Chlamydia* grows and replicates within the safe confines of a membrane-bound vacuole called the chlamydial inclusion (Fig. 1). How chlamydiae initiate these events by manipulating the host cytoskeleton, establishing the inclusion, and converting from an EB into a reticulate body (RB) is comprehensively described by Ted Hackstadt in chapter 6 ("Initial interactions of chlamydiae with the host cell"). Further details about how chlamydiae interact with the host cell and subvert a range of cellular processes are discussed in chapter 8 ("Cell biology of the chlamydial inclusion") by Raphael Valdivia and Marcela Kokes and in chapter 9 ("Protein secretion and *Chlamydia* pathogenesis") by Ken Fields. These host-pathogen interactions provide the environment within the inclusion so that RBs can replicate and eventually convert into EBs that can infect a new host cell.

This serial conversion between two specialized forms is a unique feature of chlamydial biology, and two models have emerged to account for the progression of the chlamydial developmental cycle. In chapter 7 ("Temporal gene regulation during the chlamydial developmental cycle"), Ming Tan proposes a gene regulation model in which the sequential expression of chlamydial genes in developmental classes is controlled by the temporal expression of key regulators through a domino effect. For example, soon after an EB enters a cell, there is expression of early gene products including DNA gyrase, which is an enzyme that increases DNA supercoiling. Higher global supercoiling levels, which have been shown to peak in mid-cycle, are then proposed to upregulate mid genes through their supercoiling-responsive promoters. Among the mid gene products that are expressed is σ^{28}, which subsequently directs the transcription of a subset of late genes to mediate RB-to-EB conversion. In chapter 16 ("Biomathematical modeling of *Chlamydia* infection and disease"), David Wilson describes the Type III secretion (T3S), contact-dependent model, which he has proposed together with Patrik Bavoil and colleagues. This model hypothesizes that contact between an RB and the inclusion membrane via T3S injectisomes is necessary for the RB stage and that loss of contact and associated disruption of T3S translocating activity induce RB-to-EB conversion. These two models are not mutually exclusive, however, and it is likely that gene regulation is coupled to external stimuli such as contact between an RB and the inclusion membrane and the activity state of the T3S apparatus.

As we learn more about how members of the *Chlamydiaceae* successfully infect and interact with eukaryotic cells, it is helpful and instructive to examine what aspects of this unusual biology are conserved features. Comparative genomic analysis makes clear that *Chlamydia* spp. are closely related and share a core set of 668 conserved proteins, which amounts to about two-thirds of the genome (see chapter 2). However, a number of chlamydial proteins that are proposed to have important roles in the biology and pathogenesis of *Chlamydia* are not encoded in the genomes of environmental chlamydiae that have been sequenced (see chapter 3). For example, environmental chlamydiae do not have Tarp (translocated actin recruiting phosphoprotein), an actin-nucleating chlamydial protein that is proposed to promote EB entry into a host cell. Intriguingly, they also lack both the late temporal regulator σ^{28} and its target gene *hctB*, which encodes the histone-like protein Hc2 that plays a role in the condensation of DNA in EBs. Almost all environmental chlamydiae do not have MOMP, which is the major outer membrane protein and immunodominant antigen of *Chlamydia*, or IncA, which is involved in the homotypic fusion of chlamydial inclusions. Thus, these *Chlamydia*-specific factors are not strictly necessary for the intracellular lifestyle of *Chlamydiales*, and they may represent specializations that contribute to the ability of *Chlamydia* spp. to infect vertebrate host cells and cause disease.

The host immune response is important for protection against an infection, but chlamydial diseases exemplify the role that the immune system can play in pathogenesis. In chapter 10 ("Immune recognition and host cell response during *Chlamydia* infection"), Uma Nagarajan describes a number of mechanisms by which chlamydiae are recognized by the host immune system. In chapter 11 ("*Chlamydia* immunopathogenesis"), Toni Darville and Catherine O'Connell review how chlamydiae induce and modulate host immune responses and describe how these innate and adaptive responses to chlamydiae contribute to pathology. Distinguishing between protective and pathologic immune responses is of course critical in the ongoing efforts to develop a vaccine.

The natural history as well as the hallmark of untreated chlamydial infection is a chronic infection that can lead to tissue damage and sequelae such as tubal infertility. In chapter 12 ("Chlamydial persistence redux"), Gerry Byrne and Wandy Beatty take a fresh approach to the oft-described but incompletely understood phenomenon of persistent chlamydial infection by noting the similarities and differences between persistent infections caused by *Chlamydia* and those caused by other human pathogens. In chapter 13 ("In vivo chlamydial infection"), Roger Rank discusses how animal models of chlamydial infection have been used to study chlamydial persistence and pathogenesis. These animal studies have been invaluable for learning how *Chlamydia* causes disease in humans and have a continuing role to play in the development of a vaccine and new antichlamydial agents.

A safe, effective chlamydial vaccine has been elusive. In chapter 14 ("*Chlamydia* vaccine: progress and challenges"), Ashlesh Murthy, Bernard Arulanandam, and Guangming Zhong review the considerable progress that has been made both in the selection of candidate vaccine antigens and in our understanding of the types of immune response that a vaccine must elicit. It might be sufficient if a chlamydial vaccine prevents disease rather than infection as a strategy for reducing long-term complications such as infertility in women. It might even be possible to develop a therapeutic vaccine or some other immunomodulatory approach to prevent long-term complications after the initial infection.

We appear to be at the dawn of a new age in *Chlamydia* research with the first published report of stable transformation of chlamydiae. In chapter 15, "Chlamydial genetics: decades of effort, very recent successes," Brendan Jeffrey, Tony Maurelli, and Dan Rockey describe the groundbreaking work by Yibing Wang, Simona Kahane, Ian Clarke, and colleagues, wherein EBs have been transformed with a hybrid shuttle vector constructed from the chlamydial plasmid and an *Escherichia coli* plasmid containing a penicillin resistance gene. The researchers successfully selected for penicillin-resistant *C. trachomatis* and demonstrated that they could produce green fluorescent inclusions from chlamydiae expressing green fluorescent protein. This much-awaited breakthrough was published just as this book was about to go to press and followed on the heels of three other methodologic advances in developing an experimental genetic system. In the first approach, isogenic strains have been generated by chemical mutagenesis followed by identification of strains with specific sequence mutations. In a second approach, recombinant progeny with a specific phenotype, such as tetracycline resistance, have been produced by coinfecting a host cell with two parental chlamydial strains. In the third approach, transformation of chlamydiae and allelic exchange have been accomplished by electroporating EBs with plasmid DNA. These experimental tools will have a transformative effect on *Chlamydia* research because it is hoped that they will soon allow researchers to test the function of individual chlamydial genes with genetic approaches.

The book concludes with chapter 16 ("Biomathematical modeling of *Chlamydia* infection and disease") by David Wilson, Andrew Craig, and colleagues. This is the first review of biomathematical modeling in a *Chlamydia* book. Mathematical modeling tools have been used to study and predict the behavior of viral infections, and the chapter describes how this approach is being applied to chlamydial infections with good success. Refinements of these models that take into account more parameters of the chlamydial infection and the host response will surely follow in the coming years and hold the promise of providing new insights into chlamydial biology and pathogenesis.

Of the major changes and developments in *Chlamydia* research that are described in this book, one more to mention is the new taxonomy, which amounts to a "family reunion." Within the *Chlamydia* field, the *Chlamydiaceae* are now considered to consist of only the single genus *Chlamydia*; the genus name *Chlamydophila* is no longer in use, although the species names, such as *muridarum*, *caviae*, and all other veterinary species that were introduced in 1999, have not changed (Kuo et al., 2010).

This book showcases a wide range of *Chlamydia* basic research that is being done by hundreds of individuals around the world. We have selected authors who are playing a leading role in scientific discovery and who can summarize and synthesize the latest in *Chlamydia* research. We also wish to acknowledge the many other chlamydiologists who have contributed to the book through their superb work—your continued efforts are critical, and each individual has an important part to play if we are to reduce the number of chlamydial infections and their impact on public health.

This book is intended for those who are interested in the latest in *Chlamydia* research, which includes scientists, physicians, medical students, public health professionals, epidemiologists, biocomputational scientists, and government policy makers. Because of the interdisciplinary nature of modern science, this audience also includes scientists studying other causes of sexually transmitted disease and other obligate intracellular pathogens. The esteemed chlamydiologist Dr. Gerry Byrne, in his introduction to the previous *Chlamydia* book published in 2006 (Bavoil and Wyrick, 2006), laid down a challenge to future editors: we have accepted the challenge and hope that we have faithfully represented the exciting developments in *Chlamydia* research.

REFERENCES

Bavoil, P. M., and P. B. Wyrick. 2006. Chlamydia*: Genomics and Pathogenesis.* Horizon Press, Inc., Norwich, United Kingdom.

Centers for Disease Control and Prevention. 2010. *Sexually Transmitted Diseases Surveillance 2009.* U.S. Department of Health and Human Services, Centers for Disease Control and Prevention, Atlanta, GA.

Centers for Disease Control and Prevention. 2011. Summary of notifiable diseases—United States, 2009. *MMWR Morb. Mortal. Wkly. Rep.* **58:**1–100.

Kuo, C.-C., B. Kaltenboek, P. M. Bavoil, and R. S. Stephens. 2010. Genus I. *Chlamydia*, p. 846–865. *In* N. R. Krieg, J. T. Staley, D. R. Brown, B. P. Hedlund, B. J. Paster, N. L. Ward, W. Ludwig, and W. B. Whitman (ed.), *Bergey's Manual of Systematic Bacteriology*, vol. 4. Springer Press, New York, NY.

MING TAN
PATRIK M. BAVOIL

CHLAMYDIA INFECTION AND EPIDEMIOLOGY

Byron E. Batteiger

1

INTRODUCTION

Human chlamydial disease due to *Chlamydia trachomatis* and *Chlamydia pneumoniae* is manifested by ocular trachoma, sexually transmitted urogenital infections, and respiratory infections, which collectively are major worldwide public health problems with significant morbidity affecting millions of persons. In 2003, it was estimated that worldwide 40.6 million people were suffering from active ocular trachoma, with an additional 8.2 million persons suffering from trichiasis, a condition in which eyelashes turn inward to touch the cornea as a result of trachomatous scarring, setting the stage for corneal injury, opacification, and blindness (Mariotti et al., 2009). The World Health Organization (WHO) estimated in 2002 that 1.3 million people were blind from trachoma and that an additional 1.9 million people with corneal opacity in areas of endemicity likely were blind due to trachoma (Resnikoff et al., 2004). While trachoma is largely found in poor, rural communities in low-income countries, sexually transmitted chlamydial infections are found worldwide. In 1999, WHO estimated that there were nearly 92 million new cases of chlamydial sexually transmitted infection, with the largest burden in sub-Saharan Africa and South and Southeast Asia (http://www.who.int/docstore/hiv/GRSTI/003.htm). The worldwide burden of *C. pneumoniae* infection is less well characterized. In the United States, where an estimated 2 to 5 million cases of pneumonia occur annually (http://www.cdc.gov/ncidod/dbmd/diseaseinfo/chlamydia pneumonia_t.htm), the organism is associated with 2 to 30% of community-acquired pneumonia in adults and children (Hammerschlag, 2000). By using the presence of *C. pneumoniae* antibodies as a marker for past infection, it has been found that about 50% of persons are positive by age 20 and up to 70 to 80% of the elderly are positive (Grayston, 1992), suggesting that a majority of persons have been infected with this organism at some time in their lives.

Subsequent chapters in this volume address the current state of knowledge in the basic sciences of the genus *Chlamydia* and related organisms. Researchers at every level of inquiry are in the end engaged in the task of reducing the human burden of chlamydial disease and its complications; the tools for this task arise from insights gained by basic and epidemiologic research. The intent of this introductory

Byron E. Batteiger, Department of Medicine, Division of Infectious Diseases, Indiana University School of Medicine, Indianapolis, IN 46202.

Intracellular Pathogens I: Chlamydiales
Edited by Ming Tan and Patrik M. Bavoil © 2012 ASM Press, Washington, DC
10.1128/9781555817329.ch1

chapter is to review the nature and extent of this human burden and to describe the diseases and complications that are all too familiar to everyone involved in the clinical care of patients and in the larger public health efforts to control these infections. The two main goals are to compare and contrast the clinical presentations and epidemiology of diseases caused by the major human chlamydial pathogens and to provide a clinical context for basic chlamydial research. *Chlamydia* also causes substantial and important veterinary disease, but here we will mainly consider the human pathogens *C. pneumoniae* and *C. trachomatis* and the ability of different *C. trachomatis* serovars to cause diseases as distinct as ocular trachoma and sexually transmitted infections.

HUMAN CHLAMYDIAL INFECTION: COMMON THEMES

Human chlamydial infections are associated with respiratory, ocular, and genitourinary disease syndromes. The syndromes are clinically distinct, but chlamydiae that cause human disease share a number of characteristics, including the propensity to produce (i) infections that are frequently long-lasting in the absence of treatment; (ii) repeated infections following natural clearance or antibiotic treatment of an initial infection; (iii) infections that are often asymptomatic or minimally symptomatic; and (iv) infections that can cause inflammatory and scarring complications in the absence of treatment, and often with minimal or no symptoms. These themes will be introduced here and discussed in more detail in the individual infection sections. Several of these themes are related to the idea of persistence of infection in human hosts.

Persistence: a Clinical and Epidemiological Perspective

All chlamydiae pathogenic for humans share the ability to assume what is called an in vitro persistence phenotype in response to stress. In cell culture models of infection, stressors such as cytokines (gamma interferon), antibiotics, and iron and other nutrient restriction induce morphologically aberrant noninfectious developmental forms with altered protein expression. Byrne and Beatty, in chapter 12, "Chlamydial persistence redux," discuss what has come to be known as chlamydial persistence, which they define as a reversible interruption in the productive intracellular chlamydial growth cycle mediated by environmental factors. The relationship of this in vitro phenomenon to human chlamydial disease has not been established. However, long-lasting infections, their response to treatment, and frequent repeat infections are all relevant to a clinical consideration of persistent infection.

Persistent infection from a clinical and epidemiologic standpoint refers either to a long duration of untreated infection or to infections that fail to resolve with antibiotic therapy. The most common form of clinical persistence relates to the typical natural history of chlamydial infection: prolonged asymptomatic infection in the absence of treatment despite a host immune response. Persistence can also take the form of failure of antibiotic therapy to clear an established infection, but single-dose azithromycin therapy of ocular and lower genital tract infections is generally effective >90% of the time. However, judgment of effectiveness of antibiotic therapy for *C. trachomatis*, especially among high-risk individuals with sexually transmitted infections, can be confounded by frequently occurring reinfections. Thus, there is the risk of concluding that an individual with a resolved infection, who then quickly becomes reinfected, has a "persistent" infection, especially if infrequent diagnostic sampling is done. Distinguishing treatment failure from reinfection in genital infection requires a combination of longitudinal follow-up with frequent testing, molecular strain typing, partner identification and testing, documentation of treatment, and detailed partner-specific behavioral information (e.g., coitus and condom use). When these tools were used in a sexually active high-risk population of adolescent women, over 84% of repeat genital infections were found to be reinfections; overall, antibiotic therapy was successful in 92% of infections (Batteiger et al., 2010a).

From a clinical perspective, persistent infection refers largely to the natural propensity of untreated chlamydial infections to be long-lasting and to the occasional failure of antibiotic treatment. There are far fewer data for *C. pneumoniae* infections, although small studies suggest that continued nasopharyngeal shedding may occur in as many as 20% of persons treated with antibiotics for respiratory infection (M. Hammerschlag, personal communication).

Chlamydial Infections Are Long Lasting

Untreated chlamydial infections are of variable duration but are typically long lasting. The natural history of human chlamydial genital infections is not well described because it is unethical to conduct a study to follow the course of untreated disease. Once a diagnosis of chlamydial genital infection is made by microbiological detection, antibiotic treatment is provided, since the presence of the organism is associated with risk of complications and transmission. Nevertheless, the limited available data reviewed in the subsequent sections suggest that the natural history of untreated genital infections is typically to persist for months to years. In trachoma, given the constant environmental exposure in areas of hyperendemicity, the natural history of an episode of infection is confounded by multiple opportunities for reinfection due to environmental and living conditions, but the median duration appears to be more than 4 months. Long durations of culture-proven nasopharyngeal shedding of *C. pneumoniae* of months to years have been described in individuals following an episode of acute symptomatic respiratory disease (Hammerschlag, 2000).

Repeated Chlamydial Infections Are Common

Repeated infection is common in individuals with *C. trachomatis* genital infections (Batteiger et al., 2010a) and in ocular infections among children living in areas where trachoma is hyperendemic (Hu et al., 2010). Repeated infection has been reported in individuals with *C. pneumoniae* infections as well (Hammerschlag et al., 2010). Repeated infections are thought to increase the risk of chronic inflammation and complications of *C. trachomatis* infection. Both ocular and genital infections induce at most a degree of partial immunity to reinfection, which may be manifested as reduced load of shed organisms and shorter duration of infection, but not sterilizing immunity (Batteiger et al., 2010b). This parallels the experience with animal models of experimental genital and ocular infection (Rank and Whittum-Hudson, 2010). This immunity is routinely overcome in the clinical setting as manifested by repeated infection. Vaccine development, discussed in detail in chapter 14, "*Chlamydia* vaccine: progress and challenges," is particularly challenging due to the lack of sterilizing immunity after a natural infection and concern that vaccine preparations may not be able to improve on what is observed in natural infection. As a result, a primary end point in human vaccine trials directed against genital disease might well be reduction in clinically apparent pelvic inflammatory disease (PID) rather than a reduction in infection acquisition; rate of incident infection, bacterial load, and infection duration may serve as secondary outcomes.

Chlamydial Infections Are Often Asymptomatic

C. trachomatis infections are frequently asymptomatic or minimally symptomatic. This is most notably the case in genital infections, where most individuals, both female and male, are asymptomatic. In trachoma, even when clinical examination of the eye reveals inflammatory changes, individuals are frequently asymptomatic or have only mild symptoms (Hu et al., 2010). However, even though asymptomatic, individuals who exhibit disease (signs of inflammation) are at risk for sequelae: asymptomatic *C. trachomatis* ocular infections in areas where trachoma is endemic can result in eyelid scarring, and *C. trachomatis* genital infection can result in PID, with its sequelae of tubal scarring and dysfunction leading to involuntary infertility and ectopic pregnancy. In addition, transmission of infection from asymptomatic persons frequently occurs. It is thus important to identify and treat asymptomatic ocular and genital *C. trachomatis* infections. Shedding of *C. pneumoniae* in respiratory secretions of asymptomatic children and adults is known to occur (Hyman et al., 1995) and in small studies has been shown to last for months to even years. The frequency of asymptomatic shedding, its transmissibility, how often symptomatic disease develops, and whether treatment of this condition is of any benefit are unknown. Since chronic sequelae of *C. pneumoniae* infections are much less well established than are those of *C. trachomatis* infections (Hammerschlag et al., 2009), there is no rationale for screening for and treating asymptomatic *C. pneumoniae* infections.

Infection versus Disease

In all human chlamydial infections and in many other infectious diseases as well, there are individuals who harbor the organism but have no evident disease. Since chlamydiae as intracellular organisms are intrinsically pathogenic, these types of infections cannot be considered colonization, as we would consider skin colonization with *Staphylococcus epidermidis*. Disease is manifested by symptoms or signs (like penile discharge, cervical discharge, or pneumonia), by evidence of tissue injury (like ocular inflammation in trachoma), or by sequelae such as organ dysfunction (blindness in trachoma and involuntary infertility in genital infection). In the case of trachoma, the site of disease is the conjunctiva, which is also the site of initial infection. The conjunctiva is easily sampled to assess infection by microbiological methods; disease (inflammation) and sequelae (trichiasis and corneal opacification) can be assessed by direct physical examination of the eye. Genital infections at the endocervix and male and female urethra are often asymptomatic, with or without evidence of inflammation. Urethritis in men can exhibit overt discharge, and neutrophils can be detected microscopically if discharge is absent; some infected men lack both. In women, cervical inflammation is more difficult to detect, and many infected women have no obvious physical signs. In addition, the site of sequelae in women is remote from the site of initial infection; unlike trachoma, detection of asymptomatic inflammation in the upper genital tract that may lead to sequelae is beyond the reach of current diagnostic tools.

Cause of Disease

Although chlamydiae are intrinsically pathogenic and cause injury and death of infected ocular and genital columnar epithelial cells, tissue injury is thought to be largely related to the host inflammatory response to epithelial infection. The innate and acquired immune mechanisms responsible for this inflammatory response have recently been reviewed in the context of genital infection (Darville and Hiltke, 2010) and are discussed in detail in chapter 11, "*Chlamydia* immunopathogenesis." Although hypersensitivity phenomena were previously proposed to be largely responsible for inflammation, abundant data now support the hypothesis that the inflammatory response is initiated and sustained by actively infected nonimmune host epithelial cells. Chlamydiae activate the innate immune system by inducing proinflammatory cytokines that mediate the influx and activation of neutrophils, natural killer cells, and monocytes. Both epithelial cells and recruited neutrophils release matrix metalloproteinases and other proteases that can contribute to tissue injury. The adaptive T-lymphocyte response is also recruited, and while it is key to clearance of active infection, it may also contribute to tissue injury. Once infection is resolved, these inflammatory elements abate, but in some instances they may leave in their wake tissue fibrosis that may result in end organ dysfunction and sequelae.

In the subsequent sections, the different chlamydial infections will be reviewed with these themes in mind, and the similarities and differences will be highlighted in the context of ongoing research. This chapter will focus on infections caused by *C. trachomatis* and *C. pneumoniae*, since they are the common chlamydial infections in humans.

CHLAMYDIA TRACHOMATIS INFECTIONS CAUSING TRACHOMA

Serovars A, B, Ba, and C of *C. trachomatis* cause the ocular disease trachoma; *C. trachomatis* was causally linked to trachoma shortly after the first isolation of the organism in 1957 (Hu et al., 2010). Unlike *C. pneumoniae* respiratory infections and *C. trachomatis* infections causing genital disease, disease related to trachoma can be quickly, easily, and reproducibly detected and graded by physical examination of the eye, using a simplified WHO standardized grading system (Thylefors et al., 1987), making microbiological diagnosis in areas of high endemicity unnecessary for treatment and control purposes (Hu et al., 2010). Unlike chlamydial genital and respiratory infections, for which patients are individually diagnosed and treated, the approach to trachoma in areas of hyperendemicity is mass treatment with single-dose azithromycin at the community or district level. In these areas, infection and reinfection are driven by the high prevalence of infection and occur simply through the presence of the pathogen in affected households or communities. The mass treatment approach reduces the overall chlamydial burden in the community and the risk of reinfection in individuals.

Clinical Features and Course

Trachoma, a disease recognized in antiquity, is today the leading infectious cause of blindness worldwide (Hu et al., 2010). Trachoma is characterized by a follicular chronic keratoconjunctivitis. Untreated, the course is characterized by long-lasting conjunctival infection punctuated by repeated exposure and reinfection with recurrent conjunctivitis. Most individuals ultimately clear the infection, and most often the inflammatory changes resolve without sight-threatening consequences. In a significant minority, long-lasting infection and cycles of reinfection result in chronic inflammatory changes in the tarsal conjunctiva lining the interior surface of the eyelid, which cause fibrosis. Over time, the fibrotic scar tissue contracts and the eyelid becomes distorted, resulting in entropion (in-turning of the eyelids) and trichiasis (eyelashes touching the eyeball). Trichiasis is painful and causes constant mechanical trauma to the cornea, resulting over months to years in corneal opacification, the blinding end stage of the disease. Corneal opacification is thought to result from the tissue injury resulting from trauma by the eyelashes, ocular dryness resulting from distorted eyelid anatomy, and secondary bacterial or fungal superinfection. Secondary infection by nonchlamydial pathogens may in part explain the observation that inflammation may be present when chlamydial organisms cannot be detected (Hu et al., 2010), particularly in the late scarring phases of the disease. The active chlamydial infections largely affect young children and form the basis for the scarring and its complications, which are more prevalent in older children and adults.

Epidemiology, Geographic Distribution, and Transmission

In the past, trachoma was a significant health problem in many parts of the world, including parts of Europe and North America. Now, the disease is largely found in poor, rural communities in low-income countries, especially in sub-Saharan Africa, but is also endemic in parts of south Asia, China, and Indochina (Burton and Mabey, 2009). The highest prevalence is reported from Ethiopia and Sudan, where active trachoma is found in as many as 50% of children below the age of 10 years and trichiasis in nearly 19% of adults (Burton and Mabey, 2009). Trachoma has largely disappeared from developed countries, which is thought to have resulted from overall improvements in living standards and hygiene (Hu et al., 2010). WHO estimated that in 2002, 1.3 million people worldwide were blind from trachoma (Resnikoff et al., 2004). The number of people with active trachoma worldwide is estimated to be 40 million, and the number of people with trichiasis, the immediate precursor to corneal opacification, is 8.2 million (Mariotti et al., 2009). Although global estimates of trachoma prevalence have limitations, there appears to be a downward trend in both active trachoma and blindness attributed to trachoma since 1981 (Burton and Mabey, 2009).

The transmission of trachoma is from person to person; the disease is focal and clusters within communities and households, suggesting that ongoing personal contact is required for transmission of the infection (West et al., 1991). Modes of transmission include spread from eye to eye during close physical contact, spread via infected ocular or nasal secretions on fingers, indirect spread via fomites such as soiled washcloths contaminated with infectious organisms, spread from infected nasopharyngeal secretions by aerosol, and transmission by eye-seeking flies (Hu et al., 2010). No zoonotic reservoir of infection has been found, indicating that flies serve only as passive vectors, transferring infectious secretions from person to person as they feed on eye secretions (Hu et al., 2010). Since most transmission occurs within households and is typically very efficient, failure to treat all household members during mass treatment efforts may result in rapid reinfection within the household, followed by more gradual spread into the community (Blake et al., 2009).

Disease Grading, Symptoms, and Chlamydial Infection

Grading of the clinical manifestations of trachoma is accomplished by physical examination of the eye assisted by a simple magnifying glass, which can be done by trained paramedical personnel in the field according to simplified WHO standardized criteria (Thylefors et al., 1987) (Table 1).

Active infection, associated with follicular and intense trachomatous inflammation, is most commonly found in young children, often in infants less than a year old. Active inflammation can also be associated with pannus, which is the infiltration of very small blood vessels into the upper margin of the cornea, which is normally devoid of blood vessels; however, this process rarely affects vision. Trachomatous scarring can be seen in both young children and older persons, while trichiasis and corneal opacity are largely seen in adults, representing the scarring end stage of the disease.

Discordance between symptoms reported by an individual and the stage of disease as assessed by the grading system is common. If symptoms are present, they are similar to those associated with chronic conjunctivitis due to any cause: conjunctival erythema (redness), eye discomfort, tearing, photophobia (light sensitivity), and minimal mucopurulent discharge (Hu et al., 2010). However, persons are frequently asymptomatic or minimally symptomatic even when signs of inflammation are present (Hu et al., 2010); nevertheless, this chronic but asymptomatic inflammation can result in scarring of the eyelid. This ability to produce damaging inflammation in the presence of no or minimal symptoms is characteristic of chlamydial infections in general, especially in the eye but also in the genital tract.

There is also discordance between the presence of chlamydial infection and the presence of trachomatous inflammation as measured by the grading system, especially in low-prevalence

TABLE 1 The WHO simplified system for the assessment of trachoma

Grade	Abbreviation	Description
Trachomatous inflammation—follicular	TF	Presence of five or more follicles (0.5 mm) in the upper tarsal conjunctiva
Trachomatous inflammation—intense	TI	Pronounced inflammatory thickening of the tarsal conjunctiva that obscures more than half of the deep normal vessels
Trachomatous scarring	TS	Presence of scarring in the tarsal conjunctiva
Trachomatous trichiasis	TT	At least one lash rubs on the eyeball
Corneal opacity	CO	Easily visible corneal opacity over the pupil

areas. *C. trachomatis* cannot be detected in all cases of active clinical disease, even by very sensitive nucleic acid amplification tests (NAATs) (Hu et al., 2010). Those with intense trachomatous inflammation are most likely to be infected and are more likely to have higher chlamydial burdens than those with follicular inflammation. Conversely, in some communities, the organism can be detected in individuals who do not have active disease as judged by WHO grading criteria (Hu et al., 2010). In a study from a low-prevalence area in The Gambia, only 24% of infected persons had clinically evident signs of inflammation; these persons could serve as a reservoir for reinfecting those in their households and communities (Burton et al., 2003). Possible reasons for the discrepancies between infection and disease are (i) early disease, where chlamydial infection is present but inflammation is not macroscopically evident on examination; (ii) late disease, where infection has cleared but clinical signs persist, which is commonly observed, with an estimated postinfection inflammation period of about 5.4 weeks (Grassly et al., 2008); and (iii) the limitations of the simplified WHO grading system itself, which requires five follicles in the eyelid conjunctiva to establish disease (Hu et al., 2010).

Natural History of Infection and Disease

At the population level, the prevalence of active disease is highly correlated to the prevalence of *C. trachomatis* infection; each is highest in preschool children, and each falls to low levels in adulthood. The majority of the overall chlamydial load in a community is found in children under the age of 1 year (Solomon et al., 2003); adult loads are lower than those of children. A longitudinal study of trachoma in The Gambia showed that the duration of infection and disease declines markedly with age, which is thought to be related to an acquired immune response (Bailey et al., 1999). In a more recent analysis of this longitudinal study, the estimated median duration of an episode of infection in the population overall was 17.2 weeks (95% confidence interval, 13.7 to 23.0 weeks); the median duration of disease was 21.1 weeks (95% confidence interval, 16.8 to 28.4 weeks), with the durations of infection and disease longer in the youngest group (0 to 4 years of age) (Grassly et al., 2008). In contrast, the scarring manifestations of trachoma increase with age, although in communities with very high prevalence of active disease, scarring and trichiasis can occur even in children under the age of 15 (Ngondi et al., 2006). Scarring over time is thought to be due to repeated exposure and cycles of reinfection with resulting cumulative inflammatory damage, and the earlier scarring seen in high-prevalence communities may reflect increased frequency of exposure and reinfection.

Progression of trachomatous inflammation to scarring sequelae is not uniform; a major determinant of the rate of disease progression among individuals is thought to be the burden of *C. trachomatis* infection in their community as noted above. Hu and colleagues have summarized a number of longitudinal cohort studies that examined the progression of scarring sequelae among individuals with already established scarring (Hu et al., 2010). In Tanzania, conjunctival scarring worsened in about 50% of individuals over an observation period of 5 years. Progression of conjunctival scarring to trichiasis was seen in 10% of individuals after 10 years in The Gambia and in 6% after 12 years in Tanzania. As well, progression of established trichiasis occurred in 33% of individuals after 1 year and in 37% after 4 years of observation. Finally, in The Gambia, 8% of people with established trichiasis developed new corneal scarring after 4 years. These data indicate that scarring or progression of scarring complications occurs in a substantial minority of persons but is not inevitable. The relatively slow rate of progression provides opportunities to provide mass antibiotic treatment to reduce the community burden of infection and thus the risk of reinfection and to correct eyelid deformities by surgery before end stage corneal opacity develops.

Immune Mechanisms and Genetic Predisposition

Most infectious diseases exhibit a range of severity and rates of complications that are a function of the balance between the host response and the virulence characteristics of the pathogen. Even among apparently healthy persons, genetic factors determine the effectiveness of the innate and acquired immune response to a given pathogen. For example, among humans experimentally infected with *Haemophilus ducreyi*, the bacterium that causes the sexually transmitted genital ulcer disease chancroid, some have a dysregulated innate response at the level of the dendritic cell that is ineffective in eradicating the bacterium. However, the response results in inflammation deleterious to the host, which is manifested as pustules (small skin abscesses) (Humphreys et al., 2007). Even effective responses capable of resolving the infectious process may produce inflammatory changes that can be harmful to the host. In trachoma, repeated cycles of infection, inflammation, resolution of infection, reinfection, and further inflammation are thought to be the prime drivers of overt disease and complications. The concept that the immune response can be both protective and damaging in the context of chlamydial infection is discussed in more detail in chapter 11, "*Chlamydia* immunopathogenesis," and in chapter 14, "*Chlamydia* vaccine: progress and challenges."

In trachoma, a variety of genetic polymorphisms have been associated with scarring sequelae in case-control studies, including polymorphisms of genes or promoters for tumor necrosis factor alpha, matrix metalloproteinase 9, gamma interferon, and interleukin 10 (Hammerschlag et al., 2009). Higher levels of tumor necrosis factor alpha protein or transcripts in ocular samples have been found in the eyes of those with active trachoma or scarring than in unaffected controls (Hammerschlag et al., 2009).

Control Efforts Based on the SAFE Strategy

The blindness due to corneal opacity is irreversible, but this outcome can be prevented. The Alliance for Global Elimination of Blinding Trachoma by the year 2020 was established by WHO in 1997 and recommends the SAFE strategy for trachoma control, consisting of surgery for trichiasis, antibiotics to treat *C. trachomatis* infections, facial cleanliness through personal hygiene, and environmental improvement with education and improved local economy (Hu et al., 2010).

Surgical repair of the scarred, misshapen eyelid can halt the progression to corneal opacity that can result from trichiasis. A major problem is the propensity for trichiasis to recur after surgery; recurrence rates range from 20% within 1 year to 60% after 3 years. Recurrence may be the result of ongoing inflammatory-scarring responses that are rarely due to *C. trachomatis* but rather appear to result from infection caused by other superinfecting bacteria (Hu et al., 2010). So while post hoc repair of deformity due to scarring has its role, prevention of scarring must be a high priority.

Single-dose azithromycin is effective in curing trachoma at the follicular and intense inflammation stages and works as well as the unwieldy prior recommended treatment, a 6-week course of topical tetracycline (Bailey et al., 1993). Demonstration that single-dose therapy of chlamydial infection was effective made mass treatment feasible and led to the trachoma elimination initiative. Detailed consideration of the work in this field is beyond the scope of this chapter; an excellent summary of the topic of mass treatment is presented in a systematic review by Hu and collaborators (Hu et al., 2010). In summary, use of mass treatment has been effective and has not resulted in antibiotic resistance among *C. trachomatis* isolates. There has been some increased macrolide resistance among nasopharyngeal isolates of *Streptococcus pneumoniae*; however, these have not persisted in the community longer than 12 months after a round of treatment. The deployment of mass treatment is based on the prevalence of follicular trachomatous inflammation in a given district or community; considerations regarding targeted versus universal treatment, the frequency with which mass treatment should be repeated, and the coverage of treatment in a given population are reviewed elsewhere (Hu et al., 2010).

Facial cleanliness aims to reduce transmission by removing infected secretions, which are a potential source of infection. While this is a reasonable idea and programs to improve facial cleanliness can be effective, there are no randomized controlled trials establishing that facial cleanliness adds benefit to antibiotic treatment. Control of environmental factors, largely through control of flies by building latrines or insecticide spraying, has shown little benefit over antibiotic treatment alone. A number of studies have examined the before-and-after effects of implementing the entire SAFE package and showed reductions in measures of trachoma, but interpretation of the data is difficult because there were no control groups. However, educational efforts appear to have added benefit when combined with antibiotic treatment (Hu et al., 2010).

SEXUALLY TRANSMITTED *CHLAMYDIA TRACHOMATIS* GENITAL INFECTIONS

Serovars D through K of *C. trachomatis* cause genital infections in men and women and were recognized in the 1970s as causing infections ranging in severity from asymptomatic infection to symptomatic PID. Unlike trachoma, where exposure in affected households is more or less continuous, sexually transmitted infection is episodic, driven by sexual behavior and indeed by the probability that coitus occurs with a person who is actually infected with *Chlamydia*. Additional diseases associated with this group of organisms include neonatal conjunctivitis, neonatal pneumonia, and occasionally self-limited conjunctivitis in sexually active adults and laboratory workers (Stamm and Batteiger, 2010). Serovars L1 through L3 are associated with lymphogranuloma venereum (LGV), which is characterized by transient genital ulcerations followed by inguinal lymphadenopathy; this classic syndrome is rare in developed regions. However, LGV strains have long been known to cause severe proctitis due to direct rectal inoculation from insertive anal sex (Quinn et al., 1981); in Europe, recent outbreaks of LGV-associated proctitis in men who have sex with men have been recognized (Spaargaren et al., 2005).

Epidemiology

Sexually transmitted genital infections occur worldwide. In 1999, WHO estimated there were nearly 92 million new cases of chlamydial sexually transmitted infection, with the largest burden in sub-Saharan Africa and South and Southeast Asia (http://www.who.int/docstore/hiv/GRSTI/003.htm). In the United States, over 1.2 million *C. trachomatis* cases were reported in 2009, representing the largest number of cases ever reported for any notifiable disease in the United States. The highest age-specific case rate was among women aged 15 to 19 years at 3,329 cases per 100,000; in this age group, rates were sixfold higher in non-Hispanic black women than in white women. U.S. *C. trachomatis* case rates are also high for women aged 20 to 24 (3,179.9/100,000) and women aged 25 to 29 (1,240.6/100,000) (http://www.cdc.gov/std/stats08/figures/5.htm). These data represent numbers of cases reported per 100,000 population of the given age range. However, these do not represent the population burden of infection since they are not age-specific prevalences, because testing is selective and is not representative of the population in each age group. The population prevalence of *C. trachomatis* among 15- to 25-year-old sexually experienced women in the United States is about 3 to 4% and has not changed significantly over the decade of 1999 to 2008 (Datta et al., 2012) despite recommendations to screen and treat such women annually (Meyers et al., 2007).

Diagnostic Tests

Highly sensitive and specific NAATs have largely replaced other diagnostic tests and are crucial to implementing screening strategies. NAATs can be performed with noninvasively obtained samples (urine or patient-collected vaginal swabs) without the need for pelvic examination in women or urethral swab sampling in men; samples can be obtained virtually anywhere with no requirement to attend a medical venue. NAATs have replaced cell culture methods for detection of *C. trachomatis* because of their superior sensitivity, similar specificity, and ease of sample acquisition and handling. NAATs are also preferred over direct fluorescent-antibody tests, enzyme immunoassays, and nucleic acid hybridization tests because of superiority of both sensitivity and specificity.

Clinical Features

Uncomplicated *C. trachomatis* genital infections are asymptomatic or minimally symptomatic in over 50% of men (Cecil et al., 2001) and 70% of women (Falk et al., 2005). When symptoms are present, dysuria (difficult/painful urination) may occur in both men and women; men may have scant mucopurulent urethral discharge and women undifferentiated vaginal discharge. Uncomplicated genital infections in women involve the endocervix and sometimes the urethra. The most common symptomatic presentation in men is nongonococcal urethritis (NGU), which is caused by *C. trachomatis* about 30 to 40% of the time. NGU is defined by the presence of five or more neutrophils per high-power field of a Gram's stain of endourethral contents (Martin, 2008). Asymptomatic cases of NGU occur, again illustrating the discordance of inflammation and symptoms.

Complications in Women

Complicated chlamydial infection in women involves the upper genital tract including the endometrium and fallopian tubes and can be symptomatic or asymptomatic. *C. trachomatis* is a common cause of PID, but other bacteria, including *Neisseria gonorrhoeae, Mycoplasma genitalium*, and members of the vaginal microbiome such as *Escherichia coli* and anaerobic bacteria also cause the syndrome; often, establishing a microbial etiology can be difficult. A substantial minority of untreated genital chlamydial infections detected by lower genital tract sampling lead to symptomatic PID; recent estimates based on randomized trials suggest this may occur in about 10% of patients over a 1-year follow-up (Oakeshott et al., 2010). In contrast to the reliable and convenient clinical staging of inflammation and scarring in trachoma, it is notoriously difficult to determine the presence, much less grade the clinical severity, of upper genital tract disease in women. This is due to the inaccessibility of the endometrium and fallopian tubes to noninvasive visual inspection, lack of sensitive and specific biomarkers of upper tract disease, and lack of reliable noninvasive imaging methods. An additional minority of untreated genital infections in women, the true extent of which is unknown, produce upper tract inflammation with minimal or no clinical symptoms/signs ("silent PID"), which likely resolves without complication in some, while in others it progresses to produce scarring complications of inflammatory injury.

PID can be largely asymptomatic or may manifest by symptoms of lower abdominal pain, dyspareunia (pain on coitus), and less often fever; on pelvic examination, tenderness on moving the cervix with the examining hand and on palpation of the oviducts and ovaries (adnexal tenderness) can be found. Complications of symptomatic or asymptomatic PID include involuntary infertility, defined as the inability for a couple to conceive after one year of unprotected coitus, and ectopic pregnancy, where implantation of the fertilized ovum occurs in the fallopian tube (or elsewhere in the abdominal cavity) rather than the uterine lining. Ectopic pregnancy is typically heralded by lower abdominal pain and can result in tubal rupture and potentially life-threatening hemorrhage; even when not complicated by bleeding, the condition requires extensive medical evaluation, treatment, and follow-up. Both of these complications are due to inflammatory injury and scarring of the fallopian tubes, resulting in blockage, tubal deformity, or abnormalities of ovum transport due to alterations in function or destruction of the ciliated tubal epithelium.

The concept of asymptomatic (silent) PID arose from consistent observations from multiple case-control studies of women who had either infertility of tubal origin or ectopic pregnancy. In these studies, serological evidence of past chlamydial infection is statistically more common in cases than in matched controls. The majority of the women with complications and evidence of past chlamydial infection did not have a history of symptomatic PID (Paavonen et al., 2008). Direct evidence for this sequence of events is scant, largely due to the fact that there are no reliable biomarkers for upper tract disease. Invasive procedures like laparoscopy are required to directly visualize the tubes, perform biopsies, and culture the samples to establish the presence of disease, but invasive procedures are difficult to justify in symptomless women. Since objective signs of inflammation without symptoms are commonly observed in trachoma, the analogy that silent PID occurs and leads in some cases to reproductive complications appears quite plausible. Similarly, it is likely true that upper genital tract infection and inflammation, like trachoma, often resolve uneventfully either naturally or in response to antibiotic therapy without development of complications. Human data to support these hypotheses are virtually nonexistent, but these ideas have been confirmed experimentally in an animal model of upper genital tract infection. In a vaginal challenge model where isolation of chlamydiae from the upper tract of infected guinea pigs was occasionally observed without histopathological change, the majority of animals with pathology resolved both infection and inflammation without treatment, although a substantial minority (about 20%) had upper tract inflammation at the latest time points studied (Rank and Sanders, 1992).

Complications in Men

The main consequence of chlamydial infection in men is the risk of transmission to women. However, complications of chlamydial urethritis in men may on occasion include epididymitis and a reactive arthritis syndrome. Epididymitis, a painful infection of the long, small convoluted tubules coiled on the posterior aspect of the testicle, complicates urethritis in perhaps 1 to 2% of cases (Martin, 2008). The main symptoms are testicular pain, which can be severe, and swelling. Chlamydial urethritis is also associated with reactive arthritis, of which Reiter's syndrome is the most severe form; there is a genetic predisposition for the latter syndrome in that 80% of affected individuals have the histocompatibility marker HLA-B27. Reactive arthritis is defined as an immune-mediated inflammatory process in the joints in response to primary infection at a distant mucosal site. Reiter's syndrome consists of arthritis, accompanied by conjunctivitis, urethritis, and skin lesions. It is thought that about 1% of men with NGU, of which *C. trachomatis* is a frequent cause, may develop a reactive arthritis syndrome; conversely, almost 70% of untreated men with Reiter's syndrome and urethritis harbor *C. trachomatis* at the urethral site. Chlamydial DNA, mRNA, and aberrant developmental forms have been detected in synovium of patients with reactive arthritis. These observations support the concept that chlamydial organisms, not just bacterial antigens, exist in the synovial membranes of affected patients (Carter and Hudson, 2010). Some but not all antibiotic treatment trials in patients with reactive arthritis and Reiter's syndrome show efficacy in reducing symptoms related to arthritis and reducing the risk of relapsing symptoms (Lauhio et al., 1991; Carter et al., 2010). However, chlamydiae are not a unique trigger for reactive arthritis syndromes, since an association is recognized for enteric pathogens like *Salmonella*, *Shigella*, *Campylobacter,* and *Yersinia* spp. Genetic predilection to either the localization of the organism or the resulting inflammatory response in the joints may be the major shared element in pathogenesis.

Screening for Genital Infection

Detection of infection in at-risk populations is essential to (i) reduce transmission by asymptomatic individuals, (ii) prevent complications of untreated asymptomatic infection, and (iii) decrease infection and prevalence in the population. There are no reliable objective means for assessing infection other than microbiological diagnosis; therefore, detection and subsequent treatment rely on screening of asymptomatic persons by using NAATs in groups of individuals who are at risk for genital chlamydial infection. Clinicians caring for adolescents and young adult women have an opportunity to (i) prevent *C. trachomatis* infections before they occur, (ii) identify and promptly treat symptomatic and asymptomatic chlamydial infections before complications develop, and (iii) identify symptoms suggestive of PID to provide prompt treatment. These three activities respectively represent primary, secondary, and tertiary prevention of *C. trachomatis*-associated infertility and ectopic pregnancy. The immediate goal of screening at the level of the individual is to identify and treat infected persons before they develop complications (secondary prevention) and to identify, test, and treat their sex partners to prevent reinfections. In women, the primary benefit of screening and treatment is to reduce their personal risk of reproductive sequelae. In men, who have a low risk of sequelae, the main rationale for treatment is to reduce the likelihood of reinfection of their female sex partners. The potential benefits of screening at the population level are to reduce the overall population burden and costs of reproductive sequelae and to reduce overall prevalence of *C. trachomatis* infections.

Screening of populations of sexually active women aged 25 years or less is recommended in the United States (Meyers et al., 2007) and in some other developed countries, since treatment is likely to prevent symptomatic PID in women (Scholes et al., 1996). However, to affect the population burden of disease, strategies to increase screening coverage must be designed, since the evidence to date suggests that screening coverage in the United States is low and not sufficient to affect prevalence (Heijne et al., 2010). In addition, even frequent testing and treating may not reduce prevalence in adolescent women at very high risk for reinfection (Batteiger et al., 2010a). The focus of screening in the United States is young at-risk women; screening of asymptomatic at-risk men is recommended only if resources exist. At present, it is recommended to identify and treat male partners and to retest women 3 months after treatment of an initial infection due to the risk of repeated infection (Workowski and Berman, 2010).

Natural History of Infection

The natural history of untreated infection is not well studied in humans, since an infection established by a positive diagnostic test ethically needs to be treated, given the potential risk for complications. The limited human studies examining untreated infection have been recently reviewed in detail (Geisler, 2010). These studies consistently show that humans clear chlamydial genital infections in the absence of antibiotic treatment. Two studies with women suggest that clearance after 1 year is about 45 to 55% and reaches 94% by 4 years (Morre et al., 2002; Molano et al., 2005). Although these women were in stable relationships, it is still possible that reinfection may have confounded the results. In a study of adolescent women in Indianapolis, employing weekly research samples collected over a period of 3 months, continuous shedding of chlamydial DNA in incident asymptomatic infections for up to 12 weeks was observed, the longest period of observation possible in the study (Batteiger et al., 2010a). Another study in pregnancy showed a 44% spontaneous clearance over 2 to 3 months (Sheffield et al., 2005). Two recent studies with much shorter follow-up periods, as well as a prior review, indicate that male genital infections also clear over time (Golden et al., 2000; Joyner et al., 2002; Geisler et al., 2008). The data strongly support the concept that immunity capable of resolving infection develops in humans. However, unlike in animal models, the time course to resolution in humans is measured in months to years rather than in weeks (Rank and Whittum-Hudson, 2010). Conversely, the same studies demonstrate that untreated infections may persist for many months and suggest that *C. trachomatis* is capable of evading host defenses and/or that an effective human response in some cases is slow to develop. In a clinical and epidemiological sense, these prolonged untreated infections are a common expression of persistent chlamydial infection.

Natural History of Disease: Progression to PID and Sequelae

The limited data available addressing the risk of an untreated chlamydial infection progressing to symptomatic PID are the subject of an excellent recent review (Haggerty et al., 2010). Early studies suggested rates of progression of up to 30% over a period of a few weeks. However, subsequent research based on much larger cohorts suggests that the rate of progression over 1 year is probably closer to 10% (Oakeshott et al., 2010). The rapidity of progression is also subject to much uncertainty; a summary of available data suggests that the greatest risk is within the first few weeks after diagnosis of a lower genital tract infection, with the risk lessening over time (Haggerty et al., 2010). However, since these studies generally select women with prevalent infection at entry, the true onset and thus the duration of the infections are unknown, making determination of the time course difficult. Since no reliable biomarkers of asymptomatic PID are available, the time course of this entity is entirely unknown. The contributions of the innate and adaptive immune responses in the pathogenesis and timing of PID and sequelae have been reviewed (Darville and Hiltke, 2010) and are discussed in more detail in chapter 11, "*Chlamydia* immunopathogenesis."

The mechanism by which chlamydiae, which lack intrinsic motility, ascend to the endometrium and fallopian tubes is not entirely clear. It has typically been assumed that an established endocervical infection served as the staging area for ascension. However, as noted above, careful studies in the guinea pig model of genital infection demonstrate that *Chlamydia caviae* can be recovered from the upper genital tract within just a few days after vaginal inoculation (Rank and Sanders, 1992). In humans, peristaltic motility of the upper genital tract, associated with female sexual arousal and orgasm, and the physical act of penile thrusting during coitus are capable of transporting fluids and particles throughout the uterus and oviducts. It has been hypothesized that these mechanisms serve to deliver viable chlamydiae to the upper tract very early after exposure to an infected partner (Lyons et al., 2009). It is possible that at least transient upper tract infection occurs in many infected women, which is supported by the frequency with which endometrial infection can be found in women presenting with lower genital tract infection (Jones et al., 1986). The development of established inflammation and symptomatic PID, which occurs in the minority, may then depend on the infecting inoculum or factors related to host susceptibility (Darville and Hiltke, 2010).

Excellent data indicate that symptomatic PID overall (due to *C. trachomatis* or any other organism) is associated with increased risk of involuntary infertility; estimates are that 16 to 18% of such women develop involuntary infertility after one documented episode and that repeated episodes of symptomatic disease are associated with correspondingly increased risk (Paavonen et al., 2008). In addition, it appears that repeated laboratory diagnosis of chlamydial infection is associated with increased risk of PID and ectopic pregnancy (Haggerty et al., 2010). However, most women with sequelae, even those with evidence of past chlamydial infection, do not have a history of clinically evident PID. These women are presumed to have had asymptomatic or minimally symptomatic PID, but the extent to which this occurs and its time course are unclear.

In summary, symptomatic PID is clearly related to chlamydial genital infection in women, but the rate of PID is likely lower than the initial studies suggested. However, given the very high infection case rates in women of reproductive age, even relatively low PID rates translate to a large burden of upper tract disease and sequelae.

Repeated Genital Infection, Antibiotic Efficacy, and Antibiotic Resistance

Repeated infection in at-risk individuals is a prominent epidemiologic characteristic of chlamydial genital infections. A recent longitudinal observational study in adolescent women, conducted with molecular diagnostic techniques, strain genotyping, and detailed behavioral analysis, demonstrates that repeat infection is largely due to reinfection (Batteiger et al., 2010a). Consistent with animal model studies (Rank and Whittum-Hudson, 2010), these reinfections from infected partners are related to a lack of sterilizing immunity to chlamydial genital infection (Batteiger et al., 2010b). In a mouse model of genital infection, treatment early in the infection course blunted the partial immunity to rechallenge observed in that model (Su et al., 1999). It has been hypothesized that increased case rates observed in some areas after implementation of aggressive screening and treatment may be related to increased reinfection due to population reductions in levels of protective immunity (the so-called "arrested immunity" hypothesis) (Brunham et al., 2005), although epidemiologic and case ascertainment factors are likely at play as well (Miller, 2008).

In a meta-analysis of clinical trials of lower genital tract infection, antibiotic treatment with azithromycin given as a single dose or doxycycline given as a 7-day course was effective in eradicating *C. trachomatis* in 97 and 98% of study participants, respectively (Lau and Qureshi, 2002). However, these studies used cell culture as the primary microbiological outcome, rather than the more sensitive NAATs, where eradication rates may be lower (Stamm et al., 2007). In two observational studies, the effectiveness of azithromycin therapy with long follow-up was about 92% (Golden et al., 2005; Batteiger et al., 2010a). Thus, current treatment for chlamydial genital infection is not as effective as antibiotic treatment for gonorrhea, which is typically >99% effective.

Antibiotic resistance does not appear to account for most treatment failures, and in many studies, early reinfection is difficult to exclude. However, one report clearly links clinical and microbiological treatment failure of *C. trachomatis* lower genital tract infection to in vitro resistance to azithromycin and fluoroquinolones (Somani et al., 2000). In addition, strains of *C. trachomatis* have been documented to harbor mutations in a 23S rRNA gene associated with macrolide resistance (Misyurina et al., 2004). In the latter report, the organisms with the resistance mutation were less viable in cell culture. Despite these reports, frequent treatment failures related to antibiotic resistance are not evident clinically. However, susceptibility testing for *C. trachomatis* is labor-intensive and not well standardized, and thus, systematic surveillance for emergence of resistance has not been implemented.

Lymphogranuloma Venereum

LGV is a sexually transmitted infection caused by *C. trachomatis* L1, L2, and L3 serovars and their variants. LGV strains generally cause more severe and invasive disease than the genital serovars D through K, with prominent involvement of lymph channels and regional lymph nodes with suppurative (pus-forming) and fibrosing consequences (Stamm, 2008). LGV syndromes are sporadic in North America, Europe, South America, and many parts of Asia. It is endemic in East and West Africa, India, Southeast Asia, parts of South America, and the Caribbean. LGV typically begins as a primary lesion at the point of inoculation where there is a break in the genital skin. The primary lesion can manifest as a papule (raised, red skin lesion); an ulcer or area of skin or mucosal erosion; shallow, clustered ulcers resembling herpes simplex; or urethritis. The organism enters the lymphatic channels, causing lymphangitis, and then moves to regional lymph nodes, causing lymphadenitis, sometimes progressing to suppurative necrosis with drainage via sinus tracts to the skin. Such enlarged nodes are termed buboes, and in the inguinal syndrome in men they are located above and below the inguinal ligament at the top of the thigh. The anogenitorectal syndrome and proctocolitis can occur in both men and women and is characterized by severe inflammatory changes in the distal colon and rectal mucosa. Fibrotic sequelae can result in rectal stricture. Rarely, a process termed esthiomene occurs in women with vulvar lymphedema due to sclerosing fibrosis, which can result in massive enlargement of the vulva (elephantiasis) with resulting skin and mucosal ulcerations. Occasionally, autoinoculation to the eye can occur with conjunctivitis and regional adenopathy. The recommended antibiotic treatment is 21 days of doxycycline (Workowski and Berman, 2010).

C. trachomatis and Cervical Cancer

Virtually all cases of cervical cancer are caused by high-risk human papillomavirus (HPV) types, but only a small proportion of high-risk HPV-positive women have persistence of high-risk HPV and go on to develop cervical cancer. Cigarette smoking and chlamydial infection have been proposed as cofactors that contribute to the progression to cervical cancer. Results of epidemiologic studies have been mixed in demonstrating an association between chlamydial infection and cervical dysplasia (cervical intraepithelial neoplasia [CIN]) and invasive cancer. CIN grades 1, 2, and 3 refer to dysplastic changes confined to the epithelial layers with increasing degrees of severity, while invasive cancer extends beyond the basement membrane, where it can invade locally, spread via lymphatics to regional lymph nodes, and metastasize to distant sites. One example of several positive studies is a registry study in Finland demonstrating an association between serological evidence of past chlamydial infection and invasive cervical carcinoma (Anttila et al., 2001). Using data from participants in placebo arms of recent HPV vaccine trials, a retrospective cohort study demonstrated an association between detection of *C. trachomatis* at the cervical site and CIN grade 2, but the association disappeared when CIN grade 3 (high-grade dysplasia) was included in the analysis (Lehtinen et al., 2011). Another recent study found no association between *C. trachomatis* detected at the cervical site and the risk of incident CIN after controlling for HPV-positive status (Safaeian et al., 2010). The design of epidemiologic studies to confirm the possible relationship is difficult, and virtually all published studies have methodologic design issues (Miller and Ko, 2011). Future studies must consider the timing and duration of chlamydial infection, which is made more difficult by the obligation to treat chlamydial infection and stages of CIN once found (Miller and Ko, 2011).

Nevertheless, there is biologic plausibility for an interaction between HPV and *C. trachomatis* infection in the development or progression of CIN. *C. trachomatis* may (i) increase the risk of acquisition of HPV by altering conditions at the endocervix, (ii) increase the rate of transformation to CIN, and/or (iii) increase the likelihood that CIN leads to invasive cervical cancer. Persistence of HPV infection for longer periods is associated with increased risk of cervical cancer, and several studies have shown that concomitant *C. trachomatis* infection increases the duration of HPV persistence (Shew et al., 2006). More recently, it has been demonstrated that *C. trachomatis* infection in cell culture models causes centrosomal number defects, multipolar spindles, and chromosomal segregation defects (Grieshaber et al., 2006) and dysregulates the centrosome duplication pathway (Johnson et al., 2009). Mitotic spindle pole defects have subsequently been shown to be induced by chlamydial infection, independent of the effect on centrosomes (Knowlton et al., 2011). The hypothesis is that these combined defects may contribute to genetic instability, providing a possible mechanism for *C. trachomatis* to serve as a cofactor in the development of cervical cancer.

More research is needed to confirm the epidemiologic relationship between *C. trachomatis* and cervical cancer to determine if interventions directed toward chlamydial infections add benefit to the known efficacy of the available HPV vaccines in preventing CIN.

RELATIONSHIPS AMONG *C. TRACHOMATIS* SEROVARS, DISEASE MANIFESTATIONS, AND TISSUE TROPISM

Classically, there are three disease-associated groups of *C. trachomatis*: an ocular group that typically produces trachoma (serovars A, B, Ba, and C), a genital group typically associated with genital disease (D through K), and a third group associated with LGV syndromes (L1, L2, and L3). The ocular and genital strains infect columnar epithelial cells, which are found in the conjunctiva, urethra, endocervix, endometrium, endosalpinx, and rectum. LGV strains infect columnar epithelia at these sites and in addition can productively infect macrophages, accounting for the invasive lymphangitis/lymphadenitis disease associations seen with these agents. However, there are some exceptions to these general rules.

Serovars A and C appear to be strictly associated with trachoma (serovars B and Ba cause trachoma but also cause genital disease) and are tropic for conjunctival epithelium. These ocular strains differ from genital strains in that they lack an intact *trpA* gene, which encodes a subunit of tryptophan synthase, and thus cannot utilize indole to synthesize tryptophan (Caldwell et al., 2003). Other differences are found in genes encoding polymorphic membrane proteins (Stothard et al., 2003), cytotoxin (Carlson et al., 2004), and the translocated actin-recruiting phosphoprotein (Tarp), a type III secretion effector (Carlson et al., 2005), but the relevance of these differences to pathogenesis is currently unclear.

However, both genital and LGV strains also infect the conjunctivum of the eye; conjunctivitis may occur in as many as 1 to 2% of men with chlamydial urethritis (Martin, 2008), with autoinoculation from the genital site being the mechanism. Strain identifications of chlamydiae from patients with trachoma have occasionally yielded genital strains such as serovar E; in addition, TW183, the first example of *C. pneumoniae*, was isolated from the eye of a participant in a trachoma vaccine trial. The role of these nontraditional isolates in producing or perpetuating trachoma is not known. Conversely, serovars B and Ba, typically associated with trachoma, are also found with some reasonable frequency in genital infections (Batteiger et al., 2010a).

The relationship between disease expression and potential virulence molecules responsible for differences among the strains has recently been reviewed (Byrne, 2010). Candidate molecules include the major outer membrane protein, polymorphic outer membrane proteins, type III secretion effectors, putative cytotoxin, stress response proteins, lipopolysaccharides and other glycolipids, and gene products of the chlamydial virulence plasmid (Byrne, 2010). It is clear that the major outer membrane protein (encoded by *ompA*), which is the variable surface protein whose epitopes define individual serovars and the antigenically related serogroups B group (B, Ba, D, E, L1, and L2), C group (C, H, I, Ia, J, K, L3), and intermediate group (F and G), is not related to the three disease-associated groups. Certain polymorphic outer membrane proteins, which are a family of autotransporters in the chlamydial outer membrane, have evolved in parallel with the disease-causing groups (Stothard et al., 2003), but it is not known if these proteins are responsible for the differences in disease manifestations.

C. PNEUMONIAE INFECTIONS

Chlamydia pneumoniae was originally isolated from the conjunctiva of a child without active trachoma who was enrolled in a trachoma vaccine trial in Taiwan in 1965 (Kuo et al., 1986); a similar organism was isolated from a college student with pneumonia in Seattle (Grayston, 1992). Initially dubbed "TWAR" and thought to represent a new strain of *C. psittaci*, it was shown subsequently to be morphologically unique and to be genetically distinct from *C. trachomatis* and *C. psittaci* based on comparative genome sequencing, and it was assigned its own species (Grayston et al., 1989). Unlike *C. psittaci* or *C. abortus* organisms, which have zoonotic niches and only occasionally cause human infection, *C. pneumoniae* appears to be primarily a human respiratory pathogen with evidence of human-to-human transmission (Saikku et al., 1985; Kleemola et al., 1988). Based on seroepidemiologic studies, it is very common worldwide (Grayston, 1992; Hammerschlag, 2000); the seroprevalence, based on the reference microimmunofluorescence (MIF) test, reaches 50% by age 20 and can exceed 70 to 80% in the elderly, indicating widespread ongoing exposure to the organism during adulthood (Grayston, 1992). *C. pneumoniae* causes community-acquired pneumonia (Grayston et al., 1986), which is a common but serious lower respiratory tract infection that occurs in children and adults outside the health care setting. Depending on location, patient population, and diagnostic methods used, *C. pneumoniae* causes 0 to 44.2% of community-acquired pneumonias, with the majority of studies suggesting a range of 6 to 22% (Hammerschlag, 2000; Kumar and Hammerschlag, 2007).

Several chronic diseases have been putatively linked to *C. pneumoniae*, including asthma and atherosclerosis; these associations are considered below. Full characterization of *C. pneumoniae* clinical manifestations, chronic disease associations, and frequency and consequences of asymptomatic infection has been seriously hampered by the lack of sensitive, specific, reproducible, and validated diagnostic methods (Dowell et al., 2001; Kumar and Hammerschlag, 2007).

Diagnostic Tests

Initially the organism was distinguished from other chlamydial species and linked to respiratory disease based on the MIF serology of Wang (Grayston, 1992). This test uses suspensions of purified elementary bodies of chlamydial strains that are adhered to a glass slide, incubated with dilutions of human serum, and then visualized with fluorescein-labeled secondary antibodies specific for human immunoglobulin G (IgG) or IgM. Per joint recommendations of the U.S. Centers for Disease Control and Prevention and the Canadian Laboratory Centre for Disease Control (Dowell et al., 2001), the MIF is the recommended serological test for diagnostic use and seroepidemiologic studies. However, it is subjective and operator dependent, and there is relatively low agreement even among laboratories expert in its performance (Littman et al., 2004). Some have suggested that the presence of IgA antibodies indicate chronic infection, because the relatively short half-life of IgA in the serum suggests ongoing antigenic stimulation, but this link has not been validated (Kumar and Hammerschlag, 2007). Proper use and interpretation of the MIF test for diagnosis of *C. pneumoniae* infection include obtaining and testing paired serum samples 4 to 8 weeks apart. Acute infection is defined serologically as a single IgM of ≥1:16 or a fourfold increase in IgG in paired samples; possible acute infection is defined as a single IgG of ≥1:512; and presumed past infection is defined by a single IgG of ≥1:16 (Dowell et al., 2001). The sensitivity and specificity of the MIF are unknown since there is no diagnostic gold standard and there is relatively poor correlation between MIF data and detection of the organism by nucleic acid amplification methods or culture (Kumar and Hammerschlag, 2007). Cases of culture-positive pneumonia with negative MIF have been reported particularly among children, where 50% or more with culture-positive pneumonia have a negative MIF test (Hammerschlag et al., 2009). Finally, in initial (primary) infections, IgM antibodies appear 2 to 3 weeks and IgG antibodies appear 6 to 8 weeks following infection, so that MIF is not useful in planning treatment of an acute respiratory infection (Kumar and Hammerschlag, 2007).

C. pneumoniae is a fastidious slow-growing obligate intracellular pathogen that can be grown in cell culture, but the diagnostic utility of culture is limited, since the method is not widely available and is sufficiently time-consuming that the results are not useful for acute management of pneumonia. In addition, the most commonly available diagnostic sample in cases of pneumonia is sputum, which is directly toxic to cell culture and contaminated with colonizing oral and upper airway bacteria as well (Hammerschlag et al., 2010). Standards for performance and interpretation of cell culture for *C. pneumoniae* have been established (Dowell et al., 2001).

Although a variety of DNA amplification methods for *C. pneumoniae* have been published, most are "in-house" assays; only a few have been studied well enough to be validated (Kumar and Hammerschlag, 2007), and to date there is no standardized FDA-approved amplification test for diagnostic use. Likely impediments to development and marketing of commercially available tests are the unknown clinical significance of detection of *C. pneumoniae* in asymptomatic persons and the routine use of treatment guidelines in the management of pneumonia. *C. pneumoniae* and *Mycoplasma pneumoniae* are considered common causes of pneumonia in nonhospitalized adults and children; however, clinicians rarely make a microbiological diagnosis of these causes of community-acquired pneumonia. Rather, antibiotic treatment is based on published guidelines that are designed to provide coverage of the most likely pathogens (Mandell et al., 2007). Since these guidelines recommend antibiotics effective against *C. pneumoniae* and antibiotic resistance has not been an issue, a diagnostic test would not affect clinical decision making. However, a standardized, validated, sensitive, and specific amplification test could have a major impact on better defining the epidemiology of acute and chronic conditions related to the organism.

Clinical Manifestations

The most convincing disease association with *C. pneumoniae* is pneumonia, which is typically relatively mild, often self-limited (in that symptoms resolve without specific antibiotic treatment), and very difficult to distinguish from other forms of pneumonia. It is often cited as one of the causes of "atypical pneumonia," resembling that caused by *Mycoplasma pneumoniae*. In truth, it is clinically difficult to distinguish these pneumonias from those due to classic pathogens such as *Streptococcus pneumoniae*. Given the lack of FDA-approved validated DNA amplification tests, the fact that culture is relatively insensitive and not widely available, and that serodiagnosis requires paired acute- and convalescent-phase sera 6 to 8 weeks apart, specific diagnosis of *C. pneumoniae* pneumonia is not typically attempted. Rather, these pneumonias are treated empirically based on the most likely pathogens encountered in a given clinical setting according to guidelines as noted above. Antibiotic therapy for community-acquired pneumonia includes agents active against *C. pneumoniae* and *M. pneumoniae*, namely, macrolides (azithromycin and clarithromycin) with or without a beta-lactam antibiotic directed to *S. pneumoniae*, or with selected fluoroquinolone antibiotics (levofloxacin or moxifloxacin) alone. Antibiotic resistance of *C. pneumoniae* to these agents has not been noted (Hammerschlag et al., 2010).

Although *C. pneumoniae* has been cultured from the pharynx of patients with sore throat, it is unclear how frequently *C. pneumoniae* is associated with symptomatic sore throat because of diagnostic test limitations. Currently, antibiotic treatment is only recommended for *Streptococcus pyogenes* (group A streptococcal) pharyngitis, and treatment is mainly to prevent development of serious postinfectious syndromes like rheumatic fever. There is no clinical indication at this time to seek or empirically treat *C. pneumoniae* in patients with sore throat.

Epidemiology

Based on MIF and culture studies, *C. pneumoniae* appears to be a common human respiratory pathogen. The presumed mode of transmission is from person to person via respiratory secretions. Droplet transmission has been described in a laboratory accident (Hammerschlag, 2000), but *C. trachomatis* likewise has caused lung infection via infectious aerosols during similar laboratory accidents, so this example has little relevance to natural transmission. Genomic variants of *C. pneumoniae* cause infection in some animals, most notably koalas (Wardrop et al., 1999), but there is no evidence to suggest that a zoonotic reservoir is important in human disease. The most compelling case for human-to-human transmission is found in the evaluation of outbreaks in closed populations of schoolchildren (Saikku et al., 1985) and military recruits (Kleemola et al., 1988) in Finland.

In Vitro Persistence and Chronicity of Human Infections

The ability to assume an in vitro persistence phenotype in response to stress is seen with *C. pneumoniae*. In addition, in one cell culture model of *C. pneumoniae* infection, continuous infected cell cultures were maintained for over 4 years in the absence of stressors (Kutlin et al., 2001). In this model, most inclusions were typical but a minority (about 10%) contained aberrant forms consistent with the persistence phenotype. In this cell culture model, organisms exhibiting the persistence phenotype resist bactericidal action of antibiotics (Kutlin et al., 2002). The role of this phenomenon in human disease is not known.

However, long durations of culture-proven nasopharyngeal shedding of *C. pneumoniae* have been described in individuals after an episode of acute symptomatic respiratory disease. In one of the few studies with long follow-up, five individuals shed for up to 11 months; two of these five persons shed for 7 to 9 years (Hammerschlag et al., 1992; Hammerschlag, 2000). This nasopharyngeal shedding continued despite repeated courses of antibiotics; however, the individuals were largely asymptomatic. No evidence of antibiotic resistance was found among the isolates tested. Although derived from a small number of patients, these data indicate that in certain instances, prolonged respiratory shedding of *C. pneumoniae* can occur. Among persons with symptomatic respiratory infection, there are limited data on the proportion of individuals that go on to chronically shed *C. pneumoniae*, since systematic follow-up of such cohorts has not been reported. A small study involving children with asthma and *C. pneumoniae* infection suggests that perhaps as many as 20% have nasopharyngeal shedding after treatment (Emre et al., 1994).

Potential Role in Chronic Diseases

C. pneumoniae has been linked to chronic diseases characterized by elements of chronic inflammation. The initial associations were established by seroepidemiologic surveys, the first with atherosclerotic cardiovascular disease and the second with asthma. In 1988, Saikku and coworkers reported an association with chronic coronary heart disease and acute myocardial infarction (Saikku et al., 1988). In 1991, Hahn and coworkers reported an association between serologic evidence of acute *C. pneumoniae* infection and adult-onset asthma (Hahn et al., 1991). An excellent and detailed review of these topics with extensive literature citations has been recently published (Hammerschlag et al., 2009). The goal here is to provide a clinical perspective on the state of these associations at this time.

Atherosclerosis

Inflammation is a known element of atherosclerosis, and a number of infectious agents, including *C. pneumoniae*, have been proposed to contribute to atherosclerotic disease. After the initial seroepidemiological association was published, a variety of in vitro experimental studies provided biological plausibility for the link. *C. pneumoniae* is able to infect and replicate within monocytes, macrophages, vascular endothelial cells, and smooth muscle cells, all of which are components of atherosclerotic plaque. In addition, in vitro infection resulted in oxidation of cellular low-density lipoprotein and production of proinflammatory cytokines. Further, studies of mouse and rabbit models of atherosclerosis showed that infection could induce or enhance atherosclerosis and that treatment of infection can blunt this process (Hoymans et al., 2007).

However, subsequent studies with humans proved to be inconsistent; reviews of this extensive literature have recently been published (Hoymans et al., 2007; Watson and Alp, 2008; Hammerschlag et al., 2009). Subsequent seroepidemiologic studies conducted in a prospective rather than cross-sectional manner and adequately controlling for cardiovascular risk factors showed weak or no associations between serological evidence of *C. pneumoniae* infection and atherosclerosis (Hoymans et al., 2007). Reports of bacterial isolation and detection by immunohistochemistry and PCR suggested the presence of the organism in atherosclerotic plaques, but the specificity of these nonstandardized techniques has not been uniformly borne out by subsequent studies. Similarly, evidence supporting the presence of *C. pneumoniae* in circulating blood mononuclear cells, which are presumed to be the vehicle for translocating the organisms from the respiratory system to the vessel wall, has not been convincingly and reproducibly demonstrated in humans (Hoymans et al., 2007; Hammerschlag et al., 2009).

Initial reports of small clinical experiences and case-control studies suggested a reduction in subsequent coronary events among seropositive individuals treated with antibiotics active in vitro against *C. pneumoniae* (Hammerschlag et al., 2009). These clinical studies, together with the treatment results in the animal models, led to a number of large, prospective, controlled clinical treatment trials. The results of these studies have been summarized in meta-analyses, where data from comparable clinical trials are pooled to provide more statistical power to detect the direction and magnitude of a treatment effect (Andraws et al., 2005). The results were consistent in showing no effect of antibiotic treatment on total mortality, no effect on subsequent cardiac events including myocardial infarction and unstable angina, and no relationship between outcome and *C. pneumoniae* serologic status (Hammerschlag et al., 2009). These conclusions are tempered by the fact that patient selection for the trials was hampered by the lack of diagnostic markers for endovascular infection. This is important, because it is possible that only a minority of seropositive trial participants had true endovascular infection. If the remaining majority of participants had only past, resolved respiratory infection with no current active endovascular infection, it could explain why these treatment trials showed no benefit.

At present, the preponderance of evidence suggests that *C. pneumoniae* in and of itself is neither necessary nor sufficient to cause atherosclerosis or its complications in humans. Furthermore, available clinical trials fail to demonstrate an overall benefit of antibiotic therapy in reducing mortality or cardiovascular events in patients with coronary heart disease. The evidence does not exclude a role for *C. pneumoniae* as a possible modifiable risk factor for development or progression of atherosclerosis and its complications in humans (Watson and Alp, 2008), although definition of the extent and importance of its role will require clinical studies using improved diagnostic methods capable of establishing the presence of infection at the endovascular site.

Asthma

The relationship of *C. pneumoniae* infection to asthma and the occurrence of chronic *C. pneumoniae* shedding in asymptomatic and symptomatic asthmatics have been recently reviewed in detail (Hammerschlag et al., 2009). Although evidence supports the contention that acute *C. pneumoniae* infections can be associated with exacerbations of asthma, most exacerbations are not thought to be related to bacterial infection. Current treatment guidelines for acute exacerbations of asthma in adolescents and adults do not include the use of antibiotics, unless there is clinical evidence of pneumonia. If pneumonia is present, treatment appropriate for community-acquired pneumonia is provided, which includes antibiotics active against *C. pneumoniae*. While shedding of *C. pneumoniae* has been documented for up to 6 months in asymptomatic and symptomatic asthmatics, the relationship to symptoms is uncertain. There are also insufficient data to support the hypothesis that *C. pneumoniae* is an initiator of asthma. Antibiotic treatment trials for asthma have shown inconsistent results, related in part to differing diagnostic criteria for *C. pneumoniae* infection and failure to include infection end points in some studies. Another confounding factor is that antibiotics commonly used to treat *C. pneumoniae* infection, such as macrolides, tetracyclines, and fluoroquinolones, also have immunomodulatory anti-inflammatory effects that can result in improvements in pulmonary function independent of infection status (Hammerschlag et al., 2009).

Other Conditions

A role for *C. pneumoniae* in the causation of other chronic illnesses such as macular degeneration, Alzheimer's dementia, and multiple sclerosis is highly uncertain, again due to the lack of standardized diagnostic tools. Initial reports of PCR amplification of chlamydial DNA from cerebrospinal fluid samples from patients with multiple sclerosis are now questionable because the primers that were used were not specific for *C. pneumoniae* (Hammerschlag et al., 2009).

CURRENT AND FUTURE CHALLENGES

The major challenges to progress toward worldwide public health goals of reducing the burden of sequelae caused by *C. trachomatis* infections relate to the common clinical themes of prolonged infection, repeated infection, scarring complications of infection, and their underlying biological mechanisms. The themes are interrelated and result in continued spread and persistence of chlamydial infection in the population but also suggest points of intervention that could serve to reduce the impact of infection and disease.

Long-lasting asymptomatic infections ensure a long duration of infectivity and accompanying risk of transmission. Why do effective immune responses capable of resolving existing infection take so long to develop? If chlamydial virulence factors are capable of subverting and delaying the development of an effective response, can discovering and targeting these factors with drugs or vaccines shorten infection duration and thus transmissibility? Until such tools are available, which practical, affordable population-based screening strategies are best suited to identify and treat the largest possible numbers of prevalent infections?

Repeated genital infections are most often reinfections and are a major reason (aside from inadequate implementation) that screening and treatment strategies have not reduced population prevalence of infection. Protective immunity in natural infection is at best partial, resulting in reduced chlamydial shedding and duration of infection; treatment may further impair development of protective immunity. Will vaccines, even if not capable of producing a high rate of sterilizing immunity, be able to reduce transmission by inducing a similar (or greater?) degree of protective immunity compared to natural infection and thus serve to reduce population prevalence if sufficient numbers of persons are vaccinated? Can newer antibiotics with greater activity and longer pharmacologic half-lives reduce treatment failures and perhaps even prevent early reinfections, and at least partly address the important issue of reinfection? Currently, preventing repeated genital infection depends on notifying sex partners; in practice, it is extraordinarily difficult to identify and treat the majority of partners. Mass treatment of trachoma reduces the burden of active disease in a community or district and in so doing reduces the probability of reinfection. Under the right circumstances, could properly focused mass treatment strategies provide another tool to aid in reducing or eliminating genital chlamydial infections in a population?

Chlamydial infections cause scarring sequelae in a substantial minority of infected persons. Can reliable genetic or immunological biomarkers be identified to allow us to best focus our screening and treating efforts on these more vulnerable individuals? Can biomarkers that identify women with asymptomatic PID be found and allow us to intervene earlier with antibiotics? Can vaccines, even if not able to induce sterilizing immunity in the majority of recipients, reduce the risk of symptomatic PID and scarring complications even if infection does occur?

The eventual control of chlamydial infections depends on continued progress of both basic and epidemiological research and the productive interplay between the two, whereby clinical observations drive and focus the direction of basic science and the tools and insights from basic science inform and enhance epidemiologic study and clinical care.

REFERENCES

Andraws, R., J. S. Berger, and D. L. Brown. 2005. Effects of antibiotic therapy on outcomes of patients with coronary artery disease. *JAMA* **293:**2641–2647.

Anttila, T., P. Saikku, P. Koskela, A. Bloigu, J. Dillner, I. Ikäheimo, E. Jellum, M. Lehtinen, P. Lenner, T. Hakulinen, A. Närvänen, E. Pukkala, S. Thoresen, L. Youngman, and J. Paavonen. 2001. Serotypes of *Chlamydia trachomatis* and risk for development of cervical squamous cell carcinoma. *JAMA* **285:**47–51.

Bailey, R., T. Duong, R. Carpenter, H. Whittle, and D. Mabey. 1999. The duration of human ocular *Chlamydia trachomatis* infection is age dependent. *Epidemiol. Infect.* **123:**479–486.

Bailey, R. L., P. Arullendran, H. C. Whittle, and D. C. Mabey. 1993. Randomized controlled trial of single-dose azithromycin in treatment of trachoma. *Lancet* **342:**453–456.

Batteiger, B. E., W. Tu, S. Ofner, B. Van Der Pol, D. R. Stothard, D. P. Orr, B. P. Katz, and J. D. Fortenberry. 2010a. Repeated *Chlamydia trachomatis* genital infections in adolescent women. *J. Infect. Dis.* **201:**42–51.

Batteiger, B. E., F. Xu, R. E. Johnson, and M. L. Rekart. 2010b. Protective immunity to *Chlamydia trachomatis* genital infection: evidence from human studies. *J. Infect. Dis.* **201:**S178–S189.

Blake, I. M., M. J. Burton, R. L. Bailey, A. W. Solomon, S. West, B. Muñoz, M. J. Holland, D. C. W. Mabey, M. Gambhir, M.-G. Basáñez, and N. C. Grassly. 2009. Estimating household and community transmission of ocular *Chlamydia trachomatis*. *PLoS Negl. Trop. Dis.* **3:** e401.

Brunham, R. C., B. Pourbohloul, S. Mak, R. White, and M. L. Rekart. 2005. The unexpected impact of a *Chlamydia trachomatis* infection control program on susceptibility to reinfection. *J. Infect. Dis.* **192:**1836–1844.

Burton, M. J., M. J. Holland, N. Faal, E. A. N. Aryee, N. D. E. Alexander, M. Bah, H. Faal, S. K. West, A. Foster, G. J. Johnson, D. C. W. Mabey, and R. L. Bailey. 2003. Which members of a community need antibiotics to control trachoma? Conjunctival *Chlamydia trachomatis* infection load in Gambian villages. *Investig. Ophthalmol. Vis. Sci.* **44:**4215–4222.

Burton, M. J., and D. C. W. Mabey. 2009. The global burden of trachoma: a review. *PLoS Negl. Trop. Dis.* **3:**e460.

Byrne, G. I. 2010. *Chlamydia trachomatis* strains and virulence: rethinking links to infection prevalence and disease severity. *J. Infect. Dis.* **201:**S126–S133.

Caldwell, H. D., H. Wood, D. D. Crane, R. Bailey, R. B. Jones, D. Mabey, I. Maclean, Z. Mohammed, R. W. Peeling, C. Roshick, J. Schacter, A. W. Solomon, W. E. Stamm, R. J. Suchland, L. Taylor, S. K. West, T. C. Quinn, R. J. Belland, and G. McClarty. 2003. Polymorphisms in *Chlamydia trachomatis* tryptophan synthase genes differentiate between genital and ocular isolates. *J. Clin. Investig.* **111:**1757–1769.

Carlson, J. H., S. Hughes, D. Hogan, G. Cieplak, D. E. Sturdevant, G. McClarty, H. D. Caldwell, and R. J. Belland. 2004. Polymorphisms in the *Chlamydia trachomatis* cytotoxin locus associated with ocular and genital isolates. *Infect. Immun.* **72:**7063–7072.

Carlson, J. H., S. F. Porcella, G. McClarty, and H. D. Caldwell. 2005. Comparative genomic analysis of *Chlamydia trachomatis* oculotropic and genitotropic strains. *Infect. Immun.* **73:**6407–6418.

Carter, J. D., L. R. Espinoza, R. D. Inman, K. B. Sneed, L. R. Ricca, F. B. Vasey, J. Valeriano, J. A. Stanich, C. Oszust, H. C. Gerard, and A. P. Hudson. 2010. Combination antibiotics as a treatment for chronic *Chlamydia*-induced reactive arthritis: a double-blind, placebo-controlled, prospective trial. *Arthritis Rheum.* **62:**1298–1307.

Carter, J. D., and A. P. Hudson. 2010. The evolving story of *Chlamydia*-induced reactive arthritis. *Curr. Opin. Rheumatol.* **22:**424–430.

Cecil, J. A., M. R. Howell, J. J. Tawes, J. C. Gaydos, K. T. J. McKee, T. C. Quinn, and C. A. Gaydos. 2001. Features of *Chlamydia trachomatis* and *Neisseria gonorrhoeae* infection in male Army recruits. *J. Infect. Dis.* **184:**1216–1219.

Darville, T., and T. J. Hiltke. 2010. Pathogenesis of genital tract disease due to *Chlamydia trachomatis*. *J. Infect. Dis.* **201:**S114–S125.

Datta, S. B., E. Terrone, D. Kruszon-Movan, S. Berman, R. Johnson, C. L. Satterwhite, J. Papp, and H. Weinstock. 2012. *Chlamydia trachomatis* trends in the United States among persons 14 to 39 years of age, 1999–2008. *Sex. Transm. Dis.* **39:**92–96.

Dowell, S. F., R. W. Peeling, J. Boman, G. M. Carlone, B. S. Fields, J. Guarner, M. R. Hammerschlag, L. A. Jackson, C.-C. Kuo, M. Maass, T. O. Messmer, D. F. Talkington, M. L. Tondella, S. R. Zaki, et al. 2001. Standardizing *Chlamydia pneumoniae* assays: recommendations from the Centers for Disease Control and Prevention (USA) and the Laboratory Centre for Disease Control (Canada). *Clin. Infect. Dis.* **33:**492–503.

Emre, U., P. M. Roblin, M. Gelling, W. Dumornay, M. Rao, M. R. Hammerschlag, and J. Schachter. 1994. The association of *Chlamydia pneumoniae* infection and reactive airway disease in children. *Arch. Pediatr. Adolesc. Med.* **148:**727–732.

Falk, L., H. Fredlund, and J. S. Jensen. 2005. Signs and symptoms of urethritis and cervicitis among women with or without *Mycoplasma genitalium* or *Chlamydia trachomatis* infection. *Sex. Transm. Infect.* **81:**73–78.

Geisler, W. M. 2010. Duration of untreated, uncomplicated *Chlamydia trachomatis* genital infection and factors associated with chlamydia resolution: a review of human studies. *J. Infect. Dis.* **201**(Suppl. 2)**:**S104–S113.

Geisler, W. M., C. Wang, S. G. Morrison, C. M. Black, C. I. Bandea, and E. W. I. Hook. 2008. The natural history of untreated *Chlamydia trachomatis* infection in the interval between screening and returning for treatment. *Sex. Transm. Dis.* **35:**119–123.

Golden, M. R., J. A. Schillinger, L. Markowitz, and M. E. St. Louis. 2000. Duration of untreated genital infections with *Chlamydia trachomatis*: a review of the literature. *Sex. Transm. Dis.* **27:**329–337.

Golden, M. R., W. L. H. Whittington, H. H. Handsfield, J. P. Hughes, W. E. Stamm, M. Hogben, A. Clark, C. Malinski, J. R. L. Helmers, K. K. Thomas, and K. K. Holmes. 2005. Effect of expedited treatment of sex partners on recurrent or persistent gonorrhea or chlamydial infection. *N. Engl. J. Med.* **352:**676–685.

Grassly, N. C., M. E. Ward, S. Ferris, D. C. Mabey, and R. L. Bailey. 2008. The natural history of trachoma infection and disease in a Gambian cohort with frequent follow-up. *PLoS Negl. Trop. Dis.* **2:**e341.

Grayston, J. T. 1992. Infections caused by *Chlamydia pneumoniae* strain TWAR. *Clin. Infect. Dis.* **15:**757–763.

Grayston, J. T., C.-C. Kuo, L. A. Campbell, and S.-P. Wang. 1989. *Chlamydia pneumoniae* sp. nov. for *Chlamydia* sp. strain TWAR. *Int. J. Syst. Bacteriol.* **39:**88–90.

Grayston, J. T., C.-C. Kuo, S.-P. Wang, and J. Altman. 1986. A new *Chlamydia psittaci* strain, TWAR, isolated in acute respiratory tract infections. *N. Engl. J. Med.* **315:**161–168.

Grieshaber, S. S., N. A. Grieshaber, N. Miller, and T. Hackstadt. 2006. *Chlamydia trachomatis* causes centrosomal defects resulting in chromosomal segregation abnormalities. *Traffic* **7:**940–949.

Haggerty, C. L., S. L. Gottlieb, B. D. Taylor, N. Low, F. Xu, and R. B. Ness. 2010. Risk of sequelae after *Chlamydia trachomatis* genital infection in women. *J. Infect. Dis.* **201:**S134–S155.

Hahn, D. L., R. W. Dodge, and R. Golubjatnikov. 1991. Association of *Chlamydia pneumoniae* (strain TWAR) infection with wheezing, asthmatic bronchitis, and adult-onset asthma. *JAMA* **266:**225–230.

Hammerschlag, M. 2000. *Chlamydia pneumoniae* and the lung. *Eur. Respir. J.* **16:**1001–1007.

Hammerschlag, M. R., K. Chirgwin, P. M. Roblin, M. Gelling, W. Dumornay, L. Mandel, P. Smith, and J. Schachter. 1992. Persistent infection with *Chlamydia pneumoniae* following acute respiratory illness. *Clin. Infect. Dis.* **14:**178–182.

Hammerschlag, M. R., S. A. Kohlhoff, and P. M. Apfalter. 2010. *Chlamydiophila (Chlamydia) pneumoniae*, p. 2467–2475. *In* G. L. Mandell, J. E. Bennett, and R. Dolin (ed.), *Principles and Practice of Infectious Diseases*, 7th ed., vol. 2. Churchill Livingstone Elsevier, Philadelphia, PA.

Hammerschlag, M. R., S. A. Kohlhoff, and T. Darville. 2009. *Chlamydia pneumoniae* and *Chlamydia trachomatis*, p. 27–52. *In* P. M. Fratamico, J. L. Smith, and K. A. Brogden (ed.), *Sequelae and Long-Term Consequences of Infectious Diseases*. ASM Press, Washington, DC.

Heijne, J. C. M., G. Tao, C. K. Kent, and N. Low. 2010. Uptake of regular *Chlamydia* testing by U.S. women: a longitudinal study. *Am. J. Prev. Med.* **39:**243–250.

Hoymans, V. Y., J. M. Bosmans, M. M. Ieven, and C. J. Vrints. 2007. *Chlamydia pneumonia*-based atherosclerosis: a smoking gun. *Acta Cardiol.* **62:**565–571.

Hu, V. H., E. M. Harding-Esch, M. J. Burton, R. L. Bailey, J. Kadimpeul, and D. C. W. Mabey. 2010. Epidemiology and control of trachoma: systematic review. *Trop. Med. Int. Health* **15:**673–691.

Humphreys, T. L., L. Li, X. Li, D. M. Janowicz, K. R. Fortney, Q. Zhao, W. Li, J. McClintick, B. P. Katz, D. S. Wilkes, H. J. Edenberg, and S. M. Spinola. 2007. Dysregulated immune profiles for skin and dendritic cells are associated with increased host susceptibility to *Haemophilus ducreyi* infection in human volunteers. *Infect. Immun.* **75:**5686–5697.

Hyman, C. L., P. M. Roblin, C. A. Gaydos, T. C. Quinn, J. Schachter, and M. R. Hammerschlag. 1995. Prevalence of asymptomatic carriage of *Chlamydia pneumoniae* in subjectively healthy adults: assessment by polymerase chain reaction-enzyme immunoassay and culture. *Clin. Infect. Dis.* **20:**1174–1178.

Johnson, K. A., M. Tan, and C. Sutterlin. 2009. Centrosome abnormalities during a *Chlamydia trachomatis* infection are caused by dysregulation of the normal duplication pathway. *Cell. Microbiol.* **11:**1064–1073.

Jones, R. B., J. B. Mammel, M. K. Shepard, and R. R. Fisher. 1986. Recovery of *Chlamydia trachomatis* from the endometrium of women at risk for chlamydial infection. *Am. J. Obstet. Gynecol.* **155:**35–39.

Joyner, J. L., J. M. J. Douglas, M. Foster, and F. N. Judson. 2002. Persistence of *Chlamydia trachomatis* infection detected by polymerase chain reaction in untreated patients. *Sex. Transm. Dis.* **29:**196–200.

Kleemola, M., P. Saikku, R. Visakorpi, S.-P. Wang, and J. T. Grayston. 1988. Epidemics of pneumonia caused by TWAR, a new *Chlamydia* organism, in military trainees in Finland. *J. Infect. Dis.* **157:**230–236.

Knowlton, A. E., H. M. Brown, T. S. Richards, L. A. Andreolas, R. K. Patel, and S. S. Grieshaber. 2011. *Chlamydia trachomatis* infection causes mitotic spindle pole defects independently from its effects on centrosome amplification. *Traffic* **12:**854–866.

Kumar, S., and M. R. Hammerschlag. 2007. Acute respiratory infection due to *Chlamydia pneumoniae*: current status of diagnostic methods. *Clin. Infect. Dis.* **44:**568–576.

Kuo, C.-C., H. H. Chen, S.-P. Wang, and J. T. Grayston. 1986. Characterization of TWAR strains, a new group of *Chlamydia psittaci*, p. 321–324. *In* D. Oriel, G. Ridgway, J. Schachter, D. Taylor-Robinson, and M. Ward (ed.), *Chlamydial Infections: Proceedings of the Sixth International Symposium on Human Chlamydial Infections*. Cambridge University Press, Cambridge, England.

Kutlin, A., C. Flegg, D. Stenzel, T. Reznik, P. M. Roblin, S. Mathews, P. Timms, and M. R. Hammerschlag. 2001. Ultrastructural study of *Chlamydia pneumoniae* in a continuous infection model. *J. Clin. Microbiol.* **39:**3721–3723.

Kutlin, A., P. M. Roblin, and M. R. Hammerschlag. 2002. Effect of prolonged treatment with azithromycin, clarithromycin, and levofloxacin on *Chlamydia pneumoniae* in a continuous infection model. *Antimicrob. Agents Chemother.* **46:**409–412.

Lau, C.-Y., and A. K. Qureshi. 2002. Azithromycin versus doxycycline for genital chlamydial infections. *Sex. Transm. Dis.* **29:**497–502.

Lauhio, A., M. Leirisalo-Repo, J. Lähdevirta, P. Saikku, and H. Repo. 1991. Double-blind, placebo-controlled study of three-month treatment with lymecycline in reactive arthritis, with special reference to *Chlamydia* arthritis. *Arthritis Rheum.* **34:**6–14.

Lehtinen, M., K. A. Ault, E. Lyytikainen, J. Dillner, S. M. Garland, D. G. Ferris, L. A. Koutsky, H. L. Sings, S. Lu, R. M. Haupt, J. Paavonen, et al. 2011. *Chlamydia trachomatis* infection and risk of cervical intraepithelial neoplasia. *Sex. Transm. Infect.* **87:**372–376.

Littman, A. J., L. A. Jackson, E. White, M. D. Thornquist, C. A. Gaydos, and T. L. Vaughan. 2004. Interlaboratory reliability of microimmunofluorescence test for measurement of *Chlamydia pneumoniae*-specific immunoglobulin A and G antibody titers. *Clin. Diagn. Lab. Immunol.* **11:**615–617.

Lyons, J. M., S. A. Morre, and J. A. Land. 2009. Aspects of reproductive biology that influence the distribution and spread of *Chlamydia trachomatis* within the female genital tract: a new paradigm. *Drugs Today* **45:**119–124.

Mandell, L. A., R. G. Wunderink, A. Anzueto, J. G. Bartlett, G. D. Campbell, N. C. Dean, S. F. Dowell, T. M. File, D. M. Musher, M. S. Niederman, A. Torres, and C. G. Whitney. 2007. Infectious Diseases Society of America/American Thoracic society consensus guidelines on the management of community-acquired pneumonia in adults. *Clin. Infect. Dis.* **44:**S27–S72.

Mariotti, S. P., D. Pascolini, and J. Rose-Nussbaumer. 2009. Trachoma: global magnitude of a preventable cause of blindness. *Br. J. Ophthalmol.* **93:**563–568.

Martin, D. H. 2008. Urethritis in males, p. 1107–1126. *In* K. K. Holmes, P. F. Sparling, W. E. Stamm, P. Piot, J. N. Wasserheit, L. Corey, M. S. Cohen, and D. H. Watts (ed.), *Sexually Transmitted Diseases*, 4th ed. McGraw Hill Medical, New York, NY.

Meyers, D. S., H. Halvorson, and S. Luckhaupt. 2007. Screening for chlamydial infection: an evidence update for the U.S. Preventive Services Task Force. *Ann. Intern. Med.* **147:**135–142.

Miller, W. C. 2008. Epidemiology of chlamydial infection: are we losing ground? *Sex. Transm. Infect.* **84:**82–86.

Miller, W. C., and E. M. Ko. 2011. Chlamydial infection and risk of cervical neoplasia: the challenge is in the study design. *Sex. Transm. Infect.* **87:**366–367.

Misyurina, O. Y., E. V. Chipitsyna, Y. P. Finashutina, V. N. Lazarev, T. A. Akopian, A. M. Savicheva, and V. M. Govorun. 2004. Mutations in a 23S rRNA gene of *Chlamydia trachomatis* associated with resistance to macrolides. *Antimicrob. Agents Chemother.* **48:**1347–1349.

Molano, M., C. J. L. M. Meijer, E. Weiderpass, A. Arslan, H. Posso, S. Franceschi, M. Ronderos, N. Munoz, and A. J. C. van den Brule. 2005. The natural course of *Chlamydia* *trachomatis* infection in asymptomatic Colombian women: a 5-year follow-up study. *J. Infect. Dis.* **191:**907–916.

Morre, S. A., A. J. C. van den Brule, L. Rozendaal, A. J. P. Boeke, F. J. Voorhorst, S. de Blok, and C. J. L. M. Meijer. 2002. The natural course of asymptomatic *Chlamydia trachomatis* infections: 45% clearance and no development of clinical PID after one-year follow-up. *Int. J. STD AIDS* **13:**12–18.

Ngondi, J., F. Ole-Sempele, A. Onsarigo, I. Matende, S. Baba, M. Reacher, F. Matthews, C. Brayne, and P. Emerson. 2006. Blinding trachoma in postconflict southern Sudan. *PLoS Med.* **3:**e478.

Oakeshott, P., S. Kerry, A. Aghaizu, H. Atherton, S. Hay, D. Taylor-Robinson, I. Simms, and P. Hay. 2010. Randomised controlled trial of screening for *Chlamydia trachomatis* to prevent pelvic inflammatory disease: the POPI (prevention of pelvic infection) trial. *BMJ* **340:**c1642.

Paavonen, J., L. Westrom, and D. Eschenbach. 2008. Pelvic inflammatory disease, p. 1017–1050. *In* K. K. Holmes, P. F. Sparling, W. E. Stamm, P. Piot, J. N. Wasserheit, L. Corey, M. S. Cohen, and D. H. Watts (ed.), *Sexually Transmitted Diseases*, 4th ed. McGraw Hill, New York, NY.

Quinn, T. C., S. E. Goodell, E. Mkrtichian, M. D. Schuffler, S.-P. Wang, W. E. Stamm, and K. K. Holmes. 1981. *Chlamydia trachomatis* proctitis. *N. Engl. J. Med.* **305:**195–200.

Rank, R. G., and M. M. Sanders. 1992. Pathogenesis of endometritis and salpingitis in a guinea pig model of chlamydial genital infection. *Am. J. Pathol.* **140:**927–936.

Rank, R. G., and J. A. Whittum-Hudson. 2010. Protective immunity to chlamydial genital infection: evidence from animal studies. *J. Infect. Dis.* **201:**S168–S177.

Resnikoff, S., D. Pascolini, D. Etya'ale, I. Kocur, R. Pararajasegaram, G. P. Pokharel, and S. P. Mariotti. 2004. Global data on visual impairment in the year 2002. *Bull. W. H. O.* **82:**844–851.

Safaeian, M., K. Quint, M. Schiffman, A. C. Rodriguez, S. Wacholder, R. Herrero, A. Hildesheim, R. P. Viscidi, W. Quint, and R. D. Burk. 2010. Chlamydia trachomatis and risk of prevalent and incident cervical premalignancy in a population-based cohort. *JNCI* **102:**1794–1804.

Saikku, P., M. Leinonen, K. Mattila, M. R. Ekman, M. S. Nieminen, P. H. Makela, J. K. Huttunen, and V. Valtonen. 1988. Serological evidence of an association of a novel *Chlamydia*, TWAR, with chronic coronary heart disease and acute myocardial infarction. *Lancet* **ii:**983–986.

Saikku, P., S.-P. Wang, M. Kleemola, E. Brander, E. Rusanen, and J. T. Grayston. 1985. An epidemic of mild pneumonia due to an unusual strain of *Chlamydia psittaci*. *J. Infect. Dis.* **151:**832–839.

Scholes, D., A. Stergachis, F. E. Heidrich, H. Andrilla, K. K. Holmes, and W. E. Stamm. 1996. Prevention of pelvic inflammatory disease by screening for cervical chlamydial infections. *N. Engl. J. Med.* **334:**1362–1366.

Sheffield, J. S., W. W. Andrews, M. A. Klebanoff, C. MacPherson, J. C. Carey, J. M. Ernest, R. J. Wapner, W. Trout, A. Moawad, M. Miodovnik, B. Sibai, M. W. Varner, S. N. Caritis, M. Dombrowski, O. Langer, M. J. O'Sullivan, et al. 2005. Spontaneous resolution of asymptomatic *Chlamydia trachomatis* in pregnancy. *Obstet. Gynecol.* **105:**557–562.

Shew, M. L., J. D. Fortenberry, W. Tu, B. E. Juliar, B. E. Batteiger, B. Qadadri, and D. R. Brown. 2006. Association of condom use, sexual behaviors and sexually transmitted infections with the duration of genital human papillomavirus infection among adolescent women. *Arch. Pediatr. Adolesc. Med.* **160:**151–156.

Solomon, A. W., M. J. Holland, M. J. Burton, S. K. West, N. D. E. Alexander, A. Aguirre, P. A. Massae, H. Mkocha, B. Muñoz, G. J. Johnson, R. W. Peeling, R. L. Bailey, A. Foster, and D. C. W. Mabey. 2003. Strategies for control of trachoma: observational study with quantitative PCR. *Lancet* **362:**198–204.

Somani, J., V. B. Bhullar, K. A. Workowski, C. E. Farshy, and C. M. Black. 2000. Multiple drug-resistant *Chlamydia trachomatis* associated with clinical treatment failure. *J. Infect. Dis.* **181:**1421–1427.

Spaargaren, J., H. A. S. Fennema, and S. A. Morre. 2005. New lymphogranuloma venereum *Chlamydia trachomatis* variant Amsterdam. *Emerg. Infect. Dis.* **11:**1090–1094.

Stamm, W. E. 2008. Lymphogranuloma venereum, p. 595–605. *In* K. K. Holmes, P. F. Sparling, W. E. Stamm, P. Piot, J. N. Wasserheit, L. Corey, M. S. Cohen, and D. H. Watts (ed.), *Sexually Transmitted Diseases*, 4th ed. McGraw-Hill Medical, New York, NY.

Stamm, W. E., and B. E. Batteiger. 2010. *Chlamydia trachomatis* (trachoma, perinatal infections, lymphogranuloma venereum, and other genital infections), p. 2443–2461. *In* G. L. Mandell, J. E. Bennett, and R. Dolin (ed.), *Principles and Practice of Infectious Diseases*, 7th ed., vol. 2. Churchill Livingstone Elsevier, Philadelphia, PA.

Stamm, W. E., B. E. Batteiger, W. M. McCormack, P. A. Totten, A. Sternlicht, N. M. Kivel, et al. 2007. A randomized, double-blind study comparing single-dose rifalazil with single-dose azithromycin for the empirical treatment of nongonococcal urethritis in men. *Sex. Transm. Dis.* **34:**545–552.

Stothard, D. R., G. A. Toth, and B. E. Batteiger. 2003. Polymorphic membrane protein H has evolved in parallel with the three disease-causing groups of *Chlamydia trachomatis*. *Infect. Immun.* **71:**1200–1208.

Su, H., R. Morrison, R. Messer, W. Whitmire, S. Hughes, and H. D. Caldwell. 1999. The effect of doxycycline treatment on the development of protective immunity in a murine model of chlamydial genital infection. *J. Infect. Dis.* **180:**1252–1258.

Thylefors, B., C. R. Dawson, B. R. Jones, S. K. West, and H. R. Taylor. 1987. A simple system for the assessment of trachoma and its complications. *Bull. W.H.O.* **65:**477–483.

Wardrop, S., A. Fowler, P. O'Callaghan, P. Giffard, and P. Timms. 1999. Characterization of the koala biovar of *Chlamydia pneumoniae* at four gene loci—*ompA*VD4, *ompB*, 16S rRNA, *groESL* spacer region. *Syst. Appl. Microbiol.* **22:**22–27.

Watson, C., and N. J. Alp. 2008. Role of *Chlamydia pneumoniae* in atherosclerosis. *Clin. Sci.* **114:**509–531.

West, S. K., B. Munoz, V. M. Turner, B. B. O. Mmbaga, and H. R. Taylor. 1991. The epidemiology of trachoma in central Tanzania. *Int. J. Epidemiol.* **20:**1088–1092.

Workowski, K. A., and S. Berman. 2010. Sexually transmitted diseases treatment guidelines, 2010. *MMWR Morb. Mortal. Wkly. Rep.* **59:**1–111.

DEEP AND WIDE: COMPARATIVE GENOMICS OF *CHLAMYDIA*

Garry S. A. Myers, Jonathan Crabtree, and Heather Huot Creasy

2

INTRODUCTION

Chlamydial biology continues to be poorly understood, particularly when compared to other major bacterial pathogens. This is a direct result of two limiting factors that still constrain *Chlamydia* research: chlamydiae cannot be grown outside the host cell, and there is no general method of genetic transformation for laboratory manipulation. These limitations continue to render *Chlamydia* intractable to most molecular biology techniques. In contrast, the small genome size (~1 Mbp), unremarkable G+C% content (39 to 41%), and relatively few repeat elements make the chlamydial chromosome and plasmid typically straightforward to sequence and assemble using whole-genome shotgun (WGS) sequencing. In this context, WGS sequencing has become a key methodology for the field. Indeed, the data from high-throughput sequencing-based approaches have been the underpinning of many of the key advances in *Chlamydia* research since the publication of the first chlamydial genome sequence in 1998 (Stephens et al., 1998).

Following the initial publication of the genome sequence of *Chlamydia trachomatis* serovar D in 1998, global sequencing efforts soon produced genomes for *Chlamydia pneumoniae* (Kalman et al., 1999; Read et al., 2000), *Chlamydia muridarum* (Read et al., 2000), *Chlamydia caviae* (Read et al., 2003), *Chlamydia abortus* (Thomson et al., 2005), and *Chlamydia felis* (Azuma et al., 2006). These sequences were generated by the WGS sequencing approach using Sanger dideoxy terminator chemistry, as pioneered by The Institute for Genomic Research (TIGR) (Fleischmann et al., 1995). In addition, genome sequences from members of the broader order *Chlamydiales* (i.e., including the environmental chlamydiae) have been published in recent years (Horn et al., 2004; Bertelli et al., 2010; Collingro et al., 2011), lending broad insight into this evolutionarily distinct order. Here we will focus on comparative genomics of *Chlamydia* spp., which are well-known human and animal pathogens that constitute the family *Chlamydiaceae*.

Second-generation sequencing platforms have now superseded the original sequencing technology based on Sanger chemistry, particularly for high-throughput applications such as WGS sequencing. There are three second-generation sequencing platforms in broad usage

Garry S. A. Myers, Jonathan Crabtree, and Heather Huot Creasy, Institute for Genome Sciences, University of Maryland School of Medicine, Baltimore, MD 21201.

Intracellular Pathogens I: Chlamydiales
Edited by Ming Tan and Patrik M. Bavoil © 2012 ASM Press, Washington, DC
10.1128/9781555817329.ch2

at this time: the Roche 454 Titanium instrument, Illumina GAIIx or HiSeq2000 instruments, and the ABI SOLiD instrument. There are advantages and disadvantages associated with each of these platforms (see Table 1 for a comparison of a selection of such instruments); in general, error rates in the second-generation instruments are typically higher than for Sanger chemistry. Read lengths are significantly reduced, which has required changes in the assembly algorithms and approaches developed for Sanger reads. However, the key advantage of all new sequencing technologies is the large number of sequence reads obtainable per run, enabling substantial coverage depth and resulting in a large reduction of sequencing costs. This allows small bacterial genomes, such as the 1-Mbp chromosome of *Chlamydia*, to be determined faster and less expensively than ever. Third-generation and future sequencing technologies should continue this trend, with even greater sequence coverage, longer sequence reads, and reduction of error rates, and will continue to make WGS sequencing more accessible to research groups.

The depth of sequence read coverage and the cost advantage conferred by second- and third-generation instruments come with a downside. Manual closure processes that allowed validation of genome assemblies and associated insertions and deletions (indels), rearrangements, and single nucleotide polymorphisms (SNPs) were an integral part of the original WGSS paradigm developed using longer Sanger sequence reads. These time-consuming and expensive but highly useful processes (Fraser et al., 2002) are typically no longer performed and have been traded instead for the speed, sequencing depth, and significant cost savings offered by second- and third-generation sequencing pipelines. Thus, in the absence of such manual closure and validation processes and irrespective of coverage depth, genome sequences produced on second- and third-generation instruments should be considered incomplete draft assemblies.

For many genome-scale applications, this is a minor issue that is offset by the gains in throughput, coverage, and cost. However, comparative analyses of very similar strains, such as the highly syntenic *Chlamydia*, often coalesce around collections of SNPs and indels. Many of the SNPs and indels identified within draft assemblies are often artifacts or errors inherent to the sequencing chemistry or the assembly algorithm used, combined with characteristics inherent to the sequence (for example, homopolymeric regions are often "collapsed" by 454 pyrosequencing). Furthermore, there is increasing evidence that in vivo- and in vitro-grown *Chlamydia* populations are not genetically homogenous, existing as pools of variants. Sequencing of such population pools using standard consensus-based assembly approaches summarizes valid genetic variation by majority rule into a final genomic

TABLE 1 Key parameters of selected second- and third-generation sequencing technologies, compared to first-generation Sanger sequencing

Platform name	Generation	Primary error type	Raw error rate (%)	Run time	Avg read length (bp)	No. of reads/run	Mbp/run	$/Mbp (raw)
Sanger (ABI 3730xl)	1	Balanced	0.001	2 h	~1,000	96	0.1	500–2,500
454 Titanium	2	Indel	0.5	10 h	>400	>1 M	300–500	15–30
Illumina GAIIx	2	Substitution	1	3–12 days	36–100	100–200 M	5,000–40,000	<0.50
Illumina HiSeq	2	Substitution	1.5–2.0	8+days	50–100	1.5–2.5B	200,000–300,000	<0.20
PacBio RS (beta)	3	Indel	~16	40 min	~1,000	~50 K	>30	5–10

TABLE 2 Genome features of current publicly available *Chlamydiaceae* genomes

Organism	GenBank accession no.	Chromosome size (bp)	No. of CDSs	Avg gene size (bp)	%GC	Plasmid reported?
Chlamydia trachomatis						
C. trachomatis 434/Bu	AM884176.1	1,038,842	874	351	41	No
C. trachomatis A/HAR-13	CP000051.1	1,044,459	919	345	41	Yes
C. trachomatis B/Jali20/OT	FM872308.1	1,044,352	875	351	41	Yes
C. trachomatis B/TZ1A828/OT	FM872307.1	1,044,282	880	351	41	No
C. trachomatis D-EC	CP002052.1	1,042,522	878	358	41	Yes
C. trachomatis D-LC	CP002054.1	1,042,520	878	358	41	Yes
C. trachomatis D/UW-3/CX	AE001273.1	1,042,519	894	350	41	No
C. trachomatis E/11023	CP001890.1	1,043,025	926	337	41	No
C. trachomatis E/150	CP001886.1	1,042,996	927	336	41	No
C. trachomatis G/11074	CP001889.1	1,042,875	919	340	41	No
C. trachomatis G/11222	CP001888.1	1,042,354	927	336	41	No
C. trachomatis G/9301	CP001930.1	1,042,811	921	340	41	No
C. trachomatis G/9768	CP001887.1	1,042,810	920	340	41	No
C. trachomatis L2b/UCH-1/ proctitis	AM884177.1	1,038,869	874	352	41	No
C. trachomatis Sweden2	FN652779.2	1,042,839	875	353	41	No
C. trachomatis 6276	NZ_ABYD01000001.1	1,043,181	920	340	41	No
C. trachomatis 6276s	NZ_ABYE01000001.1	1,043,182	922	339	41	No
C. trachomatis 70	NZ_ABYF01000001.1	1,048,006	923	337	41	No
C. trachomatis 70s	NZ_ABYG01000001.1	1,046,064	919	337	41	No
C. trachomatis D(s)2923	NZ_ACFJ01000001.1	1,042,757	932	335	41	No
Chlamydia pneumoniae						
C. pneumoniae AR39	AE002161.1	1,229,853	1,116	321	40	No
C. pneumoniae CWL029	AE001363.1	1,230,230	1,052	332	40	No
C. pneumoniae J138	BA000008.3	1,226,565	1,069	344	40	No
C. pneumoniae LPCoLN	CP001713.1	1,241,020	1,105	329	40	Yes
C. pneumoniae TW-183	AE009440.1	1,225,935	1,113	323	40	No
Chlamydia psittaci						
C. psittaci 6BC	CP002586.1	1,171,667	1,009	351	39	Yes
C. psittaci RD1	FQ482149.1	1,164,076	951	371	39	Yes
C. psittaci Cal10	AEZD00000000.1	1,169,283	1,005	350	39	Yes
Chlamydia pecorum						
C. pecorum E58	CP002608.1	1,106,197	988	344	40	No
Chlamydia caviae						
C. caviae GPIC	AE015925.1	1,173,390	1,005	352	39	Yes
Chlamydia abortus						
C. abortus S26/3	CR848038.1	1,144,377	932	358	39	No
Chlamydia felis						
C. felis FEC56	AP006861.1	1,166,239	1,013.00	353	39	Yes
Chlamydia muridarum						
C. muridarum Nigg	AE002160.2	1,072,950	911	358	40	Yes

TABLE 3 Pangenome analysis of the *Chlamydiaceae* across representative species of all sequenced *Chlamydiaceae* and within each *Chlamydiaceae* species that is represented by multiple genome sequences[a]

Organism (no. of predicted core peptides)	No. of predicted CDSs	No. of variable peptides	No. of unique peptides
Chlamydiaceae[b] (668)			
Chlamydia abortus S26/3	932	246	8
Chlamydia felis FEC56	1,013	259	48
Chlamydia muridarum Nigg	911	184	59
Chlamydia trachomatis D/UW-3/CX (reference)	894	187	41
Chlamydia pneumoniae AR39	1,116	201	282
Chlamydia psittaci 6BC	1,009	282	28
Chlamydia pecorum E58	988	171	138
Chlamydia caviae GPIC	1,005	271	44
Chlamydia trachomatis[b] (816)			
Chlamydia trachomatis 434/Bu	874	61	1
Chlamydia trachomatis A/HAR-13	919	91	3
Chlamydia trachomatis B/Jali20/OT	875	73	1
Chlamydia trachomatis B/TZ1A828/OT	880	76	0
Chlamydia trachomatis D-EC	878	56	0
Chlamydia trachomatis D-LC	878	55	1
Chlamydia trachomatis D/UW-3/CX (reference)	894	80	0
Chlam*ydia trachomatis* E/11023	926	104	0
Chlamydia trachomatis E/150	927	106	0
Chlamydia trachomatis G/11074	919	101	0
Chlamydia trachomatis G/11222	927	107	1
Chlamydia trachomatis G/9301	921	103	0
Chlamydia trachomatis G/9768	920	102	0
Chlamydia trachomatis L2b/UCH-1/proctitis	874	62	0
Chlamydia trachomatis Sweden2	875	73	2
Chlamydia trachomatis 6276	920	101	0
Chlamydia trachomatis 6276s	922	103	0
Chlamydia trachomatis 70	923	100	1
Chlamydia trachomatis 70s	919	98	0
Chlamydia trachomatis D(s)2923	932	108	1
Chlamydia pneumoniae (1,010)			
Chlamydia pneumoniae AR39 (reference)	1,116	75	36
Chlamydia pneumoniae CWL029	1,052	75	24
Chlamydia pneumoniae J138	1,069	47	3
Chlamydia pneumoniae LPCoLN	1,105	60	97
Chlamydia pneumoniae TW-183	1,113	74	73
Chlamydia psittaci and closely related species			
Chlamydia psittaci 6BC (reference) (771)	1,009	219	16
Chlamydia psittaci RD1 (771)	951	178	5
Chlamydia psittaci Cal10 (771)	1,005	205	5
Chlamydia pecorum E58[c] (314)	988	511	163
Chlamydia caviae GPIC[c] (315)	1,005	638	62

sequence that actually obscures the underlying diversity.

These effects can often be mitigated by a variety of approaches: by using a closed and validated genome assembly as the reference for comparison; by eliminating subsets of SNPs/indels based on knowledge of the error type and rate of the sequencing chemistry (Table 1); by using a mixture of reads obtained from different platforms to offset the limitations of each; or by validating SNPs/indels through targeted resequencing or other assays. Nevertheless, as with all sequence data in public databases, the limitations incurred by the incomplete nature of draft genome sequences should be kept in mind.

When this chapter was prepared, there were 33 chlamydial genome sequences produced by WGS sequencing that are publicly available as complete or draft genomes in GenBank (Table 2). Here we present the first comparative analysis of all 33 genomes, encapsulating the currently known genomic diversity of *Chlamydia*. From this analysis, we suggest priorities for further wide and deep sequencing. We define wide sequencing as exhaustive WGS sequencing of the diversity across chlamydial species and isolates. Conversely, deep sequencing is the application of WGS sequencing to a defined genomic region, resulting in extremely high sequence read coverage. This high level of coverage allows the detection of lower frequency mutations that would not be seen or would be ignored when generating a consensus sequence. In this case, the genomic region is the entire chlamydial chromosome, and low-frequency mutations would represent variants within a putative heterogeneous chlamydial population. Both in vivo and in vitro chlamydial population structures could be discerned by deep sequencing. Finally we identify future applications of high-throughput sequencing that will lend further insight into this unique family of bacterial pathogens.

THE PANGENOME OF THE *CHLAMYDIACEAE*

The available genomes represent all but one of the known *Chlamydiaceae* species. Only *Chlamydia suis* is not yet represented by a publicly available genome sequence, although *C. suis* is highly similar to *C. trachomatis* and *C. muridarum* (G. S. A. Myers and M. Donati, unpublished data). Taking advantage of the high degree of genomic synteny, we did a BLAST score ratio (BSR) analysis (Rasko et al., 2005) across the *Chlamydiaceae*. This approach uses the complete proteomes predicted from the genomes to examine the level of conservation among the current set of sequenced isolates. This is the first pangenome-scale view of the pathogenic *Chlamydiaceae*, identifying the core, variable, and unique predicted peptides for each sequenced strain (Table 3).

We first examined the *Chlamydiaceae* pangenome by a BSR analysis using a single representative of each of the eight sequenced species (Table 3). In accordance with the observed genome synteny, nearly two-thirds of all chlamydial proteins (n = 668) are shared over the diversity of the *Chlamydiaceae* and represent the core predicted peptides for the family. As noted elsewhere (Read et al., 2000), this degree of conservation may be explained by evolutionary restrictions enforced by the intracellular niche and the conserved chlamydial developmental cycle. As is observed across virtually all bacterial genomes, a significant number of predicted coding sequences (CDSs) encode products with no known or assignable function. Many of the predicted proteins within the core set are annotated as hypothetical; these likely represent novel

---

[a]The BSR approach was used to identify the core (>0.8 BSR), variable (between 0.8 and 0.4 BSR), and unique (<0.4 BSR) predicted peptides.

[b]For these organisms, the core predicted peptides were calculated using *C. trachomatis* D/UW-3/CX as the reference genome. Variable and unique predicted peptides were calculated using each genome as the reference against all others in the set (thus, core plus unique plus variable do not add up to the total predicted).

[c]The count of core predicted peptides for *C. pecorum* and *C. caviae* was calculated using *C. pecorum* and *C. caviae* genomes as the respective reference against all others in the set rather than *C. psittaci*.

biological functionality that is conserved across the *Chlamydiaceae*. Perhaps more significantly, hypothetical proteins (n = 139) are prominent within the variable and unique sets of predicted peptides in the *Chlamydiaceae* pangenome. Many of these uncharacterized proteins are likely to be key differentiators of the species within the *Chlamydiaceae* and are thus targets for further characterization of host and tissue tropisms and disease outcomes.

A range of chlamydial CDSs encode proteins that are known virulence factors; many of these are represented in the variable BSR categories. Using 63 key virulence factors (Collingro et al., 2011), we retrieved the BSR metrics for each factor across the *Chlamydiaceae*. These metrics were subjected to hierarchical clustering using MeV (Pearson correlation with average linkage) (Saeed et al., 2006) to generate a profile of the known *Chlamydiaceae* virulence proteins, based on protein similarity relative to *C. trachomatis* D (Fig. 1). Despite the genomic synteny and overall similarity exhibited by the *Chlamydiaceae*, this global analysis highlights the degree of sequence variation seen within these well-known virulence factors. The species clustering is consistent with established phylogeny. The Inc family of proteins is particularly variable across the *Chlamydiaceae*, with many Inc orthologs distinctly divergent or absent outside the *C. trachomatis* genome set. Indeed, virtually all of the secreted effectors identified in *C. trachomatis* are more divergent in other species than are the other selected virulence factors. As much of the experimental work on pathogenesis and virulence has been performed using *C. trachomatis*, this may be an artifact of selection bias. This selection bias is likely to be exacerbated by the high proportion of *C. trachomatis* genomes (20 of 33) used in this analysis. Investigation of species-specific virulence factors within the variable and unique protein sets is likely to be very informative.

We next examined the pangenome of each chlamydial species for which multiple genome sequences are currently publicly available: *C. trachomatis* (n = 20), *C. pneumoniae* (n = 5), and *C. psittaci* (n = 3) (Table 3). We also included *C. pecorum* and *C. caviae* in the *C. psittaci* analysis, as these are phylogenetically close to *C. psittaci*. As expected by the increase in genomic synteny when comparing within a species, the *C. trachomatis* core predicted peptides represent nearly 90% of all *C. trachomatis* peptides with very few unique peptides. These unique peptides (n = 11 over the 20 genomes) are encoded by pseudogenes or small hypothetical CDSs that may represent minor variations in CDS calling algorithms. The remaining ~10% of *C. trachomatis* peptides are defined as variable; a significant number of these are encoded by hypothetical CDSs with no characterized function (n = 140). Much of the phenotypic variation that is observed between *C. trachomatis* strains may be encapsulated within these uncharacterized proteins. Other variable peptides identified among these *C. trachomatis* strains are well known, including members of the Inc and Pmp families, or are proteins involved in metabolic processes, including tryptophan synthesis and ubiquinone reduction through succinate and fumarate oxidation. The now-familiar pattern of high representation of proteins encoded by uncharacterized hypothetical CDSs within the variable and unique sets of the pangenome is repeated in the *C. pneumoniae* (n = 445) and *C. psittaci* (n = 158) comparisons. Thus, much novel genetic diversity that is likely to differentiate host and cell tropism and pathogenic processes remains to be characterized at all taxonomic levels of the *Chlamydiaceae*.

THE CHLAMYDIAL PLASTICITY ZONE

Early comparative genomic analyses of available *Chlamydia* genomes demonstrated the high degree of similarity that is now recognized as a hallmark of the genus (Kalman et al., 1999; Read et al., 2000, 2003; Shirai et al., 2000; Thomson et al., 2005; Azuma et al., 2006). This similarity is expressed both in terms of genomic synteny, whereby the order of the genes along the chromosome is conserved, and in terms of gene content. The degree of synteny and

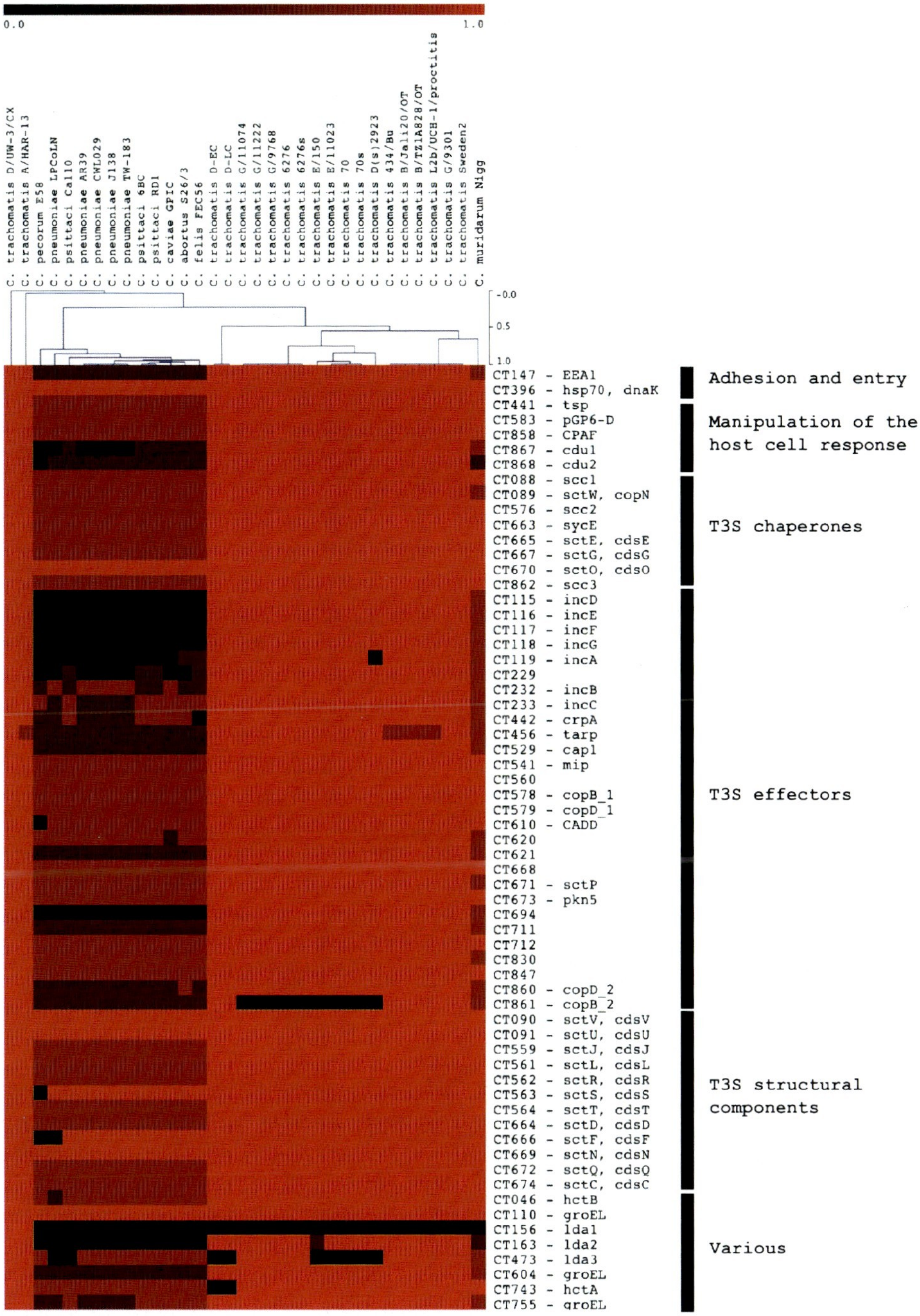

FIGURE 1 Blast Score Ratio similarity profile of selected known virulence factors across the *Chlamydiaceae*. The BSR for each protein was scored from 0.0 (black; least similar) to 1.0 (red; most similar) and hierarchically clustered by species (Pearson correlation with average linkage). doi:10.1128/9781555817329.ch2.f1

content also seems to correlate to the extent of evolutionary divergence observed. A key motivation for additional *Chlamydia* genome sequencing has been to understand how a pathogen with such a conserved and syntenic chromosome is able to infect a broad range of hosts, tissues, and cell types, causing a diversity of disease outcomes. As described in the BSR analysis above, the accumulation of genome sequences is now starting to allow insight into the complexities of chlamydial pathogenesis and host and niche specificity by enabling the exact genetic differences between all of the isolates to be discovered.

A significant amount of the variation across *Chlamydia* species is found within the plasticity zone (PZ) (Read et al., 2000) at the replication terminus of the genome (Fig. 2). Large-scale rearrangements around the origin and terminus of replications are a common feature of many bacterial genomes (Eisen et al., 2000). In *Chlamydia*, this heterogeneous region includes an array of putative chlamydial virulence factors that may play a role in host tropism or niche specificity. The size of the PZ varies from 18 to 81 kb between different *Chlamydia* species, with the *C. abortus* PZ being smallest and the *C. muridarum* PZ being the largest, with associated variability in content (Fig. 2). The PZ is typically bounded by *accBC* and *guaBA*, which encode components of acetyl coenzyme A carboxylase and GMP synthesis, respectively. Both *guaB* and *guaA* are truncated in *C. abortus* and absent in the koala isolate of *C. pneumoniae*. *guaB* is truncated in the human isolates of *C. pneumoniae*.

Another PZ denizen with significant length polymorphism across all chlamydial species is a CDS containing a domain with significant similarity to the membrane attack complex/perforin protein (MACPF) (Ponting, 1999). Proteins encoding the MACPF domain in *Toxoplasma* and *Plasmodium* have been shown to form pores that enable parasitic lysis of the host cell. The chlamydial MACPF is encoded by the koala isolate of *C. pneumoniae* (LPCoLN), *C. psittaci* 6BC, *C. pecorum*, and *C. trachomatis* serovar D (Fig. 2). Truncated forms of MACPF are encoded by *C. psittaci* 6BC, which thus has two copies of MACPF, *C. abortus*, and the human *C. pneumoniae* isolates. MACPF is absent from the PZ of *C. caviae* and *C. felis*. The precise role of MACPF in chlamydial pathogenesis has not yet been delineated but is likely to interact with membranes (Taylor et al., 2010). In addition, MACPF has been shown to be expressed in *C. trachomatis*, is subjected to proteolytic processing, and has been localized to subsets of RBs within the inclusion (Taylor et al., 2010). The degree of heterogeneity observed here across the *Chlamydiaceae* further supports MACPF as a key mediator of niche specificity between chlamydial species.

Other key virulence-associated features found within the PZ of some chlamydial species are the tryptophan (*trp*) operon and phospholipase D (*pld*). Tryptophan synthesis has been proposed to be important for the virulence of *C. trachomatis* genital strains but not the ocular strains (Caldwell et al., 2003). *C. trachomatis* strains possess a truncated *trp* operon (*trpRBA*) located within the PZ (McClarty et al., 2007) (Fig. 2). A tryptophan operon is absent in *C. muridarum* and *C. abortus* and is also missing in all sequenced strains of *C. pneumoniae*. In contrast, a full tryptophan synthesis repertoire (*trpRBACD*) is found in *C. caviae*, *C. felis*, and *C. pecorum*, although only *C. caviae* has the *trp* operon within the PZ. Phospholipase D, linked to MACPF function, is proposed to play a role in chlamydial survival late in the developmental cycle through involvement in lipid modification or metabolism and has been shown to localize to lipid-rich structures near the chlamydial inclusion (Taylor et al., 2010). *C. trachomatis* serovar D possesses three copies of *pld* within the PZ.

One of the most significant features of the chlamydial PZ revealed by this comparative analysis is the heterogeneity of the chlamydial cytotoxin (*tox*) across the *Chlamydiaceae* (Fig. 2). The chlamydial cytotoxin is a protein with high sequence similarity to the large clostridial cytotoxins (LCTs) (Belland et al., 2001; Read et al., 2003). No orthologs with similarity to

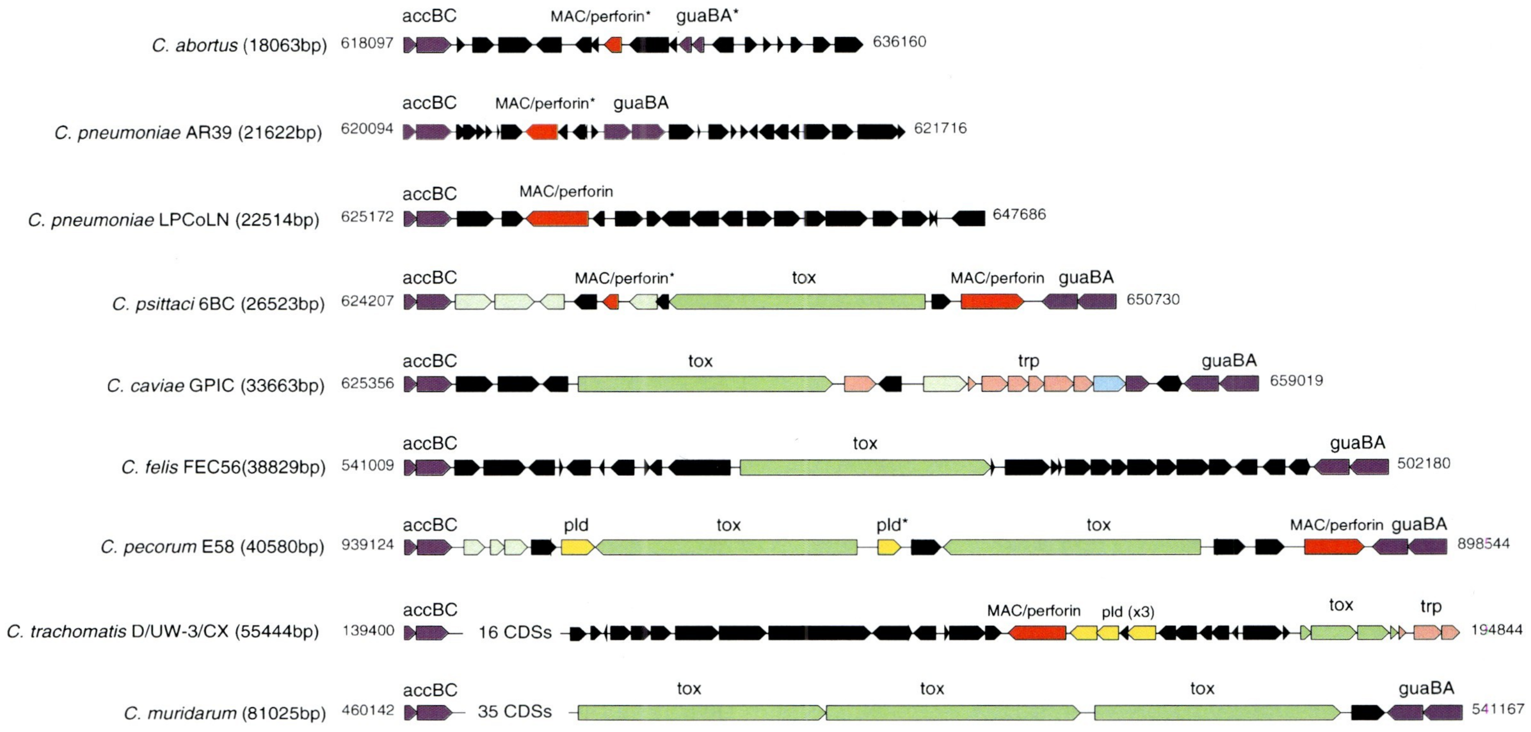

FIGURE 2 Comparison of the chlamydial plasticity zone across nine representative genomes. Regions are ordered by size, with selected coding sequences highlighted in color. Asterisks denote selected genes with evidence of truncation or decay. doi:10.1128/9781555817329.ch2.f2

tox are found in *C. abortus* or any of the *C. pneumoniae* genomes. A single full-length copy is found in the PZ of *C. psittaci*, *C. caviae*, and *C. felis*. The PZ of *C. pecorum* contains two full-length copies, while *C. muridarum* contains three. Truncated CDSs matching portions of the N- and C-terminal regions of *tox* are found within *C. trachomatis* and are transcribed (Belland et al., 2001; Carlson et al., 2004).

The full-length *tox* orthologs and one of the truncated *tox* fragments (*ct166*) from *C. trachomatis* serovar D contain glycosyltransferase and UDP-glucose binding domains that are present in LCTs (Belland et al., 2001). The glycosyltransferase activity of the LCT monoglycosylates and thus functionally inactivates Rho GTPases, leading to disassembly of the actin cytoskeleton with distinct associated cytopathic effects. A central hydrophobic region in LCTs, identified as a transmembrane segment likely involved in translocation into the cytosol, is also present in the full-length *tox* genes (Read et al., 2003). In contrast, the chlamydial *tox* genes do not contain the C-terminal repeats of LCTs that are involved with cell surface receptor binding (Belland et al., 2001; Read et al., 2003; Carlson et al., 2004). Analysis of the cytotoxicity of *C. muridarum* and *C. trachomatis* has previously demonstrated cytopathic effects in epithelial cells indistinguishable from that of LCTs from *Clostridium* and thus possibly attributable to the *tox* genes (Belland et al., 2001). Interaction with an intact host actin cytoskeleton is a known requirement for entry for chlamydial invasion, with substantial reorganization and remodeling of the cytoskeleton at the point of entry. The *tox* orthologs may play a role in these invasion-related actin-remodeling events or in immune evasion (Read et al., 2003; McClarty et al., 2007).

The *tox* products have also been identified as members of the YopT effector family of type III secreted cysteine proteases (Shao et al., 2002). Members of this family are present in bacterial pathogens of plant or animal cells. YopT induces a cytotoxic effect in mammalian cells, characterized in vitro by disruption of the actin cytoskeleton and rounding up of cells (Iriarte and Cornelis, 1998). Shao et al. (2002) demonstrate that this effect arises from YopT-mediated proteolytic C-terminal cleavage of Rho GTPases. This cleavage leads to the release of the GTPases from the cell membrane with concomitant disruption of the actin cytoskeleton. The cysteine protease domain with the invariant catalytic triad (CHD) that defines this family is present in all full-length chlamydial *tox* orthologs, although absent from *C. trachomatis tox* fragments (Read et al., 2003). More recently, Thalmann et al. (2010) found that the CT166 *tox* fragment functioned as an effector protein during entry, interfering with Rac-dependent cytoskeletal interactions.

Thus, the proteins encoded by the full-length chlamydial *tox* orthologs possess two distinct domains that inhibit actin polymerization. The presence of an actin-disrupting toxin, albeit fragmented, is a key difference among *C. trachomatis* biovars (Carlson et al., 2004). It has been postulated that this difference may account for the different sites of infection between biovars by mediating trafficking within early endosomes and thus affecting the degree of systemic dissemination by localizing chlamydial egress to an apical point of entry, rather than transiting the cell (Belland et al., 2001; McClarty et al., 2007).

The observation that *tox* interacts and interferes with the actin cytoskeleton (Thalmann et al., 2010), combined with the heterogeneity of copy number and fragmentation observed in Fig. 2, suggests the *tox* gene product is a major factor that differentiates chlamydial species. It is tempting to correlate this heterogeneity with known phenotypic variables, such as mucosal restriction or host diversity: is the ability of a particular chlamydial species to disseminate within the host organism or perhaps even between host species mediated to some degree by the nature or copy number of *tox*? Does loss of *tox* functionality enable an isolate to switch host species, or does loss of function through gene decay occur after a new host species has been successfully invaded? A plausible mechanism for *tox*-mediated host switching is not yet apparent; furthermore, it is unlikely that all chlamydial host diversity and niche specificity will be the

result of a single gene activity. Nevertheless, the *tox* variation identified here is a prime example of the utility of comparative genomics for identifying novel interspecies genetic variation for further exploration, and it focuses further attention on the processes of gene decay and loss that are shaping the *Chlamydiaceae*.

C. PNEUMONIAE AS A FORMER ZOONOTIC PATHOGEN: THE CASE FOR SEQUENCING WIDELY

C. pneumoniae was described as a distinct species only in 1988 (Cox et al., 1988). It causes widespread acute respiratory disease in humans (Saikku, 1992) and has been linked to several major human chronic diseases such as asthma (Sutherland and Martin, 2007), atherosclerosis (Watson and Alp, 2008), stroke (Elkind and Cole, 2006), and late-onset Alzheimer's disease (Balin et al., 2008). Four epidemiologically distinct human *C. pneumoniae* isolates were sequenced soon after the original *C. trachomatis* genome sequence (Kalman et al., 1999; Read et al., 2000; Shirai et al., 2000), rendering *C. pneumoniae* the most sequenced member species of the *Chlamydia* genus for several years. This highlights the degree of interest in this recently described human pathogen. However, comparative genomic analysis showed that these four isolates were not only syntenic, as typically observed with chlamydial species, but also essentially identical, with fewer than 500 SNPs or indels (Table 4) scattered around the ~1-Mbp chromosome in no discernible pattern (Myers et al., 2009). Such an extreme degree of similarity between temporally and geographically disparate isolates was interpreted as evidence for a relatively recent clonal expansion of human *C. pneumoniae* (Read et al., 2000). However, these virtually identical genomes allowed little insight into evolutionary origins or disease mechanisms.

C. pneumoniae was originally described as a human pathogen; however, it has since been found in many animals, through both sequence-based surveys and isolation of the organism, and is now considered the most cosmopolitan of all the *Chlamydiaceae* (Bodetti et al., 2002). *C. pneumoniae* is found to cause disease in both warm and cold-blooded animals including horses, marsupials, amphibians, and reptiles (Bodetti et al., 2002). No clinical evidence exists for zoonotic transfer of *C. pneumoniae* from animals to humans; however, both *C. psittaci* and *C. abortus* are well-known zoonotic pathogens.

Populations of the Australian koala (*Phascolarctos cinereus*) are notable for being infected with both *C. pneumoniae* and *C. pecorum*, with severe ongoing consequences for threatened populations (Bodetti et al., 2002). As with human infection, *C. pneumoniae* infections are commonly found in the koala respiratory tract and are linked to overt symptoms of respiratory disease (Wardrop et al., 1999). Comparative genome analysis of a koala *C. pneumoniae* isolate (LPCoLN) against the extant human *C. pneumoniae* genome sequences shows that the LPCoLN isolate is also highly similar, exhibiting the characteristic conservation of both genomic synteny and gene content (Myers et al., 2009). The human isolates all lack the virulence-associated 7.5-kb chlamydial plasmid, which is present in the LPCoLN isolate. However, a larger number of SNPs were identified within the koala isolate (Table 4); a phylogenetic analysis of synonymous SNPs showed the koala isolate is basal to the human isolates (Fig. 3), indicating that this contemporary marsupial isolate is closer to the ancestor of human *C. pneumoniae* (Myers et al., 2009), supporting earlier observations using a subset of SNPs (Rattei et al., 2007).

In contrast to the human isolates, the koala strain has accumulated SNPs in genomic hot spots. The hot spots identified by these SNP and BSR plots overlap and represent likely regions of varying evolutionary rates between the human and koala isolates. Significantly, comparative analysis of the gene regions delineated by these hot spots showed that many of the CDSs in the human isolates are truncated or fragmented relative to LPCoLN. This pattern is directional: at every hot spot, only the human isolates exhibit CDS truncation or fragmentation, whereas the koala isolate does not. These hot spots, which include the PZ,

TABLE 4 Breakdown of SNP/indels per *Chlamydiaceae* genome, where multiple genomes are available[a]

Species/strain[b]	SNPs/indels							
	Location (SNP + indel)			Strand (SNP + indel)		Type (SNPs alone)		
	Total	Coding (%)	Intergenic (%)	+ (%)	− (%)	Total	Synonymous (%)	Nonsynonymous (%)
C. trachomatis								
D/UW-3/CX								
434/Bu	9,289	9,251 (99.6)	38 (0.4)	928 (10.2)	8,342 (89.8)	7,884	2,370 (30.1)	5,514 (69.9)
A/HAR-13	3,925	3,908 (99.6)	17 (0.4)	477 (12.2)	3,445 (87.8)	3,391	959 (28.3)	2,432 (71.7)
B/Jali20/OT	4,084	4,065 (99.5)	19 (0.5)	472 (11.6)	3,609 (88.4)	3,418	971(28.4)	2,447 (71.6)
B/TZ1A828/OT	4,090	4,065 (99.4)	25 (0.6)	465 (11.4)	3,621 (88.5)	3,504	968 (27.6)	2,536 (72.4)
D-EC	19	19 (100)	0 (0)	0 (0)	19 (100)	14	3 (21.4)	11 (78.6)
D-LC	19	19 (100)	0 (0)	0 (0)	19 (100)	14	3 (21.4)	11 (78.6)
E/11023	6,700	6,693 (99.9)	7 (0.1)	689 (10.3)	6,008 (89.7)	5,887	1,758 (29.9)	4,129 (70.1)
E/150	6,727	6,720 (99.9)	7 (0.1)	657 (9.8)	6,068 (90.2)	5,879	1,755 (29.9)	4,124 (70.1)
G/11074	1,997	1,990 (99.6)	7 (0.4)	140 (7.1)	1,855 (92.9)	1,717	473 (27.5)	1,244 (72.5)
G/11222	1,717	1,717 (100)	0 (0)	88 (5.1)	1,629 (94.9)	1,449	411 (28.4)	1,038 (71.6)
G/9301	2,000	1,993 (99.7)	7 (0.4)	138 (7.0)	1,860 (93.0)	1,716	467 (27.2)	1,249 (72.8)
G/9768	2,000	1,993 (99.7)	7 (0.4)	138 (7.0)	1,860 (93.0)	1,717	467 (27.2)	1,250 (72.8)
L2b/UCH-1/proctitis	9,348	9,312 (99.6)	36 (0.4)	941 (10.3)	8,389 (89.7)	7,916	2,391 (30.2)	5,525 (69.8)
Sweden2	6,410	6,410 (100)	0 (0)	672 (10.5)	5,738 (89.5)	5,692	1,726 (30.3)	3,966 (69.7)
6276	1,976	1,973 (99.8)	3 (0.2)	98 (5.1)	1,875 (94.9)	1,658	425 (25.6)	1,233 (74.4)
6276s	1,978	1,975 (99.8)	3 (0.2)	101 (5.3)	1,874 (94.7)	1,661	432 (26.0)	1,229 (74.0)
70	6,761	6,749 (99.8)	12 (0.2)	683 (10.2)	6,071 (89.8)	5,871	1,762 (30.0)	4,109 (70.0)
70s	6,739	6,727 (99.8)	12 (0.2)	683 (10.2)	6,049 (89.8)	5,855	1,758 (30.0)	4,097 (70.0)
D(s)2923	6,298	6,291 (99.9)	7 (0.1)	665 (10.6)	5,630 (89.4)	5,540	1,653 (29.8)	3,887 (70.2)
C. pneumoniae								
AR39								
CWL029	470	380 (80.9)	90 (19.1)	181 (38.5)	289 (61.5)	299	111 (37.1)	188 (62.9)
J138	478	356 (74.5)	12 (25.5)	205 (42.9)	273 (57.1)	304	121 (39.8)	183 (60.2)
TW-183	435	351 (77.4)	84 (22.6)	200 (46.0)	235 (54.0)	304	100 (32.9)	204 (67.1)
LPCoLN	10,863	8,413 (80.7)	2,450 (19.3)	5,368 (49.4)	5,494 (50.6)	7,304	3,313 (45.4)	3,991 (54.6)
C. psittaci								
6BC	35	27 (77.1)	8 (22.9)	33 (94.3)	-	14	6 (42.9)	8 (57.1)
RD1	186	98 (52.7)	88 (47.3)	121 (65.1)	65 (34.9)	69	21(30.4)	48 (69.6)
Cal10	140	99 (70.7)	41 (29.3)	69 (49.3)	68 (48.6)	61	19 (31.1)	42 (68.9)

[a]SNPs and indels for each species were identified by comparing the selected reference genome against each set of query genomes using a custom pipeline built on the Nucmer package (Delcher et al., 2003) developed at the Institute for Genome Science, University of Maryland, Baltimore.

[b]Boldface indicates that the genome sequence was used as the reference.

also encode several virulence or metabolic factors that vary between strains and species and include the polymorphic membrane protein (Pmp) family, secreted proteins, and proteins involved in the biosynthesis of chorismate, a precursor of aromatic amino acids (Fig. 4A). Furthermore, several regions, such as CDS clusters encoding members of the Pmp fam-

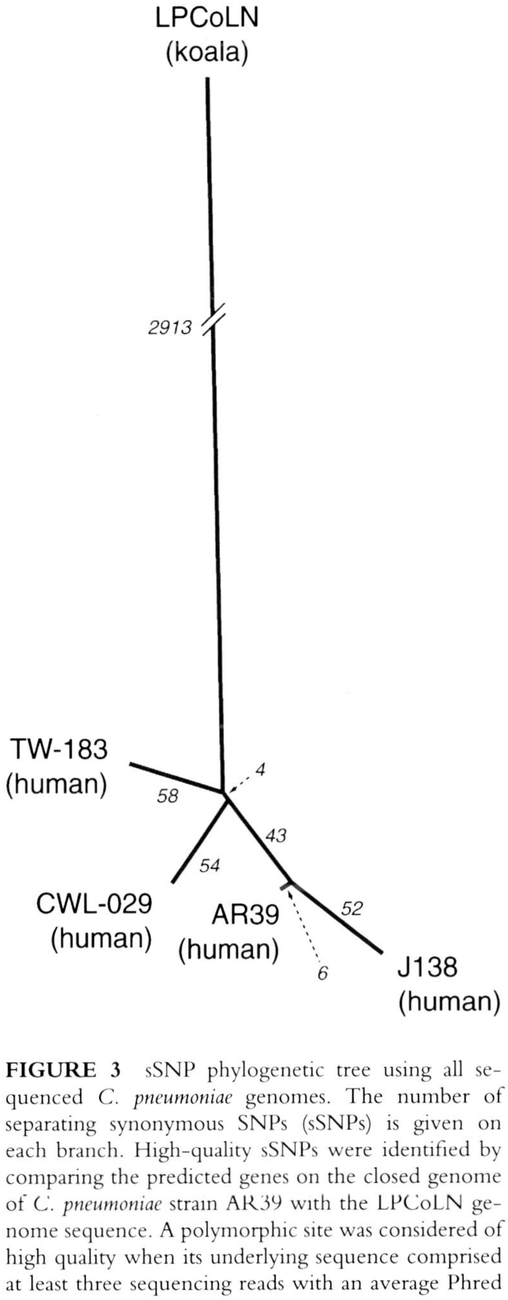

FIGURE 3 sSNP phylogenetic tree using all sequenced *C. pneumoniae* genomes. The number of separating synonymous SNPs (sSNPs) is given on each branch. High-quality sSNPs were identified by comparing the predicted genes on the closed genome of *C. pneumoniae* strain AR39 with the LPCoLN genome sequence. A polymorphic site was considered of high quality when its underlying sequence comprised at least three sequencing reads with an average Phred quality score greater than 30. sSNPs in CWL029, TW183, and J138 were similarly identified, although no assessment of quality could be made, as quality scores are not available for these genomes. Concatenated sSNPs for the individual *C. pneumoniae* isolates were further analyzed by the HKY85 method with 200 bootstrap replicates, and the results were used to generate an unrooted phylogenetic tree according to the PhyLM algorithms.
doi:10.1128/9781555817329.ch2.f3

ily, show truncation or fragmentation differences between the human isolates themselves (Fig. 4B). These observations suggest that evolutionary processes are ongoing within the human isolates of *C. pneumoniae* (Myers et al., 2009).

This directional pattern of gene fragmentation not only supports the predating of koala *C. pneumoniae* LPCoLN relative to the human isolates but also suggests that animals were the original hosts of *C. pneumoniae*. This has led to the proposal that *C. pneumoniae* is a former zoonotic pathogen that has successfully crossed the species barrier several times, adapting to humans through reductive processes of gene decay and loss, to the point where it no longer requires the animal host for transmission to humans (Myers et al., 2009). Subsequent phylogenetic examination of 23 gene loci across a diversity of animal and human isolates showed significant diversity within the animal isolates, permitting grouping of *C. pneumoniae* isolates into 5 distinct genotypes (A to E) (Mitchell et al., 2010). This phylogenetic analysis also indicated that the human isolates derived from an amphibian or reptile *C. pneumoniae* lineage and that an animal-to-human transmission has occurred at least twice (Mitchell et al., 2010).

These novel findings were made possible only by the completion of the koala *C. pneumoniae* sequence and stand in contrast to the four existing *C. pneumoniae* genome sequences, which showed virtually no difference between epidemiologically distinct human isolates. The LPCoLN genome sequence enabled the directionality of plasmid loss and gene fragmentation patterns to be discerned: the human isolates by themselves showed no indication that the gene fragments were anything but small CDSs that varied slightly between different isolates and were likely annotation artifacts.

This demonstrates why it is necessary to sequence widely across the diversity of extant chlamydial isolates, both within existing culture collections and for recent clinical isolates from humans and animals. In the continuing absence of general tools for genetic manipulation, systematic sampling of the natural

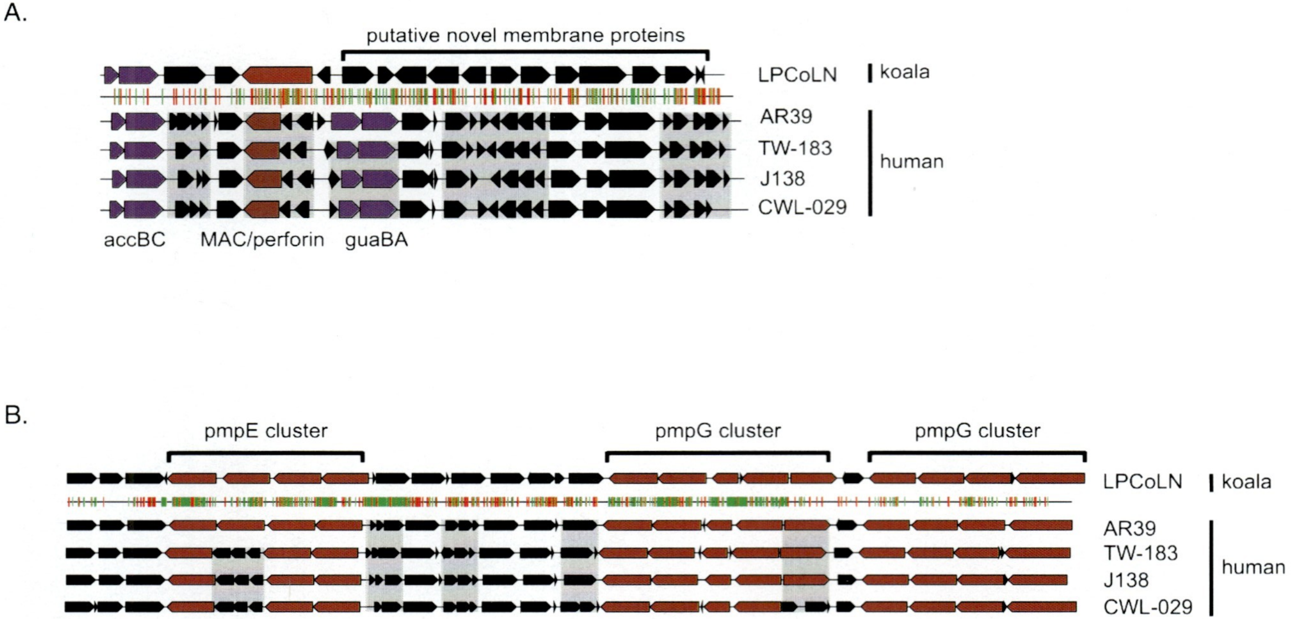

FIGURE 4 Comparison of two regions of SNP accumulation in *C. pneumoniae*, with SNP location and type (synonymous, green; nonsynonymous, red). Grey highlighting shows SNP-associated CDS fragmentation. (A) The plasticity zone. LPCoLN gene region, ORF00689 to ORF00665; AR39 gene region, CP_0585 to CP_0622. (B) Pmp cluster. LPCoLN gene region, ORF00989 to ORF00956; AR39 gene region, CP_0280 to CP_0309. doi:10.1128/9781555817329.ch2.f4

diversity of *Chlamydia* through genome sequencing is an efficient path to better understanding of these organisms.

ANIMAL *CHLAMYDIA* SPP. ARE UNDERREPRESENTED

As described above, genomic analysis of diverse isolates of *C. pneumoniae* provided evidence for a zoonotic origin of a major human pathogen. Most available genome sequences derive from human isolates of either *C. trachomatis* or *C. pneumoniae*, yet zoonotic infections from wildlife are considered to be, of all the emerging infectious diseases, the greatest threat to global health (Jones et al., 2008). More attention needs to be paid to the animal chlamydial pathogens, as these can lend significant insight into human disease. The known chlamydial zoonotic pathogens, particularly *C. psittaci*, require greater consideration in light of a putative zoonotic origin for *C. pneumoniae*. Among the other animal pathogens, there is a single genome for *C. felis* (Azuma et al., 2006), and more recently a single genome sequence for *C. pecorum* has been published (Mojica et al., 2011). No genome has yet been released for *C. suis*.

C. psittaci is perhaps the archetypal chlamydial zoonotic pathogen and is classified as a category B bioterrorism agent (CDC, 2007) due to its potential to cause fatal disease, its high infectivity, and the ease of dissemination via the respiratory route. Chlamydiosis in birds and other mammals is a disease with a global distribution; transmission of *C. psittaci* from birds to humans is reported regularly, particularly in high-risk individuals such as veterinarians, farm workers, bird owners, and animal shopkeepers (Beeckman and Vanrompay, 2009; Harkinezhad et al., 2009).

Psittacosis and the Founding of the National Institutes of Health

> "On January 6, 1930, Dr. Willis P. Martin, of Annapolis, observed three cases of illness in an Annapolis family in which a parrot had died. He suggested psittacosis, the disease having come to his attention through a newspaper article, which appeared at the time. The parrot had been purchased from a pet shop in Baltimore on December 14, 1929." (Ellicott and Halliday, 1931)

In the Maryland outbreak described by Ellicott and Halliday (1931), 36 cases of psittacosis were identified, with 5 fatalities. Three of the cases were Dr. W. R. Stokes, Director of the investigating bacteriological laboratory at the University of Maryland School of Medicine in Baltimore, Dr. D. S. Hatfield, Director of the Maryland Bureau of Communicable Diseases, and Dr. C. Armstrong, lead outbreak investigator at the Hygienic Laboratory in Washington, DC. Dr. Stokes died of the disease, while Dr. Hatfield and Dr. Armstrong recovered following injection of immune sera from recovered patients. Nationwide, the psittacosis outbreak constituted 169 cases with 33 fatalities. In May 1930, two months following the outbreak, Congress recognized and rewarded the Hygienic Laboratory for its role in identifying and controlling the outbreak by expanding the Laboratory and granting it a new name: the National Institute of Health (Lepore, 2009).

Most *C. psittaci* infections occur through inhalation of an infectious aerosol when handling an infected animal and/or contaminated material. Contaminated feces and feathers are assumed to play an essential role in zoonotic transmission to humans. The symptoms of human disease (termed psittacosis when originating from a psittacine species and ornithosis when originating from poultry) are variable, ranging from no clinical signs at all to severe systemic disease. Acute disease is characterized by a flu-like onset that progresses to atypical pneumonia and cardiovascular disease if left untreated, yet determinants of the severity of human infection are poorly understood. The potential severity of acute zoonotic disease caused by *C. psittaci* is already of concern. This concern becomes even higher if *C. psittaci* loses the need for the animal reservoir and is able to disseminate by human-to-human transmission, as is hypothesized for *C. pneumoniae*, which fortunately causes a relatively mild acute respiratory disease. A *C. psittaci* isolate able to circulate within human populations by direct transmission would be a major public health concern.

The genome sequences of four strains have been completed (Table 2), including two variants of the type strain *C. psittaci* 6BC, isolated from a parrot during the 1929-1930 psittacosis pandemic and subsequently used extensively as a standard laboratory strain. The sequence of one of these *C. psittaci* 6BC isolates (CP002586) was manually finished, and it is used as the reference genome for comparative analysis here (Grinblat-Huse et al., 2011). Draft genomes have also been published for *C. psittaci* RD1 (Seth-Smith et al., 2011), recovered from a mixed culture with *C. trachomatis* serovar L2b and *C. psittaci* Cal10 (Grinblat-Huse et al., 2011), originally termed the meningopneumonitis virus, which was isolated from ferrets inoculated with throat washings from human cases of an influenza-like respiratory infection.

As observed above for *C. pneumoniae*, the completed and draft *C. psittaci* genomes are virtually identical, exhibiting conservation of chromosome size and gene count (Table 2), synteny, and gene content. There are fewer than 200 SNPs/indels observed between *C. psittaci* 6BC, RD1, and Cal10 (Table 4), making these isolates more similar to each other than what is observed for human *C. pneumoniae*. However, unlike *C. pneumoniae*, all *C. psittaci* isolates contain the *C. psittaci* plasmid (8 CDSs; 7,553 bp), a known virulence factor. Similarly, plots of SNP and BSR comparisons identify several hot spot regions of SNP and CDS variation in contrast to the SNP scattering found in the genome of human *C. pneumoniae* strains (Grinblat-Huse et al., 2011).

Hot spots of variation occur in at least four regions: one hot spot maps to the *C. psittaci* cytotoxin ortholog within the chlamydial PZ, and two other hot spots are found within distinct *pmpG* clusters (Grinblat-Huse et al., 2011). The four available *C. psittaci* genome sequences all contain a single *tox* CDS within the PZ. If loss of *tox* function through decay is a gateway to host switching, as speculated earlier, *C. psittaci* could be a potential problem. A *C. psittaci* isolate that lacks *tox* would be a major public health concern if this loss of function conferred a heightened ability to disseminate and to be efficiently transmitted among humans. Further genome sequencing of diverse *C. psittaci* isolates obtained from animals and humans with clinically characterized infections will allow further examination of this by correlating decay or absence of *tox* with systemic disease, along with other possible determinants. The fourth hot spot centers on a CDS predicted to encode phosphatidylinositol-4-phosphate 5-kinase, a lipid-modifying enzyme involved in actin remodeling that has been shown to be instrumental for entry of chlamydiae into the host cell (Balana et al., 2005). Thus, two of the *C. psittaci* hot spots center on CDSs encoding proteins predicted to interact with the actin cytoskeleton (Grinblat-Huse et al., 2011).

Serological typing of *C. psittaci* allows division into at least 6 serotypes (A to F), with certain serotypes often isolated from specific hosts (Andersen, 1991; Vanrompay et al., 1993; Andersen, 1997; Geens et al., 2005). A recent comprehensive multilocus sequence typing (MLST) analysis of *C. psittaci* showed an association between the different MLST genotypes and the host species (Pannekoek et al., 2010). A key motivation for sequencing multiple *Chlamydia* genomes has been to understand why a pathogen with such a conserved and syntenic chromosome is able to infect a diversity of hosts, tissues, and cell types and cause a diversity of disease outcomes. The low SNP and CDS variation identified here for *C. psittaci* suggests that extensive genome sequencing deep within extant collections and fresh clinical isolates of *C. psittaci*, with strain selection directed by MLST and other typing schemes, will allow even finer correlation of gene, SNP, and indel variation with niche specificity. In addition, the low genetic variability observed in the four sequenced strains of *C. psittaci* is reminiscent of the low variability that was seen in the human isolates of *C. pneumoniae* prior to sequencing of the koala strain. Thus, the low genomic variation observed so far in *C. psittaci* may simply represent insufficient sampling of its true diversity.

C. pecorum is found in cattle and other ruminants, swine, koalas, and other marsupials and causes a wide diversity of disease with significant economic impact. Strains of *C. pecorum* were considered members of the *C. psittaci* species until separated in 1992 on the basis of DNA-DNA hybridization and immunological data (Fukushi and Hirai, 1992). *C. pecorum*-associated diseases in sheep, goats, cattle, horses, and pigs include polyarthritis, pneumonia, urogenital tract infections, spontaneous abortion, conjunctivitis, mastitis, encephalomyelitis, enteritis, pleuritis, and pericarditis; in koalas, *C. pecorum* causes conjunctivitis and infertility (Yousef Mohamad et al., 2008). Only the type strain, *C. pecorum* E58 (McNutt and Waller, 1940), has been completed, limiting comparative analysis to other species (Table 2). However, the *C. pecorum* E58 genome is a finished and closed sequence that will be a reliable scaffold for future comparative analyses within the species.

The *C. pecorum* E58 genome is 1,106,197 bp, encoding 988 CDSs (Table 2) (Mojica et al., 2011). As expected, it is highly similar to *C. psittaci* in both gene content and genomic synteny, although the evolutionary divergence of *C. pecorum* from *C. psittaci* is evident with both synteny and overall gene similarity decreasing relative to *C. psittaci*. No plasmid was recovered during sequencing. *C. pecorum* E58 possesses two full-length copies of the chlamydial *tox* arranged in tandem within the PZ (Fig. 2). Three copies of phospholipase D, which is suggested to play a role in chlamydial lipid modification or metabolism (Taylor et al., 2010), are found in the *C. pecorum* PZ. Similarly, the virulence-associated MACPF gene, which is either truncated or absent in most chlamydial genomes, is present as a single copy.

Limited gene sequencing and serological studies have suggested that there is significant strain diversity within *C. pecorum*, consistent with the observed diverse spectrum of hosts and diseases (Yousef Mohamad et al., 2008). A single genome is thus obviously insufficient; additional genome sequencing of recent clinical isolates from animals with defined symptoms is required to explore this strain diversity.

CHLAMYDIA TRACHOMATIS—MORE, MORE, MORE

C. trachomatis is represented by 20 of the 33 genomes currently available, as befitting the most recognizable pathogen of the family *Chlamydiaceae*, responsible for significant sexually transmitted disease morbidity and infectious blindness worldwide. The extant genome sequences span isolates from the mucosally restricted trachoma and genital tract biovars to the more aggressive and systemic lymphogranuloma venereum (LGV) biovars (Carlson et al., 2005; Thomson et al., 2008; Jeffrey et al., 2010; Somboonna et al., 2011).

Much of the recent literature dealing with *C. trachomatis* genomics has focused on the emergence of a Swedish variant strain or on the accumulating evidence for recombination and lateral gene transfer. The most significant difference of the *C. trachomatis* Swedish variant was shown to be a deletion of 377 bp within CDS1 of the chlamydial plasmid (Thomson and Clarke, 2010). This deletion effectively removed the target of two PCR-based laboratory tests. The subsequent failure of these tests led to undetected spread of *C. trachomatis* infections, with significant public health consequences (Thomson and Clarke, 2010). This episode highlighted both the danger of using single-target assays and the utility of whole-genome sequencing as a public health tool.

In previous years, sequence analysis of selected genes has indicated that recombination and lateral gene transfer can occur and has occurred in *C. trachomatis* (Gomes et al., 2004, 2006, 2007; Binet and Maurelli, 2009); moreover, lateral gene transfer can be induced in vitro (DeMars et al., 2007; DeMars and Weinfurter, 2008; Binet and Maurelli, 2009; Suchland et al., 2009). More recently, this has been extended by whole-genome sequence analysis of new and existing *C. trachomatis* genome sequences (Jeffrey et al., 2010; Joseph et

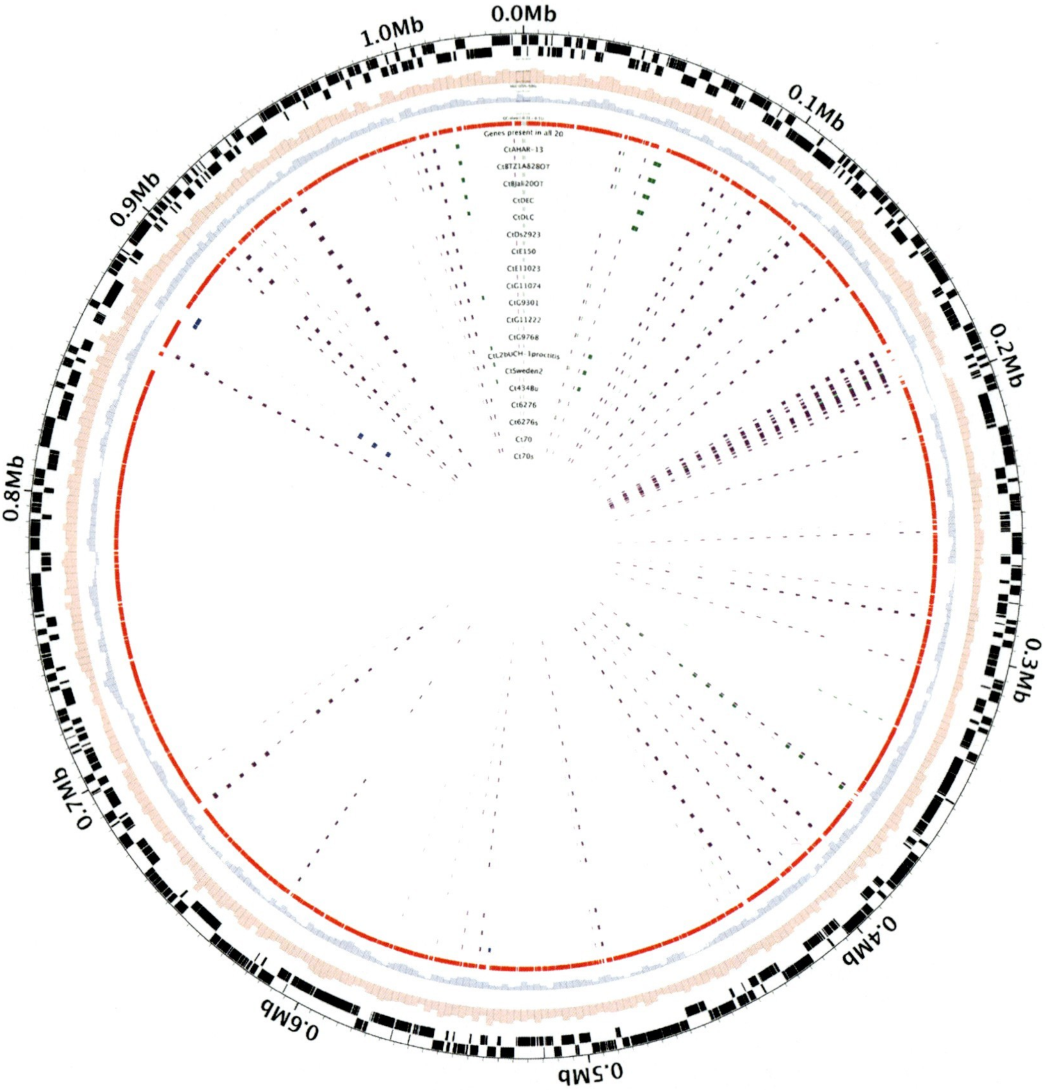

FIGURE 5 Circular representation of *C. trachomatis* proteomic similarity across 19 genomes, relative to *C. trachomatis* D, showing hot spots of gene variability. Data are from outermost circle to innermost. In the first two outermost circles, black tick marks represent predicted CDSs on the plus strand of *C. trachomatis* 6BC and the minus strand, respectively. The following two circles plot %GC and GC skew as histograms. The following circle plots the positions of proteins that are present and highly conserved (red; BSR, >0.8) across all genomes. Each subsequent circle shows the positions of variable or unique proteins only for each genome as labeled. Color coding is as follows: purple, the protein is present in <19 genomes (including the reference); green, protein is present in ≤10 genomes; blue, protein is present in ≤5 genomes; orange, protein is present only in the reference; grey, protein is absent in the reference genome. *C. trachomatis* strains, from the outermost circle moving toward the center, are as follows: A-HAR-13, B-TZA828OT, B-Jali20OT, D-EC, D-LC, D-s2923, E-150, E-1103, G-11074, G-9301, G-11222, G-9768, L2b-UCH-1proctitis, Sweden2, 434Bu, 6276, 6276s, 70, and 70s. doi:10.1128/9781555817329.ch2.f5

al., 2011; Somboonna et al., 2011). Notably, evidence has been presented for recombination between an invasive LGV isolate and a noninvasive urogenital strain of *C. trachomatis* (Somboonna et al., 2011). While the mechanism for recombination remains to be determined, these observations suggest and support the idea that *Chlamydia* exists as a mixture of variants circulating within host populations and may play a role in fitness in vivo. *Chlamydia* as a mixed population has substantial ramifications for clinical outcomes. One manifestation could be that chlamydial infection is a mixed infection with variants that can influence disease progression or resolution; the influence of host selection in a mixed population becomes a much greater question. A mixed population may also serve as a source of genomic sequence for recombination, possibly leading to emergence of resistance or to increased virulence as reported by Somboonna et al. (2011). As noted earlier, whole-genome sequencing typically creates a consensus sequence. This has the effect of obscuring population variants by "collapsing" base calls by majority rule. There is significant scope for applying the tools of deep, whole-genome sequencing to the question of chlamydial variants circulating within human and animal populations and the associated effect on disease.

Genome synteny analysis shows that all sequenced publicly available *C. trachomatis* strains are highly syntenic with very high gene content similarity, as expected for the *Chlamydiaceae*. Using *C. trachomatis* D/UW-3/CX as the reference genome, SNPs and indels were identified across the 19 remaining *C. trachomatis* genomes (Table 3). In contrast to the four human *C. pneumoniae* isolates and the four *C. psittaci* isolates, *C. trachomatis* exhibits significantly more variation between isolates at the SNP/indel level. Plotting these SNPs and indels by genomic position for each *C. trachomatis* genome shows numerous hot spots of SNP/indel accumulation (data not shown). Similarly, BSR analysis of the predicted proteomes of all 20 isolates shows hot spots of variation at the level of protein similarity. Several of these hot spots correlate to known regions of variation that are also found in *C. pneumoniae* and *C. psittaci* (see above), including the Pmp and Inc families as well as the PZ. However, several of these regions are novel to *C. trachomatis*. In addition, a significant number of the variable peptides found by this BSR analysis have no known function (n = 140; annotated as hypothetical); many of these hypothetical proteins are encoded by genes located in hot spots of BSR variation (Fig. 5). This subset of hypothetical genes differentiates the various *C. trachomatis* strains; characterization of these gene products is likely to be informative for understanding tropism and disease processes. Supporting this, genome sequence analysis of three recent clinical isolates of *C. trachomatis* (serovars E, G, and J) combined with subsequent genotyping assays on a larger set of clinical isolates identified several loci that were linked to tissue tropism and regions of recombination (Jeffrey et al., 2010). Several of these regions, including the PZ and Pmp genes as well as hypothetical genes, overlap with the regions identified by this BSR analysis.

BEYOND REPRESENTATIVE GENOMES—WHAT'S NEXT?

Wider . . . More

"More sequencing" is the usual answer to the question of what is next for chlamydial genomics. This might seem a glib response, particularly from groups that utilize high-throughput genomics as a standard analysis tool. Nevertheless, as outlined in this chapter, WGS sequencing has had an outsized effect on *Chlamydia* research, with at least one genome representative for nearly all extant species. However, the current "representative" strains may not be the best examples for each species: the sequenced strains were typically selected on the basis of convenience and availability rather than biology. In addition, most chlamydial genomes to date are derived from laboratory strains that have been passaged in vitro many times, with unknown selective pressures possibly altering the genome. Evolutionary processes that lead to gene decay and loss of function appear to be the primary force shaping and differentiating the *Chlamydiaceae*; continued wide sequencing will further define these processes.

This chapter highlights some of the insights that are possible from sequencing genomes that are not the typical laboratory strains. This provides strong justification for wide sequencing of isolates spanning the spectrum of chlamydial infections from human and, particularly, animal sources. Similarly, wide sequencing within species that are either underrepresented (for example, *C. psittaci* and *C. pecorum*) or are major human pathogens (*C. trachomatis* and *C. pneumoniae*) is an efficient method to correlate genotype to phenotype in an experimentally difficult system. The work described by Jeffrey et al. (2010), in which WGS sequencing of recent *C. trachomatis* clinical strains was used to explore tissue tropism, highlights the power of this approach.

The diversity of *C. trachomatis* has traditionally been determined through serological typing using a panel of antibodies against certain epitopes of MOMP (major outer membrane protein), encoded by *ompA*. The serotype structure derived from this typing approach is still in common usage; however, the degree of *ompA* recombination and chlamydial lateral gene transfer being discovered suggests that a typing scheme based on epitopes of a single protein is unlikely to capture the true extent of chlamydial diversity and indeed may obscure it. More modern genotyping approaches have variously targeted *ompA* or 16S rRNA or have utilized the MLST approach with subsets of conserved genes. MLST-style schemes in particular are inexpensive and effective tools for rapid assignment of phylogenetic position; several schemes have been described to better study chlamydial diversity beyond *C. trachomatis* and used to great effect (for examples, see Laroucau et al., 2008; Pannekoek et al., 2008; and Christerson et al., 2011). However, the ever-decreasing costs of sequencing, combined with the large amount of additional information contained within even just a draft genome sequence, suggest that insights from whole-genome sequencing will replace the antiquated *C. trachomatis* serological typing schema. This information will greatly expand evolutionary understanding of *Chlamydia* in a broader perspective. In lieu of fundamental laboratory tools for genetically manipulating chlamydiae, the natural diversity of *Chlamydia* can be exploited in this way to better understand chlamydial biology.

Deeper: Exploring Chlamydial Population Variation

Current and future genomes are a fundamental starting point for modern *Chlamydia* research, with many more genomes pending and needed. However, the "unit" of a genome is based on the assumption of a clonal population. As noted above, the increasing number of studies describing recombination and lateral gene transfer challenges this assumption in vivo, and even under controlled in vitro conditions. Genome assemblies, whether complete or draft genomes, are snapshots that "freeze" the genetic state of the organism at a particular point in time. Such a snapshot captures a consensus sequence that is undoubtedly useful but is essentially a summary that obscures a complex and dynamic population. Sequencing approaches that look past this consensus mode of genomics and instead treat minority polymorphisms and indels as a census of the total chlamydial population are likely to shed more light on this complexity both in vivo and in vitro.

The Host Cell: the Elephant in the Room

The wealth of chlamydial genome sequences has already enabled substantial biological insight through comparative genomic and proteomic analyses and has provided the building blocks for genome-scale tools used for transcriptomic analyses, such as whole-genome microarrays (Belland et al., 2003) and mapping cDNA transcripts and transcription start sites onto chlamydial genome annotation (Albrecht et al., 2010). Nevertheless, the true scale of modern high-throughput sequencing has yet to be fully and effectively applied to chlamydial biology—notably, the host cell response to chlamydial infection remains uncharted by genome-scale tools.

Chlamydiae are obligate intracellular pathogens whose biology is thus dynamically and intimately intertwined with that of their host cells. Analyses of the chlamydial transcriptome by microarrays or by pathogen-specific cDNA sequencing capture only a fraction of the transcriptional activity in the immediate environment of the infected cell, ignoring the host itself. It is self-evident that a full understanding of disease pathogenesis cannot be achieved by only focusing on the pathogen. The global host response side of the equation has largely been ignored so far simply because the size of the mammalian genome (and thus the dynamic mammalian transcriptome) has made genome-scale investigation cost-prohibitive. Indeed host cell DNA and RNA have historically been considered "contamination" at sequencing centers, representing something to be minimized when sequencing a chlamydial genome or transcriptome.

Sequencing of cDNA libraries to explore the transcriptional landscape has been used for over 20 years. However, such early approaches were limited by cost, insufficient sequencing depth, and experimental biases. The shift from Sanger chemistry to next-generation cDNA sequencing, popularly termed RNA-Seq, enables affordable, comprehensive sequencing of cDNA libraries with minimal bias, generating accurate gene expression profiles of the source cells. RNA-Seq also provides insights into the transcriptome beyond just sequence, yielding splicing and expression level information as well. Now, however, the availability of cheaper sequencing and greater sequencing depth makes *Chlamydia* an ideal candidate to be at the forefront of the next major sequencing frontier for infectious disease research, building on the numerous genome sequences to decipher the gene expression dynamics of *Chlamydia* and their host cells through deep and comprehensive transcriptional profiling.

REFERENCES

Albrecht, M., C. M. Sharma, R. Reinhardt, J. Vogel, and T. Rudel. 2010. Deep sequencing-based discovery of the *Chlamydia trachomatis* transcriptome. *Nucleic Acids Res.* **38:**868–877.

Andersen, A. A. 1991. Serotyping of *Chlamydia psittaci* isolates using serovar-specific monoclonal antibodies with the microimmunofluorescence test. *J. Clin. Microbiol.* **29:**707–711.

Andersen, A. A. 1997. Two new serovars of *Chlamydia psittaci* from North American birds. *J. Vet. Diagn. Investig.* **9:**159–164.

Azuma, Y., H. Hirakawa, A. Yamashita, Y. Cai, M. A. Rahman, H. Suzuki, S. Mitaku, H. Toh, S. Goto, T. Murakami, K. Sugi, H. Hayashi, H. Fukushi, M. Hattori, S. Kuhara, and M. Shirai. 2006. Genome sequence of the cat pathogen, *Chlamydophila felis*. *DNA Res.* **13:**15–23.

Balana, M. E., F. Niedergang, A. Subtil, A. Alcover, P. Chavrier, and A. Dautry-Varsat. 2005. ARF6 GTPase controls bacterial invasion by actin remodelling. *J. Cell Sci.* **118:**2201–2210.

Balin, B. J., C. S. Little, C. J. Hammond, D. M. Appelt, J. A. Whittum-Hudson, H. C. Gerard, and A. P. Hudson. 2008. *Chlamydophila pneumoniae* and the etiology of late-onset Alzheimer's disease. *J. Alzheimer's Dis.* **13:**371–380.

Beeckman, D. S., and D. C. Vanrompay. 2009. Zoonotic *Chlamydophila psittaci* infections from a clinical perspective. *Clin. Microbiol. Infect.* **15:**11–17. .

Belland, R. J., M. A. Scidmore, D. D. Crane, D. M. Hogan, W. Whitmire, G. McClarty, and H. D. Caldwell. 2001. *Chlamydia trachomatis* cytotoxicity associated with complete and partial cytotoxin genes. *Proc. Natl. Acad. Sci. USA* **98:**13984–13989.

Belland, R. J., G. Zhong, D. D. Crane, D. Hogan, D. Sturdevant, J. Sharma, W. L. Beatty, and H. D. Caldwell. 2003. Genomic transcriptional profiling of the developmental cycle of *Chlamydia trachomatis*. *Proc. Natl. Acad. Sci. USA* **100:**8478–8483.

Bertelli, C., F. Collyn, A. Croxatto, C. Ruckert, A. Polkinghorne, C. Kebbi-Beghdadi, A. Goesmann, L. Vaughan, and G. Greub. 2010. The *Waddlia* genome: a window into chlamydial biology. *PloS One* **5:**e10890.

Binet, R., and A. T. Maurelli. 2009. Transformation and isolation of allelic exchange mutants of *Chlamydia psittaci* using recombinant DNA introduced by electroporation. *Proc. Natl. Acad. Sci. USA* **106:**292–297.

Bodetti, T. J., E. Jacobson, C. Wan, L. Hafner, A. Pospischil, K. Rose, and P. Timms. 2002. Molecular evidence to support the expansion of the host range of *Chlamydophila pneumoniae* to include reptiles as well as humans, horses, koalas and amphibians. *Syst. Appl. Microbiol.* **25:**146–152.

Caldwell, H. D., H. Wood, D. Crane, R. Bailey, R. B. Jones, D. Mabey, I. Maclean, Z. Mohammed, R. Peeling, C. Roshick, J. Schachter, A. W. Solomon, W. E. Stamm, R. J. Suchland, L. Taylor, S. K. West, T. C. Quinn, R. J. Belland, and G. McClarty. 2003. Polymorphisms in *Chlamydia trachomatis* tryptophan synthase genes differentiate between genital and ocular isolates. *J. Clin. Investig.* **111:**1757–1769.

Carlson, J. H., S. Hughes, D. Hogan, G. Cieplak, D. E. Sturdevant, G. McClarty, H. D. Caldwell, and R. J. Belland. 2004. Polymorphisms in the *Chlamydia trachomatis* cytotoxin locus associated with ocular and genital isolates. *Infect. Immun.* **72:**7063–7072.

Carlson, J. H., S. F. Porcella, G. McClarty, and H. D. Caldwell. 2005. Comparative genomic analysis of *Chlamydia trachomatis* oculotropic and genitotropic strains. *Infect. Immun.* **73:**6407–6418.

CDC. 2007. *Biosafety in Microbiological and Biomedical Laboratories (BMBL)*, 5th ed. U.S. Department of Health and Human Services, Washington, DC.

Christerson, L., A. Ruettger, K. Gravningen, R. Ehricht, K. Sachse, and B. Herrmann. 2011. High-resolution genotyping of *Chlamydia trachomatis* by use of a novel multilocus typing DNA microarray. *J. Clin. Microbiol.* **49:**2838–2843.

Collingro, A., P. Tischler, T. Weinmaier, T. Penz, E. Heinz, R. C. Brunham, T. D. Read, P. M. Bavoil, K. Sachse, S. Kahane, M. G. Friedman, T. Rattei, G. S. Myers, and M. Horn. 2011. Unity in variety—the pan-genome of the Chlamydiae. *Mol. Biol. Evol.* doi: 10.1093/molbev/msr161.

Cox, R. L., C.-c. Kuo, J. T. Grayston, and L. Campbell. 1988. Deoxyribonucleic acid relatedness of *Chlamydia* sp. strain TWAR to *Chlamydia trachomatis* and *Chlamydia psittaci*. *Int. J. Syst. Bacteriol.* **38:**265–268.

Delcher, A. L., S. L. Salzberg, and A. M. Phillippy. 2003. Using MUMmer to identify similar regions in large sequence sets. *Curr. Protocols Bioinform.* **Chapter 10:**Unit 10 3.

DeMars, R., and J. Weinfurter. 2008. Interstrain gene transfer in *Chlamydia trachomatis in vitro*: mechanism and significance. *J. Bacteriol.* **190:**1605–1614.

DeMars, R., J. Weinfurter, E. Guex, J. Lin, and Y. Potucek. 2007. Lateral gene transfer in vitro in the intracellular pathogen *Chlamydia trachomatis*. *J. Bacteriol.* **189:**991–1003.

Eisen, J. A., J. F. Heidelberg, O. White, and S. L. Salzberg. 2000. Evidence for symmetric chromosomal inversions around the replication origin in bacteria. *Genome Biol.* **1:**RESEARCH0011.

Elkind, M. S., and J. W. Cole. 2006. Do common infections cause stroke? *Semin. Neurol.* **26:**88–99.

Ellicott, V. L., and C. H. Halliday. 1931. The psittacosis outbreak in Maryland, December, 1929, and January, 1930. *Public Health Rep.* **46:**843–850.

Fleischmann, R. D., M. D. Adams, O. White, R. A. Clayton, E. F. Kirkness, A. R. Kerlavage, C. J. Bult, J. F. Tomb, B. A. Dougherty, J. M. Merrick, et al. 1995. Whole-genome random sequencing and assembly of *Haemophilus influenzae* Rd. *Science* **269:**496–512.

Fraser, C. M., J. A. Eisen, K. E. Nelson, I. T. Paulsen, and S. L. Salzberg. 2002. The value of complete microbial genome sequencing (you get what you pay for). *J. Bacteriol.* **184:**6403–6405.

Fukushi, H., and K. Hirai. 1992. Proposal of *Chlamydia pecorum* sp. nov. for *Chlamydia* strains derived from ruminants. *Int. J. Syst. Bacteriol.* **42:**306–308.

Geens, T., A. Desplanques, M. Van Loock, B. M. Bonner, E. F. Kaleta, S. Magnino, A. A. Andersen, K. D. Everett, and D. Vanrompay. 2005. Sequencing of the *Chlamydophila psittaci ompA* gene reveals a new genotype, E/B, and the need for a rapid discriminatory genotyping method. *J. Clin. Microbiol.* **43:**2456–2461.

Gomes, J. P., W. J. Bruno, M. J. Borrego, and D. Dean. 2004. Recombination in the genome of *Chlamydia trachomatis* involving the polymorphic membrane protein C gene relative to *ompA* and evidence for horizontal gene transfer. *J. Bacteriol.* **186:**4295–4306.

Gomes, J. P., W. J. Bruno, A. Nunes, N. Santos, C. Florindo, M. J. Borrego, and D. Dean. 2007. Evolution of *Chlamydia trachomatis* diversity occurs by widespread interstrain recombination involving hotspots. *Genome Res.* **17:**50–60.

Gomes, J. P., A. Nunes, W. J. Bruno, M. J. Borrego, C. Florindo, and D. Dean. 2006. Polymorphisms in the nine polymorphic membrane proteins of *Chlamydia trachomatis* across all serovars: evidence for serovar Da recombination and correlation with tissue tropism. *J. Bacteriol.* **188:**275–286.

Grinblat-Huse, V., E. F. Drabek, H. H. Creasy, S. C. Daugherty, K. M. Jones, I. Santana-Cruz, L. J. Tallon, T. D. Read, T. P. Hatch, P. Bavoil, and G. S. Myers. 2011. Genome sequences of the zoonotic pathogens *Chlamydia psittaci* 6BC and Cal10. *J. Bacteriol.* **193:**4039–4040.

Harkinezhad, T., T. Geens, and D. Vanrompay. 2009. *Chlamydophila psittaci* infections in birds: a review with emphasis on zoonotic consequences. *Vet. Microbiol.* **135:**68–77.

Horn, M., A. Collingro, S. Schmitz-Esser, C. L. Beier, U. Purkhold, B. Fartmann, P. Brandt, G. J. Nyakatura, M. Droege, D. Frishman, T. Rattei, H. W. Mewes, and M. Wagner.

2004. Illuminating the evolutionary history of chlamydiae. *Science* **304:**728–730.

Iriarte, M., and G. R. Cornelis. 1998. YopT, a new *Yersinia* Yop effector protein, affects the cytoskeleton of host cells. *Mol. Microbiol.* **29:**915–929.

Jeffrey, B. M., R. J. Suchland, K. L. Quinn, J. R. Davidson, W. E. Stamm, and D. D. Rockey. 2010. Genome sequencing of recent clinical *Chlamydia trachomatis* strains identifies loci associated with tissue tropism and regions of apparent recombination. *Infect. Immun.* **78:**2544–2553.

Jones, K. E., N. G. Patel, M. A. Levy, A. Storeygard, D. Balk, J. L. Gittleman, and P. Daszak. 2008. Global trends in emerging infectious diseases. *Nature* **451:**990–993.

Joseph, S. J., X. Didelot, K. Gandhi, D. Dean, and T. D. Read. 2011. Interplay of recombination and selection in the genomes of *Chlamydia trachomatis*. *Biol. Direct.* **6:**28.

Kalman, S., W. Mitchell, R. Marathe, C. Lammel, J. Fan, R. W. Hyman, L. Olinger, J. Grimwood, R. W. Davis, and R. S. Stephens. 1999. Comparative genomes of *Chlamydia pneumoniae* and *C. trachomatis*. *Nat. Genet.* **21:**385–389.

Laroucau, K., S. Thierry, F. Vorimore, K. Blanco, E. Kaleta, R. Hoop, S. Magnino, D. Vanrompay, K. Sachse, G. S. Myers, P. M. Bavoil, G. Vergnaud, and C. Pourcel. 2008. High resolution typing of *Chlamydophila psittaci* by multilocus VNTR analysis (MLVA). *Infect. Genet. Evol.* **8:**171–181.

Lepore, J. June 1 2009. It's spreading: outbreaks, media scares, and the parrot panic of 1930. *The New Yorker* **19:**46–50.

McClarty, G., H. D. Caldwell, and D. E. Nelson. 2007. Chlamydial interferon gamma immune evasion influences infection tropism. *Curr. Opin. Microbiol.* **10:**47–51.

McNutt, S. H., and E. F. Waller. 1940. Sporadic bovine encephalomyelitis (Buss disease). *Cornell Vet.* **30:**437–448.

Mitchell, C. M., S. Hutton, G. S. Myers, R. Brunham, and P. Timms. 2010. *Chlamydia pneumoniae* is genetically diverse in animals and appears to have crossed the host barrier to humans on (at least) two occasions. *PLoS Pathog.* **6:**e1000903.

Mojica, S., H. Huot Creasy, S. Daugherty, T. D. Read, T. Kim, B. Kaltenboeck, P. Bavoil, and G. S. Myers. 2011. Genome sequence of the obligate intracellular animal pathogen *Chlamydia pecorum* E58. *J. Bacteriol.* **193:**3690.

Myers, G. S., S. A. Mathews, M. Eppinger, C. Mitchell, K. K. O'Brien, O. R. White, F. Benahmed, R. C. Brunham, T. D. Read, J. Ravel, P. M. Bavoil, and P. Timms. 2009. Evidence that human *Chlamydia pneumoniae* was zoonotically acquired. *J. Bacteriol.* **191:**7225–7233.

Pannekoek, Y., V. Dickx, D. S. Beeckman, K. A. Jolley, W. C. Keijzers, E. Vretou, M. C. Maiden, D. Vanrompay, and A. van der Ende. 2010. Multi locus sequence typing of *Chlamydia* reveals an association between *Chlamydia psittaci* genotypes and host species. *PloS One* **5:**e14179.

Pannekoek, Y., G. Morelli, B. Kusecek, S. A. Morre, J. M. Ossewaarde, A. A. Langerak, and A. van der Ende. 2008. Multi locus sequence typing of Chlamydiales: clonal groupings within the obligate intracellular bacteria *Chlamydia trachomatis*. *BMC Microbiol.* **8:**42.

Ponting, C. P. 1999. Chlamydial homologues of the MACPF (MAC/perforin) domain. *Curr. Biol.* **9:** R911–R913.

Rasko, D. A., G. S. Myers, and J. Ravel. 2005. Visualization of comparative genomic analyses by BLAST score ratio. *BMC Bioinform.* **6:**2.

Rattei, T., S. Ott, M. Gutacker, J. Rupp, M. Maass, S. Schreiber, W. Solbach, T. Wirth, and J. Gieffers. 2007. Genetic diversity of the obligate intracellular bacterium *Chlamydophila pneumoniae* by genome-wide analysis of single nucleotide polymorphisms: evidence for highly clonal population structure. *BMC Genomics* **8:**355.

Read, T. D., R. C. Brunham, C. Shen, S. R. Gill, J. F. Heidelberg, O. White, E. K. Hickey, J. Peterson, T. Utterback, K. Berry, S. Bass, K. Linher, J. Weidman, H. Khouri, B. Craven, C. Bowman, R. Dodson, M. Gwinn, W. Nelson, R. DeBoy, J. Kolonay, G. McClarty, S. L. Salzberg, J. Eisen, and C. M. Fraser. 2000. Genome sequences of *Chlamydia trachomatis* MoPn and *Chlamydia pneumoniae* AR39. *Nucleic Acids Res.* **28:**1397–1406.

Read, T. D., G. S. Myers, R. C. Brunham, W. C. Nelson, I. T. Paulsen, J. Heidelberg, E. Holtzapple, H. Khouri, N. B. Federova, H. A. Carty, L. A. Umayam, D. H. Haft, J. Peterson, M. J. Beanan, O. White, S. L. Salzberg, R. C. Hsia, G. McClarty, R. G. Rank, P. M. Bavoil, and C. M. Fraser. 2003. Genome sequence of *Chlamydophila caviae* (*Chlamydia psittaci* GPIC): examining the role of niche-specific genes in the evolution of the *Chlamydiaceae*. *Nucleic Acids Res.* **31:**2134–2147.

Saeed, A. I., N. K. Bhagabati, J. C. Braisted, W. Liang, V. Sharov, E. A. Howe, J. Li, M. Thiagarajan, J. A. White, and J. Quackenbush. 2006. TM4 microarray software suite. *Methods Enzymol.* **411:**134–193.

Saikku, P. 1992. The epidemiology and significance of *Chlamydia pneumoniae*. *J. Infect.* **25**(Suppl. 1): 27–34.

Seth-Smith, H. M., S. R. Harris, R. Rance, A. P. West, J. A. Severin, J. M. Ossewaarde, L. T. Cutcliffe, R. J. Skilton, P. Marsh, J.

Parkhill, I. N. Clarke, and N. R. Thomson. 2011. Genome sequence of the zoonotic pathogen *Chlamydophila psittaci*. *J. Bacteriol.* **193:**1282–1283.

Shao, F., P. M. Merritt, Z. Bao, R. W. Innes, and J. E. Dixon. 2002. A *Yersinia* effector and a *Pseudomonas* avirulence protein define a family of cysteine proteases functioning in bacterial pathogenesis. *Cell* **109:**575–588.

Shirai, M., H. Hirakawa, M. Kimoto, M. Tabuchi, F. Kishi, K. Ouchi, T. Shiba, K. Ishii, M. Hattori, S. Kuhara, and T. Nakazawa. 2000. Comparison of whole genome sequences of *Chlamydia pneumoniae* J138 from Japan and CWL029 from USA. *Nucleic Acids Res.* **28:**2311–2314.

Somboonna, N., R. Wan, D. M. Ojcius, M. A. Pettengill, S. J. Joseph, A. Chang, R. Hsu, T. D. Read, and D. Dean. 3 May 2011. Hypervirulent *Chlamydia trachomatis* clinical strain is a recombinant between lymphogranuloma venereum (L2) and D lineages. *mBio* **2:**e00045-11.

Stephens, R. S., S. Kalman, C. Lammel, J. Fan, R. Marathe, L. Aravind, W. Mitchell, L. Olinger, R. L. Tatusov, Q. Zhao, E. V. Koonin, and R. W. Davis. 1998. Genome sequence of an obligate intracellular pathogen of humans: *Chlamydia trachomatis*. *Science* **282:**754–759.

Suchland, R. J., K. M. Sandoz, B. M. Jeffrey, W. E. Stamm, and D. D. Rockey. 2009. Horizontal transfer of tetracycline resistance among *Chlamydia* spp. in vitro. *Antimicrob. Agents Chemother.* **53:**4604–4611.

Sutherland, E. R., and R. J. Martin. 2007. Asthma and atypical bacterial infection. *Chest* **132:**1962–1966.

Taylor, L. D., D. E. Nelson, D. W. Dorward, W. M. Whitmire, and H. D. Caldwell. 2010. Biological characterization of *Chlamydia trachomatis* plasticity zone MACPF domain family protein CT153. *Infect. Immun.* **78:**2691–2699.

Thalmann, J., K. Janik, M. May, K. Sommer, J. Ebeling, F. Hofmann, H. Genth, and A. Klos. 2010. Actin re-organization induced by *Chlamydia trachomatis* serovar D—evidence for a critical role of the effector protein CT166 targeting Rec. *PLoS ONE* **5**(3):e9887. doi:10.1371/journal.pone.0009887.

Thomson, N. R., and I. N. Clarke. 2010. *Chlamydia trachomatis*: small genome, big challenges. *Future Microbiol.* **5:**555–561.

Thomson, N. R., M. T. Holden, C. Carder, N. Lennard, S. J. Lockey, P. Marsh, P. Skipp, C. D. O'Connor, I. Goodhead, H. Norbertzcak, B. Harris, D. Ormond, R. Rance, M. A. Quail, J. Parkhill, R. S. Stephens, and I. N. Clarke. 2008. *Chlamydia trachomatis*: genome sequence analysis of lymphogranuloma venereum isolates. *Genome Res.* **18:**161–171.

Thomson, N. R., C. Yeats, K. Bell, M. T. Holden, S. D. Bentley, M. Livingstone, A. M. Cerdeno-Tarraga, B. Harris, J. Doggett, D. Ormond, K. Mungall, K. Clarke, T. Feltwell, Z. Hance, M. Sanders, M. A. Quail, C. Price, B. G. Barrell, J. Parkhill, and D. Longbottom. 2005. The *Chlamydophila abortus* genome sequence reveals an array of variable proteins that contribute to interspecies variation. *Genome Res.* **15:**629–640.

Vanrompay, D., A. A. Andersen, R. Ducatelle, and F. Haesebrouck. 1993. Serotyping of European isolates of *Chlamydia psittaci* from poultry and other birds. *J. Clin. Microbiol.* **31:**134–137.

Wardrop, S., A. Fowler, P. O'Callaghan, P. Giffard, and P. Timms. 1999. Characterization of the koala biovar of *Chlamydia pneumoniae* at four gene loci—*ompA*VD4, *ompB*, 16S rRNA, *groESL* spacer region. *Syst. Appl. Microbiol.* **22:**22–27.

Watson, C., and N. J. Alp. 2008. Role of *Chlamydia pneumoniae* in atherosclerosis. *Clin. Sci.* **114:**509–531.

Yousef Mohamad, K., S. M. Roche, G. Myers, P. M. Bavoil, K. Laroucau, S. Magnino, S. Laurent, D. Rasschaert, and A. Rodolakis. 2008. Preliminary phylogenetic identification of virulent *Chlamydophila pecorum* strains. *Infect. Genet. Evol.* **8:**764–771.

LESSONS FROM ENVIRONMENTAL CHLAMYDIAE

Alexander Siegl and Matthias Horn

3

PROLOGUE

As a curiosity in the history of *Chlamydia* research, the discovery of chlamydiae and their initial description as protozoan parasites tremendously boosted the further progression and reputation of protozoology (Halberstädter and Prowazek, 1907). (The terms "chlamydiae" and "chlamydial" refer to all members of the class *Chlamydiae*, including the *Chlamydiaceae* as well as other, more recently identified families.) It was not until 1964 that the chlamydiae were finally, after an intensive debate with respect to their taxonomic affiliation, assigned to the domain *Bacteria* (Moulder, 1964). Exactly 90 years after the first description of chlamydiae, protozoans came into play once again, this time in the form of amoebae as natural hosts of chlamydiae in the environment (Amann et al., 1997). Prior to 1997, chlamydiae were exclusively perceived as pathogens of humans and animals, and our knowledge about their diversity and biology was restricted to members of the family *Chlamydiaceae*, including the human pathogens *Chlamydia trachomatis* and *Chlamydia pneumoniae*. In this context it is interesting that already in the beginning of the 20th century, Stanislaus von Prowazek speculated about a certain diversity within the "*Chlamydozoa*" based on the observation of different diseases caused by this group of infectious agents (von Prowazek, 1912). von Prowazek was right, and the true diversity of these bacteria, which constitute the class *Chlamydiae*, is larger than ever thought before. Many of the more recently discovered chlamydiae exist in various hosts in the environment. To separate them from the pathogenic members of the *Chlamydiaceae*, they are therefore collectively referred to as environmental chlamydiae (or as *Chlamydia*-like bacteria). This chapter summarizes work on environmental chlamydiae performed primarily between 2008 and early 2011. For a more comprehensive overview of earlier studies, the reader is referred to respective book chapters and review articles (Friedman et al., 2003; Horn et al., 2006; Horn, 2008; Greub, 2009; Lamoth and Greub, 2010).

THE UNEXPECTED DIVERSITY OF CHLAMYDIAE

Despite the advances of the past two decades, the research field dealing with environmental chlamydiae can still be regarded as being in its infancy. The first genome sequences from the

Alexander Siegl and Matthias Horn, Department of Microbial Ecology, University of Vienna, 1090 Vienna, Austria.

Intracellular Pathogens I: Chlamydiales
Edited by Ming Tan and Patrik M. Bavoil © 2012 ASM Press, Washington, DC
10.1128/9781555817329.ch3

Chlamydiaceae were already published when the tremendous diversity of chlamydiae in the environment became apparent. Even with the recent increase in the number of recognized species, genera, and even families within the *Chlamydiales*, molecular diversity surveys suggest that additional members of this class exist and await description.

The first studies on environmental chlamydiae reported the discoveries of *Waddlia chondrophila*, which was isolated from an aborted bovine fetus (Dilbeck et al., 1990; Rurangirwa et al., 1999), and the cell culture contaminant *Simkania negevensis* (Kahane et al., 1993; Kahane and Friedman, 1995). However, much of the basic research with respect to environmental chlamydiae has so far been performed with chlamydiae that are associated with free-living amoebae. Members of the genus *Acanthamoeba* are well known to frequently contain bacterial endosymbionts. In an early survey, approximately 5% of *Acanthamoeba* isolates investigated were shown to harbor chlamydiae (Fritsche et al., 1993, 2000). Acanthamoebae are ubiquitous; they have been found in diverse habitats, and they are abundant in many terrestrial and aquatic environments. Interestingly, *Acanthamoeba* species are also considered to act as reservoirs or environmental vectors for a number of bacteria that cause disease in humans, including *Legionella pneumophila*, *Escherichia coli*, *Salmonella enterica* serovar Typhimurium, *Helicobacter pylori*, and *Mycobacterium* spp., just to name a few (reviewed by Salah et al. [2009]). Acanthamoebae are well-suited model organisms for the analysis of environmental chlamydiae for several reasons: (i) they are natural hosts of environmental chlamydiae, (ii) they can be cultured and handled easily in the lab, and (iii) they can serve as artificial host systems for chlamydiae with unknown natural hosts (Collingro et al., 2005a; Thomas et al., 2006; Corsaro et al., 2009; Corsaro and Venditti, 2009).

Most members of the family *Parachlamydiaceae* were identified as intracellular symbionts of amoebae, a finding which is often reflected in the chlamydial genus or species name (Fig. 1). For instance, *Protochlamydia amoebophila* UWE25 is an endosymbiont of an *Acanthamoeba* isolate obtained from soil (Collingro et al., 2005b). In this context, it should be noted that although the species name *amoebophila* (from the Greek adjective *philos*, meaning loving) indicates a mutualistic relationship, we define symbiosis within the scope of this chapter in the sense of Anton De Bary (De Bary, 1879). This interpretation of the term symbiosis neutrally describes any intimate association between organisms including mutualism, parasitism, and commensalism. Which of the three subcategories is appropriate needs to be evaluated individually for each bacterium-host association. The genus *Protochlamydia* includes another amoebal symbiont, *P. naegleriophila*, whose host is a *Naegleria* species, as reflected in the species name (Casson et al., 2008b). Members of at least two additional genera within the family *Parachlamydiaceae* were isolated from amoebae: *Parachlamydia acanthamoebae* Bn_9 was found in a human *Acanthamoeba* isolate (Michel et al., 1994), and *Neochlamydia hartmannellae* A_1Hsp was detected in the aquatic amoeba *Hartmannella vermiformis* (Horn et al., 2000).

Cocultivation of acanthamoebae with environmental samples has proven useful for the recovery of environmental chlamydiae (Collingro et al., 2005a). Using this approach, several novel species including *Criblamydia sequanensis*, *Estrella lausannensis*, and *Estrella* sp. CRIB 31 were recently recovered from samples of a water treatment plant. These environmental chlamydiae constitute the novel family *Criblamydiaceae* (Thomas et al., 2006; Corsaro et al., 2009) (Fig. 1). In addition, two isolates (CRIB 32 and E6) that are only distantly related to all other members of the *Chlamydiales* were obtained, and evidence from mixed eukaryotic cultures suggests that amoebae might not be the natural hosts of these bacteria (Corsaro and Venditti, 2009). Taken together, these findings illustrate that amoebae can serve as surrogate hosts for diverse chlamydiae in the environment, but there is also evidence that amoeba coculture is

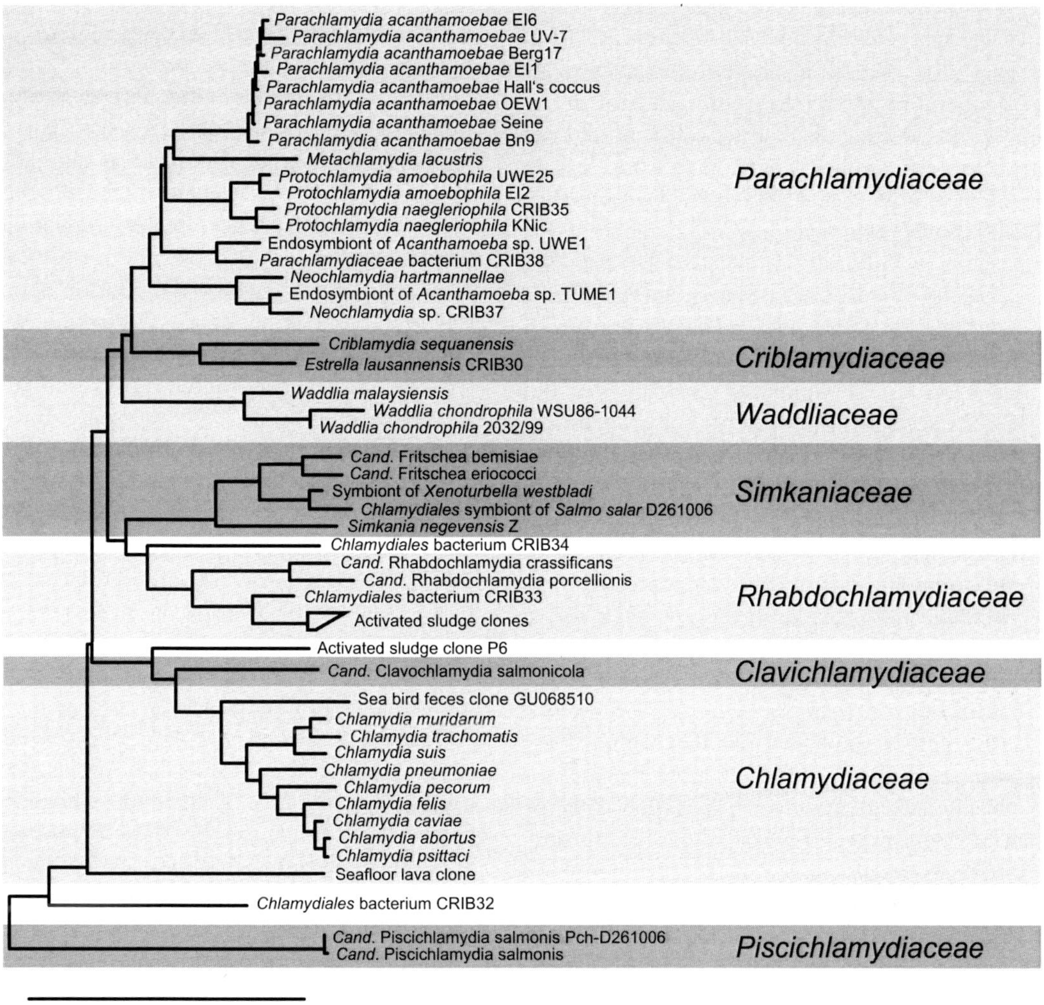

FIGURE 1 Phylogenetic relationships in the class *Chlamydiae*. The neighbor-joining tree shows known members of families within the *Chlamydiales*. Bar, 10% estimated evolutionary distance.
doi:10.1128/9781555817329.ch3.f1

strongly selective and seems to favor predominantly the recovery of *Parachlamydiaceae*. It is important to note that most other protozoa are more difficult (or impossible) to obtain in pure culture than *Acanthamoeba* spp. These protozoa are thus inherently difficult to study with respect to bacterial symbionts. We predict that much of the chlamydial diversity in the environment seen with molecular methods represents symbionts of protozoa or other members of the microfauna that are not as easily accessible as acanthamoebae. The development of new techniques suitable for the isolation, cultivation, and axenization of protozoa will help to further expand our view of environmental chlamydiae as symbionts of unicellular eukaryotes.

Within the class *Chlamydiae*, there is only one order, *Chlamydiales*, which is composed of eight known families (16S rRNA divergence, >10%) (Horn, 2008; Kuo et al., 2011) (Fig. 1): the *Chlamydiaceae*, *Parachlamydiaceae*, *Criblamydiaceae*,

Waddliaceae, *Simkaniaceae*, *Rhabdochlamydiaceae*, *Clavichlamydiaceae*, and *Piscichlamydiaceae*. However, the detection of yet uncharacterized, divergent sequence clades in molecular diversity surveys suggests that the actual number of chlamydial families is even higher. Just within the past three years (2008 to 2011), more than 200 novel 16S rRNA gene sequences that are phylogenetically affiliated with the class *Chlamydiae* but not assigned to the family *Chlamydiaceae* have been added to the NCBI nucleotide database. Intriguingly, this is almost exactly the number of all chlamydial 16S rRNA gene sequences (>1,000 bp) including the *Chlamydiaceae* that were deposited in public databases by the year 2007 (Horn, 2008).

The variety of aquatic, terrestrial, and engineered habitats in which 16S rRNA genes of environmental chlamydiae were detected is remarkable and provides further evidence for an association of these bacteria with a great range of eukaryotic host organisms including vertebrates and invertebrates (reviewed by Corsaro et al. [2003] and Horn [2008]). Some of the more recently described putative habitats for chlamydiae are noteworthy, as they include harsh environments such as permafrost soil, activated sludge, anoxic marine sediment, and hydrothermal vent fluid. rRNA gene sequences affiliated with known environmental chlamydiae were also reported from a hot spring microbial mat (Nishio et al., 2010) as well as from a tailing pond (Ramos-Padron et al., 2011), underlining the ubiquity of these organisms. In addition, sequences only distantly related to known environmental chlamydiae were recovered from anoxic water layers of Lake Tanganyika in Africa (Schubert et al., 2006) and from microbial mats on lava tube walls on Hawaii and the Azores (J. J. Hathaway, M. G. Garcia, M. Moya, M. N. Spilde, F. D. Stone, L. M. Dapkevicius, and D. E. Northup, unpublished data). Ongoing metagenomic studies and diversity surveys using next-generation sequencing methods, such as pyrosequencing, will continue to reveal an even greater diversity of chlamydiae. As these methods rely solely on the detection of DNA, the challenge to identify their natural hosts remains.

THE BROAD HOST RANGE OF ENVIRONMENTAL CHLAMYDIAE

Sponges are considered as being the oldest living metazoans, which evolved more than 600 million years ago and played a significant role in the emergence of multicellular animals. It is thus interesting that chlamydial 16S rRNA gene sequences have been recovered from different members of this ancient animal lineage. First reports came from the marine sponge *Suberites zeteki* (Zhu et al., 2008) and were complemented by similar findings from the sponges *Prostylyssa foetida* (G. B. Purushottama, S. Malathi, M. N. Venugopal, and I. Karunasagar, unpublished data) and *Suberites carnosus* (G. B. Purushottama, S. Malathi, T. Suresh, K. H. Harishchandra, I. Karunasagar, and I. Karunasagar, unpublished data). Since only inner tissue was used for DNA extraction in the case of *Suberites zeteki*, it cannot be ruled out that the detected chlamydiae served as food particles for the filter-feeding sponge rather than being permanently associated with it. Massive parallel 16S rRNA gene tag pyrosequencing of three Australian sponges also revealed the presence of chlamydiae, but at relatively low abundances (≤0.03%) (Webster et al., 2009), indicating that the class *Chlamydiae* is not an abundant part of the microbial community associated with these sponge species.

Chlamydiae and plants share a common evolutionary history (Huang and Gogarten, 2007; Becker et al., 2008; Moustafa et al., 2008; Collingro et al., 2011). However, although experimental evidence is, at least to our knowledge, lacking, it is commonly accepted that extant chlamydial species are not capable of infecting higher plants. Nevertheless, chlamydial sequences were obtained from the inner tissue of ginseng (F. Qiu and W. Song, unpublished data) and from soybean roots depicted as being an endophyte (Zhang et al., 2011). The first published report of chlamydiae associated with plant leaves described *Simkaniaceae*-related sequences obtained from citrus leaves (Sagaram et al., 2009). As the plant surface was sterilized with bleach and 70% ethanol prior to DNA extraction, the authors speculate that the bacteria detected can be regarded as true endophytes.

These findings are interesting, but ethanol treatment does not remove DNA efficiently, and thus, it is possible that the environmental chlamydiae were harbored in amoebal cysts, which were associated with the plant surface and resistant to sterilization. So far, experimental evidence such as infection studies or the detection of chlamydiae in plant tissue, e.g., by fluorescence in situ hybridization (FISH), is lacking. Whether chlamydiae are indeed associated with extant plants thus remains an open question.

Some years ago, the enigmatic marine invertebrates *Xenoturbella bocki* and *Xenoturbella westbladi* were shown to serve as hosts for chlamydial symbionts related to *Fritschea* and *Simkania* within gastrodermal cells (Israelsson, 2007). No cytopathological impact on the *Xenoturbella* host cells was observed, suggesting that these bacteria do not harm their hosts and represent mutualistic symbionts (Israelsson, 2007). However, recent FISH studies indicated that chlamydiae are not the predominant symbionts in *X. bocki*, since they are clearly outnumbered by another endosymbiont affiliated with the *Gammaproteobacteria* (Kjeldsen et al., 2010). Remarkably, symbionts were also detected in spermatid clusters of worms, enabling vertical transmission of the bacteria to the next generation and thus granting a long-term association.

The deepest-branching family represented by "*Candidatus* Piscichlamydia salmonis" has been found in marine-stage Atlantic salmon (*Salmo salar*) and in freshwater Arctic charr (*Salvelinus alpinus*) (Draghi et al., 2004, 2010). The "*Candidatus*" status is a taxonomic term used for provisional classification of microbes that have been identified solely based on 16S rRNA genes (Murray and Stackebrandt, 1995). The closest relative of the *Chlamydiaceae*, "*Candidatus* Clavochlamydia salmonicola," was reported from freshwater stage Atlantic salmon and wild brown trout (*Salmo trutta*) (Karlsen et al., 2008; Mitchell et al., 2010). In addition, a putative novel lineage within the order *Chlamydiales* was reported from a leopard shark, *Triakis semifasciata*, suffering from epitheliocystis, an infection of the gills (Polkinghorne et al., 2010). Moreover, epitheliocystis in sea bass (*Dicentrarchus labrax*) is also thought to be associated with a distinct clade of chlamydiae (A. Polkinghorne, A. Lehner, S. Crespo, and L. Vaughan, unpublished data). Molecular evidence for a novel chlamydial organism in fecal samples of seabirds was recently described (Christerson et al., 2010). The recovered 16S rRNA gene shows 92% sequence identity with *Chlamydiaceae*, thus representing a putative novel genus within this family. Its natural host is currently unknown. However, as the bird species investigated in this study feeds on fish, it is tempting to speculate that this novel bacterium is yet another fish symbiont.

Within the last years, it became evident that certain environmental chlamydiae are widespread among eukaryotes while others seem to be restricted to single host species. *Waddlia*-related sequences were recovered originally from mammals and marsupials. *W. chondrophila* was detected in humans, cattle, and the marsupial Gilbert's potoroo (Rurangirwa et al., 1999; Bodetti et al., 2003; Haider et al., 2008), whereas *Waddlia malaysiensis* seems to be restricted to fruit bats (Chua et al., 2005). Recently, a *W. chondrophila* 16S rRNA gene sequence (100% identity) was also obtained from chicken (Robertson et al., 2010), suggesting that the host range of this chlamydial organism is broader than previously thought. Other chlamydiae appear to be strictly host specific. One such case is the recently discovered "*Candidatus* Metachlamydia lacustris," a member of the *Parachlamydiaceae*, which shares its natural host *Saccamoeba lacustris* with additional symbionts belonging to the *Betaproteobacteria*. Surprisingly, "*Candidatus* Metachlamydia lacustris" is lytic to its natural host and is not capable of infecting other amoeba species tested (Corsaro et al., 2010).

Whiteflies are a well-studied model system for multipartner associations between insects and bacteria (Clark et al., 1992; Costa et al., 1995). The sweet potato whitefly *Bemisia tabaci* harbors the primary obligate symbiont *Portiera aleyrodidarum* and several facultative secondary symbionts including *Rickettsia*, *Wolbachia*, and *Cardinium* species, which are together vertically transmitted to the next generation. Environmental chlamydiae, later named "*Candidatus* Fritschea bemisiae," were also shown to be secondary symbionts of *B. tabaci* (Thao et al., 2003; Everett et al., 2005). However, subsequent studies showed that "*Candidatus* Fritschea bemisiae" seems to be restricted to certain *B. tabaci* biotypes. While it is present in biotype A from the United States (Thao et al., 2003), *Fritschea* could be detected neither in biotype Q from Croatia (Skaljac et al., 2010) nor in biotypes B and Q from China (Chu et al., 2011). The fact that the greenhouse whitefly *Trialeurodes vaporariorum* possesses the same primary symbiont as the sweet potato whitefly but lacks "*Candidatus* Fritschea bemisiae" further supports the strict host specificity of these chlamydial organisms. Conversely, the closely related species "*Candidatus* Fritschea eriococci" is so far only known from the scale insect *Eriococcus spurius* (Thao et al., 2003). Taken together, there is evidence that some environmental chlamydiae show a very narrow host range, but for many chlamydiae this might just reflect the current state of knowledge, and future studies will likely discover novel hosts for many of the known chlamydiae.

THE DEVELOPMENTAL CYCLE OF ENVIRONMENTAL CHLAMYDIAE

The presence of different developmental stages was observed already when the "*Chlamydozoa*," known now as *C. trachomatis*, were discovered (Halberstädter, 1912). Reticulate bodies (RBs) were originally designated "initial bodies," whereas the term "elementary bodies" (EBs) was already proposed in the early days of *Chlamydia* research and still holds today. Since the availability of electron microscopy, the existence of a biphasic developmental cycle with distinct morphological stages has been confirmed for all members of the *Chlamydiaceae* (Abdelrahman and Belland, 2005). All environmental chlamydiae described so far also have a similar developmental cycle, with EBs and RBs that are in general ultrastructurally similar to their counterparts from the *Chlamydiaceae* (Fig. 2A and C) (reviewed by Horn [2008]). However, some exceptions to this rule are noteworthy, and they shed light onto the morphological and biological diversity of the *Chlamydiae*.

Unusual Morphotypes of Environmental Chlamydiae

The recently identified "*Candidatus* Clavochlamydia salmonicola" forms characteristic EBs with a head-and-tail shape and a length of approximately 1 μm (Karlsen et al., 2008) (Fig. 2D). Transitional stages between these unusual EBs, intermediate bodies (IBs) resembling the IBs of the *Chlamydiaceae*, and large irregularly shaped RBs were observed. To date, virtually nothing is known about the function of this unique EB morphology. Bacteria with a similar head-and-tail shape have been observed in marine and freshwater fish on several occasions (Rourke et al., 1984; Bradley et al., 1988; Draghi et al., 2004). However, as those bacteria were not identified by 16S rRNA gene sequencing, it is unclear whether they were related to "*Candidatus* Clavochlamydia salmonicola."

Another striking variation in EB morphology can be observed for members of the *Criblamydiaceae*. Star-shaped EBs were found via electron microscopy for *Criblamydia sequanensis* (Thomas et al., 2006) as well as the related *Estrella lausannensis* strain CRIB30 (Corsaro et al., 2009), which both infect *Acanthamoeba castellanii* (Fig. 2E). Star-shaped EBs contain an elongated, intracellular structure and are surrounded by a multilayered cell wall, which was described to consist of five layers. It exhibits several branches or knobs reminiscent of envelope spikes from certain viruses (Thomas et al., 2006). Multilayered cell walls and one or more intracellular electron-translucent structures were also described for rod-shaped EBs from members of the *Rhabdochlamydiaceae*: "*Candidatus* Rhabdochlamydia porcellionis," associated with isopods (Kostanjsek et al., 2004), and "*Candidatus* Rhabdochlamydia cr assificans," found in cockroaches (Corsaro et al., 2007) (Fig. 2F). Currently, the molecular composition of the individual cell wall layers

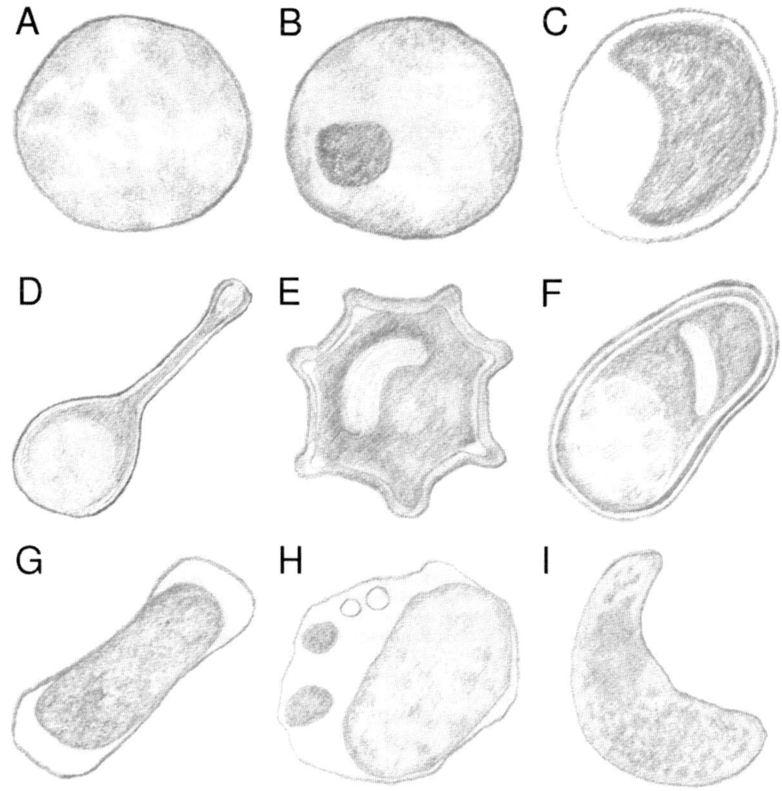

FIGURE 2 Morphological diversity of chlamydial developmental stages. Drawings show an RB (A), an IB (B), and an EB (C) as they are typically observed for members of both the *Chlamydiaceae* and environmental chlamydiae; unusual EBs of the *Clavichlamydiaceae* (D), the *Criblamydiaceae* (E), and the *Rhabdochlamydiaceae* (F); putative host-dependent morphology of a *Simkaniaceae* EB in *Acanthamoeba* species and *Xenoturbella* worms (G); a *Waddlia* aberrant body (H); and a crescent body (I). See the text for further explanations. Drawings represent simplified illustrations based on published electron micrographs and are not to scale. doi:10.1128/9781555817329.ch3.f2

remains unknown, and neither the composition nor the function of the subcellular structures within these EBs is known, but the latter might represent storage compounds such as glycogen or possibly phages.

The morphology of EBs of the *Simkaniaceae* may vary depending on the host cell. Usually these developmental forms appear oval, but elongated or even trapezoid forms were reported in several *Acanthamoeba* species (Kahane et al., 2001; Michel et al., 2005; Henning et al., 2007) (Fig. 2G). Similar observations were reported for the *Simkania*-related symbiont of the marine invertebrate *Xenoturbella bocki*, in which EBs display a rectangular shape with a five-layered cell wall (Israelsson, 2007). Interestingly, EB morphotypes reminiscent of *Simkaniaceae* have also been reported for *C. pneumoniae* strains TW-183 and AR39 (Chi et al., 1987). These chlamydial EBs were pear-shaped with a large periplasmic space. Such morphotypes could be the result of alterations in cell wall composition that affect cell wall stability resulting in irregular cell shapes and collapse during the transformation from RBs to EBs. One should also keep in mind that fixation conditions for electron microscopy can produce artifactual changes in the ultrastructure of the cell envelope of gram-negative bacteria (Silva and Sousa, 1973). It therefore remains unclear whether the observed five-layered cell walls of *Rhabdochlamydiaceae*, *Criblamydiaceae*, and *Simkaniaceae* are indeed specific adaptations of these groups of bacteria.

Another possibly host-specific developmental stage was described for *W. chondrophila*. Amorphous *Waddlia* cells named "aberrant bodies" were observed in human endometrial cells but not in any other cell lines analyzed so far (Kebbi-Beghdadi et al., 2011) (Fig. 2H). These atypical forms with diameters of more than 5 μm emerged from 72 hours postinfection onward and contained several electron-dense spots, probably indicating chromatin condensation events similar to the IBs of the *Chlamydiaceae* (Fig. 2B). The formation of aberrant bodies is reversible under laboratory conditions by growth medium renewal. For that reason, it was suggested that *Waddlia* aberrant bodies resemble a starvation-induced form similar to the in vitro persistent aberrant RBs observed for members of the *Chlamydiaceae* (Beatty et al., 1994).

One of the most thought-provoking findings with respect to the diversity of chlamydial developmental stages was the description in *P. acanthamoeba* of a second infectious form in addition to EBs (Greub and Raoult, 2002). The so-called crescent bodies of *P. acanthamoebae* show a lunate appearance and were found both inside and outside amoeba host cells (Fig. 2I). Similar forms have also been reported for other environmental chlamydiae (Greub, 2009). A very similar crescent morphology has been described for the planctomycete *Gemmata obscuriglobus* but was demonstrated to be caused by the (standard) fixative used to prepare the cells for electron microscopy (Lindsay et al., 1995). Since a similar fixative was used for the analysis of *P. acanthamoebae*, the possibility of artifacts due to the fixation procedure required for electron microscopy cannot completely be ruled out. In addition, a recent study found crescent-shaped bodies of *Protochlamydia* only in amoeba cysts with tightened cell walls but not in trophozoites (Nakamura et al., 2010), which does not fit well with the notion of crescent bodies as a distinct developmental stage. Modern fixation methods such as cryofixation, which better preserve biological structures and are less prone to artifacts, should provide more detailed insights into the ultrastructure of environmental chlamydiae.

In summary, the ultrastructural analysis of environmental chlamydiae illustrates that the developmental cycle *per se* is conserved among all known chlamydiae but their phylogenetic diversity is accompanied by a number of morphological differences. It will be a challenging task for future studies to elucidate the biological significance of these specialized structures.

Extracellular Activity of Chlamydial EBs

Traditionally, chlamydial EBs are regarded as spore-like forms which are metabolically inert. A recent study in which the metabolic activity of *P. amoebophila* was investigated at the single-cell level challenged this long-standing dogma (Haider et al., 2010). Raman microspectroscopy, which provides fingerprints of the molecular composition of individual bacterial cells, was used to differentiate between RBs and EBs and to monitor intracellular and extracellular uptake of isotope-labeled phenylalanine. Surprisingly, there was host-free uptake of phenylalanine by extracellular *Protochlamydia* EBs, which was reversibly blocked by adding the uncoupler carbonylcyanide-3-chlorophenylhydrazone (CCCP). This result shows that *Protochlamydia* uses proton or sodium-dependent transporters for uptake of this amino acid and that extracellular EBs are able to reenergize their membrane after perturbation by an uncoupler. Congruent to the Raman microspectroscopy-based findings, the same study demonstrated that extracellular *Protochlamydia* EBs are capable of *de novo* protein biosynthesis. This novel finding is not unique to *Protochlamydia*, since EBs from *C. trachomatis* also showed host-free uptake of phenylalanine and protein synthesis (Haider et al., 2010). The extracellular activity of chlamydial EBs was dependent on the incubation medium used, which may explain why EBs have not been previously shown to be metabolically active. Remarkably, EBs of

Protochlamydia remained metabolically active and infectious for at least 3 weeks outside their host cell. This long-lasting resistance might represent an adaptation trait of environmentally transmitted chlamydiae.

Independent evidence for the extracellular metabolic activity of *Protochlamydia* EBs was provided by a recent comprehensive proteomic analysis, in which 472 proteins from this infectious stage were identified (Sixt et al., 2011). Remarkably, almost all functional categories including virulence factors were recovered with overrepresentation of proteins involved in transcription, translation, energy production, and nucleotide metabolism. EBs from *Chlamydia* spp. also express a virtually complete set of proteins required for transcription, protein synthesis, and energy generation (Vandahl et al., 2001; Shaw et al., 2002). The traditional view has been that these proteins are required for immediate use upon infection of a new eukaryotic host cell. However, the presence of these proteins is consistent with the recent detection of at least a limited metabolic activity in extracellular EBs and provides evidence that some biosynthetic processes may take place in EBs.

Long-Term Persistence of Environmental Chlamydiae in Amoeba Cysts

A key feature of the natural host of many environmental chlamydiae, *Acanthamoeba* spp., is its ability to form cysts under adverse environmental conditions. *Acanthamoeba* cysts are extremely stable and show increased resistance against drought, chlorine, and antibiotics. They are thus being discussed as vehicles of dispersal for amoeba-associated bacteria (Barker and Brown, 1994; Kahane et al., 2001). *Protochlamydia* was originally described in *Acanthamoeba* trophozoites and cysts (Fritsche et al., 2000), leading to the hypothesis that in the cyst stage, both host and symbiont can survive adverse environmental conditions for extended time periods. In a recent study, the presence of *Protochlamydia* sp. in acanthamoebae was monitored by transmission electron microscopy and FISH from the trophozoite stage through encystation and after cysts had reverted to trophozoites (Nakamura et al., 2010). The predominant developmental forms inside amoeba cysts were EBs, some of which were presumably not present within inclusions. Trophozoites recovered from cysts harbored both EBs and dividing RBs, confirming the previous assumption that *Protochlamydia* spp. survive encystation of its amoeba host. Intriguingly, several studies indicate that amoebae infected with *P. acanthamoeba* cannot encyst, at least under laboratory conditions (Michel et al., 1994; Greub and Raoult, 2002; Nakamura et al., 2010). A recent proteomics-based approach showed that *P. acanthamoeba* blocks amoeba encystment at an early stage and that actin isoforms required for cyst formation are missing in infected cells (Leitsch et al., 2010). The biological advantage of a strategy to prevent host cell encystation, which results in the loss of a long-term niche for persistence, remains unclear.

HOST CELL INTERACTIONS

For obligate intracellular bacteria like the chlamydiae, the term "host cell interactions" can be subdivided into the following temporally and spatially separated stages: microbe-host recognition, internalization, replicative phase with host cell exploitation, and finally persistence within or release from the host cell to start another infectious cycle. These processes are only poorly understood for environmental chlamydiae, and our knowledge about the infection process is fragmentary.

For the naturally occurring symbiosis between *Acanthamoeba* spp. and *P. acanthamoeba* or *P. amoebophila*, the interaction is initiated by phagocytosis (Greub and Raoult, 2002). However, most studies so far have focused on infection of mammalian cells rather than on the natural amoeba hosts. For *W. chondrophila*, it was shown that internalization by human monocyte-derived macrophages occurs passively, both for living and heat-inactivated chlamydial cells (Goy et al., 2008). As deduced from studies with knockout mice, the uptake of *P. acanthamoeba* in bone marrow-derived macrophages is also mediated by phagocytosis but independent of Toll-like receptors (Roger et al., 2010). Consistent with members of the *Chlamydiaceae*, early stages of infection seem to rely on the presence of the type III secretion (T3S) system, as both structural components and associated chaperones were identified in the proteome of *P. amoebophila* UWE25 EBs (Sixt et al., 2011; more details about the chlamydial T3S system are reviewed in chapter 9, "Protein secretion and *Chlamydia* pathogenesis"). In contrast, T4S system components, which are shared by several environmental chlamydiae but are absent in the *Chlamydiaceae*, were not identified in the *P. amoebophila* proteome. This suggests that the T4S system is not required during early stages of infection.

Inclusion membrane proteins (Incs) are a characteristic feature of the *Chlamydiaceae*, and they represent key mediators of host cell interaction such as membrane fusion and the interaction with a number of host cell proteins (Rockey et al., 2002). The presence of similar proteins in environmental chlamydiae was confirmed recently for *P. amoebophila* UWE25 (Heinz et al., 2010). An *in silico* motif search led to the identification of 23 candidate Incs, and immunofluorescence analysis confirmed the expression of five Incs during infection of amoebae. In addition, immunotransmission electron microscopy verified the location of four *Protochlamydia* Inc proteins in the inclusion membrane. The findings from this study are a prime example of common themes (Incs) used by pathogenic *Chlamydiaceae* and amoeba-associated chlamydial symbionts alike.

The mechanism of establishing the intracellular niche in mammalian host cells is probably best understood for *P. acanthamoeba* (Greub et al., 2005b) and *W. chondrophila* (Croxatto and Greub, 2010; Kebbi-Beghdadi et al., 2011), which show similarities with, and differences from, *Chlamydia* spp. Both *P. acanthamoeba* and *W. chondrophila* are found inside the early endosome of their host cell, regardless of whether the infection is performed with live or heat-inactivated organisms. However, *W. chondrophila* manages to evade further steps of the phagocytic pathway and instead colocalizes with both mitochondria and the endoplasmic reticulum, possibly in order to obtain ATP and lipids from the host cell. In contrast, *Chlamydia* spp. modulate Golgi trafficking in order to scavenge host lipids (Hackstadt et al., 1995). The microtubule-dependent recruitment of mitochondria described for *Chlamydia psittaci* (Matsumoto et al., 1991) is mediated in *W. chondrophila* by both microtubules and actin filaments (Croxatto and Greub, 2010). In a remarkable finding, *S. negevensis* strain H-2 was detected not only in cytoplasmic inclusions but also in the karyoplasm of the nucleus of African green monkey kidney cells 7 days postinfection (Henning et al., 2007). The inflated appearance of the nucleus as well as the presence of condensed chromatin indicated severe cell damage or ongoing cell death. This unique observation requires further investigation, as it might imply a novel lifestyle of *Simkania* spp. within their host cells.

In summary, our understanding of how environmental chlamydiae infect and interact with their host cells is at an early stage because of the recent discovery of these bacteria, their large diversity, and the diversity of their hosts. The field would benefit from the development and establishment of model systems so that different studies can be compared and from standardization in the use of specific bacterial strains and host cells. However, the general theme that has already emerged is that there are many similarities between environmental chlamydiae and the *Chlamydiaceae* as well as pronounced differences in their intracellular lifestyles.

ENVIRONMENTAL CHLAMYDIAE AND PATHOGENICITY

As environmental chlamydiae, like their pathogenic relatives of the *Chlamydiaceae* family, are obligate intracellular bacteria of eukaryotic hosts, they have been suspected of being pathogens of humans and animals since the early days of their discovery (Birtles et al., 1997). In the environment they occur in ubiquitous soil amoebae (Fritsche et al., 1993). Thus, given the wide distribution and abundance of environmental chlamydiae, any pathogenic potential could have a major impact on public and veterinary health. Various aspects of infections caused by *Chlamydia* spp. and their implications for public health are reviewed in chapter 1, "*Chlamydia* infection and epidemiology." In this section, we will summarize the latest findings and discuss the evidence for and against potential health threats caused by environmental chlamydiae.

Of course, Koch's postulates can hardly be fulfilled to prove a causal relationship with disease for bacteria that cannot be cultivated in cell-free media, but modified versions of the original postulates that utilize molecular and immunological testing might help to clarify their role as potential pathogens of humans (Fredricks and Relman, 1996). Along these lines, a potential pathogenicity of environmental chlamydiae can be assessed by detection of bacterial DNA by PCR, by detection of the organism by immunohistochemistry, by detection of host response by serology, and by direct isolation of the organism. In these approaches the organisms should be present in diseased tissue while absent in uninvolved tissue. PCR provides only very weak evidence because it merely shows the presence of DNA but not of the organism. But even serological data and isolation of the organisms provide no direct causal evidence and have to be interpreted with caution for organisms as ubiquitous as environmental chlamydiae. Ideally, there would be an animal model available that can be used to monitor potential infections and pathologic changes upon challenge with environmental chlamydiae. Finally, the forms of evidence for microbial causation should be shown to be reproducible, and large clinical studies including healthy control groups are required.

Waddlia chondrophila and Miscarriage in Humans

W. chondrophila was originally isolated from an aborted bovine fetus and was thus proposed as a causative agent of abortion in cattle (Henning et al., 2002; Dilbeck-Robertson et al., 2003). More recently, *W. chondrophila* has been associated with lower respiratory tract disease and miscarriage in humans in several studies, as will be discussed.

Under laboratory conditions, there is accumulating evidence that *W. chondrophila* is capable of infecting diverse human cell lines (Goy et al., 2008; Kebbi-Beghdadi et al., 2011). Studies based on confocal and electron microscopy revealed that this chlamydial organism enters human monocyte-derived macrophages (Goy et al., 2008). While heat-inactivated organisms were internalized but did not have not any deleterious effects, live *W. chondrophila* multiplied logarithmically, which resulted in severe cytopathic effects and macrophage lysis. *W. chondrophila* is thus able to resist the antimicrobial activity of human macrophages. This finding led the authors to propose that macrophages serve as vectors for dissemination and establishment of *W. chondrophila* infection in humans. In addition to macrophages, *W. chondrophila* is also able to infect a human lung epithelial cell line (A549 pneumocytes) in vitro and produces similar cytopathic effects (Kebbi-Beghdadi et al., 2011). The same study showed that Ishikawa endometrial cells derived from the human uterus were susceptible to and supported the growth of *W. chondrophila*. In this case, however, no lytic behavior was detected by *Waddlia*-specific quantitative PCR, immunofluorescence, and electron microscopy. Instead, *W. chondrophila* stopped growing 48 hours postinfection and aberrant bodies were observed, suggesting persistence of *W. chondrophila* in this cell line. Remarkably, this persistent stage was reversed by the renewal of the culture medium, resulting in complete host cell lysis. Taken together, these studies demonstrate that *W. chondrophila* is capable of entering and establishing an infection in a variety of human cell lines. This finding distinguishes *W. chondrophila* from other environmental chlamydiae, such as members of the *Parachlamydiaceae*, whose primary hosts are free-living amoebae. Since *W. chondrophila* is able to efficiently infect human respiratory and endometrial cells and macrophages in cell culture, it has the potential to be a human pathogen.

There is only limited clinical evidence that *Waddlia* causes human respiratory disease. In a PCR-based assay, DNA of environmental chlamydiae was detected in approximately 1% of respiratory samples from patients (n = 387) suffering from community-acquired pneumonia (Haider et al., 2008). Within this data set, just a single sequence originating from sputum could be assigned phylogenetically to *W. chondrophila*, which does not support a considerable involvement of *W. chondrophila* in community-acquired pneumonia. Similar findings were obtained from a study in which the application of a *W. chondrophila*-specific real-time PCR assay resulted in the detection of DNA from this organism in 3 of 32 clinical samples obtained from children suffering from bronchiolitis (Goy et al., 2009). However, the limitation of both studies is that neither had a healthy control group for comparison, and thus it is unknown if this PCR-based detection bears any relationship to the respiratory infection. In addition, the mere presence of DNA does not establish a causal link between the organism and disease. All known environmental chlamydiae, including *W. chondrophila*, can use ubiquitous amoebae as host. Thus, PCR-based detection of these bacteria, particularly in respiratory samples, may not be clinically significant and may simply reflect our daily exposure to environmental chlamydiae and their amoeba hosts.

A role for *W. chondrophila* in human miscarriage was originally proposed on the basis of serological data (Baud et al., 2007). Immunoglobulin G antibody titers against *W. chondrophila* were higher in patients who had suffered a miscarriage (n = 69) than in a control group with uneventful pregnancies (n = 169). Data analysis suggested that the presence of *Waddlia*-specific antibodies correlated with contact with animals, which led to the hypothesis that *W. chondrophila* infections can be acquired zoonotically. Recently, *Waddlia* was directly detected for the first time in placental tissue from a patient who had a miscarriage (Baud et al., 2011). In this case, the identification of *W. chondrophila* was based not only on serology and PCR but also on immunohistochemistry. Since this patient did not have contact with animals, the authors could only speculate about putative modes of transmission other than zoonotic acquisition. However, to date, evidence for either foodborne or human-to-human transmission of any environmental chlamydiae does not exist.

Taken together, reports of an association between *W. chondrophila* and miscarriage in humans are limited in nature, and a causal relationship has not yet been established. Clinical data on putative infections with *W. chondrophila* are still scarce, and there are no intervention studies and no data about treatment. Such studies, including independent studies in different laboratories and hospitals, will be required to determine if there is a causal link between *W. chondrophila* infections and human miscarriage.

S. negevensis and Respiratory Disease in Humans

The original host of *S. negevensis* is currently unknown. Initially, this chlamydial organism was discovered serendipitously as a human cell line contaminant (Kahane and Friedman, 1995). Since then, its possible involvement in a variety of respiratory diseases, such as bronchiolitis (Kahane et al., 1998; Greenberg et al., 2003) and community-acquired pneumonia (Lieberman et al., 1997), has been suggested based on serological and molecular evidence. In the laboratory, *S. negevensis* can be grown not only in amoebae but also in a variety of human cell lines (Kahane et al., 2007a). Interestingly, *S. negevensis* can spread from *Acanthamoeba polyphaga* to the monocyte/macrophage cell line U937 and from there to other cell types (Kahane et al., 2008), providing a possible mode of transmission from the environment, e.g., from amoebae in drinking water to animals and humans (Kahane et al., 2007b). Remarkably, some surveys succeeded in isolating and culturing the organism directly from patients suffering from respiratory diseases (Kumar et al., 2005; Friedman et al.,

2006; Kahane et al., 2007b) or lung transplant rejection (Husain et al., 2007), thus fulfilling at least part of the second postulate originally formulated by Koch. The high seroprevalence for *S. negevensis* in unaffected individuals is notable. Seropositivity rates of up to 80% for healthy individuals were, for example, observed in Israel (Friedman et al., 1999), suggesting a ubiquitous occurrence of *S. negevensis* and frequent exposure of humans to this organism. Further research is required to determine the pathogenic potential of *S. negevensis* and to further explore its possible association with respiratory disease in humans.

Parachlamydiaceae and Human Disease

Despite the fact that the amoeba symbiont *P. acanthamoebae* grows poorly or not at all in several human cell lines tested in laboratory experiments (Maurin et al., 2002; Casson et al., 2006; Hayashi et al., 2010), these bacteria have been linked to various human diseases ranging from respiratory tract infections to miscarriage, urogenital tract and eye infections, and atherosclerosis (comprehensively reviewed by Greub [2009]). In vitro, *P. acanthamoebae* induced apoptosis (Greub et al., 2003) but prevented the release of most cytokines and the production of reactive oxygen species (Greub et al., 2005a) by human macrophages. These effects could represent a strategy to avoid detection by the host immune system, although at the expense of the host cell. A mouse model for pneumonia caused by *P. acanthamoebae* is available (Casson et al., 2008a). Infection resulted in pneumonia accompanied by dramatic weight loss and a mortality rate of 50%. *P. acanthamoebae* was detected in lesions, and living organisms were recovered from the lungs of infected mice by using amoebal coculture. Interestingly, *P. acanthamoeba* induced the expression of the cytokines interleukin 1-β (IL1-β), IL12, tumor necrosis factor alpha, and IL6 in this in vivo mouse model (Roger et al., 2010). However, intratracheal inoculation, i.e., inoculation of the surgically exposed trachea, and a dose of 10^8 bacteria were required to induce pneumonia. It is questionable whether this represents a realistic scenario for infections occurring in nature. The majority of the studies on the association of *P. acanthamoebae* with human disease rely solely on PCR-based detection. Neither immunohistochemistry data from affected tissue nor isolates from patient material are available to date. Even less is known about the putative role of other members of the *Parachlamydiaceae* in human disease. DNA of *Protochlamydia naegleriophila* and *P. amoebophila* have so far been detected only in single cases of pneumonia (Casson et al., 2008b; Haider et al., 2008). In summary, the evidence for an association of the *Parachlamydiaceae* with human diseases is currently relatively weak.

Environmental Chlamydiae and Disease in Animals

Environmental chlamydiae have been detected in a wide variety of vertebrate and invertebrate hosts, and some members of this group of bacteria have been proposed to cause disease in animals. Like *W. chondrophila* (Rurangirwa et al., 1999; Henning et al., 2002; Dilbeck-Robertson et al., 2003), *Parachlamydia* spp. have been suspected to cause bovine abortion (Borel et al., 2007). Whereas *W. chondrophila* has been isolated originally from an aborted fetus, the evidence for an association of *Parachlamydia* spp. with abortion in cattle is so far based on PCR and immunohistochemistry only. In a PCR-based study on cats with corneal diseases, *P. acanthamoeba*-like 16S rRNA gene sequences were found in both diseased and unaffected cats, showing that cats are exposed to these chlamydiae (Richter et al., 2010). As the natural *Acanthamoeba* host of *P. acanthamoebae* could not be detected by PCR in this study, it was proposed that *P. acanthamoebae* is able to persist in corneal tissue of cats (Richter et al., 2010).

Recently discovered environmental chlamydiae have been proposed to cause epitheliocystis, which is an infectious disease that affects the gills of several fish species (Nowak and LaPatra, 2006; Mitchell and Rodger, 2011). Epitheliocystis has a large economic impact because it is associated with heavy mortality in farmed fish and reduced growth of the surviving fish. Chlamydiae were suspected as possible causative agents almost 3 decades ago based on ultrastructural evidence of chlamydial forms in diseased tissue (Molnar and Boros, 1981). Two members of the *Chlamydiae* and their association with epitheliocystis in Atlantic salmon have so far been studied in some detail: "*Candidatus* Piscichlamydia salmonis" (Draghi et al., 2004) and "*Candidatus* Clavochlamydia salmonicola" (Karlsen et al., 2008). Neither organism has been cultivated in the laboratory, as indicated by the "*Candidatus*" status. Epidemiological data based on a quantitative PCR assay and histopathological analysis of samples from four different freshwater sampling sites indicated high prevalence rates of up to 100% for "*Candidatus* Clavochlamydia salmonicola" (Mitchell et al., 2010). However, "*Candidatus* Clavochlamydia salmonicola" has not been shown to cause disease and mortality in fish, since fish populations were apparently healthy and showed no other significant pathological changes at three sampling sites despite the presence of bacteria-containing cysts characteristic of epitheliocystis. Typically, epitheliocystis spreads rapidly within fish populations, especially when the density is high. "*Candidatus* Clavochlamydia salmonicola" is primarily found in freshwater fish, and the organism disappeared from the salmon gills 4 to 6 weeks after transfer to seawater (Mitchell et al., 2010).

Taken together, evidence is emerging that some environmental chlamydiae may be associated with disease in animals, but no causal relationships have been established so far.

GENOMICS AND BEYOND

Genome sequencing and analysis have been highly important during the past decade and have greatly improved our understanding of the *Chlamydiae*. Whereas full genome sequences for numerous strains and even serovars from all known *Chlamydiaceae* species (with the exception of *Chlamydia suis*) are available (see chapter 2, "Deep and wide: comparative genomics of *Chlamydia*"), genomic data on environmental chlamydiae is still scarce. First insights into the genetic repertoire of environmental chlamydiae were obtained from *P. amoebophila* UWE25, which was discovered in the year 2000 (Fritsche et al., 2000; Collingro et al., 2005b) and whose full genome sequence was published shortly thereafter (Horn et al., 2004). The genome of this amoeba symbiont and its implications for the evolution for the *Chlamydiae* have been summarized extensively (Horn et al., 2006; Horn, 2008) and are therefore just briefly repeated here. One of the first surprises was the observed size of the genome of *P. amoebophila* UWE25. Encompassing more than 2 million base pairs, the genome is twice the size of the genomes of the *Chlamydiaceae* and much larger than that of other obligate symbionts, such as *Buchnera aphidicola* strain APS (641 kb) or *Carsonella ruddii* (160 kb). Metabolic pathways that are heavily truncated in the *Chlamydiaceae* are less so in *P. amoebophila*. Much of the additional genetic repertoire includes uncharacterized genes, and a recent comprehensive proteomic analysis demonstrated that many of these additional genes are expressed (Sixt et al., 2011). *P. amoebophila* encodes many proteins associated with chlamydial virulence, including a T3S system (Horn et al., 2004) and inclusion membrane proteins (Heinz et al., 2010), suggesting that some of these proteins were already present in the last common ancestor of the *Parachlamydiaceae* and *Chlamydiaceae*, which lived around 700 million years ago (Horn et al., 2004). *Chlamydiae* are therefore likely to have evolved their intracellular lifestyle in the Precambrian era, probably during interaction with ancient unicellular eukaryotes.

Since the publication of the *P. amoebophila* genome, additional genome sequences have become available for two environmental chlamydiae, namely, *P. acanthamoebae* strain Hall's coccus (Greub et al., 2009) and *W. chondrophila* WSU 86-1044 (Bertelli et al., 2010). While the genome sequence of the former represents a draft genome used for the identification of immunogenic proteins by proteomics (described as "dirty genome approach" [Greub et al., 2009]), a combined second-generation sequencing approach yielded the full genome sequence of *W. chondrophila*. This enabled a genome-wide comparison between different environmental chlamydiae to be performed for the first time (Bertelli et al., 2010). Similar to *P. amoebophila*, *W. chondrophila* possesses a genome of around 2 Mb, encodes many key virulence-associated proteins known from the *Chlamydiaceae*, and shows greater metabolic capabilities. The lack of a *P. amoebophila*-like NAD^+ translocase (Haferkamp et al., 2004) appears to be compensated for by the capacity for *de novo* biosynthesis of NAD^+ from quinolinate or nicotinamide in *W. chondrophila*. Moreover, *W. chondrophila* is the first member of the *Chlamydiae* for which a pyrimidine *de novo* biosynthesis pathway could be reconstructed. In contrast to *P. amoebophila*, *W. chondrophila* harbors a plasmid that is 15 kb in size and, based on sequence homologies between plasmid and chromosome, was speculated to be capable of integrating into the chromosome (Bertelli et al., 2010).

A comparative pangenome analysis based on genome sequences from four families of the *Chlamydiae* was performed just recently (Collingro et al., 2011). We will therefore focus only on selected findings from this comprehensive phylogenomic comparison. This study included three novel genome sequences from *W. chondrophila* 2032/99, *P. acanthamoeba* UV-7, and *S. negevensis* Z, which together with all available complete chlamydial genome sequences allowed an analysis of genetic similarities and differences between two species of the *Parachlamydiaceae* and between this family of amoeba symbionts and members of the *Chlamydiaceae*, the *Waddliaceae*, and the *Simkaniaceae*. This analysis confirmed that the genomes of all environmental chlamydiae are two- to threefold larger than those of the *Chlamydiaceae* with little to no genomic synteny between members of the different families. An unexpectedly high number of genes (n = 560) was found in all sequenced *Chlamydiales* and can therefore be regarded as the core chlamydial gene set, which includes not only housekeeping genes but also some T3S system components and the plasmid-encoded protein pGP6-D (CT583) (Fig. 3). Only 171 orthologs are shared by *Simkania*, *Waddlia*, and the *Parachlamydiaceae* but absent from the *Chlamydiaceae*. Remarkably, environmental chlamydiae each show a large number of accessory genes (>700) not found in *Chlamydia* spp. These genes probably contribute to niche differentiation and host specificity, although many of them are so far uncharacterized. The adaptation of *Chlamydiae* to different niches is reflected also in the differences in the outer membrane protein composition, such as the absence of *ompA* (*ct681*) in the genomes of all environmental chlamydiae analyzed so far, with the exception of *S. negevensis* (Fig. 3). Several key virulence proteins such as inclusion membrane proteins, the macrophage infectivity potentiator protein Mip (CT541), and the T3S system are well conserved among all *Chlamydiae* (Fig. 3). However, the *Chlamydia* protease/proteasome-like activity factor CPAF (CT858) (Zhong et al., 2001) is not encoded by *S. negevensis*, even though it can be identified from the genomes of other sequenced environmental chlamydiae (Fig. 3). It is unclear if the function of CPAF is unnecessary for *S. negevensis* or if it can be replaced by another factor. This comparative genomic analysis shows that members of the different chlamydial families have many general similarities with respect to their intracellular lifestyle, although not all virulence strategies may be conserved.

This class-level genomic analysis has also provided new insights into the evolutionary history of the chlamydial plasmid, which is regarded as an important virulence factor in the *Chlamydiaceae*, even if it is not universally present. Most *Chlamydia* spp. contain a 7.5-kb plasmid (Carlson et al., 2008), but *W. chondrophila* WSU 86-1044 harbors a 15-kb plasmid whose genes are, with the exception of an integrase gene homologous to pGP8-D, unrelated to the *Chlamydiaceae* plasmid genes (Bertelli et al., 2010). Plasmids have not been detected in the *Parachlamydiaceae* so far.

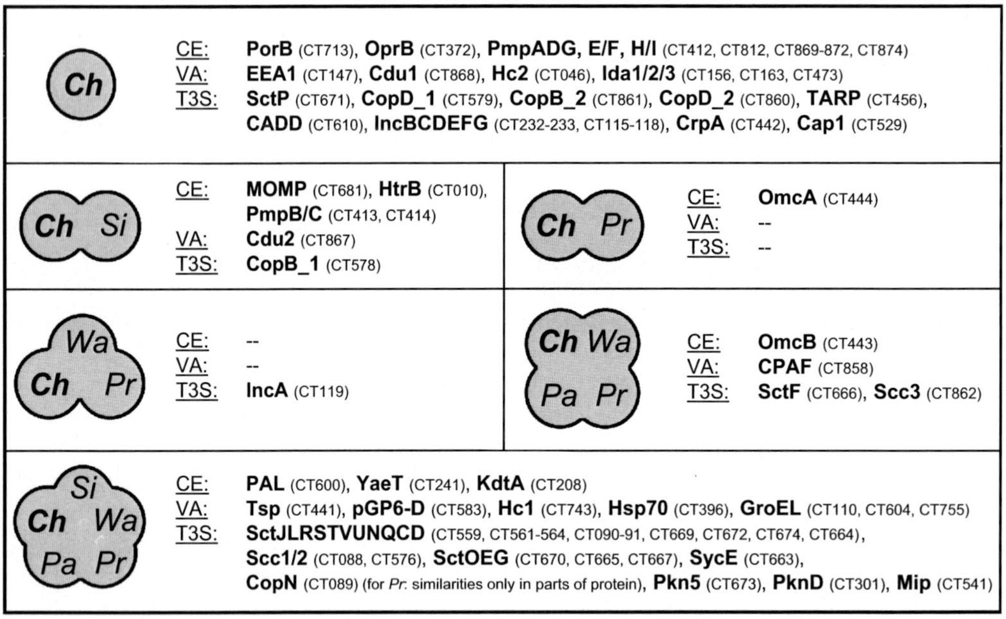

FIGURE 3 Presence/absence of selected proteins of the *Chlamydiaceae* in environmental chlamydiae. Proteins restricted to the *Chlamydiaceae* (upper panel) or shared between the *Chlamydiaceae* and different environmental chlamydiae (lower panels) are indicated. Findings are taken from the work of Collingro et al. (2011) and are based on the analysis of clusters of orthologous groups of proteins, which were determined using SIMAP (similarity matrix of proteins database) and bidirectional best BLAST hits with a cutoff value of 1E-10 (Rattei et al., 2010). Locus tags are given as for *C. trachomatis* D/UW-3/CX. Abbreviations: Ch, *Chlamydiaceae*; Si, *Simkania negevensis* Z; Wa, *Waddlia chondrophila* 2032/99; Pa, *Parachlamydia acanthamoeba* UV-7; Pr, *Protochlamydia amoebophila* UWE25; CE, cell envelope-associated genes; VA, virulence-associated genes; T3S, T3S system components and effectors. doi:10.1128/9781555817329.ch3.f3

S. negevensis is unique among the *Chlamydiae* in the fact that it contains an F-type conjugative plasmid with a size of 132 kb. Based on phylogenetic analysis, it has been suggested that this plasmid may resemble the plasmid initially acquired by the chlamydial ancestor. It has been proposed that in the course of evolution, the plasmid has been reduced in size, due to gene deletion (e.g., in the *Chlamydiaceae*) or transfer of genes to the genome, or completely lost (e.g., in the *Parachlamydiaceae*). The T4S system encoded by the *S. negevensis* plasmid and by the *Parachlamydiaceae* chromosomes may have originally been involved in conjugation, but its present-day role is unclear (Greub et al., 2004; Horn et al., 2004). This comparative analysis of all available chlamydial genomes also supports the hypothesis that ancient chlamydiae (similar to extant environmental chlamydiae) facilitated the evolution of higher plants (Huang and Gogarten, 2007; Becker et al., 2008; Moustafa et al., 2008). For example, phylogenetic analysis indicates that genes involved in carbohydrate and lipid metabolisms as well as energy production may have been horizontally transferred from a chlamydial progenitor to the ancestral primary photosynthetic eukaryote.

FUTURE LESSONS FROM ENVIRONMENTAL CHLAMYDIAE

The rapid development and affordability of novel sequencing technologies has made possible the genomic characterization of as-yet-uncultured bacteria (Handelsman, 2004). Moreover, recent improvements in genome amplification procedures based on phi29 polymerase-mediated multiple displacement amplification enable the analysis of bacteria in situations where only limited biomass or even single microbial cells are available (Binga et al., 2008; Silander and Saarela, 2008). We have recently used a metagenomic approach to sequence the genome of "*Candidatus* Clavochlamydia salmonicola" directly from tissue samples, i.e., without prior cultivation from infected fish gills (our unpublished data). By further adopting novel cultivation-independent techniques, future genome sequencing projects will characterize as-yet-uncultured environmental chlamydiae. This will add to our knowledge about the genomic diversity of the *Chlamydiae* and may provide further insights into chlamydial biology and evolution. In addition, it will facilitate the investigation of genome dynamics and processes of evolution within natural chlamydial populations—a field of research that is largely unexplored with respect to the *Chlamydiae.*

Despite this technological progress, the gold standard in microbiology has and will continue to be the host-free cultivation of microbes. Remarkably, genomic data themselves can serve as a blueprint for the design of novel media for the cultivation of yet-uncultured bacteria. Such a strategy has been successfully pursued for *Coxiella burnetii*, the causative agent of Q fever (Omsland et al., 2009). Reconstruction of metabolic requirements, using genome analysis supported by expression microarrays and metabolite typing, has facilitated the development of a growth medium suitable for the axenic culturing of this obligate intracellular pathogen. Likewise, RNA-seq data obtained from the gut microbiota of the medicinal leech *Hirudo verbana* have been used for the subsequent cultivation of the most abundant symbiont, a *Rikenella*-like bacterium (Bomar et al., 2011). Genomic comparison of environmental chlamydiae has revealed that *Waddlia* and *Parachlamydia* may be the most suitable chlamydial candidates for host-free cultivation, since these organisms have the most versatile biosynthetic capabilities among the *Chlamydiae* (Bertelli et al., 2010; Collingro et al., 2011). The synergy of the "wet lab" and *in silico* studies will certainly continue to expand our knowledge of the biology of environmental chlamydiae.

One century ago, Ludwig Halberstädter noted in one of the earliest review articles on chlamydiae, "*daß die hier in Betracht kommenden Fragen noch keineswegs abgeschlossen, vielmehr die Dinge noch sehr im Fluß sind*" ("the questions that come into consideration are in no way completed, things are rather still in flux") (Halberstädter, 1912). Ironically, the same is still true today with respect to the "true" diversity of chlamydiae, their function in the environment, and their contributions to disease. During the past decade, the discovery and analysis of environmental chlamydiae have opened new horizons with respect to the phylogenetic diversity of the chlamydiae, their occurrence in the environment, and their unique evolutionary history, and there is much more to come!

ACKNOWLEDGMENTS

Alexander Siegl is supported by a grant in the framework of the ERANET project Pathomics (I291-B09). Work in the lab of Matthias Horn is supported by grants from the Austrian Science Funds and the University of Vienna (Y277-B03).

REFERENCES

Abdelrahman, Y. M., and R. J. Belland. 2005. The chlamydial developmental cycle. *FEMS Microbiol. Rev.* **29:**949–959.

Amann, R., N. Springer, W. Schonhuber, W. Ludwig, E. N. Schmid, K. D. Muller, and R. Michel. 1997. Obligate intracellular bacterial parasites of acanthamoebae related to *Chlamydia* spp. *Appl. Environ. Microbiol.* **63:**115–121.

Barker, J., and M. R. W. Brown. 1994. Trojan horses of the microbial world: protozoa and the survival of bacterial pathogens in the environment. *Microbiology* **140:**1253–1259.

Baud, D., G. Goy, M. C. Osterheld, N. Borel, Y. Vial, A. Pospischil, and G. Greub. 2011. *Waddlia chondrophila*: from bovine abortion to human miscarriage. *Clin. Infect. Dis.* **52:**1469–1471.

Baud, D., V. Thomas, A. Arafa, L. Regan, and G. Greub. 2007. *Waddlia chondrophila*, a potential agent of human fetal death. *Emerg. Infect. Dis.* **13:**1239–1243.

Beatty, W. L., R. P. Morrison, and G. I. Byrne. 1994. Persistent chlamydiae: from cell culture to a paradigm for chlamydial pathogenesis. *Microbiol. Rev.* **58:**686–699.

Becker, B., K. Hoef-Emden, and M. Melkonian. 2008. Chlamydial genes shed light on the evolution of photoautotrophic eukaryotes. *BMC Evol. Biol.* **8:**203.

Bertelli, C., F. Collyn, A. Croxatto, C. Ruckert, A. Polkinghorne, C. Kebbi-Beghdadi, A. Goesmann, L. Vaughan, and G. Greub. 2010. The *Waddlia* genome: a window into chlamydial biology. *PLoS One* **5:**e10890.

Binga, E. K., R. S. Lasken, and J. D. Neufeld. 2008. Something from (almost) nothing: the impact of multiple displacement amplification on microbial ecology. *ISME J.* **2:**233–241.

Birtles, R. J., T. J. Rowbotham, C. Storey, T. J. Marrie, and D. Raoult. 1997. *Chlamydia*-like obligate parasite of free-living amoebae. *Lancet* **349:**925–926.

Bodetti, T. J., K. Viggers, K. Warren, R. Swan, S. Conaghty, C. Sims, and P. Timms. 2003. Wide range of *Chlamydiales* types detected in native Australian mammals. *Vet. Microbiol.* **96:**177–187.

Bomar, L., M. Maltz, S. Colston, and J. Graf. 2011. Directed culturing of microorganisms using metatranscriptomics. *mBio* **2:**e00012-11.

Borel, N., S. Ruhl, N. Casson, C. Kaiser, A. Pospischil, and G. Greub. 2007. *Parachlamydia* spp. and related *Chlamydia*-like organisms and bovine abortion. *Emerg. Infect. Dis.* **13:**1904–1907.

Bradley, T. M., C. E. Newcomer, and K. O. Maxwell. 1988. Epitheliocystis associated with massive mortalities of cultured lake trout *Salvelinus namaycush*. *Dis. Aquat. Organ.* **4:**9–17.

Carlson, J. H., W. M. Whitmire, D. D. Crane, L. Wicke, K. Virtaneva, D. E. Sturdevant, J. J. Kupko III, S. F. Porcella, N. Martinez-Orengo, R. A. Heinzen, L. Kari, and H. D. Caldwell. 2008. The *Chlamydia trachomatis* plasmid is a transcriptional regulator of chromosomal genes and a virulence factor. *Infect. Immun.* **76:**2273–2283.

Casson, N., J. M. Entenza, N. Borel, A. Pospischil, and G. Greub. 2008a. Murine model of pneumonia caused by *Parachlamydia acanthamoebae*. *Microb. Pathog.* **45:**92–97.

Casson, N., N. Medico, J. Bille, and G. Greub. 2006. *Parachlamydia acanthamoebae* enters and multiplies within pneumocytes and lung fibroblasts. *Microbes Infect.* **8:**1294–1300.

Casson, N., R. Michel, K.-D. Müller, J. D. Aubert, and G. Greub. 2008b. *Protochlamydia naegleriophila* as etiologic agent of pneumonia. *Emerg. Infect. Dis.* **14:**168–172.

Chi, E. Y., C. C. Kuo, and J. T. Grayston. 1987. Unique ultrastructure in the elementary body of *Chlamydia* sp. strain TWAR. *J. Bacteriol.* **169:**3757–3763.

Christerson, L., M. Blomqvist, K. Grannas, M. Thollesson, K. Laroucau, J. Waldenstrom, I. Eliasson, B. Olsen, and B. Herrmann. 2010. A novel *Chlamydiaceae*-like bacterium found in faecal specimens from sea birds from the Bering Sea. *Environ. Microbiol. Rep.* **2:**605–610.

Chu, D., C. S. Gao, P. De Barro, Y. J. Zhang, F. H. Wan, and I. A. Khan. 2011. Further insights into the strange role of bacterial endosymbionts in whitefly, *Bemisia tabaci*: comparison of secondary symbionts from biotypes B and Q in China. *Bull. Entomol. Res.* **18:**1–10.

Chua, P. K., J. E. Corkill, P. S. Hooi, S. C. Cheng, C. Winstanley, and C. A. Hart. 2005. Isolation of *Waddlia malaysiensis*, a novel intracellular bacterium, from fruit bat (*Eonycteris spelaea*). *Emerg. Infect. Dis.* **11:**271–277.

Clark, M. A., L. Baumann, M. A. Munson, P. Baumann, B. C. Campbell, J. E. Duffus, L. S. Osborne, and N. A. Moran. 1992. The eubacterial endosymbionts of whiteflies (Homoptera, Aleyrodoidea) constitute a lineage distinct from the endosymbionts of aphids and mealybugs. *Curr. Microbiol.* **25:**119–123.

Collingro, A., S. Poppert, E. Heinz, S. Schmitz-Esser, A. Essig, M. Schweikert, M. Wagner, and M. Horn. 2005a. Recovery of an environmental chlamydia strain from activated sludge by co-cultivation with *Acanthamoeba* sp. *Microbiology* **151:**301–309.

Collingro, A., P. Tischler, T. Weinmaier, T. Penz, E. Heinz, R. C. Brunham, T. D. Read, P. M. Bavoil, K. Sachse, S. Kahane, M. G. Friedman, T. Rattei, G. S. A. Myers, and M. Horn. 2011. Unity in variety—the pangenome of the *Chlamydiae*. *Mol. Biol. Evol.* Epub ahead of print.

Collingro, A., E. R. Toenshoff, M. W. Taylor, T. R. Fritsche, M. Wagner, and M. Horn. 2005b. '*Candidatus* Protochlamydia amoebophila,' an endosymbiont of *Acanthamoeba* spp. *Int. J. Syst. Evol. Microbiol.* **55:**1863–1866.

Corsaro, D., V. Feroldi, G. Saucedo, F. Ribas, J. F. Loret, and G. Greub. 2009. Novel *Chlamydiales* strains isolated from a water treatment plant. *Environ. Microbiol.* **11:**188–200.

Corsaro, D., R. Michel, J. Walochnik, K. D. Muller, and G. Greub. 2010. *Saccamoeba lacustris*, sp. nov. (Amoebozoa: Lobosea: Hartmannellidae), a new lobose amoeba, parasitized by the novel chlamydia '*Candidatus* Metachlamydia lacustris' (Chlamydiae: *Parachlamydiaceae*). *Eur. J. Protistol.* **46:**86–95.

Corsaro, D., V. Thomas, G. Goy, D. Venditti, R. Radek, and G. Greub. 2007. '*Candidatus* Rhabdochlamydia crassificans,' an intracellular bacterial pathogen of the cockroach *Blatta orientalis* (Insecta: Blattodea). *Syst. Appl. Microbiol.* **30:**221–228.

Corsaro, D., M. Valassina, and D. Venditti. 2003. Increasing diversity within *Chlamydiae*. *Crit. Rev. Microbiol.* **29:**37–78.

Corsaro, D., and D. Venditti. 2009. Detection of *Chlamydiae* from freshwater environments by PCR, amoeba coculture and mixed coculture. *Res. Microbiol.* **160:**547–552.

Costa, H. S., D. M. Westcot, D. E. Ullman, R. Rosell, J. K. Brown, and M. W. Johnson. 1995. Morphological variation in *Bemisia* endosymbionts. *Protoplasma* **189:**194–202.

Croxatto, A., and G. Greub. 2010. Early intracellular trafficking of *Waddlia chondrophila* in human macrophages. *Microbiology* **156:**340–355.

De Bary, A. 1879. *Die Erscheinung der Symbiose.* Verlag von Karl J. Trubner, Strassburg, Austria.

Dilbeck, P. M., J. F. Evermann, T. B. Crawford, A. C. Ward, C. W. Leathers, C. J. Holland, C. A. Mebus, L. L. Logan, F. R. Rurangirwa, and T. C. McGuire. 1990. Isolation of a previously undescribed rickettsia from an aborted bovine fetus. *J. Clin. Microbiol.* **28:**814–816.

Dilbeck-Robertson, P., M. M. McAllister, D. Bradway, and J. F. Evermann. 2003. Results of a new serologic test suggest an association of *Waddlia chondrophila* with bovine abortion. *J. Vet. Diagn. Investig.* **15:**568–569.

Draghi, A., II, J. Bebak, S. Daniels, E. R. Tulman, S. J. Geary, A. B. West, V. L. Popov, and S. Frasca, Jr. 2010. Identification of '*Candidatus* Piscichlamydia salmonis' in Arctic charr *Salvelinus alpinus* during a survey of charr production facilities in North America. *Dis. Aquat. Organ.* **89:**39–49.

Draghi, A., 2nd, V. L. Popov, M. M. Kahl, J. B. Stanton, C. C. Brown, G. J. Tsongalis, A. B. West, and S. Frasca, Jr. 2004. Characterization of "*Candidatus* piscichlamydia salmonis" (order *Chlamydiales*), a chlamydia-like bacterium associated with epitheliocystis in farmed Atlantic salmon (*Salmo salar*). *J. Clin. Microbiol.* **42:**5286–5297.

Everett, K. D., M. Thao, M. Horn, G. E. Dyszynski, and P. Baumann. 2005. Novel chlamydiae in whiteflies and scale insects: endosymbionts '*Candidatus* Fritschea bemisiae' strain Falk and '*Candidatus* Fritschea eriococci' strain Elm. *Int. J. Syst. Evol. Microbiol.* **55:**1581–1587.

Fredricks, D. N., and D. A. Relman. 1996. Sequence-based identification of microbial pathogens: a reconsideration of Koch's postulates. *Clin. Microbiol. Rev.* **9:**18–33.

Friedman, M. G., B. Dvoskin, and S. Kahane. 2003. Infections with the chlamydia-like microorganism *Simkania negevensis*, a possible emerging pathogen. *Microbes Infect.* **5:**1013–1021.

Friedman, M. G., A. Galil, S. Greenberg, and S. Kahane. 1999. Seroprevalence of IgG antibodies to the chlamydia-like microorganism '*Simkania* Z' by ELISA. *Epidemiol. Infect.* **122:**117–123.

Friedman, M. G., S. Kahane, B. Dvoskin, and J. W. Hartley. 2006. Detection of *Simkania negevensis* by culture, PCR, and serology in respiratory tract infection in Cornwall, UK. *J. Clin. Pathol.* **59:**331–333.

Fritsche, T. R., R. K. Gautom, S. Seyedirashti, D. L. Bergeron, and T. D. Lindquist. 1993. Occurrence of bacteria endosymbionts in *Acanthamoeba* spp. isolated from corneal and environmental specimens and contact lenses. *J. Clin. Microbiol.* **31:**1122–1126.

Fritsche, T. R., M. Horn, M. Wagner, R. P. Herwig, K. H. Schleifer, and R. K. Gautom. 2000. Phylogenetic diversity among geographically dispersed *Chlamydiales* endosymbionts recovered from clinical and environmental isolates of *Acanthamoeba* spp. *Appl. Environ. Microbiol.* **66:**2613–2619.

Goy, G., A. Croxatto, and G. Greub. 2008. *Waddlia chondrophila* enters and multiplies within human macrophages. *Microbes Infect.* **10:**556–562.

Goy, G., A. Croxatto, K. M. Posfay-Barbe, A. Gervaix, and G. Greub. 2009. Development of a real-time PCR for the specific detection of *Waddlia chondrophila* in clinical samples. *Eur. J. Clin. Microbiol. Infect. Dis.* **28:**1483–1486.

Greenberg, D., A. Banerji, M. G. Friedman, C. H. Chiu, and S. Kahane. 2003. High rate of *Simkania negevensis* among Canadian Inuit infants hospitalized with lower respiratory tract infections. *Scand. J. Infect. Dis.* **35:**506–508.

Greub, G. 2009. *Parachlamydia acanthamoebae*, an emerging agent of pneumonia. *Clin. Microbiol. Infect.* **15:**18–28.

Greub, G., F. Collyn, L. Guy, and C. A. Roten. 2004. A genomic island present along the bacterial chromosome of the *Parachlamydiaceae* UWE25, an obligate amoebal endosymbiont, encodes a potentially functional F-like conjugative DNA transfer system. *BMC Microbiol.* **4:**48.

Greub, G., B. Desnues, D. Raoult, and J. L. Mege. 2005a. Lack of microbicidal response in human macrophages infected with *Parachlamydia acanthamoebae*. *Microbes Infect.* **7:**714–719.

Greub, G., C. Kebbi-Beghdadi, C. Bertelli, F. Collyn, B. M. Riederer, C. Yersin, A. Croxatto, and D. Raoult. 2009. High throughput sequencing and proteomics to identify immunogenic proteins of a new pathogen: the dirty genome approach. *PLoS One* **4:**e8423.

Greub, G., J.-L. Mege, and D. Raoult. 2003. *Parachlamydia acanthamoebae* enters and multiplies within human macrophages and induces their apoptosis. *Infect. Immun.* **71:**5979–5985.

Greub, G., J. L. Mege, J. P. Gorvel, D. Raoult, and S. Meresse. 2005b. Intracellular trafficking of *Parachlamydia acanthamoebae. Cell. Microbiol.* **7:**581–589.

Greub, G., and D. Raoult. 2002. Crescent bodies of *Parachlamydia acanthamoebae* and its life cycle within *Acanthamoeba polyphaga*: an electron micrograph study. *Appl. Environ. Microbiol.* **68:**3076–3084.

Hackstadt, T., M. A. Scidmore, and D. D. Rockey. 1995. Lipid metabolism in *Chlamydia trachomatis*-infected cells: directed trafficking of Golgi-derived sphingolipids to the chlamydial inclusion. *Proc. Natl. Acad. Sci. USA* **92:**4877–4881.

Haferkamp, I., S. Schmitz-Esser, N. Linka, C. Urbany, A. Collingro, M. Wagner, M. Horn, and H. E. Neuhaus. 2004. A candidate NAD^+ transporter in an intracellular bacterial symbiont related to *Chlamydiae. Nature* **432:**622–625.

Haider, S., A. Collingro, J. Walochnik, M. Wagner, and M. Horn. 2008. Chlamydia-like bacteria in respiratory samples of community-acquired pneumonia patients. *FEMS Microbiol. Lett.* **281:**198–202.

Haider, S., M. Wagner, M. C. Schmid, B. S. Sixt, J. G. Christian, G. Hacker, P. Pichler, K. Mechtler, A. Muller, C. Baranyi, E. R. Toenshoff, J. Montanaro, and M. Horn. 2010. Raman microspectroscopy reveals long-term extracellular activity of chlamydiae. *Mol. Microbiol.* **77:**687–700.

Halberstädter, L. 1912. Trachom und Chlamydozoenerkrankung der Schleimhäute, p. 172–195. *In* S. von Prowazek (ed.), *Handbuch der Pathogenen Protozoen.* Verlag von Johann Ambrosius Barth, Leipzig, Germany.

Halberstädter, L., and S. von Prowazek. 1907. Über Zelleinschlüsse parasitärer Natur beim Trachom. *Arbeiten aus dem Kaiserlichen Gesundheitsamte, Berlin* **26:**44–47.

Handelsman, J. 2004. Metagenomics: application of genomics to uncultured microorganisms. *Microbiol. Mol. Biol. Rev.* **68:**669–685.

Hayashi, Y., S. Nakamura, J. Matsuo, T. Fukumoto, M. Yoshida, K. Takahashi, Y. Mizutani, T. Yao, and H. Yamaguchi. 2010. Host range of obligate intracellular bacterium *Parachlamydia acanthamoebae. Microbiol. Immunol.* **54:**707–713.

Heinz, E., D. D. Rockey, J. Montanaro, K. Aistleitner, M. Wagner, and M. Horn. 2010. Inclusion membrane proteins of *Protochlamydia amoebophila* UWE25 reveal a conserved mechanism for host cell interaction among the *Chlamydiae. J. Bacteriol.* **192:**5093–5102.

Henning, K., G. Schares, H. Granzow, U. Polster, M. Hartmann, H. Hotzel, K. Sachse, M. Peters, and M. Rauser. 2002. *Neospora caninum* and *Waddlia chondrophila* strain 2032/99 in a septic stillborn calf. *Vet. Microbiol.* **85:**285–292.

Henning, K., L. Zöller, B. Hauröder, H. Hotzel, and R. Michel. 2007. *Hartmannella vermiformis* (Hartmannellidae) harboured a hidden chlamydia-like endosymbiont. *Endocytobiosis Cell Res.* **18:**1–10.

Horn, M. 2008. *Chlamydiae* as symbionts in eukaryotes. *Annu. Rev. Microbiol.* **62:**113–131.

Horn, M., A. Collingro, S. Schmitz-Esser, C. L. Beier, U. Purkhold, B. Fartmann, P. Brandt, G. J. Nyakatura, M. Droege, D. Frishman, T. Rattei, H. W. Mewes, and M. Wagner. 2004. Illuminating the evolutionary history of chlamydiae. *Science* **304:**728–730.

Horn, M., A. Collingro, S. Schmitz-Esser, and M. Wagner. 2006. Environmental chlamydia genomics, p. 25–44. *In* P. M. Bavoil and P. B. Wyrick (ed.), Chlamydia*: Genomics and Pathogenesis.* Horizon Bioscience, Norfolk, United Kingdom.

Horn, M., M. Wagner, K. D. Muller, E. N. Schmid, T. R. Fritsche, K. H. Schleifer, and R. Michel. 2000. *Neochlamydia hartmannellae* gen. nov., sp. nov. (*Parachlamydiaceae*), an endoparasite of the amoeba *Hartmannella vermiformis. Microbiology* **146:**1231–1239.

Huang, J., and J. Gogarten. 2007. Did an ancient chlamydial endosymbiosis facilitate the establishment of primary plastids? *Genome Biol.* **8:**R99.

Husain, S., S. Kahane, M. G. Friedman, D. L. Paterson, S. Studer, K. R. McCurry, D. G. Wolf, A. Zeevi, J. Pilewski, and D. Greenberg. 2007. *Simkania negevensis* in bronchoalveolar lavage of lung transplant recipients: a possible association with acute rejection. *Transplantation* **83:**138–143.

Israelsson, O. 2007. Chlamydial symbionts in the enigmatic *Xenoturbella* (Deuterostomia). *J. Invertebr. Pathol.* **96:**213–220.

Kahane, S., B. Dvoskin, and M. G. Friedman. 2008. The role of monocyte/macrophages as vehicles of dissemination of *Simkania negevensis*: an in vitro simulation model. *FEMS Immunol. Med. Microbiol.* **52:**219–227.

Kahane, S., B. Dvoskin, M. Mathias, and M. G. Friedman. 2001. Infection of *Acanthamoeba polyphaga* with *Simkania negevensis* and *S. negevensis* survival within amoebal cysts. *Appl. Environ. Microbiol.* **67:**4789–4795.

Kahane, S., D. Fruchter, B. Dvoskin, and M. G. Friedman. 2007a. Versatility of *Simkania negevensis* infection in vitro and induction of host cell inflammatory cytokine response. *J. Infect.* **55:**e13–e21.

Kahane, S., R. Gonen, C. Sayada, J. Elion, and M. G. Friedman. 1993. Description and partial characterization of a new chlamydia-like microorganism. *FEMS Microbiol. Lett.* **109:**329–334.

Kahane, S., D. Greenberg, M. G. Friedman, H. Haikin, and R. Dagan. 1998. High prevalence of "*Simkania* Z," a novel *Chlamydia*-like bacterium, in infants with acute bronchiolitis. *J. Infect. Dis.* **177:**1425–1429.

Kahane, S., D. Greenberg, N. Newman, B. Dvoskin, and M. G. Friedman. 2007b. Domestic water supplies as a possible source of infection with *Simkania. J. Infect.* **54:**75–81.

Kahane, S. E. M., and M. G. Friedman. 1995. Evidence that the novel microorganism 'Z' may belong to a new genus in the family *Chlamydiaceae. FEMS Microbiol. Lett.* **126:**203–208.

Karlsen, M., A. Nylund, K. Watanabe, J. V. Helvik, S. Nylund, and H. Plarre. 2008. Characterization of '*Candidatus* Clavochlamydia salmonicola': an intracellular bacterium infecting salmonid fish. *Environ. Microbiol.* **10:**208–218.

Kebbi-Beghdadi, C., O. Cisse, and G. Greub. 2011. Permissivity of Vero cells, human pneumocytes and human endometrial cells to *Waddlia chondrophila. Microbes Infect.* **13:**566–574.

Kjeldsen, K. U., M. Obst, H. Nakano, P. Funch, and A. Schramm. 2010. Two types of endosymbiotic bacteria in the enigmatic marine worm *Xenoturbella bocki. Appl. Environ. Microbiol.* **76:**2657–2662.

Kostanjsek, R., J. Strus, D. Drobne, and G. Avgustin. 2004. '*Candidatus* Rhabdochlamydia porcellionis,' an intracellular bacterium from the hepatopancreas of the terrestrial isopod *Porcellio scaber* (Crustacea: Isopoda). *Int. J. Syst. Evol. Microbiol.* **54:**543–549.

Kumar, S., S. A. Kohlhoff, M. Gelling, P. M. Roblin, A. Kutlin, S. Kahane, M. G. Friedman, and M. R. Hammerschlag. 2005. Infection with *Simkania negevensis* in Brooklyn, New York. *Pediatr. Infect. Dis. J.* **24:**989–992.

Kuo, C.-C., M. Horn, and R. S. Stephens. 2011. The order *Chlamydiales*, p. 844–845. *In* B. Hedlund, N. R. Krieg, W. Ludwig, B. J. Paster, J. T. Staley, N. Ward, and W. B. Whitman (ed.), *Bergey's Manual of Systematic Bacteriology—The Planctomycetes, Spriochaetes, Fibrobacteres, Bacteriodetes and Fusobacteria*, 2nd ed., vol. 4. Springer, New York, NY.

Lamoth, F., and G. Greub. 2010. Amoebal pathogens as emerging causal agents of pneumonia. *FEMS Microbiol. Rev.* **34:**260–280.

Leitsch, D., M. Kohsler, M. Marchetti-Deschmann, A. Deutsch, G. Allmaier, L. Konig, B. S. Sixt, M. Duchene, and J. Walochnik. 2010. Proteomic aspects of *Parachlamydia acanthamoebae* infection in *Acanthamoeba* spp. *ISME J.* **4:**1366–1374.

Lieberman, D., S. Kahane, D. Lieberman, and M. G. Friedman. 1997. Pneumonia with serological evidence of acute infection with the *Chlamydia*-like microorganism 'Z.' *Am. J. Respir. Crit. Care Med.* **156:**578–582.

Lindsay, M. R., R. I. Webb, H. M. Hosmer, and J. A. Fuerst. 1995. Effects of fixative and buffer on morphology and ultrastructure of a freshwater planctomycete, *Gemmata obscuriglobus. J. Microbiol. Methods* **21:**45–54.

Matsumoto, A., H. Bessho, K. Uehira, and T. Suda. 1991. Morphological studies of the association of mitochondria with chlamydial inclusions and the fusion of chlamydial inclusions. *J. Electron. Microsc.* (Tokyo) **40:**356–363.

Maurin, M., A. Bryskier, and D. Raoult. 2002. Antibiotic susceptibilities of *Parachlamydia acanthamoeba* in amoebae. *Antimicrob. Agents Chemother.* **46:**3065–3067.

Michel, R., B. Hauröder-Philippczyk, K.-D. Müller, and I. Weishaar. 1994. *Acanthamoeba* from human nasal mucosa infected with an obligate intracellular parasite. *Eur. J. Protistol.* **30:**104–110.

Michel, R., K. D. Müller, L. Zöller, J. Walochnik, M. Hartmann, and E. N. Schmid. 2005. Free-living amoebae serve as a host for the *Chlamydia*-like bacterium *Simkania negevensis. Acta Protozool.* **44:**113–121.

Mitchell, S. O., and H. D. Rodger. 2011. A review of infectious gill disease in marine salmonid fish. *J. Fish Dis.* **34:**411–432.

Mitchell, S. O., T. Steinum, H. Rodger, C. Holland, K. Falk, and D. J. Colquhoun. 2010. Epitheliocystis in Atlantic salmon, *Salmo salar* L., farmed in fresh water in Ireland is associated with '*Candidatus* Clavochlamydia salmonicola' infection. *J. Fish Dis.* **33:**665–673.

Molnar, K., and G. Boros. 1981. A light and electron microscopic study of the agent of carp mucophilosis. *J. Fish Dis.* **4:**325–334.

Moulder, J. W. 1964. *The Psittacosis Group as Bacteria.* John Wiley, New York, NY.

Moustafa, A., A. Reyes-Prieto, and D. Bhattacharya. 2008. Chlamydiae has contributed at least 55 genes to Plantae with predominantly plastid functions. *PLoS One* **3:**e2205.

Murray, R. G. E., and E. Stackebrandt. 1995. Taxonomic note: implementation of the provisional status Candidatus for incompletely described procaryotes. *Int. J. Syst. Bacteriol.* **45:**186–187.

Nakamura, S., J. Matsuo, Y. Hayashi, K. Kawaguchi, M. Yoshida, K. Takahashi, Y. Mizutani, T. Yao, and H. Yamaguchi. 2010. Endosymbiotic bacterium *Protochlamydia* can survive in acanthamoebae following encystation. *Environ. Microbiol. Rep.* **2:**611–618.

Nishio, K., K. Hashimoto, and K. Watanabe. 2010. Light/electricity conversion by a self-organized photosynthetic biofilm in a single-chamber reactor. *Appl. Microbiol. Biotechnol.* **86:**957–964.

Nowak, B. F., and S. E. LaPatra. 2006. Epitheliocystis in fish. *J. Fish Dis.* **29:**573–588.

Omsland, A., D. C. Cockrell, D. Howe, E. R. Fischer, K. Virtaneva, D. E. Sturdevant, S. F. Porcella, and R. A. Heinzen. 2009. Host cell-free growth of the Q fever bacterium *Coxiella burnetii*. *Proc. Natl. Acad. Sci. USA* **106:**4430–4434.

Polkinghorne, A., H. Schmidt-Posthaus, A. Meijer, A. Lehner, and L. Vaughan. 2010. Novel *Chlamydiales* associated with epitheliocystis in a leopard shark *Triakis semifasciata*. *Dis. Aquat. Organ.* **91:**75–81.

Ramos-Padron, E., S. Bordenave, S. Lin, I. M. Bhaskar, X. Dong, C. W. Sensen, J. Fournier, G. Voordouw, and L. M. Gieg. 2011. Carbon and sulfur cycling by microbial communities in a gypsum-treated oil sands tailings pond. *Environ. Sci. Technol.* **45:**439–446.

Rattei, T., P. Tischler, S. Gotz, M. A. Jehl, J. Hoser, R. Arnold, A. Conesa, and H. W. Mewes. 2010. SIMAP—a comprehensive database of pre-calculated protein sequence similarities, domains, annotations and clusters. *Nucleic Acids Res.* **38:**D223–D226.

Richter, M., F. Matheis, E. Gonczi, S. Aeby, B. Spiess, and G. Greub. 2010. *Parachlamydia acanthamoebae* in domestic cats with and without corneal disease. *Vet. Ophthalmol.* **13:**235–237.

Robertson, T., S. Bibby, D. O'Rourke, T. Belfiore, R. Agnew-Crumpton, and A. H. Noormohammadi. 2010. Identification of chlamydial species in crocodiles and chickens by PCR-HRM curve analysis. *Vet. Microbiol.* **145:**373–379.

Rockey, D. D., M. A. Scidmore, J. P. Bannantine, and W. J. Brown. 2002. Proteins in the chlamydial inclusion membrane. *Microbes Infect.* **4:**333–340.

Roger, T., N. Casson, A. Croxatto, J. M. Entenza, M. Pusztaszeri, S. Akira, M. K. Reymond, D. Le Roy, T. Calandra, and G. Greub. 2010. Role of MyD88 and Toll-like receptors 2 and 4 in the sensing of *Parachlamydia acanthamoebae*. *Infect. Immun.* **78:**5195–5201.

Rourke, A. W., R. W. Davis, and T. M. Bradley. 1984. A light and electron microscope study of epitheliocystis in juvenile steelhead trout, *Salmo gairdneri* Richardson. *J. Fish Dis.* **7:**301–309.

Rurangirwa, F. R., P. M. Dilbeck, T. B. Crawford, T. C. McGuire, and T. F. McElwain. 1999. Analysis of the 16S rRNA gene of microorganism WSU 86-1044 from an aborted bovine foetus reveals that it is a member of the order *Chlamydiales*: proposal of *Waddliaceae* fam. nov., *Waddlia chondrophila* gen. nov., sp. nov. *Int. J. Syst. Bacteriol.* **49:**577–581.

Sagaram, U. S., K. M. DeAngelis, P. Trivedi, G. L. Andersen, S. E. Lu, and N. Wang. 2009. Bacterial diversity analysis of Huanglongbing pathogen-infected citrus, using PhyloChip arrays and 16S rRNA gene clone library sequencing. *Appl. Environ. Microbiol.* **75:**1566–1574.

Salah, I. B., E. Ghigo, and M. Drancourt. 2009. Free-living amoebae, a training field for macrophage resistance of mycobacteria. *Clin. Microbiol. Infect.* **15:**894–905.

Schubert, C. J., E. Durisch-Kaiser, B. Wehrli, B. Thamdrup, P. Lam, and M. M. Kuypers. 2006. Anaerobic ammonium oxidation in a tropical freshwater system (Lake Tanganyika). *Environ. Microbiol.* **8:**1857–1863.

Shaw, A. C., K. Gevaert, H. Demol, B. Hoorelbeke, J. Vandekerckhove, M. R. Larsen, P. Roepstorff, A. Holm, G. Christiansen, and S. Birkelund. 2002. Comparative proteome analysis of *Chlamydia trachomatis* serovar A, D and L2. *Proteomics* **2:**164–186.

Silander, K., and J. Saarela. 2008. Whole genome amplification with Phi29 DNA polymerase to enable genetic or genomic analysis of samples of low DNA yield. *Methods Mol. Biol.* **439:**1–18.

Silva, M. T., and J. C. F. Sousa. 1973. Ultrastructure of cell-wall and cytoplasmic membrane of gram-negative bacteria with different fixation techniques. *J. Bacteriol.* **113:**953–962.

Sixt, B. S., C. Heinz, P. Pichler, E. Heinz, J. Montanaro, H. J. Op den Camp, G. Ammerer, K. Mechtler, M. Wagner, and M. Horn. 2011. Proteomic analysis reveals a virtually complete set of proteins for translation and energy generation in elementary bodies of the amoeba symbiont *Protochlamydia amoebophila*. *Proteomics* **11:**1868–1892.

Skaljac, M., K. Zanic, S. G. Ban, S. Kontsedalov, and M. Ghanim. 2010. Co-infection and localization of secondary symbionts in two whitefly species. *BMC Microbiol.* **10:**142.

Thao, M. L., L. Baumann, J. M. Hess, B. W. Falk, J. C. Ng, P. J. Gullan, and P. Baumann. 2003. Phylogenetic evidence for two new insect-associated *Chlamydia* of the family *Simkaniaceae*. *Curr. Microbiol.* **47:**46–50.

Thomas, V., N. Casson, and G. Greub. 2006. *Criblamydia sequanensis*, a new intracellular *Chlamydiales* isolated from Seine river water using amoebal co-culture. *Environ. Microbiol.* **8:**2125–2135.

Vandahl, B. B., S. Birkelund, H. Demol, B. Hoorelbeke, G. Christiansen, J. Vandekerckhove, and K. Gevaert. 2001. Proteome analysis of the *Chlamydia pneumoniae* elementary body. *Electrophoresis* **22:**1204–1223.

von Prowazek, S. 1912. Chlamydozoen (Allgemeines), p. 119–121. *In* S. von Prowazek (ed.), *Handbuch der Pathogenen Protozoen*. Verlag von Johann Ambrosius Barth, Leipzig, Germany.

Webster, N. S., M. W. Taylor, F. Behnam, S. Lucker, T. Rattei, S. Whalan, M. Horn, and M. Wagner. 2009. Deep sequencing reveals exceptional diversity and modes of transmission for bacterial sponge symbionts. *Environ. Microbiol.* **12:**2070–2082.

Zhang, Y. Z., E. T. Wang, M. Li, Q. Q. Li, Y. M. Zhang, S. J. Zhao, X. L. Jia, L. H. Zhang, W. F. Chen, and W. X. Chen. 2011. Effects of rhizobial inoculation, cropping systems and growth stages on endophytic bacterial community of soybean roots. *Plant Soil* **347:**147–161.

Zhong, G., P. Fan, H. Ji, F. Dong, and Y. Huang. 2001. Identification of a chlamydial protease-like activity factor responsible for the degradation of host transcription factors. *J. Exp. Med.* **193:**935–942.

Zhu, P., Q. Li, and G. Wang. 2008. Unique microbial signatures of the alien Hawaiian marine sponge *Suberites zeteki*. *Microb. Ecol.* **55:**406–414.

THE CHLAMYDIAL CELL ENVELOPE

David E. Nelson

4

INTRODUCTION

Molecules on the surfaces of chlamydiae mediate many aspects of the extracellular and intracellular biology of *Chlamydia* spp. Chlamydial development is punctuated by changes in protein-protein interactions on elementary body (EB) and reticulate body (RB) surfaces. For example, disulfide cross-linked protein complexes render an EB metabolically inactive and resistant to osmotic stress outside host cells. Reduction of disulfide cross-links in the chlamydial outer membrane complex (COMC) concomitant with attachment and entry of the EB is rapidly followed by transition to the fragile RB, which is specialized for acquisition of nutrients during chlamydial growth and differentiation. Here we review knowledge about this progression starting with the structure of the EB envelope in the extracellular environment and the way in which this surface interacts with, and is altered during, the process of chlamydial attachment, entry, development, and exit from host cells.

David E. Nelson, Department of Biology, Indiana University, Bloomington, IN 47405.

COMPOSITION OF THE CHLAMYDIAL SURFACE

Ultrastructure of the Chlamydial Surface

Methods for purification (Moulder, 1962; Moulder et al., 1963) and subfractionation (Manire, 1966; Manire and Tamura, 1967; Tamura and Manire, 1968) of EBs were pioneered almost 50 years ago, and detailed descriptions of chlamydial ultrastructure began to appear in the literature shortly thereafter. Early studies demonstrated that *Chlamydia*-infected cells contained a mixture of small EB (~0.3 µm) and larger RB (~1.0 µm) developmental forms (Moulder et al., 1963), that RBs were more fragile than EBs (Tamura and Manire, 1967), that EBs and RBs were the infectious and vegetative chlamydial particles, respectively, and that antichlamydial activity of penicillin was associated with a block in binary fission of RBs (Higashi, 1965; Tamura and Manire, 1968; Matsumoto and Manire, 1970a, 1970b). The presence of gram-negative double membranes was confirmed by early transmission electron microscopy (TEM) studies of RBs and EBs (Tamura et al., 1971), but challenges in purification and fractionation of RB membranes shifted emphasis toward EB membranes

Intracellular Pathogens I: Chlamydiales
Edited by Ming Tan and Patrik M. Bavoil © 2012 ASM Press, Washington, DC
10.1128/9781555817329.ch4

in subsequent studies. The same technical challenges continue to impede our study and understanding of the RB surface today.

An early TEM study of purified *Chlamydia psittaci* meningopneumonitis (Cal10) particles identified the first subcellular structure associated with chlamydial surfaces. Regularly spaced hexagonal lattices were observed in negatively stained EB envelope preparations (Manire, 1966). These lattices, located between the outer and cytoplasmic EB membranes, and similar structures were present in multiple *Chlamydia* species. It was speculated that this lattice might fulfill the function of a surrogate cell wall to explain the paradox of the stability of EBs in the absence of detectable peptidoglycan (Manire and Tamura, 1967; Garrett et al., 1974).

A second class of characteristic structures was also present on the outer membrane (OM) of chlamydiae (Matsumoto, 1973, 1982; Matsumoto et al., 1973). In contrast to the lattice-like structure in EBs, these "rosettes" were larger and were present on both RBs and EBs. Disagreements concerning their sizes and arrangements did not immediately clarify if these rosettes corresponded to one or multiple classes of structures (Gregory et al., 1979). At least one class distinctly localized to a single hemisphere of EBs. Freeze-fracture and TEM analyses of purified EBs, RBs, and infected cells also showed that these structures sometimes spanned inner and outer membranes and in RBs were associated with needle-like projections that crossed the chlamydial and inclusion membranes (Matsumoto, 1981). Matsumoto posed the insightful hypothesis that *Chlamydia* might use these to communicate with host cells (Matsumoto, 1982). More recently, the rosette-like structures have been proposed to be type III secretion (T3S) complexes that *Chlamydia* uses to deliver effector proteins to host cells (Bavoil and Hsia, 1998; Fields et al., 2003). Oligomers of a chlamydial autotransporter protein accumulate on OMs of RBs and EBs and also form structures reminiscent of those observed by Matsumoto, which may explain the ambiguity concerning sizes and locations of rosettes in the early literature (discussed below) (Crane et al., 2006; Swanson et al., 2009).

Isolated EB Envelopes Retain Functions of Intact EBs

Biochemical studies of EB envelopes were facilitated by the development of means for their gentle separation from whole cells (Tamura et al., 1971; Narita et al., 1976). These envelopes were powerful immunogens and contained antigens recognized by patient convalescent-phase serum (Caldwell et al., 1975a, 1975b; Caldwell and Kuo, 1977a, 1977b). Other results indicated EB envelopes maintained characteristics of intact EBs. For example, Levy and colleagues demonstrated that envelopes attach to and invade nonphagocytic cells (Levy and Moulder, 1982). Phagocytosed envelopes (gently treated and nondenatured) also resisted fusion with lysosomes (Eissenberg et al., 1983). Thus, these results showed that envelope components are key players in early chlamydial development and are involved in the host immune responses to infection.

The COMC: a Three-Dimensional Lattice of OmpA, OmcA, and OmcB

Serological studies established that chlamydial infections elicited a combination of genus- and serovar-specific antibodies and that the corresponding antigens were abundant in EBs and purified envelopes (Schachter and Grossman, 1981). Advances in polyacrylamide gel electrophoresis (PAGE) and the realization that proteins in the COMC bound one another by intra- and interprotein disulfide cross-links were key to elucidation of the major components, structure, and serovar-specific antigens of EB envelopes.

Caldwell and colleagues initially reported partial purification of the most abundant *Chlamydia trachomatis* envelope component, a ~39-kDa protein, which they called MOMP, the major outer membrane protein (Caldwell et al., 1981). A critical advance of their approach was the use of detergent extraction to separate COMC proteins from other EB components; MOMP accounted for ~60% of insoluble protein after Sarkosyl extraction of EBs. Other studies indicated that MOMP orthologs were present in all *Chlamydia* species and were also major proteins in the EB OM (Hatch et al., 1981; Salari and Ward, 1981). Monoclonal antibodies against purified *C. trachomatis* MOMP neutralized EB infectivity in HeLa 229 cells, showing that at least some of the antigenic portions of this protein were exposed on the EB OM (Caldwell and Perry, 1982). That there was substantial intraserovar and intraspecies heterogeneity in surface-exposed MOMP epitopes was revealed by observations that monoclonal antibodies that bound MOMP from one isolate often failed to recognize MOMP of closely related strains (Caldwell and Perry, 1982; Stephens et al., 1982). In contrast, *Chlamydia pneumoniae* MOMP did not elicit strong neutralizing antibody responses (Campbell et al., 1989). Later studies confirmed that immunogenic MOMP epitopes were displayed on the surface of *C. trachomatis* EBs (Caldwell and Judd, 1982) but were not necessarily required for EB infectivity (Hackstadt and Caldwell, 1985) and corresponded to antigens recognized in the course of human and animal chlamydial disease (Newhall et al., 1982). In summary, these data showed that MOMP is both the major component of chlamydial envelopes and a critical determinant of serovar- and species-specific neutralizing antibody responses.

Hatch and others resolved two highly cysteine-rich proteins, the 12-kDa OmcA (Omp3) and 60-kDa OmcB (Omp2), which corresponded to the other major proteins in COMC and EB envelopes (Hatch et al., 1981, 1984; Newhall and Jones, 1983). Unlike MOMP, the amino acid sequences of OmcA and OmcB were highly conserved in diverse *Chlamydia* species (Clarke et al., 1988; Watson et al., 1989; Allen et al., 1990). Formation of the COMC was absolutely cysteine- and disulfide bond-dependent, which suggested that these proteins are components of the EB cell wall (Hatch et al., 1984; Allan et al., 1985). OmcA was subsequently shown to be an OM lipoprotein that is expressed in tandem with OmcB (Allen et al., 1990; Lambden et al., 1990; Everett and Hatch, 1991).

Current understanding of the structure of the EB cell wall is based upon a series of elegant experiments that demonstrated that the COMC is a disulfide cross-linked network of MOMP, OmcA, and OmcB. MOMP, OmcA, and OmcB are present in characteristic molar ratios (5:2:1) in COMCs (Everett and Hatch, 1991). In the absence of reducing agents, insoluble polymers of these proteins do not resolve by PAGE. Barbour and colleagues identified penicillin-binding proteins in *C. trachomatis* and proposed that EB protein components of envelopes might be held together by interactions between abundant OMPs and peptidoglycan precursors, but they did not identify such precursors in whole EBs or Sarkosyl-soluble EB envelope fractions (Barbour et al., 1982). Based upon reports that *C. psittaci* envelopes were efficiently dissolved in sodium dodecyl sulfate (SDS) in the presence of β-mercaptoethanol (Hatch et al., 1981), Newhall and Jones used one-dimensional and two-dimensional PAGE to analyze the effects of reducing agents on protein interactions in EBs (Newhall and Jones, 1983). They showed that multimers (dimers, trimers, and tetramers) of MOMP and the 12-kDa OmcA and 60-kDa OmcB proteins in SDS-soluble EB fractions were resolved to monomers if they were prereduced with dithiothreitol (DTT) or β-mercaptoethanol and irreversibly alkylated to prevent oxidation. Later studies confirmed that these observations were applicable to other *Chlamydia* species, that disulfide cross-linked complexes of MOMP, OmcA, and a large periplasmic domain of OmcB are present in EBs, and that these periplasmic proteinaceous complexes anchor MOMP in the OM (Hatch et al., 1984; Newhall, 1987; Melgosa et al., 1993; Everett and Hatch, 1995); reviews of this topic by Hatch and Raulston remain critical reading (Raulston, 1995; Hatch, 1996). Thus, MOMP, OmcA, and OmcB form an interconnected protein network that stabilizes the OM of extracellular EB.

Definition of the COMC

The vast majority of individual proteins in the COMC are MOMP, OmcA, and OmcB. However, other proteins, but not all proteins in the EB OM, are also cross-linked to the COMC. We note that "COMC" originally referred to the specific complex of Sarkosyl-insoluble proteins extracted from whole EBs under exacting conditions (Caldwell et al., 1981). As reviewed in detail by Hatch (Hatch, 1996), this procedure is exquisitely sensitive to alterations; different detergents and inclusion or exclusion of chelators and redox reagents alter the results achieved. A second caveat is that even under otherwise identical conditions, the purity of EB preparations impacts the composition of the COMC fraction. This has been a particularly vexing problem in proteomic analysis of the COMC and RBs. Finally, the display of proteins on the surface of EBs and during transitions between RBs and EBs is impacted by redox status (discussed below). Here we define the COMC as a disulfide cross-linked polymer of MOMP, OmcA, OmcB, and additional proteins attached to these polymers by disulfide bonds in native EBs.

Polymorphic OMPs

Genome sequencing only recently revealed the surprising diversity of Type V autotransporters, called polymorphic outer membrane proteins (Pmps), which are present in *Chlamydia* species (Grimwood and Stephens, 1999). Pmps contain an N-terminal sec signal, which mediates their secretion across the inner membrane, and a central passenger domain that contains repeats of the sequences, GG[A/L/V/I][I/L/V/Y] and FXXN (Henderson and Lam, 2001). The C terminus of a Pmp encodes a β-barrel transporter domain that forms a pore in the OM and mediates translocation of the passenger domain to the cell surface. Passenger domains from some Pmps are processed from the transporter domain, following translocation, and form oligomers on EB and RB surfaces (Vandahl et al., 2001, 2002; Vretou et al., 2003; Wehrl et al., 2004; Kiselev et al., 2007; Swanson et al., 2009).

The *C. trachomatis* genome encodes 9 Pmps, and *C. pneumoniae* encodes 21 Pmps (Stephens et al., 1998; Read et al., 2000a; Henderson and Lam, 2001). Generally, the passenger domains of Pmps share a low degree of primary amino acid sequence conservation (Grimwood and Stephens, 1999). This could reflect diversification in function or indicate that variation in Pmp coding sequences is under strong selective pressure (Grimwood and Stephens, 1999; Henderson and Lam, 2001). Consistent with this hypothesis, anti-Pmp antibodies are abundant in convalescent-phase serum from patients infected by *C. trachomatis* (Crane et al., 2006; Tan et al., 2009). Substantial strain-to-strain variation in the expression of different Pmps has been observed (Stephens and Lammel, 2001). Remarkably, the expression of individual *C. trachomatis* Pmp proteins may be switched off and on at high frequencies in vitro (Tan et al., 2010). Taken together, these observations suggest that variation in expression of Pmps may correspond to a chlamydial immune evasion mechanism. However, other than a subset of *C. pneumoniae* Pmps and *C. trachomatis* PmpD that likely function as adhesins (Wehrl et al., 2004; Molleken et al., 2010), the cellular functions of these proteins remain mysterious. The role in adherence of the Pmps is discussed in more detail in chapter 5, "Chlamydial adhesion and adhesins."

It now seems likely that many of the high-molecular-weight, less abundant OMPs observed in envelopes and COMCs of *C. trachomatis* (Caldwell et al., 1981), *C. psittaci* (Hatch et al., 1984), and *C. pneumoniae* (Melgosa et al., 1993) correspond to Pmps. Separate reports of novel COMC-associated and highly immunogenic OMPs of 89 kDa in *C. abortus* (Cevenini et al., 1989, 1991) and 98 kDa in *C. pneumoniae* (Campbell et al., 1990) provided the first direct experimental evidence that a novel family of abundant and surface-exposed OMPs were present on EBs. Longbottom and colleagues first recognized that *C. abortus* encoded and expressed multiple Pmps, originally called OMP90 family proteins (Longbottom et al., 1996). This study and subsequent ones also established that Pmps were EB envelope proteins (Griffiths et al., 1992; Longbottom et al., 1996, 1998a, 1998b; Giannikopoulou et al., 1997). However, these results failed to capture the imagination of the wider chlamydial research community until genome sequencing revealed that genes encoding OMP90 family proteins were abundant and widely distributed in *Chlamydia* species (Stephens et al., 1998; Read et al., 2000b; Carlson et al., 2005).

Our understanding of Pmp localization in EB was established by work that set high thresholds for determining if OMPs were exposed on the EB surface and were cross-linked to the COMC (Tanzer and Hatch, 2001; Tanzer et al., 2001). Using a combination of trypsin sensitivity assays, EB surface protein labeling, reducing agent treatments, and mass spectrometry, Tanzer and Hatch demonstrated that domains of multiple *C. psittaci* Pmps were surface exposed on EB (Tanzer et al., 2001). A second study involving *C. trachomatis* serovar L2 identified PmpE, G, and H as COMC-associated OMPs, showed that some Pmps were bound to the COMC, but not to one another, by disulfide cross-links, and identified a novel OMP (CT623) of unknown function (Tanzer and Hatch, 2001).

Domains of additional *C. trachomatis* Pmps locate to the EB OM, and some of these are cross-linked to the COMC. In addition to results mentioned above for PmpE, G, and H, neutralization and immunolabeling experiments have confirmed that PmpD is exposed on EB and RB OMs in *C. trachomatis* and *C. pneumoniae* (Wehrl et al., 2004; Crane et al., 2006; Kiselev et al., 2007, 2009; Swanson et al., 2009). PmpD is not disulfide cross-linked to the COMC and can be extracted with gentle detergents (Swanson et al., 2009). This observation suggests that Pmps associate with the EB OM in multiple ways. Recent studies also support the idea that domains of most, if not all, *C. trachomatis* Pmps locate to the surface of the EB OM (Birkelund et al., 2009; Liu et al., 2010; Tan et al., 2010). All *C. trachomatis* Pmps have now been identified in EBs and/or purified COMC in one or more studies. An immuno-TEM study also found evidence that passenger domains of multiple *C. trachomatis* Pmps are displayed on the EB surface (Tan et al., 2010).

The association of Pmps with the COMC and EB OM by multiple mechanisms is supported by experimental evidence. In a study of whole EB, COMC, Sarkosyl-soluble EB fractions, and peptides from β-barrels of PmpE, G, and H, but not passenger domains of these proteins, were comparatively enriched in trypsin digests of purified COMCs (Liu et al., 2010). Peptides from PmpB, C, and F were enriched in the COMC, but similar strict segregation of β-barrel peptides of these proteins was not apparent. Finally, PmpI and PmpD were associated with Sarkosyl-soluble EB fractions. Nelson and colleagues have proposed that these results indicate that different strategies are used to associate Pmps with the EB envelope and that the reason for this could be to regulate temporal and spatial display of different Pmp passenger domains during EB attachment and entry (Liu et al., 2010). A complete panel of antibodies to *C. trachomatis* Pmps has now been constructed, so that tools needed to test this idea are available (Tan et al., 2010). In any case, the sheer number and diversity of these autotransporter proteins in EBs suggest that they mediate critical roles in early chlamydial development.

Porins

MOMP is the most abundant EB OMP and is an integral component of the COMC. Envelope preparations enriched in native MOMP (Bavoil et al., 1984) and synthetic membranes containing recombinant MOMP (rMOMP) (Wyllie et al., 1998; Kubo and Stephens, 2000) promote diffusion of small molecules. These results and others suggest that MOMP is a general-purpose porin whose activation may be linked to the redox status of the COMC (Bavoil et al., 1984; Sun et al., 2007).

C. trachomatis EBs have been shown to have a second COMC-associated porin. PorB was suspected to be a porin based upon its high proportion of cysteine residues and weak similarity to MOMP (Kubo and Stephens, 2000; Stephens and Lammel, 2001). PorB copurifies with the COMC and localizes to the OMs of RBs and EBs, and anti-PorB serum neutralizes EB infectivity (Kubo and Stephens, 2000). Like MOMP, rPorB porin activity was confirmed in liposome swelling assays (Kubo and Stephens, 2000). However, it was not evident why an EB would need a second, less abundant porin until it was demonstrated that rPorB preferentially promotes diffusion of carboxylates (Kubo and Stephens, 2001). While *C. trachomatis* encodes other enzymes of the TCA cycle, it cannot make 2-oxoglutarate (Stephens et al., 1998) and utilizes exogenous 2-oxoglutarate as a carbon source (Iliffe-Lee and McClarty, 2000). Thus, PorB presumably mediates the diffusion of this substrate so that *Chlamydia* can complete the TCA pathway (Kubo and Stephens, 2001).

OprB and Omp85 are two additional *C. trachomatis* OMPs that may be porins. These less abundant EB OMPs might have specialized substrate preferences, similar to PorB. OprB, an EB OMP, shares similarity with carbohydrate-selective porins and is strongly enriched in COMC fractions (Birkelund et al., 2009; Liu et al., 2010). Anti-OprB antibody neutralizes EB infectivity, confirming that this protein is exposed on the EB surface (Birkelund et al., 2009). However, porin activity was not detected when rOprB was evaluated in a liposome-swelling assay (Stephens and Lammel, 2001). Omp85 is more distantly related to known porins and may be involved in OMP export (Liu et al., 2010) and/or Pmp autotransport (Oomen et al., 2004). Nonetheless, anti-Omp85 antibodies can neutralize *C. trachomatis* infectivity (Stephens and Lammel, 2001), and Omp85 is cysteine rich and is highly enriched in the COMC (Liu et al., 2010).

Other EB Envelope Proteins

A large body of data now confirms that EBs have most or all of the components of a T3S system (Fields et al., 2003). The CdsC secretin is an abundant OMP and is probably disulfide cross-linked to the COMC (Betts et al., 2008; Birkelund et al., 2009; Liu et al., 2010). Interestingly, the *C. trachomatis* T3S needle protein CdsF contains two cysteines, which are rarely found in needle proteins of other bacteria (Betts et al., 2008). Based on this observation, it has been proposed that activation of the EB T3S machine might be linked to reduction of the COMC during EB attachment and entry (Betts et al., 2008). The chlamydial T3S is discussed in more detail in chapter 9, "Protein secretion and *Chlamydia* pathogenesis."

Varying degrees of evidence suggest that additional EB OMPs may exist. CTL0541, CTL0645, CTL0887, and peptidoglycan-associated lipoprotein (Pal) have been identified in whole EBs, EB envelopes, and COMC fractions from *C. trachomatis* (Mygind et al., 2000; Shaw et al., 2002; Birkelund et al., 2009; Liu et al., 2010). Little is known about CTL0541 or CTL0645, although the latter protein is predicted to have a secretion signal (Liu et al., 2010). CTL0887 in lymphogranuloma venereum (LGV) corresponds to ORF623 (CT623) in *C. trachomatis* serovar D, described by Tanzer and Hatch (Tanzer and Hatch, 2001), and an EB surface-exposed 76-kDa protein of unknown function in *C. pneumoniae* (Melgosa et al., 1993). We observed that CTL0887 has a low proportion of cysteine residues and was poorly enriched in COMC versus soluble EB fractions (Liu et al., 2010), which suggests that this protein may not be attached to the COMC by disulfide cross-links.

Identifying EB OMPs is especially challenging, and many putative chlamydial OMPs may not be bona fide OMPs (Raulston, 1995; Hatch, 1996; Liu et al., 2010). Raulston and colleagues discussed three reasons why a chlamydial protein could be misassigned as an OMP (Raulston et al., 2002). First, some of these predictions were based solely upon observations of EB neutralization by crude antiserum that could not subsequently be reproduced with monoclonal antibodies. Second, factors such as redox state, divalent cations, and detergents can cause the apparent cofractionation of periplasmic EB proteins with the EB OM. Finally, it can be difficult to assess whether the presence of abundant EB proteins, such as heat shock proteins, ribosomal proteins, and chaperones, in OMP and COMC fractions indicates that they are true OMPs. Hatch suggested that the best way to ascertain if a protein is associated with the OM of an EB or RB is to have evidence from multiple, nonredundant methods (Hatch, 1996). Recent proteomic studies suggest that the EB may have additional OMPs (Birkelund et al., 2009; Liu et al., 2010). Identification and clarifications of the roles of these proteins will be key to understanding early interactions between chlamydiae and the host cell.

Lipids in EBs and RBs

Chlamydial membranes contain lipids found in other gram-negative bacteria including a modified lipopolysaccharide (LPS), which chlamydiae can synthesize de novo, and eukaryotic lipids that chlamydiae acquire, and sometimes modify, from the host cell. Chlamydiae can synthesize phosphatidylethanolamine, phosphatidylglycerol, and phosphatidylserine de novo (Wylie et al., 1997), although they prefer to construct these by SN-2 substitution of their own branched-chain lipids onto glycerophospholipids salvaged from host cells (Su et al., 2004).

Chlamydiae intercept lipids from a variety of host cell sources (Hackstadt et al., 1995; Scidmore et al., 1996a, 1996b; Carabeo et al., 2003; Beatty, 2006, 2008; Cocchiaro et al., 2008; Heuer et al., 2009), and these lipids include cholesterol, sphingomyelin, phosphatidylcholine, phosphatidylinositol, and cardiolipin, all of which are incorporated into EB and RB membranes (Newhall, 1988; Wylie et al., 1997). Somewhat surprisingly, *C. trachomatis* robustly infects cell lines that have moderate to severe defects in lipid metabolism or cells that have been treated with brefeldin A to interrupt lipid traffic (Wylie et al., 1997). Alterations in phospholipid composition are tolerated by EBs, and membranes of EBs and RBs reflect the lipid composition of their host cell (Hatch and McClarty, 1998). Thus, host cell lipid requirements of chlamydial membranes appear to be somewhat plastic, though this has not been studied extensively in species other than *C. trachomatis*. In contrast, LPS biosynthesis is essential for production of infectious EBs (Nguyen et al., 2011). A recent report that higher proportions of cholesterol contribute to the rigidity and lower permeability of an EB, compared to an RB, is intriguing and suggests new approaches to investigate these questions (Lim and Klauda, 2011).

More pertinent here is the composition of the outer membrane of the EB. As in other gram-negative bacteria, chlamydial OMs contain high proportions of LPS and are stabilized by divalent cations. All *Chlamydia* spp. synthesize similar deep-rough type LPS molecules that contain lipid A and a core oligosaccharide, and which lacks or only has minor remnants of O antigens (Nurminen et al., 1983; Caldwell and Hitchcock, 1984; Nurminen et al., 1985). LPS appears to play a key role in maturation of the chlamydial envelope, as inhibitors of lipid A biosynthesis block RB-to-EB transition and result in the accumulation of intermediate bodies (Nguyen et al., 2011). Studies of chlamydial LPS structure were facilitated by groups who pioneered methods for expression of these molecules in *Salmonella* and *Escherichia coli* (Brade et al., 1985, 1987; Nano et al., 1985; Nurminen et al., 1985; Belunis et al., 1992). Rund and colleagues eventually proposed a complete structure for *C. trachomatis* LPS following their determination that the genus-specific oligosaccharide was a 3-deoxy-D-manno-oct-2-ulopyranosonic acid (KDO). *C. psittaci* also produces branched LPS variants (Rund et al., 1999), and a monoclonal antibody specific for one of these can differentiate *C. psittaci* from *C. pneumoniae*, *Chlamydia pecorum*, and *C. trachomatis* (Muller-Loennies et al., 2006). It has been suggested that *C. trachomatis* and *C. psittaci* LPS might undergo smooth-rough phase variation (Lukacova et al., 1994), but enzymes that could mediate this change were not detected in the *C. trachomatis* genome (Stephens et al., 1998). However, there are many open reading frames (ORFs) of unknown function in chlamydial genomes, and absence of evidence in this and other cases does not constitute evidence of absence.

Anti-LPS responses are elicited during the course of natural chlamydial infections in humans and animals (Brade et al., 1990), but it is unclear how well these antibodies recognize native LPS in the OM of EBs. *C. trachomatis* LPS is tightly associated with the EB OM (Caldwell and Hitchcock, 1984), and cross-linking experiments indicate that much of the LPS in the EB OM is associated with, or is in close proximity to, MOMP (Birkelund et al., 1988, 1989). Interestingly, an antibody that reacted with chemically cross-linked LPS and MOMP from the EB OM did not react with RBs (Birkelund et al., 1988). A genus-specific antibody also bound LPS on *C. trachomatis* RBs but not EBs in another study (Collett et al., 1989). The authors of both these studies noted that fixation protocols differentially impacted antibody binding to EBs and RBs. Peterson and colleagues later characterized a genus-specific LPS monoclonal antibody that reacted with purified LPS of *C. trachomatis*, *C. psittaci*, and *C. pneumoniae* but determined that it only neutralized infectivity of the *C. pneumoniae* TW-183 strain in vitro (Peterson et al., 1998). Antibody responses to LPS elicited during natural chlamydial infection are limited, and chlamydial LPS has low endotoxin activity (Nurminen et al., 1983). However, an adoptively transferred monoclonal anti-LPS antibody protected against reinfection with *C. muridarum* in the mouse genital tract (Morrison and Morrison, 2005). These results suggest that LPS molecules are differentially oriented in, or associated with, the OM in EBs and RBs, which may be a strategy used by an extracellular EB to reduce LPS immunogenicity.

Glycoproteins and Exoglycolipids

In addition to carbohydrate moieties of LPS, other polysaccharides, possibly derived from chlamydiae or the host cell, are displayed on EBs. Kuo and colleagues reported that *C. trachomatis* MOMP is a glycoprotein based upon observations that MOMP isolated from whole EBs (i) showed faster migration on a protein gel following treatment with the *N*-glycosidase F, (ii) was weakly stained with *p*-phenylenediamine, (iii) was metabolically labeled by tritiated glucosamine or galactose, and (iv) bound various lectins (Swanson and Kuo, 1991). The polysaccharide was later shown to be an N-linked high-mannose type oligosaccharide that could bind HeLa cells and inhibit EB infectivity (Swanson and Kuo, 1994). Removal of this polysaccharide by glycanase treatment moderately inhibited the infectivity of *C. pneumoniae*, *C. trachomatis*, and *C. psittaci* EBs for select cell lines (Kuo et al., 2004). Two additional EB glycoproteins of approximately 18 kDa and 32 kDa have been proposed based upon reactivity with lectins (Hackstadt, 1986). It now seems likely that these proteins correspond to the abundant and similarly sized chlamydial histone-like proteins Hc1 and Hc2 (Hackstadt, 1991; Swanson and Kuo, 1991, 1994; Kuo et al., 1996).

The exoglycolipid antigen (GLXA) was identified in supernatants of *Chlamydia*-infected cells as an antigen that reacted with patient convalescent-phase serum (Stuart and Macdonald, 1989). GLXA does not appear to be related to chlamydial LPS and has been associated with EB and inclusion membranes and with supernatants from cultured cells infected with *C. trachomatis*, *C. pneumoniae*, or *C. psittaci* (Stuart et al., 1991; Vora and Stuart, 2003). Purified GLXA augments *C. trachomatis* infectivity in vitro and in mice (Vora and Stuart, 2003). The chemical composition of GLXA and whether it is produced by *Chlamydia* or host cells are unknown (Webley et al., 2004), and genes which could mediate this molecule's assembly have not been identified in published chlamydial genomes.

Glycosaminoglycans

Glycosaminoglycans (GAGs) can modulate chlamydial attachment and infectivity in vitro, although the specificity and in vivo significance of this observation remains a subject of debate. Exogenous heparin and heparan sulfate (HS) bind EBs and can competitively inhibit their attachment to host cells, presumably by preventing interactions between EB and host cells. The extent of this inhibition varies among *Chlamydia* strains and species (Becker et al., 1969; Kuo and Grayston, 1976; Chen and Stephens, 1994; Chen et al., 1996). Heparin-mediated inhibition of EB infectivity is sensitive to the degree of GAG sulfation and is specific for the type of GAG because chondroitin sulfate, keratin sulfate, and hyaluronic acid have much less of an effect (Zhang and Stephens, 1992). Treatment of EBs with heparin sulfate lyase similarly inhibits EB attachment, which is reversed by addition of exogenous heparin or HS (Zhang and Stephens, 1992). Davis and Wyrick suggested that the different tropisms of *C. trachomatis* LGV and trachoma biovar strains might be related to their different affinities for, or possibly the differential display of, HS on apical and basolateral surfaces of host cells (Davis and Wyrick, 1997). OmcB proteins from *C. trachomatis* LGV and trachoma biovars have different affinities for HS, which provides support for this idea (Stephens et al., 2001; Fadel and Eley, 2007). However, host HS is dispensable for EB attachment and infectivity. *C. trachomatis* LGV serovar L2 infectivity was only marginally reduced in an HS-deficient CHO cell line compared to parental cells (Rasmussen-Lathrop et al., 2000; Stephens et al., 2006). A possible explanation is that *C. trachomatis* synthesizes an HS-like molecule de novo or from unrecognized precursors in host cells (Rasmussen-Lathrop et al., 2000). Enzymes that might be involved in construction of a chlamydial HS-like molecule in the *C. trachomatis* genome were identified by weak sequence homology but have not been confirmed to mediate chlamydial HS construction (Rasmussen-Lathrop et al., 2000).

MOMP and OmcB have been proposed to be chlamydial ligands of HS. Both MOMP and OmcB bind exogenous HS in vitro, and it has been suggested that interactions between host HS and these proteins are critical for attachment of EBs to host cells (Su et al., 1996; Stephens et al., 2001; Fadel and Eley, 2007; Moelleken and Hegemann, 2008). rMOMP attaches to, but does not enter, HeLa cells and competitively inhibits binding of EBs (Su et al., 1996). Heparin and HS but not chondroitin sulfate also specifically inhibited rMOMP binding to cells. These data led Su and colleagues to propose that MOMP is an adhesin (Su et al., 1996). In contrast, OmcB was the only major protein observed when Stephens and colleagues separated EB envelopes by SDS-PAGE and probed them with labeled HS (Stephens et al., 2001). As with MOMP, binding of labeled HS by OmcB was specifically inhibited by unlabeled HS, but not by less sulfated GAGs. The putative HS binding motif in OmcB was identified, and sera against this peptide reacted with the surface of intact EB, although previous studies suggested that most of OmcB locates to the periplasm (Everett and Hatch, 1995; Mygind et al., 1998). The HS binding motif was found to be at the N terminus of OmcB (Stephens et al., 2001), consistent with a report that OmcB from *C. caviae* GPIC is an adhesin whose function is abrogated by trypsin cleavage of a small, OM-exposed peptide (Ting et al., 1995). *C. pneumoniae* OmcB also has HS binding and adhesin functions (Moelleken and Hegemann, 2008). Intriguingly, mutation of a single amino acid residue that differs between OmcB from *C. trachomatis* serovars E and L1 was sufficient to abrogate binding of L1 OmcB to HS in vitro (Moelleken and Hegemann, 2008). The authors proposed that OmcB is a determinant of tropism in *C. trachomatis*, but it is unclear if differences in this position are widely conserved in trachoma and LGV biovar strains.

Modeling the EB OM

Based upon results in the previous sections, I propose an updated model of the EB OM that retains many aspects of models by Hatch and Raulston (Raulston, 1995; Hatch, 1996) (Fig. 1). LPS is densely packed in the EB OM and is associated with disulfide-linked trimers of MOMP, in a conformation that masks

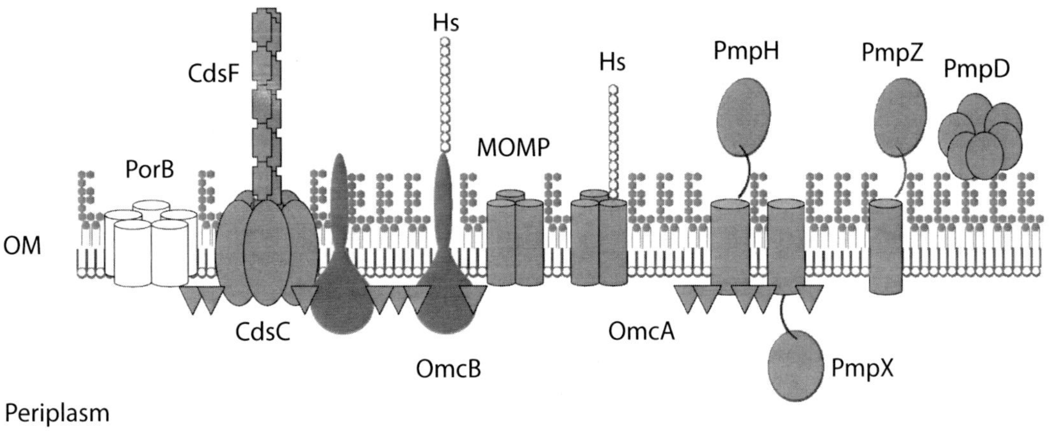

FIGURE 1 Model of the EB surface. OmcB (dumbbells) and OmcA (triangles) are located primarily in the periplasm. LPS (small gray hexagons) is primarily located in the outer leaflet of the outer membrane, while phospholipids are in the inner leaflet (small open circles). OmcA is associated with the inner leaflet of the OM by its lipid moiety (not shown). An N-terminal portion of some OmcB molecules may traverse the OM and bind HS on the EB surface (chains of open hexagons). HS may also bind unknown residues on MOMP. OmcA and OmcB are cysteine cross-linked to one another in the periplasm, and, speculatively, to other OMPs such as CdsC, PorB, and some Pmps (PmpH and hypothetical PmpX). The figure shows a range of scenarios that may occur for specific Pmps. For example, the passenger domain (shown as ovals for each of the Pmps) of PmpH has been exported to the EB surface, whereas export of the passenger domain of PmpX has not been completed. Export of the hypothetical PmpZ passenger domain has been completed, but its transporter domain is not bound to other COMC proteins. Oligomers of PmpD, which have been processed from their passenger domains, are loosely associated with the surface of the EB. Polymers of the putative T3S needle protein CdsF extend from CdsC. doi:10.1128/9781555817329.ch4.f1

immunogenic KDO epitopes (not shown). MOMP is an integral membrane protein that is anchored in the OM. OmcA and OmcB form a disulfide cross-linked heteropolymer in the periplasm. The N terminus of OmcB and a surface-exposed region of MOMP are displayed on the OM surface, where they may be bound by HS. Additional porins, such as PorB, and perhaps OprB and Omp85 (not shown), are present in the OM and may be linked laterally to MOMP, to OmcA, or to the periplasmic portion of OmcB. Polymers of the T3S needle protein CdsF extend from the EB surface and are attached to the T3S OM secretin CdsC, which is disulfide cross-linked to other COMC proteins and localizes to a single hemisphere of the EB. As shown in Fig. 1, transport and processing of the PmpD passenger domain has been completed and oligomers of this domain are peripherally associated with the EB OM. Translocation of the PmpH passenger domain is also complete, but it has not been separated from its autotransporter domain, which is disulfide cross-linked to the COMC. Based upon limited data (Buendia et al., 1997; Tanzer and Hatch, 2001; Tanzer et al., 2001; Liu et al., 2010), passenger domains of other Pmps, whose autotransporter domains are in the OM, may be retained in the periplasm, regardless of whether they are disulfide cross-linked to COMC proteins. Other proteins including CTL0645 and CTL0887 (not shown) are on the EB surface and may be disulfide cross-linked to COMC proteins. Key implications of this model are that reduction of disulfide bonds of OM proteins concomitant with chlamydial entry activates porins in the OM, the T3S system, and temporal delivery of Pmp passenger domains to the chlamydial OM.

CHANGES IN THE CHLAMYDIAL SURFACE DURING THE DEVELOPMENTAL CYCLE

EB Attachment and Entry into Host Cells

A short overview of chlamydial attachment and entry is presented here to provide background, but these critical early steps in the infection are discussed in more detail in chapter 5, "Chlamydial adhesion and adhesins," and in chapter 6, "Initial interactions of chlamydiae with the host cell." Key studies in mutant CHO cells lines by the Stephens and Hackstadt groups over the last 10 years has revealed that chlamydial attachment and entry can be separated into discrete steps (Carabeo and Hackstadt, 2001; Fudyk et al., 2002; Stephens et al., 2006; Conant and Stephens, 2007; Abromaitis and Stephens, 2009). Briefly, loose attachment of EBs to the host cell is proposed to be mediated by interactions between HS, unidentified host proteins, and EB proteins including MOMP, OmcB, and PmpD. This initial interaction is followed by tight attachment of an unknown EB ligand(s) to a host cell protein complex that includes protein disulfide isomerase (PDI), which is required for chlamydial attachment and entry (Davis et al., 2002; Conant and Stephens, 2007; Abromaitis and Stephens, 2009). Subsequently, the EB T3S system is activated and secretes tyrosine actin-recruiting protein (Tarp) into host cells (Clifton et al., 2004). Tarp elicits cystoskeletal rearrangements and pathogen-mediated endocytosis of the EB (Clifton et al., 2004). EB entry and subsequent escape from the endocytic pathway are insensitive to inhibitors of bacterial transcription/translation, whereas trafficking to the peri-Golgi region and interception of host cell lipids are not (Fields et al., 2003). Differentiating which of these processes are mediated by EB OMPs and T3S effectors is difficult because EB T3S is largely (Muschiol et al., 2006) or completely (Wolf et al., 2006) insensitive to inhibitors.

Reduction of the COMC and Chlamydial Entry

Upon entry, EB infectivity is rapidly lost, and decondensation of the EB genome and an increase in permeability of the particle are initiated independently of de novo chlamydial or host protein synthesis (Moulder, 1991). Tamura and Manire provided the first evidence that the EB decondensation and the early transition to RB might be related to disulfide bonding when they found that RBs, but not EBs, lack cysteine and methionine (Tamura and Manire, 1967). Hatch and colleagues and Newhall and Jones established that solubility of major components of EB envelopes could be increased by treatment with -mercaptoethanol and/or DTT and that reformation of protein heteropolymers in envelopes was prevented by irreversible alkylation of exposed sulfhydryls (Newhall and Jones, 1983; Hatch et al., 1984). The role of disulfide bonds in EB impermeability to small molecules was demonstrated by using native MOMP in an in vitro assay. Unreduced MOMP promoted limited diffusion of small molecules in liposome assays, but this activity was dramatically increased if the reconstituted protein was prereduced and then alkylated to prevent oxidation (Bavoil et al., 1984). An extensive study of the effects of DTT on EBs established that loss of EB infectivity and early stages of the EB-to-RB transition were dependent upon the redox status of EB cysteines (Hackstadt et al., 1985). EBs treated with DTT showed reduced infectivity, were permeable to Machiavelo red and sensitive to osmotic lysis, and could oxidize glutamate in a cell-free system. However, exposure to DTT did not detectably alter the ultrastructure of EBs, indicating that additional steps were required to complete the EB-to-RB transition (Hackstadt et al., 1985). One such step is the change from the tightly compacted DNA in an EB to the more dispersed DNA of an RB. This DNA decondensation is mediated by a metabolite of the chlamydial methylerythritol-4-phosphate pathway of isoprenoid biosynthesis, which disrupts binding of histone-like proteins to DNA (Grieshaber et al., 2004). In summary, these data show that disulfide bonds are important for maintaining the physical stability of extracellular EBs and that reduction of these cross-links plays a role in EB-to-RB conversion.

Davis and colleagues reported the first evidence that reduction of EB disulfide cross-links could be mediated by specific host cell protein(s) (Davis et al., 2002). Mass spectrometry was used to identify apical proteins from HEC-1B cells that coimmunoprecipitated with EBs. The most abundant was PDI, a protein known to catalyze thiol-disulfide interchange reactions and mediate entry of some viruses and toxins. EB infectivity was decreased when host cells were treated with a membrane impermeable thiol-alkylating reagent or bacitracin, a PDI inhibitor, which indicated that PDI-mediated disulfide exchange was critical for EB infectivity (Davis et al., 2002). Consistent with these findings, expression of wild-type PDI restored susceptibility of a profoundly *Chlamydia*-resistant CHO cell line to diverse *Chlamydia* species (Conant and Stephens, 2007). Other results showed that EBs were not directly binding PDI but that PDI was essential for EB attachment and entry (Abromaitis and Stephens, 2009). Intriguingly, mutant PDI that could not catalyze disulfide exchange supported EB attachment but not entry (Abromaitis and Stephens, 2009). More recent findings suggest that additional proteins that regulate the redox balance in host cells are involved in this process (Lazarev et al., 2010), but whether PDI is sufficient to reduce EB disulfide cross-links is unclear.

A Black Box—Changes in the EB Surface after Entry

Dissecting the changes that occur in the EB upon entry into the host cell is challenging because of the mixed population of attached and internalized EBs in the early stages of infection. However, some data suggest that MOMP and the cysteine-rich proteins OmcA and OmcB have different fates early in development. Hatch reported that within 1 hour of infection the majority of *C. psittaci* MOMP was reduced to monomers, whereas this process was slower in *C. trachomatis* (Hatch et al., 1986). Cross-linked complexes containing OmcA and OmcB from EBs may persist well into the developmental cycle (Hatch et al., 1986). However, these findings may have been an artifact caused by oxidation of these proteins during COMC isolation (Newhall, 1987), and slight disagreement between the results of these studies may reflect differences in *C. psittaci* and *C. trachomatis* biology. In any case, OmcB is not associated with membrane fractions following full reduction of the COMC, and OmcA is only weakly associated via its lipid moiety, which suggested that OmcA and OmcB were periplasmic proteins (Everett and Hatch, 1995). Little is known about disulfide bonding of other COMC-associated OMPs early in development, but the reduction of porins and autotransported proteins is presumed to be similar to the major envelope proteins. Pulse-chase experiments with radiolabeled metabolites, such as 2-oxoglutarate, in mutant cell lines with defects in attachment and entry (Carabeo and Hackstadt, 2001; Fudyk et al., 2002) could illuminate temporal aspects of porin activation and provide a promising area for future work.

Changes in the Composition of the Chlamydial Envelope during the Developmental Cycle

Studying the RB proteome at all stages of development has been especially challenging because RBs are extremely fragile and it is difficult to effectively separate RBs from EBs and intermediate bodies. There have only been a handful of studies of RB proteomes, and these studies have yet to identify OMPs exclusively found in RB membranes (Shaw et al., 2002; Vandahl et al., 2004; Skipp et al., 2005). Thus, most of our understanding of RB surfaces has been gleaned from transcriptional studies and Western blotting.

Studies of the chlamydial transcriptome revealed that characteristic classes of genes are expressed in early, middle, and late stages of development. Wichlan and Hatch provided early evidence that chlamydial development was punctuated by temporal expression of different classes of genes when they showed that EUO, a DNA binding protein absent from EBs, was transcribed and translated almost immediately upon EB entry (Wichlan and Hatch, 1993). Shaw and colleagues subsequently used reverse transcription-PCR to determine when approximately 8% of the genes in the *C. trachomatis* genome were first expressed during the developmental cycle (Shaw et al., 2000). Their analysis revealed that three classes of genes were turned on sequentially during chlamydial development, and genes whose expression was initiated soon after chlamydial entry, during RB proliferation, and during RB-to-EB differentiation, could be differentiated (Shaw et al., 2000). Following this, Belland and colleagues used a combination of DNA microarrays and quantitative reverse transcription-PCR to measure transcript levels for all annotated *C. trachomatis* coding sequences during the normal developmental cycle (Belland et al., 2003b). The results from these two studies are broadly similar, and data from the more comprehensive Belland study will primarily be considered in the following section.

Early Stages of the EB-to-RB Transition

Belland detected expression of 29 chlamydial immediate early transcripts, which can be divided into four general classes (Belland et al., 2003b). First, the transcripts for *incD-G* and *ct228-229*, which encode Inc and Inc-like proteins, were detected. Some of these Inc proteins have been shown to be secreted by T3S in surrogate systems and presumably are secreted from an early RB into the inclusion membrane, where they mediate inclusion biogenesis and early interactions with host cells. A second class of early transcripts corresponds to membrane and membrane-associated proteins with potential roles in nutrient acquisition. Included in this group are an oligonucleotide permease (*ct480*), D-alanine/glycine permease (*ct735*), an ADP/ATP translocase (*npt1*), and a nucleoside phosphate transporter (*npt2*). Additional proteins associated with oligopeptide transport were well represented by 3 hours postinfection, indicating that one of the first steps in EB-to-RB transition is initiating the acquisition of peptides, ATP, and nucleoside triphosphates from host cells (Tjaden et al., 1999). Third, chaperones and enzymes associated with protein processing and folding were detected, including *groEL*, *groES*, and a putative methionine aminopeptidase (*ct851*). Finally, transcripts of conserved chlamydial genes of unknown function were well represented.

Growth and Division of RBs: Mid-to-Late Developmental Cycle

Expression of most RB genes begins by 8 to 16 h postinfection. The expression of additional Incs, components of the T3S system, some Pmps, and genes associated with general metabolism is initiated during this phase of development. Interestingly, genes for peptidoglycan synthesis are expressed in RBs, and many of the corresponding enzymes have now been confirmed to be active in vitro (McCoy and Maurelli, 2006). The presence and composition of peptidoglycan in RBs have not been confirmed, but these observations and the presence of a penicillin-sensitive septum-like structure in RBs (Brown and Rockey, 2000) suggest that some form of murein is present and involved in RB division. Presumably, EBs lack peptidoglycan so that they can avoid activation of cytosolic pattern recognition receptors (Welter-Stahl et al., 2006). Small amounts of peptidoglycan in RBs, sequestered within the inclusion, may also go undetected by the host innate immune surveillance. Synthesis of some Pmps and MOMP is also initiated during this stage. Most of these Pmps are exposed on the RB and are not in the inclusion membrane, indicating that Pmps play roles on both RB and EB surfaces (Tan et al., 2010). In addition to a variety of amino acid permeases and ABC transporters, a group of enzymes in the phospholipase D superfamily (Pld) that are likely to mediate RB nutrient acquisition is expressed during this phase. These enzymes are interesting because they are expanded and diversified in *C. trachomatis* and close relatives (Stephens et al., 1998; Read et al., 2000b) and are differentially regulated during gamma interferon-induced persistence in *C. trachomatis* (Belland et al., 2003a; Nelson et al., 2006). Pld might mediate RB lipid metabolism, and some of these proteins locate to the face of the RB that is proximal to the inclusion membrane (Nelson et al., 2006; Taylor et al., 2010). At least one Pld family enzyme (Lda3) associates with lipid droplets in yeast and is secreted from the inclusion into host cells (Cocchiaro et al., 2008). It has been suggested that Lda3 transports liquid droplets into the inclusion. Thus, this could be a strategy *Chlamydia* uses to capture high-energy metabolites during the phase of rapid RB replication. However, the enzymatic activities of the Lda and Pld family proteins have not been confirmed (Cocchiaro et al., 2008).

Intimate association of RBs with the inner surface of the inclusion is one of the most striking ultrastructural features of this phase of development. Bavoil and colleagues have proposed that chlamydial T3S injectisomes mediate this association and that as long as this association is maintained RB-to-EB differentiation cannot occur (Wilson et al., 2006, 2009). Enlarged RBs, elicited by various triggers of chlamydial persistence, are still intimately associated with the inclusion membrane (Wilson et al., 2009). However, T3S inhibitors prevent association of RBs with the inclusion membrane (Muschiol et al., 2006). The factor(s) that tethers RBs to the inclusion membrane has not been conclusively identified, and it is unclear if physical detachment and coupled disruption of T3S are the trigger for, or the consequence of, RB-to-EB transition.

Mid-to-Late Gene Expression and the RB-to-EB Transition

Late in the developmental cycle, individual RBs convert into EBs and detach from the inclusion membrane. It is notable that many of the genes expressed during this phase are not expressed during persistent growth (Belland et al., 2003a). In *C. trachomatis*, synthesis of the histone-like proteins Hc1 and Hc2 initiates late in development (Hackstadt et al., 1991; Tao et al., 1991). Whether histones drive the RB-to-EB transition is unclear, but expression of *C. trachomatis* Hc1 is sufficient to mediate the formation of condensed chromosomes in *E. coli* similar to those observed in EBs (Barry et al., 1992). OmcA and OmcB synthesis initiates late in infection. At least two classes of transcripts produced at this same time encode proteins suspected to be involved in integration of OmcA and OmcB with MOMP into the COMC and EB envelope. CT780 and CT783 are putative thioredoxin disulfide isomerases that are predicted to oxidize cysteine residues on OmcA, OmcB, and MOMP. MtpA and MtpB are putative membrane thiol proteases that may mediate processing of these OMPs into the mature forms of these complexes (Belland et al., 2003b). PmpG and PmpH are late genes in *C. trachomatis* serovar D (Belland et al., 2003b), and more recent data suggest that PmpE-H are encoded by a single operon that is transcribed late in the developmental cycle of *C. trachomatis* serovar E (Carrasco et al., 2011). These late Pmps may have a specialized function that is different from other Pmps that are expressed earlier in development (Belland et al., 2003b). PmpG and PmpH are among the most abundant COMC-associated Pmps (Liu et al., 2010). Thus, expression of these late Pmps appears to be coordinated with the enzymes that mediate COMC assembly.

Re-formation of the COMC and EB Exit

The regularity of the hexagonal lattices originally observed by Manire suggests that reassembly of the COMC is an orderly and well-regulated process (Manire, 1966); however, we still know little about the regulation or spatial placement of individual proteins in the COMC. Re-formation of the COMC appears to be absolutely dependent on bacterial protein synthesis, cysteine, and late gene expression since inhibitors of bacterial translation, cysteine-deficient medium, and stimuli that trigger persistence all inhibit disulfide bond formation in the COMC (Newhall and Jones, 1983; Allan et al., 1985; Hatch et al., 1986; Newhall, 1987). Additional oxidation can occur following release of EBs, and there may be slight differences in this process in different *Chlamydia* species (Hatch et al., 1986; Newhall, 1987). In any case, COMC oxidation must be sufficient upon release of EBs during lysis of cells to stabilize them in the extracellular environment. Interestingly, a recent study showed that EBs can exit cells by pathogen-driven lysis of the inclusion and the host plasma membrane and by an extrusion process that requires host cell proteins (Hybiske and Stephens, 2007). The latter process has been observed with a variety of chlamydial species in different cell lines and may or may not occur in vivo (Doughri et al., 1972; Rank et al., 2011). Comparisons of the COMC in EBs released via the lysis and extrusion pathways could be useful for studying the relative contribution of host and *Chlamydia* redox proteins to EB maturation and reassembly of the COMC.

SUMMARY

The envelopes of the extracellular EB and intracellular RB differ markedly and reflect requirements for survival of chlamydiae in two unique environments. Outside the cell, the oxidized COMC provides osmotic protection, restricts loss of metabolites across the OM, and may mask immunogenic determinants that could alert host cells to the presence of these pathogens. During attachment and entry, this protective shell must be reduced and reconfigured so that proteins in the OM can begin to interact with host cells. In contrast to the impermeable EB, the OM of the intracellular RB incorporates a range of individual pores and larger macromolecular machines, such as the secretin ring of the T3S injectisome. These transporters and secretion machineries facilitate acquisition of nutrients from, and delivery of effector proteins to, host cells from inside the RB's protected niche and across the inclusion membrane. Finally, in response to a putative signal such as crowding or nutrient depletion, RBs sense and orchestrate de novo production of EB-specific proteins and the orderly reduction and assembly of EB surface proteins into the COMC. The processes that mediate each of these steps remain poorly understood, but it is likely that future studies of the chlamydial surface could reveal fascinating insights into these pathogens' unique biology and provide new targets for antichlamydial drugs and vaccines.

CONCLUSION AND FUTURE PERSPECTIVES: DISSECTING CHLAMYDIAL SURFACES IN THE EMERGING ERA OF CHLAMYDIAL GENETICS

The recent discovery (DeMars et al., 2007; DeMars and Weinfurter, 2008) and development of experimental means for lateral gene transfer (Suchland et al., 2009) are encouraging and mean that it may soon be possible to map chlamydial alleles that mediate strain-specific differences in attachment and entry. Strategies recently developed by Laszlo Kari and colleagues for mutagenesis of *C. trachomatis* and isolation of mutants with disruptions in specific genes are especially exciting because it should now be possible to directly inactivate and test the functions of some envelope proteins (Kari et al., 2011). These developments in the ability to genetically manipulate chlamydiae are discussed in chapter 15, "Chlamydial genetics: decades of efforts, very recent successes." These approaches should be especially useful in dissecting the roles of strain- and species-variable gene families associated with chlamydial surfaces, such as the chlamydial Pmp and Pld proteins. Presumably, the variability in these gene families indicates that they fulfill niche-specific functions and some may be dispensable in cell culture. More intriguingly, similar strategies might be used to generate strains with conditional mutations in essential genes. A conditional T3S mutant could provide answers to many questions concerning the roles of EB surface proteins and T3S effectors in early stages of chlamydial development.

On a separate note, developments in isotopic labeling and quantitative mass spectrometry should now make it feasible to more accurately define the proteomes of RBs and EBs. Various methods, including stable isotope labeling with amino acids in cell culture, isotope-coded affinity tags, and isobaric tagging for relative and absolute quantitation, could be employed to determine if proteins originate from RBs or EBs. It may also be time to revisit methods for RB isolation that have not been substantially refined in almost 30 years. It seems likely that approaches such as flow cytometry could be integrated into existing RB purification procedures to reduce levels of contaminating intermediate bodies and EBs.

ACKNOWLEDGMENTS

I dedicate this chapter to my outstanding postdoctoral mentor, Harlan Caldwell. Separately, I thank Thomas Hatch for critical and insightful review of this chapter and sharing his extensive insights into the chlamydial literature with me. I also thank Evelyn Toh for assistance in editing this chapter. Finally, I note that all omissions and errors of interpretation of previous studies are the fault of the author, and I especially apologize to anyone whose efforts I have overlooked.

REFERENCES

Abromaitis, S., and R. S. Stephens. 2009. Attachment and entry of *Chlamydia* have distinct requirements for host protein disulfide isomerase. *PLoS Pathog.* **5:**e1000357.

Allan, I., T. P. Hatch, and J. H. Pearce. 1985. Influence of cysteine deprivation on chlamydial differentiation from reproductive to infective life-cycle forms. *J. Gen. Microbiol.* **131:**3171–3177.

Allen, J. E., M. C. Cerrone, P. R. Beatty, and R. S. Stephens. 1990. Cysteine-rich outer membrane proteins of *Chlamydia trachomatis* display compensatory sequence changes between biovariants. *Mol. Microbiol.* **4:**1543–1550.

Barbour, A. G., K. Amano, T. Hackstadt, L. Perry, and H. D. Caldwell. 1982. *Chlamydia trachomatis* has penicillin-binding proteins but not detectable muramic acid. *J. Bacteriol.* **151:**420–428.

Barry, C. E., III, S. F. Hayes, and T. Hackstadt. 1992. Nucleoid condensation in *Escherichia coli* that express a chlamydial histone homolog. *Science* **256:**377–379.

Bavoil, P., A. Ohlin, and J. Schachter. 1984. Role of disulfide bonding in outer membrane structure and permeability in *Chlamydia trachomatis*. *Infect. Immun.* **44:**479–485.

Bavoil, P. M., and R. C. Hsia. 1998. Type III secretion in *Chlamydia*: a case of déjà vu? *Mol. Microbiol.* **28:**860–862.

Beatty, W. L. 2006. Trafficking from CD63-positive late endocytic multivesicular bodies is essential for intracellular development of *Chlamydia trachomatis*. *J. Cell Sci.* **119:**350–359.

Beatty, W. L. 2008. Late endocytic multivesicular bodies intersect the chlamydial inclusion in the absence of CD63. *Infect. Immun.* **76:**2872–2881.

Becker, Y., E. Hochberg, and Z. Zakay-Rones. 1969. Interaction of trachoma elementary bodies with host cells. *Isr. J. Med. Sci.* **5:**121–124.

Belland, R. J., D. E. Nelson, D. Virok, D. D. Crane, D. Hogan, D. Sturdevant, W. L. Beatty, and H. D. Caldwell. 2003a. Transcriptome analysis of chlamydial growth during IFN-gamma-mediated persistence and reactivation. *Proc. Natl. Acad. Sci. USA* **100:**15971–15976.

Belland, R. J., G. Zhong, D. D. Crane, D. Hogan, D. Sturdevant, J. Sharma, W. L. Beatty, and H. D. Caldwell. 2003b. Genomic transcriptional profiling of the developmental cycle of *Chlamydia trachomatis*. *Proc. Natl. Acad. Sci. USA* **100:**8478–8483.

Belunis, C. J., K. E. Mdluli, C. R. Raetz, and F. E. Nano. 1992. A novel 3-deoxy-D-manno-octulosonic acid transferase from *Chlamydia trachomatis* required for expression of the genus-specific epitope. *J. Biol. Chem.* **267:**18702–18707.

Betts, H. J., L. E. Twiggs, M. S. Sal, P. B. Wyrick, and K. A. Fields. 2008. Bioinformatic and biochemical evidence for the identification of the type III secretion system needle protein of *Chlamydia trachomatis*. *J. Bacteriol.* **190:**1680–1690.

Birkelund, S., A. G. Lundemose, and G. Christiansen. 1988. Chemical cross-linking of *Chlamydia trachomatis*. *Infect. Immun.* **56:**654–659.

Birkelund, S., A. G. Lundemose, and G. Christiansen. 1989. Immunoelectron microscopy of lipopolysaccharide in *Chlamydia trachomatis*. *Infect. Immun.* **57:**3250–3253.

Birkelund, S., M. Morgan-Fisher, E. Timmerman, K. Gevaert, A. C. Shaw, and G. Christiansen. 2009. Analysis of proteins in *Chlamydia trachomatis* L2 outer membrane complex, COMC. *FEMS Immunol. Med. Microbiol.* **55:**187–195.

Brade, H., L. Brade, and F. E. Nano. 1987. Chemical and serological investigations on the genus-specific lipopolysaccharide epitope of *Chlamydia*. *Proc. Natl. Acad. Sci. USA* **84:**2508–2512.

Brade, L., O. Holst, P. Kosma, Y. X. Zhang, H. Paulsen, R. Krausse, and H. Brade. 1990. Characterization of murine monoclonal and murine, rabbit, and human polyclonal antibodies against chlamydial lipopolysaccharide. *Infect. Immun.* **58:**205–213.

Brade, L., M. Nurminen, P. H. Makela, and H. Brade. 1985. Antigenic properties of *Chlamydia trachomatis* lipopolysaccharide. *Infect. Immun.* **48:**569–572.

Brown, W. J., and D. D. Rockey. 2000. Identification of an antigen localized to an apparent septum within dividing chlamydiae. *Infect. Immun.* **68:**708–715.

Buendia, A. J., J. Salinas, J. Sanchez, M. C. Gallego, A. Rodolakis, and F. Cuello. 1997. Localization by immunoelectron microscopy of antigens of *Chlamydia psittaci* suitable for diagnosis or vaccine development. *FEMS Microbiol. Lett.* **150:**113–119.

Caldwell, H. D., and P. J. Hitchcock. 1984. Monoclonal antibody against a genus-specific antigen of *Chlamydia* species: location of the epitope on chlamydial lipopolysaccharide. *Infect. Immun.* **44:**306–314.

Caldwell, H. D., and R. C. Judd. 1982. Structural analysis of chlamydial major outer membrane proteins. *Infect. Immun.* **38:**960–968.

Caldwell, H. D., J. Kromhout, and J. Schachter. 1981. Purification and partial characterization of the major outer membrane protein of *Chlamydia trachomatis*. *Infect. Immun.* **31:**1161–1176.

Caldwell, H. D., and C. C. Kuo. 1977a. Purification of a *Chlamydia trachomatis*-specific antigen by immunoadsorption with monospecific antibody. *J. Immunol.* **118:**437–441.

Caldwell, H. D., and C. C. Kuo. 1977b. Serologic diagnosis of lymphogranuloma venereum by counterimmunoelectrophoresis with a *Chlamydia trachomatis* protein antigen. *J. Immunol.* **118:**442–445.

Caldwell, H. D., C. C. Kuo, and G. E. Kenny. 1975a. Antigenic analysis of Chlamydiae by two-dimensional immunoelectrophoresis. I. Antigenic heterogeneity between *C. trachomatis* and *C. psittaci*. *J. Immunol.* **115:**963–968.

Caldwell, H. D., C. C. Kuo, and G. E. Kenny. 1975b. Antigenic analysis of chlamydiae by two-dimensional immunoelectrophoresis. II. A trachoma-LGV-specific antigen. *J. Immunol.* **115:**969–975.

Caldwell, H. D., and L. J. Perry. 1982. Neutralization of *Chlamydia trachomatis* infectivity with antibodies to the major outer membrane protein. *Infect. Immun.* **38:**745–754.

Campbell, L. A., C. C. Kuo, and J. T. Grayston. 1990. Structural and antigenic analysis of *Chlamydia pneumoniae*. *Infect. Immun.* **58:**93–97.

Campbell, L. A., C. C. Kuo, R. W. Thissen, and J. T. Grayston. 1989. Isolation of a gene encoding a *Chlamydia* sp. strain TWAR protein that is recognized during infection of humans. *Infect. Immun.* **57:**71–75.

Carabeo, R. A., and T. Hackstadt. 2001. Isolation and characterization of a mutant Chinese hamster ovary cell line that is resistant to *Chlamydia trachomatis* infection at a novel step in the attachment process. *Infect. Immun.* **69:**5899–5904.

Carabeo, R. A., D. J. Mead, and T. Hackstadt. 2003. Golgi-dependent transport of cholesterol to the *Chlamydia trachomatis* inclusion. *Proc. Natl. Acad. Sci. USA* **100:**6771–6776.

Carlson, J. H., S. F. Porcella, G. McClarty, and H. D. Caldwell. 2005. Comparative genomic analysis of *Chlamydia trachomatis* oculotropic and genitotropic strains. *Infect. Immun.* **73:**6407–6418.

Carrasco, J. A., C. Tan, R. G. Rank, R. C. Hsia, and P. M. Bavoil. 2011. Altered developmental expression of polymorphic membrane proteins in penicillin-stressed *Chlamydia trachomatis*. *Cell. Microbiol.* **13:**1014–1025.

Cevenini, R., M. Donati, E. Brocchi, F. De Simone, and M. La Placa. 1991. Partial characterization of an 89-kDa highly immunoreactive protein from *Chlamydia psittaci* A/22 causing ovine abortion. *FEMS Microbiol. Lett.* **65:**111–115.

Cevenini, R., A. Moroni, V. Sambri, S. Perini, and M. La Placa. 1989. Serological response to chlamydial infection in sheep, studied by enzyme-linked immunosorbent assay and immunoblotting. *FEMS Microbiol. Immunol.* **1:**459–464.

Chen, J. C., and R. S. Stephens. 1994. Trachoma and LGV biovars of *Chlamydia trachomatis* share the same glycosaminoglycan-dependent mechanism for infection of eukaryotic cells. *Mol. Microbiol.* **11:**501–507.

Chen, J. C., J. P. Zhang, and R. S. Stephens. 1996. Structural requirements of heparin binding to *Chlamydia trachomatis*. *J. Biol. Chem.* **271:**11134–11140.

Clarke, I. N., M. E. Ward, and P. R. Lambden. 1988. Molecular cloning and sequence analysis of a developmentally regulated cysteine-rich outer membrane protein from *Chlamydia trachomatis*. *Gene* **71:**307–314.

Clifton, D. R., K. A. Fields, S. S. Grieshaber, C. A. Dooley, E. R. Fischer, D. J. Mead, R. A. Carabeo, and T. Hackstadt. 2004. A chlamydial type III translocated protein is tyrosine-phosphorylated at the site of entry and associated with recruitment of actin. *Proc. Natl. Acad. Sci. USA* **101:**10166–10171.

Cocchiaro, J. L., Y. Kumar, E. R. Fischer, T. Hackstadt, and R. H. Valdivia. 2008. Cytoplasmic lipid droplets are translocated into the lumen of the *Chlamydia trachomatis* parasitophorous vacuole. *Proc. Natl. Acad. Sci. USA* **105:**9379–9384.

Collett, B. A., W. J. Newhall, R. A. Jersild, Jr., and R. B. Jones. 1989. Detection of surface-exposed epitopes on *Chlamydia trachomatis* by immune electron microscopy. *J. Gen. Microbiol.* **135:**85–94.

Conant, C. G., and R. S. Stephens. 2007. Chlamydia attachment to mammalian cells requires protein disulfide isomerase. *Cell. Microbiol.* **9:**222–232.

Crane, D. D., J. H. Carlson, E. R. Fischer, P. Bavoil, R. C. Hsia, C. Tan, C. C. Kuo, and H. D. Caldwell. 2006. *Chlamydia trachomatis* polymorphic membrane protein D is a species-common pan-neutralizing antigen. *Proc. Natl. Acad. Sci. USA* **103:**1894–1899.

Davis, C. H., J. E. Raulston, and P. B. Wyrick. 2002. Protein disulfide isomerase, a component of the estrogen receptor complex, is associated with *Chlamydia trachomatis* serovar E attached to human endometrial epithelial cells. *Infect. Immun.* **70:**3413–3418.

Davis, C. H., and P. B. Wyrick. 1997. Differences in the association of *Chlamydia trachomatis* serovar E and serovar L2 with epithelial cells in vitro may reflect biological differences *in vivo*. *Infect. Immun.* **65:**2914–2924.

DeMars, R., and J. Weinfurter. 2008. Interstrain gene transfer in *Chlamydia trachomatis* in vitro: mechanism and significance. *J. Bacteriol.* **190:**1605–1614.

Demars, R., J. Weinfurter, E. Guex, J. Lin, and Y. Potucek. 2007. Lateral gene transfer in vitro in the intracellular pathogen *Chlamydia trachomatis*. *J. Bacteriol.* **189:**991–1003.

Doughri, A. M., J. Storz, and K. P. Altera. 1972. Mode of entry and release of chlamydiae in infections of intestinal epithelial cells. *J. Infect. Dis.* **126:**652–657.

Eissenberg, L. G., P. B. Wyrick, C. H. Davis, and J. W. Rumpp. 1983. *Chlamydia psittaci* elementary body envelopes: ingestion and inhibition of phagolysosome fusion. *Infect. Immun.* **40:**741–751.

Everett, K. D., and T. P. Hatch. 1991. Sequence analysis and lipid modification of the cysteine-rich envelope proteins of *Chlamydia psittaci* 6BC. *J. Bacteriol.* **173:**3821–3830.

Everett, K. D., and T. P. Hatch. 1995. Architecture of the cell envelope of *Chlamydia psittaci* 6BC. *J. Bacteriol.* **177:**877–882.

Fadel, S., and A. Eley. 2007. *Chlamydia trachomatis* OmcB protein is a surface-exposed glycosaminoglycan-dependent adhesin. *J. Med. Microbiol.* **56:**15–22.

Fields, K. A., D. J. Mead, C. A. Dooley, and T. Hackstadt. 2003. *Chlamydia trachomatis* type III secretion: evidence for a functional apparatus during early-cycle development. *Mol. Microbiol.* **48:**671–683.

Fudyk, T., L. Olinger, and R. S. Stephens. 2002. Selection of mutant cell lines resistant to infection by *Chlamydia* spp. *Infect. Immun.* **70:**6444–6447.

Garrett, A. J., M. J. Harrison, and G. P. Manire. 1974. A search for the bacterial mucopeptide component, muramic acid, in *Chlamydia*. *J. Gen. Microbiol.* **80:**315–318.

Giannikopoulou, P., L. Bini, P. D. Simitsek, V. Pallini, and E. Vretou. 1997. Two-dimensional electrophoretic analysis of the protein family at 90 kDa of abortifacient *Chlamydia psittaci*. *Electrophoresis* **18:**2104–2108.

Gregory, W. W., M. Gardner, G. I. Byrne, and J. W. Moulder. 1979. Arrays of hemispheric surface projections on *Chlamydia psittaci* and *Chlamydia trachomatis* observed by scanning electron microscopy. *J. Bacteriol.* **138:**241–244.

Grieshaber, N. A., E. R. Fischer, D. J. Mead, C. A. Dooley, and T. Hackstadt. 2004. Chlamydial histone-DNA interactions are disrupted by a metabolite in the methylerythritol phosphate pathway of isoprenoid biosynthesis. *Proc. Natl. Acad. Sci. USA* **101:**7451–7456.

Griffiths, P. C., H. L. Philips, M. Dawson, and M. J. Clarkson. 1992. Antigenic and morphological differentiation of placental and intestinal isolates of *Chlamydia psittaci* of ovine origin. *Vet. Microbiol.* **30:**165–177.

Grimwood, J., and R. S. Stephens. 1999. Computational analysis of the polymorphic membrane protein superfamily of *Chlamydia trachomatis* and *Chlamydia pneumoniae*. *Microb. Comp. Genomics* **4:**187–201.

Hackstadt, T. 1986. Identification and properties of chlamydial polypeptides that bind eucaryotic cell surface components. *J. Bacteriol.* **165:**13–20.

Hackstadt, T. 1991. Purification and N-terminal amino acid sequences of *Chlamydia trachomatis* histone analogs. *J. Bacteriol.* **173:**7046–7049.

Hackstadt, T., W. Baehr, and Y. Ying. 1991. *Chlamydia trachomatis* developmentally regulated protein is homologous to eukaryotic histone H1. *Proc. Natl. Acad. Sci. USA* **88:**3937–3941.

Hackstadt, T., and H. D. Caldwell. 1985. Effect of proteolytic cleavage of surface-exposed proteins on infectivity of *Chlamydia trachomatis*. *Infect. Immun.* **48:**546–551.

Hackstadt, T., M. A. Scidmore, and D. D. Rockey. 1995. Lipid metabolism in *Chlamydia trachomatis*-infected cells: directed trafficking of Golgi-derived sphingolipids to the chlamydial inclusion. *Proc. Natl. Acad. Sci. USA* **92:**4877–4881.

Hackstadt, T., W. J. Todd, and H. D. Caldwell. 1985. Disulfide-mediated interactions of the chlamydial major outer membrane protein: role in the differentiation of chlamydiae? *J. Bacteriol.* **161:**25–31.

Hatch, G. M., and G. McClarty. 1998. Phospholipid composition of purified *Chlamydia trachomatis* mimics that of the eucaryotic host cell. *Infect. Immun.* **66:**3727–3735.

Hatch, T. P. 1996. Disulfide cross-linked envelope proteins: the functional equivalent of peptidoglycan in chlamydiae? *J. Bacteriol.* **178:**1–5.

Hatch, T. P., I. Allan, and J. H. Pearce. 1984. Structural and polypeptide differences between envelopes of infective and reproductive life cycle forms of *Chlamydia* spp. *J. Bacteriol.* **157:**13–20.

Hatch, T. P., M. Miceli, and J. E. Sublett. 1986. Synthesis of disulfide-bonded outer membrane proteins during the developmental cycle of *Chlamydia psittaci* and *Chlamydia trachomatis*. *J. Bacteriol.* **165:**379–385.

Hatch, T. P., D. W. Vance, Jr., and E. Al-Hossainy. 1981. Identification of a major envelope protein in *Chlamydia* spp. *J. Bacteriol.* **146:**426–429.

Henderson, I. R., and A. C. Lam. 2001. Polymorphic proteins of *Chlamydia* spp.—autotransporters beyond the Proteobacteria. *Trends Microbiol.* **9:**573–578.

Heuer, D., A. Rejman Lipinski, N. Machuy, A. Karlas, A. Wehrens, F. Siedler, V. Brinkmann, and T. F. Meyer. 2009. *Chlamydia* causes fragmentation of the Golgi compartment to ensure reproduction. *Nature* **457:**731–735.

Higashi, N. 1965. Electron microscopic studies on the mode of reproduction of trachoma virus and psittacosis virus in cell cultures. *Exp. Mol. Pathol.* **76:**24–39.

Hybiske, K., and R. S. Stephens. 2007. Mechanisms of host cell exit by the intracellular bacterium *Chlamydia*. *Proc. Natl. Acad. Sci. USA* **104:**11430–11435.

Iliffe-Lee, E. R., and G. McClarty. 2000. Regulation of carbon metabolism in *Chlamydia trachomatis*. *Mol. Microbiol.* **38:**20–30.

Kari, L., M. M. Goheen, L. B. Randall, L. D. Taylor, J. H. Carlson, W. M. Whitmire, D. Virok, K. Rajaram, V. Endresz, G. McClarty, D. E. Nelson, and H. D. Caldwell. 2011. Generation of targeted *Chlamydia trachomatis* null mutants. *Proc. Natl. Acad. Sci. USA* **108:**7189–7193.

Kiselev, A. O., M. C. Skinner, and M. F. Lampe. 2009. Analysis of *pmpD* expression and PmpD post-translational processing during the life cycle of *Chlamydia trachomatis* serovars A, D, and L2. *PLoS One* **4:**e5191.

Kiselev, A. O., W. E. Stamm, J. R. Yates, and M. F. Lampe. 2007. Expression, processing, and localization of PmpD of *Chlamydia trachomatis* serovar L2 during the chlamydial developmental cycle. *PLoS One* **2:**e568.

Kubo, A., and R. S. Stephens. 2000. Characterization and functional analysis of PorB, a *Chlamydia* porin and neutralizing target. *Mol. Microbiol.* **38:**772–780.

Kubo, A., and R. S. Stephens. 2001. Substrate-specific diffusion of select dicarboxylates through *Chlamydia trachomatis* PorB. *Microbiology* **147:**3135–3140.

Kuo, C., N. Takahashi, A. F. Swanson, Y. Ozeki, and S. Hakomori. 1996. An N-linked high-mannose type oligosaccharide, expressed at the major outer membrane protein of *Chlamydia trachomatis*, mediates attachment and infectivity of the microorganism to HeLa cells. *J. Clin. Investig.* **98:**2813–2818.

Kuo, C. C., and T. Grayston. 1976. Interaction of *Chlamydia trachomatis* organisms and HeLa 229 cells. *Infect. Immun.* **13:**1103–1109.

Kuo, C. C., A. Lee, and L. A. Campbell. 2004. Cleavage of the N-linked oligosaccharide from the surfaces of *Chlamydia* species affects attachment and infectivity of the organisms in human epithelial and endothelial cells. *Infect. Immun.* **72:**6699–6701.

Lambden, P. R., J. S. Everson, M. E. Ward, and I. N. Clarke. 1990. Sulfur-rich proteins of *Chlamydia trachomatis*: developmentally regulated transcription of polycistronic mRNA from tandem promoters. *Gene* **87:**105–112.

Lazarev, V. N., G. G. Borisenko, M. M. Shkarupeta, I. A. Demina, M. V. Serebryakova, M. A. Galyamina, S. A. Levitskiy, and V. M. Govorun. 2010. The role of intracellular glutathione in the progression of *Chlamydia trachomatis* infection. *Free Radic. Biol. Med.* **49:**1947–1955.

Levy, N. J., and J. W. Moulder. 1982. Attachment of cell walls of *Chlamydia psittaci* to mouse fibroblasts (L cells). *Infect. Immun.* **37:**1059–1065.

Lim, J. B., and J. B. Klauda. 2011. Lipid chain branching at the iso- and anteiso-positions in complex *Chlamydia* membranes: a molecular dynamics study. *Biochim. Biophys. Acta* **1808:**323–331.

Liu, X., M. Afrane, D. E. Clemmer, G. Zhong, and D. E. Nelson. 2010. Identification of *Chlamydia trachomatis* outer membrane complex proteins by differential proteomics. *J. Bacteriol.* **192:**2852–2860.

Longbottom, D., J. Findlay, E. Vretou, and S. M. Dunbar. 1998a. Immunoelectron microscopic localisation of the OMP90 family on the outer membrane surface of *Chlamydia psittaci*. *FEMS Microbiol. Lett.* **164:**111–117.

Longbottom, D., M. Russell, S. M. Dunbar, G. E. Jones, and A. J. Herring. 1998b. Molecular cloning and characterization of the genes coding for the highly immunogenic cluster of 90-kilodalton envelope proteins from the *Chlamydia psittaci* subtype that causes abortion in sheep. *Infect. Immun.* **66:**1317–1324.

Longbottom, D., M. Russell, G. E. Jones, F. A. Lainson, and A. J. Herring. 1996. Identification of a multigene family coding for the 90 kDa proteins of the ovine abortion subtype of *Chlamydia psittaci*. *FEMS Microbiol. Lett.* **142:**277–281.

Lukacova, M., M. Baumann, L. Brade, U. Mamat, and H. Brade. 1994. Lipopolysaccharide smooth-rough phase variation in bacteria of the genus *Chlamydia. Infect. Immun.* **62:**2270–2276.

Manire, G. P. 1966. Structure of purified cell walls of dense forms of meningopneumonitis organisms. *J. Bacteriol.* **91:**409–413.

Manire, G. P., and A. Tamura. 1967. Preparation and chemical composition of the cell walls of mature infectious dense forms of meningopneumonitis organisms. *J. Bacteriol.* **94:**1178–1183.

Matsumoto, A. 1973. Fine structures of cell envelopes of *Chlamydia* organisms as revealed by freeze-etching and negative staining techniques. *J. Bacteriol.* **116:**1355–1363.

Matsumoto, A. 1981. Electron microscopic observations of surface projections and related intracellular structures of *Chlamydia* organisms. *J. Electron Microsc.* (Tokyo) **30:**315–320.

Matsumoto, A. 1982. Electron microscopic observations of surface projections on *Chlamydia psittaci* reticulate bodies. *J. Bacteriol.* **150:**358–364.

Matsumoto, A., N. Higashi, and A. Tamura. 1973. Electron microscope observations on the effects of polymixin B sulfate on cell walls of *Chlamydia psittaci. J. Bacteriol.* **113:**357–364.

Matsumoto, A., and G. P. Manire. 1970a. Electron microscopic observations on the effects of penicillin on the morphology of *Chlamydia psittaci. J. Bacteriol.* **101:**278–285.

Matsumoto, A., and G. P. Manire. 1970b. Electron microscopic observations on the fine structure of cell walls of *Chlamydia psittaci. J. Bacteriol.* **104:**1332–1337.

McCoy, A. J., and A. T. Maurelli. 2006. Building the invisible wall: updating the chlamydial peptidoglycan anomaly. *Trends Microbiol.* **14:**70–77.

Melgosa, M. P., C. C. Kuo, and L. A. Campbell. 1993. Outer membrane complex proteins of *Chlamydia pneumoniae. FEMS Microbiol. Lett.* **112:**199–204.

Moelleken, K., and J. H. Hegemann. 2008. The *Chlamydia* outer membrane protein OmcB is required for adhesion and exhibits biovar-specific differences in glycosaminoglycan binding. *Mol. Microbiol.* **67:**403–419.

Molleken, K., E. Schmidt, and J. H. Hegemann. 2010. Members of the Pmp protein family of *Chlamydia pneumoniae* mediate adhesion to human cells via short repetitive peptide motifs. *Mol. Microbiol.* **78:**1004–1017.

Morrison, S. G., and R. P. Morrison. 2005. A predominant role for antibody in acquired immunity to chlamydial genital tract reinfection. *J. Immunol.* **175:**7536–7542.

Moulder, J. W. 1962. Some basic properties of the psittacosis-lymphogranuloma venereum group of agents. Structure and chemical composition of isolated particles. *Ann. N. Y. Acad. Sci.* **98:**92–99.

Moulder, J. W. 1991. Interaction of chlamydiae and host cells *in vitro. Microbiol. Rev.* **55:**143–190.

Moulder, J. W., D. L. Novosel, and J. E. Officer. 1963. Inhibition of the growth of agents of the psittacosis group by D-cycloserine and its specific reversal by D-alanine. *J. Bacteriol.* **85:**707–711.

Muller-Loennies, S., S. Gronow, L. Brade, R. MacKenzie, P. Kosma, and H. Brade. 2006. A monoclonal antibody against a carbohydrate epitope in lipopolysaccharide differentiates *Chlamydophila psittaci* from *Chlamydophila pecorum, Chlamydophila pneumoniae,* and *Chlamydia trachomatis. Glycobiology* **16:**184–196.

Muschiol, S., L. Bailey, A. Gylfe, C. Sundin, K. Hultenby, S. Bergstrom, M. Elofsson, H. Wolf-Watz, S. Normark, and B. Henriques-Normark. 2006. A small-molecule inhibitor of type III secretion inhibits different stages of the infectious cycle of *Chlamydia trachomatis. Proc. Natl. Acad. Sci. USA* **103:**14566–14571.

Mygind, P., G. Christiansen, and S. Birkelund. 1998. Topological analysis of *Chlamydia trachomatis* L2 outer membrane protein 2. *J. Bacteriol.* **180:**5784–5787.

Mygind, P. H., G. Christiansen, P. Roepstorff, and S. Birkelund. 2000. Membrane proteins PmpG and PmpH are major constituents of *Chlamydia trachomatis* L2 outer membrane complex. *FEMS Microbiol. Lett.* **186:**163–169.

Nano, F. E., P. A. Barstad, L. W. Mayer, J. E. Coligan, and H. D. Caldwell. 1985. Partial amino acid sequence and molecular cloning of the encoding gene for the major outer membrane protein of *Chlamydia trachomatis. Infect. Immun.* **48:**372–377.

Narita, T., P. B. Wyrick, and G. P. Manire. 1976. Effect of alkali on the structure of cell envelopes of *Chlamydia psittaci* elementary bodies. *J. Bacteriol.* **125:**300–307.

Nelson, D. E., D. D. Crane, L. D. Taylor, D. W. Dorward, M. M. Goheen, and H. D. Caldwell. 2006. Inhibition of chlamydiae by primary alcohols correlates with the strain-specific complement of plasticity zone phospholipase D genes. *Infect. Immun.* **74:**73–80.

Newhall, W. J. 1987. Biosynthesis and disulfide cross-linking of outer membrane components during the growth cycle of *Chlamydia trachomatis. Infect. Immun.* **55:**162–168.

Newhall, W. J. 1988. Macromolecular and antigenic composition of chlamydiae, p. 47–70. *In* A. L. Barron (ed.), *Microbiology of* Chlamydia. CRC Press, Boca Raton, FL.

Newhall, W. J., B. Batteiger, and R. B. Jones. 1982. Analysis of the human serological response to proteins of *Chlamydia trachomatis*. *Infect. Immun.* **38:**1181–1189.

Newhall, W. J., and R. B. Jones. 1983. Disulfide-linked oligomers of the major outer membrane protein of chlamydiae. *J. Bacteriol.* **154:**998–1001.

Nguyen, B. D., D. Cunningham, X. Liang, X. Chen, E. J. Toone, C. R. Raetz, P. Zhou, and R. H. Valdivia. 2011. Lipooligosaccharide is required for the generation of infectious elementary bodies in *Chlamydia trachomatis*. *Proc. Natl. Acad. Sci. USA* **108:**10284–10289.

Nurminen, M., M. Leinonen, P. Saikku, and P. H. Makela. 1983. The genus-specific antigen of *Chlamydia*: resemblance to the lipopolysaccharide of enteric bacteria. *Science* **220:**1279–1281.

Nurminen, M., E. T. Rietschel, and H. Brade. 1985. Chemical characterization of *Chlamydia trachomatis* lipopolysaccharide. *Infect. Immun.* **48:**573–575.

Oomen, C. J., P. van Ulsen, P. van Gelder, M. Feijen, J. Tommassen, and P. Gros. 2004. Structure of the translocator domain of a bacterial autotransporter. *EMBO J.* **23:**1257–1266.

Peterson, E. M., L. M. de la Maza, L. Brade, and H. Brade. 1998. Characterization of a neutralizing monoclonal antibody directed at the lipopolysaccharide of *Chlamydia pneumoniae*. *Infect. Immun.* **66:**3848–3855.

Rank, R. G., J. Whittimore, A. K. Bowlin, and P. B. Wyrick. 2011. In vivo ultrastructural analysis of the intimate relationship between polymorphonuclear leukocytes and the chlamydial developmental cycle. *Infect. Immun.* **79:**3291–3301.

Rasmussen-Lathrop, S. J., K. Koshiyama, N. Phillips, and R. S. Stephens. 2000. *Chlamydia*-dependent biosynthesis of a heparan sulphate-like compound in eukaryotic cells. *Cell. Microbiol.* **2:**137–144.

Raulston, J. E. 1995. Chlamydial envelope components and pathogen-host cell interactions. *Mol. Microbiol.* **15:**607–616.

Raulston, J. E., C. H. Davis, T. R. Paul, J. D. Hobbs, and P. B. Wyrick. 2002. Surface accessibility of the 70-kilodalton *Chlamydia trachomatis* heat shock protein following reduction of outer membrane protein disulfide bonds. *Infect. Immun.* **70:**535–543.

Read, T. D., R. C. Brunham, C. Shen, S. R. Gill, J. F. Heidelberg, O. White, E. K. Hickey, J. Peterson, T. Utterback, K. Berry, S. Bass, K. Linher, J. Weidman, H. Khouri, B. Craven, C. Bowman, R. Dodson, M. Gwinn, W. Nelson, R. DeBoy, J. Kolonay, G. McClarty, S. L. Salzberg, J. Eisen, and C. M. Fraser. 2000a. Genome sequences of *Chlamydia trachomatis* MoPn and *Chlamydia pneumoniae* AR39. *Nucleic Acids Res.* **28:**1397–1406.

Read, T. D., C. M. Fraser, R. C. Hsia, and P. M. Bavoil. 2000b. Comparative analysis of *Chlamydia* bacteriophages reveals variation localized to a putative receptor binding domain. *Microb. Comp. Genomics* **5:**223–231.

Rund, S., B. Lindner, H. Brade, and O. Holst. 1999. Structural analysis of the lipopolysaccharide from *Chlamydia trachomatis* serotype L2. *J. Biol. Chem.* **274:**16819–16824.

Salari, S. H., and M. E. Ward. 1981. Polypeptide composition of *Chlamydia trachomatis*. *J. Gen. Microbiol.* **123:**197–207.

Schachter, J., and M. Grossman. 1981. Chlamydial infections. *Annu. Rev. Med.* **32:**45–61.

Scidmore, M. A., E. R. Fischer, and T. Hackstadt. 1996a. Sphingolipids and glycoproteins are differentially trafficked to the *Chlamydia trachomatis* inclusion. *J. Cell Biol.* **134:**363–374.

Scidmore, M. A., D. D. Rockey, E. R. Fischer, R. A. Heinzen, and T. Hackstadt. 1996b. Vesicular interactions of the *Chlamydia trachomatis* inclusion are determined by chlamydial early protein synthesis rather than route of entry. *Infect. Immun.* **64:**5366–5372.

Shaw, A. C., K. Gevaert, H. Demol, B. Hoorelbeke, J. Vandekerckhove, M. R. Larsen, P. Roepstorff, A. Holm, G. Christiansen, and S. Birkelund. 2002. Comparative proteome analysis of *Chlamydia trachomatis* serovar A, D and L2. *Proteomics* **2:**164–186.

Shaw, E. I., C. A. Dooley, E. R. Fischer, M. A. Scidmore, K. A. Fields, and T. Hackstadt. 2000. Three temporal classes of gene expression during the *Chlamydia trachomatis* developmental cycle. *Mol. Microbiol.* **37:**913–925.

Skipp, P., J. Robinson, C. D. O'Connor, and I. N. Clarke. 2005. Shotgun proteomic analysis of *Chlamydia trachomatis*. *Proteomics* **5:**1558–1573.

Stephens, R. S., S. Kalman, C. Lammel, J. Fan, R. Marathe, L. Aravind, W. Mitchell, L. Olinger, R. L. Tatusov, Q. Zhao, E. V. Koonin, and R. W. Davis. 1998. Genome sequence of an obligate intracellular pathogen of humans: *Chlamydia trachomatis*. *Science* **282:**754–759.

Stephens, R. S., K. Koshiyama, E. Lewis, and A. Kubo. 2001. Heparin-binding outer membrane protein of chlamydiae. *Mol. Microbiol.* **40:**691–699.

Stephens, R. S., and C. J. Lammel. 2001. Chlamydia outer membrane protein discovery using genomics. *Curr. Opin. Microbiol.* **4:**16–20.

Stephens, R. S., J. M. Poteralski, and L. Olinger. 2006. Interaction of *Chlamydia trachomatis* with mammalian cells is independent of host cell surface heparan sulfate glycosaminoglycans. *Infect. Immun.* **74:**1795–1799.

Stephens, R. S., M. R. Tam, C. C. Kuo, and R. C. Nowinski. 1982. Monoclonal antibodies to *Chlamydia trachomatis*: antibody specificities and antigen characterization. *J. Immunol.* **128:**1083–1089.

Stuart, E. S., and A. B. Macdonald. 1989. Some characteristics of a secreted chlamydial antigen recognized by IgG from *C. trachomatis* patient sera. *Immunology* **68:**469–473.

Stuart, E. S., P. B. Wyrick, J. Choong, S. B. Stoler, and A. B. MacDonald. 1991. Examination of chlamydial glycolipid with monoclonal antibodies: cellular distribution and epitope binding. *Immunology* **74:**740–747.

Su, H., G. McClarty, F. Dong, G. M. Hatch, Z. K. Pan, and G. Zhong. 2004. Activation of Raf/MEK/ERK/cPLA2 signaling pathway is essential for chlamydial acquisition of host glycerophospholipids. *J. Biol. Chem.* **279:**9409–9416.

Su, H., L. Raymond, D. D. Rockey, E. Fischer, T. Hackstadt, and H. D. Caldwell. 1996. A recombinant *Chlamydia trachomatis* major outer membrane protein binds to heparan sulfate receptors on epithelial cells. *Proc. Natl. Acad. Sci. USA* **93:**11143–11148.

Suchland, R. J., K. M. Sandoz, B. M. Jeffrey, W. E. Stamm, and D. D. Rockey. 2009. Horizontal transfer of tetracycline resistance among *Chlamydia* spp. in vitro. *Antimicrob. Agents Chemother.* **53:**4604–4611.

Sun, G., S. Pal, A. K. Sarcon, S. Kim, E. Sugawara, H. Nikaido, M. J. Cocco, E. M. Peterson, and L. M. de la Maza. 2007. Structural and functional analyses of the major outer membrane protein of *Chlamydia trachomatis*. *J. Bacteriol.* **189:**6222–6235.

Swanson, A. F., and C. C. Kuo. 1991. Evidence that the major outer membrane protein of *Chlamydia trachomatis* is glycosylated. *Infect. Immun.* **59:**2120–2125.

Swanson, A. F., and C. C. Kuo. 1994. Binding of the glycan of the major outer membrane protein of *Chlamydia trachomatis* to HeLa cells. *Infect. Immun.* **62:**24–28.

Swanson, K. A., L. D. Taylor, S. D. Frank, G. L. Sturdevant, E. R. Fischer, J. H. Carlson, W. M. Whitmire, and H. D. Caldwell. 2009. *Chlamydia trachomatis* polymorphic membrane protein D is an oligomeric autotransporter with a higher-order structure. *Infect. Immun.* **77:**508–516.

Tamura, A., and G. P. Manire. 1967. Preparation and chemical composition of the cell membranes of developmental reticulate forms of meningopneumonitis organisms. *J. Bacteriol.* **94:**1184–1188.

Tamura, A., and G. P. Manire. 1968. Effect of penicillin on the multiplication of meningopneumonitis organisms (*Chlamydia psittaci*). *J. Bacteriol.* **96:**875–880.

Tamura, A., A. Matsumoto, G. P. Manire, and N. Higashi. 1971. Electron microscopic observations on the structure of the envelopes of mature elementary bodies and developmental reticulate forms of *Chlamydia psittaci*. *J. Bacteriol.* **105:**355–360.

Tan, C., R. C. Hsia, H. Shou, J. A. Carrasco, R. G. Rank, and P. M. Bavoil. 2010. Variable expression of surface-exposed polymorphic membrane proteins in *in vitro*-grown *Chlamydia trachomatis*. *Cell. Microbiol.* **12:**174–187.

Tan, C., R. C. Hsia, H. Shou, C. L. Haggerty, R. B. Ness, C. A. Gaydos, D. Dean, A. M. Scurlock, D. P. Wilson, and P. M. Bavoil. 2009. *Chlamydia trachomatis*-infected patients display variable antibody profiles against the nine-member polymorphic membrane protein family. *Infect. Immun.* **77:**3218–3226.

Tanzer, R. J., and T. P. Hatch. 2001. Characterization of outer membrane proteins in *Chlamydia trachomatis* LGV serovar L2. *J. Bacteriol.* **183:**2686–2690.

Tanzer, R. J., D. Longbottom, and T. P. Hatch. 2001. Identification of polymorphic outer membrane proteins of *Chlamydia psittaci* 6BC. *Infect. Immun.* **69:**2428–2434.

Tao, S., R. Kaul, and W. M. Wenman. 1991. Identification and nucleotide sequence of a developmentally regulated gene encoding a eukaryotic histone H1-like protein from *Chlamydia trachomatis*. *J. Bacteriol.* **173:**2818–2822.

Taylor, L. D., D. E. Nelson, D. W. Dorward, W. M. Whitmire, and H. D. Caldwell. 2010. Biological characterization of *Chlamydia trachomatis* plasticity zone MACPF domain family protein CT153. *Infect. Immun.* **78:**2691–2699.

Ting, L. M., R. C. Hsia, C. G. Haidaris, and P. M. Bavoil. 1995. Interaction of outer envelope proteins of *Chlamydia psittaci* GPIC with the HeLa cell surface. *Infect. Immun.* **63:**3600–3608.

Tjaden, J., H. H. Winkler, C. Schwoppe, M. Van Der Laan, T. Mohlmann, and H. E. Neuhaus. 1999. Two nucleotide transport proteins in *Chlamydia trachomatis*, one for net nucleoside triphosphate uptake and the other for transport of energy. *J. Bacteriol.* **181:**1196–1202.

Vandahl, B. B., S. Birkelund, and G. Christiansen. 2004. Genome and proteome analysis of *Chlamydia*. *Proteomics* **4:**2831–2842.

Vandahl, B. B., K. Gevaert, H. Demol, B. Hoorelbeke, A. Holm, J. Vandekerckhove, G. Christiansen, and S. Birkelund. 2001. Time-dependent expression and processing of a hypothetical protein of possible importance for regulation of the *Chlamydia pneumoniae* developmental cycle. *Electrophoresis* **22:**1697–1704.

Vandahl, B. B., A. S. Pedersen, K. Gevaert, A. Holm, J. Vandekerckhove, G. Christiansen, and S. Birkelund. 2002. The expression, processing and localization of polymorphic membrane proteins in *Chlamydia pneumoniae* strain CWL029. *BMC Microbiol.* **2:**36.

Vora, G. J., and E. S. Stuart. 2003. A role for the glycolipid exoantigen (GLXA) in chlamydial infectivity. *Curr. Microbiol.* **46:**217–223.

Vretou, E., P. Giannikopoulou, D. Longbottom, and E. Psarrou. 2003. Antigenic organization of the N-terminal part of the polymorphic outer membrane proteins 90, 91A, and 91B of *Chlamydophila abortus*. *Infect. Immun.* **71:**3240–3250.

Watson, M. W., P. R. Lambden, M. E. Ward, and I. N. Clarke. 1989. *Chlamydia trachomatis* 60 kDa cysteine rich outer membrane protein: sequence homology between trachoma and LGV biovars. *FEMS Microbiol. Lett.* **53:**293–297.

Webley, W. C., G. J. Vora, and E. S. Stuart. 2004. Cell surface display of the chlamydial glycolipid exoantigen (GLXA) demonstrated by antibody-dependent complement-mediated cytotoxicity. *Curr. Microbiol.* **49:**13–21.

Wehrl, W., V. Brinkmann, P. R. Jungblut, T. F. Meyer, and A. J. Szczepek. 2004. From the inside out—processing of the chlamydial autotransporter PmpD and its role in bacterial adhesion and activation of human host cells. *Mol. Microbiol.* **51:**319–334.

Welter-Stahl, L., D. M. Ojcius, J. Viala, S. Girardin, W. Liu, C. Delarbre, D. Philpott, K. A. Kelly, and T. Darville. 2006. Stimulation of the cytosolic receptor for peptidoglycan, Nod1, by infection with *Chlamydia trachomatis* or *Chlamydia muridarum*. *Cell. Microbiol.* **8:**1047–1057.

Wichlan, D. G., and T. P. Hatch. 1993. Identification of an early-stage gene of *Chlamydia psittaci* 6BC. *J. Bacteriol.* **175:**2936–2942.

Wilson, D. P., P. Timms, D. L. McElwain, and P. M. Bavoil. 2006. Type III secretion, contact-dependent model for the intracellular development of *Chlamydia*. *Bull. Math. Biol.* **68:**161–178.

Wilson, D. P., J. A. Whittum-Hudson, P. Timms, and P. M. Bavoil. 2009. Kinematics of intracellular chlamydiae provide evidence for contact-dependent development. *J. Bacteriol.* **191:**5734–5742.

Wolf, K., H. J. Betts, B. Chellas-Gery, S. Hower, C. N. Linton, and K. A. Fields. 2006. Treatment of *Chlamydia trachomatis* with a small molecule inhibitor of the *Yersinia* type III secretion system disrupts progression of the chlamydial developmental cycle. *Mol. Microbiol.* **61:**1543–1555.

Wylie, J. L., G. M. Hatch, and G. McClarty. 1997. Host cell phospholipids are trafficked to and then modified by *Chlamydia trachomatis*. *J. Bacteriol.* **179:**7233–7242.

Wyllie, S., R. H. Ashley, D. Longbottom, and A. J. Herring. 1998. The major outer membrane protein of *Chlamydia psittaci* functions as a porin-like ion channel. *Infect. Immun.* **66:**5202–5207.

Zhang, J. P., and R. S. Stephens. 1992. Mechanism of *C. trachomatis* attachment to eukaryotic host cells. *Cell* **69:**861–869.

CHLAMYDIAL ADHESION AND ADHESINS

Johannes H. Hegemann and Katja Moelleken

5

INTRODUCTION

As the first direct contact between pathogen and target, adhesion is a prerequisite for subsequent steps in the infection process and is the basis for host cell specificity for most if not all pathogens. Adhesion is usually a multifactorial process, involving various microbial and host proteins (Falkow, 1991; Finlay and Falkow, 1997; Pizarro-Cerda and Cossart, 2006; Nobbs et al., 2009; Bardiau et al., 2010).

Obligate intracellular pathogens, in particular, rely on specialized surface structures to mediate specific interactions with eukaryotic host cells (Kline et al., 2009). Thus, successful invasion by chlamydiae requires initial adhesion of the infectious elementary body (EB) to the host cell as a prelude to uptake and initiation of the intracellular developmental cycle (reviewed by Hackstadt [1999]; Dautry-Varsat et al. [2005]; and Campbell and Kuo [2006]). This chapter focuses on recent advances in understanding the attachment of chlamydial EBs to target cells. Mechanisms of cell entry are reviewed in chapter 6, "Initial interactions of chlamydiae with the host cell."

The natural targets of most chlamydial infections are mucosal and submucosal surfaces. However, the bacteria can productively infect a variety of phagocytic and nonphagocytic cells in cell culture. The molecular mechanisms underlying attachment and entry are still not well understood. Early work on attachment of *Chlamydia caviae* EBs to mouse fibroblasts (L cells) (Byrne, 1976; Byrne and Moulder, 1978; Bose and Paul, 1982) and adhesion of *Chlamydia trachomatis* to HeLa and McCoy cells (Vretou et al., 1989) established the involvement of thermolabile proteins, but the identity of the chlamydial adhesins and their host cell receptors remained unclear.

A number of factors, such as GroEL-1, the envelope proteins MOMP (major outer membrane protein) and OmcB, and EB-associated heparan sulfate (HS)-like glycosaminoglycans, have been implicated in the attachment process (summarized in Table 1). On susceptible eukaryotic cells, surface proteins such as receptors for mannose, mannose-6-phosphate (M6P), and estrogen as well as protein disulfide isomerase (PDI) have been proposed to mediate chlamydial adherence (Table 2). These data have been comprehensively summarized recently in

Johannes H. Hegemann and Katja Moelleken, Institut für Funktionelle Genomforschung der Mikroorganismen, Heinrich-Heine-Universität Düsseldorf, Universitätsstr. 1, Geb. 25.02. U1, 40225 Düsseldorf, Germany.

Intracellular Pathogens I: Chlamydiales
Edited by Ming Tan and Patrik M. Bavoil © 2012 ASM Press, Washington, DC
10.1128/9781555817329.ch5

TABLE 1 EB cell surface components associated with adhesion and/or infection

Chlamydial protein/structure	Host target	Host cell type(s)	Chlamydial species, strain/serovar	Reference(s)
ArtJ	Unknown	LLC-MK2	*C. trachomatis* D strain D/UW-3/CX *C. pneumoniae* FB/96	Soriani et al., 2010
EB surface protein	Soluble GAGs and/ or proteoglycans	HEp-2; HeLa229; BEC HUVE CHO-K1; pgsA-745; pgsD-677	*C. pneumoniae* TW-183; A-03 *C. trachomatis* L2/434/Bu; E (UW5-CX)	Beswick et al., 2003
		HEp-2; HeLa229 CHO-K1; pgsA-745; pgsD-677	*C. pneumoniae* GiD *C. trachomatis* L2/434/Bu	Wuppermann et al., 2001
		McCoy; HeLa229; Hec-1B	*C. trachomatis* E (UW5-CX); L2/434/Bu	Davis and Wyrick, 1997
		HeLa229	*C. caviae*	Gutiérrez-Martín et al., 1997
		L929 mouse fibroblasts HeLa229	*C. trachomatis* L2/434/Bu; B/TW-5/OT; C/TW-3/OT; G/UW-57/CX	Chen and Stephens, 1997
		HL	*C. pneumoniae* CWL-029; Kajaani 7; Parola *C. trachomatis* L2; E	Yan et al., 2006
		L929 mouse fibroblasts HeLa229	*C. trachomatis* L2/434/Bu; B/TW-5/OT	Chen and Stephens, 1996
		HeLa229 L929 mouse fibroblasts CHO-K1; CHO-761	*C. trachomatis* L2/434/Bu	Zhang and Stephens, 1992
		HeLa 229 CHO-677; F-17	*C. trachomatis* L2	Yabushita et al., 2002
GroEL-1	Unknown	HUVE HEp-2	*C. pneumoniae* GiD	Wuppermann et al., 2008
LPS	Unknown	HEp-2; HeLa229	*C. pneumoniae* TW-183; CWL-029; 2043; 1497; CM-1 *C. trachomatis* L1 (440), L3 (404), A (G-17), B (HAR-36), C (TW-3), D (IC-Cal), E (Boor), I (UW-12), J (UW-36), K (UW-31) *C. tr* MoPn (Nigg II)	Peterson et al., 1998
		McCoy; Hec-1B	*C. psittaci* (Texas turkey) *C. trachomatis* E (E/UW-5/CX); LGV1	Fadel and Eley, 2008b

MOMP/N-linked oligosaccharides	M6P/IGF2 receptor?	HL; HMEC-1	*C. trachomatis* L2/434/Bu; E/UW-5/Cx *C. pneumoniae* AR-39 *C. psittaci* 6BC	Kuo et al., 2004
		HeLa229	*C. trachomatis* L2/434/Bu *C. pneumoniae* AR-39 *C. psittaci* 6BC	Kuo et al., 1996
		HeLa229 CHO-K1; pgsA-745; pgsD-677	*C. trachomatis* MoPn	Su et al., 1996
		HeLa229	*C. trachomatis* L2/434/Bu	Swanson and Kuo, 1994
		HeLa229	*C. trachomatis* B (B/TW-5/OT); A (A/Har-13); C (C/TW-3/OT); E (E/Bour); F (F/IC-Cal-13); G (F-UW-57/Cx); H (H/UW-4/Cx); I (I/UW-12/Ur); L2 (L2-434)	Su et al., 1990
OmcB	Soluble GAGs and/or proteoglycans	HEp-2 HUVE CHO-K1; pgsA-745; pgsD-677	*C. pneumoniae* GiD *C. trachomatis* L1/440/Bu; E (DK-20)	Moelleken and Hegemann, 2008
		Hec-1B CHO-K1; pgsA-745; pgsD-677	*C. trachomatis* E (E/UW-5/CX); L1	Fadel and Eley, 2008a
		Hec-1B; HeLa229 CHO-K1; pgsA-745; pgsD-677	*C. trachomatis* L1	Fadel and Eley, 2007
		L929 mouse fibroblasts	*C. trachomatis* L2/434/Bu	Stephens et al., 2001
		HeLa229	*C. caviae*	Ting et al., 2005
Pmp6, Pmp20, Pmp21	Unknown	HEp-2	*C. pneumoniae* GiD	Moelleken et al., 2010
Pmp21		HEp-2	*C.pneumoniae* CWL029	Wehrl et al., 2004
PmpD		HeLa229 HaK	*C. trachomatis* A/HAR-13; B/TW-5/OT; Ba/Ap-2; C/TW-3/OT; D/UW-3/Cx; E/Bour; F/IC-Cal-3; G/UW-524/Cx; H/UW-4/Cx; I/UW-12/Ur; J/UW-36/Cx; K/UW-31/Cx; L1/LGV-440; L2/LGV-434; L3/LGV-404 *C. muridarum* (MoPn) *C. pneumoniae* (AR-39) *C. caviae*	Crane et al., 2006

TABLE 2 Host cell surface localized/soluble molecules with relevance to adhesion and/or infection

Host protein/structure	Chlamydial target	Host cell type and animal model	Chlamydial species and strain/serovar	Reference(s)
Apolipoprotein E4	EBs	HEp-2; HeLa 229 U937/THP-1 (monocytes) U-87MG/SW1088/SW1783 (astrocytes) CHME-5 (microglias)	*C. pneumoniae* AR-39 *C. trachomatis* K	Gerard et al., 2008
Collectin (mannose binding protein)	EBs (MOMP glycoprotein)	HeLa 229 cells; HL cells	*C. trachomatis* C/TW-3/OT; E/UW-5/Cx; L_2/434/Bu *C. pneumoniae* AR-39 *C. psittaci* (6BC)	Swanson et al., 1998
CFTR	LPS	HeLa 229; Calu-3 (airway); T84 (colon) Capan (pancreas); CF-PAC-1 (pancreas from CF patient)	*C. trachomatis* D	Ajonuma et al., 2010
M6P/insulin-like growth factor 2 receptor	High-mannose oligosaccharide	HL; HMEC-1	*C. pneumoniae* AR-39	Puolakkainen et al., 2008
		HMEC-1	*C. pneumoniae* AR-39	
Mannose receptor		Mouse L cells	*C. trachomatis* UW-5 (E/UW-5/Cx); L2 (L2/434/Bu)	Puolakkainen et al., 2005
Ovotransferrin/ lactoferrins	Unknown	BGM; HD11 (macrophage-like)	*C. psittaci* D 92/1293	Beeckman et al., 2007
PDGFR	Unknown	HCAEC	*C. pneumoniae* A03	Wang et al., 2010
		NIH 3T3; Abl/Arg$^{-/-}$; HeLa 229; L929 mouse fibroblasts; *Drosophila* S2	*C. trachomatis* L2/434/Bu	Elwell et al., 2008
PDI/estrogen receptor α and β	EBs	Polarized endometrial epithelial carcinoma HEC-1B and IK	*C. trachomatis E* (UW5-CX)	Hall et al., 2011
		CHO K1; CHO6	*C. trachomatis* L2/434/Bu	Abromaitis and Stephens, 2009
		HeLa 229 L929 mouse fibroblasts	*C. psittaci* PF6 BC	
		CHO K1; CHO6	*C. trachomatis* L2/434/Bu *C. psittaci* PF6 BC	Conant and Stephens, 2007
				Davis et al., 2002
		McCoy HEC-1B	*C. trachomatis* E (UW5-CX)	

excellent reviews (Hackstadt, 1999; Dautry-Varsat et al., 2005; Campbell and Kuo, 2006) and will be mentioned only briefly here.

GLYCOSAMINOGLYCANS AS LIGANDS FOR CHLAMYDIAL ADHESION TO EUKARYOTIC HOST CELLS

Proteoglycans are found in the extracellular matrix and as integral membrane proteins on the surface of virtually all mammalian cells. They consist of a protein core with one or more covalently attached glycosaminoglycans (GAGs), which are linear glycoside chains made up of negatively charged disaccharide repeat units. GAGs on mammalian cells are important for interactions with many microbial pathogens and are recognized by a number of viral, bacterial, and protozoan adhesins (Rostand and Esko, 1997; Wadstrom and Ljungh, 1999; Menozzi et al., 2002; Chen et al., 2008). Analysis of mutant Chinese hamster ovary (CHO) cell lines resistant to chlamydial infection has provided evidence for a two-step attachment mechanism involving reversible binding to GAGs on the target cell (see below), followed by irreversible interactions with other surface components (Carabeo and Hackstadt, 2001).

Depending on the composition of the repeat and the overall extent of sulfation and acetylation, GAGs are classified into different species, including heparin, HS, and chondroitin sulfate A. HS proteoglycan (HSPG) is the most prevalent form of GAG on the cell surface (Bernfield et al., 1999; Esko and Selleck, 2002).

Very early observations revealed that prior exposure of host cells to heparin, a soluble, highly sulfated polymer closely related to the GAGs found in HS, inhibits chlamydial attachment and uptake (Becker et al., 1969; Kuo et al., 1973). In 1992, Zhang and Stephens provided convincing evidence that HS-like GAGs are involved in the attachment of *C. trachomatis* serovar L2 to human cells. They proposed a trimolecular mechanism with soluble HS molecules bridging the gap between chlamydiae and the target cell by binding to a chlamydial OM protein and a host cell receptor (Zhang and Stephens, 1992). Specifically, they showed that infectivity was reduced when EBs were treated with HS-specific lyases and that it could be restored by coating the EBs with heparin or HS.

This elegant work initiated subsequent studies showing that GAGs of the HS type are required for the initial attachment of EBs to their targets (Chen and Stephens, 1994, 1997; Zaretzky et al., 1995; Chen et al., 1996; Davis and Wyrick, 1997; Gutierrez-Martin et al., 1997; Wuppermann et al., 2001). In most cases, pretreatment with exogenous HS or its analog, heparin, reduced infectivity to varying degrees, depending on the chlamydial species, biovar, or isolate and on the human cell lines and experimental procedures used (Chen and Stephens, 1997; Gutierrez-Martin et al., 1997; Taraktchoglou et al., 2001; Wuppermann et al., 2001; Yabushita et al., 2002; Beswick et al., 2003; Darville et al., 2004; Yan et al., 2006). In particular, comparative studies on two *C. trachomatis* biovars, the trachoma biovar (serovars A through K) and the lymphogranuloma venereum (LGV) biovar (serovars L1, L2, and L3), which differ in their infectious characteristics and invasiveness in clinical disease, revealed significantly different requirements for GAGs (for details, see the review by Campbell and Kuo [2006]). Low concentrations of soluble HS or heparin significantly block binding and infection of human cells by the LGV biovar, but the binding of EBs of serovars B, C, D, E, and G and the infectivity of B and E are only partially inhibited even at high GAG concentrations (Kuo and Grayston, 1976; Chen and Stephens, 1994, 1997; Davis and Wyrick, 1997; Taraktchoglou et al., 2001; Yabushita et al., 2002; Beswick et al., 2003; Yan et al., 2006). It is also noteworthy that EBs of *C. trachomatis* serovars B and L2 bind heparin in a saturable fashion, suggesting that both bear a specific GAG receptor on the EB surface (Chen et al., 1996). Comparison of the results described in these different reports is complicated by the wide spectrum of species, isolates, cell lines, and protocols used.

Interestingly, the *C. trachomatis* LGV and trachoma biovars behaved differently in infection experiments using polarized epithelial cells as targets (Davis and Wyrick, 1997). Overall, the data support the idea that the LGV biovars, which are primarily submucosal pathogens, are more strongly dependent on recognition of HS, because they need to adhere to the basolateral domains of mucosal epithelia. In contrast, serovar E, as a luminal pathogen, may enter and leave cells preferentially on the apical side, which exhibits a different GAG profile (Davis and Wyrick, 1997); for a review, see Campbell and Kuo, 2006.

For the *Chlamydia pneumoniae* isolate GiD, it was shown that EB binding to HS-like GAGs on the host cell surface is required for infection (Wuppermann et al., 2001). Thus, treatment of human epithelial HEp-2 cells, but not *C. pneumoniae* EBs, with HS-specific lyases reduced subsequent infection. Likewise, pretreatment of *C. pneumoniae* EBs, but not epithelial cells, with heparin reduced infection, and *C. pneumoniae* EBs bound specifically to heparin-coated microtiter plates. These data can be explained by a bimolecular model involving direct binding of EB surface proteins to HS-like GAGs on the host cell surface (Wuppermann et al., 2001). Other studies have proposed a similar mechanism for *C. trachomatis* serovars L1 and L2 (Taraktchoglou et al., 2001; Yabushita et al., 2002).

EB CELL SURFACE COMPONENTS AND THEIR RELEVANCE FOR HOST CELL ATTACHMENT

Whole EBs of *C. caviae* and the *C. trachomatis* serovars D, E, L1, and L2 exhibit similar net negative charges at neutral (physiological) pH (Kraaipoel and van Duin, 1979; Vance and Hatch, 1980; Soderlund and Kihlstrom, 1982; reviewed by Raulston [1995]). The exposed surface of the OM of chlamydiae is characterized by an outer layer of lipopolysaccharide (LPS), which is less complex than those in other gram-negative bacteria and is only weakly immunogenic (reviewed by Brade [1999]). However, antibodies directed against LPS, addition of isolated chlamydial LPS, or an inhibitor of LPS synthesis reduced infection by *C. pneumoniae* and *C. trachomatis* serovars L1 and E (Peterson et al., 1998; Fadel and Eley, 2008b). Interestingly, chlamydial LPS has recently been found in association with the cystic fibrosis transmembrane conductance regulator (CFTR) (see below) (Ajonuma et al., 2010).

The structural rigidity and osmotic stability of EBs is due to a macromolecular structure composed of OM proteins cross-linked by disulfide bonds (Hatch, 1996). The main components of this sarcosyl-insoluble chlamydial outer membrane complex (COMC) are MOMP, which accounts for around 60% of total COMC protein, and the two cysteine-rich OM complex proteins, OmcA and OmcB, which are conserved in all chlamydial species (Caldwell et al., 1981; Hatch et al., 1984; Hatch, 1999). The COMC is thought to be functionally equivalent to the peptidoglycan of other bacteria (Hackstadt et al., 1985; Hatch et al., 1986).

MOMP functions as a porin (Bavoil et al., 1984; Wyllie et al., 1998) and may act as an adhesin for host cells in some chlamydial species (Su et al., 1996). Thus, recombinant *C. trachomatis* MOMP bound HS-like GAGs on human cells and inhibited infection, while recombinant *C. pneumoniae* MOMP showed no adhesion (Su et al., 1996; Moelleken et al., 2010). In several species, MOMP itself appears to have a proteoglycan-like structure. For example, for *C. trachomatis* serovar E or *C. pneumoniae*, enzymatic removal of N-linked oligosaccharides from EBs attenuates infectivity. Thus, both *C. trachomatis* and *C. pneumoniae* carry high-mannose oligosaccharide residues, and available data suggest differences in phosphorylation of the mannose residues that might affect receptor usage. M6P, but not mannan, inhibited the infectivity of *C. pneumoniae* in human arterial endothelial cells (Puolakkainen et al., 2005). In contrast, mannan, but not M6P, inhibited the infectivity of *C. trachomatis*. These findings suggest differences in receptor usage between the two species. Specifically, *C. pneumoniae* may preferentially use the M6P/insulin-like growth factor 2 receptor, while *C. trachomatis* may use the mannose receptor, for entry and infection of endothelial cells (for details, see the review by Campbell and Kuo [2006]; see also Puolakkainen et al., 2005, 2008).

Initially both OmcA and OmcB were believed to be localized in the periplasm of *Chlamydia psittaci* 6BC (Everett et al., 1991; Everett and Hatch, 1995). However, the sensitivity of *C. caviae* OmcB to N-terminal proteolysis suggested that at least part of the protein was exposed on the EB surface (Ting et al., 1995). Similar results were obtained for OmcB from *C. trachomatis*, *C. psittaci*, and *C. pneumoniae* in studies describing surface exposure of an N-terminal segment of OmcB, while the large C-terminal domain was found to be localized at the inner surface of the COMC (Collett et al., 1989; Watson et al., 1994; Mygind et al., 1998a). Later studies using antibodies provided direct evidence for localization of some part of OmcB on the surface of *C. trachomatis* and *C. pneumoniae* EBs (Stephens et al., 2001; Montigiani et al., 2002; Vandahl et al., 2002). Moreover, OmcB induces a strong antibody response in chlamydial infections, again suggesting that it is readily accessible on the EB surface (Wagels et al., 1994; Mygind et al., 1998b; Cunningham and Ward, 2003; Klein et al., 2003; Portig et al., 2003).

It now appears that not all EB OM proteins are cross-linked by disulfide bonds within the COMC and that species-specific differences exist. Around 50% of the GroEL-1 protein of *C. pneumoniae* EBs is located on the surface and can be extracted with low-salt buffer (Wuppermann et al., 2008). Polymorphic membrane protein D (PmpD) is present on the surface of intact *C. trachomatis* EBs but can be extracted with mild detergents in the absence of reducing agents (Swanson et al., 2009). In contrast, elution of the homologous Pmp21 from the *C. pneumoniae* COMC requires reducing conditions (Wehrl et al., 2004; Moelleken et al., 2010).

Several other proteins have been found to be associated with the EB surface and/or the COMC, including PorB, OprB, PmpE, PmpG, and PmpH (Hatch et al., 1984; Kubo and Stephens, 2000; Mygind et al., 2000; Tanzer et al., 2001; Birkelund et al., 2009). A recent proteomics study on *C. trachomatis* serovar L2 (Liu et al., 2010) has identified additional COMC proteins. Comparison of sarcosyl-soluble and detergent-insoluble fractions with the proteome of whole EBs led to the specific identification of 17 low-abundance proteins in the COMC, including PmpB, PmpC, PmpE, PmpF, PmpG, and PmpH (Liu et al., 2010). More details about the chlamydial outer envelope are reviewed in chapter 4, "The chlamydial cell envelope."

OMCB IS A GAG-BINDING ADHESIN THAT IS IMPORTANT FOR INFECTION

In their original report on the role of HS GAGs in *C. trachomatis* infections, Zhang and Stephens proposed that the 60-kDa OmcB, which is rich in lysine and arginine, interacts with negatively charged GAG structures (Zhang and Stephens, 1992). Ting and colleagues later used an in vitro ligand-binding assay to enrich and identify proteins from *C. caviae* EB envelopes that bound to mammalian cells (Ting et al., 1995). They extracted EB OM proteins with detergent and demonstrated specific binding of OmcB to glutaraldehyde-fixed HeLa cell surfaces. Moreover, based on the susceptibility of OmcB on EBs to cleavage by trypsin, they proposed that OmcB on the EB participates in a specific interaction with human cells (Ting et al., 1995).

A few years later Stephens and colleagues showed, by heparin blotting of COMC protein extracts from *C. trachomatis* serovars B and L2, that the COMC possesses a distinct heparin-binding activity (Stephens et al., 2001). The only COMC protein found to bind to a heparin-agarose affinity column was OmcB, and the heparin-binding domain was mapped to its N-terminal portion (Fig. 1A). Finally, OmcB peptides derived from *C. trachomatis* and *C. pneumoniae* (amino acids [aa] 50 to 70), and including a cluster of basic amino acid residues, were shown to bind heparin (Fig. 1B) (Stephens et al., 2001). The early hints of a

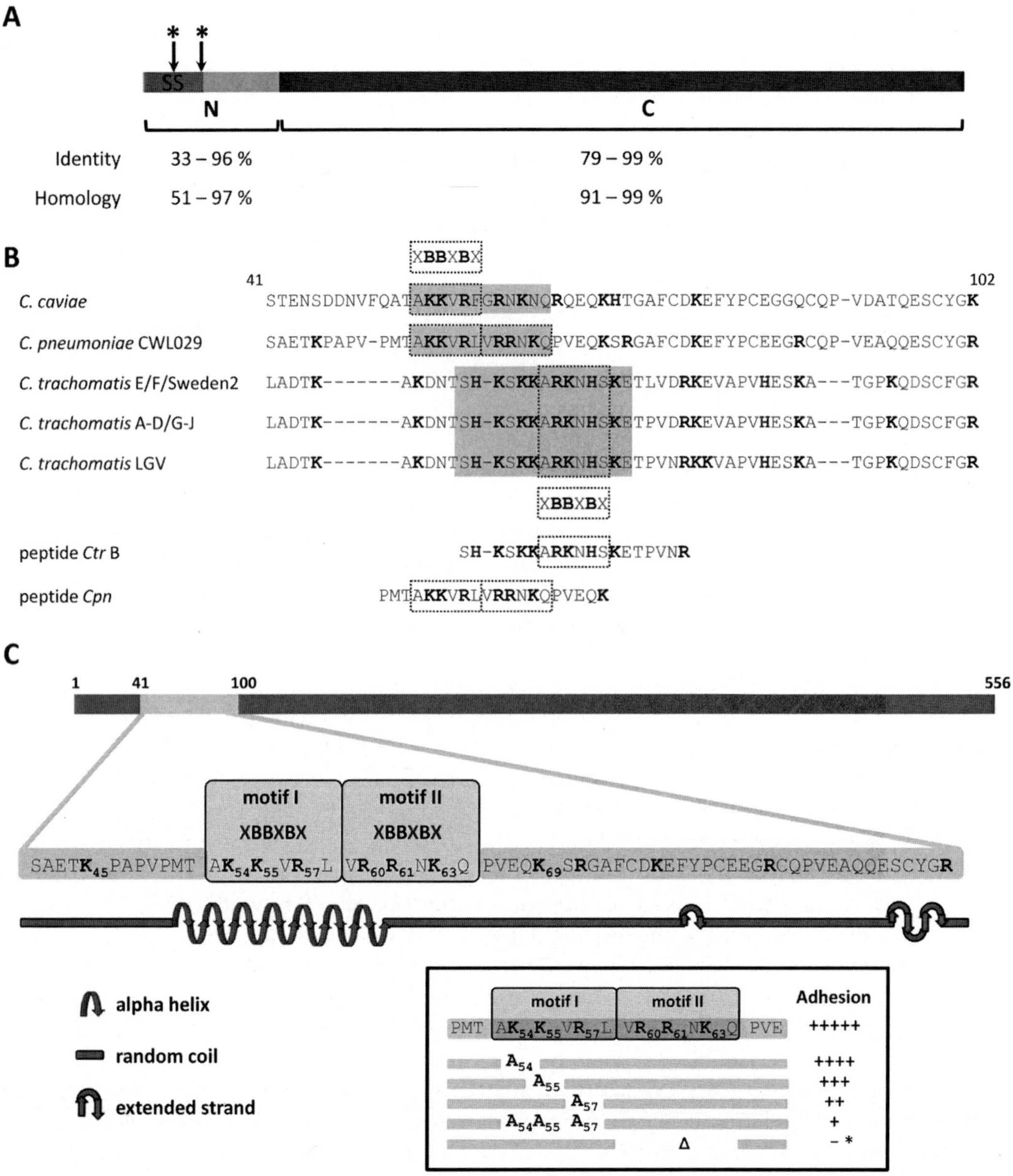
A
SS
N
C
Identity 33 – 96 % 79 – 99 %
Homology 51 – 97 % 91 – 99 %
B
XBBXBX
41
102
C. caviae STENSDDNVFQATAKKVREGRNKNQRQEQKHTGAFCDKEFYPCEGGQCQP-VDATQESCYGK
C. pneumoniae CWL029 SAETKPAPV-PMTAKKVRLVRRNKQPVEQKSRGAFCDKEFYPCEEGRCQP-VEAQQESCYGR
C. trachomatis E/F/Sweden2 LADTK-------AKDNTSH-KSKKARKNHSKETLVDRKEVAPVHESKA---TGPKQDSCFGR
C. trachomatis A-D/G-J LADTK-------AKDNTSH-KSKKARKNHSKETPVDRKEVAPVHESKA---TGPKQDSCFGR
C. trachomatis LGV LADTK-------AKDNTSH-KSKKARKNHSKETPVNRKKVAPVHESKA---TGPKQDSCFGR
XBBXBX
peptide Ctr B SH-KSKKARKNHSKETPVNR
peptide Cpn PMTAKKVRLVRRNKQPVEQK
C
1
41
100
556
motif I
XBBXBX
motif II
XBBXBX
SAETK45PAPVPMT AK54K55VR57L VR60R61NK63Q PVEQK69SRGAFCDKEFYPCEEGRCQPVEAQQESCYGR
alpha helix
random coil
extended strand
motif I
motif II
Adhesion
PMT AK54K55VR57L VR60R61NK63Q PVE +++++
A54 ++++
A55 +++
A57 ++
A54A55 A57 +
Δ – *

role for OmcB in adhesion to mammalian cells and the subsequent identification of OmcB as an EB surface protein that binds heparin were important first steps in the understanding of chlamydial adhesion.

The results mentioned above have been confirmed and extended by adhesion experiments using *Escherichia coli* and *Saccharomyces cerevisiae* as surrogate expression systems for chlamydial OmcB (Fadel and Eley, 2007; Moelleken and Hegemann, 2008). Thus, *E. coli* cells expressing OmcB from *C. trachomatis* serovar L1 acquired the capacity to bind human epithelial HeLa and Hec-1B cells, and preincubation with an OmcB-specific antibody significantly reduced attachment (Fadel and Eley, 2007). Similarly, yeast cells displaying the OmcB protein from *C. pneumoniae*, *C. caviae*, or *C. trachomatis* serovars L1 or E all showed specific binding to epithelial HEp-2 cells (Moelleken and Hegemann, 2008). Interestingly, in both surrogate expression systems, OmcB from *C. trachomatis* serovar E showed less affinity for human cells than the other chlamydial OmcB proteins analyzed. Latex beads coated with recombinant OmcB also adhered specifically to HEp-2 cells (Moelleken and Hegemann, 2008). Binding of surrogate cells expressing OmcB was dependent on GAG synthesis by mammalian cells, as both *E. coli* cells presenting *C. trachomatis* serovar L1 OmcB and yeast cells presenting *C. pneumoniae* OmcB failed to adhere to GAG-deficient mutant CHO cell lines (Fadel and Eley, 2007; Moelleken and Hegemann, 2008). Competition experiments using various GAGs revealed that heparin alone prevented adhesion of yeast displaying *C. pneumoniae* OmcB, as did treatment of HEp-2 cells with heparinase (Moelleken and Hegemann, 2008).

Furthermore, when epithelial cell lines (HeLa, Hec-1B, or HEp-2) were incubated with different amounts of recombinant OmcB from *C. trachomatis* serovar L1 or *C. pneumoniae* and subsequently challenged with the respective pathogen, a dose-dependent reduction in infection was observed. Pretreatment of EBs with the OmcB-specific antibodies likewise

FIGURE 1 A conserved N-terminal binding motif in OmcB is required for HS-dependent adhesion to HEp-2 cells. (A) The N-terminal segment (aa 41 to 84) of OmcB from *C. trachomatis* serovar L2 (chosen as generally representative of chlamydial OmcB proteins) is presumably exposed on the surface of EBs, while the rest of the protein remains in the periplasm in association with the outer membrane. The OmcB C-terminal region is highly conserved, while the N-terminal segment starting at aa 41 (the predicted site of cleavage by signal peptidase) is highly variable. Asterisks indicate the two alternative proteolytic cleavage sites identified in the *C. trachomatis* serovar L2 OmcB, which are used with the same frequency (Allen and Stephens, 1989). (B) N-terminal OmcB sequences from 16 different chlamydial species and serovars were aligned using the MultAlin tool (expasy.org). Identical sequences were grouped. Basic residues are marked in bold (R = arginine, K = lysine, H = histidine). The heparin-binding motifs XBBXBX (B = basic residue; X = hydrophatic amino acid) originally proposed by Stephens are indicated by the dashed outlines (Stephens et al., 2001). The *C. pneumoniae* OmcB has two copies of the motif. The grey-shaded box represents the extended heparin-binding region proposed here. The synthetic OmcB peptides derived from *C. pneumoniae* and *C. trachomatis* that confer heparin binding activity are shown below the alignments (Stephens et al., 2001). Dashes represent gaps in the alignment. OmcB accession nos. (obtained from NCBI): *C. caviae*, AAB61619; *C. pneumoniae* CWL029, NP_224753; *C. trachomatis* E, P23603; *C. trachomatis* F/IC-Cal3, M85196; *C. trachomatis* Sweden2, CBJ14964; *C. trachomatis* D/H/G/K, Q548P6; *C. trachomatis* A/HAR-13, YP_328263; *C. trachomatis* B/Jali20/OT, YP_002888064; *C. trachomatis* C, P26758; *C. trachomatis* I/J, Q933I7; *C. trachomatis* L1/L2/L3,P21354. (C) The *C. pneumoniae* OmcB protein harbors a duplication of the original heparin-binding motif XBBXBX (boxed with the basic residues in bold) (aa 41 to 100) (Moelleken and Hegemann, 2008). Replacement of all three basic residues in motif I by alanine residues abolishes OmcB-mediated adhesion to epithelial HEp-2 cells (Moelleken and Hegemann, 2008). Deletion of the second heparin-binding motif (motif II) also resulted in the complete loss of adhesion, indicating that the binding motif in the OmcB protein is larger than originally suggested (Fechtner et al., unpublished). Secondary-structure prediction was done with GORIV. Adhesion symbols: +++++ to −, strong adhesion to no adhesion.
doi:10.1128/9781555817329.ch5.f1

reduced infection levels in a dose-dependent manner (Fadel and Eley, 2007; Moelleken and Hegemann, 2008). Enzyme-linked immunosorbent assay and flow cytometric analyses showed that preincubation with recombinant *C. pneumoniae* OmcB dramatically decreased the incidence of EB attachment to epithelial (HEp-2) and endothelial (primary HUVE) cells (Moelleken and Hegemann, 2008). These data strongly argue that OmcB acts as an adhesin for association to human cells and is of primary importance for infection by several chlamydial species. The precise nature of the OmcB ligand(s) is not yet known. However, help may come from high-throughput technologies in the field of glycomics (Turnbull, 2010; Turnbull et al., 2010).

The N-Terminal Segment of OmcB Harbors a Heparin-Binding Domain

The OmcB protein is highly conserved among the *Chlamydia* spp. The C-terminal portion of the protein (aa 96 to aa 556) is most conserved (79 to 99% pairwise identity), while the N-terminal end (excluding the 22-aa signal sequence) is more variable (33 to 96%) (Fig. 1A). OmcB sequences from all *C. trachomatis* serovars (A-K and L1-3) are very similar, and most of the basic residues are located in the N-terminal region.

Functional dissection of the *C. pneumoniae* OmcB indeed revealed that its N-terminal half, particularly its basic domain (aa 45 to aa 78), is responsible for binding to HEp-2 cells. An N-terminal fragment neutralized *C. pneumoniae* infectivity as effectively as wild-type OmcB (Moelleken and Hegemann, 2008). Inspection of OmcB sequences from the different chlamydial species and serovars revealed that all carry at least one XBBXBX motif (where X is any hydrophobic amino acid and B is arginine, lysine, or histidine) in their basic domain. This motif has been identified as one of three major binding motifs found in heparin-binding proteins (Cardin and Weintraub, 1989; Hileman et al., 1998; Stephens et al., 2001; Capila and Linhardt, 2002). Synthetic peptides containing such heparin-binding motifs from *C. pneumoniae* OmcB (aa 50 to 70) or *C. trachomatis* serovar L2 (aa 51 to 70) bound biotinylated heparin in an enzyme-linked immunosorbent assay experiment (Stephens et al., 2001) (Fig. 1B). A recombinant *C. pneumoniae* OmcB variant without the tandem heparin-binding motifs showed none of the OmcB-associated functions (Moelleken and Hegemann, 2008). Single, double, and triple alanine substitutions within motif I caused progressive loss of adhesion, while a complete deletion of motif II also abolished adhesion, indicating that both motifs are needed for binding to human cells (Fig. 1C) (Moelleken and Hegemann, 2008) (T. Fechtner, S. Stallmann, and J. H. Hegemann, unpublished data).

The GAG Specificity of *C. trachomatis* OmcB Reflects Biovar-Specific Differences

Adhesion of the OmcB proteins from *C. pneumoniae*, *C. psittaci*, and *C. trachomatis* serovar L2, but not serovar E, to human epithelial cells is blocked by exogenous heparin, implying that OmcB from *C. trachomatis* serovar E (hereafter named E-OmcB) might bind a different ligand (Moelleken and Hegemann, 2008). *C. trachomatis* LGV biovars are strongly dependent on the presence of HS-like GAGs on their target cells for infection (see above), while infection by *C. trachomatis* serovar E is completely independent of HS-like GAGs (Fig. 2A) (Chen and Stephens, 1997; Davis and Wyrick, 1997; Taraktchoglou et al., 2001; Beswick et al., 2003; Moelleken and Hegemann, 2008).

E-OmcB differs from the OmcBs of the LGV serovars L1 and L2 (L-OmcB) at only three positions (66, 68, and 71), which lie C terminal to the heparin-binding motif within the basic domain. When presented on *E. coli* or yeast cells, all three OmcB proteins bound to human epithelial cells, and preincubation of host cells with recombinant OmcB protein prior to exposure to the parent serovar significantly reduced infection. Yet adhesion of yeast or *E. coli* cells presenting E-OmcB was not inhibited by heparin or HS (Fadel and Eley, 2008a; Moelleken and Hegemann, 2008). When the latter authors replaced the proline

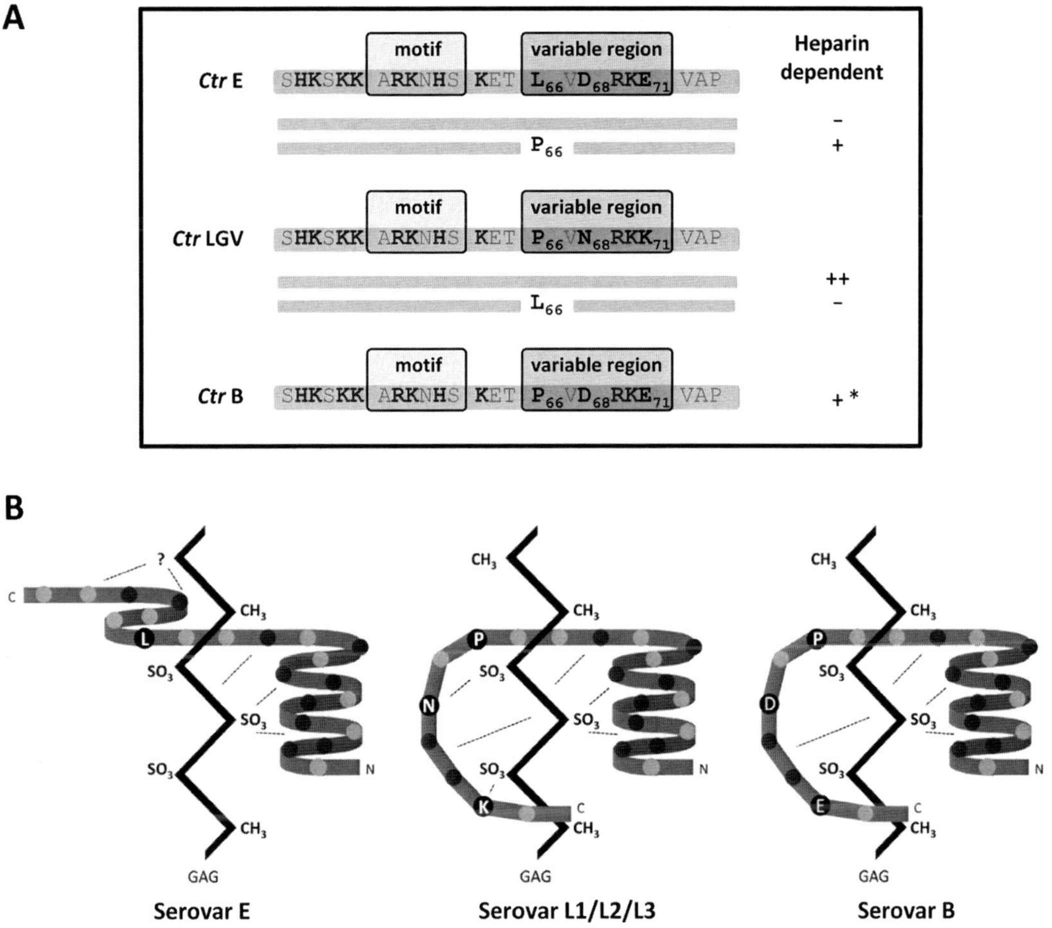

FIGURE 2 Dependence of different *C. trachomatis* OmcB proteins on the presence of heparin on target cells for successful infection. (A) The *C. trachomatis* serovars E, LGV, and B vary in their requirement for HS-like GAGs on target cells for successful infection. This variability is seen in the OmcB proteins from these serovars. The extent to which binding of these OmcBs to epithelial HEp-2 cells is inhibited by heparin ("Heparin-dependent") is indicated. A stretch of residues C terminal to the heparin-binding domain ("motif") exhibits serovar-specific variability ("variable region") at positions 66, 68, and 71 (relevant residues are shown in bold). The amino acid residue at position 66 determines whether or not infection is dependent on heparin. Thus, replacement of the proline at position 66 in OmcB from biovar LGV with a leucine corresponding to the residue at the same position in OmcB from serovar E makes infection by the former independent of heparin, and vice versa (Moelleken and Hegemann, 2008). The sequence of the serovar E $OmcB_{L66P}$ variant corresponds to that of the OmcB proteins of the trachoma and genital serovars A to D and F to K. The asterisk reflects heparin dependence of *C. trachomatis* serovar B attachment (Chen and Stephens, 1994). (B) A model depicting binding of OmcB proteins from serovars E, L2, and B to highly sulfated GAG structures like HS or heparin, incorporating secondary structure predictions based on the amino acid sequences shown in panel A. Relevant residues are marked, and serovar-specific residues are enlarged. Strong heparin dependence of OmcB from LGV could be due to a series of interactions of the basic residues in the heparin-binding motif presented in an α-helical structure, supported by additional basic amino acids and by an asparagine at position 68 (known to form hydrogen bonds to GAGs). The leucine residue at position 66 of the OmcB from serovar E results in a structural change and thereby alters GAG recognition. The moderate heparin dependence of OmcB of serovar B could be due to loss of contact sites due to different residues at positions 68 (aspartic acid, D) and 71 (glutamic acid, E). doi:10.1128/9781555817329.ch5.f2

at position 66 in L-OmcB with a leucine as found in E-OmcB at the same position, the resulting L-OmcB$_{P66L}$ mutant bound to human cells in the presence of heparin. Conversely, binding of E-OmcB$_{L66P}$ was inhibited by heparin. Hence, the residue at position 66 determines whether or not OmcB recognizes heparin-type GAGs (Fig. 2A) (Moelleken and Hegemann, 2008).

Interestingly, the sequence of the N-terminal basic domain of serovar E OmcB$_{L66P}$ mutant protein is identical to that of *C. trachomatis* serovars A to D and F to K. Intriguingly, partial dependence of adhesion on heparin has been reported for serovars B, C, and G during infection of epithelial cells. Thus, the in vitro adhesion data for recombinant OmcB protein parallel the results from infection studies (Fig. 2A) (Kuo and Grayston, 1976; Chen and Stephens, 1997; Moelleken and Hegemann, 2008). Structure prediction programs suggest that the heparin-binding motif in the basic domains of OmcB proteins from several species form part of an alpha-helix embedded in a random-coil segment (Fig. 2B). Only the E-OmcB protein harbors a second, very short helix C terminal to the first, which is lost in the E-OmcB$_{L66P}$ variant. Conversely, introduction of the leucine residue into the L-OmcB is predicted to induce formation of a second helix. Thus, the local conformation in this region may largely determine the type of GAG required for adhesion.

The nature of the GAG molecule recognized by chlamydial OmcB proteins, including E-OmcB, remains unknown, but there is good evidence to suggest that E-OmcB and L-OmcB bind related ligands (Fadel and Eley, 2008a) (K. Moelleken and J. H. Hegemann, unpublished data). It is likely that different OmcB proteins have evolved subtly different GAG specificities by incorporating single substitutions that affect the number and spatial arrangement of GAG-binding motifs (Fig. 2B). For example, OmcB from the LGV biovars shows strong adhesion to heparin-like GAG structures on epithelial HEp-2 cells, and addition of heparin markedly reduces adhesion and subsequent bacterial infection. OmcB from serovars A to K differs from L-OmcB at three positions and exhibits less affinity for heparin-like GAGs, as addition of heparin has only a moderate effect on infection. Finally E-OmcB, with only a single residue difference relative to OmcB from serovars A to K, exhibits binding to HEp-2 cells (albeit less than L-OmcB) but is completely insensitive to inhibition by heparin, suggesting that it binds a related GAG structure (Fig. 2B) (Moelleken and Hegemann, 2008). The differences correlate with the relative binding affinities of L-OmcB and E-OmcB for CHO-K1 cells: while *E. coli* cells presenting L-OmcB at their surface bind to wild-type CHO-K1 cells but not to GAG-deficient CHO cells, *E. coli* cells presenting E-OmcB show weaker binding to K1 cells, but only moderately reduced binding to CHO mutant cell lines (Fadel and Eley, 2008a). Determination of the precise chemical nature of the OmcB ligand(s) is therefore of critical importance for a molecular understanding of this aspect of adhesion of chlamydiae to mammalian cells.

Regardless of the structure of the GAG recognized by OmcB, we propose that the attachment of chlamydiae to human cells via OmcB-GAG interactions is only a first step towards successful internalization. Blocking this initial interaction between an EB and the target cell by the addition of excess soluble HS, recombinant OmcB, or anti-OmcB antibody never inhibits the infection by more than 90% (Zhang and Stephens, 1992; Wuppermann et al., 2001; Fadel and Eley, 2007; Moelleken and Hegemann, 2008). This residual infectivity points to the presence of additional adhesin-receptor interactions, and the recent identification of the Pmp protein family as a new group of chlamydial adhesins supports this concept. The combined activities of this molecular cross talk might set the stage for the next step in infection (internalization of EBs) and moreover might account in part for tissue tropism and the spread of the pathogen.

THE PMP PROTEIN FAMILY—A POTENTIAL SOURCE OF DIVERSITY IN ADHESION?

Identification and Phylogenetic Characterization of the Pmp Protein Family

By definition, adhesins must be located on the bacterial cell surface. This basic criterion is fulfilled by the family of polymorphic membrane proteins (Pmps). Its first representatives were identified by immunoblotting experiments with *C. abortus* as major immunogens of around 90 kDa (Longbottom et al., 1996, 1998b). The Pmps are composed of a C-terminal domain that is probably inserted in the outer membrane of the EBs and an external N-terminal domain (Longbottom et al., 1998a). Members of the family were also identified in *C. pneumoniae* by screening an expression library with rabbit serum directed against the COMC (Knudsen et al., 1999). The *pmp* genes account for 3 to 5% of the total coding capacity, emphasizing their functional importance (Stephens et al., 1998; Kalman et al., 1999; Read et al., 2003).

In *C. trachomatis* the *pmp* gene family consists of nine members (*pmpA* through *pmpI*), which are subdivided on phylogenetic grounds into subtypes *A* (*pmpA*), *B* (*pmpB* and *pmpC*), *D* (*pmpD*), *E* (*pmpE* and *pmpF*), *G* (*pmpG* and *pmpI*), and *H* (*pmpH*). *C. pneumoniae* CWL029 has 21 *pmp* genes (Stephens et al., 1998; Grimwood and Stephens, 1999; Rockey et al., 2000; Tan et al., 2006). This is largely due to the expansion of subtype *G*, which includes 13 members (*pmp1* through *pmp13*), but there are also four subtype *E* genes (*pmp15* through *pmp18*), while other subtypes are represented by a single gene each (Grimwood and Stephens, 1999; Kalman et al., 1999). Five of the genes among the expanded subtypes *E* and *G* (*pmp3* through *pmp5*, *pmp12*, and *pmp17*) probably do not code for functional products (Grimwood et al., 2001). Interestingly, the recently sequenced genome of *C. pneumoniae* LPCoLN, isolated from koala, has full-length coding versions of these genes, suggesting that *pmp* genes might be subject to host-specific selection (Myers et al., 2009; Mitchell et al., 2010). In the CWL029 isolate, 13 of the remaining 16 *pmp* genes code for proteins with between 842 and 979 residues (90 to 100 kDa), while Pmp6, Pmp20, and Pmp21 are significantly larger (1407, 1723, and 1609 aa respectively), and variously processed forms have been identified (see below). Since the six Pmp subtypes are conserved in different chlamydial species, it has been suggested that all six perform specific roles during chlamydial infection (Grimwood and Stephens, 1999).

Overall Structure and Expression of Pmp Proteins

Members of the Pmp families from various species show little overall similarity. However, all have multiple repeats of the tetrapeptide motifs GGA(I,L,V) and FxxN in the N-terminal two-thirds of the proteins. The Pmp proteins of *C. trachomatis* and *C. pneumoniae* contain an average of 6.5 and 5 copies of the GGA(I,L,V) motifs, respectively. The FxxN motif occurs on average 13.6 times per Pmp protein in *C. trachomatis* and 11.3 times in *C. pneumoniae* (Grimwood and Stephens, 1999). The two Pmp motifs are interspersed and often appear as doublets, as in PmpD from *C. trachomatis* and Pmp6, 20, and 21 from *C. pneumoniae* (Fig. 3). All Pmp proteins also have a significant number of cysteine residues and might therefore contribute to the structural rigidity of the EB via disulfide cross-linking.

Structurally, the Pmp proteins share the basic properties of classical autotransporters (ATs) (also called type V secretion systems). Like ATs, they have a modular organization, with (i) an N-terminal signal sequence that mediates transport across the plasma membrane, (ii) a large C-terminal domain termed the translocation unit (or β-barrel) that facilitates transit through the OM, and (iii) a central passenger domain (PD), which determines whether the protein remains surface localized or is secreted (Henderson and Lam, 2001; Dautin and Bernstein, 2007; Wells et al., 2007). PDs also display considerable sequence

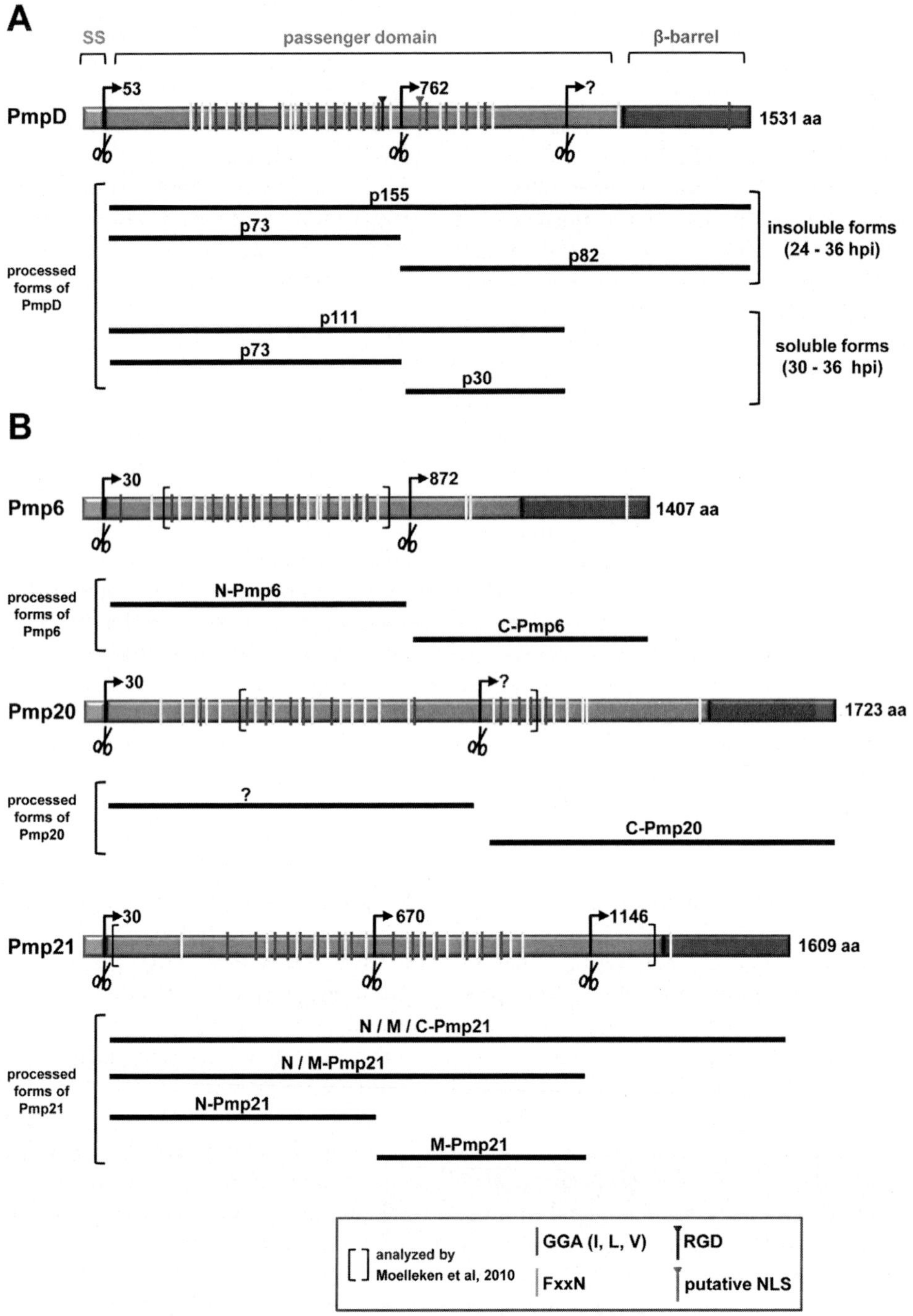

A
SS
passenger domain
β-barrel
53
762
?
PmpD
1531 aa
processed forms of PmpD
p155
p73
p82
insoluble forms (24 - 36 hpi)
p111
p73
p30
soluble forms (30 - 36 hpi)
B
30
872
Pmp6
1407 aa
processed forms of Pmp6
N-Pmp6
C-Pmp6
30
?
Pmp20
1723 aa
processed forms of Pmp20
?
C-Pmp20
30
670
1146
Pmp21
1609 aa
processed forms of Pmp21
N / M / C-Pmp21
N / M-Pmp21
N-Pmp21
M-Pmp21
analyzed by Moelleken et al, 2010
GGA (I, L, V)
FxxN
RGD
putative NLS

variation and might therefore serve many different virulence-related functions (Henderson and Nataro, 2001). For further information on AT proteins, see chapter 4 (mentioned above) and chapter 9, "Protein secretion and *Chlamydia* pathogenesis."

All *pmp* genes in *C. trachomatis* and *C. pneumoniae* are transcribed (Grimwood et al., 2001; Vandahl et al., 2002; Nunes et al., 2007; Tan et al., 2010; Carrasco et al., 2011). Moreover, all nine Pmp proteins are found in the outer membrane of *C. trachomatis* L2 and on the surface of EBs from *C. trachomatis* E (Mygind et al., 2000; Tanzer and Hatch, 2001; Shaw et al., 2002; Skipp et al., 2005; Tan et al., 2010). In *C. pneumoniae* CWL029, the 16 full-length Pmp proteins are expressed in EBs, and evidence for surface localization of Pmp6, Pmp8, Pmp10, Pmp11, Pmp20, and Pmp21 was obtained (Grimwood et al., 2001; Montigiani et al., 2002; Vandahl et al., 2002).

Patients infected with *C. trachomatis* or *C. pneumoniae* have variable titers of antibodies to various Pmp proteins, which is consistent with their exposure on the bacterial cell surface (Bunk et al., 2008; Tan et al., 2009). In line with this, it was shown that soluble recombinant Pmp20 and Pmp21 proteins from *C. pneumoniae* stimulate production of proinflammatory cytokines in human endothelial cells and monocytes (Niessner et al., 2003; Wehrl et al., 2004). Earlier work had shown that monoclonal antibodies directed against one of the 90-kDa Pmp proteins from *C. abortus* could reduce infection in vitro (Cevenini et al., 1991), and specific antibodies directed against Pmp21 from *C. pneumoniae* or its ortholog from *C. trachomatis*, PmpD, inhibit infection in vitro (Wehrl et al., 2004; Crane et al., 2006). All these data are consistent with the assumption that PmpD/Pmp21 act at the pathogen/host cell interface and thus might serve a critical function in pathogenesis (Crane et al., 2006), possibly acting as adhesins (Wehrl et al., 2004).

Pmp6, Pmp20, and Pmp21 Are *C. pneumoniae* Adhesins

Indeed, in agreement with the hypothesis that PmpD/Pmp21 act as adhesins, yeast cells or fluorescently labeled latex beads bearing the entire PD of Pmp21 from *C. pneumoniae* GiD adhered strongly to human epithelial cells (Moelleken et al., 2010). The PDs of members of two other Pmp subtypes, Pmp6 and Pmp20, also conferred comparably strong binding to HEp-2 cells. Thus, representatives of three of the six Pmp subtypes (B/C, D, and G) bind to epithelial cells, prompting the authors to postulate that Pmp6, Pmp20, and Pmp21 expressed on the EB surface might bind human cells to initiate uptake of the bacteria and subsequent infection (Moelleken et al., 2010). Indeed, pretreatment of HEp-2 cells with any of the three recombinant proteins reduced susceptibility to infection upon subsequent exposure to EBs. Hence, Pmp6, Pmp20, and Pmp21 are relevant for infection. However,

FIGURE 3 General properties of selected Pmp proteins from *C. trachomatis* and *C. pneumoniae*. Pmp proteins including PmpD from *C. trachomatis* (A) and Pmp6, Pmp20, and Pmp21 from *C. pneumoniae* (B) are presumably autotransporters, as they exhibit the typical three-domain structure with N-terminal signal sequence (ss), passenger domain, and β-barrel (Henderson and Lam, 2001). Pmp proteins are characterized by the presence of the repeat motifs GGA(I,V,L) and FxxN (motif positions are indicated by white and grey vertical bars) (Grimwood and Stephens, 1999). PmpD, Pmp6, Pmp20, and Pmp21 are processed posttranslationally (at positions indicated by scissors and the numbered N-terminal residue) to yield processed forms indicated by the black lines beneath each protein. (A) *C. trachomatis* PmpD exhibits a complex (presumably biphasic) processing pattern, yielding insoluble as well as soluble forms (drawn according to their apparent molecular weight) (Kiselev et al., 2007, 2009; Swanson et al., 2009). (B) Processing of the *C. pneumoniae* Pmp6, Pmp20, and Pmp21 proteins results in processed forms labeled for their relative positions (N, M, or C terminal) in the full-length protein (Vandahl et al., 2002; Wehrl et al., 2004; Moelleken et al., 2010). Functionally characterized protein derivatives of Pmp6, Pmp20, and Pmp21 are marked by thin brackets. NLS, nuclear localization signal; RGD, integrin binding site. Domain structure was predicted with Pfam HMM search at expasy.org. doi:10.1128/9781555817329.ch5.f3 the dose-response relationship for Pmp6 differed from that for Pmp20 and Pmp21, suggesting that binding to human cells might vary with Pmp subtype.

The ability of recombinant Pmp proteins to attenuate infectivity implies that they compete with native Pmp proteins on the EB surface for binding to sites on the host cell and thus reduce the number of adherent EBs. Indeed, pretreatment of target epithelial cells with the recombinant full-length PD of Pmp21 caused a dose-dependent reduction in the subsequent binding of fluorescently labeled, infectious *C. pneumoniae* EBs (Moelleken et al., 2010). Furthermore, specific antibodies raised against either the N-terminal or central segment of Pmp21 also neutralized infection in a concentration-dependent manner (Wehrl et al., 2004; Crane et al., 2006; Moelleken et al., 2010). Strikingly, the anti-PmpD antibody showed pan-neutralizing activity against all *C. trachomatis* serovars, suggesting an important and common function for PmpD in the pathogenesis of infection (Crane et al., 2006). Furthermore, given the degree of similarity between members of the Pmp family throughout *Chlamydia* spp., it is reasonable to assume that many, if not all, Pmp proteins exhibit the adhesion and infection-neutralizing properties seen for Pmp6, Pmp20, and Pmp21. Careful functional characterization of the entire panel of Pmp proteins from one species is therefore urgently needed.

Preliminary experimental evidence indicates that Pmp6, Pmp20, and Pmp21 act in related functional pathways. Thus, preincubation of epithelial HEp-2 cells with a mixture of recombinant Pmp6 and Pmp21 had a greater effect on infectivity than either protein alone (Moelleken et al., 2010). These data suggest that Pmp6 and Pmp21 have similar but nonredundant roles in adhesion and infection, possibly recognizing the same host cell receptor or a group of related receptors. Alternatively both proteins could form hetero-oligomeric structures (see below) with enhanced biological activity relative to the individual proteins.

The Naturally Processed Passenger Domain of Pmp21 of *C. pneumoniae* Carries Several Adhesion Domains

In all AT proteins analyzed thus far, the N-terminal signal sequence is cleaved by signal peptidase I during Sec-dependent transport through the inner membrane (Dautin and Bernstein, 2007). Moreover, many of these proteins, including some Pmps, undergo further proteolytic cleavage—a step which is probably associated with their function on or beyond the cell surface (Henderson et al., 2004; Wells et al., 2007). Thus, posttranslationally processed forms of Pmp6, Pmp20, and Pmp21 are associated with EBs (Fig. 3) (Vandahl et al., 2001, 2002). Pmp6 is cleaved once, yielding a C-terminal ~60-kDa fragment and an N-terminal 90-kDa fragment that retains the adhesion and infection-modulating activities; both can be extracted from EBs (Fig. 3B) (Vandahl et al., 2002). Processing of Pmp20 is less well characterized, but a C-terminal form of ~90 kDa is associated with EBs. The N-terminal species could not be detected; it may be degraded or released from EBs (Fig. 3B) (Vandahl et al., 2002). It is not clear whether full-length forms of Pmp6 and Pmp20 are present on EBs. Full-length Pmp21 (N/M/C-Pmp21) was found to be associated with EBs, and processed fragments consisting of the entire passenger domain (N/M-Pmp21), as well as N-Pmp21, M-Pmp21, and C-Pmp21, were detected in EB extracts (Fig. 3B) (Vandahl et al., 2002; Wehrl et al., 2004; Moelleken et al., 2010). The fully processed N-Pmp21 and M-Pmp21 forms appear to dominate during infection, as well as on EBs (Wehrl et al., 2004; Moelleken et al., 2010).

Interestingly, a processing pattern similar to that seen for Pmp21 was described for its ortholog in *C. trachomatis*, PmpD (Kiselev et al., 2007, 2009; Swanson et al., 2009). PmpD, like Pmp21, is expressed late in infection (Kiselev et al., 2009; Swanson et al., 2009). PmpD is processed into an N-terminal fragment (63 to 73 kDa) and C-terminal fragment (80 to 82 kDa), together with the entire PD (111 to 120 kDa), all in roughly similar amounts. There is, however, disagreement as to whether the different forms are soluble or insoluble and at what time in the developmental cycle they appear (Kiselev et al., 2009; Swanson et al., 2009). Moreover, the Caldwell group has proposed a biphasic cleavage process, starting with production of the insoluble forms p73 and p82 from full-length PmpD beginning at 24 h postinfection, followed by the appearance of the complete PD p111 with its processed soluble derivatives p73 and p30 from 36 h on (Swanson et al., 2009). In addition, both studies report the presence of soluble PmpD material, not associated with the bacteria, in the inclusion lumen. These fragments might perform additional functions unrelated to adhesion and might even enter the nucleus, as the soluble p111 and p30 forms carry a predicted nuclear localization signal (Swanson et al., 2009). Evidence for cleavage of several other Pmp proteins has recently been reported in lysates of purified EBs from *C. trachomatis* serovar E (Tan et al., 2010).

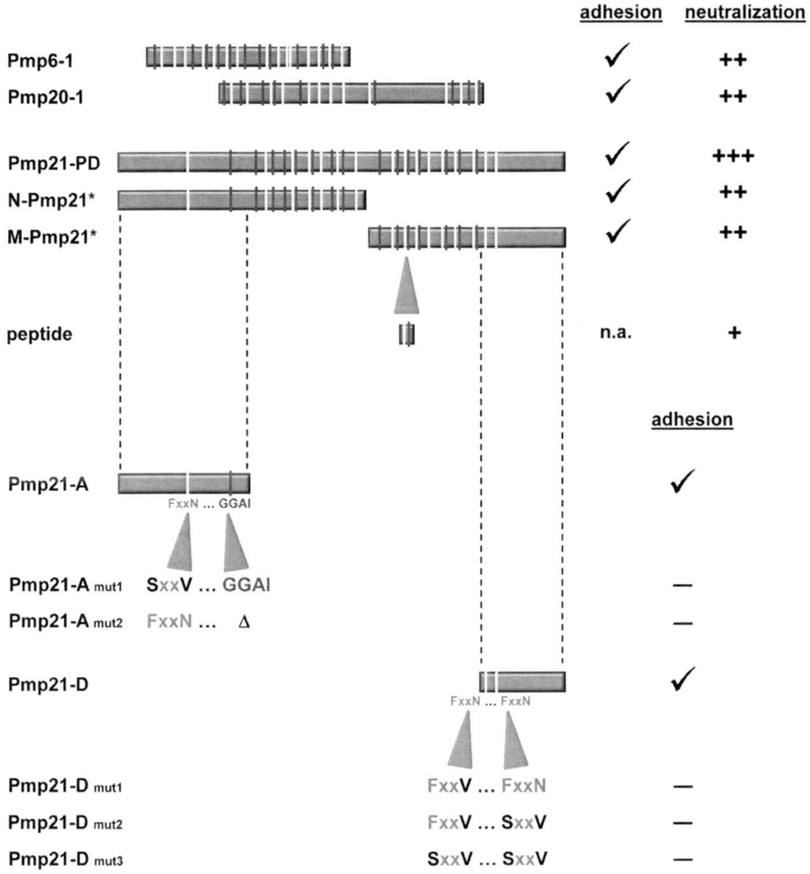

FIGURE 4 *C. pneumoniae* Pmp proteins are adhesins, and binding is mediated by repeat motifs. The PDs of the three largest Pmp proteins from *C. pneumoniae* mediate adhesion to epithelial HEp-2 cells and attenuate subsequent infection (Moelleken et al., 2010). Recombinant subdomains of N-Pmp21 and M-Pmp21, each harboring two motifs only, adhered efficiently to human cells, but point mutations in the motifs, or a 4-residue-deletion (labeled by Δ), abrogated binding. Likewise, a synthetic peptide derived from M-Pmp21 carrying a doublet of FxxN and GGAI was able to significantly reduce infection, while a scrambled peptide showed no activity. Repeat motifs GGA(I,V,L) and FxxN are labeled as in Fig. 3. doi:10.1128/9781555817329.ch5.f4

The protease(s) responsible for processing the Pmp proteins remains to be identified. Processing may occur via an autocatalytic mechanism (Wells et al., 2007) or by a separate surface-bound protease. The fact that intact Pmp PDs can be generated in yeast and purified from *E. coli* suggests that Pmp PDs lack intrinsic protease activity (Moelleken et al., 2010). Full-length PmpD containing the signal sequence predominates during infection by *C. trachomatis* in the presence of penicillin. These findings are consistent with a putative chlamydial signal protease involved in later processing events, whose activity is blocked by β-lactam compounds such as penicillin (Kiselev et al., 2007).

The fact that the complete Pmp21 PD shows adhesive properties raised two questions: (i) what part of the domain is involved in binding, and (ii) is it unmasked by processing? To clarify these issues, the recombinant N- and M-Pmp21 forms were tested. Both exhibited strong adhesive properties and significantly inhibited subsequent infection, albeit to a lesser extent than the complete PD (Fig. 4) (Moelleken et al., 2010). Moreover, when the PD was arbitrarily divided into four regions of similar size, all four retained a significant adhesion capacity. Furthermore, they exhibited a variable, region-dependent ability to block infection. Even small fragments of 50 aa in length displayed undiminished adhesion (Moelleken et al., 2010). These data show that the complete PD harbors several independent adhesion domains. However, the capacity of an isolated region of the PD to block infection is not strictly correlated with its adhesive activity. Pmp variants resembling the fully processed forms N-Pmp21 and M-Pmp21, as well as the full-length PD, were most effective, suggesting that they bind optimally to the cognate receptor(s).

The Repeat Motifs Are Essential for the Function of Pmp Proteins

All Pmp21 domains that exhibit adhesion contain the repeat motifs GGA(I,L,V) and FxxN. Further mutation studies revealed that a minimum of two copies of either motif (FxxN + GGAV or FxxN + FxxN) is required for adhesion. The relevance of the motifs for Pmp function was then stringently tested using a synthetic 32-aa peptide harboring a typical tetrapeptide motif doublet (FxxN-x_8-GGAI) derived from Pmp21. Preincubation of human epithelial cells with this peptide attenuated subsequent infection by up to 40%, suggesting that the repeat motifs play an important role in the function of Pmp21 during adhesion and initiation of infection (Fig. 4) (Moelleken et al., 2010). In general, the distribution of either peptide motif in the Pmp proteins follows no obvious pattern, but doublets with an N-terminal FxxN motif and a C-terminal GGA(I,L,V) motif separated by 4 to 18 residues are a prominent feature of most Pmp proteins (Fig. 3) (Grimwood and Stephens, 1999). However, longer Pmp21 domains harboring only two motifs (FxxN + FxxN or FxxN + GGAI) that are far apart in the primary sequence also exhibited adhesion, so the two motifs do not have to be near each another (Fig. 4). Since the number of motifs and/or doublets does not directly correlate with adhesive strength or neutralizing capacity, other factors such as additional binding sequences or protein conformation may be involved.

The different Pmp proteins may contribute to adhesion by collectively interacting with the same or related host cell surface receptors or by individually recognizing specific receptors. The Pmp family consists of six subtypes represented by at least one ortholog in both *C. trachomatis* and *C. pneumoniae*. This observation led to the original suggestion that each subtype has a specific function in *Chlamydia* biology (Grimwood and Stephens, 1999). The adhesion studies now suggest that perhaps all Pmp proteins act as adhesins. If each Pmp subtype were to bind its own host cell receptor, six adhesin-receptor pairs could contribute to cell attachment. The expansion of the *C. trachomatis* PmpG subtype to 13 paralogous members in *C. pneumoniae* could then represent an adaptation to the specific needs of this species in infecting relevant cells and tissues in vivo.

Interestingly, other bacterial cell surface proteins such as OmpA, YfaL, and FHA, from *Rickettsia rickettsii*, *E. coli*, and *Bordetella pertussis*, respectively, also carry multiple GGA(I,L,V) and FxxN motifs (Grimwood and Stephens, 1999). However, their role in protein function is unclear (Mazar and Cotter, 2006; Wells et al., 2010).

Do Pmp Proteins Form Oligomeric Adhesin Structures?

Structure predictions have suggested that a large region of the PD of Pmp6 folds into a parallel β-strand in a helical pattern with three faces that form a β-helix (Bradley et al., 2001; Vandahl et al., 2002). Analysis of over 500 ATs revealed that nearly all adopt a right-handed parallel β-helix structure, which might be required for efficient translocation across the OM and for folding (Junker et al., 2006).

Systematic analysis of all Pmp proteins from various species also revealed that most have right-handed β-helix domains (E. Becker, T. Fechtner, K. Moelleken, and J. H. Hegemann, unpublished data). The β-helix in many Pmp proteins is predicted to be very long and could provide a rigid platform on which multiple adhesive sites (e.g., such as those identified for Pmp21) can be presented.

The β-helices can associate with each other effectively to generate oligomers (Fig. 5A). Preliminary evidence for oligomerization of the PmpD protein from *C. trachomatis* has been presented recently (Swanson et al., 2009). Analysis of immunoaffinity-purified PmpD detected stable complexes of around 530 kDa and 850 kDa, which consisted of full-length p155 and the processed p73 PD plus the p82 translocator domain, all of which are associated with the bacterial surface (Fig. 3) (Swanson et al., 2009). Assuming that the processed Pmp21 forms that contain the β-barrel are C terminally inserted into the outer membrane of the EB, the elongated N-terminal β-helix probably projects from the EB surface. This would permit interactions with other processed forms of the Pmp proteins via β-helix structures or other interaction domains, resulting in homo-oligomers that could facilitate coordinated binding to sites on the host cell (Fig. 5B). The *C. trachomatis* proteins PmpA-I could form at least nine specific oligomers with identical or distinct adhesive functions. Even more intriguingly, processed forms of different Pmp proteins might form hetero-oligomers (Fig. 5B). In fact the PmpD oligomers isolated by immunoaffinity purification might well have contained additional Pmp proteins (Swanson et al., 2009). If this were the case, the range of possible oligomeric forms would be greatly expanded, which might be relevant for immune evasion or niche adaptation (see below).

Electron microscopic examination of the isolated 850-kDa PmpD oligomers revealed a homogeneous population of defined particles exhibiting symmetrical flower-like structures, which could represent physiological complexes present on the surfaces of RBs and EBs (Swanson et al., 2009). The rosettes could be related to the projections and other structures observed more than 25 years ago on the *C. psittaci* and *C. trachomatis* EB surface (Gregory et al., 1979; Chang et al., 1982; Matsumoto, 1982), which were later proposed to represent the chlamydial type III secretion apparatus (Bavoil and Hsia, 1998).

The Pmp Proteins Are Subject to Antigenic Variation

In their original description of the *pmp* gene family, Grimwood and Stephens speculated that the Pmp proteins might be subject to antigenic variation (Grimwood and Stephens, 1999). The first direct evidence for the differential expression of a Pmp protein was reported for *C. pneumoniae*, where the Pmp10 protein was found to be differentially expressed in infected epithelial HEp-2 cells (Pedersen et al., 2001). The subsequent discovery of heterogeneous immunoreactivity to recombinant PmpC in sera from patients infected with different *C. trachomatis* strains suggested a role for PmpC in antigenic variation (Gomes et al., 2005), and *C. trachomatis*-infected patients were later shown to display variable antibody responses to all nine members of the Pmp family, suggesting variable expression of the *pmp* gene family (Tan et al., 2009). Variable expression of Pmps has since been demonstrated in cells infected with *C. trachomatis*, where between

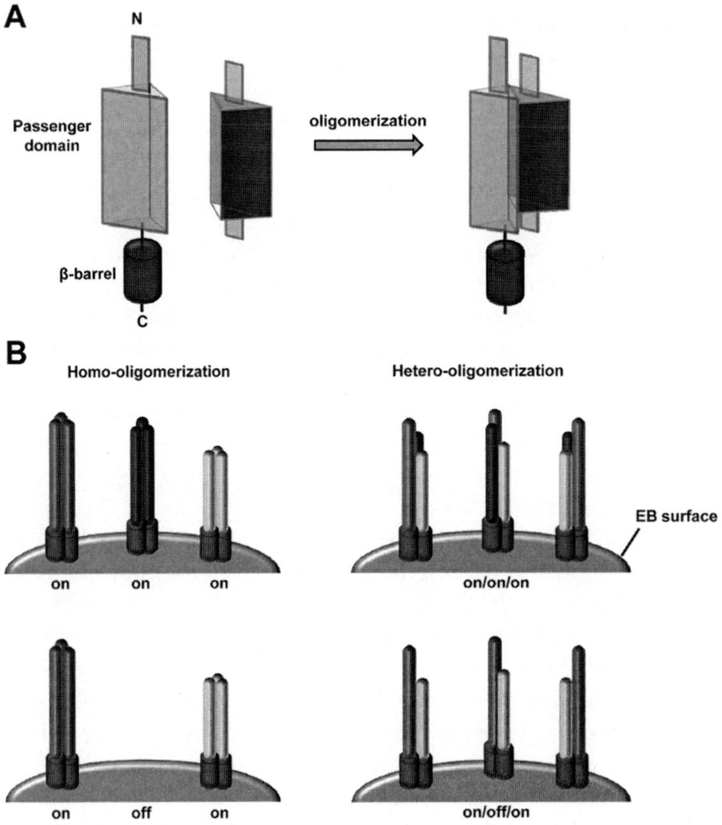

FIGURE 5 Oligomerization model for the Pmp proteins. (A) Experimental evidence for native PmpD oligomers on infectious *C. trachomatis* EBs was provided by Swanson and colleagues (Swanson et al., 2009). Structure predictions reveal three-stranded β-helix domains in most of the Pmp proteins. In other proteins these β-helices (triangular prisms) have been shown to interact with each other to form oligomeric structures. The model shows the passenger domain (light grey prism) of a Pmp protein that has a β-barrel (cylinder) interacting with a processed form of Pmp (dark grey prism) via β-helix or other interaction domains. (B) On the EB cell surface Pmp oligomers could be formed by full-length and/or processed forms of a single species of Pmp protein, forming homo-oligomers (single-shaded oligomers). Alternatively or in addition, hetero-oligomers could be formed from processed and nonprocessed forms of the same Pmp subtype or a combination of Pmp subtypes. Variable expression of individual Pmp proteins indicated by "on" and "off" might increase the diversity of Pmp complexes.
doi:10.1128/9781555817329.ch5.f5

0.1% and 10% of inclusions in a single infection did not express a given Pmp subtype as measured by immunofluorescence with subtype-specific antibodies (Tan et al., 2010). The molecular basis for these findings is unclear at present (Shirai et al., 2000; Gomes et al., 2005, 2006, 2007). However, given that Pmp proteins act as adhesins, the data may suggest that there is considerable pressure to diversify their expression, e.g., to adapt to new niches.

DO *CHLAMYDIA* SPECIES POSSESS ADDITIONAL ADHESINS?

It would be beneficial for *Chlamydia* to utilize a range of diverse adhesin-receptor interactions to bind and enter eukaryotic cells, since this pathogen only replicates intracellularly and each new developmental cycle requires infection of a new host cell. It is noteworthy that neither OmcB nor the Pmp proteins studied thus far promote internalization into human cells. Interestingly, the genome sequences of *C. caviae* and *Chlamydia suis* contain a gene (*ilp*) for an invasin-like protein (Read et al., 2003), but the available data suggest it is unlikely to encode a functional protein (Burall et al., 2007). Two other chlamydial proteins have recently been implicated in EB adherence. The arginine transporter protein ArtJ, which is exposed on *C. pneumoniae* and *C. trachomatis* EBs, may play a role in host cell adhesion during infection, as recombinant ArtJ binds to epithelial cells in vitro and both anti-ArtJ antibodies and recombinant ArtJ reduce infectivity (Soriani et al., 2010). *C. pneumoniae* GroEL-1 is another candidate adhesin. Recombinant GroEL1 adheres to human cells (unlike GroEL-2 and GroEL-3) and competitively reduces infectivity. Here again, it is not clear whether GroEL-1 is directly involved in the adhesion process (Wuppermann et al., 2008). However, it seems likely that additional chlamydial adhesins remain to be discovered.

WHAT RECEPTOR MOLECULES MIGHT CHLAMYDIAL ADHESINS RECOGNIZE ON HUMAN CELLS?

The chlamydial adhesin OmcB adheres to various epithelial cell lines (HEp-2, HeLa, and Hec-1B) and primary endothelial (HUVE) cells (Fadel and Eley, 2007; Moelleken and Hegemann, 2008), implying that the GAG structures recognized are present on a variety of cell types. However, given the diversity of GAGs present on mammalian cells, identification of the precise chemical structures bound by OmcB poses a huge challenge.

As cell adhesion receptors, integrins provide an evolutionarily conserved portal for microbial pathogen adherence and cell entry in eukaryotes (Ulanova et al., 2008), and PmpD from *C. trachomatis* contains an RGD tripeptide sequence, a characteristic recognition motif found in some bona fide integrin ligands. Moreover, since the majority of integrin ligands do not possess an RGD motif, the absence of RGD motifs in the other Pmp proteins does not preclude the possibility that they bind to integrins (Barczyk et al., 2010).

Clearly, the quest for eukaryotic receptors for Pmp proteins has only just begun. The striking similarities between the Pmp proteins from different chlamydial species suggest that they may recognize a group of related proteins, which may be cell surface localized or associated with the extracellular matrix. However, it cannot be excluded at present that the Pmp proteins of different *Chlamydia* species, which infect different host and cell types, might have evolved distinct recognition specificities.

Recently adhesion and uptake of EBs of *C. trachomatis* serovar E has been linked with the CFTR. Nonfunctional CFTR mutant cell lines or addition of a CFTR inhibitor reduced the internalization of *C. trachomatis* EBs, and homozygous mice lacking *cftr* exhibited a reduced chlamydial load (Ajonuma et al., 2010). Moreover, CFTR colocalized with, and bound to, chlamydial LPS, confirming earlier suggestions of a possible role of LPS in *C. trachomatis* infection (Fadel and Eley, 2008b; Ajonuma et al., 2010). It remains to be shown whether CFTR acts as a pattern recognition molecule that extracts chlamydial LPS from the outer membrane into epithelial cells, as has been shown for *Pseudomonas aeruginosa* (Schroeder et al., 2002). This might result in the activation of infection-promoting signal cascades.

In recent years several other human cell surface proteins have been implicated in chlamydial adhesion, but in almost all cases the molecular function and the involvement of a chlamydial adhesin remain elusive (Table 2). Apolipoprotein E4 and mannose/M6P receptors play a role in EB binding that is specific for certain *Chlamydia* spp. The chlamydial adhesins for both proteins are unknown, but theoretically MOMP glycan could be a target (Gerard et al., 2008; Puolakkainen et al., 2008). The mammalian receptor for platelet-derived growth factor (PDGFR) is activated during infection by *C. trachomatis* and is recruited to the site of EB entry. Moreover, downregulation of the PDGFR decreased EB attachment but not entry, and again no direct interaction of the receptor with EBs was shown (Elwell et al., 2008; Wang et al., 2010).

Finally the cell surface-associated PDI, which is part of the estrogen receptor complex, was identified as interacting with EBs of *C. trachomatis* serovar E (Davis et al., 2002). Interestingly, primary human endometrial epithelial cells exhibited an estrogen-dependent enhancement of EB attachment and infection (Maslow et al., 1988; Wyrick et al., 1989). Moreover, Wyrick and colleagues recently showed that antibodies against membrane-associated estrogen receptors α and β or preincubation with the estrogen receptor inhibitor tamoxifen reduced subsequent infection of polarized endometrial epithelial cells by *C. trachomatis* serovar E (Hall et al., 2011). Again a direct interaction between these receptors and the chlamydial EB has not been shown yet. Likewise, there is also evidence indicating a role of PDI in infection: (i) depletion of PDI by small interfering RNA reduces attachment of *C. trachomatis* and *C. psittaci* EBs, and attachment is not dependent on its enzymatic activity; and (ii) enzymatic PDI activity is involved in chlamydial entry (Abromaitis and Stephens, 2009). Whether PDI, possibly as part of the estrogen receptor complex, is directly relevant for EB attachment remains to be seen. However, its enzymatic activity might be relevant if restructuring of the rigid EB cell surface is required for entry-related processes.

A MODEL FOR CHLAMYDIAL ADHERENCE TO HUMAN CELLS

The hypothetical model shown in Fig. 6 summarizes our current understanding of the general processes and molecules involved in attachment and internalization of the chlamydial EB. The precise spatiotemporal order of adhesin function has not been studied in detail. Initial interac-

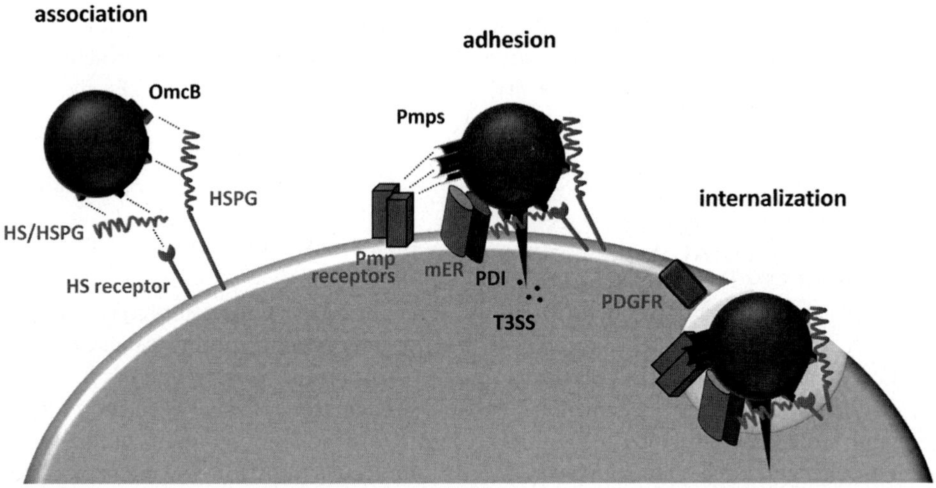

FIGURE 6 Generalized model for chlamydial adhesion to host cells. Attachment of chlamydiae to host cells is proposed to be a two-step mechanism. The initial, reversible association of the EB with the host cell is via binding of the OmcB protein to HS and/or HSPG in soluble form or associated with the host cell surface. Additional interactions may involve chlamydial high-mannose oligosaccharide structures and host cell mannose receptors (not shown). In the next step, Pmp proteins (and possibly other adhesins including invasin-like chlamydial proteins) bind to as yet unidentified receptors on the host cell surface. Surface-localized PDI and/or membrane-associated estrogen receptor (mER) may serve as a structural component of one or more of the host cell receptors. These tight interactions may then allow interaction of the type III secretion system (T3SS) of the EB with the host membrane and release of the first wave of effector proteins (including Tarp) into the host cytosol (see chapter 6). More details are provided in the text.
doi:10.1128/9781555817329.ch5.f6

tions of infectious EBs with their host cells very probably occur via glycosaminoglycan chains and are probably reversible in nature (Carabeo and Hackstadt, 2001), and their on-off dynamics might facilitate EB concentration in the vicinity of the host cell surfaces. OmcB binding to proteoglycans of the HS type anchored to the host cell surface is apparently a bimolecular reaction (Davis and Wyrick, 1997; Wuppermann et al., 2001; Fadel and Eley, 2007, 2008a; Moelleken and Hegemann, 2008). Alternatively, or in parallel, a trimolecular mechanism may occur in which soluble HS or HSPG binds to the EB and to a HS receptor, thereby attaching the EB to the host surface (Zhang and Stephens 1992). In some chlamydial species, this first round of interactions might be accompanied by MOMP-GAG interactions strengthening anchorage of the EB to the host cell surface (Su et al., 1996). Further interactions may occur via chlamydial mannose-rich moieties and host cell mannose receptors (not included in the model) (reviewed by Campbell and Kuo [2006]).

A second stage in attachment may involve more specific and irreversible adhesion of chlamydial Pmp adhesins to host cell receptors, followed by internalization of EBs. Possible candidates for the Pmp receptor include the PDI/estrogen receptor complex, the PDGFR, and the other host cell surface molecules that have been implicated in adhesion (see above). In addition, secretion of effector proteins by the type III secretion system is required to stabilize attachment and/or to initiate host cell entry.

OUTLOOK

The first chlamydial adhesin proteins have been identified and characterized, but their specific function is yet to be determined. What is the precise nature of the GAG molecules recognized by OmcB? Do all Pmp proteins act as adhesins possibly via oligomerization? Is phase variation relevant for adhesion and/or for cell type specificity? What receptors do Pmps recognize on target cells? Are there additional chlamydial adhesins? Finally, elucidation of the spatial and temporal order of the various adhesin-receptor interactions required for successful host cell entry is crucial for the understanding of EB adhesion (i.e., which comes first?). Moreover, as most of the available data have been obtained with transformed cell lines grown in vitro, attempts must be made to translate these basic observations into a more natural context. This is a particular challenge given that chlamydiae still cannot be genetically modified and grown axenically.

ACKNOWLEDGMENTS

We are grateful to our former colleagues F. Wuppermann and G. Murra and all current lab members for the creative and stimulating environment during the past several years. We thank T. Fechtner and E. Becker for their great help in preparing figures. J.H.H. expresses his great gratitude to Ursula Fleig for her endless patience and support during the preparation of this chapter. Work in our laboratory is supported by the Deutsche Forschungsgemeinschaft (SFB 590, FOR 729), the Bundesministerium für Bildung und Forschung (CHI consortium), the ERA-NET Projects ECIBUG and Pathomics, and the Manchot graduate school "Molecules of Infection—MOI."

REFERENCES

Abromaitis, S., and R. S. Stephens. 2009. Attachment and entry of *Chlamydia* have distinct requirements for host protein disulfide isomerase. *PLoS Pathog.* **5:**e1000357.

Ajonuma, L. C., K. L. Fok, L. S. Ho, P. K. Chan, P. H. Chow, L. L. Tsang, C. H. Wong, J. Chen, S. Li, D. K. Rowlands, Y. W. Chung, and H. C. Chan. 2010. CFTR is required for cellular entry and internalization of *Chlamydia trachomatis*. *Cell Biol. Int.* **34:**593–600.

Allen, J. E., and R. S. Stephens. 1989. Identification by sequence analysis of two-site posttranslational processing of the cysteine-rich outer membrane protein 2 of *Chlamydia trachomatis* serovar L2. *J. Bacteriol.* **171:**285–291.

Barczyk, M., S. Carracedo, and D. Gullberg. 2010. Integrins. *Cell Tissue Res.* **339:**269–280.

Bardiau, M., M. Szalo, and J. G. Mainil. 2010. Initial adherence of EPEC, EHEC and VTEC to host cells. *Vet. Res.* **41:**1–16.

Bavoil, P., A. Ohlin, and J. Schachter. 1984. Role of disulfide bonding in outer membrane structure and permeability in *Chlamydia trachomatis*. *Infect. Immun.* **44:**479–485.

Bavoil, P. M., and R. C. Hsia. 1998. Type III secretion in *Chlamydia*: a case of déjà vu? *Mol. Microbiol.* **28:**860–862.

Becker, Y., E. Hochberg, and Z. Zakay-Rones. 1969. Interaction of trachoma elementary bodies with host cells. *Isr. J. Med. Sci.* **5:**121–124.

Beeckman, D. S., C. M. Van Droogenbroeck, B. J. De Cock, P. Van Oostveldt, and D. C. Vanrompay. 2007. Effect of ovotransferrin and lactoferrins on *Chlamydophila psittaci* adhesion and invasion in HD11 chicken macrophages. *Vet. Res.* **38:**729–739.

Bernfield, M., M. Gotte, P. W. Park, O. Reizes, M. L. Fitzgerald, J. Lincecum, and M. Zako. 1999. Functions of cell surface heparan sulfate proteoglycans. *Annu. Rev. Biochem.* **68:**729–777.

Beswick, E. J., A. Travelstead, and M. D. Cooper. 2003. Comparative studies of glycosaminoglycan involvement in *Chlamydia pneumoniae* and *C. trachomatis* invasion of host cells. *J. Infect. Dis.* **187:**1291–1300.

Birkelund, S., M. Morgan-Fisher, E. Timmerman, K. Gevaert, A. C. Shaw, and G. Christiansen. 2009. Analysis of proteins in *Chlamydia trachomatis* L2 outer membrane complex, COMC. *FEMS Immunol. Med. Microbiol.* **55:**187–195.

Bose, S. K., and R. G. Paul. 1982. Purification of *Chlamydia trachomatis* lymphogranuloma venereum elementary bodies and their interaction with HeLa cells. *J. Gen. Microbiol.* **128:**1371–1379.

Brade, H. 1999. Chlamydial lipopolysaccharide, p. 229–242. *In* H. Brade, S. M. Opal, S. N. Vogel, and D. C. Morrison (ed.), *Endotoxin in Health and Disease.* Marcel Dekker Inc., New York, NY.

Bradley, P., L. Cowen, M. Menke, J. King, and B. Berger. 2001. BETAWRAP: successful prediction of parallel beta -helices from primary sequence reveals an association with many microbial pathogens. *Proc. Natl. Acad. Sci. USA* **98:**14819–14824.

Bunk, S., I. Susnea, J. Rupp, J. T. Summersgill, M. Maass, W. Stegmann, A. Schrattenholz, A. Wendel, M. Przybylski, and C. Hermann. 2008. Immunoproteomic identification and serological responses to novel *Chlamydia pneumoniae* antigens that are associated with persistent C. pneumoniae infections. *J. Immunol.* **180:**5490–5498.

Burall, L. S., Z. Liu, R. Rank, and P. M. Bavoil. 2007. The chlamydial invasin-like protein gene conundrum. *Microbes Infect.* **9:**873–880.

Byrne, G. I. 1976. Requirements for ingestion of *Chlamydia psittaci* by mouse fibroblasts (L cells). *Infect. Immun.* **14:**645–651.

Byrne, G. I., and J. W. Moulder. 1978. Parasite-specified phagocytosis of *Chlamydia psittaci* and *Chlamydia trachomatis* by L and HeLa cells. *Infect. Immun.* **19:**598–606.

Caldwell, H. D., J. Kromhout, and J. Schachter. 1981. Purification and partial characterization of the major outer membrane protein of *Chlamydia trachomatis. Infect. Immun.* **31:**1161–1176.

Campbell, L. A., and C. C. Kuo. 2006. Interactions of *Chlamydia* with the host cells that mediate attachment and uptake, p. 505–522. *In* P. M. Bavoil and P. B. Wyrick (ed.), *Chlamydia Genomics and Pathogenesis.* Horizon Bioscience, Norfolk, United Kingdom.

Capila, I., and R. J. Linhardt. 2002. Heparin-protein interactions. *Angew. Chem. Int. Ed. Engl.* **41:**391–412.

Carabeo, R. A., and T. Hackstadt. 2001. Isolation and characterization of a mutant Chinese hamster ovary cell line that is resistant to *Chlamydia trachomatis* infection at a novel step in the attachment process. *Infect. Immun.* **69:**5899–5904.

Cardin, A. D., and H. J. Weintraub. 1989. Molecular modeling of protein-glycosaminoglycan interactions. *Arteriosclerosis* **9:**21–32.

Carrasco, J. A., C. Tan, R. G. Rank, R. C. Hsia, and P. M. Bavoil. 2011. Altered developmental expression of polymorphic membrane proteins in penicillin-stressed *Chlamydia trachomatis. Cell Microbiol.* **13:**1014–1025.

Cevenini, R., M. Donati, E. Brocchi, F. De Simone, and M. La Placa. 1991. Partial characterization of an 89-kDa highly immunoreactive protein from *Chlamydia psittaci* A/22 causing ovine abortion. *FEMS Microbiol. Lett.* **65:**111–115.

Chang, J. J., K. Leonard, T. Arad, T. Pitt, Y. X. Zhang, and L. H. Zhang. 1982. Structural studies of the outer envelope of *Chlamydia trachomatis* by electron microscopy. *J. Mol. Biol.* **161:**579–590.

Chen, J. C., and R. S. Stephens. 1994. Trachoma and LGV biovars of *Chlamydia trachomatis* share the same glycosaminoglycan-dependent mechanism for infection of eukaryotic cells. *Mol. Microbiol.* **11:**501–507.

Chen, J. C., J. P. Zhang, and R. S. Stephens. 1996. Structural requirements of heparin binding to *Chlamydia trachomatis. J. Biol. Chem.* **271:**11134–11140.

Chen, J. C. R., and R. S. Stephens. 1997. *Chlamydia trachomatis* glycosaminoglycan dependent and independent attachment to eukaryotic cells. *Microb. Pathog.* **22:**23–30.

Chen, Y., M. Gotte, J. Liu, and P. W. Park. 2008. Microbial subversion of heparan sulfate proteoglycans. *Mol. Cells* **26:**415–426.

Collett, B. A., W. J. Newhall, R. A. Jersild, Jr., and R. B. Jones. 1989. Detection of surface-exposed epitopes on *Chlamydia trachomatis* by immune electron microscopy. *J. Gen. Microbiol.* **135:**85–94.

Conant, C. G., and R. S. Stephens. 2007. *Chlamydia* attachment to mammalian cells requires protein disulfide isomerase. *Cell. Microbiol.* **9:**222–232.

Crane, D. D., J. H. Carlson, E. R. Fischer, P. Bavoil, R. C. Hsia, C. Tan, C. C. Kuo, and

H. D. Caldwell. 2006. *Chlamydia trachomatis* polymorphic membrane protein D is a species-common pan-neutralizing antigen. *Proc. Natl. Acad. Sci. USA* **103:**1894–1899.

Cunningham, A. F., and M. E. Ward. 2003. Characterization of human humoral responses to the major outer membrane protein and OMP2 of *Chlamydophila pneumoniae. FEMS Microbiol. Lett.* **227:**73–79.

Darville, T., S. Yedgar, M. Krimsky, C. W. Andrews, Jr., T. Jungas, and D. M. Ojcius. 2004. Protection against *Chlamydia trachomatis* infection *in vitro* and modulation of inflammatory response *in vivo* by membrane-bound glycosaminoglycans. *Microbes Infect.* **6:**369–376.

Dautin, N., and H. D. Bernstein. 2007. Protein secretion in gram-negative bacteria via the autotransporter pathway. *Annu. Rev. Microbiol.* **61:**89–112.

Dautry-Varsat, A., A. Subtil, and T. Hackstadt. 2005. Recent insights into the mechanisms of *Chlamydia* entry. *Cell. Microbiol.* **7:**1714–1722.

Davis, C. H., J. E. Raulston, and P. B. Wyrick. 2002. Protein disulfide isomerase, a component of the estrogen receptor complex, is associated with *Chlamydia trachomatis* serovar E attached to human endometrial epithelial cells. *Infect. Immun.* **70:**3413–3418.

Davis, C. H., and P. B. Wyrick. 1997. Differences in the association of *Chlamydia trachomatis* serovar E and serovar L2 with epithelial cells *in vitro* may reflect biological differences *in vivo. Infect. Immun.* **65:**2914–2924.

Elwell, C. A., A. Ceesay, J. H. Kim, D. Kalman, and J. N. Engel. 2008. RNA interference screen identifies Abl kinase and PDGFR signaling in *Chlamydia trachomatis* entry. *PLoS Pathog.* **4:** e1000021.

Esko, J. D., and S. B. Selleck. 2002. Order out of chaos: assembly of ligand binding sites in heparan sulfate. *Annu. Rev. Biochem.* **71:**435–471.

Everett, K. D., A. A. Andersen, M. Plaunt, and T. P. Hatch. 1991. Cloning and sequence analysis of the major outer membrane protein gene of *Chlamydia psittaci* 6BC. *Infect. Immun.* **59:**2853–2855.

Everett, K. D., and T. P. Hatch. 1995. Architecture of the cell envelope of *Chlamydia psittaci* 6BC. *J. Bacteriol.* **177:**877–882.

Fadel, S., and A. Eley. 2007. *Chlamydia trachomatis* OmcB protein is a surface-exposed glycosaminoglycan-dependent adhesin. *J. Med. Microbiol.* **56:**15–22.

Fadel, S., and A. Eley. 2008a. Differential glycosaminoglycan binding of *Chlamydia trachomatis* OmcB protein from serovars E and LGV. *J. Med. Microbiol.* **57:**1058–1061.

Fadel, S., and A. Eley. 2008b. Is lipopolysaccharide a factor in infectivity of *Chlamydia trachomatis*? *J. Med. Microbiol.* **57:**261–266.

Falkow, S. 1991. Bacterial entry into eukaryotic cells. *Cell* **65:**1099–1102.

Finlay, B. B., and S. Falkow. 1997. Common themes in microbial pathogenicity revisited. *Microbiol. Mol. Biol. Rev.* **61:**136–169.

Gerard, H. C., E. Fomicheva, J. A. Whittum-Hudson, and A. P. Hudson. 2008. Apolipoprotein E4 enhances attachment of *Chlamydophila (Chlamydia) pneumoniae* elementary bodies to host cells. *Microb. Pathog.* **44:**279–285.

Gomes, J. P., W. J. Bruno, A. Nunes, N. Santos, C. Florindo, M. J. Borrego, and D. Dean. 2007. Evolution of *Chlamydia trachomatis* diversity occurs by widespread interstrain recombination involving hotspots. *Genome Res.* **17:**50–60.

Gomes, J. P., R. C. Hsia, S. Mead, M. J. Borrego, and D. Dean. 2005. Immunoreactivity and differential developmental expression of known and putative *Chlamydia trachomatis* membrane proteins for biologically variant serovars representing distinct disease groups. *Microbes Infect.* **7:**410–420.

Gomes, J. P., A. Nunes, W. J. Bruno, M. J. Borrego, C. Florindo, and D. Dean. 2006. Polymorphisms in the nine polymorphic membrane proteins of *Chlamydia trachomatis* across all serovars: evidence for serovar Da recombination and correlation with tissue tropism. *J. Bacteriol.* **188:**275–286.

Gregory, W. W., M. Gardner, G. I. Byrne, and J. W. Moulder. 1979. Arrays of hemispheric surface projections on *Chlamydia psittaci* and *Chlamydia trachomatis* observed by scanning electron microscopy. *J. Bacteriol.* **138:**241–244.

Grimwood, J., L. Olinger, and R. S. Stephens. 2001. Expression of *Chlamydia pneumoniae* polymorphic membrane protein family genes. *Infect. Immun.* **69:**2383–2389.

Grimwood, J., and R. S. Stephens. 1999. Computational analysis of the polymorphic membrane protein superfamily of *Chlamydia trachomatis* and *Chlamydia pneumoniae. Microb. Comp. Genomics* **4:**187–201.

Gutierrez-Martin, C. B., D. M. Ojcius, R. C. Hsia, R. Hellio, P. M. Bavoil, and A. Dautry-Varsat. 1997. Heparin mediated inhibition of *Chlamydia psittaci* adherence to HeLa cells. *Microb. Pathog.* **22:**47–57.

Hackstadt, T. 1999. Cell biology, p. 101–138. *In* R. S. Stephens (ed.), Chlamydia*: Intracellular Biology, Pathogenesis, and Immunity*. ASM Press, Washington, DC.

Hackstadt, T., W. J. Todd, and H. D. Caldwell. 1985. Disulfide-mediated interactions of the chlamydial major outer membrane protein: role in the differentiation of chlamydiae? *J. Bacteriol.* **161:**25–31.

Hall, J. V., M. Schell, S. Dessus-Babus, C. G. Moore, J. D. Whittimore, M. Sal, B. D. Dill, and P. B. Wyrick. 2011. The multifaceted role of oestrogen in enhancing *Chlamydia trachomatis* infection in polarized human endometrial epithelial cells. *Cell. Microbiol.* **13:**1183–1199.

Hatch, T. P. 1996. Disulfide cross-linked envelope proteins: the functional equivalent of peptidoglycan in chlamydiae? *J. Bacteriol.* **178:**1–5.

Hatch, T. P. 1999. Developmental biology, p. 26–67. *In* R. S. Stephens (ed.), Chlamydia*: Intracellular Biology, Pathogenesis, and Immunity*. ASM Press, Washington, DC.

Hatch, T. P., I. Allan, and J. H. Pearce. 1984. Structural and polypeptide differences between envelopes of infective and reproductive life cycle forms of *Chlamydia* spp. *J. Bacteriol.* **157:**13–20.

Hatch, T. P., M. Miceli, and J. E. Sublett. 1986. Synthesis of disulfide-bonded outer membrane proteins during the developmental cycle of *Chlamydia psittaci* and *Chlamydia trachomatis*. *J. Bacteriol.* **165:**379–385.

Henderson, I. R., and A. C. Lam. 2001. Polymorphic proteins of *Chlamydia* spp.—autotransporters beyond the *Proteobacteria*. *Trends Microbiol.* **9:**573–578.

Henderson, I. R., and J. P. Nataro. 2001. Virulence functions of autotransporter proteins. *Infect. Immun.* **69:**1231–1243.

Henderson, I. R., F. Navarro-Garcia, M. Desvaux, R. C. Fernandez, and D. Ala'Aldeen. 2004. Type V protein secretion pathway: the autotransporter story. *Microbiol. Mol. Biol. Rev.* **68:**692–744.

Hileman, R. E., J. R. Fromm, J. M. Weiler, and R. J. Linhardt. 1998. Glycosaminoglycan-protein interactions: definition of consensus sites in glycosaminoglycan binding proteins. *Bioessays* **20:**156–167.

Junker, M., C. C. Schuster, A. V. McDonnell, K. A. Sorg, M. C. Finn, B. Berger, and P. L. Clark. 2006. Pertactin beta-helix folding mechanism suggests common themes for the secretion and folding of autotransporter proteins. *Proc. Natl. Acad. Sci. USA* **103:**4918–4923.

Kalman, S., W. Mitchell, R. Marathe, C. Lammel, J. Fan, R. W. Hyman, L. Olinger, J. Grimwood, R. W. Davis, and R. S. Stephens. 1999. Comparative genomes of *Chlamydia pneumoniae* and *C. trachomatis*. *Nat. Genet.* **21:**385–389.

Kiselev, A. O., M. C. Skinner, and M. F. Lampe. 2009. Analysis of PmpD expression and PmpD post-translational processing during the life cycle of *Chlamydia trachomatis* serovars A, D, and L2. *PLoS One* **4:**e5191.

Kiselev, A. O., W. E. Stamm, J. R. Yates, and M. F. Lampe. 2007. Expression, processing, and localization of PmpD of *Chlamydia trachomatis* serovar L2 during the chlamydial developmental cycle. *PLoS ONE* **2:**e568.

Klein, M., A. Kotz, K. Bernardo, and M. Kronke. 2003. Detection of *Chlamydia pneumoniae*-specific antibodies binding to the VD2 and VD3 regions of the major outer membrane protein. *J. Clin. Microbiol.* **41:**1957–1962.

Kline, K. A., S. Falker, S. Dahlberg, S. Normark, and B. Henriques-Normark. 2009. Bacterial adhesins in host-microbe interactions. *Cell Host Microbe* **5:**580–592.

Knudsen, K., A. S. Madsen, P. Mygind, G. Christiansen, and S. Birkelund. 1999. Identification of two novel genes encoding 97- to 99-kilodalton outer membrane proteins of *Chlamydia pneumoniae*. *Infect. Immun.* **67:**375–383.

Kraaipoel, R. J., and A. M. van Duin. 1979. Isoelectric focusing of *Chlamydia trachomatis*. *Infect. Immun.* **26:**775–778.

Kubo, A., and R. S. Stephens. 2000. Characterization and functional analysis of PorB, a *Chlamydia* porin and neutralizing target. *Mol. Microbiol.* **38:**772–780.

Kuo, C. C., and T. Grayston. 1976. Interaction of *Chlamydia trachomatis* organisms and HeLa 229 cells. *Infect. Immun.* **13:**1103–1109.

Kuo, C. C., A. Lee, and L. A. Campbell. 2004. Cleavage of the N-linked oligosaccharide from the surface of Chlamydia species affects attachment and infectivity of the organism in human epithelial and endothelial cells. *Infect. Immun.* **72:**6699–6701.

Kuo, C. C., S. P. Wang, and J. T. Grayston. 1973. Effect of polycations, polyanions and neuraminidase on the infectivity of trachoma-inclusion conjunctivitis and lymphogranuloma venereum organisms HeLa cells: sialic acid residues as possible receptors for trachoma-inclusion conjunction. *Infect. Immun.* **8:**74–79.

Liu, X., M. Afrane, D. E. Clemmer, G. Zhong, and D. E. Nelson. 2010. Identification of *Chlamydia trachomatis* outer membrane complex proteins by differential proteomics. *J. Bacteriol.* **192:**2852–2860.

Longbottom, D., J. Findlay, E. Vretou, and S. M. Dunbar. 1998a. Immunoelectron microscopic localisation of the OMP90 family on the outer membrane surface of *Chlamydia psittaci*. *FEMS Microbiol. Lett.* **164:**111–117.

Longbottom, D., M. Russell, S. M. Dunbar, G. E. Jones, and A. J. Herring. 1998b. Molecular cloning and characterization of the genes coding for the highly immunogenic cluster of 90-kilodalton envelope proteins from the *Chlamydia psittaci* subtype that causes abortion in sheep. *Infect. Immun.* **66:**1317–1324.

Longbottom, D., M. Russell, G. E. Jones, F. A. Lainson, and A. J. Herring. 1996. Identification of a multigene family coding for the 90 kDa proteins of the ovine abortion subtype of *Chlamydia psittaci*. *FEMS Microbiol. Lett.* **142:**277–281.

Maslow, A. S., C. H. Davis, J. Choong, and P. B. Wyrick. 1988. Estrogen enhances attachment of *Chlamydia trachomatis* to human endometrial epithelial cells in vitro. *Am. J. Obstet. Gynecol.* **159:**1006–1014.

Matsumoto, A. 1982. Surface projections of *Chlamydia psittaci* elementary bodies as revealed by freeze-deep-etching. *J. Bacteriol.* **151:**1040–1042.

Mazar, J., and P. A. Cotter. 2006. Topology and maturation of filamentous haemagglutinin suggest a new model for two-partner secretion. *Mol. Microbiol.* **62:**641–654.

Menozzi, F. D., K. Pethe, P. Bifani, F. Soncin, M. J. Brennan, and C. Locht. 2002. Enhanced bacterial virulence through exploitation of host glycosaminoglycans. *Mol. Microbiol.* **43:**1379–1386.

Mitchell, C. M., K. M. Hovis, P. M. Bavoil, G. S. Myers, J. A. Carrasco, and P. Timms. 2010. Comparison of koala LPCoLN and human strains of *Chlamydia pneumoniae* highlights extended genetic diversity in the species. *BMC Genomics* **11:**442.

Moelleken, K., and J. H. Hegemann. 2008. The *Chlamydia* outer membrane protein OmcB is required for adhesion and exhibits biovar-specific differences in glycosaminoglycan binding. *Mol. Microbiol.* **67:**403–419.

Moelleken, K., E. Schmidt, and J. H. Hegemann. 2010. Members of the Pmp protein family of *Chlamydia pneumoniae* mediate adhesion to human cells via short repetitive peptide motifs. *Mol. Microbiol.* **78:**1004–1017.

Montigiani, S., F. Falugi, M. Scarselli, O. Finco, R. Petracca, G. Galli, M. Mariani, R. Manetti, M. Agnusdei, R. Cevenini, M. Donati, R. Nogarotto, N. Norais, I. Garaguso, S. Nuti, G. Saletti, D. Rosa, G. Ratti, and G. Grandi. 2002. Genomic approach for analysis of surface proteins in *Chlamydia pneumoniae*. *Infect. Immun.* **70:**368–379.

Myers, G. S., S. A. Mathews, M. Eppinger, C. Mitchell, K. K. O'Brien, O. R. White, F. Benahmed, R. C. Brunham, T. D. Read, J. Ravel, P. M. Bavoil, and P. Timms. 2009. Evidence that human *Chlamydia pneumoniae* was zoonotically acquired. *J. Bacteriol.* **191:**7225–7233.

Mygind, P., G. Christiansen, and S. Birkelund. 1998a. Topological analysis of *Chlamydia trachomatis* L2 outer membrane protein 2. *J. Bacteriol.* **180:**5784–5787.

Mygind, P., G. Christiansen, K. Persson, and S. Birkelund. 1998b. Analysis of the humoral immune response to *Chlamydia* outer membrane protein 2. *Clin. Diagn. Lab. Immunol.* **5:**313–318.

Mygind, P. H., G. Christiansen, P. Roepstorff, and S. Birkelund. 2000. Membrane proteins PmpG and PmpH are major constituents of *Chlamydia trachomatis* L2 outer membrane complex. *FEMS Microbiol. Lett.* **186:**163–169.

Niessner, A., C. Kaun, G. Zorn, W. Speidl, Z. Turel, G. Christiansen, A. S. Pedersen, S. Birkelund, S. Simon, A. Georgopoulos, W. Graninger, R. de Martin, J. Lipp, B. R. Binder, G. Maurer, K. Huber, and J. Wojta. 2003. Polymorphic membrane protein (PMP) 20 and PMP 21 of *Chlamydia pneumoniae* induce proinflammatory mediators in human endothelial cells in vitro by activation of the nuclear factor-kappaB pathway. *J. Infect. Dis.* **188:**108–113.

Nobbs, A. H., R. J. Lamont, and H. F. Jenkinson. 2009. *Streptococcus* adherence and colonization. *Microbiol. Mol. Biol. Rev.* **73:**407–450.

Nunes, A., J. P. Gomes, S. Mead, C. Florindo, H. Correia, M. J. Borrego, and D. Dean. 2007. Comparative expression profiling of the *Chlamydia trachomatis pmp* gene family for clinical and reference strains. *PLoS ONE* **2:**e878.

Pedersen, A. S., G. Christiansen, and S. Birkelund. 2001. Differential expression of Pmp10 in cell culture infected with *Chlamydia pneumoniae* CWL029. *FEMS Microbiol. Lett.* **203:**153–159.

Peterson, E. M., L. M. de la Maza, L. Brade, and H. Brade. 1998. Characterization of a neutralizing monoclonal antibody directed at the lipopolysaccharide of *Chlamydia pneumoniae*. *Infect. Immun.* **66:**3848–3855.

Pizarro-Cerda, J., and P. Cossart. 2006. Bacterial adhesion and entry into host cells. *Cell* **124:**715–727.

Portig, I., J. C. Goodall, R. L. Bailey, and J. S. Gaston. 2003. Characterization of the humoral immune response to *Chlamydia* outer membrane protein 2 in chlamydial infection. *Clin. Diagn. Lab. Immunol.* **10:**103–107.

Puolakkainen, M., C. C. Kuo, and L. A. Campbell. 2005. *Chlamydia pneumoniae* uses the mannose 6-phosphate/insulin-like growth factor 2 receptor for infection of endothelial cells. *Infect. Immun.* **73:**4620–4625.

Puolakkainen, M., A. Lee, T. Nosaka, H. Fukushi, C. C. Kuo, and L. A. Campbell. 2008. Retinoic acid inhibits the infectivity and growth of *Chlamydia pneumoniae* in epithelial and endothelial cells through different receptors. *Microb. Pathog.* **44:**410–416.

Raulston, J. E. 1995. Chlamydial envelope components and pathogen-host cell interactions. *Mol. Microbiol.* **15:**607–616.

Read, T. D., G. S. Myers, R. C. Brunham, W. C. Nelson, I. T. Paulsen, J. Heidelberg, E.

Holtzapple, H. Khouri, N. B. Federova, H. A. Carty, L. A. Umayam, D. H. Haft, J. Peterson, M. J. Beanan, O. White, S. L. Salzberg, R. C. Hsia, G. McClarty, R. G. Rank, P. M. Bavoil, and C. M. Fraser. 2003. Genome sequence of *Chlamydophila caviae* (*Chlamydia psittaci* GPIC): examining the role of niche-specific genes in the evolution of the *Chlamydiaceae*. *Nucleic Acids Res.* **31:**2134–2147.

Rockey, D. D., J. Lenart, and R. S. Stephens. 2000. Genome sequencing and our understanding of chlamydiae. *Infect. Immun.* **68:**5473–5479.

Rostand, K. S., and J. D. Esko. 1997. Microbial adherence to and invasion through proteoglycans. *Infect. Immun.* **65:**1–8.

Schroeder, T. H., M. M. Lee, P. W. Yacono, C. L. Cannon, A. A. Gerceker, D. E. Golan, and G. B. Pier. 2002. CFTR is a pattern recognition molecule that extracts *Pseudomonas aeruginosa* LPS from the outer membrane into epithelial cells and activates NF-kappa B translocation. *Proc. Natl. Acad. Sci. USA* **99:**6907–6912.

Shaw, A. C., K. Gevaert, H. Demol, B. Hoorelbeke, J. Vandekerckhove, M. R. Larsen, P. Roepstorff, A. Holm, G. Christiansen, and S. Birkelund. 2002. Comparative proteome analysis of *Chlamydia trachomatis* serovar A, D and L2. *Proteomics* **2:**164–186.

Shirai, M., H. Hirakawa, K. Ouchi, M. Tabuchi, F. Kishi, M. Kimoto, H. Takeuchi, J. Nishida, K. Shibata, R. Fujinaga, H. Yoneda, H. Matsushima, C. Tanaka, S. Furukawa, K. Miura, A. Nakazawa, K. Ishii, T. Shiba, M. Hattori, S. Kuhara, and T. Nakazawa. 2000. Comparison of outer membrane protein genes *omp* and *pmp* in the whole genome sequences of *Chlamydia pneumoniae* isolates from Japan and the United States. *J. Infect. Dis.* **181**(Suppl. 3)**:**S524–S527.

Skipp, P., J. Robinson, C. D. O'Connor, and I. N. Clarke. 2005. Shotgun proteomic analysis of *Chlamydia trachomatis*. *Proteomics* **5:**1558–1573.

Soderlund, G., and E. Kihlstrom. 1982. Physicochemical surface properties of elementary bodies from different serotypes of *Chlamydia trachomatis* and their interaction with mouse fibroblasts. *Infect. Immun.* **36:**893–899.

Soriani, M., P. Petit, R. Grifantini, R. Petracca, G. Gancitano, E. Frigimelica, F. Nardelli, C. Garcia, S. Spinelli, G. Scarabelli, S. Fiorucci, R. Affentranger, M. Ferrer-Navarro, M. Zacharias, G. Colombo, L. Vuillard, X. Daura, and G. Grandi. 2010. Exploiting antigenic diversity for vaccine design: the *Chlamydia* ArtJ paradigm. *J. Biol. Chem.* **285:**30126–30138.

Stephens, R. S., S. Kalman, C. Lammel, J. Fan, R. Marathe, L. Aravind, W. Mitchell, L. Olinger, R. L. Tatusov, Q. Zhao, E. V. Koonin, and R. W. Davis. 1998. Genome sequence of an obligate intracellular pathogen of humans: *Chlamydia trachomatis*. *Science* **282:**754–759.

Stephens, R. S., K. Koshiyama, E. Lewis, and A. Kubo. 2001. Heparin-binding outer membrane protein of chlamydiae. *Mol. Microbiol.* **40:**691–699.

Su, H., L. Raymond, D. D. Rockey, E. Fischer, T. Hackstadt, and H. D. Caldwell. 1996. A recombinant *Chlamydia trachomatis* major outer membrane protein binds to heparan sulfate receptors on epithelial cells. *Proc. Natl. Acad. Sci. USA* **93:**11143–11148.

Su, H., N. G. Watkins, Y. X. Zhang, and H. D. Caldwell. 1990. *Chlamydia trachomatis*-host cell interactions: role of the chlamydial major outer membrane protein as an adhesin. *Infect. Immun.* **58:**1017–1025.

Swanson, A. F., R. A. Ezekowitz, A. Lee, and C. C. Kuo. 1998. Human mannose-binding protein inhibits infection of HeLa cells by *Chlamydia trachomatis*. *Infect. Immun.* **66:**1607–1612.

Swanson, A. F., and C. C. Kuo. 1994. Binding of the glycan of the major outer membrane protein of *Chlamydia trachomatis* to HeLa cells. *Infect. Immun.* **62:**24–28.

Swanson, K. A., L. D. Taylor, S. D. Frank, G. L. Sturdevant, E. R. Fischer, J. H. Carlson, W. M. Whitmire, and H. D. Caldwell. 2009. *Chlamydia trachomatis* polymorphic membrane protein D is an oligomeric autotransporter with a higher-order structure. *Infect. Immun.* **77:**508–516.

Tan, C., R. C. Hsia, H. Shou, J. A. Carrasco, R. G. Rank, and P. M. Bavoil. 2010. Variable expression of surface-exposed polymorphic membrane proteins in *in vitro*-grown *Chlamydia trachomatis*. *Cell. Microbiol.* **12:**174–187.

Tan, C., R. C. Hsia, H. Shou, C. L. Haggerty, R. B. Ness, C. A. Gaydos, D. Dean, A. M. Scurlock, D. P. Wilson, and P. M. Bavoil. 2009. *Chlamydia trachomatis*-infected patients display variable antibody profiles against the nine-member polymorphic membrane protein family. *Infect. Immun.* **77:**3218–3226.

Tan, C., J. K. Spitznagel, H.-Z. Shou, R.-C. Hsia, and P. M. Bavoil. 2006. The polymorphic membrane protein gene family of the *Chlamydiaceae*, p. 195–218. *In* P. M. Bavoil and P. B. Wyrick (ed.), *Chlamydia Genomics and Pathogenesis*. Horizon Bioscience, Norfolk, United Kingdom.

Tanzer, R. J., and T. P. Hatch. 2001. Characterization of outer membrane proteins in *Chlamydia trachomatis* LGV serovar L2. *J. Bacteriol.* **183:**2686–2690.

Tanzer, R. J., D. Longbottom, and T. P. Hatch. 2001. Identification of polymorphic outer membrane proteins of *Chlamydia psittaci* 6BC. *Infect. Immun.* **69:**2428–2434.

Taraktchoglou, M., A. A. Pacey, J. E. Turnbull, and A. Eley. 2001. Infectivity of *Chlamydia trachomatis* serovar LGV but not E is dependent on host cell heparan sulfate. *Infect. Immun.* **69:**968–976.

Ting, L. M., R. C. Hsia, C. G. Haidaris, and P. M. Bavoil. 1995. Interaction of outer envelope proteins of *Chlamydia psittaci* GPIC with the HeLa cell surface. *Infect. Immun.* **63:**3600–3608.

Turnbull, J. E. 2010. Heparan sulfate glycomics: towards systems biology strategies. *Biochem. Soc. Trans.* **38:**1356–1360.

Turnbull, J. E., R. L. Miller, Y. Ahmed, T. M. Puvirajesinghe, and S. E. Guimond. 2010. Glycomics profiling of heparan sulfate structure and activity. *Methods Enzymol.* **480:**65–85.

Ulanova, M., S. Gravelle, and R. Barnes. 2008. The role of epithelial integrin receptors in recognition of pulmonary pathogens. *J. Innate Immun.* **1:**4–17.

Vance, D. W., Jr., and T. P. Hatch. 1980. Surface properties of *Chlamydia psittaci. Infect. Immun.* **29:**175–180.

Vandahl, B. B., S. Birkelund, H. Demol, B. Hoorelbeke, G. Christiansen, J. Vandekerckhove, and K. Gevaert. 2001. Proteome analysis of the *Chlamydia pneumoniae* elementary body. *Electrophoresis* **22:**1204–1223.

Vandahl, B. B., A. S. Pedersen, K. Gevaert, A. Holm, J. Vandekerckhove, G. Christiansen, and S. Birkelund. 2002. The expression, processing and localization of polymorphic membrane proteins in *Chlamydia pneumoniae* strain CWL029. *BMC Microbiol.* **2:**36.

Vretou, E., P. C. Goswami, and S. K. Bose. 1989. Adherence of multiple serovars of *Chlamydia trachomatis* to a common receptor on HeLa and McCoy cells is mediated by thermolabile protein(s). *J. Gen. Microbiol.* **135:**3229–3237.

Wadstrom, T., and A. Ljungh. 1999. Glycosaminoglycan-binding microbial proteins in tissue adhesion and invasion: key events in microbial pathogenicity. *J. Med. Microbiol.* **48:**223–233.

Wagels, G., S. Rasmussen, and P. Timms. 1994. Comparison of *Chlamydia pneumoniae* isolates by Western blot (immunoblot) analysis and DNA sequencing of the *omp2* gene. *J. Clin. Microbiol.* **32:**2820–2823.

Wang, A., S. C. Johnston, J. Chou, and D. Dean. 2010. A systemic network for *Chlamydia pneumoniae* entry into human cells. *J. Bacteriol.* **192:**2809–2815.

Watson, M. W., P. R. Lambden, J. S. Everson, and I. N. Clarke. 1994. Immunoreactivity of the 60 kDa cysteine-rich proteins of *Chlamydia trachomatis, Chlamydia psittaci* and *Chlamydia pneumoniae* expressed in *Escherichia coli. Microbiology* **140:**2003–2011.

Wehrl, W., V. Brinkmann, P. R. Jungblut, T. F. Meyer, and A. J. Szczepek. 2004. From the inside out—processing of the chlamydial autotransporter PmpD and its role in bacterial adhesion and activation of human host cells. *Mol. Microbiol.* **51:**319–334.

Wells, T. J., M. Totsika, and M. A. Schembri. 2010. Autotransporters of *Escherichia coli*: a sequence-based characterization. *Microbiology* **156:**2459–2469.

Wells, T. J., J. J. Tree, G. C. Ulett, and M. A. Schembri. 2007. Autotransporter proteins: novel targets at the bacterial cell surface. *FEMS Microbiol. Lett.* **274:**163–172.

Wuppermann, F. N., J. H. Hegemann, and C. A. Jantos. 2001. Heparan sulfate-like glycosaminoglycan is a cellular receptor for *Chlamydia pneumoniae. J. Infect. Dis.* **184:**181–187.

Wuppermann, F. N., K. Moelleken, M. Julien, C. A. Jantos, and J. H. Hegemann. 2008. *Chlamydia pneumoniae* GroEL1 protein is cell surface associated and required for infection of HEp-2 cells. *J. Bacteriol.* **190:**3757–3767.

Wyllie, S., R. H. Ashley, D. Longbottom, and A. J. Herring. 1998. The major outer membrane protein of *Chlamydia psittaci* functions as a porin-like ion channel. *Infect. Immun.* **66:**5202–5207.

Wyrick, P. B., J. Choong, C. H. Davis, S. T. Knight, M. O. Royal, A. S. Maslow, and C. R. Bagnell. 1989. Entry of genital *Chlamydia trachomatis* into polarized human epithelial cells. *Infect. Immun.* **57:**2378–2389.

Yabushita, H., Y. Noguchi, H. Habuchi, S. Ashikari, K. Nakabe, M. Fujita, M. Noguchi, J. D. Esko, and K. Kimata. 2002. Effects of chemically modified heparin on *Chlamydia trachomatis* serovar L2 infection of eukaryotic cells in culture. *Glycobiology* **12:**345–351.

Yan, Y., S. Silvennoinen-Kassinen, M. Leinonen, and P. Saikku. 2006. Inhibitory effect of heparan sulfate-like glycosaminoglycans on the infectivity of *Chlamydia pneumoniae* in HL cells varies between strains. *Microbes Infect.* **8:**866–872.

Zaretzky, F. R., R. Pearce-Pratt, and D. M. Phillips. 1995. Sulfated polyanions block *Chlamydia trachomatis* infection of cervix-derived human epithelia. *Infect. Immun.* **63:**3520–3526.

Zhang, J. P., and R. S. Stephens. 1992. Mechanism of *C. trachomatis* attachment to eukaryotic host cells. *Cell* **69:**861–869.

INITIAL INTERACTIONS OF CHLAMYDIAE WITH THE HOST CELL

Ted Hackstadt

6

INTRODUCTION

The initial stages of interaction between chlamydial elementary bodies (EBs) and the eukaryotic host cell are critical to the establishment of a productive infection. Within the first few minutes to hours of contact of an EB with a susceptible host cell, chlamydiae initiate multiple processes that can determine the success or failure of this parasitic interaction. Even prior to chlamydial transcription and translation, host cellular signal transduction pathways are triggered and fusion with endocytic compartments is inhibited. The supposedly metabolically dormant EBs respond to unknown environmental signals by the initiation of transcription and translation. Among the early gene products are several proteins, believed to be type III secreted, that extensively modify the inclusion membrane. Once modified by de novo-synthesized chlamydial proteins, a number of interactions with cellular organelles and compartments that effectively define the properties of the chlamydial inclusion are established.

Ted Hackstadt, Host-Parasite Interactions Section, Laboratory of Intracellular Parasites, Rocky Mountain Laboratories, NIAID, NIH, 903 South 4th Street, Hamilton, MT 59840, USA.

ENTRY

Chlamydiae are taken up by host cells so efficiently that the process has been termed "parasite-specified endocytosis" by Byrne and Moulder (Byrne and Moulder, 1978). Despite the efficiency of the process, the attachment and entry phases of chlamydial interactions with the host cell have been among the most controversial. As described in the preceding chapter (chapter 5, "Chlamydial adhesion and adhesins"), multiple EB surface ligands and cognate host cell receptors have been described. It appears that chlamydiae employ an entry mechanism that involves several distinct levels of interaction. For the purpose of this chapter, we will focus upon the active interactions of chlamydiae with the host cell and cover the stage of development up to the division of reticulate bodies (RBs).

Based upon mutant cell lines selected for resistance to chlamydial infection, attachment has been thought to occur in at least two stages, a reversible and probably electrostatic interaction followed by a subsequent, irreversible stage. Surprisingly, different mutant cell lines displayed unique patterns of susceptibility or resistance to different *C. trachomatis* serovars or to *C. pneumoniae* (Carabeo and Hackstadt, 2001; Fudyk et al., 2002). Subsequently,

Intracellular Pathogens I: Chlamydiales
Edited by Ming Tan and Patrik M. Bavoil © 2012 ASM Press, Washington, DC
10.1128/9781555817329.ch6

protein disulfide isomerase was identified as defective in one of the mutant cell lines and shown to function in both attachment, as part of a complex, and internalization, for which enzymatic activity is required, possibly for reduction of the disulfide cross-linked outer membrane complex of EBs (Davis and Wyrick, 1997; Conant and Stephens, 2007; Abromaitis and Stephens, 2009). The nature of the irreversible stage of attachment remains unclear.

Because they are obligate intracellular parasites, uptake by eukaryotic cells is essential for survival of chlamydiae. As such, there is a widely held belief that chlamydiae may utilize more than one mechanism to gain access to host cells. Indeed, despite the difficulties of comparing studies between different chlamydial species or serovars, conditions, and cell types, there is sufficient evidence to suggest that individual chlamydiae are capable of utilizing different mechanisms for entry (Moulder, 1991; Scidmore et al., 1996b; Hackstadt, 1999). In any case, active participation of the host cell in the requirement for cytoskeletal rearrangements is apparent. Recent studies have focused primarily on the role of the actin cytoskeleton, although earlier studies have provided compelling evidence for participation of a microfilament-independent uptake mechanism, which is deserving of further study (Prain and Pearce, 1989).

Tarp ACTIN NUCLEATING ACTIVITY

Many intracellular pathogenic bacteria utilize a specialized secretion mechanism known as a type III secretion (T3S) system to directly inject bacterial effector proteins into the cytosol of eukaryotic cells (Hueck, 1998). Several of these effectors interact with the host cytoskeleton to promote, or inhibit, bacterial internalization. It appears that chlamydiae also use a T3S effector protein to promote entry. *C. trachomatis* encodes a T3S protein called Tarp (acronym for translocated actin recruiting phosphoprotein), which is tyrosine phosphorylated by host cell kinase(s) and is spatially and temporally associated with the recruitment of actin at the site of EB invasion (Clifton et al., 2004). Tarp is present in all pathogenic *Chlamydia* species examined to date. However, analysis of Tarp orthologs from *Chlamydia trachomatis*, *Chlamydia muridarum*, *Chlamydia caviae*, and *Chlamydia pneumoniae* indicates that only *C. trachomatis* Tarp is phosphorylated despite all *Chlamydia* strains demonstrating the recruitment of actin to the site of entry (Clifton et al., 2005).

EBs have a preformed T3S apparatus (Fields et al., 2003) but have been thought to display little metabolic or transcriptional activity (Hatch et al., 1985; Moulder, 1991). Tarp exists in EBs in an unphosphorylated state and is not exposed on the surface of EBs. Translocation occurs upon contact with host cells even in the presence of inhibitors of bacterial transcription or translation (Clifton et al., 2004). Thus, chlamydial T3S may be activated to secrete in the absence of gene expression. EBs have an unusually high ATP content of about 40 mM (Tipples and McClarty, 1993), which is presumably necessary for early events including activation of T3S. Incubation at 4°C inhibits Tarp secretion either into cells (Clifton et al., 2004) or by cell-free inducers of T3S (Jamison and Hackstadt, 2008). Paraformaldehyde fixation of chlamydial EBs inhibits induction of Tarp secretion by in vitro stimulators of T3S as well as by cellular contact but does not appreciably reduce attachment (Jamison and Hackstadt, 2008). Invasion assays corroborated these results, as paraformaldehyde-fixed EBs did not become internalized. The data support the necessity for Tarp secretion from EBs for entry, but not for attachment.

The signals triggering Tarp secretion are unclear, although, typically, contact of a bacterium with the host cell activates T3S (Hueck, 1998). A variety of environmental conditions have been identified as capable of inducing the T3S system in gram-negative bacteria in a cell-free environment (Bahrani et al., 1997; Lee et al., 2001; Rivera-Amill et al., 2001; van der Goot et al., 2004; Hayward et al., 2005). An analysis of potential stimulatory conditions for induction of the chlamydial T3S system indicated that sphingolipid- and cholesterol-rich liposomes, fetal bovine serum, and bovine serum albumin induced secretion of Tarp from EBs in a temperature-dependent fashion (Jamison and Hackstadt, 2008). As has been suggested for *Shigella* (van der Goot et al., 2004), this may suggest involvement of ordered lipid domains in the induction of the chlamydial T3S system. Indeed, lipid rafts have been implicated in the internalization of some chlamydial strains (Stuart et al., 2003), although the specificity of the treatments used to deplete lipid rafts has been questioned (Gabel et al., 2004). A role for lipid rafts in activation of chlamydial T3S has not been directly examined. Demonstration of a requirement for lipid rafts to trigger Tarp secretion may help reconcile some of these disparate findings.

Comparison of the sequenced Tarp orthologs reveals a surprising diversity of isomers. Tarp displays at least three distinct domains including an actin binding domain with structural and primary amino acid sequence and structural similarity to Wiskott-Aldrich syndrome protein homology 2 (Wb) domain family proteins. The number of Wb domains in the Tarp orthologs varies from one to four depending upon the strain or species (Lutter et al., 2010). In addition, a single proline-rich domain is necessary to promote Tarp oligomerization and is required for Tarp-dependent nucleation of new actin filaments. *C. trachomatis* Tarp also bears one to nine tyrosine-rich repeat units, of approximately 50 amino acid residues each, which are the site of tyrosine phosphorylation. *C. pneumoniae*, *C. caviae*, and *C. muridarum* Tarp proteins lack the tyrosine-rich repeat domain and are not tyrosine phosphorylated (Clifton et al., 2005).

Both monomeric actin (G-actin) and filamentous actin (F-actin) associate with Tarp directly, and this binding occurs independently of Tarp phosphorylation. In pyrene actin polymerization assays, Tarp induces rapid and dose-dependent actin polymerization in the absence of additional host or chlamydial factors (Jewett et al., 2006). The data suggest that Tarp independently nucleates new actin filaments by a mechanism distinct from known actin nucleators (Qualmann and Kessels, 2009). The minimal actin binding domain of Tarp has been localized to an ~100-amino-acid fragment. Interestingly, this minimal actin binding domain binds G-actin but does not promote polymerization in in vitro assays. A proline-rich domain appears to be required in conjunction with the actin binding domain to allow oligomerization of Tarp and stimulation of actin nucleation (Jewett et al., 2006). This ability to oligomerize is hypothesized to bring multiple actin subunits together to nucleate a new actin filament in juxtaposition with the surface-associated EBs.

The multiple functional actin binding domains of some Tarp orthologs suggest the possibility of a hybrid actin nucleation mechanism involving multiple Wb domains on a single peptide. The presence of multiple Wb domains could potentially allow one Tarp protein to align multiple actin monomers together to initiate an actin filament. Tarp fragments harboring multiple actin binding domains nucleated actin in vitro even in the absence of the proline-rich oligomerization domain (Jewett et al., 2010). However, the proline-rich domain appears to be conserved and thus likely plays an essential role in vivo even in strains bearing multiple Wb domains (Lutter et al., 2010). The biological advantage to strains bearing different numbers of Tarp Wb domains is unclear. Intrinsic differences in Tarp structure are, perhaps, not surprising given that *Chlamydia* spp. differ both clinically and biologically, demonstrating diverse tissue tropisms and varying degrees of localized and systemic infection. Like many chlamydial functions, confirmation of an essential role for Tarp and defining the advantages of multiple Wb domains may await the ability to genetically modify chlamydiae before a definitive answer is possible.

TYROSINE PHOSPHORYLATION OF Tarp AND ACTIVATION OF CELL SIGNALING

Translocation of Tarp and its tyrosine phosphorylation appear to be one of the first means of communication with the host cell to actively subvert host processes for parasite purposes. A number of secreted effectors from pathogenic bacteria such as enteropathogenic *Escherichia coli*, *Citrobacter* spp., *Helicobacter pylori*, and *Bartonella henselae* have been shown to be phosphorylated by host cell tyrosine kinases shortly after translocation into susceptible target cells (Backert and Selbach, 2005). Dissecting the signal transduction pathways that initiate these events is crucial to our understanding of infectious diseases.

The number of tyrosine-rich repeat units in Tarp varies, with one to nine repeat units present in Tarp from *C. trachomatis* serovars representative of different diseases and tissue tropisms. A comparison of Tarp structures from laboratory strains and clinical isolates representative of 16 serovars of *C. trachomatis* taken from various geographical locations and anatomical sites showed that the greatest variation was in the number of tyrosine-rich repeat units and in the number of putative Wb domains (Lutter et al., 2010). The trachoma biovars tended toward fewer tyrosine-rich repeats and more Wb domains, while the lymphogranuloma venereum (LGV) biovar displayed a greater number of tyrosine-rich repeats and fewer Wb domains.

Each tyrosine-rich repeat unit contains at least one Src-like consensus target, and some harbor two overlapping Src-like consensus targets (Backert and Selbach, 2005; Jewett et al., 2008; Lane et al., 2008; Mehlitz et al., 2008). Mutational analysis of a single tyrosine-rich repeat harboring overlapping Src-like target sequences indicated that both tyrosine residues can be phosphorylated by HeLa extracts or purified Src kinase (Jewett et al., 2008). The significance of overlapping kinase sites remains unknown, but differential phosphorylation could potentially lead to unique interactions. Additionally, it is unclear if all potential sites are phosphorylated in vivo or if the different repeats are preferentially phosphorylated by a specific kinase family member in different cell types or developmental stages. Identification of the kinases mediating Tarp phosphorylation is an initial step in mapping the signal transduction networks initiated by *C. trachomatis* to establish residence within an intracellular niche. Purified Src, Yes, and Fyn, which are members of the Src family kinases, as well as Abl and Syk kinases are capable of phosphorylating Tarp (Elwell et al., 2008; Jewett et al., 2008; Mehlitz et al., 2008). Kinase redundancy is not unusual in microbes having tyrosine phosphorylated effectors. *Chlamydia* species other than *C. trachomatis* harbor Tarp orthologs that are not phosphorylated upon entry (Clifton et al., 2005) and therefore are presumed to initiate alternate signaling pathways.

Tarp is believed to be required for internalization, but the role of tyrosine phosphorylation in the entry process per se is more controversial. One study has proposed a model in which phosphorylated Tarp, by virtue of its Src-homology 2 (Sb) domain, directly or indirectly activates two guanine nucleotide exchange factors (GEFs), Vav2 and Sos1/Abi1/Eps8, respectively, to trigger Rac activation, which is known to be required for chlamydial internalization (Lane et al., 2008). The p85 subunit of phosphoinositide-3-kinase (PI3K) is recruited to the Sb domain of Tarp phosphotyrosine and activated to produce phosphatidylinositol-3,4,5-triphosphate [PI(3,4,5)P_3], which is in turn proposed to activate Vav2. The multiprotein complex Sos1/Abi1/Eps8 is hypothesized to be recruited via the Abi1 subunit of the complex to Tarp phosphotyrosine as an alternative means to recruit GEFs and thus activate Rac1. The redundant pathways to activate GEFs are proposed to promote the function of this signaling pathway under conditions of variable PI(3,4,5)P_3 availability (Lane et al., 2008).

Tarp appears to be somewhat promiscuous in its phosphorylation by different host kinases (Elwell et al., 2008; Jewett et al., 2008; Mehlitz et al., 2008). Because of this promiscuity, tyrosine phosphorylation of Tarp had been difficult to inhibit by standard inhibitors with broad specificity such as genistein (Clifton et al., 2004; Mehlitz et al., 2008). However, one kinase inhibitor, PP2 [4-amino-5-(4-chlorophenyl)-7-(*t*-butyl)pyrazolo-[3,4-d]pyrimidine], was found to inhibit Tarp phosphorylation in L929 fibroblasts (Jewett et al., 2008). This may be due to the fact that at high concentrations, PP2 has been shown to inhibit tyrosine kinases other than Src and Abl. PP2 reduced Tarp phosphorylation to below levels of detection without demonstrable reduction in internalization, thus indicating that Tarp phosphorylation is not required for bacterial entry (Jewett et al., 2008). A similar conclusion was reached by Mehlitz et al., using Src, Yes, Fyn-deficient cells further inhibited with specific inhibitors of Abl and Syk. In these experiments, phosphorylation of Tarp was reduced to about 20% of control levels with minimal effects on inclusion formation (Mehlitz et al., 2008). Therefore, if Tarp phosphorylation is required for *C. trachomatis* entry, it may be necessary only during specific conditions or infection of certain cell types.

What then is the function of *C. trachomatis* Tarp phosphorylation, and how are these requirements circumvented in chlamydial species lacking phosphotyrosine on Tarp? Although the specific mechanisms differ, enteropathogenic *E. coli* Tir is an appropriate analogy. Signaling for Nck recruitment to enteropathogenic *E. coli* attachment sites requires tyrosine phosphorylation of Tir, but the orthologous enterohemorrhagic *E. coli* Tir is not tyrosine phosphorylated and uses alternative signaling pathways (DeVinney et al., 2001; Gruenheid et al., 2001). Certainly, alternate signaling pathways are possible for Tarp, and there is increasing evidence for this. *C. trachomatis* Tarp binds multiple host proteins containing Sb domains with affinity for the phosphotyrosine residues on Tarp (Mehlitz et al., 2010). Notable among these is SHC1, which has been shown to play a role in early resistance to apoptosis (Mehlitz et al., 2010). Other chlamydial factors have been shown to function in the inhibition of apoptosis that occurs after about 16 h postinfection. The chlamydial protease/proteasome-like activity factor, CPAF (Zhong et al., 2001), is necessary and sufficient for the degradation of Bc only proteins, and thus likely plays a major role in the inhibition of apoptosis in *Chlamydia*-infected cells through the middle and late stages of chlamydial development (Pirbhai et al., 2006). Other chlamydial species besides *C. trachomatis* that have a phosphorylated Tarp are predicted to similarly inhibit apoptosis early, although this has not been demonstrated and would be expected to be triggered though distinct pathways. With the many Sb domain-containing proteins binding *C. trachomatis* Tarp, the redundancy of tyrosine kinases, and potential for cross talk between signaling pathways, a complete understanding of essential pathways and those elicited only under specific circumstances may be a challenge to unravel.

Analysis of *C. pneumoniae* requirements for entry have identified an involvement of the MEK/ERK1/2 pathway. A MEK-dependent phosphorylation of ERK1/2 leads to a PI3K- and FAK-dependent activity and activation of AKT. Activation of the adaptor protein SHC1 is also observed. Interestingly, SHC1 activation by *C. trachomatis* has been linked to tyrosine phosphorylation of Tarp (Mehlitz et al., 2010). Activation of SHC1 via the MEK/ERK pathway in *C. pneumoniae*-infected cells may indicate activation of an essential pathway by an alternative mechanism by a species whose Tarp is not phosphorylated. Inhibition of entry of *C. pneumoniae* EBs by specific inhibitors of MEK or PI3K suggests that these kinases have an essential role in entry. Inhibition of these pathways also blocks the transient microvillar hypertrophy associated with internalization of chlamydiae (Carabeo et al., 2002; Coombes and Mahony, 2002).

OTHER EARLY T3S EFFECTORS

It is unlikely that chlamydiae secrete only a single T3S effector during their initial contact with host cells. At least one other T3S effector, CT694, has been shown to interact with the cellular actin binding protein AHNAK (Hower et al., 2009). CT694 is present in EBs and is secreted by 1 hour postinfection in a pattern similar to that observed with Tarp, in which it appears adjacent to, but does not overlap, internalized EBs. Based upon the disruption of actin filaments in cells ectopically expressing CT694 and the known interactions of AHNAK with actin, a role in modulation of the actin cytoskeleton has been predicted (Hower et al., 2009). This could possibly function in the disruption of actin filaments during the transient Tarp-mediated actin cytoskeletal remodeling that accompanies entry. Other functions, such as a role in maintaining the actin scaffolds thought to support inclusion structure (Kumar and Valdivia, 2008), are possible. It is probable that a battery of proteins expressed later in development are retained in EBs for immediate translocation upon contact, although their identity and function remain to be determined.

TOXINS

C. muridarum encodes three full copies of a gene with similarities to the large clostridial toxins (Read et al., 2000). In contrast, *C. caviae* encodes a single copy and *C. trachomatis* encodes only a single partial copy of the cytotoxin (Read et al., 2003; Carlson et al., 2004). These chlamydial cytotoxins are among the few genes known to vary with chlamydial species and tissue tropism. *C. trachomatis* urogenital strains encode both the predicted UDP-glucose binding and glucosyltransferase domains of the toxin, while ocular strains, with the exception of serovar B, encode only the UDP-binding domain. LGV strains are deficient in both of the active sites and are the only serovars nontoxic to cells at high multiplicities of infection (Belland et al., 2001; Carlson et al., 2004). The cytotoxins are thus thought to be a factor in tissue tropism or pathogenesis. The toxin appears to play a major role in the resistance of *C. muridarum* to gamma interferon-induced p47 GTPases found in murine cells (Nelson et al., 2005; Coers et al., 2008). Although the mechanism of translocation has not yet been defined, the cytotoxin is present in EBs and translocated into cells very early during infection.

ACTIVATION OF THE HOST ACTIN NUCLEATION MACHINERY

Although Tarp clearly has independent actin-nucleating activities and may be involved in activation of signaling pathways, it is apparent that Tarp alone is insufficient to promote internalization of EBs. Entry of chlamydiae is also dependent on signal transduction cascades initiated following activation of host Rho family GTPases and culminating in the activation of the Arp2/3 complex, which is a host actin nucleator (Carabeo et al., 2004; Subtil et al., 2004). Therefore, it has been proposed that the chlamydial and cellular actin-nucleating activities function in concert to promote chlamydial invasion. Tarp nucleates the formation of linear actin filaments (Jewett et al., 2006), whereas host Arp2/3 nucleates actin filaments that grow from the side of existing linear filaments (Robinson et al., 2001), and the two mechanisms may thus function synergistically. Although the role for Tarp phosphorylation in entry and the recruitment of Arp2/3 is somewhat controversial, Tarp proteins secreted from *C. pneumoniae* and *C. caviae* are not phosphorylated and therefore activate Arp2/3 by an alternate signaling pathway. The signaling cascades initiated by the different chlamydiae appear to be species specific. *C. trachomatis* invasion requires Rac GTPase, while *C. caviae* requires both Rac and Cdc42 (Carabeo et al., 2004, 2007; Subtil et al., 2004).

The Rac/WAVE2/Abi1/Arp2/3 pathway required for *C. trachomatis* internalization is thought to function synergistically with the actin-nucleating activity of Tarp (Fig. 1). The Rac GTPase protein is central to this pathway and acts on Abi-1 and WAVE2 to activate Arp2/3-dependent actin recruitment and polymerization. However, the mechanism by which chlamydiae activate Rac is less clear. As described above, *C. trachomatis* Tarp phosphorylation may lead to GEF activation. Another possibility is that binding to a secondary receptor may be necessary for Rac activation, microvillar hypertrophy, and pedestal formation, but whether Rac activation is a direct consequence of receptor binding akin to *Listeria* InlB signaling is not clear. Identification of components upstream of Rac in this signal transduction pathway is expected to shed light on this problem and add to the

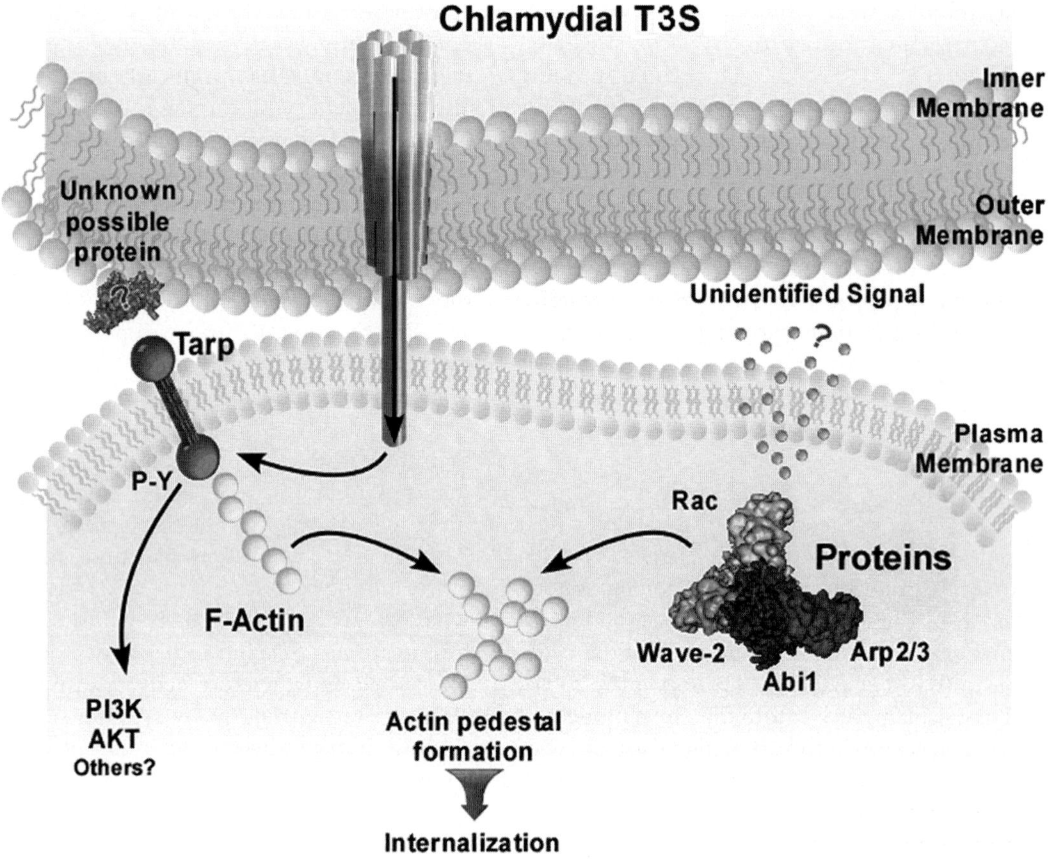

FIGURE 1 Model of cooperative cell signaling during the internalization of *C. trachomatis*. Upon contact with the eukaryotic host cell, Tarp is translocated from EBs by the chlamydial T3S system and exposed to the cytosol. *C. trachomatis* Tarp, but not that of other species, is tyrosine phosphorylated by host kinases. A variety of host proteins display affinity for tyrosine phosphorylated Tarp. These include those potentially acting as GEFs for activation of Rac as well as those functioning in cellular signaling pathways regulating apoptosis and other regulatory cascades. Tarp is believed to independently nucleate linear actin filament formation. Rac activation, and CDC42 for *C. caviae*, is required for activation of the cellular actin nucleating complex, Arp2/3. Tarp and Arp2/3 are proposed to function synergistically to promote the actin cytoskeletal rearrangements promoting chlamydial internalization. The mechanisms of Rac and/or CDC42 activation by those species whose Tarp is not phosphorylated, or when *C. trachomatis* Tarp phosphorylation is inhibited, are undefined. Tarp is not predicted to be a membrane protein but remains associated with the nascent inclusion membrane for several hours postinfection. The means for long-term retention at the inclusion membrane is unknown. doi:10.1128/9781555817329.ch6.f1

rapidly growing list of molecular mechanisms of actin cytoskeleton remodeling by intracellular pathogens.

CREATION OF AN INTRACELLULAR NICHE

Chlamydiae undergo their entire intracellular developmental cycle within a parasitophorous vacuole, termed an inclusion, that is unique among intracellular parasites (Moulder, 1991; Hackstadt, 1999, 2000; Wyrick, 2000; Fields and Hackstadt, 2002). Chlamydiae are endocytosed into a tightly membrane-bound vesicle that grows throughout the developmental cycle to accommodate an increasing number of intracellular bacteria. The chlamydial inclusion displays no markers of the endocytic or lysosomal pathway and thus had been considered among those intracellular parasites that occupy "nonfusogenic" vacuoles (Fig. 2). Rather than being strictly nonfusogenic with host vesicular trafficking pathways, the chlamydial inclusion intercepts sphingomyelin and cholesterol from an apparent exocytic pathway that delivers these lipids from the Golgi apparatus to the plasma membrane. Although all species of *Chlamydia* intersect this pathway, no other intracellular parasites have yet been found to similarly interact with this host vesicular trafficking pathway. Sequestration of chlamydiae within a vesicle that intersects an exocytic pathway is hypothesized to provide a unique, protected intracellular niche in which the chlamydiae replicate (Hackstadt et al., 1997).

TRIGGERING OF EARLY DIFFERENTIATION

Once internalized, EBs are triggered via unknown signals to initiate development. The DNA of EBs is held in a condensed state by two histone H1 homologs, Hc1 and Hc2 (Wagar and Stephens, 1988; Hackstadt et al., 1991; Tao et al., 1991; Perara et al., 1992; Brickman et al., 1993). Both *hctA* and *hctB*,

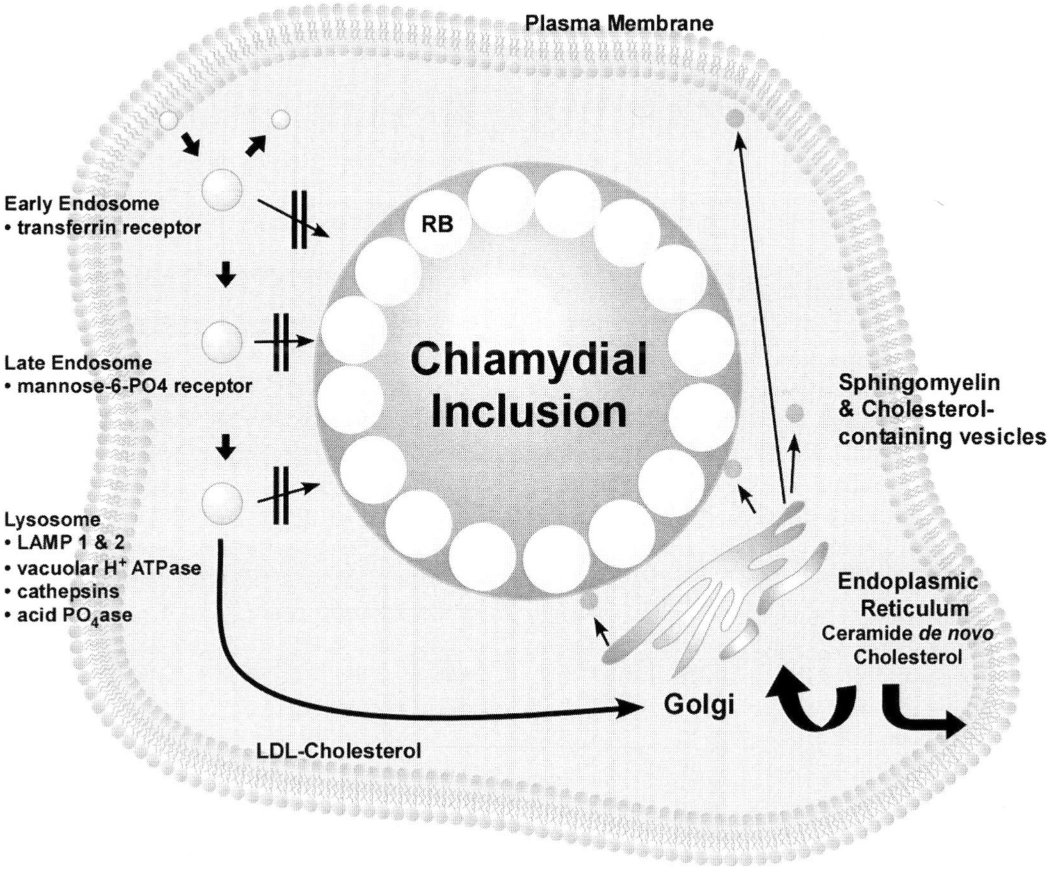

FIGURE 2 Model of the vesicular interactions of the mature chlamydial inclusion. Markers for the plasma membrane and fluid phase and early and late endosomal and lysosomal markers are absent from the inclusion membrane. Instead, chlamydiae intercept sphingolipids and cholesterol directly from the Golgi apparatus. All chlamydial species interact with cells similarly. These interactions require modification of the inclusion membrane by chlamydial protein(s) and are initiated by 2 h postinfection. doi:10.1128/9781555817329.ch6.f2

encoding Hc1 and Hc2, respectively, are transcribed late in the developmental cycle concomitant with RB differentiation back to EBs and nucleoid condensation. Although EBs had long been considered to exhibit little to no metabolic activity, they clearly have the ability to respond to unknown signals intracellularly and initiate growth.

Differentiation of EBs to RBs is accompanied by dispersal of the chromatin as chlamydiae become transcriptionally active. Paradoxically, dissociation of the nucleoid requires chlamydial transcription and translation with negligible proteolysis or degradation of Hc1 (Grieshaber et al., 2004). Expression of Hc1 in *E. coli* is effectively lethal, presumably because *E. coli* lacks the means to release the histone-DNA complex (Barry et al., 1992). Two separate loci on the chlamydial genome that rescue *E. coli* from the growth restrictions imposed by Hc1 have been identified. The first locus, CT804, encodes IspE, an intermediate enzyme of the nonmevalonate methylerythritol 4-phosphate (MEP) pathway of isoprenoid biosynthesis (Grieshaber et al., 2004). The second encodes a small regulatory RNA that acts as an additional checkpoint to negatively regulate Hc1 synthesis (Grieshaber et al., 2006). *E. coli* bacteria coexpressing IspE and Hc1 grow normally, although they express Hc1 to levels equivalent to that which condensed the chromatin of parental Hc1-expressing controls. This situation is equivalent to that of germinating EBs in which the histone remains present but no longer functions in condensation of the chromatin. Inhibition of the MEP pathway with fosmidomycin, a herbicide that targets an upstream step in the MEP pathway, abolished IspE rescue of Hc1-expressing *E. coli*. Deproteinated extract from IspE-expressing bacteria caused dispersal of purified chlamydial nucleoids, suggesting that chlamydial histone-DNA interactions are disrupted by a small metabolite within the MEP pathway rather than by direct action of IspE (Grieshaber et al., 2004). It appears that 2*C*-methyl-erythritol-2,4-cyclodiphosphate dissociates Hc1 from chlamydial chromatin.

TEMPORAL GENE EXPRESSION DURING THE DEVELOPMENTAL CYCLE

The chlamydial developmental cycle consists of a programmed series of events in which the chlamydiae are sequentially triggered to express a cascade of chlamydial gene products in response to changing intracellular microenvironments encountered throughout their developmental cycle. Maturation occurs through the interplay of host cell activities and modulation of those activities by the chlamydiae. Analysis of temporal gene expression patterns throughout the chlamydial developmental cycle have identified at least three temporal classes of developmentally expressed genes. In *C. trachomatis* L2, these have been defined as early (initiated by 2 h postinfection); midcycle (initiated by 10 to 12 h postinfection); and late (initiated by 16 h postinfection) (Shaw et al., 2000; Belland et al., 2003). The mechanisms that regulate this temporal gene expression are discussed in chapter 7, "Temporal gene regulation during the chlamydial developmental cycle."

ESTABLISHING CELLULAR INTERACTIONS OF THE INCLUSION

EBs are initially taken up into a plasma membrane-derived vesicle that is not fusogenic to a demonstrable degree with compartments of the endocytic pathway. The membrane appears to be tightly associated with the endocytosed EBs. During the first hour postinfection, there is a demonstrated absence of endocytic and lysosomal markers within the inclusion membrane or lumen of the inclusion to suggest that the nascent chlamydial inclusion is minimally interactive with endosomal compartments during this interval early in infection (Scidmore et al., 2003). Even when prevented from modifying the properties of the inclusion by incubation in the presence of protein synthesis inhibitors, vesicles containing EBs are very slow to acquire lysosomal characteristics. Whereas over 80% of endocytosed *Yersinia pseudotuberculosis* was observed in LAMP1-positive vesicles by 1 h postinfection, vesicles containing chloramphenicol-inhibited EBs showed a half-time for acquisition of this lysosomal marker of approximately 12 to 24 h.

Establishment of the vacuolar properties of the mature inclusion is dependent upon chlamydial modification of the vesicle and appears less dependent upon the route of internalization (Scidmore et al., 1996b). Within a few hours following endocytosis, the properties of the nascent chlamydial inclusion change dramatically. Endocytosed EBs are rapidly transported to the peri-Golgi region of the host cell and become fusogenic with Golgi-derived vesicles. Transport to the peri-Golgi region and acquisition of sphingomyelin are dependent on early chlamydial protein synthesis, suggesting that chlamydiae actively modify the inclusion membrane in order to intersect an exocytic pathway. In the absence of chlamydial protein synthesis, the endocytosed EBs are ultimately degraded within lysosomes (Scidmore et al., 1996b).

These results imply a two-stage mechanism for chlamydial avoidance of lysosomal fusion: (i) an initial phase of delayed maturation to lysosomes due to intrinsic activities or structural components of EBs; and (ii) an active modification of the vesicular interactions of the inclusion requiring chlamydial protein synthesis.

PROTEINS MODIFYING THE INCLUSION MEMBRANE

Chlamydial early protein synthesis is required to promote microtubule-dependent intracellular trafficking of the nascent inclusion to the peri-Golgi region of the host cell and to establish fusogenicity with sphingomyelin and cholesterol-containing vesicles (Fields and Hackstadt, 2002). Chlamydiae, like virtually all other intracellular pathogens, alter the fusion competence of only the vacuole they occupy rather than cause a global interference with vesicular trafficking (Hackstadt, 2000). A likely means of specifically inhibiting lysosomal maturation of the inclusion without interruption of lysosomal fusion throughout the remainder of the cell would be the modification of the inclusion membrane by the insertion of chlamydial products. A large number of chlamydial proteins are now known to be inserted in the inclusion membrane such that they are exposed to the cytosol, and the majority of these are synthesized early in the developmental cycle. These inclusion membrane proteins, or Incs, are thus candidates for factors controlling the cellular interactions of the nascent chlamydial inclusion.

Chlamydial proteins modifying the inclusion membrane were first identified in *C. caviae* (Rockey and Rosquist, 1994; Rockey et al., 1995; Bannantine et al., 1998). The inclusion membrane protein IncA is exposed on the cytoplasmic face of the inclusion membrane and phosphorylated by the host cell (Rockey et al., 1997). IncB and C are similarly localized to the inclusion membrane. Incs from *C. trachomatis* and other chlamydial species were soon identified. All known chlamydial Inc proteins share a common structural feature of a predicted hydrophobic domain of 40 or more amino acid residues (Bannantine et al., 2000). Despite this shared structural feature, there is little or no primary sequence similarity even within the predicted hydrophobic domains. Searches of the *C. trachomatis* genome have revealed upwards of 40 proteins bearing this characteristic feature (Bannantine et al., 2000; Shaw et al., 2000; Rockey et al., 2002; Li et al., 2008; Dehoux et al., 2011). Approximately one-half that number of such proteins have now been confirmed as localized to the inclusion membrane by the development and application of specific antibodies. Of 39 confirmed or putative *C. trachomatis* Inc proteins, approximately 70% were shown to be transcribed early in infection (Shaw et al., 2000). Several Incs that have been examined in heterologous systems were found to be secreted by T3S systems (Fields and Hackstadt, 2000; Subtil et al., 2001; Fields et al., 2003); thus, it is widely assumed that most Incs represent T3S effectors. Despite their presumed importance in controlling interactions with the host cell, functions of the Inc proteins are for the most part unknown. The Inc proteins do not share structural similarities to eukaryotic proteins, except for SNARE-like motifs, which will be discussed below, and homology searches do not provide substantial clues about their function.

INTERACTIONS WITH THE GOLGI APPARATUS

One of the initial events in infection is the expression of a chlamydial gene product(s) that effectively isolates the inclusion from the endosomal/lysosomal pathway and initiates fusion competence with a subset of exocytic vesicles, which deliver sphingolipids from the Golgi apparatus to the plasma membrane (Hackstadt et al., 1995, 1996). Although chlamydiae intercept sphingolipids in transit to the plasma membrane, the processing and export of cellular glycoproteins are not inhibited (Scidmore et al., 1996a). Thus, chlamydiae interrupt a specific cellular pathway with minimal interference of normal cellular function. Collectively, the data suggest that the inclusion occupies a site distal to the trans-Golgi apparatus with properties of an exocytic vesicle in which fusion with the plasma membrane is inhibited or delayed (Hackstadt et al., 1997). Although these lipids do not appear to be further metabolized, the incorporation of sphingolipids is important for chlamydial development (van Ooij et al., 1998; Robertson et al., 2009). Insights into the cellular interactions of the inclusion have been provided through studies of the trafficking of a fluorescent analog of ceramide (C_6-NBD-ceramide). This probe is a vital stain for the Golgi apparatus and has been used extensively in conjunction with either fluorescence or electron microscopy to study sphingolipid trafficking in viable cells (Lipsky and Pagano, 1985a, 1985b). When cells infected with *C. trachomatis* are labeled with C_6-NBD-ceramide, a substantial proportion of the sphingomyelin endogenously synthesized from the added fluorescent ceramide analog is diverted to the inclusion, where it is subsequently incorporated into the cell walls of chlamydiae (Hackstadt et al., 1995). The energy and temperature dependence of sphingomyelin delivery to intracellular chlamydiae is characteristic of the vesicular transport of sphingomyelin to the plasma membrane. C_6-NBD-sphingomyelin, introduced directly into the plasma membrane, is not delivered in significant amounts to the chlamydial inclusion, indicating that the primary route of sphingomyelin delivery to the inclusion is not from the plasma membrane but directly from the Golgi apparatus (Hackstadt et al., 1996). These findings suggest a novel mechanism by which chlamydiae may evade lysosomal fusion. By appearing as a secretory vesicle, the chlamydial inclusion is apparently not perceived by the host cell as a vesicle destined to fuse with lysosomes. No other pathogens are known to similarly interact with this pathway.

The cell wall of *C. trachomatis* contains the eukaryotic lipids cholesterol and sphingomyelin (Newhall, 1988). Chlamydiae do not encode the enzymatic capacity for synthesis of these lipids, which are not typically found in prokaryotes (Stephens et al., 1998), and thus they are likely acquired from the host cell. Because sphingomyelin and cholesterol are frequently associated in membrane microdomains and may share transport pathways from the Golgi apparatus to the plasma membrane, it is likely that these lipids might be transported via the same mechanisms. *C. trachomatis* obtains both de novo-synthesized or low density lipoprotein-derived cholesterol from the host cell. Acquisition of cholesterol from either source is inhibited by brefeldin A and nocodazole (Carabeo et al., 2003). Thus, the same pharmacological inhibitors of Golgi or microtubule function that inhibit sphingomyelin delivery to the inclusion also block cholesterol trafficking, which implies that the Golgi apparatus is a necessary intermediate in the transport of cholesterol to the inclusion. Infection of mutant CHO cell lines defective in either the low density lipoprotein or cholesterol biosynthetic pathways resulted in modest or no reduction in yields of progeny EBs. However, simultaneous inhibition of both pathways was lethal, suggesting that either cholesterol or the activity of this vesicular transport pathway is necessary for chlamydial growth, although deleterious effects on the host cell cannot be ruled out.

Fragmentation of the Golgi apparatus in *C. trachomatis*-infected cells occurs at about 16 h postinfection. This fragmentation is triggered by the proteolytic cleavage of golgin-84 by unknown proteases and is associated with enhanced sphingolipid transport to the inclusion and an acceleration of the developmental cycle to increase progeny EB production (Heuer et al., 2009). The primary effect of this phenomenon appears to be to stimulate differentiation to EBs, as sphingolipid transport is initiated well before fragmentation of the Golgi apparatus is detected. Specific Rab GTPases appear to play a role in this process as described below (Rejman Lipinski et al., 2009).

Sphingomyelin is incorporated into the chlamydial cell wall and not further metabolized; yet depletion of cellular sphingomyelin through inhibition of synthesis leads to inhibition of chlamydial replication as well as loss of inclusion membrane stability (van Ooij et al., 2000; Robertson et al., 2009). CD63-positive multivesicular bodies have also been implicated as potential sources of sphingomyelin and cholesterol for chlamydiae acting either in addition to or in conjunction with Golgi-dependent lipid trafficking (Beatty, 2008; Robertson et al., 2009).

Chlamydiae appear to have an unusually large requirement for lipids as they acquire multiple lipid species from various cellular sources (Fig. 3). *C. trachomatis* translocates at least three proteins that associate specifically with cellular

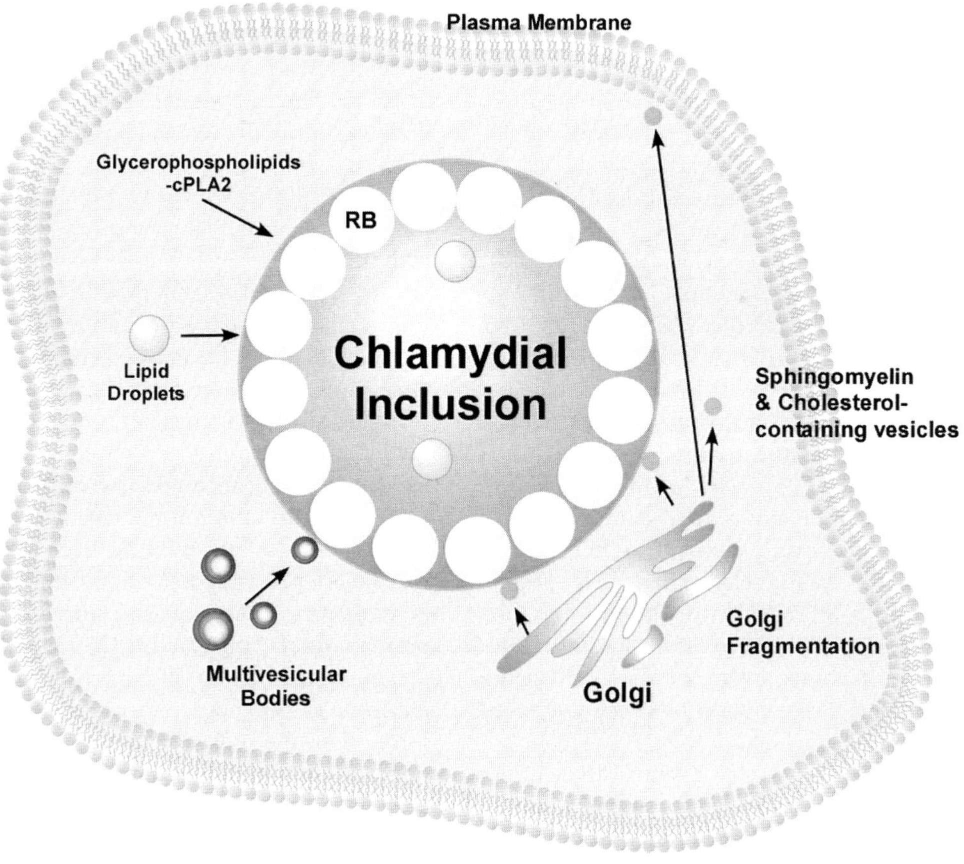

FIGURE 3 Sources of lipids for chlamydiae. Chlamydiae utilize multiple host sources for lipid acquisition. doi:10.1128/9781555817329.ch6.f3

neutral lipid storage organelles known as lipid droplets (Kumar et al., 2006). These lipid droplet-associated (Lda) proteins appear to function in the translocation of lipid droplets into the lumen of the chlamydial inclusion (Cocchiaro et al., 2008). Presumably, this translocation provides a source of nutrient lipids for chlamydiae, although metabolism of these neutral lipids by chlamydiae has not yet been demonstrated. Chlamydiae also target host phospholipids by a process that involves a pathway of Ras/Raf/MEK/ERK activation of cytoplasmic phospholipase A2 (cPLA2) to deacylate host phospholipids, which are subsequently reacylated by chlamydial branched-chain fatty acids for incorporation into chlamydial membrane (Wylie et al., 1997; Su et al., 2004).

MICROTUBULE-DEPENDENT TRAFFICKING

By 2 to 4 h postinfection, endocytosed EBs are trafficked via microtubules to a perinuclear location and early endosomal markers coalesce in close apposition to the chlamydial inclusion membrane. At an ultrastructural level, no evidence of fusion with endosomal compartments bearing Tf, TfR, or M6PR is observed (Scidmore et al., 2003). The association of tubular endosomes containing Tf with early inclusions is similar to that observed with 18-h inclusions in that there is a very close association of these vesicles with the inclusion but no evidence of Tf in the inclusion membrane itself or within the lumen of the inclusion (Scidmore et al., 1996a). Transport of EBs to the perinuclear region and the concentration of recycling endosomes adjacent to the inclusion membrane require chlamydial early protein synthesis. It is probable that the concentration of endosomes observed near the chlamydial inclusion is due to effects on the cytoskeleton that cause a redistribution of vesicles involved in receptor recycling.

Chlamydiae move toward the minus end of microtubules and aggregate at the microtubule-organizing center (MTOC), which is located in a perinuclear region of the host cells (Higashi, 1965; Campbell et al., 1989a, 1989b; McBride and Wilde, 1990; Majeed and Kihlstrom, 1991; Hackstadt et al., 1996; Scidmore et al., 1996b). In mammalian cells, the major minus-end-directed microtubule motor is cytoplasmic dynein. *C. trachomatis* actively modifies the inclusion membrane to recruit dynein and selected components of the dynactin complex for migration along microtubules in a fashion similar to host vesicular trafficking (Clausen et al., 1997; Grieshaber et al., 2003). Surprisingly, overexpression of the protein p50 dynamitin, a subunit of the dynactin complex that links vesicular cargo to the dynein motor in minus-end-directed vesicle trafficking, does not abrogate chlamydial migration, even though host vesicle transport is inhibited (Grieshaber et al., 2003). Nascent chlamydial inclusions, however, colocalize with the p150$^{(Glued)}$ dynactin subunit, thus suggesting that p150$^{(Glued)}$ may be required for dynein activation or processivity but that the cargo binding activity of dynactin, supplied by p50 dynamitin subunits and possibly other subunits, is not. Because chlamydial transcription and translation are required for this intracellular trafficking, chlamydial proteins modifying the cytoplasmic face of the inclusion membrane are likely candidates for proteins controlling the interactions of the inclusion. These findings suggest that chlamydiae circumvent the necessity for an intact dynein/dynactin motor complex in a unique manner, in which a chlamydial protein supplants a requirement for at least the dynamitin component of dynactin.

A novel structure on the chlamydial inclusion membrane has recently been described that is enriched with active Src-family kinases, as well as at least four inclusion membrane proteins (IncB, Inc101, Inc222, and Inc850) (Mital et al., 2010). Inc microdomains are localized at the point of contact of centrosomes with the inclusion membrane and may represent a complex of chlamydial and host proteins that mediates the interactions with dynein to direct migration along microtubule tracks to the MTOC. Two of the microdomain Incs, IncB and Inc850, are early gene products synthesized by 2 h postinfection. Inc101 and Inc222 are considered midcycle genes expressed first around 8 to 12 h postinfection. It is difficult to visualize when these microdomains first form because early, nascent inclusions are quite small. However, by 12 h postinfection, when RBs are beginning to divide, microdomains can be clearly observed by immunofluorescent staining. These microdomains are observed through at least 36 h postinfection. It is worth noting that centripetal migration of nascent inclusions has been initiated by 2 h postinfection (Hackstadt et al., 1996; Grieshaber et al., 2003), and thus the critical chlamydial proteins mediating interaction with dynein must be present and functional by that time.

Although these microdomains are comprised of four chlamydial inclusion membrane proteins and host Src-family kinases, additional host or chlamydial proteins may be required for intracellular interactions. Filipin staining indicates that the microdomains may be further enriched in cholesterol. One of the four Incs, Inc850, showed consistent colocalization with centrosomes when ectopically expressed in HeLa cells (Mital et al., 2010). Inc850 is expressed as early as 1 hour postinfection and thus is present at the time when trafficking to the MTOC is initiated. This suggests that Inc850 should bear increased consideration as the possible linkage of the inclusion to the dynein motor.

This specialized domain on the *C. trachomatis* inclusion membrane implies a functional interaction with the host microtubule network and the microtubule motor, dynein. These microdomains thus appear to act as a platform for interactions controlling trafficking and positioning of the chlamydial inclusion. A better understanding of this complex should promote a more complete characterization of host and chlamydial proteins involved in this domain and aid in discerning interactions with the host cell that promote infection or subsequent cellular transformation and malignancy.

REGULATION OF VESICLE TRAFFICKING

Although the chlamydial inclusion has been listed among those considered nonfusogenic (Sinai and Joiner, 1997), it is fusogenic with at least a subset of exocytic vesicles but does not appear to fuse with compartments of the endocytic pathway to any detectable degree (Taraska et al., 1996; Heinzen and Hackstadt, 1997; van Ooij et al., 1997; Scidmore et al., 2003).

Vesicle fusion in eukaryotic cells is an exquisitely regulated process meant to maintain the identity of the various compartments (Rothman and Wieland, 1996; Parlati et al., 2002). Soluble NSF attachment protein receptors (SNARE) proteins play a central role in conferring specificity to vesicle fusion. Until recently, SNAREs had been classified as v-SNAREs (vesicle-SNAREs) or t-SNAREs (target-SNAREs), but they are now termed Q-SNAREs (based upon an essential glutamine in the active site) or R-SNAREs (an essential arginine in the active site), respectively. The interaction of complementary SNARE proteins on opposing faces of vesicle and acceptor membranes forms a complex required for fusion of the lipid bilayers.

Bioinformatic screening has indicated that at least three *C. trachomatis* inclusion membrane proteins, IncA, Inc813, and Inc223, exhibit a characteristic coiled-coil secondary structure and a predicted SNARE-like motif (Delevoye et al., 2008). Orthologs of IncA are annotated in all chlamydial species, although the similarity is low. Inc813 and Inc223 are found only in *C. trachomatis* and *C. muridarum*. The SNARE-like motif of IncA is maintained in *C. trachomatis*, *C. caviae*, *C. muridarum*, *Chlamydia felis*, *Chlamydia psittaci*, and *Chlamydia abortus* but is absent from *C. pneumoniae* (Delevoye et al., 2008). IncA has been shown to interact with the SNARE proteins Vamp3, Vamp7, and Vamp8, although individual depletion of these SNAREs by small interfering RNA had no deleterious impact on chlamydial growth (Delevoye et al., 2008). Similarly, the trans-Golgi SNARE syntaxin 6 is also recruited to the inclusion membrane via interactions with unknown chlamydial protein(s). Depletion of syntaxin 6 also had no significant impact on chlamydial development (Moore et al., 2011). Subsequent studies using in vitro cell fusion assays demonstrated that IncA inhibited SNARE-mediated membrane fusion and thus may function as an inhibitory i-SNARE. Interestingly, the inhibitory activity of IncA was restricted to endocytic SNAREs but did not interrupt exocytic complexes (Paumet et al., 2009). Although IncA is not expressed until well after fusogenicity with endocytic compartments is blocked, it may function in controlling vesicle fusion at later times in the developmental cycle (Hackstadt et al., 1999).

IncA has also been shown to interact with itself on opposing membrane faces and has been proposed to play a role in the homotypic vesicle fusion (Hackstadt et al., 1999) that is characteristic of *C. trachomatis* inclusions within cells infected with multiple EBs (Ridderhof and Barnes, 1989). Interestingly, IncA of *C. caviae* is approximately 30% larger than *C. trachomatis* IncA but does not interact with itself in a yeast two-hybrid system or with *C. trachomatis* IncA (Hackstadt et al., 1999), although both share the predicted SNARE-like motif (Delevoye et al., 2008). This is consistent with the lack of fusion between *C. trachomatis* and *C. caviae* inclusions in coinfected cells. Indeed, clinical isolates of *C. trachomatis* deficient in IncA are also nonfusogenic (Suchland et al., 2000). Although IncA is thought to be essential for homotypic fusion of *C. trachomatis* inclusions, other factors, such a microtubules, appear to also be required (Ridderhof and Barnes, 1989; Schramm and Wyrick, 1995),

Also playing a role in regulation of vesicle fusion is the Rab family of small GTPases (Novick and Zerial, 1997; Schimmoller et al., 1998). A number of Rab family GTPases are recruited to the chlamydial inclusion membrane, evidently via interactions with Incs or other chlamydial proteins on the inclusion membrane. The patterns of recruitment, however, show little commonality across the genus. Rab1, Rab4, Rab11, and Rab14 are recruited to *C. trachomatis*, *C. muridarum*, and *C. pneumoniae* inclusions. Rab6, however, is recruited only to *C. trachomatis*, and Rab10 is specific to *C. muridarum* and *C. pneumoniae* inclusions (Rzomp et al., 2003; Brumell and Scidmore, 2007). Rab GTPases associated with endocytic maturation are excluded from the chlamydial inclusion. It appears that specific Inc proteins may play a role in Rab recruitment to the inclusion membrane. *C. trachomatis* Inc229 mediates recruitment of Rab4 (Rzomp et al., 2006), and *C. pneumoniae* Cpn585 displays affinity for Rab 1, Rab10, and Rab11 (Cortes et al., 2007). RNA interference screens in *Drosophila* S2 cells for host proteins essential for chlamydial infection have confirmed requirements for Rab1 and Rab14 by *C. trachomatis* (Elwell et al., 2008) and Rab11 by *C. caviae* (Derre et al., 2007). These screens also identified several additional Rab GTPases whose role in a *Chlamydia* infection remains to be determined.

More recently, Rab6 and Rab11 have been implicated in the fragmentation of the Golgi apparatus that is associated with increased rates of chlamydial propagation. Knockdown of either Rab6 or Rab11 decreased *C. trachomatis* progeny EB formation at later time points but did not inhibit inclusion formation (Rejman Lipinski et al., 2009). Although sphingomyelin transport to the inclusion was disrupted in Rab6 and Rab11 knockdown cells, the late timing of the disruption and correlation with Golgi fragmentation suggest that other factors are likely involved in controlling the delivery of sphingomyelin, which is initiated very early. Whether depletion of Rab6 or other Rabs affects chlamydial species other than *C. trachomatis* was not determined. Similarly, Rab14 has been implicated in an essential requirement for sphingolipid transport, although this too occurred later in infection (Capmany and Damiani, 2010).

The inclusion membrane-localized Rab GTPases accordingly recruit multiple Rab effectors to further modify the inclusion membrane. Oculocerebrorenal syndrome of Lowe protein 1 (OCRL1), a phosphatidylinositol-

5-phosphatase, is recruited to the inclusion membrane in a Rab-dependent fashion, where it modifies the composition of the inclusion membrane by the production of phosphatidylinositol-4-phosphate (PI4P) (Moorhead et al., 2010). Two additional host proteins, Arf1 and PI4KIIα, involved in recruitment of PI4K-binding proteins and PI4K synthesis at the Golgi, respectively, are also recruited to the inclusion membrane. Depletion of each of these factors is detrimental to chlamydial replication. These findings suggest that the phosphoinositide composition of the inclusion membrane contributes to the creation of this unique intracellular niche.

Specific Incs are also known to recruit host proteins. For example, IncG recruits the adaptor molecule 14-3-3β in a species-specific fashion (Scidmore and Hackstadt, 2001). This molecule has been shown to sequester phosphorylated BAD and thus to contribute to resistance to apoptosis in *C. trachomatis*-infected cells. Those species that do not produce IncG or recruit 14-3-3β are not similarly resistant to apoptosis (Verbeke et al., 2006). Although the mechanism of recruitment is yet unclear, the Rab6 effector Bicaudal D1 is recruited to *C. trachomatis* LGV biovar inclusions. This recruitment is Rab6-independent as well as microtubule and dynein independent (Moorhead et al., 2007). The function of Bicaudal D1 in certain *C. trachomatis* biovars is unknown.

POLARIZED CELLS TO DISTINGUISH TRAFFICKING PATHWAYS

An epithelial cell barrier is often the first line of host defense that potential pathogens must breach. Although nonpolarized cells are more commonly used to study *Chlamydia* infections in cell culture, polarized cells more closely resemble the in vivo environment encountered by chlamydiae. Thus, polarized cell models are highly relevant and have contributed extensively to our knowledge of chlamydial pathogenic processes. Studies using polarized endometrial or cervical epithelial cells have been used to analyze entry and egress, including demonstration that the more invasive LGV strains exit basolaterally to presumably promote more systemic disease (Wyrick et al., 1989, 1993; Davis and Wyrick, 1997; Davis et al., 2002). They have also provided valuable insights in the study of antibiotic delivery (Wyrick et al., 1993), effects of the hormonal cycle (Maslow et al., 1988; Guseva et al., 2003), and the effects of cytokines (Igietseme et al., 1994; Kane and Byrne, 1998; Dessus-Babus et al., 2000).

A polarized epithelial cell model has also been used to study vectorial trafficking of lipids and proteins to the inclusion. The chlamydial inclusion appears to preferentially intercept and retain sphingomyelin, suggesting a divergence in sphingomyelin and glucosylceramide (GlcCer) trafficking (Hackstadt et al., 1995). Ceramide is the immediate biosynthetic precursor of sphingomyelin and GlcCer, which are synthesized in the *cis*-Golgi (Lipsky and Pagano, 1985a, 1985b). These lipids are then believed to be transported to the plasma membrane by a vesicular process in which they are oriented toward the luminal surface of the vesicle such that they are exposed on the outer leaflet of the plasma membrane upon fusion (Lipsky and Pagano, 1985a, 1985b). In polarized cells, GlcCer and sphingomyelin are transported asymmetrically such that GlcCer is enriched on the apical surface whereas sphingomyelin is equally distributed or slightly enriched on the basolateral surface (van Meer et al., 1987; van'tHof and vanMeer, 1990). The chlamydial inclusion preferentially intercepts basolaterally trafficked sphingomyelin. Consistent with previous observations of C_6-NBD-lipid trafficking in infected cells, these changes in lipid trafficking were specific to chlamydiae (Moore et al., 2008). The interruption of a basolaterally directed vesicle trafficking pathway further demonstrates the specificity of chlamydial interactions with vesicle trafficking pathways of the host cell and implies a requirement for specific host vesicle trafficking machinery (SNAREs) to promote this process. At least one such SNARE, syntaxin 6, has been shown to be recruited to the chlamydial inclusion, although knockdown of syntaxin 6 had no negative effect on sphingomyelin trafficking to the inclusion (Moore et al., 2011).

MULTIPLICATION OF RBs AND SUBSEQUENT DIFFERENTIATION

Although chlamydiae appear to expend considerable effort in establishing the appropriate intracellular niche before initiating replication, they continue to produce and secrete additional factors throughout development that serve to further regulate interactions with the host as chlamydiae replicate. Once chlamydiae have established an appropriate intracellular habitat and have differentiated into RBs, a secondary, or midcycle, temporal class of genes are expressed (Shaw et al., 2000; Belland et al., 2003). These midcycle genes include many structural genes and those involved in secondary metabolism. During this phase of development, chlamydiae express several additional genes controlling interactions with the host. Among these, CPAF (chlamydial protease/proteasome-like activity factor) plays a role in multiple essential functions including inhibition of apoptosis and evasion of immune recognition (Zhong, 2009). *C. trachomatis* also modulates NF-κB signaling to downregulate innate immunity and inflammation. The RelA subunit of NF-κB is degraded by CPAF (Christian et al., 2010) or a chlamydial tail-specific protease, CT441, to block NF-κB activation (Lad et al., 2007). NF-κB activity may be further regulated by two deubiquitinating enzymes, ChlaDub1 and ChlaDub2, that inhibit ubiquitination of IκBα, thereby inhibiting its degradation and subsequent activation of NF-κB (Misaghi et al., 2006). *C. pneumoniae* also suppresses interleukin-17-induced NF-κB activation by sequestering an activator of NF-κB (Act1) (Wolf et al., 2009). In *C. psittaci*-infected, but not *C. trachomatis*-infected cells, recruitment of mitochondria to the inclusion membrane is initiated by 10 to 12 h postinfection (Matsumoto et al., 1991). Thus, chlamydiae modulate host activities at multiple stages of their intracellular development. These later interactions will be discussed in greater detail in subsequent chapters.

OUTLOOK

Although there has been impressive progress over the last decade in defining the cellular interactions of chlamydiae with host cells, the field remains rich with opportunity. Many of these interactions are conserved throughout the genus, yet the fundamental mechanisms responsible for the diverse diseases and tissue tropisms of the different species and serovars remain poorly understood. The high degree of sequence identity as well as synteny in gene order and content of the genomes creates a great paradox in chlamydial pathobiology. Relatively few genes have been correlated with disease outcomes or tissue tropism. Many of the observed differences in outcomes of chlamydial infection could be due to subtle differences in the cellular signal transduction pathways induced rather than the presence or absence of specific genes. As obligate intracellular parasites, the pathogenic mechanisms of chlamydiae involve communication with the host cell on many levels throughout the developmental cycle including signal transduction pathways, regulation of vesicle fusion, and acquisition of nutrients from the host cell. The molecular mechanisms of many of these interactions remain to be fully elucidated. Although the stable introduction of nucleic acids has continued to prove to be an elusive goal, novel and creative techniques that complement the natural ability of chlamydiae to exchange genetic information will undoubtedly provide new tools to dissect chlamydial pathogenesis. An improved understanding of the complex interactions of these important pathogens with the host cell should provide great potential for improved chemotherapeutic and immunoprophylactic interventions.

ACKNOWLEDGMENTS

This research was supported in part by the Intramural Research Program of the NIAID/NIH. I thank J. Mital, E. Lutter, A. Omsland, L. Bauler, and C. Dooley for critical review of the manuscript.

REFERENCES

Abromaitis, S., and R. S. Stephens. 2009. Attachment and entry of *Chlamydia* have distinct requirements for host protein disulfide isomerase. *PLoS Pathog.* **5:**e1000357.

Backert, S., and M. Selbach. 2005. Tyrosine-phosphorylated bacterial effector proteins: the enemies within. *Trends Microbiol.* **13:**476–484.

Bahrani, F. K., P. J. Sansonetti, and C. Parsot. 1997. Secretion of Ipa proteins by *Shigella flexneri*: inducer molecules and kinetics of activation. *Infect. Immun.* **65:**4005–4010.

Bannantine, J. P., R. S. Griffiths, W. Viratyosin, W. J. Brown, and D. D. Rockey. 2000. A secondary structure motif predictive of protein localization to the chlamydial inclusion membrane. *Cell. Microbiol.* **2:**35–47.

Bannantine, J. P., D. D. Rockey, and T. Hackstadt. 1998. Tandem genes of *Chlamydia psittaci* that encode proteins localized to the inclusion membrane. *Mol. Microbiol.* **28:**1017–1026.

Barry, C. E., III, S. F. Hayes, and T. Hackstadt. 1992. Nucleoid condensation in *Escherichia coli* that express a chlamydial histone homolog. *Science* **256:**377–379.

Beatty, W. L. 2008. Late endocytic multivesicular bodies intersect the chlamydial inclusion in the absence of CD63. *Infect. Immun.* **76:**2872–2881.

Belland, R. J., M. A. Scidmore, D. D. Crane, D. M. Hogan, W. Whitmire, G. McClarty, and H. D. Caldwell. 2001. *Chlamydia trachomatis* cytotoxicity associated with complete and partial cytotoxin genes. *Proc. Natl. Acad. Sci. USA* **98:**13984–13989.

Belland, R. J., G. Zhong, D. D. Crane, D. Hogan, D. Sturdevant, J. Sharma, W. L. Beatty, and H. D. Caldwell. 2003. Genomic transcriptional profiling of the developmental cycle of *Chlamydia trachomatis*. *Proc. Natl. Acad. Sci. USA* **100:**8478–8483.

Brickman, T. J., C. E. Barry III, and T. Hackstadt. 1993. Molecular cloning and expression of *hctB* encoding a strain-variant chlamydial histone-like protein with DNA-binding activity. *J. Bacteriol.* **175:**4274–4281.

Brumell, J. H., and M. A. Scidmore. 2007. Manipulation of Rab GTPase function by intracellular bacterial pathogens. *Microbiol. Mol. Biol. Rev.* **71:**636–652.

Byrne, G. I., and J. W. Moulder. 1978. Parasite-specified phagocytosis of *Chlamydia psittaci* and *Chlamydia trachomatis* by L and HeLa cells. *Infect. Immun.* **19:**598–606.

Campbell, S., S. J. Richmond, and P. Yates. 1989a. The development of *Chlamydia trachomatis* inclusions within the host eukaryotic cell during interphase and mitosis. *J. Gen. Microbiol.* **135:**1153–1165.

Campbell, S., S. J. Richmond, and P. S. Yates. 1989b. The effect of *Chlamydia trachomatis* infection on the host cell cytoskeleton and membrane compartments. *J. Gen. Microbiol.* **135:**2379–2386.

Capmany, A., and M. T. Damiani. 2010. *Chlamydia trachomatis* intercepts Golgi-derived sphingolipids through a Rab14-mediated transport required for bacterial development and replication. *PLoS One* **5:**e14084.

Carabeo, R. A., C. A. Dooley, S. S. Grieshaber, and T. Hackstadt. 2007. Rac interacts with Abi-1 and WAVE2 to promote an Arp2/3-dependent actin recruitment during chlamydial invasion. *Cell. Microbiol.* **9:**2278–2288.

Carabeo, R. A., S. Grieshaber, A. Hasenkrug, C. A. Dooley, and T. Hackstadt. 2004. Requirement for the Rac GTPase in *Chlamydia trachomatis* invasion of non-phagocytic cells. *Traffic* **5:**418–425.

Carabeo, R. A., S. S. Grieshaber, E. Fischer, and T. Hackstadt. 2002. *Chlamydia trachomatis* induces remodeling of the actin cytoskeleton during attachment and entry into HeLa cells. *Infect. Immun.* **70:**3793–3803.

Carabeo, R. A., and T. Hackstadt. 2001. Isolation and characterization of a mutant Chinese hamster ovary cell line that is resistant to *Chlamydia trachomatis* infection at a novel step in the attachment process. *Infect. Immun.* **69:**5899–5904.

Carabeo, R. A., D. J. Mead, and T. Hackstadt. 2003. Golgi-dependent transport of cholesterol to the *Chlamydia trachomatis* inclusion. *Proc. Natl. Acad. Sci. USA* **100:**6771–6776.

Carlson, J. H., S. Hughes, D. Hogan, G. Cieplak, D. Sturdevant, G. McClarty, H. D. Caldwell, and R. J. Belland. 2004. Polymorphisms in the *Chlamydia trachomatis* cytotoxin locus associated with ocular and genital isolates. *Infect. Immun.* **72:**7063–7072.

Christian, J., J. Vier, S.A. Paschen, and G. Häcker. 2010. Cleavage of the NF-κB family protein p65/RelA by the chlamydial protease-like activity factor (CPAF) impairs proinflammatory signaling in cells infected with Chlamydiae. *J. Biol. Chem.* **285:**41320–41327.

Clausen, J. D., G. Christiansen, H. U. Holst, and S. Birkelund. 1997. *Chlamydia trachomatis* utilizes the host cell microtubule network during early events of infection. *Mol. Microbiol.* **25:** 441–449.

Clifton, D. R., C. A. Dooley, S. S. Grieshaber, R. A. Carabeo, K. A. Fields, and T. Hackstadt. 2005. Tyrosine phosphorylation of chlamydial Tarp is species specific and not required for the recruitment of actin. *Infect. Immun.* **73:**3860–3868.

Clifton, D. R., K. A. Fields, S. Grieshaber, C. A. Dooley, E. Fischer, D. Mead, R. A. Carabeo, and T. Hackstadt. 2004. A chlamydial type III translocated protein is tyrosine phosphorylated at the site of entry and associated with recruitment of actin. *Proc. Natl. Acad. Sci. USA* **101:** 10166–10171.

Cocchiaro, J., Y. Kumar, E. R. Fischer, T. Hackstadt, and R. H. Valdivia. 2008. Cytoplasmic lipid droplets are translocated into the lumen of the *Chlamydia trachomatis* parasitophorous vacuole. *Proc. Natl. Acad. Sci. USA* **105:**9379–9384.

Coers, J., I. Bernstein-Hanley, D. Grotsky, I. Parvanova, J. C. Howard, G. A. Taylor, W. F. Dietrich, and M. N. Starnbach. 2008. *Chlamydia muridarum* evades growth restriction by the IFN-gamma-inducible host resistance factor Irgb10. *J. Immunol.* **180:**6237–6245.

Conant, C. G., and R. S. Stephens. 2007. Chlamydia attachment to mammalian cells requires protein disulfide isomerase. *Cell. Microbiol.* **9:** 222–232.

Coombes, B. K., and J. B. Mahony. 2002. Identification of MEK- and phosphoinositide 3-kinase-dependent signalling as essential events during *Chlamydia pneumoniae* invasion of HEp2 cells. *Cell. Microbiol.* **4:**447–460.

Cortes, C., K. A. Rzomp, A. Tvinnereim, M. A. Scidmore, and B. Wizel. 2007. *Chlamydia pneumoniae* inclusion membrane protein Cpn0585 interacts with multiple Rab GTPases. *Infect. Immun.* **75:**5586–5596.

Davis, C. H., J. E. Raulston, and P. B. Wyrick. 2002. Protein disulfide isomerase, a component of the estrogen receptor complex, is associated with *Chlamydia trachomatis* serovar E attached to human endometrial epithelial cells. *Infect. Immun.* **70:**3413–3418.

Davis, C. H., and P. B. Wyrick. 1997. Differences in the association of *Chlamydia trachomatis* serovar E and serovar L2 with epithelial cells in vitro may reflect biological differences in vivo. *Infect. Immun.* **65:**2914–2924.

Dehoux, P., R. Flores, C. Dauga, G. Zhong, and A. Subtil. 2011. Multi-genome identification and characterization of chlamydiae-specific type III secretion substrates: the Inc proteins. *BMC Genomics* **12:**109.

Delevoye, C., M. Nilges, P. Dehoux, F. Paumet, S. Perrinet, A. Dautry-Varsat, and A. Subtil. 2008. SNARE protein mimicry by an intracellular bacterium. *PLoS Pathog.* **4:**e1000022.

Derre, I., M. Pypaert, A. Dautry-Varsat, and H. Agaisse. 2007. RNAi screen in *Drosophila* cells reveals the involvement of the Tom complex in *Chlamydia* infection. *PLoS Pathog.* **3:**1446–1458.

Dessus-Babus, S., S. T. Knight, and P. B. Wyrick. 2000. Chlamydial infection of polarized HeLa cells induces PMN chemotaxis but the cytokine profile varies between disseminating and nondisseminating strains. *Cell. Microbiol.* **2:**317–327.

DeVinney, R., J. L. Puente, A. Gauthier, D. Goosney, and B. Finlay. 2001. Enterohaemorrhagic and enteropathogenic *Escherichia coli* use different Tir-based mechanism for pedestal formation. *Mol. Microbiol.* **41:**1445–1458.

Elwell, C. A., A. Ceesay, J. H. Kim, D. Kalman, and J. N. Engel. 2008. RNA interference screen identifies Abl kinase and PDGFR signaling in *Chlamydia trachomatis* entry. *PLoS Pathog.* **4:** e1000021.

Fields, K. A., and T. Hackstadt. 2000. Evidence for the secretion of *Chlamydia trachomatis* CopN by a type III secretion mechanism. *Mol. Microbiol.* **38:**1048–1060.

Fields, K. A., and T. Hackstadt. 2002. The chlamydial inclusion: escape from the endocytic pathway. *Annu. Rev. Cell Dev. Biol.* **18:**221–245.

Fields, K. A., D. Mead, C. A. Dooley, and T. Hackstadt. 2003. *Chlamydia trachomatis* type III secretion: evidence for a functional apparatus during early-cycle development. *Mol. Microbiol.* **48:**671–683.

Fudyk, T., L. Olinger, and R. S. Stephens. 2002. Selection of mutant cell lines resistant to infection by *Chlamydia trachomatis* and *Chlamydia pneumoniae*. *Infect. Immun.* **70:**6444–6447.

Gabel, B. R., S. C. Ijzendoorn, and J. N. Engel. 2004. Lipid raft-mediated entry is not required for *Chlamydia trachomatis* infection of cultured epithelial cells. *Infect. Immun.* **72:**7367–7373.

Grieshaber, N., E. Fischer, D. Mead, C. A. Dooley, and T. Hackstadt. 2004. Chlamydial histone-DNA interactions are disrupted by a metabolite in the methylerythritol phosphate pathway of isoprenoid biosynthesis. *Proc. Natl. Acad. Sci. USA* **101:**7451–7456.

Grieshaber, N. A., S. S. Grieshaber, E. R. Fischer, and T. Hackstadt. 2006. A small RNA inhibits translation of the histone-like protein Hc1 in *Chlamydia trachomatis*. *Mol. Microbiol.* **59:** 541–550.

Grieshaber, S., N. Grieshaber, and T. Hackstadt. 2003. *Chlamydia trachomatis* uses host cell dynein to traffic to the microtube organizing center in a p50 dynamitin independent process. *J. Cell Biol.* **116:**3793–3802.

Gruenheid, S., R. DeVinney, F. Bladt, D. Goosney, S. Gelkp, G. D. Gish, T. Pawson, and B. Finlay. 2001. Enteropathogenic *E. coli* Tir binds Nck to initiate actin pedestal formation in host cells. *Nat. Cell Biol.* **3:**856–859.

Guseva, N. V., S. T. Knight, J. D. Whittimore, and P. B. Wyrick. 2003. Primary cultures of female swine genital epithelial cells in vitro: a new approach for the study of hormonal modulation of chlamydia infection. *Infect. Immun.* **71:** 4700–4710.

Hackstadt, T. 1999. Cell biology, p. 101–138. *In* R. S. Stephens (ed.), Chlamydia*: Intracellular Biology, Pathogenesis, and Immunity*. ASM Press, Washington, DC.

Hackstadt, T. 2000. Redirection of host vesicle trafficking pathways by intracellular parasites. *Traffic* **1:**93–99.

Hackstadt, T., W. Baehr, and Y. Ying. 1991. *Chlamydia trachomatis* developmentally regulated protein is homologous to eukaryotic histone a. *Proc. Natl. Acad. Sci. USA* **88:**3937–3941.

Hackstadt, T., E. R. Fischer, M. A. Scidmore, D. D. Rockey, and R. A. Heinzen. 1997. Origins and functions of the chlamydial inclusion. *Trends Microbiol.* **5:**288–293.

Hackstadt, T., D. D. Rockey, R. A. Heinzen, and M. A. Scidmore. 1996. *Chlamydia trachomatis* interrupts an exocytic pathway to acquire endogenously synthesized sphingomyelin in transit from the Golgi apparatus to the plasma membrane. *EMBO J.* **15:**964–977.

Hackstadt, T., M. A. Scidmore-Carlson, E. I. Shaw, and E. R. Fischer. 1999. The *Chlamydia trachomatis* IncA protein is required for homotypic vesicle fusion. *Cell. Microbiol.* **1:**119–130.

Hackstadt, T., M. A. Scidmore, and D. D. Rockey. 1995. Lipid metabolism in *Chlamydia trachomatis*-infected cells: directed trafficking of Golgi-derived sphingolipids to the chlamydial inclusion. *Proc. Natl. Acad. Sci. USA* **92:**4877–4881.

Hatch, T. P., M. Miceli, and J. A. Silverman. 1985. Synthesis of protein in host-free reticulate bodies of *Chlamydia psittaci* and *Chlamydia trachomatis*. *J. Bacteriol.* **162:**938–942.

Hayward, R. D., R. J. Cain, E. J. McGhie, N. Phillips, M. J. Garner, and V. Koronakis. 2005. Cholesterol binding by the bacterial type III translocon is essential for virulence effector delivery into mammalian cells. *Mol. Microbiol.* **56:**590–603.

Heinzen, R. A., and T. Hackstadt. 1997. The *Chlamydia trachomatis* parasitophorous vacuolar membrane is not passively permeable to low-molecular-weight compounds. *Infect. Immun.* **65:**1088–1094.

Heuer, D., A. Rejman Lipinski, N. Machuy, A. Karlas, A. Wehrens, F. Siedler, V. Brinkmann, and T. F. Meyer. 2009. *Chlamydia* causes fragmentation of the Golgi compartment to ensure reproduction. *Nature* **457:**731–735.

Higashi, N. 1965. Electron microscopic studies on the mode of reproduction of trachoma virus and psittacosis virus in cell cultures. *Exp. Mol. Pathol.* **76:**24–39.

Hower, S., K. Wolf, and K. A. Fields. 2009. Evidence that CT694 is a novel *Chlamydia trachomatis* T3S substrate capable of functioning during invasion or early cycle development. *Mol. Microbiol.* **72:**1423–1437.

Hueck, C. J. 1998. Type III protein secretion systems in bacterial pathogens of animals and plants. *Microbiol. Mol. Biol. Rev.* **62:**379–433.

Igietseme, J. U., P. B. Wyrick, D. Goyeau, and R. G. Rank. 1994. An in vitro model for immune control of chlamydial growth in polarized epithelial cells. *Infect. Immun.* **62:**3528–3535.

Jamison, W. P., and T. Hackstadt. 2008. Induction of type III secretion by cell-free *Chlamydia trachomatis* elementary bodies. *Microb. Pathog.* **45:**435–440.

Jewett, T. J., C. A. Dooley, D. J. Mead, and T. Hackstadt. 2008. *Chlamydia trachomatis* Tarp is phosphorylated by Src family tyrosine kinases. *Biochem. Biophys. Res. Commun.* **371:**339–344.

Jewett, T. J., E. R. Fischer, D. J. Mead, and T. Hackstadt. 2006. Chlamydial TARP is a bacterial nucleator of actin. *Proc. Natl. Acad. Sci. USA* **103:**15599–15604.

Jewett, T. J., N. J. Miller, C. A. Dooley, and T. Hackstadt. 2010. The conserved Tarp actin binding domain is important for chlamydial invasion. *PLoS Pathog.* **6:**e1000997.

Kane, C. D., and G. I. Byrne. 1998. Differential effects of γ interferon on *Chlamydia trachomatis* growth in polarized and nonpolarized human epithelial cells in culture. *Infect. Immun.* **66:** 2349–2351.

Kumar, Y., J. Cocchiaro, and R. H. Valdivia. 2006. The obligate intracellular pathogen *Chlamydia trachomatis* targets host lipid droplets. *Curr. Biol.* **16:**1646–1651.

Kumar, Y., and R. H. Valdivia. 2008. Actin and intermediate filaments stabilize the *Chlamydia trachomatis* vacuole by forming dynamic structural scaffolds. *Cell Host Microbe* **4:**159–169.

Lad, S. P., J. Li, J. da Silva Correia, Q. Pan, S. Gadwal, R. J. Ulevitch, and E. Li. 2007. Cleavage of p65/RelA of the NF-kappaB pathway by *Chlamydia*. *Proc. Natl. Acad. Sci. USA* **104:**2933–2938.

Lane, B. J., C. Mutchier, S. Al Khodor, S. S. Grieshaber, and R. A. Carabeo. 2008. Chlamydial entry involves TARP binding of guanine nucleotide exchange factors. *PLoS Pathog.* **4:** e1000014.

Lee, V. T., S. K. Mazmanian, and O. Schneewind. 2001. A program of *Yersinia enterocolitica* type III secretion reactions is activated by specific signals. *J. Bacteriol.* **183:**4970–4978.

Li, Z., C. Chen, D. Chen, Y. Wu, Y. Zhong, and G. Zhong. 2008. Characterization of fifty putative inclusion membrane proteins encoded in the *Chlamydia trachomatis* genome. *Infect. Immun.* **76:**2746–2757.

Lipsky, N. G., and R. E. Pagano. 1985a. Intracellular translocation of fluorescent sphingolipids in cultured fibroblasts: endogenously synthesized sphingomyelin and glucocerebroside analogues pass through the Golgi apparatus en route to the plasma membrane. *J. Cell Biol.* **100:**27–34.

Lipsky, N. G., and R. E. Pagano. 1985b. A vital stain for the Golgi apparatus. *Science* **228:**745–747.

Lutter, E. I., C. Bonner, M. Holland, R. J. Suchland, W. E. Stamm, T. J. Jewett, G. McClarty, and T. Hackstadt. 2010. Phylogenetic analysis of *Chlamydia trachomatis* Tarp and correlation with clinical phenotype. *Infect. Immun.* **78:**3678–3688.

Majeed, M., and E. Kihlstrom. 1991. Mobilization of F-actin and clathrin during redistribution of *Chlamydia trachomatis* to an intracellular site in eucaryotic cells. *Infect. Immun.* **59:**4465–4472.

Maslow, A. S., C. H. Davis, J. Choong, and P. B. Wyrick. 1988. Estrogen enhances attachment of *Chlamydia trachomatis* to human endometrial epithelial cells in vitro. *Am. J. Obstet. Gynecol.* **159:**1006–1014.

Matsumoto, A., H. Bessho, K. Uehira, and T. Suda. 1991. Morphological studies of the association of mitochondria with chlamydial inclusions and the fusion of chlamydial inclusions. *J. Electron Microsc.* **40:**356–363.

McBride, T., and E. Wilde III. 1990. Intracellular translocation of *Chlamydia trachomatis*, p. 36–39. *In* W. R. Bowie, H. D. Caldwell, R. P. Jones, et al. (ed.), *Chlamydial Infections.* Cambridge University Press, Cambridge, United Kingdom.

Mehlitz, A., S. Banhart, S. Hess, M. Selbach, and T. F. Meyer. 2008. Complex kinase requirements for *Chlamydia trachomatis* Tarp phosphorylation. *FEMS Microbiol. Lett.* **289:**233–240.

Mehlitz, A., S. Banhart, A. P. Maurer, A. Kaushansky, A. G. Gordus, J. Zielecki, G. Macbeath, and T. F. Meyer. 2010. Tarp regulates early *Chlamydia*-induced host cell survival through interactions with the human adaptor protein SHC1. *J. Cell Biol.* **190:**143–157.

Misaghi, S., Z. R. Balsara, A. Catic, E. Spooner, H. L. Ploegh, and M. N. Starnbach. 2006. *Chlamydia trachomatis*-derived deubiquitinating enzymes in mammalian cells during infection. *Mol. Microbiol.* **61:**142–150.

Mital, J., N. J. Miller, E. R. Fischer, and T. Hackstadt. 2010. Specific chlamydial inclusion membrane proteins associate with active Src family kinases in microdomains that interact with the host microtubule network. *Cell. Microbiol.* **12:** 1235–1249.

Moore, E. R., E. R. Fischer, D. J. Mead, and T. Hackstadt. 2008. The chlamydial inclusion preferentially intercepts basolaterally directed sphingomyelin-containing exocytic vacuoles. *Traffic* **9:**2130–2140.

Moore, E. R., D. J. Mead, C. A. Dooley, J. Sager, and T. Hackstadt. 2011. The trans-Golgi SNARE syntaxin 6 is recruited to the chlamydial inclusion membrane. *Microbiology* **157:**830–838.

Moorhead, A. M., J. Y. Jung, A. Smirnov, S. Kaufer, and M. A. Scidmore. 2010. Multiple host proteins that function in phosphatidylinositol-4-phosphate metabolism are recruited to the chlamydial inclusion. *Infect. Immun.* **78:**1990–2007.

Moorhead, A. R., K. A. Rzomp, and M. A. Scidmore. 2007. The Rab6 effector Bicaudal D1 associates with *Chlamydia trachomatis* inclusions in a biovar-specific manner. *Infect. Immun.* **75:** 781–791.

Moulder, J. W. 1991. Interaction of chlamydiae and host cells in vitro. *Microbiol. Rev.* **55:**143–190.

Nelson, D. E., et al. 2005. Chlamydial interferon gamma immune evasion is linked to host infection tropism. *Proc. Natl. Acad. Sci. USA* **102:** 10658–10663.

Newhall, W. J. 1988. Macromolecular and antigenic composition of chlamydiae, p. 47–70. *In* A. L. Barron (ed.), *Microbiology of Chlamydia.* CRC Press, Boca Raton, FL.

Novick, P., and M. Zerial. 1997. The diversity of Rab proteins in vesicle transport. *Curr. Opin. Cell Biol.* **9:**496–504.

Parlati, F., O. Varlamov, K. Paz, J. A. McNew, D. Hurtado, T. H. Sollner, and J. E. Rothman. 2002. Distinct SNARE complexes mediating membrane fusion in Golgi transport based on combinatorial specificity. *Proc. Natl. Acad. Sci. USA* **99:**5424–5429.

Paumet, F., J. Wesolowski, A. Garcia-Diaz, C. Delevoye, N. Aulner, H. A. Shuman, A. Subtil, and J. E. Rothman. 2009. Intracellular bacteria encode inhibitory SNARE-like proteins. *PLoS ONE* **4:**e7375.

Perara, E., D. Ganem, and J. N. Engel. 1992. A developmentally regulated chlamydial gene with apparent homology to eukaryotic histone a. *Proc. Natl. Acad. Sci. USA* **89:**2125–2129.

Pirbhai, M., F. Dong, Y. Zhong, K. Z. Pan, and G. Zhong. 2006. The secreted protease factor CPAF is responsible for degrading pro-apoptotic BC-only proteins in *Chlamydia trachomatis*-infected cells. *J. Biol. Chem.* **281:**31495–31501.

Prain, C. J., and J. H. Pearce. 1989. Ultrastructural studies on the intracellular fate of *Chlamydia psittaci* (strain guinea pig inclusion conjunctivitis) and *Chlamydia trachomatis* (strain lymphogranuloma venereum 434): modulation of intracellular events and relationship with endocytic mechanism. *J. Gen. Microbiol.* **135:**2107–2123.

Qualmann, B., and M. M. Kessels. 2009. New players in actin polymerization—Wb-domain-containing actin nucleators. *Trends Cell Biol.* **19:**276–285.

Read, T. D., et al. 2000. Genome sequences of *Chlamydia trachomatis* MoPn and *Chlamydia pneumoniae* AR39. *Nucleic Acids Res.* **28:**1397–1406.

Read, T. D., et al. 2003. Genome sequence of *Chlamydiophila caviae* (*Chlamydia psittaci* GPIC): examining the role of niche-specific genes in the evolution of the Chlamydiaceae. *Nucleic Acids Res.* **31:**2134–2147.

Rejman Lipinski, A., J. Heymann, C. Meissner, A. Karlas, V. Brinkmann, T. F. Meyer, and D. Heuer. 2009. Rab6 and Rab11 regulate *Chlamydia trachomatis* development and golgin-84-dependent Golgi fragmentation. *PLoS Pathog.* **5:** e1000615.

Ridderhof, J. C., and R. C. Barnes. 1989. Fusion of inclusions following superinfection of HeLa cells by two serovars of *Chlamydia trachomatis. Infect. Immun.* **57:**3189–3193.

Rivera-Amill, V., B. J. Kim, J. Seshu, and M. E. Konkel. 2001. Secretion of the virulence-associated *Campylobacter* invasion antigens from *Campylobacter jejuni* requires a stimulatory signal. *J. Infect. Dis.* **183:**1607–1616.

Robertson, D. K., L. Gu, R. K. Rowe, and W. L. Beatty. 2009. Inclusion biogenesis and reactivation of persistent *Chlamydia trachomatis* requires host cell sphingolipid biosynthesis. *PLoS Pathog.* **5:**e1000664.

Robinson, R. C., K. Turbedsky, D. A. Kaiser, J. B. Marchand, H. N. Higgs, S. Choe, and T. D. Pollard. 2001. Crystal structure of Arp2/3 complex. *Science* **294:**1679–1684.

Rockey, D. D., D. Grosenbach, D. E. Hruby, M. G. Peacock, R. A. Heinzen, and T. Hackstadt. 1997. *Chlamydia psittaci* IncA is phosphorylated by the host cell and is exposed on the cytoplasmic face of the developing inclusion. *Mol. Microbiol.* **24:**217–228.

Rockey, D. D., R. A. Heinzen, and T. Hackstadt. 1995. Cloning and characterization of a *Chlamydia psittaci* gene coding for a protein localized in the inclusion membrane of infected cells. *Mol. Microbiol.* **15:**617–626.

Rockey, D. D., and J. L. Rosquist. 1994. Protein antigens of *Chlamydia psittaci* present in infected cells but not detected in the infectious elementary body. *Infect. Immun.* **62:**106–112.

Rockey, D. D., M. A. Scidmore, J. P. Bannantine, and W. J. Brown. 2002. Proteins in the chlamydial inclusion membrane. *Microbes Infect.* **4:**333–340.

Rothman, J. E., and F. T. Wieland. 1996. Protein sorting by transport vesicles. *Science* **272:**227–234.

Rzomp, K. A., A. R. Moorhead, and M. A. Scidmore. 2006. The GTPase Rab4 interacts with *Chlamydia trachomatis* inclusion membrane protein CT229. *Infect. Immun.* **74:**5362–5373.

Rzomp, K. A., L. D. Scholtes, B. J. Briggs, G. R. Whittaker, and M. A. Scidmore. 2003. Rab GTPases are recruited to chlamydial inclusions in both a species-dependent and species-independent manner. *Infect. Immun.* **71:**5855–5870.

Schimmoller, F., I. Simon, and S. R. Pfeffer. 1998. Rab GTPases, directors of vesicle docking. *J. Biol. Chem.* **273:**22161–22164.

Schramm, N., and P. B. Wyrick. 1995. Cytoskeletal requirements in *Chlamydia trachomatis* infection of host cells. *Infect. Immun.* **63:**324–332.

Scidmore, M. A., E. Fischer, and T. Hackstadt. 2003. Restricted fusion of *Chlamydia trachomatis* vesicles with endocytic compartments during the initial stages of infection. *Infect. Immun.* **71:**973–984.

Scidmore, M. A., E. R. Fischer, and T. Hackstadt. 1996a. Sphingolipids and glycoproteins are differentially trafficked to the *Chlamydia trachomatis* inclusion. *J. Cell Biol.* **134:**363–374.

Scidmore, M. A., and T. Hackstadt. 2001. Mammalian 14-3-3beta associates with the *Chlamydia trachomatis* inclusion membrane via its interaction with IncG. *Mol. Microbiol.* **39:**1638–1650.

Scidmore, M. A., D. D. Rockey, E. R. Fischer, R. A. Heinzen, and T. Hackstadt. 1996b. Vesicular interactions of the *Chlamydia trachomatis* inclusion are determined by chlamydial early protein synthesis rather than route of entry. *Infect. Immun.* **64:**5366–5372.

Shaw, E. I., C. A. Dooley, E. R. Fischer, M. A. Scidmore, K. A. Fields, and T. Hackstadt. 2000. Three temporal classes of gene expression during the *Chlamydia trachomatis* developmental cycle. *Mol. Microbiol.* **37:**913–925.

Sinai, A. P., and K. A. Joiner. 1997. Safe haven: the cell biology of nonfusogenic pathogen vacuoles. *Annu. Rev. Microbiol.* **51:**415–462.

Stephens, R. S., et al. 1998. Genome sequence of an obligate intracellular pathogen of humans: *Chlamydia trachomatis. Science* **282:**754–759.

Stuart, E. S., W. C. Webley, and L. C. Norkin. 2003. Lipid rafts, caveolae, caveolin-1, and entry by chlamydiae into host cells. *Exp. Cell. Res.* **287:** 67–78.

Su, H., G. McClarty, F. Dong, G. M. Hatch, Z. Pan, K,, and G. Zhong. 2004. Activation of RAf/MEK/ERK/cPLA2 signaling pathway is essential for chlamydial acquisition of host glycerophospholipids. *J. Biol. Chem.* **279:**9409–9416.

Subtil, A., C. Parsot, and A. Dautry-Varsat. 2001. Secretion of predicted Inc proteins of *Chlamydia pneumoniae* by a heterologous type III machinery. *Mol. Microbiol.* **39:**792–800.

Subtil, A., B. Wyplosz, M. E. Balana, and A. Dautry-Varsat. 2004. Analysis of *Chlamydia caviae* entry sites and involvement of Cdc42 and Rac activity. *J. Cell Sci.* **117:**3923–3933.

Suchland, R. J., D. D. Rockey, J. P. Bannantine, and W. E. Stamm. 2000. Isolates of *Chlamydia trachomatis* that occupy nonfusogenic inclusions lack IncA, a protein localized to the inclusion membrane. *Infect. Immun.* **68:**360–367.

Tao, S., R. Kaul, and W. M. Wenman. 1991. Identification and nucleotide sequence of a developmentally regulated gene encoding a eukaryotic histone a-like protein from *Chlamydia trachomatis*. *J. Bacteriol.* **173:**2818–2822.

Taraska, T., D. M. Ward, R. S. Ajioka, P. B. Wyrick, S. R. Davis-Kaplan, C. H. Davis, and J. Kaplan. 1996. The late chlamydial inclusion membrane is not derived from the endocytic pathway and is relatively deficient in host proteins. *Infect. Immun.* **64:**3713–3727.

Tipples, G., and G. McClarty. 1993. The obligate intracellular bacterium *Chlamydia trachomatis* is auxotrophic for three of the four ribonucleoside triphosphates. *Mol. Microbiol.* **8:**1105–1114.

van der Goot, F. G., G. Tran van Nhieu, A. Allaoui, P. Sansonetti, and F. Lafont. 2004. Rafts can trigger contact-mediated secretion of bacterial effectors via a lipid-based mechanism. *J. Biol. Chem.* **279:**47792–47798.

van Meer, G., E. H. K. Stelzer, R. W. Winjnaendts-van-Resandt, and K. Simons. 1987. Sorting of sphingolipids in epithelial (Madin-Darby Canine Kidney) cells. *J. Cell Biol.* **105:**1623–1635.

van Ooij, C., G. Apodaca, and J. Engel. 1997. Characterization of the *Chlamydia trachomatis* vacuole and its interaction with the host endocytic pathway in HeLa cells. *Infect. Immun.* **65:** 758–766.

van Ooij, C., L. Kalman, S. van Ijzendoorn, M. Nishijima, K. Hanada, K. Mostov, and J. N. Engel. 2000. Host cell-derived sphingolipids are required for the intracellular growth of *Chlamydia trachomatis*. *Cell. Microbiol.* **2:**627–637.

van'tHof, W., and G. vanMeer. 1990. Generation of lipid polarity in intestinal epithelial (Caco-2) cells: sphingolipid synthesis in the Golgi complex and sorting before vesicular traffic to the plasma membrane. *J. Cell Biol.* **111:**977–986.

Verbeke, P., L. Welter-Stahl, S. Ying, J. Hansen, G. Hacker, T. Darville, and D. M. Ojcius. 2006. Recruitment of BAD by the *Chlamydia trachomatis* vacuole correlates with host-cell survival. *PLoS Pathog.* **2:**e45.

Wagar, E. A., and R. S. Stephens. 1988. Developmental-form-specific DNA-binding proteins in *Chlamydia* spp. *Infect. Immun.* **56:**1678–1684.

Wolf, K., G. V. Plano, and K. A. Fields. 2009. A protein secreted by the respiratory pathogen *Chlamydia pneumoniae* impairs IL-17 signaling via interaction with human Act1. *Cell. Microbiol.* **11:**767–779.

Wylie, J. L., G. M. Hatch, and G. McClarty. 1997. Host cell phospholipids are trafficked to and then modified by *Chlamydia trachomatis*. *J. Bacteriol.* **179:**7233–7242.

Wyrick, P. B. 2000. Intracellular survival by *Chlamydia*. *Cell. Microbiol.* **2:**275–282.

Wyrick, P. B., J. Choong, C. H. Davis, S. T. Knight, M. O. Royal, A. S. Maslow, and C. R. Bagnell. 1989. Entry of genital *Chlamydia trachomatis* into polarized human epithelial cells. *Infect. Immun.* **57:**2378–2389.

Wyrick, P. B., C. H. Davis, S. T. Knight, J. Choong, J. E. Raulston, and N. Schramm. 1993. An in vitro human epithelial cell culture system for studying the pathogenesis of *Chlamydia trachomatis*. *Sex. Transm. Dis.* **20:**248–256.

Zhong, G. 2009. Killing me softly: chlamydial use of proteolysis for evading host defenses. *Trends Microbiol.* **17:**467–474.

Zhong, G., P. Fan, H. Ji, F. Dong, and Y. Huang. 2001. Identification of a chlamydial protease-like activity factor responsible for the degradation of host transcription factors. *J. Exp. Med.* **193:**935–942.

TEMPORAL GENE REGULATION DURING THE CHLAMYDIAL DEVELOPMENTAL CYCLE

Ming Tan

7

THE CHLAMYDIAL DEVELOPMENTAL CYCLE

The *Chlamydiales* are obligate intracellular bacteria that grow and proliferate only inside a eukaryotic host cell (reviewed by Moulder [1991] and Hackstadt [1999]). Replication occurs within a membrane-bound cytoplasmic inclusion, with the cycle from entry to release of progeny taking 48 to 72 h depending on the species and strain. A defining feature of this infectious cycle is the sequential conversion between two morphologically distinct forms within the confines of the chlamydial inclusion (AbdelRahman and Belland, 2005).

These two developmental forms of chlamydiae have separate roles that are specialized for either extracellular survival or intracellular growth and replication (Schachter, 1988). The elementary body (EB) is the infectious form that binds and enters the host cell. By electron microscopy, the EB begins its striking change into the larger reticulate body (RB) at 2 to 3 hours postinfection (hpi), and conversion is complete by 6 to 8 hpi (Ward, 1988; Belland et al., 2003b). Known steps in this process include dispersion of the EB's condensed chromatin and reduction of the disulfide-linked outer membrane complex that forms its protective coat. The developmental cycle then transitions to a replicative phase during which the metabolically active RB divides repeatedly by binary fission, yielding up to 1,000 organisms per infected cell. Starting at around 18 to 24 hpi, individual RBs convert into EBs with condensation of genomic DNA and heavy disulfide cross-linking of outer membrane proteins. This late conversion step is asynchronous, with some RBs undergoing terminal differentiation into EBs while others continue to replicate to produce more RBs. For *C. trachomatis*, the developmental cycle is completed by approximately 48 hpi, and chlamydiae are released by lysis of the host cell or extrusion of the inclusion (Hybiske and Stephens, 2007). EBs can then begin a new developmental cycle by infecting a neighboring host cell.

The rate of *C. trachomatis* replication during the developmental cycle has been measured by quantitative PCR. Timms and colleagues measured cDNA for 16S rRNA, which allowed them to calculate a chlamydial doubling time of 3 hours, in the period between 8 and 40 hpi, and a yield of 200 to 300 progeny per cell (Mathews et al., 1999). Hackstadt and colleagues quantified genome copy number

Ming Tan, Departments of Microbiology & Molecular Genetics, and Medicine, University of California, Irvine, CA 92697-4025.

Intracellular Pathogens I: Chlamydiales
Edited by Ming Tan and Patrik M. Bavoil © 2012 ASM Press, Washington, DC
10.1128/9781555817329.ch7

to derive an average doubling time of 2 hours (Shaw et al., 2000). They measured a maximal doubling rate of once every 1.45 hours and a yield of just over 1,000 organisms per infected cell. As a rough guide, these findings are consistent with 8 to 10 rounds of RB replication during the developmental cycle.

REGULATION OF GENE EXPRESSION DURING THE DEVELOPMENTAL CYCLE

Chlamydial genes are transcribed as three main temporal classes, which correspond to the major stages of the developmental cycle (Shaw et al., 2000; Belland et al., 2003b; Nicholson et al., 2003). Early genes are a group of genes that are first transcribed during EB-to-RB conversion, within the first few hours after chlamydial entry. Midcycle genes represent the largest temporal group by far and are first expressed during the RB stage. Late genes are another small group of genes that are upregulated at the end of the developmental cycle, when RBs are converting into EBs. Put another way, midgenes are expressed during chlamydial growth and replication, while early and late genes appear to be specialized subsets of genes involved in the establishment and conclusion of the intracellular infection, respectively.

The first indications that chlamydial genes are differentially transcribed during the course of the developmental cycle came from expression studies of individual genes. These studies detected chlamydial RNA with Northern blots, which favored the identification of highly expressed genes. Transcripts for *ompA*, the gene coding for the major outer membrane protein (MOMP), were first detected between 4 and 8 hours post infection (hpi) and were expressed at high levels by 12 hpi in several *Chlamydia* spp. (Stephens et al., 1988; Yuan et al., 1990). In contrast, chlamydial rRNA was present as early as 2 hpi (Engel and Ganem, 1990a), while *omcAB* transcripts were not detected until 24 hpi (Lambden et al., 1990). Thus, it was already apparent from these early studies that gene expression during the developmental cycle is temporally regulated.

To identify additional early and late genes, Hatch and colleagues screened *C. trachomatis* genomic libraries for clones that preferentially hybridized to early or late chlamydial RNA. They synthesized these developmental RNA samples by isolating chlamydiae from infected cells at different times postinfection and incubating the host-free chlamydiae with radiolabeled nucleotides. Using this approach, they identified an early gene, *euo*, which was expressed by 1.5 hpi (Wichlan and Hatch, 1993). They also identified two new late genes, *ltuA* and *ltuB*, that were only expressed late in the developmental cycle (Fahr et al., 1995).

In the most ambitious study prior to the use of DNA microarrays, Hackstadt and colleagues determined the transcriptional patterns for 8% of *C. trachomatis* genes (Shaw et al., 2000). In a quantitative reverse transcription-PCR analysis of 70 genes, they found that these genes grouped into three temporal classes, which they classified as early, midcycle, and late genes. Significantly, these temporal classes corresponded to the three main stages of the chlamydial developmental cycle. This landmark study demonstrated that differential temporal expression, which had been previously noted for individual genes, is a prominent feature of the developmental cycle.

The three main temporal groups of chlamydial genes were confirmed in two groundbreaking microarray studies that measured transcript levels for almost all the genes in the *C. trachomatis* genome at a number of times during the developmental cycle. These studies analyzed the transcriptional profiles for the approximately 900 chromosomal open reading frames in serovars D (Belland et al., 2003b) and L2 (Nicholson et al., 2003). The Belland microarray study also examined the eight open reading frames located on the serovar L2 plasmid. A similar microarray analysis found that genes were also expressed as three main temporal classes in the related species *C. pneumoniae* (Maurer et al., 2007). Smaller subsets of genes at very early and late times have been proposed, although it is not known if they are distinct from early and late genes, respectively. These include 29 immediate early genes whose transcripts were detected by 1 hpi (Belland et al., 2003b) and very late genes that were transcribed at their highest levels at the end of the developmental cycle (Nicholson et al., 2003; Maurer et al., 2007). In the *C. pneumoniae* study, these very late genes, which were called tardy genes, showed an association with mRNAs found in EBs (Maurer et al., 2007). In these microarray studies, transcripts for almost all genes were detectable at the end of the developmental cycle, indicating that the entire chlamydial genome is expressed during normal growth and replication.

These global transcriptional profiles provide a wealth of clues about the potential function and regulation of chlamydial genes. For example, the expression of immediate early genes within an hour of chlamydial uptake suggests that these genes share a related function in establishing the intracellular infection (Belland et al., 2003b). In addition, their selective expression indicates that these genes are likely to be coregulated by a mechanism that is different from later temporal classes of genes. On a larger scale, these studies demonstrate that temporal regulation of gene expression during the developmental cycle is common to *Chlamydia* spp.

The three microarray studies showed general agreement, but not all orthologous genes were assigned to the same main temporal group in the different studies. These discrepancies may reflect strain differences or the way in which the data from the different times were normalized so they could be compared. The results also depend on technical issues such as the sensitivity of the hybridization method used to detect transcript levels and the ability of the data analysis to resolve genes into separate groups. It is also possible that chlamydial genes may not divide neatly into temporal groups that can be easily distinguished. For example, temporal groups could overlap in their expression patterns, or individual genes within a group of coregulated genes may vary in the timing of their expression. A single temporal group of chlamydial genes could actually be composed of subsets that are regulated by separate mechanisms, which appears to be the case for late genes, as will be discussed below. Finally, the identification of late genes is complicated because not all RBs convert into EBs at the same time. This late asynchrony is a hallmark of the developmental cycle, and there are no practical ways to separately analyze the subpopulation of chlamydiae that are undergoing RB-to-EB conversion.

An important caveat about the microarray studies, like the earlier Northern blot and reverse transcription-PCR analyses, is that they only measured steady-state levels of chlamydial transcripts. Different transcriptional profiles have been ascribed to differential initiation of transcription, but they could instead be due to disparities in message stability. Similarly, any shutoff in transcription of a specific gene will not be apparent if stable transcripts continue to be present. Transcriptional profiles have often been used to extrapolate when a protein is present, but it is important to remember that transcript and protein levels may not correlate because of posttranscriptional and translational control mechanisms.

The timing of chlamydial gene expression reported in different studies can be difficult to compare because the length of the developmental cycle is often not the same. For example, it has been observed that the developmental cycle in cell culture is accelerated at a higher multiplicity of infection (Fan et al., 1998). Also, chlamydial species and strains differ in the length of a productive intracellular infection, and it is well known that it takes longer for *C. pneumoniae* (about 72 h) than *C. trachomatis* (about 48 h). Regardless of the strain or experimental condition used, a productive chlamydial infection undergoes the three main stages of the developmental cycle. It is reasonable to assume that the general control mechanisms that regulate these sequential steps in the production of infectious progeny have also been preserved during evolution of the *Chlamydiales*.

Taken together, these studies raised the question of whether the progression of the developmental cycle is controlled by the coordinated expression of chlamydial genes as three temporal groups. This developmental gene expression could be programmed so that it follows a defined course once set in motion by entry of the bacterium into a host cell. Alternatively, it could respond to external cues such as environmental and metabolic signals, as has been shown for chlamydial transcription factors that are regulated by cofactors such as amino acids, nucleotides, and metal ions (Wyllie and Raulston, 2001; Akers and Tan, 2006; Carlson et al., 2006; Schaumburg and Tan, 2006; Akers et al., 2011; Case et al., 2011). The factors that control this temporal expression of chlamydial genes could do so in both a programmed and a responsive manner and have the potential to be master regulators of the developmental cycle.

In considering how gene expression during the developmental cycle is regulated, we will focus on the control of transcription initiation by RNA polymerase, which is a critical step in the regulation of bacterial gene expression (Browning and Busby, 2004). The activity of RNA polymerase can be modulated by transcription factors that control the expression of specific target genes. Fewer than 10 transcription factors have been identified in *Chlamydia*, however, and the ones that have been studied mainly have specific roles in the homeostatic regulation of amino acids, nucleotides, and metal ions (Wyllie and Raulston, 2001; Akers and Tan, 2006; Carlson et al., 2006; Schaumburg and Tan, 2006; Akers et al., 2011; Case et al., 2011). This paucity of transcription factors is a general feature of obligate intracellular bacteria (Madan Babu et al., 2006), presumably because they encounter a controlled environment within an infected cell. Bacteria also regulate transcription by using alternative forms of RNA polymerase to transcribe specific subsets of genes. An alternative RNA polymerase has a promoter specificity different from that of the main form of RNA polymerase because it contains an alternative sigma factor instead of the major sigma factor. *Chlamydia* appears to use alternative RNA polymerases as a mechanism for regulating transcription, since all *Chlamydia* spp. encode two alternative sigma factors, σ^{28} and σ^{54}, in addition to the major chlamydial sigma factor, σ^{66} (Stephens et al., 1998; Tan, 2006).

We will review the mechanisms of temporal gene regulation utilized during the three main stages of the developmental cycle, as well as how transcription is silenced in EBs. We will begin with midcycle because the majority of chlamydial genes are regulated during this stage of RB growth and replication. The developmental cycle can be thought of as a circle, however, and events at the end of one developmental cycle prepare chlamydiae for the beginning of the next cycle in another host cell. Thus, we will discuss the transcriptional regulation of midcycle genes, late genes, EBs, and early genes in sequence, mirroring the events that occur during replication, conversion of an RB into an EB, and then back again into an RB at the start of a new round of infection.

TRANSCRIPTIONAL REGULATION OF MIDCYCLE GENES

Midcycle genes make up the largest temporal group of chlamydial genes. These genes are not expressed during the early stage of infection and are first transcribed around the time when EB-to-RB conversion has been completed. In the Belland microarray study, more than 500 genes that were not expressed at 3 hpi were transcribed at 8 hpi (Belland et al., 2003b), which suggests that *Chlamydia* utilizes one or more mechanisms to coordinately upregulate the expression of hundreds of genes at the start of midcycle. These midcycle genes are expressed during active chlamydial growth and replication, when RBs are dividing by binary fission, and include housekeeping genes, such as enzymes of glycolysis and gluconeogenesis. They also include MOMP and the inclusion membrane protein IncA, which is involved in homotypic fusion of chlamydial inclusions.

Regulation of Midcycle Genes by DNA Supercoiling

There is accumulating evidence that an increase in the level of chlamydial DNA supercoiling is utilized to upregulate genes in midcycle. There is precedent in other bacteria for the level of DNA supercoiling to function as a global regulator of gene expression (reviewed by Dorman [2006]). In fact, DNA supercoiling has been proposed to be the highest level in the hierarchy of prokaryotic gene regulation because of its ability to coordinately regulate a large number of genes (Hatfield and Benham, 2002). In contrast, a regulon directly controlled by a bacterial transcription factor is much smaller, and only a few global regulators in *Escherichia coli* control more than 100 target genes (Martinez-Antonio and Collado-Vides, 2003).

DNA supercoiling regulates bacterial gene expression by modulating promoter activity, thereby controlling the level of transcription from individual genes. Changes in DNA supercoiling can modulate promoter activity directly, by altering DNA structure and melting energy, or indirectly, by affecting the binding of transcription factors that regulate promoter activity (Bae et al., 2006). Not all bacterial promoters are affected by DNA supercoiling levels, but DNA microarray studies in *E. coli* and *Haemophilus influenzae* have demonstrated that changes in supercoiling alter the expression of hundreds of genes in vivo (Gmuender et al., 2001; Peter et al., 2004).

DNA supercoiling varies during the developmental cycle and is highest in midcycle, when RBs are actively growing and dividing. The supercoiling level of the chlamydial plasmid has been measured as a marker of chlamydial DNA supercoiling (Solbrig et al., 1990; Barry et al., 1993) and was highest at 18, 24, and 28 hpi and lower (more relaxed) at earlier (2 and 6 hpi) and late (40 and 46 hpi) times (Niehus et al., 2008). The superhelical density, which was estimated by resolving the chlamydial plasmid on two-dimensional gels, changed from a range of −0.063 to −0.077 at 24 hpi to −0.028 to −0.035 at 46 hpi (the numbers are negative because bacterial DNA is negatively supercoiled). These values are similar to those for *E. coli*, where the superhelical density of a reporter plasmid was approximately −0.05 during logarithmic growth and −0.03 in stationary phase (Hatfield and Benham, 2002).

Intriguingly, these higher levels of in vivo supercoiling in the midstage of the developmental cycle correlate with the upregulation of midcycle but not late promoters by increased DNA supercoiling levels in vitro (Niehus et al., 2008). Chlamydial promoters were cloned on a transcription plasmid, and topoisomers of each plasmid, differing only in their supercoiling level, were transcribed by chlamydial RNA polymerase in vitro. In this supercoiling sensitivity transcription assay, promoters for five midcycle genes were upregulated >4-fold in response to increased DNA supercoiling levels (Niehus et al., 2008; Case et al., 2010). The effect was greatest with the *ompA* promoter, which was upregulated >50-fold. In contrast, promoters for six late genes were insensitive to changes in DNA supercoiling (Niehus et al., 2008; Case et al., 2010).

Together, these results provide evidence for a novel use of DNA supercoiling as a developmental regulator of gene expression. It is not yet known whether increased DNA supercoiling is the general mechanism for regulating midcycle gene expression since only a small number of midcycle genes have been tested and shown to have supercoiling-responsive promoters. Nevertheless, this regulatory mechanism is likely to be important for virulence since midcycle genes with supercoiling-responsive promoters include the immunodominant MOMP and three operons encoding components of the type III secretion system for secreting chlamydial proteins into the infected host cell (Niehus et al., 2008; Case et al., 2010). DNA supercoiling provides an elegant way for chlamydiae to sense when they are in the RB stage, since supercoiling levels are higher during log-phase bacterial growth (Hatfield and Benham, 2002). In addition, DNA supercoiling provides a global mechanism to coordinately transcribe several hundred midcycle genes.

The enzymes that cause DNA supercoiling levels to increase during midcycle represent a new class of temporal regulators in *Chlamydia*. The supercoiling levels of bacterial DNA are regulated by the counterbalancing action of topoisomerases (Drlica, 1992).

For example, DNA gyrase introduces negative supercoils into DNA, which increases DNA superhelicity, while DNA topoisomerase I removes negative supercoils to relax DNA. All *Chlamydia* spp. are unusual in encoding two DNA gyrases, although there is only a single DNA topoisomerase I (Stephens et al., 1998). A DNA gyrase encoded as two subunits by *C. trachomatis gyrA_1* and *gyrB_1* is transcribed from early times (Belland et al., 2003a). This early gyrase is likely to cause the increase in DNA supercoiling that peaks in midcycle and is therefore a predicted regulator of midcycle genes and potentially the master regulator of gene expression in midcycle. The second gyrase (*gyrA_2* and *gyrB_2*) has a late temporal pattern (Belland et al., 2003a), which suggests that it has a specialized role in modulating DNA topology during late stages in the developmental cycle. Whether it also has a role in temporal gene regulation remains to be seen.

A role for DNA supercoiling as a mechanism for the temporal regulation of the developmental cycle has implications for antichlamydial therapy. Fluoroquinolones have been widely used as antibiotics to treat a number of bacterial infections including those caused by *Chlamydia*. Their mechanism of action is the inhibition of DNA gyrase and a related enzyme, topoisomerase IV, which results in a block in bacterial DNA replication (Hooper, 2000). However, fluoroquinolones may have additional *Chlamydia*-specific effects by preventing DNA supercoiling levels from increasing in midcycle, thereby inhibiting expression of midcycle genes and progression of the developmental cycle.

Regulation of Midcycle Genes by ChxR

Expression of midcycle genes in *Chlamydia* has also been proposed to be regulated by a transcription factor called ChxR (Koo and Stephens, 2003). ChxR is an ortholog of the OmpR/PhoB subfamily of bacterial transcription factors. These transcription factors are called response regulators and are usually paired with a sensor kinase to form a two-component signal transduction system that is used by a bacterium to respond to an environmental signal (Gao and Stock, 2009). In this signaling pathway, the environmental signal typically induces autophosphorylation of the membrane-bound sensor histidine kinase. The sensor kinase then transfers this phosphate group to its cognate response regulator, which promotes the ability of the transcription factor to homodimerize and regulate transcription.

ChxR has been demonstrated to be an atypical OmpR/PhoB response regulator because it does not require phosphorylation to be an active transcription factor. Similar phosphorylation-independent response regulators have been described in a number of bacteria. ChxR lacks the conserved aspartate residue at its active site that would be phosphorylated by a sensor kinase (Koo and Stephens, 2003). In addition, *Chlamydia* does not appear to encode a partner sensor kinase for ChxR (Stephens et al., 1998). Hefty and colleagues have provided experimental evidence that ChxR does not require phosphorylation for activity by showing that unphosphorylated ChxR homodimerizes and is able to bind to a target promoter (Hickey et al., 2011).

ChxR is the first chlamydial transcriptional activator that has been shown to be functionally active. Intriguingly, the other chlamydial transcription factors that have been examined in functional studies are all repressors, which negatively regulate target genes (Wilson and Tan, 2002; Akers and Tan, 2006; Schaumburg and Tan, 2006; Akers et al., 2011; Case et al., 2011). Using an *E. coli* in vitro transcription assay, five target promoters of ChxR have been identified. One of these targets is its own promoter, suggesting that ChxR regulates its own expression (Koo and Stephens, 2003). Autoregulation of ChxR expression was verified in a chromatin immunoprecipitation assay in which ChxR was found to recognize its own promoter in vivo (Hickey et al., 2011). A total of six binding sites upstream of the *chxR* promoter have been characterized, which has allowed the identification of a ChxR DNA recognition motif (Hickey et al., 2011).

The main rationale for proposing ChxR as a temporal regulator is that it is only expressed during midcycle and late times in the developmental cycle (Koo and Stephens, 2003; Hickey et al., 2011). Certainly, this temporal expression of ChxR indicates that its activity as an activator is restricted to the RB stage of the developmental cycle. However, this feature is not unusual, since other chlamydial regulators such as TrpR and NrdR are also only expressed during the RB stage (Akers and Tan, 2006; Case et al., 2011). It is perhaps no surprise that a number of chlamydial regulators would function when chlamydiae are metabolically active. One difference, however, is that these other regulators each regulate a small number of target genes, whereas ChxR has the potential to control a larger regulon. The precise role of ChxR in regulating chlamydial gene expression during the developmental cycle remains to be determined.

TRANSCRIPTIONAL REGULATION OF LATE GENES

Late genes represent a specialized group of genes that are upregulated late in the developmental cycle when RBs are converting into EBs. Belland et al. identified 26 late genes in their microarray study (Belland et al., 2003b). Several late genes have obvious functions in an EB, such as *omcAB*, which encodes two EB-specific cysteine-rich outer membrane proteins (Clarke et al., 1988; Hatch, 1996), and *hctA* and *hctB*, which encode the histone-like proteins Hc1 and Hc2, which mediate DNA condensation (Barry et al., 1992; Brickman et al., 1993). Several late genes are predicted to encode thioredoxin disulfide isomerases and membrane thiol proteases, which may have a role in the formation of the highly cross-linked outer membrane complex of the EB (Belland et al., 2003b).

Regulation of Late Gene Expression by σ^{28} RNA Polymerase

We described the first temporal regulator of gene expression in *Chlamydia* when we showed that the alternative sigma factor σ^{28} recognizes promoters for a subset of late genes (Yu and Tan, 2003; Yu et al., 2006b). A role for alternative sigma factors in the temporal regulation of chlamydial gene expression was proposed even before these transcriptional regulators were identified in *Chlamydia* (Plaunt and Hatch, 1988; Engel and Ganem, 1990b; Fahr et al., 1995). This prediction was based on the example of *Bacillus subtilis*, where a cascade of alternative sigma factors regulates developmental gene expression during sporulation (Stragier and Losick, 1990). Alternative sigma factors also have important roles in the regulation of virulence gene expression in many bacteria (Fang, 2005; Kazmierczak et al., 2005).

When its genome was sequenced, *C. trachomatis* was shown to encode two potential alternative sigma factors with sequence similarity to σ^{28} and σ^{54} in other bacteria (Stephens et al., 1998). σ^{28} and σ^{54} orthologs have since been identified in all the sequenced genomes of *Chlamydia*, although σ^{28} does not appear to be encoded by *Protochlamydiaceae* (Horn et al., 2004). In other bacteria, σ^{28} regulates the expression of genes involved in flagellar synthesis, chemotaxis, and motility (Haldenwang, 1995). Chlamydiae, however, are nonmotile organisms that do not have flagella, and therefore the role of σ^{28} in the regulation of chlamydial gene expression was not immediately obvious.

The identification of σ^{28} target genes required the development of a σ^{28} RNA polymerase transcription assay, which in turn required active σ^{28} RNA polymerase. Chlamydial σ^{28} RNA polymerase was reconstituted by adding recombinant *C. trachomatis* σ^{28} to native core enzyme that had been biochemically purified from *C. trachomatis* RBs (Yu and Tan, 2003). In vitro and in vivo transcription studies have also been performed with a hybrid σ^{28} RNA polymerase consisting of recombinant *C. trachomatis* σ^{28} and *E. coli* core enzyme (Shen et al., 2004, 2006).

The first clue that σ^{28} has a role in temporal gene regulation came when chlamydial σ^{28} RNA polymerase was shown to transcribe the *C. trachomatis hctB* promoter in an in vitro transcription assay (Yu and Tan,

2003). *hctB* encodes the histone-like protein Hc2, a known late gene that mediates DNA condensation during RB-to-EB conversion (Brickman et al., 1993). This finding was consistent with, and provided an explanation for, results from previous studies by Hatch and colleagues showing that promoters for four late genes, but not *hctB*, were transcribed by σ^{66} RNA polymerase (Mathews et al., 1993; Fahr et al., 1995). Intriguingly, one of these σ^{66}-dependent late genes is *hctA*, which encodes the other chlamydial histone-like protein Hc1. Together, these findings indicate that a single temporal class of chlamydial genes can be regulated by more than one mechanism (Yu and Tan, 2003). Furthermore, regulation of the two histone-like proteins by different forms of chlamydial RNA polymerase suggests that RB-to-EB conversion may be controlled by more than one signal or pathway.

The role of σ^{28} as a late regulator has been supported by the identification of additional σ^{28}-dependent late genes through a combined bioinformatics and functional approach. We delineated the preferred sequence recognized by *C. trachomatis* σ^{28} RNA polymerase (Yu et al., 2006a) and searched for sequences in the *C. trachomatis* genome that resemble this sequence and the sequence of σ^{28}-dependent promoters from other bacteria. Candidate promoters were then tested for transcription by chlamydial σ^{28} RNA polymerase. Six σ^{28} target genes were identified, and three have late expression patterns (Yu et al., 2006b). Besides *hctB*, the σ^{28}-transcribed late genes are *tsp*, the gene for the tail-specific protease Tsp (Lad et al., 2007), and *tlyC_1*, which encodes a predicted hemolysin (Stephens et al., 1998). Two other σ^{28} target genes, *dnaK* and *pgk*, do not have a late transcription pattern (Belland et al., 2003b), but they are each transcribed from tandem σ^{66} and σ^{28} promoters (Yu et al., 2006b). Thus, their pattern of expression by σ^{28} RNA polymerase cannot be discerned from the overall transcriptional profile, which is a composite of transcription by σ^{28} and σ^{66} RNA polymerases (Yu et al., 2006b).

These studies showing that σ^{28} is a regulator of late gene expression revealed a number of features about temporal regulation in *Chlamydia*. They demonstrated that late genes are transcribed as two classes, since separate subsets of late genes are transcribed by σ^{28} and σ^{66} RNA polymerases. They also showed that an alternative form of RNA polymerase can be used as a mechanism for temporal regulation during the developmental cycle. However, this novel role for σ^{28} as a temporal regulator also raised more questions about how σ^{28} itself is regulated so that it transcribes late genes at late times.

Regulation of σ^{28} by an Anti-Sigma Factor

σ^{28} RNA polymerase could be regulated by alternative mechanisms in which either σ^{28} protein or σ^{28}-dependent activity is absent until late developmental times. There is evidence, however, that σ^{28} protein is expressed during midcycle (Douglas and Hatch, 2000; Shen et al., 2004), before σ^{28}-dependent late genes are transcribed. These observations suggest that it is σ^{28} activity that is regulated to prevent premature transcription of late genes, such as *hctB*, that encode proteins involved in RB-to-EB conversion.

There is precedent from other bacteria for the regulation of RNA polymerase activity by an anti-sigma factor that binds and inactivates the sigma subunit. For example in *B. subtilis*, RsbW binds σ and sequesters σ^{B}, so that σ^{B} RNA polymerase is inhibited (Price, 2002). This control mechanism has been called a partner-switching mechanism because RsbW has an alternative binding partner, RsbV, which serves as an antagonist or anti-anti-sigma factor (Price, 2002). When RsbV binds RsbW, it causes σ^{B} to be released, and as a consequence, σ^{B} RNA polymerase is active.

The partner-switching mechanism is actually a signaling pathway involving additional components that regulate the ability of RsbV to bind RsbW. RsbV-RsbW binding only occurs when RsbV is in an unphosphorylated state, and RsbW actively prevents this binding since it is the kinase for RsbV. An additional upstream regulator, RsbU, is a phosphatase that dephosphorylates RsbV. Thus, in *Bacillus subtilis*, σ^B RNA polymerase is activated by a signaling pathway in which RsbU dephosphorylates RsbV, allowing RsbV to bind RsbW and releasing σ^B (Price, 2002).

Chlamydia appears to express all the components of this partner-switching mechanism, but until recently there were conflicting data about its potential role in regulating σ^{28} RNA polymerase activity. σ^{28} and σ^B are related sigma factors (Gruber and Gross, 2003), and all *Chlamydia* spp. encode a predicted RsbW ortholog, two proteins, RsbV1 and RsbV2, which have sequence similarity to RsbV, and two potential RsbU orthologs, RsbU and CT589 (Stephens et al., 1998; Hua et al., 2006). *C. trachomatis* RsbW bound σ^{28} in a pull-down assay (Douglas and Hatch, 2006). However, in a separate study, RsbW did not interact with σ^{28} in a yeast two-hybrid assay and did not alter σ^{28}-dependent transcription (Hua et al., 2006). *C. trachomatis* RsbW was able to bind and phosphorylate RsbV1 and RsbV2, however, demonstrating that the kinase activity of chlamydial RsbW was conserved. In a third study that used *Salmonella* as a heterologous genetic system, *C. trachomatis* RsbW did not inhibit chlamydial σ^{28} (Karlinsey and Hughes, 2006).

Recently, we have produced experimental evidence to support a role for the RsbW partner-switching mechanism in regulating chlamydial σ^{28} RNA polymerase activity (C. J. Rosario and M. Tan, unpublished data). We demonstrated that RsbW bound σ^{28} in an in vitro pull-down assay. Moreover, RsbW inhibited σ^{28}-dependent transcription of the *hctB* promoter but not transcription by σ^{66} RNA polymerase. Based on these findings, we propose a model in which RsbW binds and inhibits σ^{28} prior to late times in the developmental cycle. This inhibition is then counteracted at the end of midcycle by RsbV1 and RsbV2, which bind RsbW, allowing σ^{28} RNA polymerase to transcribe its late genes targets.

The relative expression patterns of σ^{28} and its proposed regulators provide support for this model. σ^{28} protein is expressed during midcycle at 16 hpi (Shen et al., 2004). However, its inhibitor RsbW is expressed as an early gene from 3 hpi (Belland et al., 2003b), ensuring that RsbW is available to bind σ^{28} and prevent premature σ^{28}-dependent transcription. RsbV1 and RsbV2 are expressed as midcycle genes (Belland et al., 2003b), which would allow them to counteract RsbW at the end of midcycle to initiate transcription of σ^{28}-regulated late genes.

This signaling pathway provides a potential means for external stimuli to regulate σ^{28}-dependent transcription of late genes. It is not clear why there are two RsbV proteins, but they could represent parallel pathways for σ^{28} to be controlled by separate input signals (Hua et al., 2006). The predicted serine phosphatases RsbU and CT589 have not been studied, but each is predicted to be a transmembrane protein with an extracytoplasmic region that could sense external signals (Hua et al., 2006).

Regulation of σ^{66}-Dependent Late Genes by a Transcriptional Repressor

The σ^{66}-dependent late genes have been a group of target genes in search of a regulatory mechanism. There must be a separate mechanism to control the expression of late genes, such as *omcAB* and *hctA* (Fahr et al., 1995), so that they are not transcribed by σ^{66} RNA polymerase together with early and midcycle genes. An anti-sigma factor is unlikely to be the mechanism for regulating σ^{66}-dependent late genes because it would interfere with transcription of early and midcycle genes by σ^{66} RNA polymerase. This class of late genes also does not appear to be regulated by DNA supercoiling because transcription of representative σ^{66}-dependent late promoters was not altered by changes in supercoiling levels (Niehus et al., 2008; Case et al., 2010).

A potential mechanism for the regulation of σ^{66}-dependent late genes is a transcription factor that selectively modulates the activity of late promoters. For example, a transcriptional activator that is only available at late time points could bind upstream of late promoters and upregulate their activity. However, late promoters can be transcribed by purified *E. coli* RNA polymerase in vitro, which indicates that an activator is not necessary (Fahr et al., 1995; Schaumburg and Tan, 2000). In addition, sequences upstream of late promoters, where an activator would typically bind, can be deleted without altering promoter activity (J. C. Akers and M. Tan, unpublished results). Alternatively, late genes could be regulated by a transcriptional repressor that prevents transcription of late promoters during early and mid-times, until repression is relieved late in the developmental cycle.

A repressor of σ^{66}-dependent late genes would be predicted to have a number of properties. It should be expressed as an early gene, so that it is present from early times to provide continuous repression of specific target genes until late in the developmental cycle. It should be a DNA-binding protein because bacterial repressors typically block RNA polymerase-promoter interactions by binding to an operator sequence in the vicinity of the promoter. Its target late genes, but not early and midcycle genes, should each contain this operator near the promoter so that the putative repressor can specifically repress late genes. In addition, a mechanism must exist to relieve this late repressor so that σ^{66}-dependent late promoters can be transcribed late in the developmental cycle.

There is evidence that EUO is this predicted repressor of late genes. Hatch and colleagues, who discovered EUO, demonstrated that it is transcribed as early as 1 hpi and that it bound the promoter region for the late operon, *omcAB* (Wichlan and Hatch, 1993; Zhang et al., 1998). They showed that EUO preferentially binds AT-rich sequences and proposed a DNA binding sequence based on binding assays with *C. psittaci* genomic fragments (Zhang et al., 1998, 2000). We have recently found that EUO is a selective repressor of late promoters (Rosario and Tan, 2012). With in vitro binding assays, recombinant EUO bound promoters for late genes but not early or midcycle genes. EUO also selectively repressed transcription of late promoters but not early or midcycle promoters in vitro.

How is repression of σ^{66}-dependent late genes by EUO relieved at late times so that these genes can be transcribed? It is possible that late in the developmental cycle, metabolic changes within an RB serve as a signal that is sensed by EUO. For example, EUO-dependent repression could require a corepressor that is depleted by late times, leading to derepression of its target genes. Alternatively, repression by EUO could be relieved by an inducer that accumulates in an RB that is about to convert into an EB. A cofactor model could explain the observed asynchrony in RB-to-EB conversion, since only RBs with cofactor levels above a threshold would express σ^{66}-dependent late genes and initiate the conversion process. However, we have found no evidence that EUO-mediated repression involves a cofactor (Rosario and Tan, unpublished data).

Repression by EUO could instead be relieved as a consequence of the decrease in EUO protein levels that has been measured late in the developmental cycle. In *C. psittaci*, EUO transcripts were only detected during early and midcycle, and EUO protein was only present between 1 and 20 hpi and peaked at 15 hpi (Zhang et al., 1998). EUO protein was not detected at later times or in EBs. This temporal regulation of EUO levels could account for the relief of EUO-mediated repression at late time points, but it is not known how EUO levels are regulated.

TRANSCRIPTIONAL REGULATION IN EBs

Histone-Like Proteins

EBs have long been considered to be transcriptionally silent. This lack of gene expression is not due to the absence of the transcriptional machinery, since σ^{66} RNA polymerase has been detected in EBs (Shaw et al., 2002; Skipp et al., 2005; Sixt et al., 2011). Instead, it is likely that RNA polymerase is unable to access and transcribe genomic DNA because the genetic material in an EB is condensed into a prominent nucleoid. mRNA has been detected in EBs, but it has been attributed to transcripts that were synthesized in an RB before it converted into an EB (Belland et al., 2003b). In support of this model, an association has been demonstrated in *C. pneumoniae* between genes that are transcribed at very late times and mRNAs found in EBs (Maurer et al., 2007). It is not clear if these transcripts are used as templates for translation, but protein synthesis in host-free EBs has recently been described (Haider et al., 2010).

Chromatin condensation in EBs is mediated by two histone-like proteins that are encoded by all *Chlamydia* spp. Hc1 and Hc2 are *Chlamydia*-specific proteins that resemble eukaryotic histones (Hackstadt, 1991; Tao et al., 1991; Perara et al., 1992). These nucleoid-associated proteins bind DNA without known sequence specificity and induce the formation of a condensed nucleoid (Barry et al., 1992, 1993; Brickman et al., 1993). Both proteins are only expressed at late times, with the gene encoding Hc1 transcribed by σ^{66} RNA polymerase (Fahr et al., 1995) and the gene for Hc2 transcribed by σ^{28} RNA polymerase (Yu and Tan, 2003). By causing chromatin condensation, this late expression of the histone-like proteins is believed to be one of the forces that drive RB-to-EB conversion at the end of the developmental cycle (Barry et al., 1992).

The histone-like proteins have been proposed to function as silencers of overall chlamydial gene expression in EBs by causing chromatin condensation. In studies that used *E. coli* as a heterologous system, expression of either Hc1 or Hc2 inhibited overall transcription and translation in vivo (Barry et al., 1993; Pedersen et al., 1994, 1996). Each histone-like protein also inhibited transcription by T7 RNA polymerase in vitro (Pedersen et al., 1994, 1996). The two histone-like proteins are not equivalent, however, and may have different effects on chlamydial gene expression. For instance, Hc1 produces a more compact nucleoid than Hc2 (Barry et al., 1992; Brickman et al., 1993). Additionally, Hc1 preferentially binds supercoiled DNA, while Hc2 has a higher affinity for RNA and linearized DNA (Pedersen et al., 1996). Intriguingly, Hc2 is unique to the *Chlamydiaceae* and is not encoded by the environmental chlamydiae (Collingro et al., 2011). This observation makes it likely that Hc1 is the primary histone-like protein for condensing chlamydial DNA and silencing gene expression in EBs and that Hc2 has a specialized role.

The histone-like proteins may also regulate gene expression by altering DNA supercoiling. Hc1 expression in *E. coli* decreased the supercoiling level of a reporter plasmid and modulated the activity of supercoiling-dependent promoters (Barry et al., 1993). Thus, the histone-like proteins have the potential to regulate the transcription of genes with supercoiling-responsive promoters (Barry et al., 1993; Niehus et al., 2008).

Novel Regulator Scc4

Shen and colleagues have shown that Scc4 is a novel regulator of σ^{66} RNA polymerase (Rao et al., 2009). Scc4 was identified in a bacterial two-hybrid screen for chlamydial proteins that bind to σ^{66} RNA polymerase. Scc4 interacts with both β and σ subunits of σ^{66} RNA polymerase, thereby preventing contacts with the −35 promoter element. Scc4 inhibited transcription of three *C. trachomatis* σ^{66} promoters by *E. coli* σ^{70} RNA polymerase or a hybrid polymerase composed of *E. coli* core enzyme and a recombinant sigma factor containing both *C. trachomatis* σ^{66} and *E. coli* σ^{70} sequences. However, Scc4 did not inhibit all promoters tested, demonstrating that RNA polymerase bound by Scc4 is still transcriptionally active. This promoter-specific resistance to Scc4-mediated inhibition has been proposed to be due to the presence of an extended −10 element, which is a promoter feature that provides additional contacts with RNA polymerase to compensate for a suboptimal −35 promoter element (Rao et al., 2009).

Scc4 has the potential to be a temporal regulator of chlamydial transcription. Its expression is developmentally regulated, and Scc4 transcript and protein levels increase late in the developmental cycle (Rao et al., 2009). Shen and colleagues have proposed that Scc4 may facilitate σ^{28}-dependent transcription late in the developmental cycle because it inhibits σ^{66} RNA polymerase but not σ^{28} RNA polymerase (Rao et al., 2009). Scc4 was also more abundant in purified EBs than RBs (Rao et al., 2009), which raises the possibility that it is an inhibitor of σ^{66}-dependent transcription in EBs. Intriguingly, Scc4 is also proposed to be a chaperone for the type III secretion system (Betts-Hampikian and Fields, 2010) (see chapter 9, "Protein secretion and *Chlamydia* pathogenesis"), and it interacted with another type III chaperone, SccI, in a yeast two-hybrid assay (Spaeth et al., 2009). Thus, Scc4 has the potential to serve dual roles that link secretion of protein effectors and gene expression in *Chlamydia* (Rao et al., 2009).

TRANSCRIPTIONAL REGULATION OF EARLY GENES

Early genes are transcribed within 3 hours of an EB entering its host cell, when the large majority of other chlamydial genes are transcriptionally silent (Shaw et al., 2000; Belland et al., 2003b). This selective transcription of early genes indicates that they are likely to have important roles in establishing the chlamydial inclusion and in EB-to-RB conversion. Early genes encode subunits of RNA polymerase, 16S rRNA, ribosomal proteins, and DNA polymerase, which are the basic machinery for transcription, translation, and DNA replication. They also include the genes for the major chlamydial heat shock proteins, DnaK, GroEL, and GroES. Early genes also encode proteins involved in nutrient and energy acquisition, such as an ADP/ATP translocase, and a large number of inclusion membrane proteins. Some early genes encode proteins that have roles in regulating later temporal classes of genes. Among them are DNA gyrase, which is predicted to upregulate midcycle gene expression by increasing supercoiling levels in midcycle (Niehus et al., 2008), and EUO (Rosario and Tan, 2012) and RsbW, which are repressors of late genes.

The selective transcription of early genes at the start of an intracellular chlamydial infection indicates that these early genes must be regulated differently from later temporal classes. One possibility is that there are specific mechanisms to directly regulate the expression of early genes, which are only known to be transcribed by σ^{66} RNA polymerase. Alternatively, early genes may be constitutively active, and they may appear to be selectively expressed because transcription of midcycle and late genes is regulated to prevent their expression until later in the developmental cycle. In addition, there must be a mechanism to allow an EB to become transcriptionally competent when it enters a host cell. This relief of transcriptional silencing and the selective transcription of early genes may be accomplished together in a single step or as separate mechanisms.

Disruption of Histone-DNA Binding and Chromatin Decondensation

An obvious mechanism for reversing the general silencing of transcription in EBs is to relieve histone-mediated chromatin condensation. Grieshaber and Hackstadt provided experimental evidence that chromatin decondensation can be induced by disrupting the binding of histone-like proteins to DNA (Grieshaber et al., 2004, 2006). Using a genetic selection in *E. coli*, they demonstrated that binding of Hc1 and Hc2 to DNA was reversible. Furthermore, they showed that histone-DNA binding was disrupted by a small metabolite from the nonmevalonate methylerythritol 4-phosphate (MEP) pathway of isoprenoid biosynthesis.

This mechanism of chromatin decondensation is likely to be important for EB-to-RB conversion, but it may not be the initial switch that turns on chlamydial transcription at the very beginning of the developmental cycle. The Belland microarray study showed that 29 immediate early genes were transcribed by 1 hpi when intracellular chlamydiae were still at the EB stage by electron microscopy. Furthermore, 200 early genes were transcribed by 3 hpi when transitional forms in the process of EB-to-RB conversion were present and undergoing chromatin decondensation (Belland et al., 2003b). Thus, chlamydial transcription precedes full chromatin decondensation, which is completed by 6 to 8 hpi. There is experimental evidence that chlamydial transcription and translation are, in fact, necessary for chromatin decondensation (Grieshaber et al., 2004). Intriguingly, the seven genes that make up the MEP pathway in *Chlamydia* are all transcribed as early genes within 2 hpi (Grieshaber et al., 2004). These findings support a model in which the MEP pathway metabolite is synthesized as a consequence of early gene expression, leading to disruption of histone-DNA binding and chromatin decondensation. However, unless there is a presynthesized pool of the metabolite that is released upon EB uptake into the host cell, it cannot be the mechanism that initiates chlamydial transcription at the start of the developmental cycle.

Regulation of Early Genes by DNA Supercoiling

The effect of DNA supercoiling on early genes has been examined in light of the important role of DNA topology for temporal regulation in *Chlamydia*. As discussed already, expression of midcycle genes has been proposed to be upregulated by increased supercoiling levels in midcycle (Niehus et al., 2008). At early times in the developmental cycle, however, supercoiling levels are lower, and early genes but not midcycle genes are transcribed. This difference in expression pattern could result if early promoters are relatively insensitive to changes in DNA supercoiling, allowing them to be transcribed at supercoiling levels that are below the threshold required to upregulate midcycle genes.

We have used a supercoiling sensitivity transcription assay to determine if early promoters are supercoiling independent compared to midcycle promoters (E. Cheng and M. Tan, unpublished data). We tested promoters for 11 early genes located at various positions on the *C. trachomatis* chromosome and plasmid. For seven early promoters, there was little change in the amount of in vitro transcription over a range of DNA superhelicities. These findings support a model in which early genes are selectively transcribed at early times, when supercoiling levels are low because they have supercoiling-insensitive promoters.

This study also showed that a subset of early genes resemble midcycle genes in having promoters that are supercoiling responsive. Four early promoters were transcribed at higher levels from a more supercoiled DNA template, and their genes had higher in vivo transcript levels in midcycle (Cheng and Tan, unpublished). In contrast, early genes with supercoiling-insensitive promoters had transcript levels that remained the same between early times and midcycle. These findings support a broadened role of supercoiling as a general mechanism to upregulate the expression of the large group of midcycle genes as well as a subset of early genes in the mid-stage of the developmental cycle.

REGULATION BY SMALL RNAs

There is emerging evidence that gene expression during the chlamydial developmental cycle is also regulated by small RNAs (sRNAs). As has been discussed, studies on chlamydial gene regulation have focused on RNA polymerase and how transcription initiation can be modulated by DNA topology and transcription factors (Tan, 2006; Niehus et al., 2008). There is ample precedent from other bacteria, however, for additional control mechanisms after genes have been transcribed. In particular, there has been growing interest in the regulation of gene expression by sRNAs, which are short, noncoding RNAs of about 50 to 500 nucleotides (nt) in length that target transcripts of specific genes.

sRNAs can regulate gene expression by a number of different mechanisms (Gottesman and Storz, 2010). *cis*-encoded sRNAs are antisense sRNAs that regulate gene expression at the posttranscriptional level by base pairing with their respective target mRNAs and altering message stability. Their targets are the easiest to identify because the *cis*-encoded sRNA and its target mRNA are synthesized from the same gene, albeit from complementary DNA strands. In contrast, *trans*-encoded RNAs modulate translation initiation by base pairing their target mRNAs with limited complementarity. For example, a *trans*-encoded sRNA can inhibit translation by binding and blocking the ribosome-binding site. Alternatively, a *trans*-encoded sRNA can bind its target mRNA and stimulate translation initiation by altering the structure of the ribosome-binding site and facilitating access by the translational machinery. The target genes of *trans*-encoded sRNAs can be difficult to identify because they do not share the same genome location as their sRNA. In addition, the limited sequence complementarity makes them difficult to predict by computational methods. A third class of sRNAs has been shown to modify the activity of a protein by competing with the natural RNA or DNA targets for binding to the protein.

The first evidence that sRNAs have a role in chlamydial gene regulation came when a *trans*-encoded RNA called IhtA was shown to inhibit translation of the histone-like protein Hc1. Grieshaber, Hackstadt, and colleagues identified IhtA in a clever screen for chlamydial genes that rescued the lethal effect of Hc1 expression in *E. coli* (Grieshaber et al., 2004). This screen implicated genes encoding the MEP pathway in the disruption of Hc1-DNA binding, as has already been discussed. However, its forward genetic design allowed it to reveal a second locus that did not correspond to a known coding sequence. Instead, it was located in an intergenic region that was found to encode a 120-nt sRNA, which was named IhtA. In a heterologous expression assay in *E. coli*, IhtA inhibited translation of Hc1, but not Hc2, and did not affect Hc1 transcript levels (Grieshaber et al., 2005, 2006). IhtA and Hc1 have opposite expression patterns in *Chlamydia* since IhtA is transcribed from early times until its levels decrease late in the developmental cycle, while Hc1 protein levels are lowest during the RB stage. These findings support a model in which IhtA prevents the translation of Hc1 in RBs (Grieshaber et al., 2005). IhtA therefore represents a new class of chlamydial temporal regulators that are sRNAs which downregulate translation of specific target genes.

Two complementary genome-wide studies have since identified many more sRNAs in *C. trachomatis*. A deep sequencing study in serovar L2b provided experimental evidence for 16 *trans*-encoded sRNA and 25 *cis*-encoded sRNAs ranging in size from 90 to 400 nt (Albrecht et al., 2010). An intergenic microarray study identified 31 *trans*-encoded sRNAs and 3 *cis*-encoded sRNAs from serovar D (Abdelrahman et al., 2011). *Chlamydia* spp. are likely to encode more sRNAs because neither study examined chlamydial RNA before 24 hpi, and both would therefore have missed sRNAs that are only present at earlier times. It is not known whether sRNAs are conserved among *Chlamydia* spp. or if there are species and strain-specific sRNAs that might account for the range of clinical manifestations caused by different chlamydial isolates.

So far only one other chlamydial sRNA besides IhtA has been shown to regulate the expression of a specific target gene. CTIG270 is a *cis*-encoded sRNA that is transcribed from the strand opposite to *ftsI*, which encodes a protein involved in peptidoglycan synthesis and cell division in other bacteria. CTIG270 is only expressed late in the developmental cycle, at the same time that *ftsI* is downregulated, and it decreased *ftsI* transcripts to undetectable levels in an *E. coli* expression assay (Abdelrahman et al., 2011). These data support a role for CTIG270 as an antisense sRNA that reduces *ftsI* message stability at late time points.

How likely is it that additional chlamydial sRNAs are involved in temporal gene regulation during the developmental cycle? The examples of IhtA and CTIG270 illustrate that an sRNA can function as a temporal regulator if its own expression pattern is developmentally regulated. In other words, the temporal expression of IhtA is necessary for the developmental regulation of its target gene, *hctA*. A number of sRNAs with temporal expression patterns were identified in the intergenic microarray study (Abdelrahman et al., 2011), which makes them candidate temporal regulators. sRNAs were also differentially expressed in response to gamma interferon treatment (Abdelrahman et al., 2011), which suggests that sRNAs may regulate gene expression during chlamydial persistence (see chapter 12, "Chlamydial persistence redux"). The expression pattern of an sRNA could be controlled at the level of sRNA stability. However, it is reasonable to predict that sRNA levels will be controlled by the same mechanisms of transcriptional regulation that govern the expression of chlamydial mRNA. Thus, this new class of chlamydial temporal regulators is likely to function downstream of transcriptional regulation, both in controlling gene expression after transcription initiation and in being dependent on transcriptional mechanisms for their own temporal regulation.

PERSPECTIVE: PUTTING IT ALL TOGETHER

The following model for temporal gene expression during the chlamydial developmental cycle integrates the regulatory mechanisms that have been discussed in this chapter (Fig. 1) (Table 1). In some instances, the proposed mechanism for a temporal class of genes has been extrapolated from studies on representative genes. For example, the supercoiling response has been studied in vitro for a small number of early, midcycle, and late promoters. Additional genome-wide and in vivo studies would provide stronger support for the model.

At the beginning of the intracellular infection, when the transcriptionally silent EB enters a host cell, an undefined switch makes the EB transcriptionally competent, and immediate early genes are transcribed within an hour (Belland et al., 2003b). Next, early genes, which are important for establishing the intracellular infection, are transcribed while the EB is converting into an RB (Shaw et al., 2000; Belland et al., 2003b). A possible reason for this selective transcription is that promoters for early, but not midcycle, genes can be transcribed at the low supercoiling levels that are present early in the developmental cycle (Niehus et al., 2008; Case et al., 2010). Although promoters for late genes can also be

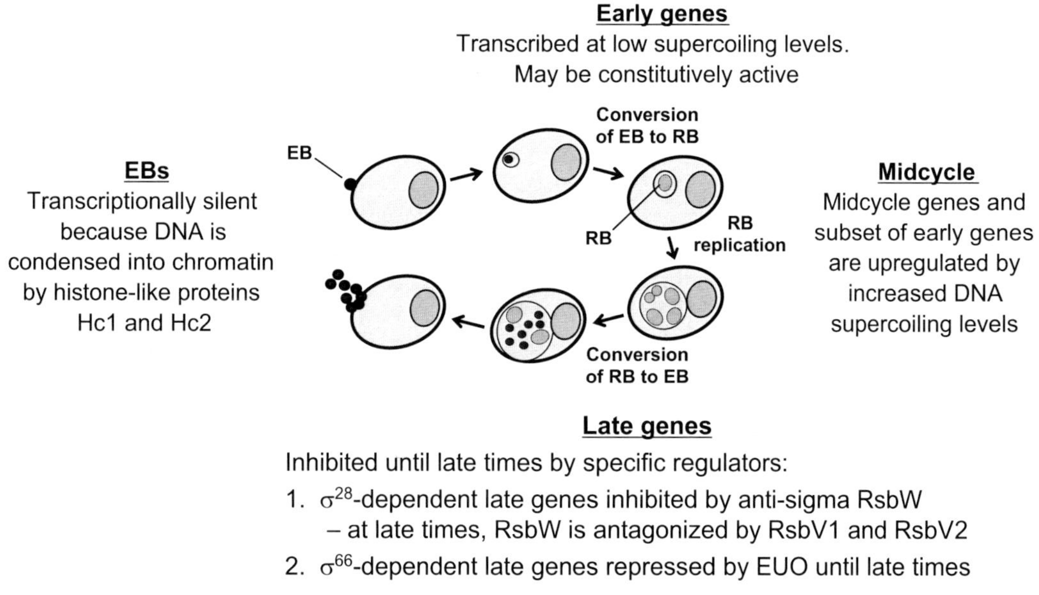
Early genes
Transcribed at low supercoiling levels.
May be constitutively active
Conversion of EB to RB
EB
EBs
Transcriptionally silent because DNA is condensed into chromatin by histone-like proteins Hc1 and Hc2
RB
RB replication
Midcycle
Midcycle genes and subset of early genes are upregulated by increased DNA supercoiling levels
Conversion of RB to EB
Late genes
Inhibited until late times by specific regulators:
1. σ28-dependent late genes inhibited by anti-sigma RsbW
– at late times, RsbW is antagonized by RsbV1 and RsbV2
2. σ66-dependent late genes repressed by EUO until late times

FIGURE 1 Proposed mechanisms for temporal regulation of gene expression during the chlamydial developmental cycle. doi:10.1128/9781555817329.ch7.f1

TABLE 1 Chlamydial factors for which there are data to support a role in regulating temporal gene expression during the developmental cycle

Expressed as:	Temporal regulator	Function	Regulatory role	Reference(s)
Early genes	MEP pathway genes	Produce metabolite that disrupts binding of Hc1 and Hc2 to DNA	Prevent Hc1- and Hc2-mediated DNA condensation in RBs	Grieshaber et al., 2004, 2006
	IhtA	sRNA that inhibits Hc1 translation	Prevents Hc1-mediated DNA condensation in RBs	Grieshaber et al., 2005
	DNA gyrase (encoded by *gyrA_1* and *gyrB_1*)	Increases global DNA supercoiling levels	Upregulates genes in midcycle (midcycle genes and subset of early genes)	Niehus et al., 2008; Cheng and Tan, 2012
	EUO	Represses promoters of late genes transcribed by σ^{66} RNA polymerase	Prevents premature expression of σ^{66}-dependent late genes	Rosario and Tan, 2012
	RsbW	Antagonist of late regulator σ^{28}	Prevents premature expression of σ^{28}-dependent late genes	Rosario and Tan, unpublished
Midcycle genes	σ^{28}	Alternative sigma factor that is present in σ^{28} RNA polymerase	Regulates subset of late genes	Yu et al., 2006b
Late genes	Hc1 and Hc2	Histone-like proteins that bind and condense DNA	Global silencing of transcription in EBs	Barry et al., 1992; Brickman et al., 1993
	CTIG270	Antisense sRNA that downregulates *ftsI* transcript levels	Prevents RB cell division?	AbdelRahman et al., 2011
	Scc4 (CT663)	Inhibitor of σ^{66} RNA polymerase	Unknown	Rao et al., 2009

transcribed at low supercoiling levels (Niehus et al., 2008; Case et al., 2010), they are inhibited by the repressor EUO, which is expressed from very early times in the developmental cycle (Zhang et al., 1998).

This early stage of the developmental cycle is defined by the conversion of the EB into an RB. Although a number of processes may be involved in this conversion, a prominent step is the decondensation of chromatin. This decondensation is mediated by the production of a small metabolite by the MEP pathway, which disrupts binding of the histone-like proteins Hc1 and Hc2 to chlamydial DNA (Grieshaber et al., 2004, 2006). In addition, the sRNA IhtA prevents new translation of Hc1 until late in the developmental cycle (Grieshaber et al., 2005).

The expression of midcycle genes begins when EB-to-RB conversion has been completed and the developmental cycle enters the RB replication phase. This stage is characterized by increased supercoiling levels, which peak in midcycle. The increase in supercoiling is likely caused by DNA gyrase, which is encoded by two early genes (Belland et al., 2003b). Since Hc1 appears to decrease supercoiling levels (Barry et al., 1993), the disassociation of this histone-like protein from DNA may also contribute to increased supercoiling. This higher level of DNA supercoiling is proposed to be the driving force that upregulates midcycle genes (Niehus et al., 2008; Case et al., 2010), as well as a subset of early genes that are transcribed at higher levels in midcycle (Cheng and Tan, 2012).

Late genes consist of two subsets that are each transcribed by a different form of RNA polymerase and negatively regulated by a specific inhibitor to prevent their premature expression until late in the developmental cycle. These genes are a specialized group of genes involved in RB-to-EB conversion and other late events. One subset of late genes are transcribed by σ^{66} RNA polymerase but repressed by EUO until late in the developmental cycle (Rosario and Tan, 2012). This mechanism leads to specific repression of σ^{66}-dependent late genes without preventing the expression of early and midcycle genes. The mechanism of derepression that allows late expression of these genes has not been determined. However, it may involve the regulation of EUO protein levels, since the transcription of σ^{66}-dependent late genes coincides with the decrease in EUO protein and transcript levels at late times (Zhang et al., 1998). A second subset of late genes are transcribed by an alternative form of RNA polymerase containing σ^{28} (Yu and Tan, 2003; Yu et al., 2006b). This alternative sigma factor is proposed to be bound and sequestered by an anti-sigma factor, RsbW, until late time points when RsbW binds to one or both RsbV proteins and releases σ^{28}. σ^{28} RNA polymerase is then active and transcribes its target late genes. RsbW, RsbV1, and RsbV2 are part of a signaling pathway (Hua et al., 2006) that may allow RBs to sense and respond to an environmental signal by activating σ^{28}-dependent late gene expression.

These parallel control mechanisms may allow late gene expression to respond to more than one regulatory signal and provide safeguards to ensure that conversion into an EB is tightly controlled. This regulation of late gene expression is likely to determine whether an RB continues to replicate or if it terminally differentiates into an EB, thus controlling the balance between the production of more progeny and the proportion of these progeny that are infectious. Since RB-to-EB conversion is asynchronous, the regulatory inputs are predicted to act at the level of individual RBs and not to affect all RBs in a similar manner at the same time.

The expression of specific genes during the developmental cycle may also be regulated by sRNAs. For example, late expression of the sRNA CTIG270 is proposed to downregulate *ftsI* (Abdelrahman et al., 2011), which may favor RB-to-EB conversion over RB replication since *ftsI* may have a role in RB cell division. sRNAs can also provide additional safety mechanisms to regulate expression of specific genes. For example, Hc1 expression prior to late time points is prevented not only by EUO, which represses its transcription (Rosario and Tan, 2012), but also by the sRNA IhtA, which prevents its translation (Grieshaber et al., 2005).

Expression of late genes mediates the conversion of an RB into an EB. Transcription of the genes for Hc1 by σ^{66} RNA polymerase (Fahr et al., 1995) and Hc2 by σ^{28} RNA polymerase (Yu and Tan, 2003) lead to condensation of chlamydial DNA (Barry et al., 1992; Brickman et al., 1993). Other σ^{66}-dependent late genes encode EB-specific cysteine-rich outer membrane proteins (Clarke et al., 1988; Mathews et al., 1993) and putative thioredoxin disulfide isomerases and membrane thiol proteases (Belland et al., 2003b). These genes may be involved in formation of the outer membrane complex that protects the EB from the extracellular environment upon release from the host cell.

An individual EB represents both the end of a developmental cycle in one infected cell and the potential of beginning another cycle in a new host cell. EBs contain σ^{66} RNA polymerase (Shaw et al., 2002; Skipp et al., 2005; Sixt et al., 2011), but their DNA is condensed by the histone-like proteins, which prevents transcription (Barry et al., 1993; Pedersen et al., 1994, 1996). EBs also contain an inhibitor of RNA polymerase activity, called Scc4 (Rao et al., 2009). EBs are thus transcriptionally silent, but they carry both the transcriptional machinery and genetic material to initiate gene expression in the next developmental cycle.

FINAL WORDS

The regulators of temporal gene expression in *Chlamydia* have a critical role in the successful progression and completion of the developmental cycle. *Chlamydia* orchestrates this temporal expression with a wide range of mechanisms of prokaryotic gene regulation, including different forms of RNA polymerase, transcription factors, enzymes that modulate DNA supercoiling, nucleoid-associated proteins, and sRNAs (Table 1). In considering how these mechanisms relate to each other, it becomes apparent that a regulator for a given temporal class of genes is usually expressed in a prior temporal class. For example, DNA gyrase, a regulator of midcycle gene expression, and EUO, a late repressor, are both expressed as early genes. Thus, the regulatory mechanisms for the different temporal classes of genes are interconnected, and each stage of the developmental cycle prepares the way for the next in an ongoing cycle. The mechanisms that control the temporal expression of a chlamydial protein are not the only regulators of its function, since its activity may depend on subsequent processing, covalent modification, or secretion into the host cytosol. Nonetheless, the temporal regulators of the developmental cycle are important for chlamydial growth and replication and the production of infectious progeny and may be attractive therapeutic targets for novel antichlamydial strategies.

ACKNOWLEDGMENTS

I thank Christine Suetterlin, Patrik Bavoil, Johnny Akers, Allan Chen, Eric Cheng, and Chris Rosario for critical reviews and suggestions that improved the manuscript. This work was supported by a grant from the NIH (AI 44198).

REFERENCES

Abdelrahman, Y. M., and R. J. Belland. 2005. The chlamydial developmental cycle. *FEMS Microbiol. Rev.* **29:**949–959.

Abdelrahman, Y. M., L. A. Rose, and R. J. Belland. 2011. Developmental expression of non-coding RNAs in *Chlamydia trachomatis* during normal and persistent growth. *Nucleic Acids Res.* **39:**1843–1854.

Akers, J. C., H. HoDac, R. H. Lathrop, and M. Tan. 2011. Identification and functional analysis of CT069 as a novel transcriptional regulator in *Chlamydia. J. Bacteriol.* **193:**6123–6131.

Akers, J. C., and M. Tan. 2006. Molecular mechanism of tryptophan-dependent transcriptional regulation in *Chlamydia trachomatis. J. Bacteriol.* **188:**4236–4243.

Albrecht, M., C. M. Sharma, R. Reinhardt, J. Vogel, and T. Rudel. 2010. Deep sequencing-based discovery of the *Chlamydia trachomatis* transcriptome. *Nucleic Acids Res.* **38:**868–877.

Bae, S. H., S. H. Yun, D. Sun, H. M. Lim, and B. S. Choi. 2006. Structural and dynamic basis of a supercoiling-responsive DNA element. *Nucleic Acids Res.* **34:**254–261.

Barry, C., S. Hayes, and T. Hackstadt. 1992. Nucleoid condensation in *Escherichia coli* that express a chlamydial histone homolog. *Science* **256:**377–379.

Barry, C. E., III, T. J. Brickman, and T. Hackstadt. 1993. Hc1-mediated effects on DNA structure: a potential regulator of chlamydial development. *Mol. Microbiol.* **9:**273–283.

Belland, R. J., D. E. Nelson, D. Virok, D. D. Crane, D. Hogan, D. Sturdevant, W. L. Beatty, and H. D. Caldwell. 2003a. Transcriptome analysis of chlamydial growth during IFN-gamma-mediated persistence and reactivation. *Proc. Natl. Acad. Sci. USA* **100:**15971–15976.

Belland, R. J., G. Zhong, D. D. Crane, D. Hogan, D. Sturdevant, J. Sharma, W. L. Beatty, and H. D. Caldwell. 2003b. Genomic transcriptional profiling of the developmental cycle of *Chlamydia trachomatis. Proc. Natl. Acad. Sci. USA* **100:**8478–8483.

Betts-Hampikian, H. J., and K. A. Fields. 2010. The chlamydial type III secretion mechanism: revealing cracks in a tough nut. *Frontiers Microbiol.* doi: 10.3389/fmicb.2010.00114.

Brickman, T. J., C. E. Barry III, and T. Hackstadt. 1993. Molecular cloning and expression of *hctB* encoding a strain-variant chlamydial histone-like protein with DNA-binding activity. *J. Bacteriol.* **175:**4274–4281.

Browning, D. F., and S. J. Busby. 2004. The regulation of bacterial transcription initiation. *Nat. Rev. Microbiol.* **2:**57–65.

Carlson, J. H., H. Wood, C. Roshick, H. D. Caldwell, and G. McClarty. 2006. *In vivo* and *in vitro* studies of *Chlamydia trachomatis* TrpR:DNA interactions. *Mol. Microbiol.* **59:**1678–1691.

Case, E. D., J. C. Akers, and M. Tan. 2011. CT406 encodes a chlamydial ortholog of NrdR, a repressor of ribonucleotide reductase. *J. Bacteriol.* **193:**4396–4404.

Case, E. D., E. M. Peterson, and M. Tan. 2010. Promoters for *Chlamydia* type III secretion genes show a differential response to DNA supercoiling that correlates with temporal expression pattern. *J. Bacteriol.* **192:**2569–2574.

Cheng, E., and M. Tan. 2012. Differential effects of DNA supercoiling on *Chlamydia* early promoters correlate with expression patterns in midcycle. *J. Bacteriol.* **194:**3109–3115.

Clarke, I. N., M. E. Ward, and P. R. Lambden. 1988. Molecular cloning and sequence analysis of a developmentally regulated cysteine-rich outer membrane protein from *Chlamydia trachomatis*. *Gene* **71:**307–314.

Collingro, A., P. Tischler, T. Weinmaier, T. Penz, E. Heinz, R. C. Brunham, T. D. Read, P. M. Bavoil, K. Sachse, S. Kahane, M. G. Friedman, T. Rattei, G. S. Myers, and M. Horn. 2011. Unity in variety—the pan-genome of the *Chlamydiae*. *Mol. Biol. Evol.* **28:**3253–3270.

Dorman, C. J. 2006. DNA supercoiling and bacterial gene expression. *Sci. Prog.* **89:**151–166.

Douglas, A., and T. P. Hatch. 2006. The phosphorelay system in *Chlamydia trachomatis*, p. 177–180. *In* M. Chernesky (ed.), *Chlamydial Infections: Proceedings of the Eleventh International Symposium on Human Chlamydial Infections*. Niagara-on-the-Lake, Canada.

Douglas, A. L., and T. P. Hatch. 2000. Expression of the transcripts of the sigma factors and putative sigma factor regulators of *Chlamydia trachomatis* L2. *Gene* **247:**209–214.

Drlica, K. 1992. Control of bacterial DNA supercoiling. *Mol. Microbiol.* **6:**425–433.

Engel, J., and D. Ganem. 1990a. Identification and comparison of putative chlamydial promoter elements, p. 245–260. *In* L. Van der Ploeg (ed.), *Immune Recognition and Evasion: Molecular Aspects of Host Parasite Interaction*. Academic Press Inc., San Diego, CA.

Engel, J., and D. Ganem. 1990b. A PCR-based approach to cloning sigma factors from eubacteria and its application to the isolation of a sigma70 homolog from *Chlamydia trachomatis*. *J. Bacteriol.* **172:**2447–2455.

Fahr, M. J., A. L. Douglas, W. Xia, and T. P. Hatch. 1995. Characterization of late gene promoters of *Chlamydia trachomatis*. *J. Bacteriol.* **177:**4252–4260.

Fan, T., H. Lu, H. Hu, L. Shi, G. A. McClarty, D. M. Nance, A. H. Greenberg, and G. Zhong. 1998. Inhibition of apoptosis in *Chlamydia*-infected cells: blockade of mitochondrial cytochrome c release and caspase activation. *J. Exp. Med.* **187:**487–496.

Fang, F. C. 2005. Sigma cascades in prokaryotic regulatory networks. *Proc. Natl. Acad. Sci. USA* **102:**4933–4934.

Gao, R., and A. M. Stock. 2009. Biological insights from structures of two-component proteins. *Annu. Rev. Microbiol.* **63:**133–154.

Gmuender, H., K. Kuratli, K. Di Padova, C. P. Gray, W. Keck, and S. Evers. 2001. Gene expression changes triggered by exposure of *Haemophilus influenzae* to novobiocin or ciprofloxacin: combined transcription and translation analysis. *Genome Res.* **11:**28–42.

Gottesman, S., and G. Storz. 27 October 2010. Bacterial small RNA regulators: versatile roles and rapidly evolving variations. *Cold Spring Harb. Perspect. Biol.* doi: 10.1101/cshperspect.a003798

Grieshaber, N. A., E. R. Fischer, D. J. Mead, C. A. Dooley, and T. Hackstadt. 2004. Chlamydial histone-DNA interactions are disrupted by a metabolite in the methylerythritol phosphate pathway of isoprenoid biosynthesis. *Proc. Natl. Acad. Sci. USA* **101:**7451–7456.

Grieshaber, N. A., S. S. Grieshaber, E. R. Fischer, and T. Hackstadt. 2005. A small RNA inhibits translation of the histone-like protein Hc1 in *Chlamydia trachomatis*. *Mol. Microbiol.* **59:**541–550.

Grieshaber, N. A., J. B. Sager, C. A. Dooley, S. F. Hayes, and T. Hackstadt. 2006. Regulation of the *Chlamydia trachomatis* histone H1-like protein Hc2 is IspE dependent and IhtA independent. *J. Bacteriol.* **188:**5289–5292.

Gruber, T. M., and C. A. Gross. 2003. Multiple sigma subunits and the partitioning of bacterial transcription space. *Annu. Rev. Microbiol.* **57:**441–466.

Hackstadt, T. 1991. Purification and N-terminal amino acid sequences of *Chlamydia trachomatis* histone analogs. *J. Bacteriol.* **173:**7046–7049.

Hackstadt, T. 1999. Cell biology, p. 101–138. *In* R. S. Stephens (ed.), *Chlamydia: Intracellular Biology, Pathogenesis, and Immunity*. ASM Press, Washington, DC.

Haider, S., M. Wagner, M. C. Schmid, B. S. Sixt, J. G. Christian, G. Hacker, P. Pichler, K. Mechtler, A. Muller, C. Baranyi, E. R. Toenshoff, J. Montanaro, and M. Horn. 2010. Raman microspectroscopy reveals long-term extracellular activity of chlamydiae. *Mol. Microbiol.* **77:**687–700.

Haldenwang, W. G. 1995. The sigma factors of *Bacillus subtilis*. *Microbiol. Rev.* **59:**1–30.

Hatch, T. P. 1996. Disulfide cross-linked envelope proteins: the functional equivalent of peptidoglycan in chlamydiae? *J. Bacteriol.* **178:**1–5.

Hatfield, G. W., and C. J. Benham. 2002. DNA topology-mediated control of global gene expression in *Escherichia coli*. *Annu. Rev. Genet.* **36:**175–203.

Hickey, J. M., L. Weldon, and P. S. Hefty. 2011. The atypical OmpR/PhoB response regulator ChxR from *Chlamydia trachomatis* forms homodimers in vivo and binds a direct repeat of nucleotide sequences. *J. Bacteriol.* **193:**389–398.

Hooper, D. C. 2000. Mechanisms of action and resistance of older and newer fluoroquinolones. *Clin. Infect. Dis.* **31**(Suppl. 2)**:**S24–S28.

Horn, M., A. Collingro, S. Schmitz-Esser, C. L. Beier, U. Purkhold, B. Fartmann, P. Brandt, G. J. Nyakatura, M. Droege, D. Frishman, T. Rattei, H. W. Mewes, and M. Wagner. 2004. Illuminating the evolutionary history of chlamydiae. *Science* **304:**728–730.

Hua, L., P. S. Hefty, Y. J. Lee, Y. M. Lee, R. S. Stephens, and C. W. Price. 2006. Core of the partner switching signalling mechanism is conserved in the obligate intracellular pathogen *Chlamydia trachomatis*. *Mol. Microbiol.* **59:**623–636.

Hybiske, K., and R. S. Stephens. 2007. Mechanisms of host cell exit by the intracellular bacterium *Chlamydia*. *Proc. Natl. Acad. Sci. USA* **104:**11430–11435.

Karlinsey, J. E., and K. T. Hughes. 2006. Genetic transplantation: *Salmonella enterica* serovar Typhimurium as a host to study sigma factor and anti-sigma factor interactions in genetically intractable systems. *J. Bacteriol.* **188:**103–114.

Kazmierczak, M. J., M. Wiedmann, and K. J. Boor. 2005. Alternative sigma factors and their roles in bacterial virulence. *Microbiol. Mol. Biol. Rev.* **69:**527–543.

Koo, I. C., and R. S. Stephens. 2003. A developmentally regulated two-component signal transduction system in *Chlamydia*. *J. Biol. Chem.* **278:**17314–17319.

Lad, S. P., J. Li, J. da Silva Correia, Q. Pan, S. Gadwal, R. J. Ulevitch, and E. Li. 2007. Cleavage of p65/RelA of the NF-kappaB pathway by *Chlamydia*. *Proc. Natl. Acad. Sci. USA* **104:**2933–2938.

Lambden, P. R., J. S. Everson, M. E. Ward, and I. N. Clarke. 1990. Sulfur-rich proteins of *Chlamydia trachomatis*: developmentally regulated transcription of polycistronic mRNA from tandem promoters. *Gene* **87:**105–112.

Madan Babu, M., S. A. Teichmann, and L. Aravind. 2006. Evolutionary dynamics of prokaryotic transcriptional regulatory networks. *J. Mol. Biol.* **358:**614–633.

Martinez-Antonio, A., and J. Collado-Vides. 2003. Identifying global regulators in transcriptional regulatory networks in bacteria. *Curr. Opin. Microbiol.* **6:**482–489.

Mathews, S. A., A. Douglas, K. S. Sriprakash, and T. P. Hatch. 1993. *In vitro* transcription in *Chlamydia psittaci* and *Chlamydia trachomatis*. *Mol. Microbiol.* **7:**937–946.

Mathews, S. A., K. M. Volp, and P. Timms. 1999. Development of a quantitative gene expression assay for *Chlamydia trachomatis* identified temporal expression of sigma factors. *FEBS Lett.* **458:**354–358.

Maurer, A. P., A. Mehlitz, H. J. Mollenkopf, and T. F. Meyer. 2007. Gene expression profiles of *Chlamydophila pneumoniae* during the developmental cycle and iron depletion-mediated persistence. *PLoS Pathog.* **3:**e83.

Moulder, J. W. 1991. Interaction of chlamydiae and host cells *in vitro*. *Microbiol. Rev.* **55:**143–190.

Nicholson, T. L., L. Olinger, K. Chong, G. Schoolnik, and R. S. Stephens. 2003. Global stage-specific gene regulation during the developmental cycle of *Chlamydia trachomatis*. *J. Bacteriol.* **185:**3179–3189.

Niehus, E., E. Cheng, and M. Tan. 2008. DNA supercoiling-dependent gene regulation in *Chlamydia*. *J. Bacteriol.* **190:**6419–6427.

Pedersen, L. B., S. Birkelund, and G. Christiansen. 1994. Interaction of the *Chlamydia trachomatis* histone H1-like protein (Hc1) with DNA and RNA causes repression of transcription and translation *in vitro*. *Mol. Microbiol.* **11:**1085–1098.

Pedersen, L. B., S. Birkelund, and G. Christiansen. 1996. Purification of recombinant *Chlamydia trachomatis* histone H1-like protein Hc2, and comparative functional analysis of Hc2 and Hc1. *Mol. Microbiol.* **20:**295–311.

Perara, E., D. Ganem, and J. Engel. 1992. A developmentally regulated chlamydial gene with apparent homology to eukaryotic histone H1. *Proc. Natl. Acad. Sci. USA* **89:**2125–2129.

Peter, B. J., J. Arsuaga, A. M. Breier, A. B. Khodursky, P. O. Brown, and N. R. Cozzarelli. 2004. Genomic transcriptional response to loss of chromosomal supercoiling in *Escherichia coli*. *Genome Biol.* **5:**R87.

Plaunt, M. R., and T. P. Hatch. 1988. Protein synthesis early in the developmental cycle of *Chlamydia psittaci*. *Infect. Immun.* **56:**3021–3025.

Price, C. W. 2002. General stress response, p. 369–384. *In* A. L. Sonenshein, J. A. Hoch, and R. Losick (ed.), Bacillus subtilis *and Its Closest Relatives: from Genes to Cells*. ASM Press, Washington, DC.

Rao, X., P. Deighan, Z. Hua, X. Hu, J. Wang, M. Luo, J. Wang, Y. Liang, G. Zhong, A. Hochschild, and L. Shen. 2009. A regulator from *Chlamydia trachomatis* modulates the activity of RNA polymerase through direct interaction with the beta subunit and the primary sigma subunit. *Genes Dev.* **23:**1818–1829.

Rosario, C. J., and M. Tan. 2012. The early gene product EUO is a transcriptional repressor that selectively regulates promoters of *Chlamydia* late genes. *Mol. Microbiol.* **84:**1097–1107.

Schachter, J. 1988. The intracellular life of *Chlamydia*. *Curr. Topics Microbiol. Immunol.* **138:**109–139.

Schaumburg, C. S., and M. Tan. 2000. A positive cis-acting DNA element is required for high-level transcription in *Chlamydia*. *J. Bacteriol.* **182:**5167–5171.

Schaumburg, C. S., and M. Tan. 2006. Arginine-dependent gene regulation via the ArgR repressor is species specific in *Chlamydia*. *J. Bacteriol.* **188:**919–927.

Shaw, A. C., K. Gevaert, H. Demol, B. Hoorelbeke, J. Vandekerckhove, M. R. Larsen, P. Roepstorff, A. Holm, G. Christiansen, and S. Birkelund. 2002. Comparative proteome analysis of *Chlamydia trachomatis* serovar A, D and L2. *Proteomics* **2:**164–186.

Shaw, E. I., C. A. Dooley, E. R. Fischer, M. A. Scidmore, K. A. Fields, and T. Hackstadt. 2000. Three temporal classes of gene expression during the *Chlamydia trachomatis* developmental cycle. *Mol. Microbiol.* **37:**913–925.

Shen, L., X. Feng, Y. Yuan, X. Luo, T. P. Hatch, K. T. Hughes, J. S. Liu, and Y. X. Zhang. 2006. Selective promoter recognition by chlamydial sigma28 holoenzyme. *J. Bacteriol.* **188:**7364–7377.

Shen, L., M. Li, and Y. X. Zhang. 2004. *Chlamydia trachomatis* sigma28 recognizes the *fliC* promoter of *Escherichia coli* and responds to heat shock in chlamydiae. *Microbiology* **150:**205–215.

Sixt, B. S., C. Heinz, P. Pichler, E. Heinz, J. Montanaro, H. J. Op den Camp, G. Ammerer, K. Mechtler, M. Wagner, and M. Horn. 2011. Proteomic analysis reveals a virtually complete set of proteins for translation and energy generation in elementary bodies of the amoeba symbiont *Protochlamydia amoebophila*. *Proteomics* **11:**1868–1892.

Skipp, P., J. Robinson, C. D. O'Connor, and I. N. Clarke. 2005. Shotgun proteomic analysis of *Chlamydia trachomatis*. *Proteomics* **5:**1558–1573.

Solbrig, M. V., M. L. Wong, and R. S. Stephens. 1990. Developmental stage specific plasmid supercoiling in *Chlamydia trachomatis*. *Mol. Microbiol.* **4:**1–7.

Spaeth, K. E., Y. S. Chen, and R. H. Valdivia. 2009. The *Chlamydia* type III secretion system C-ring engages a chaperone-effector protein complex. *PLoS Pathog.* **5:**e1000579.

Stephens, R. S., S. Kalman, C. Lammel, J. Fan, R. Marathe, L. Aravind, W. Mitchell, L. Olinger, R. L. Tatusov, Q. Zhao, E. V. Koonin, and R. W. Davis. 1998. Genome sequence of an obligate intracellular pathogen of humans: *Chlamydia trachomatis*. *Science* **282:**754–759.

Stephens, R. S., E. A. Wagar, and U. Edman. 1988. Developmental regulation of tandem promoters for the major outer membrane protein of *Chlamydia trachomatis*. *J. Bacteriol.* **170:**744–750.

Stragier, P., and R. Losick. 1990. Cascades of sigma factors revisited. *Mol. Microbiol.* **4:**1801–1806.

Tan, M. 2006. Regulation of gene expression, p. 103–131. *In* P. M. Bavoil and P. B. Wyrick (ed.), *Chlamydia: Genomics and Pathogenesis*. Horizon Bioscience, Wymondham, United Kingdom.

Tao, S., R. Kaul, and W. M. Wenman. 1991. Identification and nucleotide sequence of a developmentally regulated gene encoding a eukaryotic histone H1-like protein from *Chlamydia trachomatis*. *J. Bacteriol.* **173:**2818–2822.

Ward, M. E. 1988. The chlamydial developmental cycle, p. 71–98. *In* A. L. Barron (ed.), *Microbiology of Chlamydia*. CRC Press, Boca Raton, FL.

Wichlan, D. G., and T. P. Hatch. 1993. Identification of an early-stage gene of *Chlamydia psittaci* 6BC. *J. Bacteriol.* **175:**2936–2942.

Wilson, A. C., and M. Tan. 2002. Functional analysis of the heat shock regulator HrcA of *Chlamydia trachomatis*. *J. Bacteriol.* **184:**6566–6571.

Wyllie, S., and J. E. Raulston. 2001. Identifying regulators of transcription in an obligate intracellular pathogen: a metal-dependent repressor in *Chlamydia trachomatis*. *Mol. Microbiol.* **40:** 1027–1036.

Yu, H. H. Y., E. G. Di Russo, M. A. Rounds, and M. Tan. 2006a. Mutational analysis of the promoter recognized by *Chlamydia* and *Escherichia coli* sigma 28 RNA polymerase. *J. Bacteriol.* **188:**5524–5531.

Yu, H. H. Y., D. Kibler, and M. Tan. 2006b. *In silico* prediction and functional validation of sigma 28-regulated genes in *Chlamydia* and *Escherichia coli*. *J. Bacteriol.* **188:**8206–8212.

Yu, H. H. Y., and M. Tan. 2003. Sigma 28 RNA polymerase regulates *hctB*, a late developmental gene in *Chlamydia*. *Mol. Microbiol.* **50:**577–584.

Yuan, Y., Y. X. Zhang, D. S. Manning, and H. D. Caldwell. 1990. Multiple tandem promoters of the major outer membrane protein gene (*omp1*) of *Chlamydia psittaci*. *Infect. Immun.* **58:** 2850–2855.

Zhang, L., A. L. Douglas, and T. P. Hatch. 1998. Characterization of a *Chlamydia psittaci* DNA binding protein (EUO) synthesized during the early and middle phases of the developmental cycle. *Infect. Immun.* **66:**1167–1173.

Zhang, L., M. M. Howe, and T. P. Hatch. 2000. Characterization of *in vitro* DNA binding sites of the EUO protein of *Chlamydia psittaci*. *Infect. Immun.* **68:**1337–1349.

CELL BIOLOGY OF THE CHLAMYDIAL INCLUSION

Marcela Kokes and Raphael H. Valdivia

8

INTRODUCTION

The chlamydiae are obligate intracellular pathogens that reside within a membrane vacuole in the cytoplasm termed an inclusion (Moulder, 1991). All aspects of *Chlamydia* survival are intimately linked to the cell biology of its host. The inclusion must avoid degradation by lysosomal compartments and limit detection by innate immune surveillance systems that may activate cell autonomous defense mechanisms or recruit immune cells with potent antimicrobial functions. At the same time, *Chlamydia* must ensure the acquisition of nutrients to support robust bacterial replication and prepare it for eventual exit and dissemination.

Not unexpectedly, *Chlamydia* is equipped with a vast arsenal of proteins to modulate cellular pathways important for invasion, inclusion remodeling, innate immune manipulation, and nutrient acquisition (Betts et al., 2009; Cocchiaro and Valdivia, 2009). Despite the lack of genetic tools to identify these chlamydial factors by bacterial mutational analysis, significant progress has been made to link chlamydial proteins to unique events during the infectious cycle. Effector proteins translocated early in infection have been defined as central to cytoskeletal rearrangements that mediate bacterial invasion (Clifton et al., 2004; Jewett et al., 2006; Lane et al., 2008). Similarly, inclusion membrane proteins expressed early after entry have been associated with the recruitment of Rab GTPases during inclusion biogenesis (Rzomp et al., 2006; Cortes et al., 2007).

Recent advances in cell biological techniques and new tools to perform loss-of-function experiments in mammalian cells have accelerated our understanding of the extent to which *Chlamydia* manipulates the host. For instance, new markers of functionally distinct subcellular compartments and fluorescent reporter proteins now permit the tracking of cellular events in real time in live cells. These tools have revealed an underappreciated complexity of the interactions between *Chlamydia* and its host cell. Similarly, RNA interference (RNAi)-mediated silencing of mammalian gene expression and the application of unbiased genome-wide screens for host factors required for chlamydial replication have revealed new pathways that this pathogen must modulate for efficient replication (Derré et al., 2007; Elwell et al., 2008; Gurumurthy et al.,

Marcela Kokes and Raphael H. Valdivia, Department of Molecular Genetics and Microbiology and Center for Microbial Pathogenesis, Durham, NC 27710.

Intracellular Pathogens I: Chlamydiales
Edited by Ming Tan and Patrik M. Bavoil © 2012 ASM Press, Washington, DC
10.1128/9781555817329.ch8

2010). On the bacterial side, the availability of genome sequences has allowed the implementation of genome-wide functional approaches to identify proteins likely involved in the co-option of its host. These include targeted screens for chlamydial type III secreted proteins based on their ability to be recognized by type III secretion systems of enteric bacteria (Ho and Starnbach, 2005; Subtil et al., 2005), screens for chlamydial proteins that disrupt basic eukaryotic cellular processes upon ectopic expression (Sisko et al., 2006), and the generation of comprehensive libraries of antibodies to chlamydial proteins that can be screened to identify proteins translocated into the host cell at different stages in infection (Li et al., 2008). Overall, these approaches suggest that ~10% of *Chlamydia* open reading frames encode proteins with the potential to access the host cytoplasm or inclusion membranes (Valdivia, 2008). Achieving a molecular understanding of the functions of these proteins, their host targets, and the pathways they target will be a daunting task, especially when one considers that these proteins can be highly variable among closely related *Chlamydia* species and/or have no homology to proteins with known functions. Nonetheless, this is a fertile area of study that will not only reveal unique pathogenic strategies used by the *Chlamydiales* but also further our understanding of the basic cell biology of eukaryotic cells.

SETTING UP FOR SUCCESS—EARLY EVENTS IN NASCENT INCLUSION BIOGENESIS

Trafficking through the mammalian endomembrane system involves multiple protein and lipid determinants. These include small GTPases that initiate vesicle formation, coat proteins, lipid-modifying enzymes, lipid subdomains, tethering factors, and membrane fusogenic proteins (reviewed by Lippincott-Schwartz and Phair [2010]). In particular, Arf and Rab GTPases (reviewed by Stenmark [2009]) function as molecular switches that alternate between a GTP-bound "on" conformation and a GDP-bound "off" state to recruit coat proteins, generate vesicular carriers, and eventually dock these vesicles to the appropriate target membrane. In addition, phosphatidylinositol is differentially phosphorylated to define distinct subcellular membrane compartments that recruit proteins important in membrane trafficking events (reviewed by Behnia and Munro [2005]). In the following sections we discuss how *Chlamydia* co-opts these various determinants of cellular membrane traffic (Fig. 1).

The Nascent Inclusion Is Rapidly Segregated from Classical Early Endocytic Pathways

Shortly after invasion, *Chlamydia*-containing endosomes lose plasma membrane and early endocytic markers (Scidmore et al., 2003). Some of this rapid remodeling of the nascent endosome is likely mediated by the chlamydial proteins synthesized and translocated early after invasion (Scidmore et al., 1996). In addition, physical properties unique to the elementary body (EB)—or possibly its method of entry—lead to a significant delay in the fusion of EB-containing endosomes with lysosomes. For instance, heat-killed EBs fuse with endosomes at rates significantly slower than those of other invasive bacteria (Scidmore et al., 2003). Overall, there is a concerted effort by intracellular chlamydiae to dissociate their inclusions from endolysosomal trafficking pathways.

Rab GTPases associated with classical endolysosomal transport (Rab5, Rab7, and Rab9) are not recruited to inclusions (Rzomp et al., 2003). Similarly, the phosphoinositides PI3P and $PI(4,5)P_2$, which mark endosomal and the plasma membranes, respectively, are absent from inclusions (Moorhead et al., 2010). As a result, the inclusion was originally viewed as being completely dissociated from all endocytic traffic (reviewed by Fields and Hackstadt [2002]). However, with the application of newer cell biological tools, there is a growing appreciation that the inclusion is not completely segregated from membrane transport

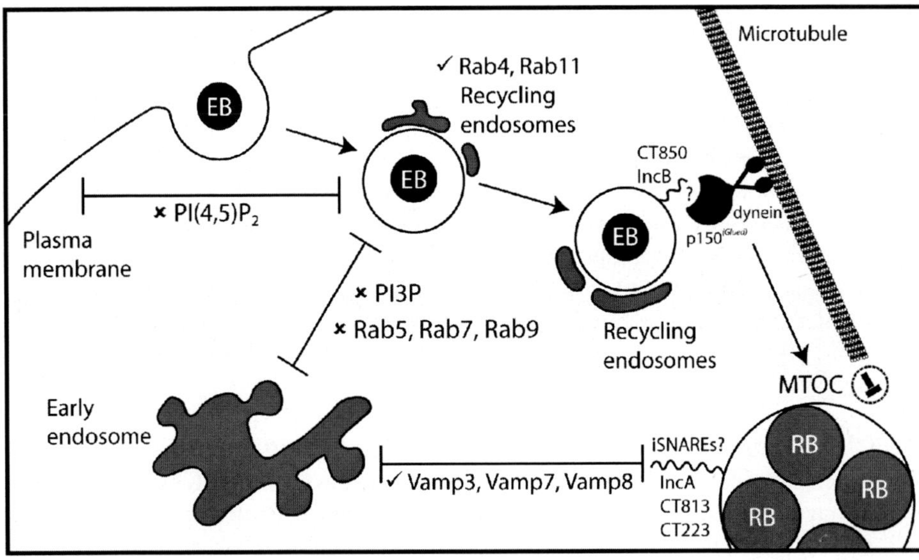

FIGURE 1 Early events in nascent inclusion biogenesis. Soon after chlamydial entry, the nascent inclusion rapidly loses plasma membrane and early endocytic and classic endolysosomal markers (Scidmore et al., 2003), including the phosphoinositides PI(4,5)P_2 and PI3P and endolysosomal Rabs (Rab5, Rab7, and Rab9) (Moorhead et al., 2010; Rzomp et al., 2003). Recycling endosomes and their associated Rabs localize to early inclusions (Rzomp et al., 2006, 2003) and may facilitate migration to the MTOC, in a microtubule- and dynein-dependent manner (Clausen et al., 1997). p150$^{(Glued)}$, the component of the dynactin protein complex linking vesicular cargo to dynein, is required for migration to the MTOC, although p50 dynamitin is not (Grieshaber et al., 2003). *Chlamydia* inclusion membrane proteins IncB and Ct850 are postulated to play a role in this interaction to promote association of the nascent inclusion with the microtubule-organizing center (Mital et al., 2010). Endocytic pathway-associated SNAREs, including Vamp3, -7, and -8, localize around the inclusion in a fusion-inhibited state potentially as a result of interactions with *Chlamydia* Incs (IncA, CT813, and CT223) (Delevoye et al., 2008; Paumet et al., 2009). These Inc proteins have been proposed to function as inhibitory SNARE (iSNARE) mimics (Paumet et al., 2009). doi:10.1128/9781555817329.ch8.f1

pathways and that some proteins that normally function in Golgi apparatus-mediated transport and on recycling/sorting endosomes are recruited to inclusions (reviewed by Cocchiaro and Valdivia [2009]).

The inclusion membrane is an important interface between *Chlamydia* and the host cell. The limiting membrane of the inclusion is heavily modified with *Chlamydia*-encoded integral membrane proteins. These proteins are collectively termed Incs, for inclusion membrane proteins (Rockey et al., 2002). A set of Incs is prepackaged into EBs (Saka et al., 2011), and another set is expressed within 1 h after bacterial invasion (Belland et al., 2003). This suggests that modification of the nascent inclusion by Inc proteins begins immediately after bacterial entry.

Different Incs mediate interactions with specific host subcellular compartments. Based on tertiary structural predictions, three *C. trachomatis* Incs (IncA/CT119, CT223, and CT813) display soluble *N*-ethylmaleimide-sensitive factor (SNARE)-like motifs (Delevoye et al., 2008). SNAREs are integral membrane proteins that are central to vesicle-mediated membrane transport (reviewed by Chen and Scheller [2001] and Martens and McMahon [2008]). Vesicular SNAREs (v-SNAREs) pair with specific target SNAREs (t-SNAREs) on acceptor membranes to provide specificity to vesicle targeting and the energy required to fuse two membranes

(Chen and Scheller, 2001). IncA interacts through its SNARE-like domain with a subset of host SNAREs that are enriched at the inclusion (Delevoye et al., 2008). These three SNAREs, Vamp3, Vamp7, and Vamp8, are often associated with membrane fusion events in the endocytic pathway. SNAREs that are not recruited to the inclusion, Vamp4 and Sec22, are associated with exocytosis and endoplasmic reticulum (ER)-to-Golgi apparatus transport (Delevoye et al., 2008).

The recruitment of these SNAREs to the inclusion, however, should not be seen as evidence that the inclusion fuses with these compartments. IncA, in reconstituted SNARE-mediated fusion reactions with purified protein and liposomes, inhibits rather than promotes membrane fusion (Paumet et al., 2009). IncA disrupts in vitro fusion between t-SNARE (syntaxin7/syntaxin8/Vti1b) and v-SNARE (Vamp8) in liposomes, as well as t-SNARE (syntaxin4/SNAP23) and v-SNARE-mediated membrane transport in in vivo cellular assays (Paumet et al., 2009). Therefore, some Incs may have evolved to mimic SNARE-like motifs that act as inhibitory SNAREs (iSNAREs) to limit fusion with host endocytic compartments (Paumet et al., 2009). Consistent with this, Vamp8 at the inclusion, as assessed by immunoelectron microscopy, is largely due to the accumulation of Vamp8-positive compartments intimately associated with the inclusion rather than on the inclusion membrane itself (Delevoye et al., 2008). Similarly, clinical strains that are deficient in the production of IncA fail to efficiently recruit Vamp8 to the inclusion membrane (Delevoye et al., 2008). Nonetheless, these strains are still viable and presumably segregated from endolysosomal transport (Suchland et al., 2000), indicating that there is an IncA-independent mechanism for *Chlamydia* avoidance of lysosomal fusion.

In contrast to a potential role in inhibiting fusion with host vesicles, IncA is required for homotypic fusion between *C. trachomatis* inclusions. Upon infection with a high dose of EBs, several nonfused inclusions will persist within an infected cell until approximately 10 hours postinfection (hpi). At this stage, inclusions begin to fuse and form one larger inclusion. Blocking IncA function by microinjection with anti-IncA antibodies results in multiple inclusions (Hackstadt et al., 1999). Similarly, infections performed at lower temperature (e.g., 32°C), which limits IncA export (Fields et al., 2002), also lead to the formation of multiple inclusions (Van Ooij et al., 1998), and clinical strains lacking IncA display fragmented inclusions (Suchland et al., 2000). Interestingly, even when cells are infected by *C. trachomatis* with a very low multiplicity of infection, microinjection of anti-IncA antibodies results in multilobed, fragmented inclusions (Hackstadt et al., 1999), suggesting that IncA may somehow function to prevent septation or fission of an established inclusion in addition to facilitating fusion between originally separate inclusions. Because IncA forms oligomers, it was originally thought that SNARE-like pairings would facilitate homotypic fusion of adjacent inclusions (Delevoye et al., 2004). However, IncA cannot drive membrane fusion reactions in vitro (Paumet et al., 2009), suggesting that there are additional host or bacterial factors mediating IncA-dependent inclusion fusion. Homotypic fusion of *C. trachomatis* inclusions could serve to consolidate resources and reduce competition among multiple growing inclusions. However, this function is dispensable for viability in cells, as bacterial replication is largely unaffected by pharmacological agents that disrupt inclusion fusion (Schramm and Wyrick, 1995), and not all *Chlamydia* species and strains display fusogenic vacuoles (Rockey et al., 1996). Nonetheless, mathematical models of reticulate body (RB)-to-EB transition postulate that large, fused inclusions lead to the greatest EB yield (Wilson et al., 2006); the slow growth rates that have been observed in *Chlamydia* species with multiple or fragmented inclusions corroborate this prediction.

It is clear that Incs on early inclusions likely play important roles in remodeling the nascent inclusion to segregate from the endolysosomal pathway and maintain single inclusion morphology in fusogenic *Chlamydia* species.

However, the role played by soluble effectors secreted early or even during entry should not be discounted when considering early interactions with host cell biology.

The Nascent Inclusion Migrates to the Host Centrosome

In tissue culture cells, the nascent inclusion migrates to a perinuclear region of the cell corresponding to the microtubule-organizing center (MTOC) or centrosome (Clausen et al., 1997). Many secretory organelles reside near the MTOC, and inclusion migration to this site presumably facilitates interactions with these lipid- and nutrient-rich compartments. This migration occurs along microtubules and requires de novo chlamydial protein synthesis (Scidmore et al., 1996; Clausen et al., 1997), indicating that a bacterial protein, most likely an Inc protein, coordinates transport to the MTOC.

How the inclusion gets to the MTOC is not well understood. Because inclusions intimately associate with recycling endosomes early during infection (van Ooij et al., 1997; Scidmore et al., 2003) and recycling endosomes localize pericentriolarly (Ullrich et al., 1996), *Chlamydia* could potentially use this association to migrate to the MTOC. Consistent with this, two Rab GTPases (Rab4 and Rab11) present on recycling endosomes are also recruited to the inclusion membrane early in infection (Rzomp et al., 2003, 2006). Rabs interact with molecular motors via adaptor proteins to regulate vesicular trafficking (Stenmark, 2009) and thus could influence inclusion migration to the MTOC. Rab6 and its effector BICD1, which interacts with the microtubule-dependent motor dynein (Matanis et al., 2002), localize to the inclusion and may mediate dynein-mediated inclusion migration (Moorhead et al., 2007). Consistent with this, inclusion migration is dependent on host microtubules and dynein, both of which are recruited to early inclusions (Clausen et al., 1997; Grieshaber et al., 2003). In addition, the p150$^{(Glued)}$ component of the dynactin protein complex, which links vesicular cargo to the dynein motor, is required for nascent inclusion migration. Unexpectedly, the cargo recruitment component p50 dynamitin is not necessary (Grieshaber et al., 2003), indicating that its function is performed by an unknown host cargo receptor adaptor or a *Chlamydia* protein on the inclusion membrane.

Additional observations support the notion that Inc proteins may participate in dynein recruitment to the inclusion and inclusion migration to the centrosome. Host centrosomes intimately associate with the inclusion in a dynein-dependent dynactin-independent manner (Grieshaber et al., 2006). Four Incs (IncB/CT232, CT101, CT222, and CT850) localize to patches on the inclusion surface that associate with host centrosomes (Mital et al., 2010). CT222 and CT850 stably interact with each other, and ectopically expressed CT850 localizes to host centrosomes (Mital et al., 2010). IncB and CT850 are expressed by 2 hpi (Shaw et al., 2000; Belland et al., 2003) and are thus good candidates for mediating early interactions with dynein and facilitating inclusion migration (Mital et al., 2010).

TAKING CONTROL—MANIPULATION OF HOST TRAFFICKING AND ORGANELLES FOR NUTRIENT ACQUISITION

Chlamydia relies on host cells for obtaining vital nutrients, including amino acids, nucleotides, and lipids. Consistent with their obligate intracellular nature, chlamydiae have lost the ability to synthesize many of these essential metabolic building blocks (Stephens et al., 1998). To acquire nutrients, the inclusion must interact with numerous intracellular trafficking pathways and organelles (Fig. 2).

The Inclusion Membrane Recruits Critical Mediators of Membrane Traffic

In addition to Rab4 and Rab11, *Chlamydia* recruits Rab1 in many species, Rab6 and Rab10 in select species, and Rab14 in at least one species to inclusion membranes (Rzomp et al., 2003; Capmany and Damiani, 2010). This recruitment of Rabs is independent of

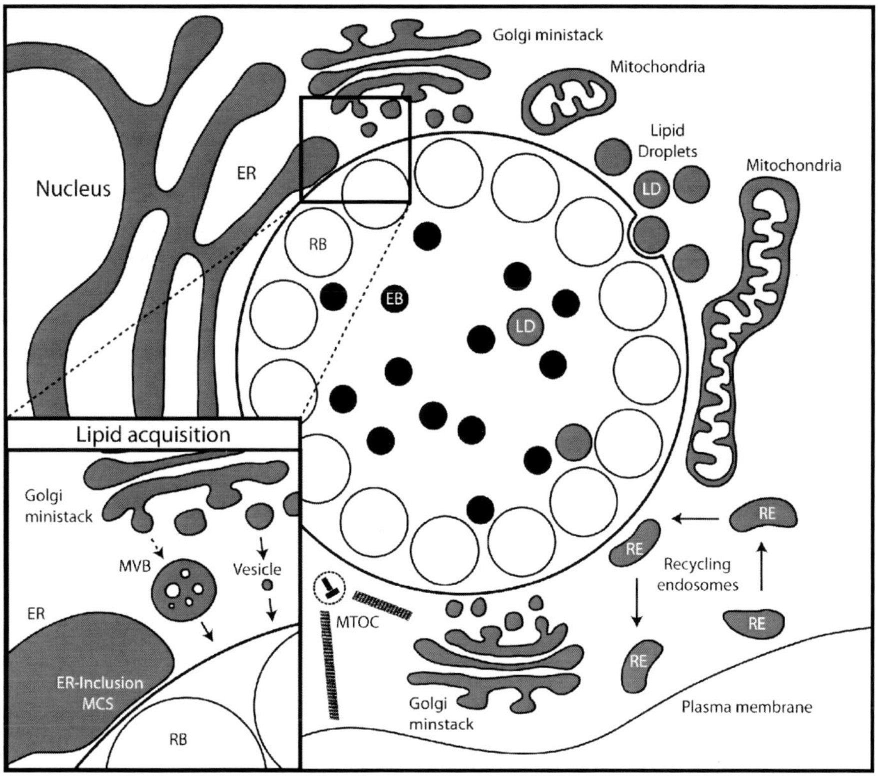

FIGURE 2 The inclusion interacts with multiple subcellular compartments. Fragmented Golgi apparatus ministacks, the ER, LDs, mitochondria, and recycling endosomes (RE) closely associate with the inclusion (Heuer et al., 2009; Peterson and de la Maza, 1988; Kumar et al., 2006; Matsumoto et al., 1991; van Ooij et al., 1997; Scidmore et al., 1996). These interactions may facilitate nutrient acquisition directly from these organelles. Golgi apparatus fragmentation enhances sphingolipid uptake (Heuer et al., 2009), and lipid droplets translocate into the lumen of the inclusion (Cocchiaro et al., 2008). Additional pathways for lipid delivery (inset) include vesicular transport of Golgi apparatus-derived exocytic vesicles (Hackstadt et al., 1996; Carabeo et al., 2003), MVBs (Beatty, 2008, 2006), and transfer at membrane contact sites (MCS) between the ER and inclusion membranes (Derré et al., 2011; Elwell et al., 2011). The inclusion remains in close association with centrosomes at the MTOC throughout intracellular infection (Grieshaber et al., 2006). doi:10.1128/9781555817329.ch8.f2

microtubules or the Golgi apparatus, indicating that these proteins do not traffic through membrane intermediates en route to the inclusion. Rab association with the inclusion may occur via a direct binding to Inc proteins or indirectly by Rab effectors and regulators that associate with the inclusion. For instance, the Inc protein CT229 interacts with Rab4 in vitro, and EGFP-Rab4 expressed in *Chlamydia*-infected cells colocalizes with endogenous CT229 around the inclusion in vivo (Rzomp et al., 2006). Interestingly, CT229 interacts only with Rab4-GTP, suggesting that CT229 may mimic Rab4 effectors by binding Rab4-GTP and recruiting it to the inclusion (Rzomp et al., 2006). Similarly, the *Chlamydia pneumoniae* Inc protein Cpn0585 interacts with several Rabs in a GTP-dependent manner (Cortes et al., 2007). In contrast, the Rab6 effector BICD1 localizes to the inclusion even in the presence of a dominant-negative version of Rab6, suggesting that BICD1 is recruited to the inclusion independently of active Rab6 (Moorhead et al., 2007). Overall, current evidence predominantly supports a model in which Inc proteins directly recruit active Rab proteins and their bound effectors to the inclusion.

Rab proteins mediate vesicular traffic between specific subcellular compartments, and thus it is perhaps not surprising that *Chlamydia* would co-opt these proteins to gain access to essential factors within vesicular carriers. Indeed, RNAi-based depletion and dominant-negative expression experiments indicate that Rab1, Rab6, Rab11, Rab14, and Rab4/Rab11 play important roles in *Chlamydia* replication within host cells (Derré et al., 2007; Elwell et al., 2008; Rejman Lipinski et al., 2009; Capmany and Damiani, 2010; Ouellette and Carabeo, 2010).

Normal Rab functions provide clues as to how Rab recruitment to the inclusion may influence interactions with the host. Rabs specify organelle identity in part by regulating the recruitment of lipid kinases and phosphatases that control the levels of phosphoinositides on membranes (Behnia and Munro, 2005). OCRL1 (oculocerebrorenal syndrome of Lowe protein 1) is a PI(4,5)P_2 5-phosphatase, binds to Rab1, Rab6, and Rab14, and is recruited to the inclusion (Moorhead et al., 2010). The PI-OH(4)-kinase PI4KIIα also localizes to the inclusion, suggesting that host enzymes are in place at the inclusion to generate phosphatidylinositol 4-phosphate (PI4P). Consistent with these observations, the pleckstrin homology (PH) domains of OSBP and GPBP, which specifically bind to PI4P, prominently decorate the inclusion membrane (Moorhead et al., 2010). Similarly, Arf1, a small GTPase that can recruit PI-OH(4)-kinases and PI4P-binding proteins, also localizes to the inclusion and may enhance the recruitment of these PI4P-generating enzymes. Small interfering RNA (siRNA)-mediated depletion of OCRL, PI4KIIα, or Arf1 reduces infectivity and inclusion formation, indicating that the generation of PI4P pools at the inclusion membrane plays an important role in bacterial replication and survival (Moorhead et al., 2010).

Chlamydia Acquires Host Lipids

Even though *Chlamydia* has the capacity to synthesize its own lipids, it will preferentially acquire host-derived lipids. As a result, the lipid composition of *C. trachomatis*, which includes cholesterol, sphingolipids (SLs), and glycerophospholipids, mimics that of its host cell (Wylie et al., 1997; Hatch and McClarty, 1998b). Glycerophospholipids are acquired as lysophospholipid precursors through a process that may involve a host calcium-dependent cytosolic phospholipase A_2 (cPLA_2) (Su et al., 2004). The lysophospholipids are reacylated with a *Chlamydia*-derived branched fatty acid prior to their incorporation into bacterial membranes (Wylie et al., 1997).

SLs are composed of sphingosine with both a nonpolar tail of fatty acid chains and a head group of either phosphocholine (sphingomyelin) or different sugars (glycosphingolipids) (reviewed by Lippincott-Schwartz and Phair [2010]). SLs are synthesized in the ER as ceramide precursors, processed in the Golgi apparatus, and accumulate at the plasma membrane (reviewed by Perry and Ridgway [2005]). SLs are required for chlamydial replication (van Ooij et al., 2000; Robertson et al., 2009), and treatment of infected cells with myriocin, a potent inhibitor of serine palmitoyltransferase—the initial step in SL biosynthesis—leads to loss of inclusion membrane integrity, RB-to-EB differentiation defects, and early EB release (Robertson et al., 2009). Furthermore, homotypic fusion between inclusions and the ability to reactivate from a persistent state are inhibited (Robertson et al., 2009), indicating that SLs are important for the function of the inclusion membrane. The fatty acid chains of SLs are typically saturated, and SLs can thus pack closely in lipid bilayers. The impact of SL depletion on inclusion membrane stability could be consistent with the role of this lipid in imparting order and rigidity to membranes (Lippincott-Schwartz and Phair, 2010).

Chlamydia Acquires Host Lipids Partly via Golgi Apparatus-Dependent Vesicular Trafficking Mechanisms

Tracking of fluorescently labeled ceramide analogues in live cells indicate that SL transport to the inclusion shares features of canonical Golgi apparatus-to-plasma membrane vesicular transport (Hackstadt et al., 1995; Wolf and Hackstadt, 2001). In addition, consistent with the inclusion's segregation from endocytic traffic, plasma membrane SLs are not trafficked to the inclusion (Hackstadt et al., 1996). SL-containing vesicles, seemingly in the process of fusing with the inclusion, have been observed by electron microscopy (Hackstadt et al., 1996). Of host SLs, only sphingomyelin, and not glucosylceramide, is delivered to the chlamydial inclusion (Moore et al., 2008), suggesting highly specific interactions with host pathways. Furthermore, in polarized epithelial cells, *Chlamydia* preferentially intercepts basolaterally targeted SLs (Moore et al., 2008), indicating that *Chlamydia* may interact with specific branches of exocytic pathways. Indeed, the trans-Golgi apparatus-associated SNARE syntaxin 6, required for basolaterally directed exocytosis, localizes to the inclusion of multiple *Chlamydia* species (Moore et al., 2010). However, syntaxin 6 depletion does not significantly block SL delivery to the inclusion or inclusion development (Moore et al., 2010), suggesting that any role in facilitating fusion of Golgi apparatus-derived SL-containing vesicles may be compensated by other SNAREs.

Experiments using the pharmacological agent brefeldin A (BFA) were important in revealing the importance of Golgi apparatus-dependent vesicular trafficking in SL acquisition by *Chlamydia* (Hackstadt et al., 1996; Carabeo et al., 2003). BFA treatment causes Golgi apparatus dispersion (Lippincott-Schwartz et al., 1989) by inhibiting guanine nucleotide exchange factors that activate Arf1, a small GTPase critical in vesicle formation (reviewed by D'Souza-Schorey and Chavrier, 2006). These BFA-targeted guanine nucleotide exchange factors are the *cis*-Golgi-localized GBF1, which is required for Golgi apparatus stack assembly and maintenance, and *trans*-Golgi apparatus-localized BIG1/BIG2, which are required for *trans*-Golgi network maintenance (Claude et al., 1999; Mansour et al., 1999; Yamaji et al., 2000; Manolea et al., 2008). A recent study reveals that only GBF1 participates in SL transport to the inclusion (Elwell et al., 2011). siRNA-mediated depletion of GBF1 reduced SL delivery to the inclusion by ~60%, while BIG1 and/or BIG2 depletion had no effect. This decrease in SL transport did not affect the production of infectious progeny, but it did impact the integrity of the inclusion membrane with GBF1 depletion, resulting in loss of inclusion integrity and bacterial release into the cytosol of the host cell (Elwell et al., 2011), as has been observed with myriocin treatment (Robertson et al., 2009). These data suggest that *Chlamydia* acquire SLs via GBF1-dependent and -independent mechanisms, and blocking the former vesicular pathway is insufficient to impact *Chlamydia* replication.

As with SLs, de novo-synthesized and extracellular cholesterol is found in EBs and the inclusion membrane (Carabeo et al., 2003). In eukaryotic cells, SLs and cholesterol are often cotransported in membrane vesicles, and cholesterol delivery to the inclusion resembles SL vesicular delivery in that it is partially dependent on microtubules and is BFA sensitive (Carabeo et al., 2003).

Chlamydia Acquires Host Lipids via Nonvesicular Pathways

Early ultrastructural studies noted the presence of thin ER tubules closely apposed with *C. trachomatis* inclusion membranes (Peterson and de la Maza, 1988; Giles and Wyrick, 2008), and ER proteins (SERCA2, IP3-R, and calreticulin) are enriched at the periphery of the inclusion (Majeed et al., 1999). The significance of this association with ER tubules was unknown. There is a growing appreciation that interorganelle lipid transport relies extensively on membrane contact sites (MCSs) between organelles, especially with the ER (reviewed by Levine and Loewen [2006]). For example, ceramide was originally thought to be mainly transported from the ER to the Golgi apparatus via vesicular carriers. However, it is now apparent that the cytosolic lipid transfer protein (CERT) is responsible for the bulk of this transport by transferring ceramide directly at ER-Golgi body MCSs (Hanada et al., 2003; Levine and Loewen, 2006).

A pair of recent studies provide evidence that the inclusion forms MCSs with the ER where CERT transfers SL precursors from the ER to be converted to sphingomyelin by host proteins at the inclusion membrane (Derré et al., 2011; Elwell et al., 2011). Endogenous CERT associates with patches on the *C. trachomatis* inclusion, where it colocalizes with VAPA and VAPB (Derré et al., 2011; Elwell et al., 2011). VAPA and VAPB are ER resident proteins that recruit FFAT motif-containing proteins, such as CERT, to the cytoplasmic face of the ER. This binding is required for CERT to transfer ceramide from the ER to the Golgi apparatus (Kawano et al., 2006). Cryoimmunogold electron micrographs suggest that CERT localizes to the inclusion membrane at sites in close apposition to VAPB-positive ER tubules (Derré et al., 2011). The distance between membranes at these apposition sites is ~10 nm, and they ultrastructurally resemble interorganelle MCSs (Derré et al., 2011).

Although the PH domain of CERT binds PI4P on Golgi apparatus membranes and is necessary and sufficient for CERT binding to the inclusion (Derré et al., 2011), a mutation that disrupts PI4P binding and Golgi apparatus localization has no effect on CERT recruitment to the inclusion (Elwell et al., 2011), and CERT remains on the inclusion upon inhibition of PI4P synthesis (Derré et al., 2011). This suggests that other factors on the inclusion membrane are required for CERT recruitment. An attractive candidate for CERT recruitment is IncD (CT115), an inclusion membrane protein that interacts with the PH domain of CERT and colocalizes with CERT in patches on the inclusion membrane (Derré et al., 2011). *Chlamydia caviae* lacks IncD and does not recruit CERT to the inclusion (Derré et al., 2011). However, *C. caviae* inclusions do accumulate SLs (Rockey et al., 1996), suggesting that other pathways of SL acquisition may suffice in this species.

Depletion of CERT or VAPA and VAPB reduces inclusion size and the number of infectious progeny recovered from *C. trachomatis* infection (Derré et al., 2011; Elwell et al., 2011), and inhibition of CERT activity results in ~56% decrease in chlamydial SL acquisition (Elwell et al., 2011). Unlike vesicle-mediated pathways of SL acquisition, CERT-mediated pathways are important for progression through the normal infectious cycle. Consequently, it has been speculated that SLs obtained via CERT or GBF1-dependent mechanisms may be utilized differently by *Chlamydia* since their disruption has different consequences on the infectious cycle (i.e., decreased replication versus inclusion membrane disruption) (Elwell et al., 2011). Ceramide transported from the ER to the *trans*-Golgi apparatus via CERT is converted to sphingomyelin through the activity of sphingomyelin synthases SMS1 and SMS2 (Tafesse et al., 2007). Interestingly, SMS2 localizes to the inclusion and partially overlaps with CERT, while SMS1 remains associated with the Golgi apparatus during infection (Elwell et al., 2011). Depletion of SMS1 or SMS2 results in decreased recovery of infectious progeny (Elwell et al., 2011). Together, these data suggest that SMS2 at the inclusion may convert CERT-transferred ceramide to sphingomyelin for bacterial use.

The finding that *Chlamydia* forms MCSs with the ER, where ceramide is directly transferred between these membranes, highlights the complexity of *Chlamydia* control of host cells and raises the critical question of whether nonvesicular communication between the inclusion and subcellular components is more widespread. As advances are made in cellular interorganelle communication, our ability to address these questions will expand in parallel.

Chlamydia Induces Golgi Apparatus Fragmentation

The Golgi apparatus of mammalian cells consists of stacks of flattened cisternae that assemble at the MTOC (Sütterlin and Colanzi, 2010). During *Chlamydia* infection, the Golgi apparatus disperses into several distinct ministacks that envelop the inclusion (Heuer et al., 2009). Golgi apparatus fragmentation correlates with cleavage of the matrix protein golgin-84, and pharmacological inhibitors of this cleavage inhibit Golgi body fragmentation and decrease infectivity of *Chlamydia* approximately twofold. Although inflammatory caspases and calpains are likely responsible for golgin-84 cleavage, this does not preclude the possibility that a chlamydial protease may also participate in this processing. Golgi apparatus fragmentation upon golgin-84 cleavage is postulated to enhance lipid delivery to the inclusion and thus *Chlamydia* replication. Consistent with this prediction, Golgi body fragmentation induced by RNAi-based silencing of giantin, GPP130, or golgin-84 enhances bacterial replication (Heuer et al., 2009).

Rab6 and Rab11 modulate *Chlamydia*-induced Golgi apparatus fragmentation (Rejman Lipinski et al., 2009). Rab6 or Rab11 depletion inhibits Golgi apparatus fragmentation in infected cells and reduces infectivity, EB formation, and SL delivery to the inclusion but does not affect golgin-84 cleavage. Thus, Rab6 and Rab11 appear to act downstream of golgin-84 cleavage to destabilize the Golgi apparatus. While Rab6 or Rab11 depletion blocks golgin-84-cleavage-dependent Golgi apparatus fragmentation, fragmentation induced by RNAi of the Golgi apparatus-tethering factor p115 is unaffected by Rab6 or Rab11 depletion, implying independent pathways (Rejman Lipinski et al., 2009). Significantly, Golgi apparatus fragmentation by p115 depletion rescues defects in SL transport to the inclusion in cells lacking Rab6 or Rab11, implying that Rab6 and Rab11 function can be bypassed by alternative mechanisms of Golgi apparatus fragmentation (Rejman Lipinski et al., 2009). Indeed, Golgi apparatus fragmentation itself appears to be a greater determinant of efficient SL delivery to the *Chlamydia* inclusion than other functions of Rab6 and Rab11, possibly by facilitating SL delivery to the inclusion by increasing Golgi apparatus proximity and association with the inclusion.

Chlamydia Interactions with Multivesicular Bodies

Multivesicular bodies (MVBs) are late endocytic compartments in which the limiting membrane of endosomes has invaginated into the lumen to form intraluminal vesicles containing membrane proteins destined for degradation (reviewed by Piper and Katzmann [2007]). MVBs can also fuse with the plasma membrane, resulting in release of specialized intraluminal vesicles termed exosomes (reviewed by van Niel et al. [2006]). Recent studies suggest an interaction between *Chlamydia* and MVBs as a potential intermediate in post-Golgi apparatus SL acquisition (Beatty, 2006, 2008; Robertson et al., 2009). Several markers of MVBs, including CD63, MLN64, and lysobisphosphatidic acid (LBPA), localize to the periphery and within inclusions in fixed cells (Beatty, 2006). Immunoelectron microscopy revealed CD63-positive and LBPA-positive structures adjacent to the inclusion, small CD63-positive and LBPA-positive vesicles inside the inclusion lumen, and CD63 at the inclusion membrane and within the inclusion (Beatty, 2006, 2008). Additionally, pharmacological inhibitors of MVBs disrupt inclusion growth and infectivity (Beatty, 2006). Intraluminal vesicles of MVBs are enriched in cholesterol and SLs (Piper and Katzmann, 2007), and treatment with U18666A, a cell-permeable amphiphilic aminosteroid that inhibits cholesterol transport from late endosomes, disrupts SL and cholesterol delivery to the inclusion (Beatty, 2006, 2008). However, the exact contribution of MVB function in *Chlamydia* host lipid acquisition is unclear since some inhibitors used to interfere with MVBs also inhibit autophagy (Al-Younes et al., 2004). Additionally, recent genome-wide RNAi screens indicated that neither MVB nor autophagy pathways are required for *C. caviae* or *C. trachomatis* replication (Derré et al., 2007; Elwell et al., 2008), potentially due to redundant mechanisms of lipid acquisition.

Other studies did not observe CD63 within the inclusion lumen or any effect of U18666A on SL acquisition in live cells (Ouellette and Carabeo, 2010). Further studies will be required to address these discrepancies.

Chlamydia Targets Lipid Droplets

Chlamydia acquires host neutral lipids presumably via interactions with lipid droplets (LDs) (Kumar et al., 2006; Cocchiaro et al., 2008). LDs are ubiquitous among eukaryotic cells with crucial roles in lipid homeostasis and energy metabolism (reviewed by Farese and Walther [2009]; Guo et al. [2009]; and Walther and Farese [2009]). LDs have a unique structure comprising a hydrophobic lipid core of sterol esters and triacylgycerols encased in a single phospholipid monolayer. LD formation can be inhibited pharmacologically with the long-chain acyl coenzyme A synthase inhibitor triacsin C, which specifically blocks cholesterol ester and triacylglycerol synthesis (Igal et al., 1997). Treatment of cells with triacsin C during infection interferes with *Chlamydia* infectivity, suggesting a potential role for LDs in chlamydial pathogenesis (Kumar et al., 2006). However, caution should be exercised when considering the implications of inhibiting lipid biosynthesis since various aspects of membrane traffic may be simultaneously impaired. For example, inhibitors of acyl coenzyme A: cholesterol acyltransferase (ACAT), which functions in cholesterol ester synthesis, and lysophospholipid acyltransferase (LPAT) have antichlamydial effects (Ouellette and Carabeo, 2010). These effects have been reported to arise from the disruption of endosome recycling rather than LD formation (Ouellette and Carabeo, 2010).

LDs accumulate at the *Chlamydia* inclusion and translocate into the inclusion lumen (Kumar et al., 2006; Cocchiaro et al., 2008). The chlamydial protein Lda3 (CT473), which binds to LDs and to the inclusion membrane when expressed ectopically, may serve as the link between LDs and the inclusion. Within the inclusion lumen, LDs associate with RBs, allowing the possibility of direct lipid transport between the two. Alternatively, lipases present on LDs may become uncoupled from host regulatory signals upon translocation into the inclusion lumen and generate biosynthetic precursors for bacterial membrane biosynthesis (Cocchiaro et al., 2008). In addition, because LDs have been implicated in vesicular trafficking, signaling, and inflammatory responses (Guo et al., 2009), it is possible that bacterial manipulation of these organelles may serve functions beyond the acquisition of nutrients. Recently, a membrane attack complex/perforin-like chlamydial protein (CT153) was reported to localize to a subpopulation of RBs, some of which may be closely apposed to intraluminal LDs (Taylor et al., 2010). CT153 lies immediately upstream of plasticity zone phospholipase D-like proteins, including Lda1(CT156), which localizes to LDs when ectopically expressed (Kumar et al., 2006), raising the possibility that the functions of these proteins are related (Taylor et al., 2010).

Recycling Endosomes Intimately Associate with Inclusions

Transferrin (Tf) is an iron-binding protein that delivers iron to cells through receptor-mediated endocytosis (reviewed by Richardson and Ponka [1997]). Tf-containing recycling endosomes associate with the inclusion (Scidmore et al., 1996; van Ooij et al., 1997), where they are found closely juxtaposed to the inclusion membrane without evidence of fusion (Scidmore et al., 2003). Inhibitors of LPAT interfere with one of the Tf recycling pathways and decrease *Chlamydia* replication (Ouellette and Carabeo, 2010). Similarly, Rab4 and Rab11, which are required for Tf receptor recycling, are enriched at the inclusion (Rzomp et al., 2003), and their disruption results in smaller inclusions and increased retention of Tf around the inclusion (Ouellette and Carabeo, 2010). These findings are consistent with a role for recycling endosomes in promoting *Chlamydia* replication. However, removal of the Tf-containing fraction from cell serum rescues LPAT inhibitor-induced defects on bacterial replication (Ouellette and Carabeo, 2010).

Although iron is an essential element for normal completion of the infectious cycle (Thompson and Carabeo, 2011), these data suggest that overavailability of iron as a result of increased retention of Tf-containing recycling endosomes around the inclusion interferes with replication (Ouellette and Carabeo, 2010).

Chlamydia Species Interact with Mitochondria

Early electron microscopy studies indicated that mitochondria closely associate with chlamydial inclusions, especially *Chlamydia psittaci* inclusions (Peterson and de la Maza, 1988; Matsumoto et al., 1991). More recently, in a genome-wide RNAi screen for host factors necessary for *C. caviae* infection, the Tim-Tom mitochondrial protein import complex was identified as important for *Chlamydia* infection (Derré et al., 2007). Recognition and import of nucleus-encoded mitochondrial proteins is disrupted upon Tim-Tom depletion and presumably results in some general mitochondrial dysfunction, although ATP levels remained comparable, indicating that energy production is not dramatically altered (Derré et al., 2007). siRNA-mediated depletion of essential Tom components resulted in smaller inclusions and decreased numbers of infectious progeny in cells infected with *C. caviae*. In contrast, depletion of Tom components did not affect the amount of infectious progeny recovered with *C. trachomatis* infection. These data suggest that *C. caviae* may be particularly susceptible to alterations in mitochondrial function. *C. caviae* and *C. psittaci* are closely related (reviewed by Stephens et al. [2009]), and mitochondrial interactions could be more critical in these species than other *Chlamydia* spp.—indeed, the same early ultrastructural studies that described mitochondria surrounding *C. psittaci* inclusions described an apparent lack of mitochondria around both *C. trachomatis* and *C. pneumoniae* inclusions (Matsumoto et al., 1991).

Upon expression of *Chlamydia*-specific genes in eukaryotic cells, at least one phospholipase D-like protein (CT084) displayed tropism for eukaryotic mitochondria (Sisko et al., 2006), raising the possibility that one or more secreted chlamydial proteins may target host mitochondria during infection, as has been described for other bacterial effectors (Nagai et al., 2005; Papatheodorou et al., 2006). The consequence of these interactions and how they benefit the bacteria remain to be determined, although it is not difficult to envision scenarios where mitochondria may enhance acquisition of nutrients, including energy metabolites, or dampen proapoptotic signals. Indeed, increased rates of mitochondrial respiration are observed during the chlamydial infectious cycle (Hatch and McClarty, 1998a), which may provide increased levels of ATP for replicating chlamydiae (Ojcius et al., 1998).

Impact of *Chlamydia* Infections on the Cell Cycle

Chlamydia also targets several host functions that are not immediately linked to the manipulation of membrane transport. For example, *Chlamydia* infection leads to extra, or supernumerary, centrosomes, which may contribute to abnormal mitotic spindles during mitosis and mistakes in chromosome segregation (Grieshaber et al., 2006). Interestingly, these defects persist in cells even after *Chlamydia* infection has been eliminated, suggesting a potential oncogenic mechanism to explain the observed epidemiological connection between *Chlamydia* infection and certain cancers (reviewed by Littman et al. [2005]; Verma et al. [2008]; and Simonetti et al. [2009]).

Supernumerary centrosomes in *Chlamydia*-infected cells are likely the result of dysregulation of the host centrosome duplication pathway. Production of these extra centrosomes requires progression through S phase and the host kinases Cdk2 and Plk4, both of which are important for normal centrosome duplication (Johnson et al., 2009). Despite the extra centrosomes in *Chlamydia*-infected cells, there was only a modest increase in the number of mature centrioles (core components of centrosomes) per cell, which is consistent with an overproduction of immature centrioles in infected cells (Johnson et al., 2009).

While containing immature centrioles, all centrosomes in infected cells are functional as MTOCs (Knowlton et al., 2011). *Chlamydia*-infected cells also display defects in cytokinesis (Greene and Zhong, 2003), which makes a small contribution to the production of extra centrosomes since both centrosomes may be retained in one daughter cell (Johnson et al., 2009). Ectopic expression of three chlamydial Incs (CT223, CT224, and CT225) disrupts cytokinesis in the absence of infection, suggesting that *Chlamydia* effectors may contribute to inhibition of cytokinesis in infected cells (Alzhanov et al., 2009).

Chlamydia can also cause mitotic spindle defects independently of its ability to cause supernumerary centrosomes (Knowlton et al., 2011). Mammalian cells can cluster and suppress extra centrosomes to form two active poles in a mitotic spindle (Quintyne et al., 2005). *Chlamydia* disrupts centrosome clustering even in cell types that characteristically display numerous centrosomes (Knowlton et al., 2011). This increased spread of centrosomes has been postulated to be a result of centrosome interactions with the inclusion (Grieshaber et al., 2006; Knowlton et al., 2011). During mitosis, the spindle assembly checkpoint (SAC) monitors microtubule attachment to kinetochores, allowing delay until proper spindle organization is achieved (Basto et al., 2008). Once the SAC is passed in uninfected cells, ubiquitination and degradation of cyclin B1 and securin lead to the separation of sister chromatids and progression to anaphase (reviewed by Musacchio and Salmon [2007]). *Chlamydia*-infected cells prematurely exit metaphase and presumably segregate chromosomes before they are properly aligned on the metaphase plate, a defect that is normally prevented by the SAC (Knowlton et al., 2011). This bypass of the SAC is achieved in part by premature degradation of both cyclin B1 (Balsara et al., 2006) and securin, which permits completion of mitosis without proper suppression of extra centrosomes (Knowlton et al., 2011).

Several stages of the cell cycle may be altered by proteolysis in *Chlamydia*-infected cells. DNA synthesis occurring in S phase proceeds normally (Bose and Liebhaber, 1979; Greene et al., 2004), but cellular proteins regulating the G_2/M transition are processed (Balsara et al., 2006). Cdk1 protein levels decrease over the course of infection, and cyclin B1 is N-terminally truncated (Balsara et al., 2006) by the secreted chlamydial protease/proteasome-like activity factor (CPAF) (Paschen et al., 2008). These changes could result in cell cycle delay after growth but prior to division, and it has been speculated that this could benefit *Chlamydia* by providing greater access to host cellular resources (Balsara et al., 2006). The overall picture may be more complex than simply blocking cell cycle progression, since the type III secretion effector protein CT847 interacts with the host Grap2 cyclin D-interacting protein (GCIP), another target of proteolysis during infection (Chellas-Géry et al., 2007). Degradation of GCIP would stimulate transition from G_1 to S phase and bring terminally differentiated infected epithelial cells back into a productive growth phase (Chellas-Géry et al., 2007). It has been postulated that efficient *Chlamydia* replication may require levels of nutrients only found in actively replicating cells. Consistent with this prediction, siRNA-mediated GCIP depletion prior to infection enhanced *Chlamydia* development (Chellas-Géry et al., 2007).

STRATEGIES TO COMBAT INNATE IMMUNITY—THE CELL BIOLOGY ANGLE

Chlamydia defends itself from innate immune defense mechanisms designed to clear intracellular pathogens by evading detection and suppressing antimicrobial clearance strategies. Much of the focus of investigations into chlamydial anti-immune strategies has centered on the interruption of innate immune signaling pathways. However, there is also a growing appreciation that manipulation of the host cell's internal architecture may influence the outcome of antichlamydial responses, as signaling from distinct subcellular compartments within a cell can have a major impact on innate immune outcomes (reviewed by Barton and Kagan [2009]).

Maintenance of Inclusion Integrity

Receptors within the cytoplasm of mammalian cells constantly survey the environment for the presence of pathogens (reviewed by Medzhitov [2001]). By residing in a modified vacuole, *Chlamydia* can limit the availability of microbial products in the host cytoplasm and evade detection (reviewed by Kumar and Valdivia [2009]). As the inclusion expands, however, *Chlamydia* faces the challenge of maintaining the inclusion membrane barrier intact. The stability of the inclusion is partly dependent on a scaffold of filamentous actin (F-actin) and intermediate filaments (IFs) that surround the *Chlamydia* inclusion (Kumar and Valdivia, 2008). Perturbation of this cytoskeletal cage results in inclusion membrane disruption and spillage of bacteria into the host cell cytosol, which in turn leads to a hyperactive immune response, as evidenced by an increase in interleukin-8 activation (Kumar and Valdivia, 2008). Rho-family GTPases are central regulators of F-actin (reviewed by Etienne-Manneville and Hall [2002]) and are required for F-actin assembly around the inclusion. The clostridial C3-transferase toxin, which inhibits RhoA-C, results in reduced F-actin assembly at the inclusion and a loss of inclusion integrity (Kumar and Valdivia, 2008). The assembly of IFs and F-actin at the inclusion are interdependent since F-actin depolymerization disrupts the IF cage and the F-actin cage is poorly assembled in cells lacking IFs (Kumar and Valdivia, 2008). IFs are relatively static cytoskeletal scaffolding structures but appear to become more flexible as a result of proteolytic processing by the *Chlamydia* protease CPAF (Dong et al., 2004; Kumar and Valdivia, 2008). Removal of the head domain of IFs hinders their ability to form extensive filaments, making them less rigid while still maintaining some of their structural properties. As a result, it has been speculated that F-actin and IFs collaborate to maintain structural integrity of the inclusion (Kumar and Valdivia, 2008). These data support the model that CPAF, which accumulates in the cytoplasm through the infectious cycle, progressively modifies filaments at the inclusion periphery to facilitate enhanced flexibility and to permit expansion of the inclusion to accommodate replicating bacteria (Kumar and Valdivia, 2008). Interestingly, treatment of infected cells with a cell-permeable CPAF-specific inhibitory peptide resulted in a loss of inclusion membrane integrity with bacteria observed directly in the cytoplasm (Jorgensen et al., 2011). Whether this collapse is due to the lack of IF modifications or interference with other CPAF functions remains to be determined.

Cellular Mechanisms To Prevent Apoptosis

In addition to evading immune detection, pathogens also suppress host cellular protection systems. For example, apoptosis is a common antimicrobial defense mechanism that is actively blocked in *Chlamydia*-infected cells (reviewed by Sharma and Rudel [2009]). A prominent mechanism underlying this anti-apoptotic state likely includes the degradation of proapoptotic BH3-only proteins (Fischer et al., 2004; Dong et al., 2005; Paschen et al., 2008). Other mechanisms to block apoptosis in *Chlamydia*-infected cells have been described. For example, 14-3-3β is recruited to the inclusion membrane by IncG (CT118) (Scidmore and Hackstadt, 2001). Normally, 14-3-3β sequesters the phosphorylated pro-apoptotic BH3-only protein BAD (Bcl2 antagonist of cell death) in the cytoplasm and prevents BAD recruitment to mitochondria, where it promotes cytochrome *c* release and apoptosis (reviewed by Brazil et al. [2002]). BAD is phosphorylated during chlamydial infection by AKT/PKB and colocalizes with 14-3-3β at the inclusion, where it is presumably incapable of executing its proapoptotic functions (Verbeke et al., 2006). A similar strategy is employed to control the proapoptotic regulator protein kinase C δ (PKCδ) (Tse et al., 2005). PKCδ also exerts its proapoptotic functions in the mitochondria and nucleus (reviewed by Yoshida [2007]). PKCδ binds to diacylglycerol at the inclusion membrane, effectively sequestering the kinase from its normal targets and protecting the infected cell from apoptosis (Tse et al., 2005).

Cell Autonomous Immunity and Autophagy

C. trachomatis and *Chlamydia muridarum* are closely related (Read et al., 2000); however, mice are refractory to disease by *C. trachomatis* but not *C. muridarum* (Morrison and Caldwell, 2002). This species-specific resistance is largely mediated by gamma interferon (IFN-γ)-induced cell autonomous immune responses. In murine cells, IFN-γ induces the expression of immunity-related GTPases (IRGs) (reviewed by Taylor [2007]). Effective suppression of *C. trachomatis* infection requires several IRG proteins, including Irgm1, Irgm3, and Irgb10 (Coers et al., 2008). The antimicrobial activity of these proteins likely involves their recruitment to the pathogen-containing vacuole, as has been exemplified for *Toxoplasma gondii* (Taylor, 2007). The mechanisms underlying *Chlamydia* clearance by IRGs is not known, but, like in the case of *Toxoplasma*, a subset of IRGs is recruited to the inclusion (Coers et al., 2008; Al-Zeer et al., 2009). Irgb10 localizes to *C. trachomatis* inclusions, and its ectopic expression is sufficient to selectively reduce *C. trachomatis* replication (Coers et al., 2008). In contrast, *C. muridarum* evades Irgb10 deposition on inclusion membranes (Coers et al., 2008). The mechanism underlying this resistance to IRGs is unknown, although it has been postulated that protease-like activities of the large toxin present in *C. muridarum* may disarm IRGs (Nelson et al., 2005).

Autophagy is a bulk degradation pathway in which cytosolic materials, such as protein aggregates, whole organelles, and intracellular pathogens, are enveloped by endomembranes into an autophagosome that subsequently fuses with lysosomes to eliminate captured cargo (Deretic, 2011). Autophagy is important in starvation responses, turnover of dysfunctional organelles, and clearance of large cytosolic aggregates. Recent studies of IRG-mediated clearance of intracellular pathogens reveal an important a role for autophagy in cell autonomous defense mechanisms to infection (reviewed by Deretic [2011]). Not surprisingly, autophagy appears to be involved in IRG-mediated clearance of *C. trachomatis* in IFN-γ-primed murine cells (Al-Zeer et al., 2009). While autophagosomes do not appear to fuse with *C. trachomatis* inclusions during infection in untreated human cells (Al-Younes et al., 2004), *C. trachomatis* inclusions fuse with autophagosomes in IFN-γ-treated murine cells, and autophagy-deficient IFN-γ-treated murine cells are more permissive to *C. trachomatis* infection (Al-Zeer et al., 2009). Thus, it appears that IRG-mediated clearance of *C. trachomatis* in murine cells at least partly resembles *T. gondii* clearance: in both cases, IRGs localize to the parasitophorous vacuole and the pathogen is ultimately eliminated by autophagic mechanisms.

While IRGs are largely absent in humans, a different family of IFN-induced proteins, the human guanylate binding proteins (hGBPs), appears to contribute to the control of *Chlamydia* replication in IFN-γ-treated human cells (Tietzel et al., 2009). hGBP1 and hGBP2 localize to the inclusion, and their depletion partially alleviates the anti-*Chlamydia* effects of IFN-γ-primed cells. These proteins also have mild anti-*Chlamydia* effects when overexpressed, and these effects are enhanced by addition of subinhibitory concentrations of IFN-γ, suggesting that hGBPs may act as potentiators of IFN-γ for the clearance of *Chlamydia* infections (Tietzel et al., 2009).

GETTING OUT—*CHLAMYDIA* EXITS HOST CELLS BY LYSIS OR EXTRUSION

At the end of the infectious cycle, the exit of EBs from the infected cells involves both extrusion of the inclusion and cell lysis (Hybiske and Stephens, 2007). As assessed by live video microscopy, approximately one-half of exit events involve the extrusion of a portion of the intact inclusion, leaving the infected cell intact. Extrusion is a relatively slow process, with the inclusion continuously encased by plasma membrane as it protrudes out of the cell. Inhibitors of actin polymerization and myosin II function, which is important for the contractile property of actomyosin fibers, completely inhibit extrusion. RhoA-C GTPases are also required for the pinching and release of extruded inclusions, implying two sequential actomyosin-mediated steps in extrusion. In contrast, the lysis exit pathway is characterized by rapid sequential rupture of the inclusion, organellar, and plasma membranes and requires the activity of cysteine proteases (Hybiske and Stephens, 2007). This suggests a programmed set of lytic events rather than disordered rupture due to mechanical forces exerted by the inclusion (Hybiske and Stephens, 2007).

Different exit strategies could potentially be favored at different stages of infection or in response to diverse host environments. This could allow for fine control over local inflammatory responses by controlling the extent of host cell lysis, although it should be noted that these extrusion events have not yet been documented in infected animals. *C. trachomatis* serovars D and L2 and *C. caviae* all display similar lysis and extrusion mechanisms (Hybiske and Stephens, 2007), but some differences may exist among serovars and strains since *C. trachomatis* serovar E exit is characterized by plasma membrane disruption prior to inclusion membrane permeability (Beatty, 2007). Fusion of lysosomes with the plasma membrane repairs any rupture, and some bacteria remain in the cell after exit (Beatty, 2007). Whether extrusion and membrane-repaired exit of EBs enable persistent infections allowing bacteria to hide within viable cells remains to be determined.

CONCLUSIONS AND FUTURE PERSPECTIVES

As *Chlamydia* can only replicate within the confines of a eukaryotic cell, it is not surprising that it has evolved an intricate arsenal of effector proteins with which to carry out such extensive alterations to the cell biology of its host as has been explored in this chapter. As we continue to learn about the function of these effector proteins, a few general mechanistic strategies come to light. One strategy is the modification of host factors by proteolysis and/or degradation to alter the cell cycle, inhibit apoptosis, and suppress innate immune responses (Balsara et al., 2006; Chellas-Géry et al., 2007; Cocchiaro and Valdivia, 2009). In addition, protein modification by effectors like NUE (CT737), a histone methyltransferase, may silence genes involved in antimicrobial responses (Pennini et al., 2010). Conversely, removal of protein modifications by effectors like the *Chla*Dub proteins, which display deubiquitinating and deneddylating activity (Misaghi et al., 2006), may assist in silencing immune-related signaling (Le Negrate et al., 2008).

Another set of effectors are restricted to the inclusion membrane. These include Inc proteins that act to sequester host proteins to the inclusion membrane, thus interfering with innate defense processes such as apoptosis (Tse et al., 2005; Verbeke et al., 2006). Inc proteins also interact extensively with components of the endolysosomal system, including SNAREs and Rab proteins (Rzomp et al., 2003, 2006; Delevoye et al., 2008; Paumet et al., 2009), presumably to avoid lysosomal degradation. Finally, Inc proteins may facilitate interactions with other organelles such as the ER to facilitate lipid acquisition (Derré et al., 2011; Elwell et al., 2011) or centrosomes for correct positioning of the inclusion within an infected cell (Mital et al., 2010).

There is also a growing appreciation that *Chlamydia* effectors may work cooperatively and be subjected to different levels of post-secretion regulation. For instance, CPAF cleaves a number of early effectors to regulate their activity and may play a role in niche protection by limiting superinfection of cells (Jorgensen et al., 2011). Similarly, Incs can form multiprotein complexes with other Incs to form centrosome-associated subdomains on the inclusion membrane (Mital et al., 2010). The regulation of effectors by other effectors and their ability to form multieffector complexes may add another level of complexity and dynamism to their functions. Indeed, our inference of how effectors affect the cell biology of the host, which is largely based on biochemical and gain-of-function approaches, may not tell the whole story. This leads to the important questions of how individual effectors contribute to the many documented alterations in the host and how these changes influence pathogenesis. Definitive answers to these questions will require loss-of-function approaches. Reassuringly, exciting progress in the development of genetic tools for *Chlamydia* (reviewed in chapter 15, "Chlamydial genetics: decades of effort, very recent successes") as well as structure-based design of specific inhibitors (Jorgensen et al., 2011) should help address the function of effectors.

Given the long evolutionary history of the association of *Chlamydia* spp. with eukaryotic cells (Horn, 2008), these bacteria should reveal new insights into basic aspects of eukaryotic cell biology, primordial mechanisms of cell autonomous innate immunity, and novel pathogenic strategies.

REFERENCES

Al-Younes, H. M., V. Brinkmann, and T. F. Meyer. 2004. Interaction of *Chlamydia trachomatis* serovar L2 with the host autophagic pathway. *Infect. Immun.* **72:**4751–4762.

Al-Zeer, M. A., H. M. Al-Younes, P. R. Braun, J. Zerrahn, and T. F. Meyer. 2009. IFN-gamma-inducible Irga6 mediates host resistance against *Chlamydia trachomatis* via autophagy. *PLoS One* **4:**e4588.

Alzhanov, D. T., S. K. Weeks, J. R. Burnett, and D. D. Rockey. 2009. Cytokinesis is blocked in mammalian cells transfected with *Chlamydia trachomatis* gene CT223. *BMC Microbiol.* **9:**2.

Balsara, Z. R., S. Misaghi, J. N. Lafave, and M. N. Starnbach. 2006. *Chlamydia trachomatis* infection induces cleavage of the mitotic cyclin B1. *Infect. Immun.* **74:**5602–5608.

Barton, G. M., and J. C. Kagan. 2009. A cell biological view of Toll-like receptor function: regulation through compartmentalization. *Nat. Rev. Immunol.* **9:**535–542.

Basto, R., K. Brunk, T. Vinadogrova, N. Peel, A. Franz, A. Khodjakov, and J. W. Raff. 2008. Centrosome amplification can initiate tumorigenesis in flies. *Cell* **133:**1032–1042.

Beatty, W. L. 2006. Trafficking from CD63-positive late endocytic multivesicular bodies is essential for intracellular development of *Chlamydia trachomatis*. *J. Cell Sci.* **119:**350–359.

Beatty, W. L. 2007. Lysosome repair enables host cell survival and bacterial persistence following *Chlamydia trachomatis* infection. *Cell. Microbiol.* **9:**2141–2152.

Beatty, W. L. 2008. Late endocytic multivesicular bodies intersect the chlamydial inclusion in the absence of CD63. *Infect. Immun.* **76:**2872–2881.

Behnia, R., and S. Munro. 2005. Organelle identity and the signposts for membrane traffic. *Nature* **438:**597–604.

Belland, R. J., G. Zhong, D. D. Crane, D. Hogan, D. Sturdevant, J. Sharma, W. L. Beatty, and H. D. Caldwell. 2003. Genomic transcriptional profiling of the developmental cycle of *Chlamydia trachomatis*. *Proc. Natl. Acad. Sci. USA* **100:**8478–8483.

Betts, H. J., K. Wolf, and K. A. Fields. 2009. Effector protein modulation of host cells: examples in the *Chlamydia* spp. arsenal. *Curr. Opin. Microbiol.* **12:**81–87.

Bose, S. K., and H. Liebhaber. 1979. Deoxyribonucleic acid synthesis, cell cycle progression, and division of *Chlamydia*-infected HeLa 229 cells. *Infect. Immun.* **24:**953–957.

Brazil, D. P., J. Park, and B. A. Hemmings. 2002. PKB binding proteins. Getting in on the Akt. *Cell* **111:**293–303.

Capmany, A., and M. T. Damiani. 2010. *Chlamydia trachomatis* intercepts Golgi-derived sphingolipids through a Rab14-mediated transport required for bacterial development and replication. *PLoS One* **5:**e14084.

Carabeo, R. A., D. J. Mead, and T. Hackstadt. 2003. Golgi-dependent transport of cholesterol to the *Chlamydia trachomatis* inclusion. *Proc. Natl. Acad. Sci. USA* **100:**6771–6776.

Chellas-Géry, B., C. N. Linton, and K. A. Fields. 2007. Human GCIP interacts with CT847, a novel *Chlamydia trachomatis* type III secretion substrate, and is degraded in a tissue-culture infection model. *Cell. Microbiol.* **9:**2417–2430.

Chen, Y. A., and R. H. Scheller. 2001. SNARE-mediated membrane fusion. *Nat. Rev. Mol. Cell Biol.* **2:**98–106.

Claude, A., B. P. Zhao, C. E. Kuziemsky, S. Dahan, S. J. Berger, J. P. Yan, A. D. Armold, E. M. Sullivan, and P. Melançon. 1999. GBF1: a novel Golgi-associated BFA-resistant guanine nucleotide exchange factor that displays specificity for ADP-ribosylation factor 5. *J. Cell Biol.* **146:**71–84.

Clausen, J. D., G. Christiansen, H. U. Holst, and S. Birkelund. 1997. *Chlamydia trachomatis* utilizes the host cell microtubule network during early events of infection. *Mol. Microbiol.* **25:**441–449.

Clifton, D. R., K. A. Fields, S. S. Grieshaber, C. A. Dooley, E. R. Fischer, D. J. Mead, R. A. Carabeo, and T. Hackstadt. 2004. A chlamydial type III translocated protein is tyrosine-phosphorylated at the site of entry and associated with recruitment of actin. *Proc. Natl. Acad. Sci. USA* **101:**10166–10171.

Cocchiaro, J. L., Y. Kumar, E. R. Fischer, T. Hackstadt, and R. H. Valdivia. 2008. Cytoplasmic lipid droplets are translocated into the lumen of the *Chlamydia trachomatis* parasitophorous vacuole. *Proc. Natl. Acad. Sci. USA* **105:**9379–9384.

Cocchiaro, J. L., and R. H. Valdivia. 2009. New insights into *Chlamydia* intracellular survival mechanisms. *Cell. Microbiol.* **11:**1571–1578.

Coers, J., I. Bernstein-Hanley, D. Grotsky, I. Parvanova, J. C. Howard, G. A. Taylor, W. F. Dietrich, and M. N. Starnbach. 2008. *Chlamydia muridarum* evades growth restriction by the IFN-gamma-inducible host resistance factor Irgb10. *J. Immunol.* **180:**6237–6245.

Cortes, C., K. A. Rzomp, A. Tvinnereim, M. A. Scidmore, and B. Wizel. 2007. *Chlamydia pneumoniae* inclusion membrane protein Cpn0585 interacts with multiple Rab GTPases. *Infect. Immun.* **75:**5586–5596.

Delevoye, C., M. Nilges, A. Dautry-Varsat, and A. Subtil. 2004. Conservation of the biochemical properties of IncA from *Chlamydia trachomatis* and *Chlamydia caviae*: oligomerization of IncA mediates interaction between facing membranes. *J. Biol. Chem.* **279:**46896–46906.

Delevoye, C., M. Nilges, P. Dehoux, F. Paumet, S. Perrinet, A. Dautry-Varsat, and A. Subtil. 2008. SNARE protein mimicry by an intracellular bacterium. *PLoS Pathog.* **4:**e1000022.

Deretic, V. 2011. Autophagy in immunity and cell-autonomous defense against intracellular microbes. *Immunol. Rev.* **240:**92–104.

Derré, I., M. Pypaert, A. Dautry-Varsat, and H. Agaisse. 2007. RNAi screen in *Drosophila* cells reveals the involvement of the Tom complex in *Chlamydia* infection. *PLoS Pathog.* **3:**e155.

Derré, I., R. Swiss, and H. Agaisse. 2011. The lipid transfer protein CERT interacts with the *Chlamydia* inclusion protein IncD and participates to ER-*Chlamydia* inclusion membrane contact sites. *PLoS Pathog.* **7:**e1002092.

Dong, F., M. Pirbhai, Y. Xiao, Y. Zhong, Y. Wu, and G. Zhong. 2005. Degradation of the proapoptotic proteins Bik, Puma, and Bim with Bcl-2 domain 3 homology in *Chlamydia trachomatis*-infected cells. *Infect. Immun.* **73:**1861–1864.

Dong, F., H. Su, Y. Huang, Y. Zhong, and G. Zhong. 2004. Cleavage of host keratin 8 by a *Chlamydia*-secreted protease. *Infect. Immun.* **72:**3863–3868.

D'Souza-Schorey, C., and P. Chavrier. 2006. ARF proteins: roles in membrane traffic and beyond. *Nat. Rev. Mol. Cell Biol.* **7:**347–358.

Elwell, C. A., A. Ceesay, J. H. Kim, D. Kalman, and J. N. Engel. 2008. RNA interference screen identifies Abl kinase and PDGFR signaling in *Chlamydia trachomatis* entry. *PLoS Pathog.* **4:**e1000021.

Elwell, C. A., S. Jiang, J. H. Kim, A. Lee, T. Wittmann, K. Hanada, P. Melancon, and J. N. Engel. 2011. *Chlamydia trachomatis* co-opts GBF1 and CERT to acquire host sphingomyelin for distinct roles during intracellular development. *PLoS Pathog.* **7:**e1002198.

Etienne-Manneville, S., and A. Hall. 2002. Rho GTPases in cell biology. *Nature* **420:**629–635.

Farese, R. V., Jr., and T. C. Walther. 2009. Lipid droplets finally get a little R-E-S-P-E-C-T. *Cell* **139:**855–860.

Fields, K. A., E. Fischer, and T. Hackstadt. 2002. Inhibition of fusion of *Chlamydia trachomatis* inclusions at 32 degrees C correlates with restricted export of IncA. *Infect. Immun.* **70:**3816–3823.

Fields, K. A., and T. Hackstadt. 2002. The chlamydial inclusion: escape from the endocytic pathway. *Annu. Rev. Cell Dev. Biol.* **18:**221–245.

Fischer, S. F., J. Vier, S. Kirschnek, A. Klos, S. Hess, S. Ying, and G. Häcker. 2004. *Chlamydia* inhibits host cell apoptosis by degradation of proapoptotic BH3-only proteins. *J. Exp. Med.* **200:**905–916.

Giles, D. K., and P. B. Wyrick. 2008. Trafficking of chlamydial antigens to the endoplasmic reticulum of infected epithelial cells. *Microbes Infect.* **10:**1494–1503.

Greene, W., Y. Xiao, Y. Huang, G. McClarty, and G. Zhong. 2004. *Chlamydia*-infected cells continue to undergo mitosis and resist induction of apoptosis. *Infect. Immun.* **72:**451–460.

Greene, W., and G. Zhong. 2003. Inhibition of host cell cytokinesis by *Chlamydia trachomatis* infection. *J. Infect.* **47:**45–51.

Grieshaber, S. S., N. A. Grieshaber, and T. Hackstadt. 2003. *Chlamydia trachomatis* uses host cell dynein to traffic to the microtubule-organizing center in a p50 dynamitin-independent process. *J. Cell Sci.* **116:**3793–3802.

Grieshaber, S. S., N. A. Grieshaber, N. Miller, and T. Hackstadt. 2006. *Chlamydia trachomatis* causes centrosomal defects resulting in chromosomal segregation abnormalities. *Traffic* **7:**940–949.

Guo, Y., K. R. Cordes, R. V. Farese, and T. C. Walther. 2009. Lipid droplets at a glance. *J. Cell Sci.* **122:**749–752.

Gurumurthy, R. K., A. P. Mäurer, N. Machuy, S. Hess, K. P. Pleissner, J. Schuchhardt, T. Rudel, and T. F. Meyer. 2010. A loss-of-function screen reveals Ras- and Raf-independent MEK-ERK signaling during *Chlamydia trachomatis* infection. *Sci. Signal.* **3:**ra21.

Hackstadt, T., D. D. Rockey, R. A. Heinzen, and M. A. Scidmore. 1996. *Chlamydia trachomatis* interrupts an exocytic pathway to acquire endogenously synthesized sphingomyelin in transit from the Golgi apparatus to the plasma membrane. *EMBO J.* **15:**964–977.

Hackstadt, T., M. A. Scidmore, and D. D. Rockey. 1995. Lipid metabolism in *Chlamydia trachomatis*-infected cells: directed trafficking of Golgi-derived sphingolipids to the chlamydial inclusion. *Proc. Natl. Acad. Sci. USA* **92:**4877–4881.

Hackstadt, T., M. A. Scidmore-Carlson, E. I. Shaw, and E. R. Fischer. 1999. The *Chlamydia trachomatis* IncA protein is required for homotypic vesicle fusion. *Cell. Microbiol.* **1:**119–130.

Hanada, K., K. Kumagai, S. Yasuda, Y. Miura, M. Kawano, M. Fukasawa, and M. Nishijima. 2003. Molecular machinery for non-vesicular trafficking of ceramide. *Nature* **426:**803–809.

Hatch, G. M., and G. McClarty. 1998a. Cardiolipin remodeling in eukaryotic cells infected with *Chlamydia trachomatis* is linked to elevated mitochondrial metabolism. *Biochem. Biophys. Res. Commun.* **243:**356–360.

Hatch, G. M., and G. McClarty. 1998b. Phospholipid composition of purified *Chlamydia trachomatis* mimics that of the eucaryotic host cell. *Infect. Immun.* **66:**3727–3735.

Heuer, D., A. Rejman Lipinski, N. Machuy, A. Karlas, A. Wehrens, F. Siedler, V. Brinkmann, and T. F. Meyer. 2009. *Chlamydia* causes fragmentation of the Golgi compartment to ensure reproduction. *Nature* **457:**731–735.

Ho, T. D., and M. N. Starnbach. 2005. The *Salmonella enterica* serovar Typhimurium-encoded type III secretion systems can translocate *Chlamydia trachomatis* proteins into the cytosol of host cells. *Infect. Immun.* **73:**905–911.

Horn, M. 2008. Chlamydiae as symbionts in eukaryotes. *Annu. Rev. Microbiol.* **62:**113–131.

Hybiske, K., and R. S. Stephens. 2007. Mechanisms of host cell exit by the intracellular bacterium *Chlamydia*. *Proc. Natl. Acad. Sci. USA* **104:**11430–11435.

Igal, R. A., P. Wang, and R. A. Coleman. 1997. Triacsin C blocks de novo synthesis of glycerolipids and cholesterol esters but not recycling of fatty acid into phospholipid: evidence for functionally separate pools of acyl-CoA. *Biochem. J.* **324:**529–534.

Jewett, T. J., E. R. Fischer, D. J. Mead, and T. Hackstadt. 2006. Chlamydial TARP is a bacterial nucleator of actin. *Proc. Natl. Acad. Sci. USA* **103:**15599–15604.

Johnson, K. A., M. Tan, and C. Sütterlin. 2009. Centrosome abnormalities during a *Chlamydia trachomatis* infection are caused by dysregulation of the normal duplication pathway. *Cell. Microbiol.* **11:**1064–1073.

Jorgensen, I., M. M. Bednar, V. Amin, B. K. Davis, J. P. Y. Ting, D. G. McCafferty, and R. H. Valdivia. 2011. The *Chlamydia* protease CPAF regulates host and bacterial proteins to maintain pathogen vacuole integrity and promote virulence. *Cell Host Microbe* **10:**21–32.

Kawano, M., K. Kumagai, M. Nishijima, and K. Hanada. 2006. Efficient trafficking of ceramide from the endoplasmic reticulum to the Golgi apparatus requires a VAMP-associated protein-interacting FFAT motif of CERT. *J. Biol. Chem.* **281:**30279–30288.

Knowlton, A. E., H. M. Brown, T. S. Richards, L. A. Andreolas, R. K. Patel, and S. S. Grieshaber. 2011. *Chlamydia trachomatis* infection causes mitotic spindle pole defects independently from its effects on centrosome amplification. *Traffic* **12:**854–866.

Kumar, Y., J. Cocchiaro, and R. H. Valdivia. 2006. The obligate intracellular pathogen *Chlamydia trachomatis* targets host lipid droplets. *Curr. Biol.* **16:**1646–1651.

Kumar, Y., and R. H. Valdivia. 2009. Leading a sheltered life: intracellular pathogens and maintenance of vacuolar compartments. *Cell Host Microbe* **5:**593–601.

Kumar, Y., and R. H. Valdivia. 2008. Actin and intermediate filaments stabilize the *Chlamydia trachomatis* vacuole by forming dynamic structural scaffolds. *Cell Host Microbe* **4:**159–169.

Lane, B. J., C. Mutchler, S. Al Khodor, S. S. Grieshaber, and R. A. Carabeo. 2008. Chlamydial entry involves TARP binding of guanine nucleotide exchange factors. *PLoS Pathog.* **4:**e1000014.

Le Negrate, G., A. Krieg, B. Faustin, M. Loeffler, A. Godzik, S. Krajewski, and J. C. Reed. 2008. ChlaDub1 of *Chlamydia trachomatis* suppresses NF-kappaB activation and inhibits IkappaBalpha ubiquitination and degradation. *Cell. Microbiol.* **10:**1879–1892.

Levine, T., and C. Loewen. 2006. Inter-organelle membrane contact sites: through a glass, darkly. *Curr. Opin. Cell Biol.* **18:**371–378.

Li, Z., C. Chen, D. Chen, Y. Wu, Y. Zhong, and G. Zhong. 2008. Characterization of fifty putative inclusion membrane proteins encoded in the *Chlamydia trachomatis* genome. *Infect. Immun.* **76:**2746–2757.

Lippincott-Schwartz, J., and R. D. Phair. 2010. Lipids and cholesterol as regulators of traffic in the endomembrane system. *Annu. Rev. Biophys.* **39:**559–578.

Lippincott-Schwartz, J., L. C. Yuan, J. S. Bonifacino, and R. D. Klausner. 1989. Rapid redistribution of Golgi proteins into the ER in cells treated with brefeldin A: evidence for membrane cycling from Golgi to ER. *Cell* **56:**801–813.

Littman, A. J., L. A. Jackson, and T. L. Vaughan. 2005. *Chlamydia pneumoniae* and lung cancer: epidemiologic evidence. *Cancer Epidemiol. Biomarkers Prev.* **14:**773–778.

Majeed, M., K. H. Krause, R. A. Clark, E. Kihlström, and O. Stendahl. 1999. Localization of intracellular Ca2+ stores in HeLa cells during infection with *Chlamydia trachomatis*. *J. Cell Sci.* **112:**35–44.

Manolea, F., A. Claude, J. Chun, J. Rosas, and P. Melançon. 2008. Distinct functions for Arf guanine nucleotide exchange factors at the Golgi complex: GBF1 and BIGs are required for assembly and maintenance of the Golgi stack and trans-Golgi network, respectively. *Mol. Biol. Cell* **19:**523–535.

Mansour, S. J., J. Skaug, X. H. Zhao, J. Giordano, S. W. Scherer, and P. Melançon. 1999. p200 ARF-GEP1: a Golgi-localized guanine nucleotide exchange protein whose Sec7 domain is targeted by the drug brefeldin A. *Proc. Natl. Acad. Sci. USA* **96:**7968–7973.

Martens, S., and H. T. McMahon. 2008. Mechanisms of membrane fusion: disparate players and common principles. *Nat. Rev. Mol. Cell Biol.* **9:**543–556.

Matanis, T., A. Akhmanova, P. Wulf, E. Del Nery, T. Weide, T. Stepanova, N. Galjart, F. Grosveld, B. Goud, C. I. De Zeeuw, A. Barnekow, and C. C. Hoogenraad. 2002. Bicaudal-D regulates COPI-independent Golgi-ER transport by recruiting the dynein-dynactin motor complex. *Nat. Cell Biol.* **4:**986–992.

Matsumoto, A., H. Bessho, K. Uehira, and T. Suda. 1991. Morphological studies of the association of mitochondria with chlamydial inclusions and the fusion of chlamydial inclusions. *J. Electron Microsc.* **40:**356–363.

Medzhitov, R. 2001. Toll-like receptors and innate immunity. *Nat. Rev. Immunol.* **1:**135–145.

Misaghi, S., Z. R. Balsara, A. Catic, E. Spooner, H. L. Ploegh, and M. N. Starnbach. 2006. *Chlamydia trachomatis*-derived deubiquitinating enzymes in mammalian cells during infection. *Mol. Microbiol.* **61:**142–150.

Mital, J., N. J. Miller, E. R. Fischer, and T. Hackstadt. 2010. Specific chlamydial inclusion membrane proteins associate with active Src family kinases in microdomains that interact with the host microtubule network. *Cell. Microbiol.* **12:**1235–1249.

Moore, E. R., E. R. Fischer, D. J. Mead, and T. Hackstadt. 2008. The chlamydial inclusion preferentially intercepts basolaterally directed sphingomyelin-containing exocytic vacuoles. *Traffic* **9:**2130–2140.

Moore, L., D. Mead, C. Dooley, J. Sager, and T. Hackstadt. 2010. The trans-Golgi SNARE syntaxin 6 is recruited to the chlamydial inclusion membrane. *Microbiology* **157:**830–838.

Moorhead, A. M., J.-Y. Jung, A. Smirnov, S. Kaufer, and M. A. Scidmore. 2010. Multiple host proteins that function in phosphatidylinositol-4-phosphate metabolism are recruited to the chlamydial inclusion. *Infect. Immun.* **78:**1990–2007.

Moorhead, A. R., K. A. Rzomp, and M. A. Scidmore. 2007. The Rab6 effector Bicaudal D1 associates with *Chlamydia trachomatis* inclusions in a biovar-specific manner. *Infect. Immun.* **75:**781–791.

Morrison, R. P., and H. D. Caldwell. 2002. Immunity to murine chlamydial genital infection. *Infect. Immun.* **70:**2741–2751.

Moulder, J. W. 1991. Interaction of chlamydiae and host cells *in vitro*. *Microbiol. Rev.* **55:**143–190.

Musacchio, A., and E. D. Salmon. 2007. The spindle-assembly checkpoint in space and time. *Nat. Rev. Mol. Cell Biol.* **8:**379–393.

Nagai, T., A. Abe, and C. Sasakawa. 2005. Targeting of enteropathogenic *Escherichia coli* EspF to host mitochondria is essential for bacterial pathogenesis: critical role of the 16th leucine residue in EspF. *J. Biol. Chem.* **280:**2998–3011.

Nelson, D. E., D. P. Virok, H. Wood, C. Roshick, R. M. Johnson, W. M. Whitmire, D. D. Crane, O. Steele-Mortimer, L. Kari, G. McClarty, and H. D. Caldwell. 2005. Chlamydial IFN-gamma immune evasion is linked to host infection tropism. *Proc. Natl. Acad. Sci. USA* **102:**10658–10663.

Ojcius, D. M., H. Degani, J. Mispelter, and A. Dautry-Varsat. 1998. Enhancement of ATP levels and glucose metabolism during an infection by *Chlamydia*. NMR studies of living cells. *J. Biol. Chem.* **273:**7052–7058.

Ouellette, S., and R. A. Carabeo. 2010. A functional slow recycling pathway of transferrin is required for growth of *Chlamydia*. *Frontiers Microbiol.* **1:**112.

Papatheodorou, P., G. Domańska, M. Oxle, J. Mathieu, O. Selchow, B. Kenny, and J. Rassow. 2006. The enteropathogenic *Escherichia coli* (EPEC) Map effector is imported into the mitochondrial matrix by the TOM/Hsp70 system and alters organelle morphology. *Cell. Microbiol.* **8:**677–689.

Paschen, S. A., J. G. Christian, J. Vier, F. Schmidt, A. Walch, D. M. Ojcius, and G.

Häcker. 2008. Cytopathicity of *Chlamydia* is largely reproduced by expression of a single chlamydial protease. *J. Cell Biol.* **182:**117–127.

Paumet, F., J. Wesolowski, A. Garcia-Diaz, C. Delevoye, N. Aulner, H. A. Shuman, A. Subtil, and J. E. Rothman. 2009. Intracellular bacteria encode inhibitory SNARE-like proteins. *PLoS One* **4:**e7375.

Pennini, M. E., S. Perrinet, A. Dautry-Varsat, and A. Subtil. 2010. Histone methylation by NUE, a novel nuclear effector of the intracellular pathogen *Chlamydia trachomatis*. *PLoS Pathog.* **6:**e1000995.

Perry, R. J., and N. D. Ridgway. 2005. Molecular mechanisms and regulation of ceramide transport. *Biochim. Biophys. Acta* **1734:**220–234.

Peterson, E. M., and L. M. de la Maza. 1988. *Chlamydia* parasitism: ultrastructural characterization of the interaction between the chlamydial cell envelope and the host cell. *J. Bacteriol.* **170:**1389–1392.

Piper, R. C., and D. J. Katzmann. 2007. Biogenesis and function of multivesicular bodies. *Annu. Rev. Cell Dev. Biol.* **23:**519–547.

Quintyne, N. J., J. E. Reing, D. R. Hoffelder, S. M. Gollin, and W. S. Saunders. 2005. Spindle multipolarity is prevented by centrosomal clustering. *Science* **307:**127–129.

Read, T. D., R. C. Brunham, C. Shen, S. R. Gill, J. F. Heidelberg, O. White, E. K. Hickey, J. Peterson, T. Utterback, K. Berry, S. Bass, K. Linher, J. Weidman, H. Khouri, B. Craven, C. Bowman, R. Dodson, M. Gwinn, W. Nelson, R. DeBoy, J. Kolonay, G. McClarty, S. L. Salzberg, J. Eisen, and C. M. Fraser. 2000. Genome sequences of *Chlamydia trachomatis* MoPn and *Chlamydia pneumoniae* AR39. *Nucleic Acids Res.* **28:**1397–1406.

Rejman Lipinski, A., J. Heymann, C. Meissner, A. Karlas, V. Brinkmann, T. F. Meyer, and D. Heuer. 2009. Rab6 and Rab11 regulate *Chlamydia trachomatis* development and golgin-84-dependent Golgi fragmentation. *PLoS Pathog.* **5:**e1000615.

Richardson, D. R., and P. Ponka. 1997. The molecular mechanisms of the metabolism and transport of iron in normal and neoplastic cells. *Biochim. Biophys. Acta* **1331:**1–40.

Robertson, D. K., L. Gu, R. K. Rowe, and W. L. Beatty. 2009. Inclusion biogenesis and reactivation of persistent *Chlamydia trachomatis* requires host cell sphingolipid biosynthesis. *PLoS Pathog.* **5:**e1000664.

Rockey, D. D., E. R. Fischer, and T. Hackstadt. 1996. Temporal analysis of the developing *Chlamydia psittaci* inclusion by use of fluorescence and electron microscopy. *Infect. Immun.* **64:**4269–4278.

Rockey, D. D., M. A. Scidmore, J. P. Bannantine, and W. J. Brown. 2002. Proteins in the chlamydial inclusion membrane. *Microbes Infect.* **4:**333–340.

Rzomp, K. A., A. R. Moorhead, and M. A. Scidmore. 2006. The GTPase Rab4 interacts with *Chlamydia trachomatis* inclusion membrane protein CT229. *Infect. Immun.* **74:**5362–5373.

Rzomp, K. A., L. D. Scholtes, B. J. Briggs, G. R. Whittaker, and M. A. Scidmore. 2003. Rab GTPases are recruited to chlamydial inclusions in both a species-dependent and species-independent manner. *Infect. Immun.* **71:**5855–5870.

Saka, H. A., J. W. Thompson, Y.-S. Chen, Y. Kumar, L. G. Dubois, M. A. Moseley, and R. H. Valdivia. 2011. Quantitative proteomics reveals metabolic and pathogenic properties of *Chlamydia trachomatis* developmental forms. *Mol. Microbiol.* **82:**1185–1203.

Schramm, N., and P. B. Wyrick. 1995. Cytoskeletal requirements in *Chlamydia trachomatis* infection of host cells. *Infect. Immun.* **63:**324–332.

Scidmore, M. A., E. R. Fischer, and T. Hackstadt. 2003. Restricted fusion of *Chlamydia trachomatis* vesicles with endocytic compartments during the initial stages of infection. *Infect. Immun.* **71:**973–984.

Scidmore, M. A., and T. Hackstadt. 2001. Mammalian 14-3-3beta associates with the *Chlamydia trachomatis* inclusion membrane via its interaction with IncG. *Mol. Microbiol.* **39:**1638–1650.

Scidmore, M. A., D. D. Rockey, E. R. Fischer, R. A. Heinzen, and T. Hackstadt. 1996. Vesicular interactions of the *Chlamydia trachomatis* inclusion are determined by chlamydial early protein synthesis rather than route of entry. *Infect. Immun.* **64:**5366–5372.

Sharma, M., and T. Rudel. 2009. Apoptosis resistance in *Chlamydia*-infected cells: a fate worse than death? *FEMS Immunol. Med. Microbiol.* **55:**154–161.

Shaw, E. I., C. A. Dooley, E. R. Fischer, M. A. Scidmore, K. A. Fields, and T. Hackstadt. 2000. Three temporal classes of gene expression during the *Chlamydia trachomatis* developmental cycle. *Mol. Microbiol.* **37:**913–925.

Simonetti, A. C., J. H. de L. Melo, P. R. E. de Souza, D. Bruneska, and J. L. de Lima Filho. 2009. Immunological's host profile for HPV and *Chlamydia trachomatis*, a cervical cancer cofactor. *Microbes Infect.* **11:**435–442.

Sisko, J. L., K. Spaeth, Y. Kumar, and R. H. Valdivia. 2006. Multifunctional analysis of *Chlamydia*-specific genes in a yeast expression system. *Mol. Microbiol.* **60:**51–66.

Stenmark, H. 2009. Rab GTPases as coordinators of vesicle traffic. *Nat. Rev. Mol. Cell Biol.* **10:**513–525.

Stephens, R. S., S. Kalman, C. Lammel, J. Fan, R. Marathe, L. Aravind, W. Mitchell, L. Olinger, R. L. Tatusov, Q. Zhao, E. V. Koonin, and

R. W. Davis. 1998. Genome sequence of an obligate intracellular pathogen of humans: *Chlamydia trachomatis*. *Science* **282:**754–759.

Stephens, R. S., G. Myers, M. Eppinger, and P. M. Bavoil. 2009. Divergence without difference: phylogenetics and taxonomy of *Chlamydia* resolved. *FEMS Immunol. Med. Microbiol.* **55:**115–119.

Su, H., G. McClarty, F. Dong, G. M. Hatch, Z. K. Pan, and G. Zhong. 2004. Activation of Raf/MEK/ERK/cPLA2 signaling pathway is essential for chlamydial acquisition of host glycerophospholipids. *J. Biol. Chem.* **279:**9409–9416.

Subtil, A., C. Delevoye, M.-E. Balañá, L. Tastevin, S. Perrinet, and A. Dautry-Varsat. 2005. A directed screen for chlamydial proteins secreted by a type III mechanism identifies a translocated protein and numerous other new candidates. *Mol. Microbiol.* **56:**1636–1647.

Suchland, R. J., D. D. Rockey, J. P. Bannantine, and W. E. Stamm. 2000. Isolates of *Chlamydia trachomatis* that occupy nonfusogenic inclusions lack IncA, a protein localized to the inclusion membrane. *Infect. Immun.* **68:**360–367.

Sütterlin, C., and A. Colanzi. 2010. The Golgi and the centrosome: building a functional partnership. *J. Cell Biol.* **188:**621–628.

Tafesse, F. G., K. Huitema, M. Hermansson, S. van der Poel, J. van den Dikkenberg, A. Uphoff, P. Somerharju, and J. C. M. Holthuis. 2007. Both sphingomyelin synthases SMS1 and SMS2 are required for sphingomyelin homeostasis and growth in human HeLa cells. *J. Biol. Chem.* **282:**17537–17547.

Taylor, G. A. 2007. IRG proteins: key mediators of interferon-regulated host resistance to intracellular pathogens. *Cell. Microbiol.* **9:**1099–1107.

Taylor, L. D., D. E. Nelson, D. W. Dorward, W. M. Whitmire, and H. D. Caldwell. 2010. Biological characterization of *Chlamydia trachomatis* plasticity zone MACPF domain family protein CT153. *Infect. Immun.* **78:**2691–2699.

Thompson, C. C., and R. A. Carabeo. 2011. An optimal method of iron starvation of the obligate intracellular pathogen, *Chlamydia trachomatis*. *Frontiers Microbiol.* **2:**20.

Tietzel, I., C. El-Haibi, and R. A. Carabeo. 2009. Human guanylate binding proteins potentiate the anti-chlamydia effects of interferon-gamma. *PLoS One* **4:**e6499.

Tse, S. M. L., D. Mason, R. J. Botelho, B. Chiu, M. Reyland, K. Hanada, R. D. Inman, and S. Grinstein. 2005. Accumulation of diacylglycerol in the *Chlamydia* inclusion vacuole: possible role in the inhibition of host cell apoptosis. *J. Biol. Chem.* **280:**25210–25215.

Ullrich, O., S. Reinsch, S. Urbé, M. Zerial, and R. G. Parton. 1996. Rab11 regulates recycling through the pericentriolar recycling endosome. *J. Cell Biol.* **135:**913–924.

Valdivia, R. H. 2008. *Chlamydia* effector proteins and new insights into chlamydial cellular microbiology. *Curr. Opin. Microbiol.* **11:**53–59.

van Niel, G., I. Porto-Carreiro, S. Simoes, and G. Raposo. 2006. Exosomes: a common pathway for a specialized function. *J. Biochem.* **140:**13–21.

van Ooij, C., G. Apodaca, and J. Engel. 1997. Characterization of the *Chlamydia trachomatis* vacuole and its interaction with the host endocytic pathway in HeLa cells. *Infect. Immun.* **65:** 758–766.

van Ooij, C., E. Homola, E. Kincaid, and J. Engel. 1998. Fusion of *Chlamydia trachomatis*-containing inclusions is inhibited at low temperatures and requires bacterial protein synthesis. *Infect. Immun.* **66:**5364–5371.

van Ooij, C., L. Kalman, S. van Ijzendoorn, M. Nishijima, K. Hanada, K. Mostov, and J. N. Engel. 2000. Host cell-derived sphingolipids are required for the intracellular growth of *Chlamydia trachomatis*. *Cell. Microbiol.* **2:**627–637.

Verbeke, P., L. Welter-Stahl, S. Ying, J. Hansen, G. Häcker, T. Darville, and D. M. Ojcius. 2006. Recruitment of BAD by the *Chlamydia trachomatis* vacuole correlates with host-cell survival. *PLoS Pathog.* **2:**e45.

Verma, V., D. Shen, P. C. Sieving, and C. C. Chan. 2008. The role of infectious agents in the etiology of ocular adnexal neoplasia. *Survey Ophthalmol.* **53:**312–331.

Walther, T. C., and R. V. Farese, Jr. 2009. The life of lipid droplets. *Biochim. Biophys. Acta* **1791:**459–466.

Wilson, D. P., P. Timms, D. L. S. McElwain, and P. M. Bavoil. 2006. Type III secretion, contact-dependent model for the intracellular development of *Chlamydia*. *Bull. Math. Biol.* **68:** 161–178.

Wolf, K., and T. Hackstadt. 2001. Sphingomyelin trafficking in *Chlamydia pneumoniae*-infected cells. *Cell. Microbiol.* **3:**145–152.

Wylie, J. L., G. M. Hatch, and G. McClarty. 1997. Host cell phospholipids are trafficked to and then modified by *Chlamydia trachomatis*. *J. Bacteriol.* **179:**7233–7242.

Yamaji, R., R. Adamik, K. Takeda, A. Togawa, G. Pacheco-Rodriguez, V. J. Ferrans, J. Moss, and M. Vaughan. 2000. Identification and localization of two brefeldin A-inhibited guanine nucleotide-exchange proteins for ADP-ribosylation factors in a macromolecular complex. *Proc. Natl. Acad. Sci. USA* **97:**2567–2572.

Yoshida, K. 2007. PKCdelta signaling: mechanisms of DNA damage response and apoptosis. *Cell. Signal.* **19:**892–901.

PROTEIN SECRETION AND *CHLAMYDIA* PATHOGENESIS

Kenneth A. Fields

9

INTRODUCTION

Intuitively, it seems that an obligate intracellular microbe must possess an array of deployable, host-interactive factors that contribute to survival within an otherwise inhospitable niche. *Chlamydia* spp. certainly fit, if not define, this maxim. In the nearly 15 years following definitive evidence of a type III secretion (T3S) system in *Chlamydia* (Hsia et al., 1997), it has become clear that chlamydiae exploit multiple secretion mechanisms to accomplish a diverse array of host alterations. While significant progress has been made in unraveling the contributions of protein secretion to chlamydial pathogenesis, many pivotal open questions remain. For example, the identity and function of T3S effectors continue to emerge, yet elucidation of *Chlamydia*-specific molecular T3S mechanisms and the degree to which virulence depends on T3S remain elusive. Conversely, chlamydial protease/proteasome-like factor (CPAF) is perhaps the best functionally characterized chlamydial antihost protein, but it remains unclear exactly how this protein gains access to host substrates. This chapter will therefore endeavor to explore current knowledge regarding chlamydial protein secretion as well as point out substantial gaps in understanding. An attempt will also be made to predict where research could lead over the next 15 years to fill these noted gaps. The T3S pathway will be emphasized, given that this area has received considerable attention and fruitful research has provided significant insight regarding this secretion mechanism.

Kenneth A. Fields, Department of Microbiology and Immunology, University of Miami Miller School of Medicine, Miami, FL 33136.

SECRETION MECHANISMS EMPLOYED BY *CHLAMYDIA* SPECIES

Specialized secretion mechanisms enabling delivery of bacterium-derived, antihost exoproteins represent a common theme contributing to the virulence of bacterial pathogens. For *Chlamydia* spp., in particular, encapsulation by a parasitophorous vesicle necessitates mechanisms capable of traversing a membrane barrier. It is now apparent that chlamydiae exploit multiple secretion mechanisms to gain access to the host cytosol (Fig. 1). To remain consistent with convention, protein export from chlamydiae will be referred to as "secretion," whereas steps involved in traversing the host or vacuolar membrane will be referred to as "translocation."

Direct injection of secreted proteins is most clearly accomplished via the use of a nonflagellar T3S (NF-T3S) mechanism (Pallen

Intracellular Pathogens I: Chlamydiales
Edited by Ming Tan and Patrik M. Bavoil © 2012 ASM Press, Washington, DC
10.1128/9781555817329.ch9

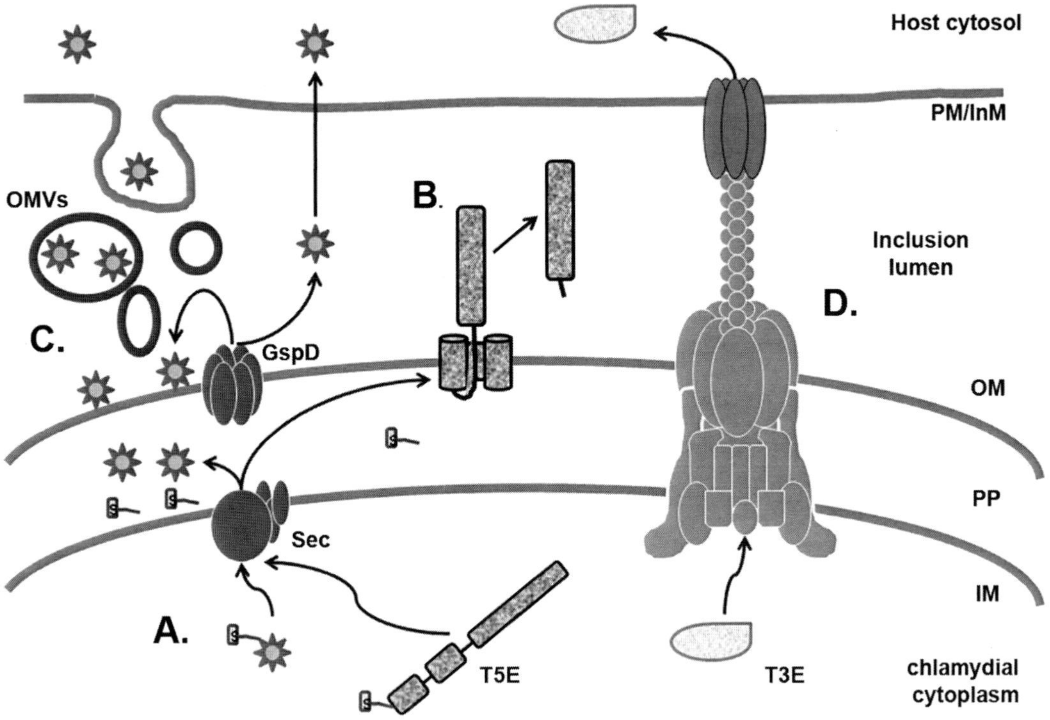

FIGURE 1 Secretion pathways employed for proteins secretion in *Chlamydia* spp. (A) The T2S system mediates export of proteins (T2E) across the chlamydial IM via the Sec protein machinery. Once in the periplasm (PP), signal peptidases can cleave secretion signals (S), and the protein is subsequently secreted across the chlamydial OM through the GspD secretin. (B) T5S substrates (T5E) also gain access to the periplasm via the IM Sec machinery, but domains within the secreted protein mediate direct insertion into the OM. Subsequent cleavage of the passenger domain would release T5S substrates into the inclusion lumen. (C) Translocation of type II and type V secreted proteins could be achieved via formation of OMVs that subsequently fuse with host plasma membrane (PM) or inclusion membranes (InM) to release effector proteins into the host cytosol. (D) The T3S system mediates single-step secretion and translocation of effector proteins directly into the cytosol of an infected cell. doi:10.1128/9781555817329.ch9.f1

et al., 2005). Chlamydial genomes contain the coding capacity for a complete apparatus (Fig. 1D) to secrete and translocate antihost proteins across bacterial and host membranes, respectively. Significant progress has been made in understanding the contributions of T3S to chlamydial pathogenesis, and it is now clear that chlamydiae employ this mechanism to translocate proteins through the host plasma membrane during invasion as well as the inclusion membrane during development. However, it is important to emphasize that T3S does not represent the entire arsenal for chlamydial protein secretion and other equally important secretion mechanisms contribute to chlamydial survival. These complementary systems warrant consideration, although comparatively little is known about them.

Chlamydial genomes (Stephens et al., 1998) encode machinery capable of mediating stepwise protein secretion via the type II secretion (T2S) system (Fig. 1A) (reviewed by Saier [2006]). An ortholog of the SecY inner membrane translocase (*Chlamydia trachomatis* CT510) is coded for by genomes, as are predicted signal peptidases (CT020 and CT408) required to cleave Sec-dependent secretion signals once substrates gain access to the periplasm. Required orthologs of the general secretory pathway (Gsp) proteins are also evident, including the outer membrane (OM) secretin GspD (CT572), to allow substrate export to the bacterial surface. Thus far, demonstration that secretion of CPAF requires T2S (Chen et al., 2010) represents the only direct evidence that this secretion mechanism is active in *Chlamydia* spp. One interesting question is how T2S substrates gain access to the host cytosol. Unlike what is seen for T3S, secretion of T2S substrates is not polarized, and this mechanism lacks the ability to support translocation of secreted proteins. Hence, T2S would be expected to release effectors into the inclusion. Indeed, detection of CPAF within the lumen of inclusions supports this notion (Heuer et al., 2003). Membrane vesicle-mediated translocation represents a potential mechanism for T2S substrates to traverse the inclusion membrane (Fig. 1C). In theory, translocation could be accomplished by encapsulation of T2S effectors in bacterial OM vesicles (OMVs) or envaginated portions of the inclusion membrane. The physiological relevance of OMVs in protein translocation remains controversial, yet evidence from other bacterial systems indicates that antihost proteins can cross host membranes via this mechanism (reviewed by Ellis and Kuehn [2010]). Membrane vesicles carrying CPAF (Heuer et al., 2003) or other chlamydial antigens (Giles et al., 2006) are apparent within the lumen of inclusions, raising the possibility that OMVs complete the deployment of T2S substrates. However, the source of this membrane remains unknown and significant work is still needed to establish a direct role of OMVs in chlamydial protein translocation.

Chlamydiae have also evolved to exploit the type V secretion (T5S) mechanism (Fig. 1B; reviewed by Saier [2006]). Also referred to as the autotransporter pathway, this mechanism requires Sec-dependent secretion of substrates to the periplasm. However, a C-terminal β-barrel domain of the effector protein mediates subsequent insertion into the OM and display of the functional passenger domain on the bacterial surface. The polymorphic OM protein (Pmp) family (Tanzer et al., 2001) exemplifies the chlamydial autotransporters (Stephens et al., 1998; Henderson and Lam, 2001). This diverse group of surface proteins has been implicated in chlamydial attachment (Wehrl et al., 2004; Crane et al., 2006) and immune evasion/niche adaptation (Tan et al., 2010). It is also possible that cleavage of passenger domains would liberate functional domains that, like T2S substrates, could gain access to the host cytosol, since evidence of processing has been documented (Vandahl et al., 2002; Wehrl et al., 2004; Kiselev et al., 2007). The Pmps are discussed in more detail in chapters 4 ("The chlamydial cell envelope") and 5 ("Chlamydial adhesion and adhesins").

It is clear that *Chlamydia* spp. lack a type IV secretion system (which is shared by several environmental chlamydiae), and there is no evidence for a twin-arginine translocation (Tat) pathway or type VI secretion. Like other pathogenic bacteria, *Chlamydia* therefore relies on a select combination of mechanisms to create and maintain a hospitable niche. Why so many? Intuitively, one answer could be that each system is adapted to achieve physiologically distinctive goals. T5S is obviously the most useful mechanism to achieve surface-anchored positioning of host-interactive proteins. T3S effectors are designed to enter host cells to neutralize a precise target during a specific time frame. Conversely, bulk transport of T2S effectors could be expected to have broader, less regulated impact. It is also possible that the comparatively high abundance of T2S effectors could have specific roles beyond targeting a single infected cell. Cell lysis would disgorge pools of these effector proteins capable of *trans* effects on neighboring cells. Regardless, protein secretion is clearly central to chlamydial survival and therefore pathogenesis.

THE CHLAMYDIAL T3S MECHANISM

The T3S system is recognized as a paradigm in gram-negative bacterial pathogenesis and is essential for virulence of human pathogens such as *Yersinia* spp., *Salmonella* spp., *Shigella flexneri*, and *Pseudomonas aeruginosa* (Hueck, 1998). Protein secretion and translocation are accomplished by an elegantly complex secretory apparatus collectively referred to as the "injectisome" (Fig. 2). The injectisome is composed of a generally conserved set of apparatus components. These include components of the multipartite core or basal secretory apparatus, the extended needle complex (NC), and the tip complex (TC), which bridges the space between the bacterial and host membranes. Finally, the translocon is required to form pores in eukaryotic membranes through which secreted substrates gain access to the host cytosol. Detailed reviews that summarize molecular mechanisms governing function of the respective components in other bacteria are available (Moraes et al., 2008; Mueller et al., 2008; Marlovits and Stebbins, 2009).

The genomes of all sequenced *Chlamydia* spp., as well as the more divergent *Parachlamydiaceae*, contain a set of genes encoding a complete and functional T3S apparatus (Stephens et al., 1998; Kalman et al., 1999; Read et al., 2000, 2003; Thomson et al., 2005; Azuma et al., 2006; Horn et al., 2004). Unlike genes for T3S systems in other bacteria that are carried on virulence plasmids or chromosomal pathogenicity islands (Hueck, 1998), chlamydial T3S system coding sequences are dispersed on the chromosome. Apparatus-specific genes

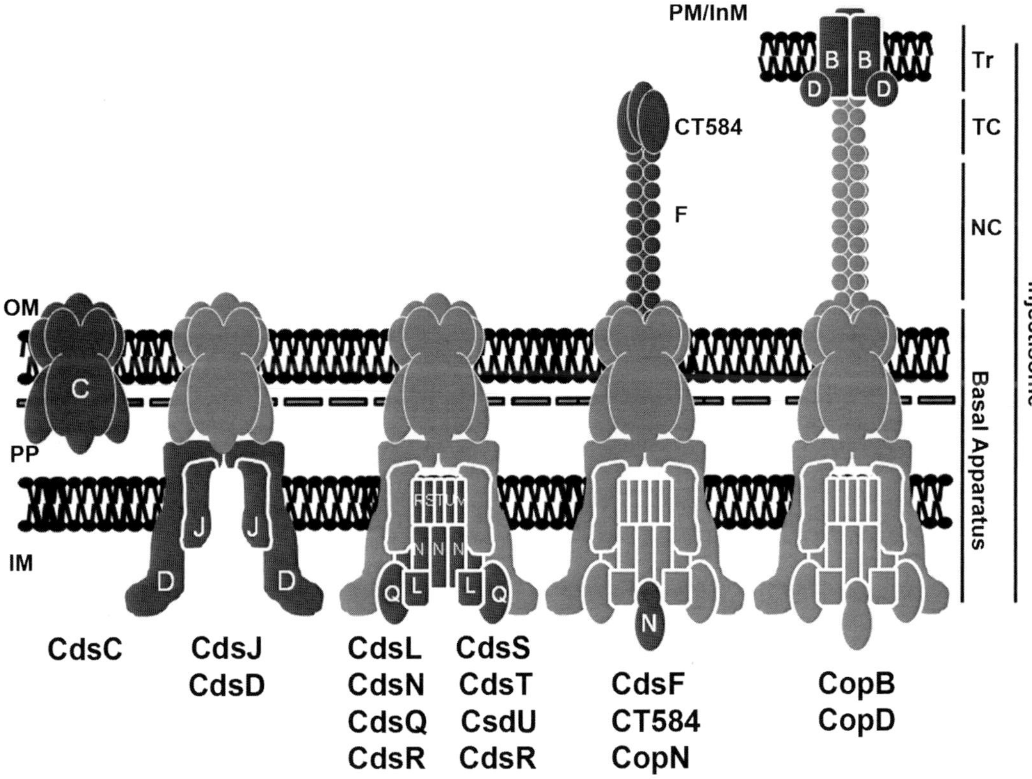

FIGURE 2 A working model for the assembly and composition of the chlamydial T3SS. Stepwise addition of proteins is indicated with newly added components shown in dark grey and previously assembled components in light grey (shown with Cds or Cop letter designation only). Schematic representations of flagellar proteins are omitted for clarity but include FlhA/CT060 (CdsV paralog), FliI/CT717 (CdsN paralog), FliH/CT718 (CdsL paralog), and FliF/CT719 (CdsJ paralog). The completed injectisome spans the bacterial IM, periplasm (PP), and OM and includes TC, NC, and basal apparatus complexes. Secreted translocon (Tr) components are shown localized to the host plasma membrane (PM) or inclusion membrane (InM). Correct stoichiometry of multimeric proteins is not indicated. This figure was adapted from previously published images (Betts-Hampikian and Fields, 2010). doi:10.1128/9781555817329.ch9.f2

are arranged in roughly 5 main loci, while chaperone and effector genes seem to be dispersed more randomly throughout the genome (Stephens et al., 1998). Apparatus genes appear to be grouped together by function. For example, the smaller *C. trachomatis ct088-091* locus encodes products seemingly involved in apparatus regulation and perhaps substrate specificity, while the *ct576-579* and *ct860-861* loci are likely involved in effector translocation. Any functional significance represented in the remaining loci is less easy to decipher, but the *ct559-ct564* locus encodes all inner membrane-associated apparatus components while the *ct663-674* locus is dominated by genes coding for more distal components functioning as periplasmic, OM, and extracellular portions of the apparatus. Thus, coding sequences are roughly grouped in locations where expression of functionally related proteins can be facilitated.

T3S genes are transcribed by the major form of chlamydial RNA polymerase, σ^{66} RNA polymerase (Case et al., 2010; Hefty and Stephens, 2007). Expression studies indicate that the bulk of apparatus components are synthesized roughly during midcycle development (reviewed by Abdelrahman and Belland [2005]). Conspicuous late developmental gene products seem to include only putative translocon and TC proteins (Belland et al., 2003). Promoters for mid, but not late, T3S operons were upregulated by increased DNA supercoiling (Case et al., 2010). Since promoters for other mid and late genes also demonstrate this differential response to supercoiling, it has been proposed that T3S genes are temporally regulated by the same mechanisms that control developmental gene expression in *Chlamydia* (Niehus et al., 2008; Case et al., 2010) (reviewed in chapter 7, "Temporal gene regulation during the chlamydial developmental cycle"). Interestingly, T3S-specific genes are carried in at least 10 operons, and internal promoters allow for independent expression of key structural components (Hefty and Stephens, 2007). The implications for these independent transcripts will be discussed below. However, it should be emphasized that genes encoding orthologs of a flagellar T3S system are found within chlamydial genomes separated from NF-T3S loci. Flagellar proteins include orthologs of flagellar FlhA, FliI, FliH, and FliF corresponding to *C. trachomatis* CT060, CT717, CT718, and CT719, respectively. Since chlamydiae are not motile and the "flagellar" homologs represent duplications of functionally significant core apparatus components, it has been postulated that these gene products could replace nonflagellar homologs to confer functionally distinct and developmentally responsive properties to the chlamydial apparatus (Betts-Hampikian and Fields, 2010).

Assembly and Function of Apparatus Components

In the years since Hsia and colleagues (Hsia et al., 1997) first discovered direct evidence that chlamydiae employ a T3S system, significant progress has been made in elucidating the molecular mechanisms involved in the assembly and function of this complex virulence determinant. Questions regarding apparatus assembly remain open, but the field of bacterial T3S has matured to the extent that it now provides a framework for this process in *Chlamydia* spp. We have recently proposed that basic aspects of T3S apparatus assembly in *Chlamydia* spp. likely mirror those found in other systems (Betts-Hampikian and Fields, 2010). In this working model, the assembly of the core apparatus begins with membrane insertion of the OM secretin CdsC followed by addition of inner membrane (IM) components CdsD and CdsJ (Fig. 2). Unfortunately, current temporal expression studies have not been performed at high enough resolution to discriminate whether these proteins are the first de novo-synthesized components of the apparatus. Some discrepancies exist among studies. For example, reverse transcription-PCR (Fields et al., 2003; Slepenkin et al., 2003) and quantitative PCR (Hefty and Stephens, 2007) analyses indicated coexpression of apparatus components, whereas a microarray study indicated that some components may be expressed before others (Belland et al., 2003). Therefore, there is a need for a detailed temporal expression analysis. Interestingly, *cdsD* and *cdsJ* are positioned at the beginning of polycistronic transcripts while an internal promoter is capable of driving expression of only *cdsC* (Hefty and Stephens, 2007). These data are consistent with enhanced or independent synthesis and insertion of these core components. All three of these proteins were easily identifiable based on homology with corresponding proteins in other T3S systems (Stephens et al., 1998). In addition, both CdsC and CdsJ primary sequences contain predicted Sec-dependent secretion signals that would be required to traverse the chlamydial IM prior to full apparatus assembly. CdsJ is also a predicted lipoprotein similar to its *Yersinia* homolog, YscJ (Silvia-Herzog et al., 2008), and CdsD contains a predicted phospholipid binding (BON) domain to mediate membrane interactions (Johnson et al., 2008).

Biochemical data also support the genome sequence-assigned functional roles in T3S. CdsC and CdsJ partition as integral membrane proteins in Triton X-114 extraction studies (Fields et al., 2003), and as expected, CdsC is detected in chlamydial OM complex preparations (Betts et al., 2008; Birkelund et al., 2009; Liu et al., 2010). While its C terminus is homologous to *Yersinia* YscD, CdsD is unique among T3S apparatus proteins. CdsD contains multiple forkhead-associated homologous domains, whereas orthologs in other T3S systems typically contain one (Pallen et al., 2005). CdsD may therefore participate in unique interactions since forkhead-associated domains are often involved in phosphorylation-dependent protein-protein interactions (Durocher et al., 2000). Interestingly, in vitro studies indicate that CdsD is phosphorylated (Johnson and Mahony, 2007) by the chlamydial eukaryotic-like serine-threonine protein kinase, PknD (Verma and Maurelli, 2003). Although these data need to be confirmed in vivo, they would be consistent with an IM localization of CdsD where the N-terminal domain is oriented within the chlamydial cytoplasm. Indeed, protein interaction studies support a role for CdsD in the chlamydial IM, since portions of CdsD have been shown to interact in vitro with IM-associated T3S proteins CdsN (Stone et al., 2008), CdsL, and CdsQ (Johnson et al., 2008).

Based on sequence similarity, CdsR/S/T/U/V constitute the complement of integral membrane components forming the IM core of the basal apparatus. Although direct investigation of these proteins remains limited, a revealing story has emerged regarding peripherally IM-associated components CdsL, CdsN, and CdsQ. Current data describing CdsL and CdsN are consistent with this pair forming the energy system required to drive the T3S process. CdsN is homologous to the β-subunit of F_0F_1 ATPases and seems to function similarly to ATPases of other T3S systems. In general, T3S ATPases are thought to couple ATP hydrolysis with release of effectors from chaperones or unfolding of secretion substrates prior to secretion (Akeda and Galan, 2005). This activity has not been directly demonstrated for CdsN, yet CdsN is capable of multimerization and ATP hydrolysis (Stone et al., 2008). In addition, protein interactions between CdsN and secretion substrates CopN (Stone et al., 2008) and CT621 (Spaeth et al., 2009) have been detected. CdsL interacts with CdsN (Stone et al., 2008), where it could function similarly to homologs by regulating CdsN activity (Blaylock et al., 2006). Indeed, in vitro studies show that CdsL is capable of inhibiting CdsN-dependent ATPase activity (Stone et al., 2008, 2011). In vitro coprecipitation (Johnson et al., 2008) and yeast two-hybrid (Spaeth et al., 2009) analyses have indicated that both CdsN and CdsL are capable of interacting with the C-ring protein CdsQ. CdsQ likely forms an interaction hub linking the basal apparatus with secretion substrates as proposed by Spaeth et al. (2009). CdsQ self-oligomerizes (Johnson et al., 2008) and is capable of interactions with basal apparatus proteins CdsD (Stone et al., 2008) and CdsS and CdsT (Spaeth et al., 2009). CdsQ is also capable of interaction with a secretion chaperone (Spaeth et al., 2009) that may usher multiple secretion substrates to the T3S apparatus (discussed below).

Completion of the basal apparatus would then allow secretion of components of the extracellular portions of the injectisome. The extracellular injectisome would be comprised of at least two proteins making up the NC and TC in elementary bodies (EBs) prior to contact with a target host cell. The needle filament is undoubtedly composed of CdsF. A multifaceted approach was used to demonstrate CdsF polymerization on the surface of chlamydiae (Betts et al., 2008). According to models for other T3S systems, the CdsF filament would extend outward from the CdsC secretin in the OM until the TC caps it. Indirect evidence implicates *C. trachomatis* CT584 as the chlamydial TC protein. Biophysical properties of CT584 are consistent with the TC family of proteins (Markham et al., 2009), and yeast two-hybrid studies identified an interaction between CdsF and CT584 (Spaeth et al., 2009). Why CT584 is encoded outside the T3S-encoding loci is unclear but interesting. At any rate, secretion and assembly of these extracellular components would complete the apparatus and allow CopN to associate with the cytoplasmic face of the basal apparatus and impose a secretion block until stimulatory signals are received (discussed below).

Many important questions remain to be addressed regarding the composition and arrangement of the chlamydial T3S apparatus. For example, there is a pressing need to confirm in vitro observations in chlamydiae and in the context of infection. Moreover, there are apparatus components that remain unaccounted for. Preliminary evidence suggests that *C. trachomatis* CT671 and CT670 could correspond to the needle ruler protein and its chaperone, respectively. CT671 is a T3S substrate (Subtil et al., 2005), while crystallography indicates that CT670 belongs to the YscO family of proteins (Lorenzini et al., 2010). In addition, the *Chlamydia pneumoniae* ortholog of CT670 is capable of dimerization (Stone et al., 2008), a property consistent with chaperone function. It is also unclear exactly how T3S injectisomes are arranged on the surface of chlamydiae and whether the arrangement differs between RBs and EBs.

Finally, the precise role of "flagellar" T3S homologs CT060, CT717, CT718, and CT719 needs to be addressed. The genes are transcribed distinctly from NF-T3S genes in both *C. trachomatis* (Belland et al., 2001) and *C. pneumoniae* (Maurer et al., 2007). Moreover, protein interaction studies have revealed that flagellar components are capable of interacting with NF-T3S proteins (Stone et al., 2010; Spaeth et al., 2009) in a manner that would be consistent with their predicted function in T3S. Further investigation is this area could reveal that the chlamydial T3S machine is not static but changes in response to secretion needs and/or developmental cues. For example, it is entirely possible that the invasion-associated apparatus present on EBs is subtly augmented to accommodate differing secretion requirements in early reticulate bodies (RBs). This is an interesting possibility given that many of the early development secretion substrates are Inc proteins. The Inc large hydrophobic domains likely interact distinctly with the secretory apparatus and may therefore require components different from those required by the translocated actin-recruiting phosphoprotein (Tarp) and CT694. Close examination of the flagellar homologs reveals that they duplicate functionally significant components of the core T3S system. CT060 and CT719 represent paralogs of the IM structural proteins CdsV and CdsJ, respectively, whereas remaining genes encode a potential alternative TS3-associated ATPase (CT717) and its regulator (CT718). Other T3S system-expressing bacteria employ multiple systems to achieve distinct goals. For example, *Salmonella* employs a *Salmonella* pathogenicity island 1-encoded T3S apparatus to gain entry into host cells and then expresses an entirely new T3S system encoded by *Salmonella* pathogenicity island 2 to enable intracellular survival (Hueck, 1998). Consistent with its minimal genome, *Chlamydia* may accomplish the same goal by merely switching central components of the T3S apparatus that are involved in recognition of secretion substrates.

Regulation of Secretion Activity

In general, T3S is regulated and stimulated in vivo upon contact with target host molecules and/or membranes and is polarized in that only secretion machines in contact with a respective stimulus are active (Hueck, 1998). By extension, one would therefore predict that the chlamydial T3S system is active so long as the bacteria are in contact with a eukaryotic membrane. Examination of invasion-related secretion substrates (discussed below) has provided direct evidence that initial activation of secretion occurs during attachment and invasion. Furthermore, continued secretion of T3S system substrates such as Inc proteins clearly indicates active secretion while chlamydiae remain associated with the inclusion membrane. Currently there is no direct indication that RBs cease T3S when they detach from the inclusion membrane. Indirectly, however, the contact-dependent model proposed by Wilson and collaborators (Wilson et al., 2006) entails coincidental (or coupled) disruption of T3S activity and physical detachment of RBs from the inclusion membrane and subsequent differentiation to EBs (this contact-dependent model is discussed in more detail in chapter 16, "Biomathematical modeling of *Chlamydia* infection and disease"). In addition, secretion activity must also be inactivated to enable accumulation of invasion-associated effector proteins prior to the cessation of metabolic capability associated with EB formation. Temporal regulation of T3S activity remains a central question that will profoundly shape our understanding of how T3S contributes to overall chlamydial biology. Given the technical complexities associated with the chlamydial developmental cycle, deciphering the molecular mechanisms involved in regulating T3S activity represents an essential step to understanding subtleties regarding when and how the secretion system is employed.

Intriguing evidence has thus far provided some tantalizing clues regarding regulation of secretion activity. Activation will be discussed here, and speculation regarding inactivation of secretion activity will be discussed in "Future Prospects" below. As with apparatus assembly, information from other T3S systems has provided a framework to generate preliminary hypotheses. According to reasonable working models from genetically tractable T3S systems, activation of secretion is contact dependent and is directly controlled by a small set of apparatus proteins involved in sensing and responding to exogenous signals. For example, oligomerized LcrV of *Yersinia* spp. represents a TC situated atop the extracellular needle filament composed of YscF (Mueller et al., 2005). In general, sensing of contact by TC proteins is thought to initiate a signal (Veenendall et al., 2007) that is transduced through the needle filament via conformational changes (Deane et al., 2006; Torruellas et al., 2005). This signal results in activation of secretion by inducing the T3S-dependent export of a negative regulatory protein, exemplified by *Yersinia* YopN, which otherwise blocks the apparatus from the intracellular face (Ferracci et al., 2005).

Since the chlamydial apparatus includes orthologs of regulatory proteins found in other T3S systems, it is reasonable to predict that some variations of these regulatory cascades may manifest themselves in *Chlamydia* spp. Chlamydial EBs contain a functional T3S system (Fields et al., 2003), and Clifton and collaborators (Clifton et al., 2004) showed that secretion activity commences during invasion by demonstrating that the *C. trachomatis* effector Tarp is detectable within host cell phagocytic pedestals associated with invading EBs. In addition, in vitro treatment of EBs with agents such as cholesterol-rich liposomes stimulates release, albeit very inefficiently, of Tarp (Jamison and Hackstadt, 2008), thereby providing additional evidence that activation of chlamydial T3S is contact dependent. If activation proceeds as in other T3S systems, then the TC protein CT584 will sense this contact. In vitro analyses indicate that recombinant CT584 is capable of multimerization and exhibits spectral transitions that are comparable to those of other TC proteins such as *Yersinia* spp. LcrV and *Shigella* spp. IpaD (Markham et al., 2009). The established CT584 association with CdsF

(Spaeth et al., 2009) is important, since this interaction would be necessary to propagate activation signals to the core T3S apparatus. Whether CdsF is capable of transducing activation signals is unclear, yet cross-linking studies revealed a fundamental difference in the migration patterns of CdsF multimers. CdsF dimerization was apparent in EB-derived material, whereas dimeric cross-linked forms were absent from bacteria treated during infection (Betts et al., 2008). These data are consistent with the conformational changes proposed to correspond to an activation status in other T3S systems (Blocker et al., 2008). However, conformational alterations are likely more complicated in the chlamydial needle due to the existence of disulfide bonding within the T3S apparatus (see "Future Prospects," below).

Characterization of CopN, the putative negative regulator of chlamydial T3S, has yielded interesting insights into the chlamydial T3S mechanism. CopN is detectable in EBs, where it would presumably impose a secretion blockage. Evidence for release of the CopN blockage during invasion through secretion has not been obtained due to sensitivity limits; yet later secretion by chlamydiae is apparent due to accumulation of detectible levels of CopN in association with the inclusion membrane (Fields and Hackstadt, 2000). As discussed below, Scc1 and CT663 comprise a heterodimeric secretion chaperone that binds the N terminus of CopN and is necessary for efficient secretion (Silvia-Herzog et al., 2011). However, CopN represents an apparent fusion protein corresponding to *Yersinia* YopN and TyeA (Pallen et al., 2005; Betts-Hampikian and Fields, 2010). This is significant since TyeA is essential for YopN to maintain blockage of the *Yersinia* T3S apparatus (Ferracci et al., 2005). In the absence of activation signals, TyeA apparently associates with both YopN and the T3S apparatus to prevent premature secretion of YopN (Joseph and Plano, 2007). Upon activation, the YopN-TyeA interaction is disrupted, allowing YopN secretion. In *Chlamydia*, the secretion chaperone Scc3 has been shown to associate with the C-terminal "TyeA-like" portion of CopN (Slepenkin et al., 2005), raising the possibility that this interaction could mediate secondary association of CopN with the T3S apparatus. While it is possible that the detected association of CopN with CdsN (Stone et al., 2008) could be involved in regulation, it is more likely that this interaction represents the general role of CdsN in initiating substrate secretion. Regardless, the data are thus far consistent with a regulatory function for CopN similar to that of its orthologs in heterologous hosts.

Role of Chaperones

Secretion chaperones also play a prominent role in facilitating and potentially regulating T3S. Chlamydial genomes contain homologs of class I, II, and III chaperones functioning as prototypical homodimeric secretion chaperones (Parsot et al., 2003), tetratricopeptide repeat-containing chaperones (Pallen et al., 2003), and heterodimeric chaperones (Quinaud et al., 2007), respectively. Database similarity searches reveal a list of obvious *C. trachomatis* T3S chaperones including CT043, CT088, CT260, CT274, CT576, CT663, and CT862, while empirical data indicate chaperone functions for CT665 and CT667. In general, it has been proposed that chaperones are used to pilot substrates to the T3S apparatus, prevent premature association of proteins that participate in interactions external to the bacteria, or maintain integral membrane proteins in a secretable conformation (Feldman and Cornelis, 2003). However, the full role of chlamydial chaperones is unclear. If these nine proteins represent the full complement of chlamydial chaperones, the sheer number of putative effector proteins makes it unlikely that all secretion substrates have chaperones. Indeed not all T3S effectors require chaperones for secretion (Page and Parsot, 2002; Feldman and Cornelis, 2003). Another possibility is that some chaperones interact with multiple secretion substrates. CT043, CT088 (designated Scc1), and CT663 belong to the CesT family of chaperones (Fields and Hackstadt, 2006). *Escherichia coli* CesT is capable of interacting with multiple effector proteins (Thomas et al., 2005), raising the possibility that CT043, CT088, or CT663 could associate with many substrates. However, although an interaction partner for CT043 has yet to be identified, CT088 and CT663 likely act together as a heterodimeric chaperone. CT088 and CT663 interact with each other in yeast two-hybrid (Spaeth et al., 2009) and coprecipitation (Silvia-Herzog et al., 2011) studies. In addition, the two together, but not alone, interact with CopN and facilitate its secretion by the heterologous *Yersinia* T3S system (Silvia-Herzog et al., 2011). Although not predicted to be a CesT family chaperone, *C. trachomatis* CT260 currently represents the only identified multicargo chaperone. CT260 interacted with secretion substrates Cap1, CT618, and CT225 in yeast two-hybrid studies (Spaeth et al., 2009) and has been designated "multiple cargo secretion chaperone" (McsC). McsC also interacted with CdsQ and may therefore function similarly to homologs in other systems that dock with the secretory apparatus via the SctQ protein (Gonzalez-Pedrajo et al., 2006; Morita-Ishihara et al., 2006).

CT274, CT576 (Scc2), and CT862 (Scc3) represent class II chaperones. All three contain tetratricopeptide repeat domains and probably function as homodimers (Fields et al., 2005). This group of chaperones is most likely directly involved in normal apparatus function since all members interact with mobile or secreted components of the T3S apparatus. CT274 was found to interact with CT668 (Spaeth et al., 2009), which is secreted via a T3S mechanism (Subtil et al., 2005). Based on the position of its coding sequence immediately downstream of CdsE/F/G, CT668 likely corresponds to *Yersinia* YopR (Blaylock et al., 2010) and therefore could govern needle polymerization. In contrast, Scc2 interacted with the translocator proteins CopD and CopB (Spaeth et al., 2009; Fields et al., 2005) and was capable of substituting for *Yersinia* SycD in mediating YopD secretion (Fields et al., 2005). Scc3 interacts with CopN, yet the impact of this interaction is not clear (see above).

Predicted structural features place CdsE (CT665) and CdsG (CT667) in the class III family of T3S chaperones (Betts et al., 2008). This group of chaperones is typically specific for extracellular needle proteins (Quinaud et al., 2005, 2007; Yip et al., 2005).Consistent with predictions, CdsE and CdsG interact to form a heteromeric complex and can interact with the needle protein CdsF (Betts et al., 2008; Spaeth et al., 2009). Moreover, the heteromeric CdsE-CdsG complex is required to stabilize CdsF prior to secretion since the half-life of ectopically expressed CdsF is greatly enhanced in the presence of CdsE and CdsG (Betts et al., 2008).

Effector Translocation

The translocation of effector proteins across the inclusion membrane represents the final step in the chlamydial T3S route, yet it remains an enigmatic process. Although this also remains true in genetically tractable systems, basic requirements are apparent. Translocon proteins, exemplified by *Yersinia* YopB and YopD, are secreted and associate with host membranes (Mueller et al., 2008). YopB family proteins intercalate into the host membrane and oligomerize to form translocation pores providing access to the host cytosol (Hakansson et al., 1996; Hume et al., 2003; Neyt and Cornelis, 1999), whereas the role of YopD is less clear. Current data are consistent with *C. trachomatis* CopB and CopD participating in the chlamydial translocon. Position-specific iterative BLAST searches reveal significant sequence similarity of CopB with YopB-family proteins, and CopD contains features consistent with translocon proteins (Pallen et al., 2005). Both proteins are secreted by chlamydiae (Fields et al., 2005; Ho and Starnbach, 2005), and CopB-specific signal is detectible in inclusion membranes during immunolocalization assays (Fields et al., 2005). CopB and CopD are unlikely to be the only chlamydial translocator proteins. De novo expression of CopB occurs during late development (Belland et al., 2003) and is detectible in EBs (Fields et al., 2003). However, CopB levels rapidly decrease during infection and are below detection by approximately 8 h postinfection for *C. trachomatis* serovar L2 (Chellas-Géry et al., 2011). Since effector proteins such as CT847 (Chellas-Géry et al., 2007) and IncA (Hackstadt et al., 1999) are translocated in the interim during which CopB is absent, a secondary translocon must exist. The best candidates for alternative translocator proteins are the apparent paralogs CopB2 and CopD2. Although endogenous CopB2 is below immunodetection outside the inclusion, ectopic expression of CopB2 in HeLa cells results in accumulation of CopB2 in a peri-inclusion localization consistent with membrane localization (Chellas-Géry et al., 2011). Importantly, CopB2 is detectible via immunoblotting at times during development when CopB is absent. While these data are consistent with a role of CopB2 in translocation, biochemical data raise doubts. For example, unlike CopB, CopB2 does not undergo partition as an integral membrane protein (Chellas-Géry et al., 2011). Therefore, effector protein translocation represents an additional area of chlamydial T3S where significant open questions remain.

EFFECTOR PROTEINS

Deployment of antihost proteins to effect alterations in host physiology represents an efficient method to promote microbial survival. The arsenal of chlamydial effectors has begun to emerge. The identification and functional characterization of novel effector proteins have constituted an exciting and productive facet of chlamydial research, yet clearly more remain to be discovered. At this writing, fewer than 20 such proteins have been identified, and even fewer have an established function. In this section, we will concentrate on the handful of antihost proteins that have been comparatively well characterized. Discussion here will be confined to emphasizing the broad role of secreted effectors in chlamydial development to underline how this system is intimately intercalated into the biology and pathogenesis of *Chlamydia* spp. The reader is also referred to several recent reviews for detailed listings of secreted proteins (Betts et al., 2009; Cocchiaro and Valdivia, 2009; Valdivia, 2008; Scidmore, 2011; Karyagina et al., 2009).

Invasion-Associated Effectors

Upon attachment to a host cell, extracellular EBs secrete a pool of effector proteins involved in gaining access to the host cell interior. Since de novo protein synthesis is not required for chlamydial invasion, this pool of effectors is synthesized by RBs and packaged into EBs for subsequent rounds of infection (Valdivia, 2008). In fact, de novo synthesis of some effectors does not occur until late development (Hower et al., 2009). Invasion-related secretion was apparent from observations that host-mediated phosphorylation of *C. trachomatis* L2 Tarp is detectible within 5 minutes of adding chlamydiae to a HeLa monolayer. Curiously, Tarp is targeted by multiple host kinases (Jewett et al., 2008; Tietzel et al., 2009; Elwell et al., 2008; Mehlitz et al., 2008, 2010), yet the functional implications of tyrosine phosphorylation remain unresolved. In addition, phosphorylation is observed only for *C. trachomatis* Tarp (Clifton et al., 2005) and occurs on tyrosine residues that are lacking in the *Chlamydia muridarum*, *Chlamydia caviae*, and *C. pneumoniae* orthologs. Like many effector proteins, Tarp is likely multifunctional. Nucleation of actin polymerization remains its best-characterized activity (Jewett et al., 2006) and is consistent with the role of T3S in remodeling the host cytoskeletal architecture to facilitate invasion. Importantly, recent evidence indicates that antibodies specific for the actin-nucleating domain of *C. trachomatis* Tarp interfere with both actin nucleation and chlamydial invasion (Jewett et al., 2010). While chlamydial invasion likely involves multiple, redundant mechanisms, functional characterization of Tarp currently represents the best evidence supporting the importance of the T3S system in this process. More details about Tarp are presented in chapter 6 ("Initial interactions of chlamydiae with the host cell").

However, it is clear that Tarp is not the sole effector contributing to cytoskeletal rearrangements. For example, CT694 is deployed via T3S during invasion and interacts with the host protein AHNAK (Hower et al., 2009). The full implications of this interaction remain to be resolved, but AHNAK has established roles in membrane cytoarchitecture (Benaud et al., 2004), modulation of Ca^{2+} channel activity (Radermacher and Crabtree, 2008), and enlargeosome formation (Cocucci et al., 2007). All of these functions would be consistent with events occurring during chlamydial invasion and early development. Inhibition of stress fiber formation hints at a role in actin depolymerization, but CT694 represents another obvious multifunctional effector, since localization of CT694 to the plasma membrane and induction of host cell toxicity occurred independent of the AHNAK binding domain. CT694 orthologs are present only in *C. trachomatis* and *C. muridarum*, again reinforcing the notion of effector divergence contributing to species-specific requirements in chlamydial invasion (Dautry-Varsat et al., 2005). Indeed, CT166 represents an additional species-specific effector protein affecting the host cytoskeleton. CT166 is encoded within the chlamydial plasticity zone of *C. caviae*, *C. muridarum*, and *C. trachomatis* genital serovars and has been correlated with immediate host cell toxicity (Belland et al., 2001). It remains unclear how CT166 is exported from chlamydiae, but the primary sequence lacks an identifiable Sec-dependent secretion signal. CT166 effects on host actin are manifested via glucosylation of Rac1 (Thalmann et al., 2010). Hence, CT694 and CT166 may function in part to reverse actin polymerization subsequent to chlamydial entry to enable host cell recovery, similar to *Salmonella* SptP (Fu and Galan, 1999).

Inclusion Biogenesis and Maintenance

After invasion, chlamydiae are contained within a membrane-bound vesicle termed the inclusion. This compartment increases in size and changes properties throughout development to accommodate accumulating bacteria. Modification of the inclusion membrane with a set of *Chlamydia*-derived integral membrane proteins represents a primary mechanism to establish and maintain inclusion integrity and maturation. The inclusion membrane protein (Inc) family consists of a diverse group of proteins first described in immunolocalization studies as showing a rim-like staining pattern in the inclusion membrane (Rockey et al., 1997). Select candidates have been shown to be secreted via T3S using heterologous systems (Fields et al., 2003; Subtil et al., 2001) or by blocking chlamydial T3S with small-molecule inhibitors (Wolf et al., 2006; Bailey et al., 2007). Since Inc proteins lack apparent Sec-dependent secretion signals, it is presumed that these proteins are secreted via the T3S mechanism. Bioinformatic analyses indicate that Inc family proteins contain a distinctive bilobed hydrophobic domain (Bannantine et al., 2000) that is typically situated toward one terminus of the protein and presumably constitutes the membrane-interactive domain. Early examples such as IncA (Rockey et al., 1997) and IncG (Scidmore and Hackstadt, 2001) established that significant, nonhydrophobic Inc domains could be exposed to the eukaryotic cytosol and be host interactive. This theme has continued, and it is now appreciated that Inc proteins contribute to essential aspects of chlamydial inclusion biology such as mobilization along microtubules and centrosome anchoring (Mital et al., 2010), interactions with an array of host proteins mediating vesicular interactions (reviewed by Betts et al. [2009]), and subversion of proinflammatory signal transduction cascades (Wolf et al., 2009). The complement of Inc proteins and even the extent of primary sequence similarity vary among chlamydial species and may account for subtle differences in infection biology.

Importantly, the vast majority of putative Incs remain functionally uncharacterized, and Inc biology represents a rich area for future investigation. Moreover, several interesting questions remain to be solved. For example, it has been proposed that membrane-interactive effector proteins require secretion chaperones to prevent spurious folding prior to secretion (Feldman and Cornelis, 2003). Spaeth et al.

(2009) recently provided evidence that the Inc Cap1 interacts with a multicargo chaperone, raising the possibility that chlamydial Incs also require chaperones. Once secreted, it is unknown how Inc proteins intercalate into the inclusion membrane. Are T3S translocon proteins required? Although elucidation of these properties in *Chlamydia* will be difficult, knowing how this process occurs seems essential to address overall T3S mechanisms that are unique to *Chlamydia* spp. Finally, it is also unclear why putative Inc proteins are so numerous. It is likely that not all directly interact with host factors and some may be involved in multiprotein complexes present within the inclusion membrane. Preliminary evidence supporting this notion is evident from protein-protein interaction studies (Spaeth et al., 2009). In addition, Mital et al. (2010) described Inc-containing microdomains involved in inclusion interactions with the microtubule network.

Targeting Host Proteins for Proteolysis

The chlamydial T2S substrate CPAF is a serine protease and represents the most thoroughly characterized secreted effector protein. Originally identified as a protease capable of degrading host transcription factors RFX5 and USF-1 (Zhong et al., 2001), it is now appreciated that CPAF plays an essential role in multiple aspects of chlamydial development and illustrates how an effector with a single enzymatic activity can accomplish diverse tasks. Host protein degradation by CPAF contributes to immune evasion, protection of infected cells from apoptosis, remodeling of the host cytoskeleton to accommodate the expanding inclusion, and dissemination (reviewed by Zhong [2009]). This impressive array of functions raises the question of what factors regulate CPAF activity. The answer(s) to this question remains elusive but is important to understand overall contributions to virulence.

Besides CPAF and Tail-specific protease (Tsp) (discussed below), other proteases could exist, since degradation of host cyclin B1 (Balsara et al., 2006) and golgin-84 (Heuer et al., 2009) has been documented. Although degradation could be directly accomplished by host factors, the *C. trachomatis* genome contains an abundance of predicted proteases (Stephens et al., 1998). In addition, it is also clear that an effector need not possess protease activity to accomplish host protein degradation. For example, the T3S effector CT847 binds human Grap2-cyclin D interacting protein (GCIP). Although a direct function for CT847 remains unresolved, GCIP is degraded by the host proteasome during chlamydial infection (Chellas-Géry et al., 2007). Overall, the degree to which proteolysis is involved in chlamydial infection is notable. Proteolysis is attractive from the pathogen standpoint for multiple reasons. First, degradation of host proteins obviously results in irreversible blockage/alteration of a given pathway until de novo synthesis replenishes the host factor. In addition, it has been suggested that large-scale degradation of proteins could have a secondary nutritional benefit by increasing the pools of peptides available for growing chlamydiae (Zhong, 2009).

Modulation of Host Gene Expression

Microarray data indicate profound alterations in host gene expression in response to chlamydial infection (Burton et al., 2011; Alvesalo et al., 2008; Ren et al., 2003). It is not surprising that secreted effector proteins may directly contribute to modulation of host cell gene expression. The eukaryotic transcription factor NF-κB represents one important target for chlamydiae to disarm proinflammatory signaling. As such, this signal transduction pathway is a prime example of a process targeted at multiple levels by distinct effector proteins. Some effectors, such as the *C. pneumoniae*-specific Inc protein CP0236, act well upstream by preventing association of signal transduction factors with proinflammatory receptors at the plasma membrane (Wolf et al., 2009). However, *Chlamydia*-infected cells are subjected to diverse stimuli capable of resulting in NF-κB activation. Interestingly, all *Chlamydia* spp. except *C. pneumoniae* are capable of preventing NF-κB activation via stabilization of the NF-κB inhibitor IκBα. Chlamydiae secrete two cysteine proteases with deubiquitinating activity (Misaghi et al., 2006) possibly via T3S. ChlaDub1 has been reported to deubiquitinate IκBα to prevent the degradation required to allow efficient activation of NF-κB (Le Negrate et al., 2008). If these mechanisms fail to neutralize signaling, NF-κB itself is directly targeted. Chlamydial Tsp (Lad et al., 2007b) and CPAF (Christian et al., 2010) have each been reported to degrade the p65 subunit of NF-κB. Like CPAF, Tsp contains an N-terminal secretion signal and would therefore be predicted to be secreted via a T2S mechanism. Once in the cytosol, Tsp and/or CPAF degrades the RelA portion of NF-κB to prevent active transcription factors from reaching the nucleus (Christian et al., 2010; Lad et al., 2007a).

Effector-dependent methylation of host histones represents an additional potential mechanism to modulate host gene expression. The *C. trachomatis* T3S substrate protein nuclear effector (NUE) localizes to the host nucleus during infection and has in vitro methyltransferase activity capable of targeting histones H2B, H3, and H4 (Pennini et al., 2010). The potential impact on host chromatin structure and gene expression remains unknown, but NUE illustrates the capability of diffusible effector proteins to gain access to host organelles. Indeed, ectopic expression studies have revealed that multiple uncharacterized chlamydial proteins are capable of nuclear localization (Sisko et al., 2006). Proteins tropic for mitochondria were also observed, and it seems reasonable to predict that other organelles are targeted by effectors given the intimate dependence of chlamydial development on host cell processes.

Proteins involved in basic secretion mechanisms are generally highly conserved among gram-negative pathogens including *Chlamydia*, whereas the complement of effector proteins in each chlamydial species differs and is likely to account for species-specific differences in chlamydial biology. The Inc proteins are perhaps the best examples of this principle since chlamydial genomes contain significant differences in Inc-coding capacity (Rockey and Alzhanov, 2006). For example, *C. trachomatis* expresses a subset of Inc proteins such as CT229 and CT813 that enable specific interactions of inclusions with Rab4 and SNAREs, respectively (reviewed by Betts et al. [2009]). Neither CT229 nor CT813 homologs are identifiable in the *C. pneumoniae* genome, and yet this species recruits Rab4 to the inclusion, presumably by an alternative mechanism. Conversely, *C. pneumoniae* expresses Cpn0585 to enable interactions with Rabs 1, 10, and 11 (Cortes et al., 2007). The recruitment of Rab10 has been reported for *C. pneumoniae* and *C. muridarum*, whereas Rab6 associates uniquely with *C. trachomatis* (Rzomp et al., 2003). These effector-mediated differences likely manifest as species-specific fusogenicity with subsets of host vesicles. Similarly, the T3S effector CT694 is present in *C. trachomatis* and *C. muridarum* but not in other chlamydial species (Hower et al., 2009). Hence, species-specific differences in invasion requirements may reflect, in part, the expression of this invasion-associated effector. Even when effector proteins such as Tarp are present in all chlamydial species, subtle differences that likely correspond to unique biology exist. *C. muridarum*, *C. psittaci*, and *C. pneumoniae* Tarps lack the tyrosine repeats of *C. trachomatis* Tarps that become phosphorylated by host kinases (Clifton et al., 2005). In addition, the precise actin-nucleating mechanisms differ subtly among chlamydial Tarp proteins (Jewett et al., 2010), raising the possibility for species-specific differences in modulation of actin networks. Additional examples of effectors unique to chlamydial species exist (reviewed by Betts et al. [2009]), and it is likely that other instances will emerge to help explain how unique pathogenic properties are manifested among the chlamydial species.

FUTURE PROSPECTS

While the field of chlamydial protein secretion is no longer in its infancy, there is plenty of opportunity for growth. Open questions abound, many of which can be addressed in the absence of a genetic system. Even without genetics, protein secretion affords opportunities to push the envelope of *Chlamydia* research. The discussion below focuses on three general areas in which protein secretion research could have a major impact on understanding chlamydial pathogenesis. The areas are by no means exhaustive and are discussed to illustrate how wide open the field currently is.

Effector Biology

Elucidation and characterization of novel effector proteins remain a major focus and represent an exciting prospect for future advances. While T5S and T2S substrates are predictable based on primary sequences, there are no such predictive sequences for T3S substrates (Cambronne and Roy, 2006). Attempts to elucidate conserved features in primary sequences that are consistent with secretion signals have met with limited success, and predictions are most often of dubious efficacy. Direct testing is therefore essential, and various experimental techniques including proteomic profiling, immunolocalization, and ectopic expression studies have proven useful in identification of novel effectors. Perhaps the most efficacious mean has been the use of heterologous secretion systems to identify secreted chlamydial proteins (Chen et al., 2010; Ho and Starnbach, 2005; Fields and Hackstadt, 2000; Subtil et al., 2005).

Once effectors have been identified, new technologies can be used to address antihost function. For example, ectopic expression in yeast provides an attractive model to examine protein function in a system amenable to genetic manipulation (Curak et al., 2009). Indeed, a recent genome-wide expression study of chlamydial proteins in *Saccharomyces cerevisiae* revealed preliminary evidence for antihost function (Sisko et al., 2006). This study included *C. trachomatis* gene products annotated as hypothetical with respect to function. The results must be taken with a grain of salt since it is likely that many of the tested proteins are not actually secreted proteins. The method does, however, provide an entry point to investigate effectors. For example, Mehlitz and collaborators (Mehlitz et al., 2010) successfully employed a yeast model to identify interactions between Tarp and the host protein SHC1. Regardless of technology, any findings must then be confirmed in the context of chlamydial infection. This step is especially important to distinguish relevant roles. Moreover, there is a need for extension of effector studies to animal models to address relevance. Several effector proteins have been shown to bind/alter specific host proteins. When available, specific murine knockout lines should be employed to ask whether the identified proteins have an impact on chlamydial infection. The utility of this approach is that some indication of relevance can be established without the need to create null mutations in chlamydial genes.

The dynamic interplay among effectors represents one question that has not yet been approached. This question is likely to be relevant since the functions of effectors in other T3S-expressing systems are often interdependent or at least intimately associated (Cain et al., 2008). Elucidation of the interplay among secreted effectors will be essential to fully appreciate chlamydial biology as well as to remind us that functional studies of a protein in isolation may not always have complete physiological relevance. Admittedly, these studies may require identification of a critical mass of chlamydial effectors. An added benefit, as more effector proteins are characterized, is that host-dependent functional nodes that will have predictive value in directing identification of still more effectors may become apparent. As discussed, NF-κB signaling represents a node that is targeted at multiple levels by chlamydial effectors. Given the extent to which NF-κB signaling is targeted, potential targets of effectors can be prioritized by first examining whether or not chlamydial infection has an impact on a given host factor in the pathway. Once a target is established, a plethora of experimental techniques exist to identify the responsible effector protein.

Finally, the issue of substrate specificity represents an area of effector biology that needs to be addressed. Is T3S substrate specificity governed merely by temporal regulation of effector gene expression? In other words, is the de novo synthesis of IncA during midcycle development the sole factor influencing when this effector is secreted? Intuitively, this seems unlikely. If it is true, for example, how would secretion of invasion-associated effectors such as CT694 be prevented prior to conversion of chlamydiae to EBs? Other factors must be involved. There is the potential that differing subsets of chaperones could be involved. In addition, chlamydiae express CdsU, which is homologous to a substrate switch mechanism that exists in the *Yersinia* T3S system (Riordan and Schneewind, 2008). As mentioned above, it is also possible that "flagellar" homologs could contribute substrate specificity to the secretion apparatus. These questions will be admittedly difficult to address in the absence of tractable genetics but do represent an interesting area that will reveal biology unique to the chlamydial system.

T3S and the Developmental Cycle

Multiple lines of evidence suggest that T3S and chlamydial development are intimately linked. Both, for example, are regarded as being dependent on contact with host membranes. EB attachment to the host plasma membrane activates T3S during invasion (Clifton et al., 2004) and also initiates a cascade resulting in decondensation of the chlamydial nucleoid (Grieshaber et al., 2004). Conversely, detachment of RBs from the inclusion membrane coincides with conversion back to infectious EBs (Abdelrahman and Belland, 2005). Intuitively, it seems that inclusion membrane detachment should also result in abolishment of T3S activity. Based on mathematical modeling, Wilson and collaborators (Wilson et al., 2006) have proposed a direct correlation between inclusion membrane attachment and T3S activity. They further describe a model in which spatial constraints increase during development and eventually limit T3S-mediated attachment to inclusion membranes (Hoare et al., 2008). In this "contact-dependent" model, progression of development leads to fewer T3S injectisomes per RB or to fewer being able to maintain contact with the inclusion membrane. This decreased association would result in decreased T3S, since T3S activity likely requires membrane contact, which is proposed to lead to developmental cues initiating RB-to-EB conversion.

If the above is true, T3S may be passively responding to the degree of attachment but actively contributing to transducing signals leading to chlamydial differentiation. Several unresolved questions must be answered to provide essential support for this working model. First, the T3S apparatus must be intimately involved in attachment of RBs to the inclusion membrane. This could certainly be possible but would represent a unique attribute of the chlamydial system, since the apparatus is not involved in the attachment of other T3S-expressing bacteria. Second, the model fails to explain the apparent existence of *C. pneumoniae* and *C. psittaci* RBs within the interior of inclusions (Wolf et al., 2000). Ultrastructural analyses indicate that these RBs may exist without apparent contact with the inclusion membrane. This discrepancy may be the result of fixation artifacts yet needs to be resolved if the contact-dependent model is to apply to all chlamydial species. Finally, the basis of the model relies on enumeration of RB-associated projections associated with the inclusion membrane (Wilson et al., 2006). Multiple groups have observed spike-like projections on the surfaces of both purified EBs and RBs (reviewed by Matsumoto [1988]) and emanating from RBs to protrude through the inclusion membrane inside a *Chlamydia*-infected cell (Nichols et al., 1985). Matsumoto (1982) was the first to report that the *C. psittaci* projections on RBs decreased in apparent abundance during an infection time course. These projections are presumed to correspond to structures isolated from EB preparations (Chang et al., 1997; Matsumoto, 1988) that strongly resemble the purified nail-like T3S injectisomes first characterized in *Salmonella* (Kubori et al., 2000), but they have not yet been proven to be the chlamydial T3S machinery. Their identification as T3S injectisomes will likely require immunogold labeling with antibodies specific for surface T3S components such as CdsF or CT584. Alternatively, high-resolution electron microscopy could reveal structural features consistent with T3S systems in other bacteria. Regardless, this issue must be resolved to support the contact-dependent model proposed by Wilson and collaborators (Wilson et al., 2006).

Interestingly, treatments that interfere with specific chlamydial effectors such as IncA (Alzhanov et al., 2004), CT229 (Rzomp et al., 2006), or CopN (Huang et al., 2008) also interfere with developmental progression. These effects on development could be indirect. However, we have recently found that disulfide bonding within surface-exposed T3S components CdsF and CdsC correlates with developmental stages. EB-localized CdsF and CdsC both contain disulfide bonds that become reduced in RB forms (Betts-Hampikian and Fields, 2011). This intriguing finding illustrates an additional link between T3S and chlamydial development, since disulfide bonding in the chlamydial envelope proteins is thought to have a role in governing differentiation of developmental forms (Hackstadt et al., 1985; Hatch et al., 1986; Newhall and Jones, 1983).

If termination of T3S activity initiates RB-to-EB conversion, it is also possible that late-expressed T3S system proteins could be direct regulators of chlamydial differentiation. For example, de novo expression of the putative TC protein CT584 may cap secretion channels, leading to inactivation of T3S (and therefore detachment in the contact-dependent model). According to working models, CopN would also have to be involved because this protein is predicted to negatively regulate T3S activity. Intriguingly, CopN secretion may also be indirectly linked to early gene expression in chlamydiae. The secretion chaperone CT663 has been implicated in association with RNA polymerase and negative regulation of σ^{66}-dependent transcription in ectopic expression studies in *E. coli* (Rao et al., 2009). Secretion of CopN during invasion would seemingly result in free CT663 prior to de novo expression of *copN* during midcycle development. CT663 could not be exerting broad inhibitory activity during this time since σ^{66} is the primary chlamydial sigma factor and is active during early development (Hatch, 1999). However, Rao and collaborators (Rao et al., 2009) proposed that CT663 levels are sufficient to inhibit σ^{66} only during late development. This work certainly needs to be confirmed in *Chlamydia* spp.; however, it indicates that a T3S chaperone could contribute to global alterations in chlamydial gene expression that correlate with developmental cues leading to initiation of EB formation. This possibility is further strengthened by consideration of established roles of chaperones linking T3S activity to gene regulation in other bacteria. For example, free pools of the *Shigella* T3S chaperone IpgC can bind and subsequently activate the AraC-like transcription factor MxiE (Marvis et al., 2002).

Therapeutics

Due to the prevalence and morbidity associated with *Chlamydia*-mediated human disease, a high priority has been placed on the development of an efficacious preventative vaccine for *Chlamydia* infections. It seems reasonable to consider select T3S proteins in formulating multivalent vaccines given the essential nature of T3S in chlamydial pathogenesis and the surface exposure of functionally significant components. Apparatus components such as CdsF, CT584, and CdsC are well conserved among chlamydial species, are presented on the surface of infectious particles, and are functionally required for T3S activity. Antigenic profiling of sera from *Chlamydia*-infected individuals indicates low-level antibody reactivity to these proteins (Wang et al., 2010; Sharma et al., 2006; Cruz-Fisher et al., 2011). In contrast, corresponding TC and NC proteins in other gram-negative bacteria are highly antigenic and represent protective antigens in infection models (Sawa et al., 1999; Espina et al., 2006; Matson et al., 2005; Lawton et al., 1963). This low-level antigenicity of the chlamydial proteins may have specifically evolved to limit presentation of epitopes. It may be possible, however, to engineer antibodies that react with T3S proteins and are effective at limiting infection. In contrast to apparatus proteins, several effector proteins represent immunodominant antigens (Sharma et al., 2006; Wang et al., 2010; Cruz-Fisher et al., 2011). Immunization with Tarp (Wang et al., 2009) or CopN (Tammiruusu et al., 2007; Sambri et al., 2004) confers modest levels of protection in infection models. It is unclear how these antibodies may be functioning, however, since the effectors have an intracellular localization.

The development of synthetic inhibitory reagents has been pursued as an additional mechanism to interfere with T3S activity. For example, initial attempts at using peptide mimetics specific for T3S apparatus components have yielded promising results in inhibiting *C. pneumoniae* infectivity (Stone et al., 2011). Perhaps the most promising results, however, have come from the development and use of small-molecule inhibitors specific for T3S or its effectors. The recent discovery that certain acylated hydrazones of salicylaldehydes called INPs are specific, small-molecule inhibitors of T3S systems (reviewed by Keyser et al. [2008]) has provided a potentially valuable tool to investigate contributions of T3S to chlamydial pathogenesis. Multiple groups have shown that treatment of *Chlamydia*-infected cultures with these INPs interferes with T3S and inhibits productive development of chlamydiae (Wolf et al., 2006; Bailey et al., 2007; Muschiol et al., 2006; Slepenkin et al., 2007). In addition, a small-molecule inhibitor of CopN interferes with *C. pneumoniae* growth in cell culture (Huang et al., 2008). Unfortunately, one complication plaguing both peptide mimetics and small-molecule inhibitors is the possibility that inhibitory activities are the result of indirect effects on host cells rather than direct effects on chlamydial T3S. This point is exemplified by the observation that chelation of iron may contribute to the ability of INPs to interfere with chlamydial development (Slepenkin et al., 2007). Regardless, refinement of these molecules holds great promise, and this approach will likely have a significant impact on research regarding basic T3S-mediated virulence mechanisms.

REFERENCES

Abdelrahman, Y. M., and R. J. Belland. 2005. The chlamydial developmental cycle. *FEMS Microbiol. Rev.* **29:**949–959.

Akeda, Y., and J. E. Galan. 2005. Chaperone release and unfolding of substrates in type III secretion. *Nature* **437:**911–915.

Alvesalo, J., D. Greco, M. Leinonen, T. Raitila, P. Vuerela, and P. Auvinen. 2008. Microarray analysis of a *Chlamydia pneumoniae*-infected human epithelial cell line by use of ontology heirarchy. *J. Infect. Dis.* **197:**156–162.

Alzhanov, D., J. Barnes, D. Hruby, and D. D. Rockey. 2004. Chlamydial development is blocked in host cells transfected with *Chlamydophila caviae incA*. *BMC Microbiol.* **4:**1–10.

Azuma, Y., H. Hirakawa, A. Yamashita, Y. Cai, M. A. Rahman, H. Suzuki, S. Mitaku, H. Toh, S. Goto, T. Murakami, K. Sugi, H. Hayashi, H. Fukushi, H. Hattori, S. Kuhara, and M. Shirai. 2006. Genome sequence of the cat pathogen, *Chlamydophila felis*. *DNA Res.* **13:**15–23.

Bailey, L., A. Gylfe, C. Sundin, S. Muschiol, M. Elofsson, P. Nordstrom, B. Henriques-Normark, R. Lugert, A. Waldenstrom, H. Wolf-Watz, and S. Bergstrom. 2007. Small molecule inhibitors of type III secretion in *Yersinia* block the *Chlamydia pneumoniae* infection cycle. *FEBS Lett.* **58:**587–595.

Balsara, Z. R., S. Misaghi, J. N. Lafave, and M. N. Starnbach. 2006. *Chlamydia trachomatis* infection induces cleavage of the mitotic cyclin B1. *Infect. Immun.* **74:**5602–5608.

Bannantine, J. P., R. S. Griffiths, W. Viratyosin, W. J. Brown, and D. D. Rockey. 2000. A secondary structure motif predictive of protein localization to the chlamydial inclusion membrane. *Cell. Microbiol.* **2:**35–47.

Belland, R. J., M. A. Scidmore, D. D. Crane, D. M. Hogan, W. Whitmire, G. McClarty, and H. D. Caldwell. 2001. *Chlamydia trachomatis* cytotoxicity associated with complete and partial cytotoxin genes. *Proc. Natl. Acad. Sci. USA* **98:**13984–13989.

Belland, R. J., G. Zhong, D. D. Crane, D. Hogan, D. Sturdevant, J. Sharma, W. L. Beatty, and H. D. Caldwell. 2003. Genomic transcriptional profiling of the developmental cycle of *Chlamydia trachomatis*. *Proc. Natl. Acad. Sci. USA* **100:**8478–8483.

Benaud, C., B. J. Gentil, N. Assard, M. Court, J. Garin, C. Delphin, and J. Baudier. 2004. AHNAK interaction with the annexin 2/S100A10 complex regulates cell membrane cytoarchitecture. *J. Cell Biol.* **164:**133–144.

Betts, H. J., L. E. Twiggs, M. S. Sal, P. B. Wyrick, and K. A. Fields. 2008. Bioinformatic and biochemical evidence for the identification of the type III secretion system needle protein of *Chlamydia trachomatis*. *J. Bacteriol.* **190:**1680–1690.

Betts, H. J., K. Wolf, and K. A. Fields. 2009. Effector protein modulation of host cells: examples in the *Chlamydia* spp. arsenal. *Curr. Opin. Microbiol.* **12:**81–87.

Betts-Hampikian, H. J., and K. A. Fields. 2010. The chlamydial type III secretion mechanism: revealing cracks in a tough nut. *Frontiers Microbiol.* **1:**114.

Betts-Hampikian, H. J., and K. A. Fields. 2011. Disulfide bonding within components of the *Chlamydia* type III secretion appratus correlates with development. *J. Bacteriol.* **193**:6950-6959.

Birkelund, S., M. Morgan-Fisher, E. Timmerman, K. Gevaert, A. C. Shaw, and G. Christiansen. 2009. Analysis of proteins in *Chlamydia trachomatis* L2 outer membrane complex, COMC. *FEMS Immunol. Med. Microbiol.* **55:**187–195.

Blaylock, B., B. J. Berube, and O. Schneewind. 2010. YopR impacts type III needle polymerization in *Yersinia* species. *Mol. Microbiol.* **75:**221–229.

Blaylock, B., K. E. Riordan, D. M. Missiakas, and O. Schneewind. 2006. Characterization of the *Yersinia enterocolitica* type III secretion ATPase YscN and its regulator, YscL. *J. Bacteriol.* **188:**3525–3534.

Blocker, A., J. E. Deane, A. K. Veenendall, P. Roversi, J. L. Hodgkinson, S. Johnson, and S. M. Lea. 2008. What's the point of the type III secretion needle? *Proc. Natl. Acad. Sci. USA* **105:**6507–6513.

Burton, M. J., S. N. Rajak, J. Bauer, H. A. Weiss, S. B. Tolbert, A. Shoo, E. Habtamu, A. Manjurano, P. M. Emerson, D. Mabey, M. J. Holland, and R. L. Bailey. 2011. Conjunctival transcriptome in scarring trachoma. *Infect. Immun.* **79:**499–511.

Cain, R. J., R. D. Hayward, and V. Koronakis. 2008. Deciphering interplay between *Salmonella* invasion effectors. *PLoS Pathog.* **4:**e1000037.

Cambronne, E. D., and C. R. Roy. 2006. Recognition and delivery of effector proteins into eukaryotic cells by bacterial secretion systems. *Traffic* **7:**929–939.

Case, E. D., E. M. Peterson, and M. Tan. 2010. Promoters for *Chlamydia* type III secretion genes show a differential response to DNA supercoiling that correlates with temporal expression pattern. *J. Bacteriol.* **192:**2569–2574.

Chang, J. J., K. R. Leonard, and Y. X. Zhang. 1997. Structural studies of the surface projections of *Chlamydia trachomatis* by electron microscopy. *J. Med. Microbiol.* **46:**1013–1018.

Chellas-Géry, B., C. N. Linton, and K. A. Fields. 2007. Human GCIP interacts with CT847, a novel *Chlamydia trachomatis* type III secretion substrate, and is degraded in a tissue-culture infection model. *Cell. Microbiol.* **9:**2417–2430.

Chellas-Géry, B., K. Wolf, J. Tisoncik, T. Hackstadt, and K. A. Fields. 2011. Biochemical and immunolocalization analyses of putative type III secretion translocator proteins CopB and CopB2 of *Chlamydia trachomatis* reveal significant distinctions. *Infect. Immun.* **79:**3035–3046.

Chen, D., L. Lei, C. Lu, R. Flores, M. P. DeLisa, T. C. Roberts, F. E. Romesberg, and G. Zhong. 2010. Secretion of the chlamydial virulence factor CPAF requires the Sec-dependent pathway. *Microbiology* **56:**3031–3040.

Christian, J., J. Vier, S. A. Paschen, and G. Häcker. 2010. Cleavage of the NF-κB family protein p65/RelA by the chlamydial protease-like activity factor (CPAF) impairs proinflammatory signaling in cells infected with chlamydiae. *J. Biol. Chem.* **285:**41320–41327.

Clifton, D. R., C. A. Dooley, S. S. Grieshaber, R. A. Carabeo, K. A. Fields, and T. Hackstadt. 2005. Tyrosine phosphorylation of the chlamydial effector protein Tarp is species specific and not required for recruitment of actin. *Infect. Immun.* **73:**3860–3868.

Clifton, D. R., K. A. Fields, S. S. Grieshaber, C. A. Dooley, E. R. Fischer, D. J. Mead, R. A. Carabeo, and T. Hackstadt. 2004. A chlamydial type III translocated protein is tyrosine-phosphorylated at the site of entry and associated with recruitment of actin. *Proc. Natl. Acad. Sci. USA* **101:**10166–10171.

Cocchiaro, J. L., and R. H. Valdivia. 2009. New insights into *Chlamydia* intracellular survival mechanisms. *Cell. Microbiol.* **11:**1571–1578.

Cocucci, E., G. Racchetti, P. Podini, and J. Meldolesi. 2007. Enlargeosome traffic: exocytosis triggered by various signals is followed by endocytosis, membrane shedding or both. *Traffic* **8:**742–757.

Cortes, C., K. A. Rzomp, A. Tvinnereim, M. A. Scidmore, and B. Wizel. 2007. *Chlamydia pneumoniae* inclusion membrane protein Cpn0585 interacts with multiple Rab GTPases. *Infect. Immun.* **75:**5586–5596.

Crane, D. D., J. H. Carlson, E. R. Fischer, P. Bavoil, R. Hsia, C. Tan, C. C. Kuo, and H. D. Caldwell. 2006. *Chlamydia trachomatis* polymorphic membrane protein D is a species-common pan-neutralizing antigen. *Proc. Natl. Acad. Sci. USA* **103:**1894–1899.

Cruz-Fisher, M. I., C. Cheng, G. Sun, S. Pal, A. Teng, D. M. Molina, M. A. Kayala, A. Vigil, P. Baldi, P. L. Felgner, X. Liang, and L. M. de la Maza. 2011. Identification of immunodominant antigens by probing a whole *Chlamydia trachomatis* open reading frame proteome microarray using sera from immunized mice. *Infect. Immun.* **79:**246–257.

Curak, J., J. Rohde, and I. Stagljar. 2009. Yeast as a tool to study bacterial effectors. *Curr. Opin. Microbiol.* **12:**18–23.

Dautry-Varsat, A., A. Subtil, and T. Hackstadt. 2005. Recent insights into the mechanisms of *Chlamydia* entry. *Cell. Microbiol.* **7:**1714–1722.

Deane, J. E., P. Roversi, F. S. Cordes, S. Johnson, R. Kenjale, S. Daniall, F. Booy, W. D. Picking, W. L. Picking, A. Blocker, and S. M. Lea. 2006. Molecular model of a type three secretion system needle: implications for host cell sensing. *Proc. Natl. Acad. Sci. USA* **103:**12529–12533.

Durocher, D., I. A. Taylor, D. Sarbassova, L. F. Haire, S. L. Westcott, S. P. Jackson, S. J. Smerdon, and M. B. Yaffe. 2000. The molecular basis of FHA domain: phosphopeptide binding specificity and implications for phospho-dependent signaling mechanisms. *Mol. Cell* **6:**1169–1182.

Ellis, T. N., and M. J. Kuehn. 2010. Virulence and immunomodulatory roles of bacterial outer membrane vesicles. *Microbiol. Mol. Biol. Rev.* **74:**81–94.

Elwell, C. A., A. Ceesay, J. H. Kim, D. Kalman, and J. N. Engel. 2008. RNA interference screen identifies Able Kinase and PDGFR signaling in *Chlamydia trachomatis* entry. *PLoS Pathog.* **4:** e1000021.

Espina, M., A. J. Olive, R. Kenjale, D. S. Moore, S. F. Ausar, C. R. Middaugh, R. W. Kaminski, E. V. Oaks, M. A. Baxter, W. D. Picking, and W. L. Picking. 2006. IpaD localizes to the tip of the type III secretion system needle of *Shigella flexneri. Infect. Immun.* **74:**4391–4400.

Feldman, M. F., and G. R. Cornelis. 2003. The multitalented type III chaperones: all you can do with 15 kDa. *FEMS Microbiol. Lett.* **219:**151–158.

Ferracci, F., F. D. Schubot, D. S. Waugh, and G. V. Plano. 2005. Selection and characterization of *Yersinia pestis* YopN mutants that constitutively block Yop secretion. *Mol. Microbiol.* **57:**970–987.

Fields, K. A., E. R. Fischer, D. J. Mead, and T. Hackstadt. 2005. Analysis of putative *Chlamydia trachomatis* chaperones Scc2 and Scc3 and their use in the identification of type III secretion substrates. *J. Bacteriol.* **187:**6466–6478.

Fields, K. A., and T. Hackstadt. 2000. Evidence for the secretion of *Chlamydia trachomatis* CopN by a type III secretion mechanism. *Mol. Microbiol.* **38:**1048–1060.

Fields, K. A., and T. Hackstadt. 2006. The *Chlamydia* type III secretion system: structure and implications for pathogenesis, p. 220–233. *In* P. Bavoil and P. B. Wyrick (ed.), Chlamydia: *Genomics and Pathogenesis.* Horizon Press, Norfolk, United Kingdom.

Fields, K. A., D. J. Mead, C. A. Dooley, and T. Hackstadt. 2003. *Chlamydia trachomatis* type III secretion: evidence for a functional apparatus during early-cycle development. *Mol. Microbiol.* **48:**671–683.

Fu, Y., and J. E. Galan. 1999. A *Salmonella* protein antagonizes Rac-1 and Cdc42 to mediate host-cell recovery after bacterial invasion. *Nature* **401:** 293–297.

Giles, D. K., J. D. Whittimore, R. W. LaRue, J. E. Raulston, and P. B. Wyrick. 2006. Ultrastructural analysis of chlamydial antigen-containing vesicles everting for the *Chlamydia trachomatis* inclusion. *Microbes Infect.* **8:**1579–1591.

Gonzalez-Pedrajo, B., T. Minamino, M. Kihara, and K. Namba. 2006. Interactions between C ring proteins and export apparatus components: a possible mechanism for facilitating type III protein export. *Mol. Microbiol.* **60:**984–998.

Grieshaber, N. A., E. R. Fischer, D. J. Mead, C. A. Dooley, and T. Hackstadt. 2004. Chlamydial histone-DNA interactions are disrupted by a metabolite in the methylerythritol phosphate pathway of isoprenoid biosynthesis. *Proc. Natl. Acad. Sci. USA* **101:**7451–7456.

Hackstadt, T., M. Scidmore-Carlson, E. Shaw, and E. Fischer. 1999. The *Chlamydia trachomatis* IncA protein is required for homotypic vesicle fusion. *Cell. Microbiol.* **1:**119–130.

Hackstadt, T., W. J. Todd, and H. D. Caldwell. 1985. Disulfide-mediated interactions of the chlamydial major outer membrane protein: role in the differentiation of chlamydiae? *J. Bacteriol.* **161:** 25–31.

Hakansson, S., K. Schesser, C. Persson, E. E. Galyov, R. Rosqvist, F. Homble, and H. Wolf-Watz. 1996. The YopB protein of *Yersinia pseudotuberculosis* is essential for the translocation of Yop effector proteins across the target cell plasma membrane and displays a contact-dependent membrane disrupting activity. *EMBO J.* **15:**5812–5823.

Hatch, T. P. 1999. Developmental biology, p. 29–67. *In* R. S. Stephens (ed.), Chlamydia: *Intracellular Biology, Pathogenesis, and Immunity.* ASM Press, Washington, DC.

Hatch, T. P., M. Miceli, and J. E. Sublett. 1986. Synthesis of disulfide-bonded outer membrane proteins during the developmental cycle of *Chlamydia psittaci* and *Chlamydia trachomatis. J. Bacteriol.* **165:**379–385.

Hefty, P. S., and R. S. Stephens. 2007. Chlamydial type III secretion system is encoded on ten operons preceded by sigma 70-like promoter elements. *J. Bacteriol.* **189:**198–206.

Henderson, I. R., and A. C. Lam. 2001. Polymorphic proteins of *Chlamydia* spp.—autotransporters beyond the proteobacteria. *Trends Microbiol.* **9:**573–578.

Heuer, D., V. Brinkmann, T. F. Meyer, and A. J. Szczepek. 2003. Expression and translocation of chlamydial protease during acute and persistent infection of the epithelial HEp-2 cells with *Chlamydophila (Chlamydia) pneumoniae. Cell. Microbiol.* **5:**315–322.

Heuer, D., L. Rejman, N. Machuy, A. Karlas, A. Wehrens, F. Siedler, V. Brinkmann, and T. F. Meyer. 2009. *Chlamydia* causes fragmentation of the Golgi compartment to ensure reproduction. *Nature* **457:**731–735.

Ho, T. D., and M. N. Starnbach. 2005. The *Salmonella enterica* serovar Typhimurium-encoded type III secretion systems can translocate *Chlamydia trachomatis* proteins into the cytosol of host cells. *Infect. Immun.* **73:**905–911.

Hoare, A., P. Timms, P. M. Bavoil, and D. P. Wilson. 2008. Spatial constraints within the chlamydial host cell inclusion predict interrupted development and persistence. *BMC Microbiol.* **8:**1–9.

Horn, M., A. Collingro, S. Schmitz-Esser, C. L. Beier, U. Purkhold, B. Fartmann, P. Brandt, G. J. Nyakatura, M. Droege, D. Frishman, T. Rattei, H. Mewes, and M. Wagner. 2004. Illuminating the evolutionary history of *Chlamydiae. Science* **304:**728–730.

Hower, S., K. Wolf, and K. A. Fields. 2009. Evidence that CT694 is a novel *Chlamydia trachomatis* T3S substrate capable of functioning during invasion or early cycle development. *Mol. Microbiol.* **72:**1423–1437.

Hsia, R. C., Y. Pannekoek, E. Ingerowski, and P. M. Bavoil. 1997. Type III secretion genes identify a putative virulence locus of *Chlamydia. Mol. Microbiol.* **25:**351–359.

Huang, J., C. F. Lesser, and S. Lory. 2008. The essential role of the CopN protein in *Chlamydia pneumoniae* intracellular growth. *Nature* **456:** 112–115.

Hueck, C. J. 1998. Type III protein secretion systems in bacterial pathogens of animals and plants. *Microbiol. Mol. Biol. Rev.* **62:**379–433.

Hume, P. J., E. J. McGhie, R. D. Hayward, and V. Koronakis. 2003. The purified *Shigella* IpaB and *Salmonella* SipB translocators share biochemical properties and membrane topology. *Mol. Microbiol.* **49:**425–439.

Jamison, W. P., and T. Hackstadt. 2008. Induction of type III secretion by cell-free *Chlamydia trachomatis* elementary bodies. *Microb. Pathog.* **45:**435–440.

Jewett, T. J., C. A. Dooley, D. J. Mead, and T. Hackstadt. 2008. *Chlamydia trachomatis* Tarp is phosphorylated by Src family tyrosine kinases. *Biochem. Biophys. Res. Commun.* **371:**339–344.

Jewett, T. J., E. R. Fischer, D. J. Mead, and T. Hackstadt. 2006. Chlamydial TARP is a bacterial nucleator of actin. *Proc. Natl. Acad. Sci. USA* **103:**15599–15604.

Jewett, T. J., N. J. Miller, C. A. Dooley, and T. Hackstadt. 2010. The conserved Tarp actin binding domain is important for chlamydial invasion. *PLoS Pathog.* **6:**e1000997.

Johnson, D. L., and J. B. Mahony. 2007. *Chlamydophila pneumoniae* PknD exhibits dual amino acid specificity and phosphorylates Cpn0712, a putative type III secretion YscD homolog. *J. Bacteriol.* **189:**7549–7555.

Johnson, D. L., C. B. Stone, and J. B. Mahony. 2008. Interactions between CdsD, CdsQ, and CdsL, three putative *Chlamydophila pneumoniae* type III secretion proteins. *J. Bacteriol.* **190:**2972–2980.

Joseph, S. S., and G. V. Plano. 2007. Identification of TyeA residues required to interact with YopN and to regulate Yop secretion. *Adv. Exp. Med. Biol.* **603:**235–245.

Kalman, S., W. Mitchell, R. Marathe, C. Lammel, J. Fan, R. W. Hyman, L. Olinger, J. Grimwood, R. W. Davis, and R. S. Stephens. 1999. Comparative genomes of *Chlamydia pneumoniae* and *C. trachomatis. Nat. Genet.* **21:** 385–389.

Karyagina, A. S., A. V. Alexeevsky, S. A. Spirin, N. A. Zigangirova, and A. L. Gintsburg. 2009. Effector proteins of chlamydiae. *Mol. Biol.* **43:**897–916.

Keyser, P., M. Elofsson, S. Rosell, and H. Wolf-Watz. 2008. Virulence blockers as alternatives to antibiotics: type III secretion inhibitors against Gram-negative bacteria. *J. Intern. Med.* **264:** 17–29.

Kiselev, A. O., W. Stamm, J. R. Yates, and M. F. Lampe. 2007. Expression, processing, and localization of PmpD of *Chlamydia trachomatis* serovar L2 during the chlamydial developmental cycle. *PLoS One* **2:**e568–e569.

Kubori, T., A. Sukhan, S. Aizawa, and J. Galan. 2000. Molecular characterization and assembly of the needle complex of the *Salmonella typhimurium* type III secretion system. *Proc. Natl. Acad. Sci. USA* **97:**10225–10230.

Lad, S. P., J. Li, J. da Silva Correia, Q. Pan, S. Gadwal, R. J. Ulevitch, and E. Li. 2007a. Cleavage of p65/RelA of the NF-κB pathway by *Chlamydia. Proc. Natl. Acad. Sci. USA* **104:**2933–2938.

Lad, S. P., G. Yang, D. A. Scott, G. Wang, P. Nair, J. Mathison, V. S. Reddy, and E. Li. 2007b. Chlamydial CT441 is a PDZ domain-containing tail-specific protease that interferes with the NF-kB pathway of immune responses. *J. Bacteriol.* **189:**6619–6625.

Lawton, W. D., R. L. Erdman, and M. J. Surgalla. 1963. Biosynthesis and purification of V and W antigens in *Pasteurella pestis J. Immunol.* **91:**179–184.

Le Negrate, G., A. Krieg, M. L. Faustin, A. Godzik, S. Krajewski, and J. C. Reed. 2008. ChlaDub1 of *Chlamydia trachomatis* suppresses NF-κB activation and inhibits I-κBα ubiquitination and degradation. *Cell. Microbiol.* **10:** 1879–1892.

Liu, X., M. Afrane, D. E. Clemmer, G. Zhong, and D. E. Nelson. 2010. Identification of *Chlamydia trachomatis* outer membrane complex proteins by differential proteomics. *J. Bacteriol.* **192:**2852–2860.

Lorenzini, E., A. Singer, B. Singh, R. Lam, T. Skarina, N. Y. Chirgadze, A. Savchenko, and R. S. Gupta. 2010. Structure and protein-protein interaction studies on *Chlamydia trachomatis* protein CT670 (YscO Homolog). *J. Bacteriol.* **192:**2746–2756.

Markham, A. P., Z. A. Jaafar, K. E. Kemege, C. R. Middaugh, and P. S. Hefty. 2009. Biophysical characterization of *Chlamydia trachomatis* CT584 supports its potential role as a type III secretion needle tip protein. *Biochemistry* **48:** 10353–10361.

Marlovits, T. C., and C. E. Stebbins. 2009. Type III secretion systems shape up as they ship out. *Curr. Opin. Microbiol.* **13:**1–6.

Marvis, M., A. L. Page, R. Tournebize, B. Demers, P. Sansonetti, and C. Parsot. 2002. Regulation of transcription by the activity of the *Shigella flexneri* type III secretion apparatus. *Mol. Microbiol.* **43:**1543–1553.

Matson, J. S., K. A. Durick, D. S. Bradley, and M. L. Nilles. 2005. Immunization of mice with YscF provides protection from *Yersinia pestis* infections. *BMC Microbiol.* **24:**38–48.

Matsumoto, A. 1982. Electron microscopic observations of surface projections on *Chlamydia psittaci* reticulate bodies. *J. Bacteriol.* **150:**358–364.

Matsumoto, A. 1988. Structural characteristics of chlamydial bodies, p. 21–45. *In* A. L. Barron (ed.), *Microbiology of* Chlamydia. CRC Press, Boca Raton, FL.

Maurer, A. P., A. Mehlitz, H. J. Mollenkopf, and T. F. Meyer. 2007. Gene expression profiles of *Chlamydophila pneumoniae* during the developmental cycle and iron depletion-mediated persistence. *PLoS Pathog.* **3:**752–769.

Mehlitz, A., S. Banhart, S. Hess, M. Selback, and T. F. Meyer. 2008. Complex kinase requirements for *Chlamydia trachomatis* Tarp phosphorylation. *FEMS Microbiol. Lett.* **289:**233–240.

Mehlitz, A., S. Banhart, A. P. Maurer, A. Kaushansky, A. G. Gordus, J. Zielecki, G. MacBeath, and T. F. Meyer. 2010. Tarp regulates early *Chlamydia*-induced host cell survival through interactions with the human adaptor protein SHC1. *J. Cell Biol.* **190:**143–157.

Misaghi, S., Z. R. Balsara, A. Catic, E. Spooner, H. L. Ploegh, and M. N. Starnbach. 2006. *Chlamydia trachomatis*-derived deubiquitinating enzymes in mammalian cells during infection. *Mol. Microbiol.* **61:**142–150.

Mital, J., N. J. Miller, E. R. Fischer, and T. Hackstadt. 2010. Specific chlamydial inclusion membrane proteins associate with active Src family kinases in microdomains that interact with the host microtubule network *Cell. Microbiol.* **12:** 1235–1249.

Moraes, T. F., T. Spreter, and N. C. Strynadka. 2008. Piecing together the type III injectisome of bacterial pathogens. *Curr. Opin. Struct. Biol.* **18:**258–266.

Morita-Ishihara, T., M. Ogawa, H. Sagara, M. Yoshida, E. Katayama, and C. Sasakawa. 2006. *Shigella* Spa33 is an essential C-ring component of type III secretion machinery. *J. Biol. Chem.* **281:**599–607.

Mueller, C. A., P. Broz, and G. Cornelis. 2008. The type III secretion system tip complex and translocon. *Mol. Microbiol.* **68:**1085–1095.

Mueller, C. A., P. Broz, S. Muller, P. Ringler, F. Erne-Brand, I. Sorg, M. Kuhn, A. Engel, and G. Cornelis. 2005. The V-antigen of *Yersinia* forms a distinct structure at the tip of injectisome needles. *Science* **310:**674–676.

Muschiol, S., L. Bailey, A. Gylfe, C. Sundin, K. Hultenby, S. Bergstrom, M. Elofsson, H. Wolf-Watz, S. Normark, and B. Henriques-Normark. 2006. A small-molecule inhibitor of type III secretion inhibits different stages of the infectious cycle of *Chlamydia trachomatis*. *Proc. Natl. Acad. Sci. USA* **103:**14566–14571.

Newhall, W. J., and R. B. Jones. 1983. Disulfide-linked oligomers of the major outer membrane protein of chlamydiae. *J. Bacteriol.* **154:**344–349.

Neyt, C., and G. R. Cornelis. 1999. Insertion of a Yop translocation pore into the macrophage plasma membrane by *Yersinia enterocolitica*: requirement for translocators YopB and YopD. *Mol. Microbiol.* **33:**971–981.

Nichols, B A., P. Y. Setzer, F. Pang, and C. R. Dawson. 1985. New view of the surface projections of *Chlamydia trachomatis*. *J. Bacteriol.* **164:** 344–349.

Niehus, E., E. Cheng, and M. Tan. 2008. DNA supercoiling-dependent gene regulation in *Chlamydia*. *J. Bacteriol.* **190:**6419-6427.

Page, A. L., and C. Parsot. 2002. Chaperones of the type III secretion pathway: jacks of all trades. *Mol. Microbiol.* **46:**1–11.

Pallen, M. J., S. A. Beatson, and C. M. Bailey. 2005. Bioinformatics, genomics and evolution of non-flagellar type III secretion systems: a Darwinian perspective. *FEMS Microbiol. Rev.* **29:** 201–229.

Pallen, M. J., M. S. Francis, and K. Futterer. 2003. Tetratricopeptide-like repeats in type III-secretion chaperones and regulators. *FEMS Microbiol. Lett.* **223:**53–60.

Parsot, C., C. Hamiaux, and A. Page. 2003. The various and varying roles of specific chaperones in type III secretion systems. *Curr. Opin. Microbiol.* **6:**7–14.

Pennini, M. E., S. Perrinet, A. Dautry-Varsat, and A. Subtil. 2010. Histone methylation by NUE, a novel nuclear effector of the intracellular pathogen *Chlamydia trachomatis. PLoS Pathog.* **6:** e10000995.

Quinaud, M., J. Chabert, E. Faudry, E. Neumann, D. Lemaire, A. Pastor, S. Elsen, A. Dessen, and I. Attree. 2005. The PscE-PscF-PscG complex controls type III secretion needle biogenesis in *Pseudomonas aeruginosa. J. Biol. Chem.* **280:**36293–36300.

Quinaud, M., S. Ple, V. Job, C. Contreras-Martel, J. P. Simorre, I. Attree, and A. Dessen. 2007. Structure of the heterotrimeric complex that regulates type III secretion needle formation. *Proc. Natl. Acad. Sci. USA* **104:**7803–7808.

Radermacher, A. N., and G. R. Crabtree. 2008. Monster protein controls calcium entry and fights infection. *Immunity* **28:**13–14.

Rao, X., P. Deighan, Z. Hua, J. Wang, M. Luo, J. Wang, Y. Liang, G. Zhong, A. Hochschild, and L. Shen. 2009. A regulator from *Chlamydia trachomatis* modulates the activity of RNA polymerase through direct interaction with the beta subunit and the primary sigma subunit. *Genes Dev.* **23:**1818–1829.

Read, T., G. Myers, R. Brunham, W. Nelson, I. Paulsen, J. Heidelberg, E. Holtzapple, H. Khouri, N. Federova, H. Carty, L. Umayam, D. Haft, J. Peterson, M. Beanan, O. White, S. Salzberg, R. Hsia, G. McClarty, R. Rank, P. Bavoil, and C. Fraser. 2003. Genome sequence of *Chlamydophila caviae* (*Chlamydia psittaci* GPIC): examining the role of niche-specific genes in the evolution of the Chlamydiaceae. *Nucleic Acids Res.* **31:**2134–2147.

Read, T. D., R. C. Brunham, C. Shen, S. R. Gill, J. F. Heidelberg, O. White, E. K. Hickey, J. Peterson, T. Utterback, K. Berry, S. Bass, K. Linher, J. Weidman, H. Khouri, B. Craven, C. Bowman, R. Dodson, M. Gwinn, W. Nelson, R. DeBoy, J. Kolonay, G. McClarty, S. L. Salzberg, J. Eisen, and C. M. Fraser. 2000. Genome sequences of *Chlamydia trachomatis* MoPn and *Chlamydia pneumoniae* AR39. *Nucleic Acids Res.* **28:**1397–1406.

Ren, Q., S. J. Robertson, D. Howe, L. F. Barrows, and R. A. Heizen. 2003. Comparative DNA microarray analysis of host cell transcriptional responses to infection by *Coxiella burnetii* or *Chlamydia trachomatis. Ann. N. Y. Acad. Sci.* **990:**701–713.

Riordan, K. E., and O. Schneewind. 2008. YscU cleavage and the assembly of *Yersinia* type III secretion machine complexes. *Mol. Microbiol.* **68:** 1485–1501.

Rockey, D. D., and D. T. Alzhanov. 2006. Proteins in the chlamydial inclusion membrane, p. 234–254. *In* P. Bavoil and P. B. Wyrick (ed.), Chlamydia*: Genomics and Pathogenesis.* Horizon Press, Norfolk, United Kingdom.

Rockey, D. D., D. Grosenbach, D. E. Hruby, M. G. Peacock, R. A. Heinzen, and T. Hackstadt. 1997. *Chlamydia psittaci* IncA is phosphorylated by the host cell and is exposed on the cytoplasmic face of the developing inclusion. *Mol. Microbiol.* **24:**217–228.

Rzomp, K. A., A. R. Moorhead, and M. A. Scidmore. 2006. The GTPase Rab4 interacts with *Chlamydia trachomatis* inclusion membrane protein CT229. *Infect. Immun.* **74:**5362–5373.

Rzomp, K. A., L. D. Scholtes, B. J. Briggs, G. R. Whittaker, and M. A. Scidmore. 2003. Rab GTPases are recruited to chlamydial inclusions in both a species-dependent and species-independent manner. *Infect. Immun.* **71:**5855–5870.

Saier, M. H. 2006. Protein secretion and membrane insertion systems in gram-negative bacteria. *J. Membr. Biol.* **214:**75–90.

Sambri, V., M. Donati, K. Storni, M. Di Leo, M. Agnisdei, R. Petracca, O. Finco, G. Grandi, G. Ratti, and C. Cevenini. 2004. Experimental infection by *Chlamydia pneumoniae* in the hamster. *Vaccine* **22:**1131–1137.

Sawa, T., T. Yahr, M. Ohara, K. Kurahashi, M. A. Gropper, J. P. Wiener-Kronish, and D. W. Frank. 1999. Active and passive immunization with the *Pseudomonas* V antigen protects against type III intoxication and lung injury. *Nat. Med.* **5:**392–398.

Scidmore, M. A. 2011. Recent advances in *Chlamydia* subversion of host cytoskeletal and membrane trafficking pathways. *Microbes Infect.* **13:**527–535.

Scidmore, M. A., and T. Hackstadt. 2001. Mammalian 14-3-3 beta associates with the *Chlamydia trachomatis* inclusion membrane via its interaction with IncG. *Mol. Microbiol.* **39:**1638–1650.

Sharma, J., Y. Zhong, F. Dong, J. M. Piper, G. Wang, and G. Zhong. 2006. Profiling of human antibody responses to *Chlamydia trachomatis* urogenital tract infection using microplates with 156 chlamydial fusion proteins. *Infect. Immun.* **74:**1490–1499.

Silvia-Herzog, E., F. Ferracci, M. W. Jackson, S. S. Joseph, and G. V. Plano. 2008. Membrane localization and topology of the *Yersinia pestis* YscJ lipoprotein. *Microbiology* **154:**593–607.

Silvia-Herzog, E., S. S. Joseph, A. Avery, J. Coba, K. Wolf, K. A. Fields, and G. V. Plano. 2011. Scc1 (CP0432) and Scc4 (CP0033) function as a type III secretion chaperone for CopN of *Chlamydia pneumoniae*. *J. Bacteriol.* **193:**3490–3496.

Sisko, J. L., K. Spaeth, Y. Kumar, and R. H. Valdivia. 2006. Multifunctional analysis of *Chlamydia*-specific genes in a yeast expression system. *Mol. Microbiol.* **60:**51–66.

Slepenkin, A., L. M. de la Maza, and E. M. Peterson. 2005. Interaction between components of the type III secretion system of *Chlamydiaceae*. *J. Bacteriol.* **187:**473–479.

Slepenkin, A., P. A. Enquist, U. Hagglund, L. M. de la Maza, M. Elofsson, and E. M. Peterson. 2007. Reversal of the antichlamydial activity of putative type III secretion inhibitors by iron. *Infect. Immun.* **75:**3478–3489.

Slepenkin, A., V. Motin, L. M. de la Maza, and E. M. Peterson. 2003. Temporal expression of type III secretion genes in *Chlamydia pneumoniae*. *Infect. Immun.* **71:**2555–2562.

Spaeth, K. E., Y. S. Chen, and R. H. Valdivia. 2009. The *Chlamydia* type III secretion system C-ring engages a chaperone-effector protein complex. *PLoS Pathog.* **5:**e1000579.

Stephens, R. S., S. Kalman, C. Lammel, J. Fan, R. Marathe, L. Aravind, W. Mitchell, L. Olinger, R. L. Tatusov, Q. Zhao, E. V. Koonin, and R. W. Davis. 1998. Genome sequence of an obligate intracellular pathogen of humans: *Chlamydia trachomatis*. *Science* **282:**754–759.

Stone, C. B., D. C. Bulir, J. D. Gilchrist, R. K. Toor, and J. B. Mahony. 2010. Interactions between flagellar and type III secretion proteins in *Chlamydia pneumoniae*. *BMC Microbiol.* **10:**18.

Stone, C. B., D. C. Bulir, C. A. Emdin, R. M. Pirie, E. A. Porfilio, J. W. Slootstra, and J. B. Mahony. 2011. *Chlamydia pneumoniae* CdsL regulates CdsN ATPase activity and disruption with a peptide mimetic prevents bacterial invasion. *Frontiers Microbiol.* doi:10.3389/fmicb.2011.00021.

Stone, C. B., D. L. Johnson, D. C. Bulir, J. D. Gilchrist, and J. B. Mahony. 2008. Characterization of the putative type III secretion ATPase CdsN (Cpn0707) of *Chlamydophila pneumoniae*. *J. Bacteriol.* **190:**6580–6588.

Subtil, A., C. Delevoye, M. E. Balana, L. Tastevin, S. Perrinet, and A. Dautry-Varsat. 2005. A directed screen for chlamydial proteins secreted by a type III mechanism identifies a translocated protein and numerous other new candidates. *Mol. Microbiol.* **56:**1636–1647.

Subtil, A., C. Parsot, and A. Dautry-Varsat. 2001. Secretion of predicted Inc proteins of *Chlamydia pneumoniae* by a heterologous type III machinery. *Mol. Microbiol.* **39:**792–800.

Tammiruusu, A., T. Penttila, R. Lahesmaa, M. Sarvas, M. Puolakkainen, and J. M. Vuola. 2007. Intranasal administration of chlamydial outer protein N (CopN) induces protection against pulmonary *Chlamydia pneumoniae* infection in a mouse model. *Vaccine* **25:**283–290.

Tan, C., R. Hsia, H. Shou, J. Carrasco, R. Rank, and P. Bavoil. 2010. Variable expression of surface-exposed polymorphic proteins in in vitro-grown *Chlamydia trachomatis*. *Cell. Microbiol.* **12:**174–187.

Tanzer, R. J., D. Longbottom, and T. P. Hatch. 2001. Identification of polymorphic outer membrane proteins of *Chlamydia psittaci* 6BC. *Infect. Immun.* **69:**2428–2434.

Thalmann, J., K. Janik, M. May, K. Sommer, J. Ebeling, K. Hofmann, H. Genth, and A. Klos. 2010. Actin re-organization induced by *Chlamydia trachomatis* serovar D—evidence for a critical role of the effector protein CT166 targeting Rac. *PLoS One* doi:10.1371/journal.pone.0009887.

Thomas, N. A., W. Deng, J. L. Puente, E. A. Frey, C. K. Yip, N. C. Strynadka, and B. B. Finlay. 2005. CesT is a multi-effector chaperone and recruitment factor required for the efficient type III secretion of both LEE- and non-LEE-encoded effectors of enteropathogenic *Escherichia coli*. *Mol. Microbiol.* **57:**1762–1779.

Thomson, N. R., C. Yeats, K. Bell, M. T. Holden, S. D. Bentley, M. Linvingstone, A. M. Cerdeno-Tarraga, B. Harris, J. Doggett, D. Ormond, K. Mungall, K. Clarke, T. Feltwell, Z. Hance, M. Sanders, M. A. Quail, C. Price, B. G. Barrell, J. Parkhill, and D. Longbottom. 2005. The *Chlamydophila abortus* genome sequence reveals an array of variable proteins that contribute to interspecies variation. *Genome Res.* **15:**629–640.

Tietzel, I., C. El-Haibi, and R. A. Carabeo. 2009. Chlamydial entry involves Tarp binding of guanine nucleotide exchange factors. *PLoS Pathog.* **4:**e1000014.

Torruellas, J., M. W. Jackson, J. W. Pennock, and G. V. Plano. 2005. The *Yersinia pestis* type III secretion needle plays a role in regulation of Yop secretion. *Mol. Microbiol.* **57:**1719–1733.

Valdivia, R. H. 2008. *Chlamydia* effector proteins and new insights into chlamydial cellular microbiology. *Curr. Opin. Microbiol.* **11:**53–59.

Vandahl, B. B., A. S. Pedersen, K. Gevaert, A. Holm, J. Vandekerckhove, J. Christensen, and S. Birkelund. 2002. The expression, processing and localization of polymorphic membrane proteins in *Chlamydia pneumoniae* strain CWL029. *BMC Microbiol.* **2:**36.

Veenendall, A. K., J. L. Hodgkinson, L. Schwarzer, D. Stabat, S. F. Zenk, and A. Blocker. 2007. The type III secretion system needle tip complex mediates host cell sensing and translocon insertion. *Mol. Microbiol.* **63:**1719–1730.

Verma, A., and A. T. Maurelli. 2003. Identification of two eukaryote-like serine/threonine kinases encoded by *Chlamydia trachomatis* serovar L2 and characterization of interacting partners of Pkn1. *Infect. Immun.* **71:**5772–5784.

Wang, J., L. Chen, F. Chen, Y. X. Zhang, J. Baseman, S. Perdue, I. T. Hyeh, R. Shain, M. J. Holland, R. L. Bailey, D. Mabey, P. Yu, and G. Zhong. 2009. A chlamydial type III-secreted effector protein (Tarp) is predominantly recognized by antibodies from humans infected with *Chlamydia trachomatis* and induces protective immunity against upper genital tract pathologies in mice. *Vaccine* **27:**2967–2980.

Wang, J., Y. X. Zhang, C. Lu, L. Lei, P. Yu, and G. Zhong. 2010. A genome-wide profiling of the humoral immune response to *Chlamydia trachomatis* infection reveals vaccine candidate antigens expressed in humans. *J. Immunol.* **185:**1670–1680.

Wehrl, W., V. Brinkmann, P. R. Jungblut, T. F. Meyer, and A. J. Szczepek. 2004. From the inside out—processing of the chlamydial autotransporter PmpD and its role in bacterial adhesion and activation of human host cells. *Mol. Microbiol.* **51:**319–334.

Wilson, D. P., P. Timms, D. L. McElwain, and P. Bavoil. 2006. Type III secretion, contact-dependent model for the intracellular development of *Chlamydia. Bull. Math. Biol.* **68:**161–178.

Wolf, K., H. J. Betts, B. Chellas-Géry, S. Hower, C. L. Linton, and K. A. Fields. 2006. Treatment of *Chlamydia trachomatis* with a small molecule inhibitor of the *Yersinia* type III secretion system disrupts progression of the chlamydial developmental cycle. *Mol. Microbiol.* **61:**1543–1555.

Wolf, K., E. R. Fischer, and T. Hackstadt. 2000. Ultrastructural analysis of developmental events in *Chlamydia pneumoniae*-infected cells. *Infect. Immun.* **68**:2379–2385.

Wolf, K., G. V. Plano, and K. A. Fields. 2009. A protein secreted by the respiratory pathogen *Chlamydia pneumoniae* impairs signaling via interaction with human Act1. *Cell. Microbiol.* **11:**769–779.

Yip, C. K., T. G. Kimbrough, H. B. Felise, M. Vuckovic, N. A. Thomas, R. A. Pfuetzner, E. A. Frey, B. B. Finlay, S. I. Miller, and N. C. Strynadka. 2005. Structural characterization of the molecular platform for type III secretion system assembly. *Nature* **435:**702–707.

Zhong, G. 2009. Killing me softly: chlamydial use of proteolysis for evading host defenses. *Trends Microbiol.* **17:**467–473.

Zhong, G., P. Fan, H. Ji, F. Dong, and Y. Huang. 2001. Identification of a chlamydial protease-like activity factor responsible for the degradation of host transcription factors. *J. Exp. Med.* **193:**935–942.

IMMUNE RECOGNITION AND HOST CELL RESPONSE DURING *CHLAMYDIA* INFECTION

Uma M. Nagarajan

10

INTRODUCTION

The hallmark of *Chlamydia trachomatis* infection in the genital tract and the conjunctiva is an acute inflammatory response. Several lines of evidence point to the major role of this response in the development of salpingitis/hydrosalpinx, a manifestation of oviduct pathology during *C. trachomatis* infection. The initiator of acute inflammatory responses is primarily the infected epithelial cell (Rasmussen et al., 1997). Cytokines/chemokines produced by the infected epithelial cells set the stage for recruitment of inflammatory cells to the site of infection. Further recognition of host "danger signals" and chlamydial factors by inflammatory cells primes them to induce optimal adaptive immunity and antibody responses, which eventually lead to clearance of infection. However, during this process, the inflammatory response can lead to pathology in the genital tract, as collateral damage. Owing to its anatomical structure, oviduct pathology leads to blockade and development of hydrosalpinx, leading to infertility. The inflammatory response also leads to cervicitis and endometritis. The primary step in this pathological response is recognition of chlamydiae by the host cell, followed by signaling cascades that lead to the cytokine response. Therefore, an understanding of the interaction of host receptors with chlamydial effectors is critical for understanding the mechanisms through which chlamydial infection leads to host pathology.

The central theme of the vertebrate immune system is to discriminate self from nonself. The definition of "nonself" has been broadened to include danger signals, which alert the immune system to the presence of infection, cell death, or inflammation. Innate immune cells possess multiple germ line-encoded receptors that have evolved to broadly distinguish between self and nonself. These receptors are called pathogen recognition receptors (PRRs), and they detect conserved structures called microbe-associated molecular patterns (MAMPs) that are present on microbes. PRRs are membrane bound or cytosolic, allowing microbial recognition in different cellular compartments. Although there is a finite number of PRRs, they are used in various combinations to recognize different intracellular pathogens, and even different species of *Chlamydia*. These variations are largely determined by specific

Uma M. Nagarajan, Division of Pediatric Infectious Diseases, Department of Pediatrics and Department of Microbiology and Immunology, University of Arkansas for Medical Sciences, Arkansas Children's Hospital Research Institute, 13 Children's Way, Little Rock, AR 72202.

Intracellular Pathogens I: Chlamydiales
Edited by Ming Tan and Patrik M. Bavoil
10.1128/9781555817329.ch10

intracellular niches occupied by the pathogen, effector molecules produced by the pathogen, the infected mucosal tissue or cell type, and sometimes the pathogen load. From the host's perspective, recognition of chlamydiae by PRRs triggers the upregulation of host cytokines and chemokines following infection. The cytokines/chemokines function to induce recruitment of inflammatory cells to the site of infection, eradicate infection, and induce effective adaptive immunity. The cytokine induction that follows recognition by a specific PRR can even benefit the pathogen more than the host, which suggests that the pathogen has learned to exploit host immune pathways. For instance, *Chlamydia* infection leads to expression of beta interferon (IFN-β) (Nagarajan et al., 2008; Qiu et al., 2008) and interleukin-10 (IL-10) (Igietseme et al., 2000) in infected cells, which facilitate chlamydial propagation while being deleterious to the host in a mouse genital infection model. The overall goal of this chapter is to review the various PRRs that recognize chlamydiae and the ensuing cellular signaling pathways that result in cytokine induction. Although the specific chlamydial effectors that interact with the recognition receptors have not been identified, candidate effectors will be discussed.

RECOGNITION OF CHLAMYDIAE

Recognition of chlamydial MAMPs by host PRRs occur at multiple levels, owing to *Chlamydia*'s obligate intracellular biphasic lifestyle. Initially, there is recognition at the surface of the host cell, but the majority of recognition occurs intracellularly. Further, chlamydial elementary bodies and reticulate bodies produce different effectors that could be recognized by different host receptors. The specific host receptors engaged in recognition of chlamydiae are discussed below.

Toll-Like Receptors

The Toll-like receptors (TLRs) are responsible for the recognition of unique MAMPs, such as bacterial lipoprotein, which initiate an immediate cytokine response upon infection (reviewed by Takeda et al. [2003]). The TLRs are type I transmembrane receptors with a unique pathogen recognition leucine-rich repeat (LRR), a cysteine-rich domain, a transmembrane domain, and an intracellular Toll/IL-1R motif (reviewed by Martin and Wesche [2002] and Jin and Lee [2008]). At least 11 different TLRs are known in humans and mice, and their respective ligands have been identified. Different TLRs are expressed on the cell surface or on intracellular membranes to recognize MAMPs at unique cellular compartments.

TLR2 and TLR4

Of the surface TLRs, the first two to be implicated in recognition of chlamydiae were TLR2 and TLR4. TLR2 typically recognizes peptidoglycan and lipoproteins, while TLR4 has been shown to bind *Escherichia coli* lipopolysaccharides (LPS) and to induce proinflammatory cytokines. One of the earliest studies using two different isolates of *Chlamydia pneumoniae* (CM-1 and TW-183) showed a predominant role for TLR2 versus TLR4 in dendritic cell activation (Prebeck et al., 2001). Subsequently, Netea and colleagues (Netea et al., 2002) also showed that the non-LPS component of *C. pneumoniae* induced cytokine production through a TLR2-dependent pathway. Likewise, a predominant role for TLR2 had been demonstrated in the recognition of *Chlamydia muridarum* (Darville et al., 2003) and *C. trachomatis* (Erridge et al., 2004), leading to induction of the cytokines tumor necrosis factor alpha (TNF-α), IL-1β, and IL-6 in inflammatory cells. The importance of TLR2 recognition in chlamydial pathogenesis was most evident during *C. muridarum* genital infection when TLR2 knockout (KO) mice were found to have significantly lower inflammatory cytokine responses, including TNF-α and MIP-2, and significantly less oviduct pathology than wild-type (WT) mice (Darville et al., 2003). Despite lower cytokine responses in TLR2 KO mice, infection was cleared at a rate similar to that of the WT mice. These data suggested a major role for TLR2 in recognition of *C. muridarum*, induction of inflammatory cytokines, and immunopathology, which are discussed in detail in chapter 11, "*Chlamydia* immunopathogenesis." Interestingly, the role for TLR2 appears to be specific to the particular mucosal site of infection. Contrary to the results of genital infection, lung infection of TLR2 KO mice with *C. muridarum* resulted in severe disease compared to WT mice (He et al., 2011), suggesting a tissue-specific role for TLR2 in disease outcome. The exact mechanism leading to the increased TLR2-mediated pathology in lungs but not in the genital tract is unclear. Nevertheless, these findings emphasize the versatility of the innate immune system in being able to mount a different response to the same pathogen depending on the anatomical site.

Independent of its recognition at mucosal surfaces, TLR2 recognition by *C. pneumoniae* in macrophages has also been shown to play a role in foam cell formation, which mediates the development of atherosclerotic lesions (Cao et al., 2007), supporting an inflammatory role for TLR2-mediated recognition. Furthermore, the ability of *C. pneumoniae* to increase cellular ATP content in mouse macrophages was found to be dependent on TLR2 expression (Yaraei et al., 2005).

Recognition of chlamydiae may also be mediated through TLR4. The AR39 strain of *C. pneumoniae* was shown to stimulate human vascular smooth muscle cell proliferation at least in part through TLR4, which was confirmed using diphosphoryl lipid A, a competitive TLR4 antagonist (Sasu et al., 2001). It is important to note that TLR2 was not expressed in these cells. These data would suggest that in the absence of TLR2 signaling, TLR4 could play a compensatory role by recognizing specific chlamydial MAMPs. We have observed that, during in vitro infection with *C. muridarum*, TLR4 KO macrophages produced higher levels of inflammatory cytokines, suggesting some degree of recognition through this receptor and a possible regulatory role. However, TLR4 KO mice do not respond any differently from WT mice to *C. muridarum* genital infection with respect to the course of the infection and pathology (Darville et al., 2003), suggesting that TLR4 does not play a major role during in vivo genital infection. It is noteworthy that all in vitro studies discussed above relate to infected epithelial cells or macrophages. During in vivo infection, the role of these receptors on uninfected inflammatory cells also comes into play, determining the overall outcome.

Overall, in the context of recognition and disease outcome, TLR2 stands out as the major innate receptor influencing *Chlamydia*-mediated immunopathology. Figure 1 summarizes recognition of chlamydiae by extracellular and intracellular TLRs and the resulting signaling cascade that leads to proinflammatory cytokine induction.

CHLAMYDIAL MAMPs FOR TLR2 AND TLR4

The identity of the chlamydial MAMPs that are recognized by TLR2 and TLR4 has been the subject of intense investigation. Gram-negative bacteria, especially commensals and pathogens such as *Salmonella* spp. and *Yersinia* spp., produce LPS with hexacylated lipid A, which is well recognized by TLR4 on the cell surface, leading to induction of a potent IFN-β response via the TRIF (Toll-IL-1 receptor domain-containing adaptor inducing IFN-β) pathway (Frendéus et al., 2001). Likewise, purified *E. coli* LPS is a potent stimulator for this pathway and is routinely used as a positive control for TLR4 activation (Tapping et al., 2000). Therefore, at the outset, chlamydial LPS was the obvious ligand for TLR4. However, *Chlamydia* spp., like *Legionella* spp. and *Francisella* spp., produce weakly stimulatory LPS (Ingalls et al., 1995), with one report even describing chlamydial LPS recognition by TLR2 (Erridge et al., 2004). Instead, recombinant Hsp60 from *C. pneumoniae* has been shown to act via TLR4 in a MyD88-dependent manner to induce TNF-α (Bulut et al., 2002), suggesting chlamydial Hsp60 as a major candidate for TLR4 stimulation.

The strongest evidence for a chlamydial TLR2 ligand comes from studies with plasmid-cured

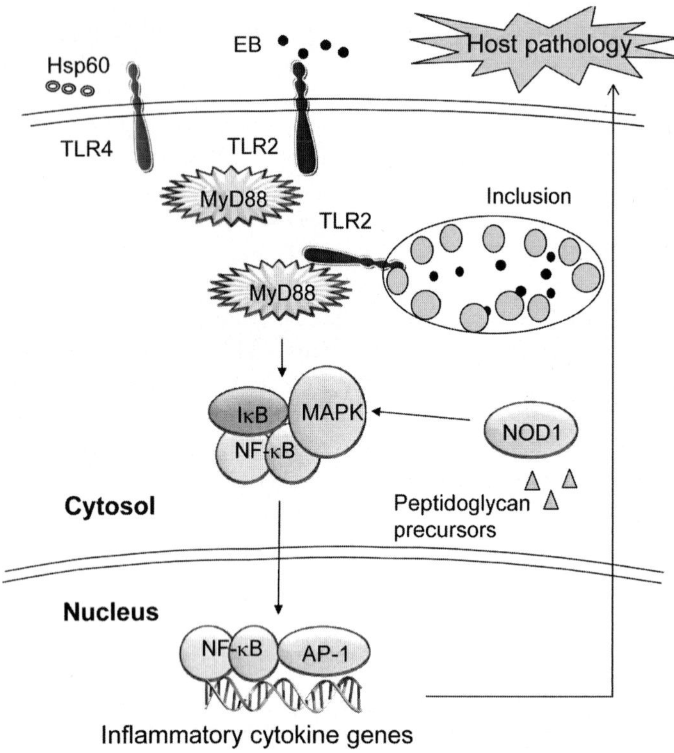

FIGURE 1 Schematic representation of the recognition of chlamydiae by TLR and NOD signaling pathways. TLR2 is the major TLR implicated in chlamydial recognition. TLR2 is normally expressed on the cell surface, but it has also been shown to localize in the vicinity of the chlamydial inclusion membrane. The specific chlamydial ligand recognized by TLR2 is not yet known. TLR4-mediated chlamydial recognition has been observed in the absence of TLR2 expression, and chlamydial Hsp60 is recognized as a TLR4 ligand. MyD88 is the common adaptor molecule for most TLRs and is necessary for TLR2/4-mediated activation of NF-kB and MAPK. This activation leads to expression of proinflammatory cytokine genes, which have been implicated in oviduct pathology during chlamydial genital infection. NOD1, an intracellular cytosolic receptor, also contributes to NF-kB activation by recognizing chlamydial peptidoglycan precursors.
doi:10.1128/9781555817329.ch10.f1

strains of *C. muridarum* (O'Connell et al., 2007) and *C. trachomatis* (O'Connell et al., 2011) that fail to induce cytokine stimulation in a TLR2-dependent manner in HEK cells transfected with TLR2. During in vivo genital infection of mice, the plasmid-deficient *C. muridarum* strain fails to induce oviduct pathology despite normal infection clearance, replicating the TLR2 KO mouse phenotype, as discussed in detail in chapter 11. However, it is not clear if the TLR2-ligand is encoded by the plasmid or from a chromosomal locus that is regulated by the plasmid. Initial microarray analysis has not provided clues, and further proteomic and lipidomic approaches may help to determine the TLR2-specific chlamydial ligand (O'Connell et al., 2011). In addition to the plasmid-coded or -regulated TLR2 ligand, a chlamydial lipoprotein called macrophage infectivity potentiator (MIP) has been shown to induce inflammatory cytokines through TLR2/TLR6 and CD14 (Bas et al., 2008). These studies involved the use of recombinant MIP because mutant chlamydial strains lacking MIP are unavailable. Therefore, the role for MIP in the context of infection needs further exploration.

As for any receptor-ligand interaction, it would be expected that the hypothetical chlamydial ligand/effector will directly interact with TLR2, leading to the question of where this interaction takes place. Using confocal microscopy, overexpressed TLR2 was shown to colocalize with the inclusion membrane (O'Connell et al., 2006). The location of TLR2 appears to be specific since TLR4 did not colocalize to the inclusion. Further, the downstream adaptor molecule MyD88 was found to colocalize with TLR2, which suggests that TLR2 localization at the inclusion membrane is functionally relevant. These data need to be validated further for endogenous TLR2 and MyD88. However, the exact location of TLR2 needs to be settled since the endoplasmic reticulum (ER) is in close proximity to the inclusion, and ER transmembrane proteins such as Sec61α have been observed in close proximity to the inclusion (Prantner et al., 2010). Confocal microscopy cannot distinguish between membranes that are adjacent, and higher-resolution imaging by electron microscopy would be helpful to distinguish the inclusion membrane from the surrounding ER.

OTHER TLRS

Inflammatory cytokine induction in TLR2-deficient macrophages was significantly reduced but not abolished in *C. muridarum*-infected macrophages (Darville et al., 2003). Similar results were also obtained from macrophages deficient for both TLR2 and TLR4 (U. M. Nagarajan and T. Darville, unpublished observation). However, macrophages deficient for MyD88, the central adaptor molecule for all the TLRs with the exception of TLR3, showed minimal TNF-α and IL-6 responses following in vitro and in vivo *C. muridarum* infection (Nagarajan et al., 2005, 2011). Similar results were also reported for *C. pneumoniae* during in vitro and in vivo infection (Naiki et al., 2005). These findings suggest the contribution of another TLR(s) and a cooperative role for TLRs in cytokine induction during infection. However, MyD88 has additional roles besides TLR signaling, which include its adaptor role in IL-1R signaling and its effect downstream of IFN-γR signaling (reviewed by Reiling et al. [2008]). Thus, the absence of cytokine induction in MyD88 KO cells and mice may not be entirely due to disruption of the TLR pathway signaling. Ideally, cells deficient in multiple TLRs should be tested to define the cooperative role for TLRs, since cells and mice deficient for individual TLRs such as TLR3, TLR7, and TLR9 show minimal changes in inflammatory cytokine levels (Nagarajan and Darville, unpublished). Among intracellular TLRs, TLR7 and TLR9 form a subgroup due to their exclusive intracellular localization in the ER and recognition of MAMPs (single-stranded RNA [ssRNA] and CpG DNA, respectively) in endosomal/lysosomal compartments (Heil et al., 2003; Latz et al., 2004). Since TLR9 does not contribute significantly to the cytokine response during *C. muridarum* infection (Nagarajan and Darville, unpublished), it is likely that chlamydial CpG DNA does not serve as a ligand for TLR9 stimulation because it is nonstimulatory or inaccessible.

Besides TLR2 and TLR4, other TLRs have not been directly demonstrated to contribute to the inflammatory response during chlamydial infection. It is likely that TLR1 and TLR6, which cooperate with TLR2 by dimerization, could contribute to TLR2-mediated recognition. More recently, TLR3 was shown to contribute to the IFN-β response in *Chlamydia*-infected oviduct epithelial cells (Derbigny et al., 2011). TLR3 is located intracellularly in the endosomal compartment in human dendritic cells (DCs) but is also expressed on the cell surface in fibroblasts (Matsumoto et al., 2003). TLR3 binds viral double-stranded RNA (dsRNA) to induce an IFN response mediated through the adaptor molecule TRIF (Toshchakov et al., 2002; Hoebe et al., 2003; Yamamoto et al., 2003; Tabeta et al., 2004). This contribution by TLR3 and TRIF to the IFN-β response was specific for mouse oviduct epithelial cells and did not occur in other cell types (Prantner et al., 2010; Derbigny et al., 2011). A direct interaction between TLR3 and the chlamydial inclusion was not examined. Therefore, it remains to be determined where this TLR3-mediated recognition is taking place and whether the chlamydial TLR3 ligand is dsRNA, as it is for viruses (Tabeta et al., 2004). Previous findings implicate the TLR-independent pathway in an IFN-β response in mouse oviduct epithelial cells (Prantner et al., 2010). It is possible that there is cross talk between the TLR3 and TLR-independent pathways or that they function in an independent but additive manner in inducing an IFN-β response. Nevertheless, it is important to note that both TLR3 KO and TRIF KO mice demonstrate no alterations in infection clearance or genital pathology during in vivo genital infection with *C. muridarum* (U. M. Nagarajan, unpublished observation), suggesting that recognition and signaling from these receptors are dispensable.

NOD Proteins

The nucleotide oligomerization domain (NOD) proteins are a group of proteins that fulfill the role of intracellular sensing of pathogens to induce an inflammatory response. The NOD-like receptor (NLR) family consists of >20 proteins that contain an LRR domain, a nucleotide-oligomerization domain, and a caspase recruitment domain (CARD). More recently, these proteins have been classified into two groups: NLRC proteins (CARD) and NLRPs (Pyrin domain, PYD).

NLRC PROTEINS

NOD1 and NOD2 are the two well-characterized members of this family. The NOD proteins are cytosolic proteins that sense bacterial peptidoglycan-derived components, iE-DAP (γ-D-glutamyl-meso-diaminopimelic acid) (Chamaillard et al., 2003) and muramyl dipeptide (Girardin et al., 2003), respectively. Muramic acid, the major component of peptidoglycan (PGN), has not been detected in *Chlamydia* spp. (Fox et al., 1990), but the genes encoding the PGN precursors are present and several indirect lines of evidence point to the presence of PGN precursors in the *Chlamydiaceae* family (McCoy and Maurelli, 2005; McCoy et al., 2006; Patin et al., 2009).

Consistent with the presence of PGN, NOD1-deficient fibroblasts had much reduced NF-κB activation following chlamydial infection compared to WT fibroblasts. Similar results were also obtained with HeLa cells when NOD1 expression was depleted by small interfering RNA (siRNA), suggesting that precursors of PGN in chlamydiae can stimulate this pathway (Welter-Stahl et al., 2006). When recognizing bacterial ligands, NOD1 and NOD2 utilize the adaptor protein Rip2 to activate NF-κB and mitogen-activated protein kinase (MAPK) (Chin et al., 2002) via TRAF2 and TRAF5 (Hasegawa et al., 2008). This contribution to NF-κB activation has been shown to be required for IL-8 induction (Buchholz and Stephens, 2008) and for optimal IFN-β induction (Prantner et al., 2010). NOD1 and NOD2 also play a predominant role in recognizing *C. pneumoniae* in endothelial cells (Opitz et al., 2005). Again, IL-8 expression was severely impaired following siRNA-mediated depletion of NOD1. An important difference between recognition of chlamydiae by TLR2 and NOD1 is that chlamydial entry was necessary for NOD1-mediated NF-κB activation while TLR2 could sense extracellular heat-killed bacteria. Taken together, it can be concluded that NOD1 plays a contributory role in the induction of inflammatory cytokines and IFN-β during a chlamydial infection. Unexpectedly, it has been determined recently that NOD2 can also recognize structurally unrelated viral ssRNA (Sabbah et al., 2009). Upon recognition of ssRNA, NOD2 can signal via viral adaptor molecule mitochondrial antiviral signaling protein (MAVS) to induce IFN-β. The role of NOD2 in recognition of chlamydiae has not been explored, but such a role is less likely because MAVS does not contribute to the IFN-β response during *C. muridarum* infection (Prantner et al., 2010).

Contrary to the results from NOD1-deficient cells, NOD1 KO mice showed no difference in cytokine induction during *C. muridarum* genital infection, with infection course and pathology similar to those of WT mice (Welter-Stahl et al., 2006). These data suggest likely compensation by the TLR pathway or that NOD1-mediated recognition plays only a contributory role.

NLRPS

The vast majority of other characterized NLRs, which have recently been called NLRPs, are engaged in activation of caspase-1 (reviewed by Ting et al. [2008]). Before discussing the role of specific NLRPs during chlamydial infection, it is important to understand the overall outcome of caspase-1 activation. Caspase-1 is the central component of the host large multiprotein complex called the "inflammasome" (reviewed by Martinon et al. [2002]) and exists as a zymogen, procaspase-1 with a long prodomain containing the CARD domain. The CARD domain is essential for homotypic protein-protein interactions that lead to the formation of the 700-kDa multiprotein oligomeric inflammasome (Martinon et al., 2002), which mediates autocatalytic cleavage of procaspase-1 to its active form (Thornberry et al., 1992). Caspase-1 has many substrates in the cell (Shao et al., 2007), but its two main targets relevant to inflammation are pro-IL-1β (Cerretti et al., 1994) and pro-IL-18 (Ghayur et al., 1997). During chlamydial lung infection, IL-18 has been shown to play a role in stimulating IFN-γ production from T cells (Robinson et al., 1997; Fantuzzi et al., 1999) and natural killer cells (Chaix et al., 2008), although its contribution during genital infection has not been tested. In contrast, IL-1β has a significant role in genital infection, as it is necessary for optimal clearance of infection but is also a major player in oviduct pathology (Prantner et al., 2009). Caspase-1 KO mice also develop less oviduct pathology (Cheng et al., 2008), emphasizing the important role of this pathway in disease outcome. Surprisingly, loss of caspase-1 also leads to decreased production of cytokines like IL-1α, IL-6, and TNF-α, even though none of these are direct caspase-1 substrates (Kuida et al., 1995).

The NLRPs constitute the major cytosolic pathogen-sensing system, which operates in parallel to classical PRR pathways and culminates in the assembly of the inflammasome and activation of caspase-1. Currently, four distinct inflammasomes have been identified (Fig. 2). All four inflammasomes include procaspase-1 and the adaptor molecule ASC (apoptosis-associated speck-like protein). However, the different inflammasomes are distinguishable from each other by the specific NLRP molecule that links the particular recognition stimulus to inflammasome assembly. NLRP proteins have a generic structure consisting of an LRR pathogen recognition domain followed by a NOD and a PYD or CARD. The PYD of the NLRP facilitates its association with the adaptor ASC protein, which contains both CARD and PYD (Srinivasula et al., 2002), bridging the PYD-containing sensor NLRs to the CARD-containing caspase-1 (Fig. 2) (Martinon and Tschopp, 2004). The ASC protein also appears to increase the efficiency of inflammasome assembly by CARD-containing NLRs, possibly by outcompeting endogenous inhibitors for binding to caspase-1. The central role of the ASC protein in inflammasome assembly is supported by the fact that ASC-deficient mice have phenotypes identical to those of caspase-1 KO mice during LPS-induced toxic shock (Mariathasan et al., 2004) and *Listeria monocytogenes* infection (Ozoren et al., 2006).

The best-studied inflammasome is of the NLRP3 variety. It is activated by diverse elements such as ATP, uric acid, fungal zymosan, bacterial RNA, influenza virus, microbial toxins, amyloid beta, and UV light (Kanneganti et al., 2006a, 2006b; Mariathasan et al., 2006; Martinon et al., 2006; Feldmeyer et al., 2007; Halle et al., 2008; Lamkanfi et al., 2009). Based on this extreme diversity, it has been proposed that NLRP3 does not recognize these stimuli directly but is a general sensor of cell stress. From a clinical perspective, hyperactivity of the NLRP3 inflammasome and the resulting excessive IL-1β production are also associated with three hereditary periodic fever syndromes: Muckle-Wells syndrome, familial cold urticaria, and chronic infantile neurological cutaneous and articular syndrome (Hoffman et al., 2001; Feldmann et al., 2002). These findings emphasize the importance of tightly regulated IL-1β production, suggesting that caspase-1 activation might have arisen as an evolutionary safeguard system to prevent deleterious inflammation. A second inflammasome featuring NLRC4 is centered on the detection of flagellin delivered by bacterial type III secretion (T3S) and type IV secretion systems (Amer et al., 2006; Miao et al., 2006, 2008; Suzuki et al., 2007). NLRC4 has also been shown to recognize the rod protein of the T3S apparatus of a number of pathogens (Miao et al., 2010). The NALP1 inflammasome reconstituted in vitro can be activated by sensing the bacterial cell wall component muramyl dipeptide (Faustin et al., 2007).

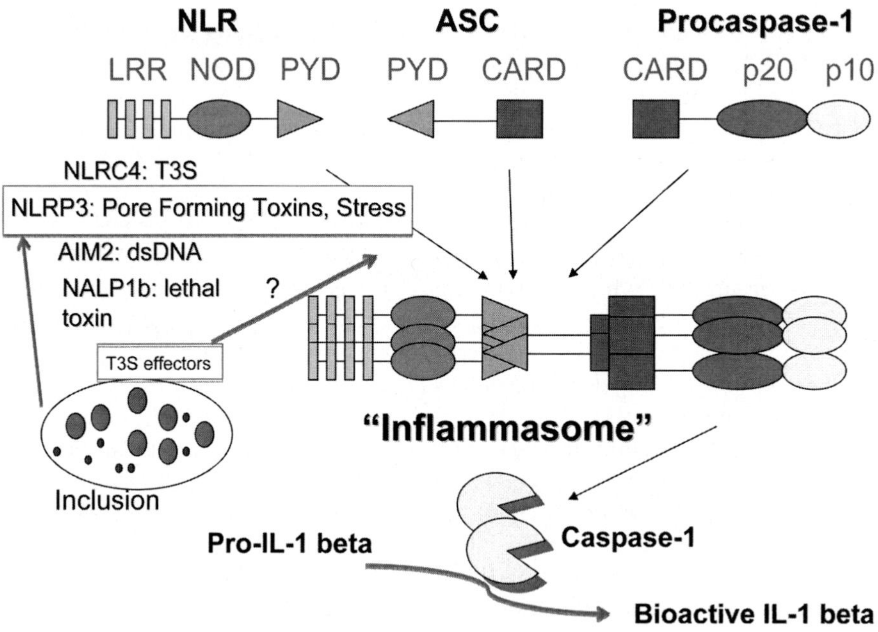

FIGURE 2 Schematic representation of inflammasome activation and caspase-1 activation during chlamydial infection. Cleavage of procaspase-1 to form active caspase-1 requires formation of the inflammasome. The inflammasome is a complex resulting from recognition of PRR or a danger signal by the NLRP proteins, the adaptor ASC and the procaspase-1. ASC is essential for the interaction of the pyrin (PYD) domain of NLRP with the CARD domain of procaspase-1. Of the NLRPs, NLRP3 recognizes ligands such as pore-forming toxins and danger signals like ATP, uric acid, or mitochondrial damage. NLRC4 recognizes the T3S apparatus and flagellin, AIM2 recognizes dsDNA, and NALP1b recognizes toxins such as Bacillus anthracis lethal toxin. During chlamydial infection, NLRP3 is involved in caspase-1 activation, but multiple NLRPs that are yet to be identified are likely activated. The chlamydial T3S apparatus contributes to caspase-1 activation, but the specific effectors and NLRP involved are not known. NLRC4 is not activated during chlamydial infection. doi:10.1128/9781555817329.ch10.f2

However, the NALP1 inflammasome has been mostly linked with activation of caspase-1 following exposure to *Bacillus anthracis* lethal toxin (Boyden and Dietrich, 2006). The most recently discovered inflammasome consists of the NLR protein AIM2, which is able to sense dsDNA (Hornung et al., 2009) that could be derived from the pathogen or host (Fig. 2).

CHLAMYDIA-INDUCED INFLAMMASOME ACTIVATION

IL-1 and caspase-1 are major players in host pathology, as evidenced from fallopian tube organ culture studies (Hvid et al., 2007) and decreased pathology in mice deficient for IL-1β (Prantner et al., 2009) and caspase-1 (Cheng et al., 2008). The mechanism of caspase-1 activation during *Chlamydia* infection is an active area of investigation. One of the earliest studies on caspase-1 activation during chlamydial infection was carried out by Brunham and colleagues, which showed that several biovars of *Chlamydia* spp. induced IL-18 secretion in epithelial cells (Lu et al., 2000). IL-18 secretion requires cleavage of pro-IL-18 by caspase-1, which was demonstrated in *Chlamydia*-infected HeLa cells. Importantly, activation of caspase-1 did not require host protein synthesis but required bacterial protein synthesis, suggesting that the process is dependent on one or more chlamydial factors. Subsequent studies in HeLa cells infected with *C. trachomatis* confirmed these findings, and siRNA knockdown studies demonstrated a role for the NLRP3 inflammasome in caspase-1 activation (Abdul-Sater et al., 2009).

Although *C. muridarum* infection is restricted to epithelial cells during in vivo genital infection in mice, it was shown that macrophages and neutrophils, rather than epithelial cells, are the major producers of IL-1β (Prantner et al., 2009). The behavior of macrophages during in vitro chlamydial infection varies significantly depending on the biovar used. *C. pneumoniae* infection results in high levels of IL-1β induction in resting macrophages, a significant portion of which is NLRP3 and ASC dependent (He et al., 2010). Conversely, resting macrophages respond poorly to *C. muridarum* infection, producing low levels of IL-1β, which is caspase-1 dependent (Cheng et al., 2008; Prantner et al., 2009). Caspase-1 activation in resting macrophages is also NLRP3 and ASC dependent (Nagarajan et al., 2012). Further, macrophages prestimulated in vitro with TLR ligands respond to chlamydial infection by producing high levels of IL-1β, as observed during in vivo genital infection (Prantner et al., 2009). These findings suggest that macrophages that respond to in vivo infection are distinct from macrophages infected in vitro. One possibility is that during in vivo infection, prestimulation of macrophages by chlamydial TLR ligands occurs and is necessary to obtain high levels of mature IL-1β. Unlike previous observations in HeLa cells (Lu et al., 2000), caspase-1 activation in prestimulated macrophages did not require bacterial growth but required *C. muridarum* viability (Prantner et al., 2009). The requirement for viable chlamydiae suggests that a chlamydial effector is required for host caspase-1 activation. Furthermore, caspase-1 activation in prestimulated macrophages does not require NLRP3 and NLRC4 but requires ASC protein, suggesting that NALP1, AIM2, or an unidentified NLRP is involved in caspase-1 activation in prestimulated macrophages (Nagarajan et al., 2012). The requirement for viable bacteria but not bacterial growth for caspase-1 activation in activated macrophages suggests that elucidating the signaling cascade in uninfected immune cells exposed to chlamydiae is critical to understand chlamydial pathogenesis.

There is no question that IL-1 signaling is detrimental to the host in both the *C. pneumoniae* lung infection model (He et al., 2010) and the *C. muridarum* genital tract model (Prantner et al., 2009). Further, caspase-1 KO mice showed reduced oviduct pathology but no alteration in infection course in the *C. muridarum* genital infection model (Cheng et al., 2008). However, recent studies from our laboratory using inflammasome KO mice indicate that inflammasome components may not be critical for IL-1β secretion during in vivo genital infection. NLRP3 and ASC-deficient mice had significantly lower levels of IL-18, but not IL-1β, than did control mice. Inflammasome-independent IL-1β secretion has been previously reported in neutrophils, where an activated serine protease could serve the same function as caspase-1 (Greten et al., 2007). This putative protease could also compensate for lack of caspase-1 activation in inflammatory cells during in vivo infection. Further, ASC KO mice have a delayed infection course compared to WT mice but no significant reduction in oviduct pathology (Nagarajan et al., 2012). These data suggest an additional function for ASC protein besides inflammasome activation.

CHLAMYDIAL EFFECTORS FOR CASPASE-1

Several studies have shown that caspase-1 activation during bacterial infection involves active contribution by the bacteria. In particular, it was demonstrated that a functional T3S apparatus is required for caspase-1 activation during infection with *Salmonella* spp. (Miao et al., 2006), *Yersinia* spp. (Daffis et al., 2007), *Shigella* spp. (Suzuki et al., 2007), *Pseudomonas* spp. (Sutterwala et al., 2007;

Suzuki et al., 2007; Galle et al., 2008; Miao et al., 2008), and *Burkholderia* spp. (Kagami et al., 2005). In certain instances, this activation has been shown to be due to a particular protein, such as *Yersinia pestis* YopJ (Lilo et al., 2008), *Pseudomonas aeruginosa* flagellin (Miao et al., 2008), or *Salmonella enterica* flagellin (Franchi et al., 2006). While these bacterial proteins are secreted via T3S, only YopJ would be considered a classic T3S effector protein. Separate studies have demonstrated that the T3S translocator proteins IpaB of *Shigella* (Chen et al., 1996; Thirumalai et al., 1997; Hilbi et al., 1998) and SipB of *Salmonella* (Hersh et al., 1999; Dreher et al., 2002) colocalize with caspase-1 and are necessary and sufficient for its activation. The NLRC4 (IPAF) inflammasome has been proposed as a general T3S sensor that induces caspase-1 activation, although it is unclear how it senses the T3S effectors or translocators. While the T3S apparatus and T3S translocator proteins are relatively conserved, T3S effectors are more specific to the needs of the individual pathogen and are not conserved. Therefore, it is unlikely that *Chlamydia* spp. would possess a structural homologue to YopJ, though one cannot rule out the possibility of a functional homologue. Additionally, *Chlamydia* spp. are nonmotile and do not express flagellin. By analogy to *Yersinia enterocolitica*, in which caspase-1 activation is dependent on the T3S translocator proteins, YopB and YopD, the chlamydial T3S translocators, CopB and CopD, are predicted to be important for *Chlamydia*-dependent caspase-1 activation via NLRC4. However, NLRC4 KO cells can activate caspase-1 effectively during chlamydial infection, suggesting alternative effectors or pathways for *Chlamydia*.

Additional non-T3S bacterial proteins have been also linked to caspase-1 activation. This list features many toxins including *L. monocytogenes* listeriolysin O, *Streptococcus pyogenes* streptolysin O, *Staphylococcus aureus* alpha-hemolysin, and *Bacillus anthracis* lethal toxin. The first three toxins uniformly activate the NLRP3 inflammasome (Meixenberger et al., 2010), while lethal toxin activates the NLRP1 inflammasome (Terra et al., 2010). Additionally, the NLRP1 inflammasome can be activated in conjunction with NOD2 in response to the bacterial cell wall component muramyl dipeptide. Considering the role of NLRP3 in *Chlamydia*-induced inflammasome activation, it is possible that non-T3S chlamydial effectors or PGN precursors could be used by *Chlamydia* spp. to induce caspase-1 activation.

Overall, caspase-1 activation during chlamydial infection appears to require both bacterial and host factors. The requirement of viable chlamydiae for caspase-1 activation demonstrates that a chlamydial factor is necessary (Lu et al., 2000; Prantner et al., 2009). A recent study also shows that an inhibitor of CPAF (chlamydial protease/proteasome-like factor) leads to caspase-1-dependent death of infected cells (Jorgensen et al., 2011). These data indicate that chlamydial growth and CPAF production actively inhibit caspase-1 dependent cell death and also likely inhibit inflammasome activation in the infected cell. This interesting possibility may not apply to inflammatory cells, which are the major cytokine producers that are recruited to the site of infection, since they activate caspase-1 in response to infected epithelial cells. Two independent studies (Wolf et al., 2006; Prantner and Nagarajan, 2009) point to the role of the T3S apparatus in caspase-1 activation and mature IL-1β production in resting macrophages. However, these studies relied on inhibitors of the T3S apparatus, which could have had other effects on the infected cell. Further studies with mutant chlamydiae that fail to specifically induce caspase-1 activation would be ideal to discern the role of the T3S apparatus. Since bacterial growth is not essential for caspase-1 activation in prestimulated immune cells, it could be speculated that a host-derived factor is also involved in driving activation of caspase-1 after it has been triggered by a chlamydial effector(s). This factor could be host stress, mitochondrial dysfunction, or extracellular nucleotides, all of which are NLRP3-dependent factors (reviewed by Vance [2010]) or NLRP3-independent factors.

Nucleic Acids and Nucleotide Sensors

In addition to the inflammatory pathways discussed above, innate immune cells and epithelial cells possess a parallel set of PRRs, called the Rig-like receptors (RLRs), and DNA sensors that reside in the cytosol. These receptors recognize MAMPs such as viral dsRNA, ssRNA, and dsDNA and induce inflammatory cytokine and a type I IFN response. Many of these MAMPs are identical or analogous to the structures recognized by TLRs, suggesting that RLRs, like NLRs, might have evolved to complement TLR surveillance by sensing MAMPs in locations inaccessible to TLRs. It has also been proposed that RLR/NLRs may be needed for redundancy signaling when TLRs become refractory to their ligands after initial MAMP exposure (Kim et al., 2008). The evolution of this pathway appears to be driven by viruses as the host cell incorporates multiple means of inducing IFN-β to limit viral replication. Nevertheless, MAMPs from intracellular bacteria are also recognized by these receptors.

Chlamydia infection induces a type I IFN response, with IFN-β as the predominant cytokine, suggesting recognition of chlamydial MAMP by nucleic acid sensors (Yao et al., 2001; Johnson, 2004; Nagarajan et al., 2005). However, the outcome of this recognition is not beneficial to the host as observed by improved outcome of disease during *C. muridarum* genital and lung infection in mice deficient for type I IFN signaling (Nagarajan et al., 2008; Qiu et al., 2008). These data would suggest that chlamydiae manipulate this viral defense pathway to their benefit. In this section, the discussion will be focused mainly on the recognition of *Chlamydia* spp. by this pathway that leads to induction of IFN-β.

RLRs

RLRs consist of the RNA helicases Rig-I and MDA5. These helicases have some overlapping specificity, since both are able to respond to dsRNA (Kato et al., 2008), yet they have distinct roles in sensing actual viruses (Kato et al., 2006). Uncapped ssRNA can also be sensed by Rig-I (Hornung et al., 2006), suggesting that this protein might be involved in sensing bacterial RNA in addition to viral RNA. Similar to TLRs, RLRs also exhibit signaling convergence with Rig-I and MDA-5 using the adaptor protein MAVS (Seth et al., 2005) linked via homotypic CARD domain interactions.

Initial studies showed that *C. muridarum*-induced IFN-β is independent of TLR2 and TLR4, and some contribution from the MyD88 pathway was suggested (Nagarajan et al., 2005). However, when TLR4-MyD88 double-KO macrophages were generated, the contribution from the MyD88 pathway disappeared, suggesting a predominant role for intracellular receptors (Prantner et al., 2010). Further, when the contribution of RLRs was tested during *C. muridarum* infection in HeLa cells, recognition through these receptors was not observed (Prantner et al., 2010). In contrast, during *C. pneumoniae* infection in HUVEC cells, signaling through MAVS was found to be essential for IRF3 activation (Buss et al., 2010). MAVS associates with TRAF3, leading to activation of IRF transcription factors and IFN-β expression (Saha et al., 2006). In addition to TRAF3, a host protein called STING is downstream of the MAVS signaling cascade (Zhong et al., 2008). STING is a unique ER transmembrane protein, which is now considered a central molecule through which all DNA/RNA sensors function. In our studies, STING was found to be essential for IFN-β induction during *C. muridarum* infection in both mouse and human epithelial cells (Prantner et al., 2010). STING was found to localize in close proximity to the inclusion (Prantner et al., 2010). Independent of these findings, it was also reported that TLR3 contributes to IFN-β induction in mouse oviduct epithelial cells during *C. muridarum* infection (Derbigny et al., 2011). Taken together, these data are contradictory, suggesting the use of different host receptors for IFN-β induction during chlamydial infection. These observed differences could be due to (i) the overall complexity of pathways induced for IFN-β production and their ability to compensate for each other, (ii) differences between *C. muridarum* and *C. pneumoniae* infections, (iii) use of different pathways in different cell types, and/or (iv) infection dose. The use of multiple receptors to induce the expression of the same cytokine may not be unique to *Chlamydia* spp., as multiple receptors have been suggested to play a role in IFN-β induction during *L. monocytogenes* infection (Soulat et al., 2006; Crimmins et al., 2008; Ishikawa et al., 2009; Yang et al., 2010).

DNA SENSORS

The ambiguity regarding RLRs and NLRs underscores the point that the field of cytosolic receptors is still in relative infancy with new pathways still being delineated. For example, delivery of either dsDNA (Stetson and Medzhitov, 2006) or the bacterial regulator cyclic di-GMP (McWhirter et al., 2009) or cyclic di-AMP into the cytosol triggers PRR pathways leading to IFN-β expression. The details of these pathways are being investigated, but the host protein STING plays a crucial role in dsDNA recognition (Ishikawa and Barber, 2008). It has been suggested that there are multiple cytosolic DNA sensors, with DAIs (DNA-dependent activator of IFN-regulatory factors) and RNA polymerase III being the first two to be characterized (Takaoka et al., 2007; Ablasser et al., 2009). This characterization has been followed by the discovery of IFI16/IFI204, which is a major DNA-sensor that recognizes dsDNA of variable length in human and mouse cells (Unterholzner et al., 2010). A more recently discovered intracellular DNA sensor is LRRFIP1, a unique sensor that initiates IFN-β induction through the β-catenin pathway (Yang et al., 2010). Additionally, since STING is downstream of MAVS, its involvement in dsDNA sensing suggests that it may be the major adaptor protein for many cytosolic pathways, analogous to the role of MyD88 for the TLRs.

The study of DNA sensors during chlamydial infection has just begun, and data from our laboratory do not support a role for DAI, RNA polymerase III, or IFI16 in IFN-β induction in mouse epithelial cells and HeLa cells (L. Yeruva, D. Prantner, and U. M. Nagarajan, unpublished observation). Additional data indicate that cyclic dinucleotides generated by chlamydiae could be inducing IFN-β response. Cyclic-di GMP and cyclic di-AMP are bacterial second messengers that can induce a type I IFN response through an unknown receptor (McWhirter et al., 2009). Using mutants that overexpressed multidrug transporters and hyperinduced IFN-β, Woodward et al. showed that secreted cyclic-di-AMP made by *L. monocytogenes* plays a major role in inducing IFN-β (Woodward et al., 2010). The specific sensor/receptor for this cyclic dinucleotide has not yet been identified, although this pathway functions through the downstream adaptor molecule STING. We have observed that macrophages expressing bacterial phosphodiesterase, which cleaves cyclic-di-GMP, induce a significantly lower IFN-β response during *C. muridarum* infection than control macrophages expressing the vector alone (Yeruva, Prantner, and Nagarajan, unpublished). These data suggest that either cyclic-di-GMP or cyclic-di-AMP is produced during infection to induce IFN-β. Further experiments are needed to confirm the specific dinucleotide inducing the response and identification of the receptor that binds to the cyclic di-nucleotides. The receptors and pathways involved in IFN-β induction during chlamydial infection are summarized in Fig. 3.

A role for DNA/RNA sensors in recognizing intracellular chlamydiae raises the question of where the chlamydial DNA/RNA MAMPs would be detected, since these sensors typically recognize viral DNA/RNA molecules in the host cytosol. All *Chlamydia* spp. possess a T3S apparatus that appears to be essential for IFN-β-induced CXCL10 expression during *C. muridarum* infection of macrophages, based on studies using T3S inhibitors (Prantner and Nagarajan, 2009). Secretion of DNA through type IV secretion system has been reported (reviewed by Christie [2001]), but there are no reports of DNA secreted via bacterial T3S. Further, the T3S inhibitors also blocked induction of other cytokines such as IL-6. Therefore, it is not clear if the inhibitors specifically block delivery of ligands through T3S or alter the T3S apparatus to affect inclusion membrane permeability and prevent leakage of bacterial products. An alternative possibility is that the DNA detected is host mitochondrial DNA that was released following damage to mitochondria in *Chlamydia*-infected cells. This argument is countered by the observation that *Chlamydia* spp. inhibit host apoptosis (Fan et al., 1998). However, in the environment of other innate receptor recognition and production of cytokines like TNF-α, it is likely that mitochondrial damage occurs during in vivo infection and may indirectly contribute to RLR/NLR/DNA sensor-mediated activation.

SIGNALING EVENTS POSTRECOGNITION

During chlamydial infection, a multitude of receptors are engaged in recognition and activation of signaling cascades, which culminate in activation of transcription factors and expression of immune response genes. In general, TLR and NOD1 signaling both lead to activation of MAPK and subsequent activation of transcription factors AP1 and NF-κB (Kawai and Akira, 2006). These transcription factors bind to response elements of genes encoding proinflammatory cytokines IL-1β (Hiscott et al., 1993) and TNF-α (Nedwin et al., 1985) and chemokines KC (Ohmori et al., 1995) and MIP-1α (Widmer et al., 1993), and they induce upregulation of these target genes. Therefore, these gene classes are universally induced when any of the TLRs detect their ligands. Different TLRs induce different expression patterns of their target genes because only some TLRs (TLR3, TLR4, and TLR7/9) can activate IRF family members, which are important transcription factors, in conjunction with NF-κB and AP-1, for expression of IFN-β (Wathelet et al., 1998) and certain chemokines such as CXCL10 (Ohmori and Hamilton, 1995). The different expression

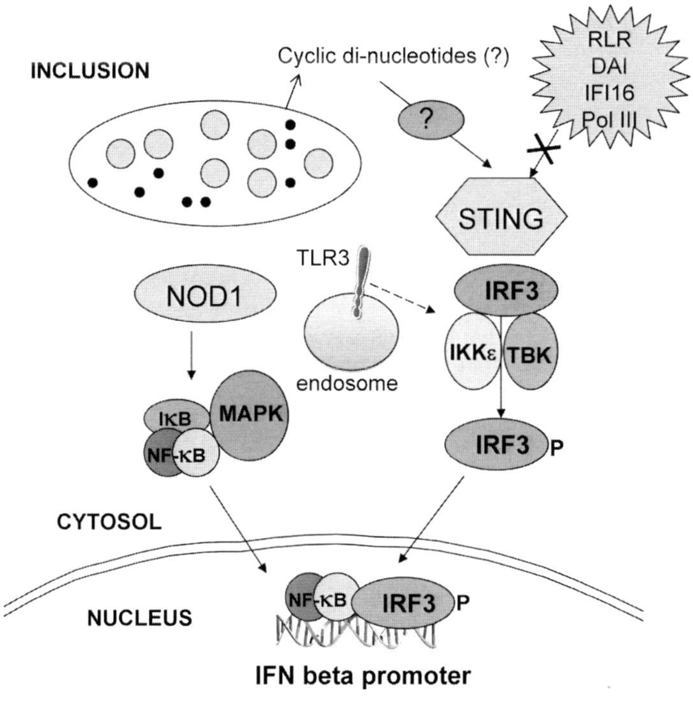

FIGURE 3 Schematic representation of receptors engaged in IFN-β induction during chlamydial infection and the resulting signaling pathway. The adaptor molecule STING is involved in chlamydially induced IFN-β induction. DNA and RNA sensors DAI, IFI16, and RLRs, which function upstream of STING, do not contribute to IFN-β induction, but detection of chlamydial dicyclic nucleotide(s) has been implicated. STING activation leads to phosphorylation of IRF3 by the kinases TBK/IKKε. Phosphorylated IRF3 translocates to the nucleus, binds the IFN-β promoter, and initiates transcription cooperatively with p65 binding. NOD1 detection and signaling lead to NF-κB activation, which contributes to IFN-β induction. TLR3- and TRIF-mediated signaling has also been shown to contribute to IFN-β induction in mouse oviduct epithelial cells (dashed line/arrow).
doi:10.1128/9781555817329.ch10.f3

patterns are determined by the differential recruitment of TRAF3, which is required for activation of IRF factors (Hacker et al., 2006) following lysine-63 ubiquitination (Tseng et al., 2010), and TRAF6, which targets NF-κB and AP-1 (Deng et al., 2000). IRF3 activation involves its phosphorylation by kinases TBK (Tank-binding kinase) or IKKε (IκB kinase ε), followed by its nuclear translocation to activate IFN-β and IFN response genes (Fitzgerald et al., 2003).

Chlamydial infection of epithelial cells leads to an overall inflammatory response that involves production of inflammatory cytokines like TNF-α and IL-1β, which have been implicated in host pathology (see chapter 11). The expression of these genes involves TLR2 activation and activation of NF-κB. Human endocervical cells express TLR2 (Herbst-Kralovetz et al., 2008), suggesting that TLR2-mediated signaling occurs during endocervical infection in humans. A number of reports demonstrate that there is activation of NF-κB and of the MAPK pathway during chlamydial infection (Buchholz and Stephens, 2007; Rodriguez et al., 2008). Likewise, activation of IRF3 has been observed during chlamydial infection (Nagarajan et al., 2005; Derbigny et al., 2007; Trumstedt et al., 2007). However, the network of signaling pathways downstream of the recognizing receptors and each receptor's specific contribution to downstream activation of NF-κB or IRF3 are not always clear when multiple receptors are involved. NOD1 signaling and TLR2 signaling both activate NF-κB induction, but only NOD1 contributes indirectly to IFN-β induction during chlamydial infection, indicating that the two pathways are not equivalent. In a similar fashion, cells treated with NOD1 siRNA show a large reduction in IL-8 and a partial reduction in IFN-β expression, while STING siRNA affects only IFN-β but not IL-8 expression (Prantner et al., 2010). These data indicate that during chlamydial infection, STING does not significantly contribute to NF-κB activation but only activates IRF3 phosphorylation (Fig. 3). Taken together, these data suggest that there is compartmentalization of NOD1- and STING-mediated recognition and TLR2 recognition. Another example of compartmentalization of intracellular events is the activation of MAPK ERK and the calcium-dependent phospholipase cPLA(2) (Vignola et al., 2010), where the two events were uncoupled from each other, during chlamydial infection. MAPK p38 was necessary for cPLA(2) activation, which is required for type I IFN and IRG induction. On the other hand, significant cross talk between signaling pathways also occurs during infection. MyD88 signaling was shown to affect iNOS induction during *C. pneumoniae* infection through activation of NF-κB and phosphorylation of the MAPK-JNK, which leads to the translocation of c-Jun, an AP-1 transcription factor. However, phosphorylation of STAT-1 and IRF-1 was delayed in MyD88 KO cells, though these are normally considered to be MyD88-independent transcription factors (Rodriguez et al., 2008). These data suggest that there is both cross talk between multiple receptors and compartmentalization of signaling pathways leading to differential cytokine induction during chlamydial infection.

There are also reports that chlamydiae can modulate host signaling by degrading activating transcription factors. Two groups demonstrated that *C. trachomatis* infection proteolytically cleaves the NF-κB subunit p65, which has been proposed to downregulate the NF-κB-mediated inflammatory response by preventing nuclear translocation of p65 (Lad et al., 2007; Christian et al., 2010). It was first shown that *C. trachomatis* protease Tsp (CT441) cleaves human p65 but not mouse p65 (Lad et al., 2007). Subsequently another protease, CPAF, has been shown to cleave both human and mouse p65 (Christian et al., 2010). In both studies, degradation of p65 occurred late (>16 to 24 h) in the chlamydial developmental cycle, while early p65 activation (3 to 8 h postinfection) in infected cells was unaffected. We have also observed enrichment of p65 in the nucleus and enhanced NF-κB-dependent promoter activation at 3 to 8 h postinfection in cells infected with either *C. muridarum* or *C. caviae* (U. M. Nagarajan, unpublished observation). Consequently, mRNA levels for the NF-κB targets, TNF-α and IL-1β, were highest at these early times, with protein levels increasing until 24 h (Darville et al., 2003; Nagarajan et al., 2005). Our data suggest that early NF-κB activation is unaltered and sufficient to induce chemokine and cytokine expression. Further, during an in vivo genital infection using *C. muridarum*, a strong NF-κB-dependent inflammatory response has been observed as early as 3 h postinfection and increases significantly at 24 h postinfection in the cervix (Rank et al., 2010). Therefore, the biological significance of p65 degradation late in the chlamydial developmental cycle is not evident at this point. Future in vivo studies with chlamydial mutants that cannot degrade p65 may provide clues about whether cleavage of the p65 subunit of NF-κB promotes the survival of chlamydiae in the host and if it is associated with chronic infection in humans. Overall, although individual signaling studies provide important clues to host-pathogen interactions, it is important to understand their significance in primary human endocervical cells and during in vivo infection in animal models.

HOST IMMUNE RECOGNITION: SUMMARY AND FUTURE DIRECTIONS

Based on the diverse repertoire of host receptors, pathogens will generally be sensed by multiple pathways leading to activation of PRRs and caspase-1. The change in gene expression and subsequent cytokine secretion is a cumulative result of all these interactions and is crucial for the development of a protective or pathological immune response. For instance, *Pseudomonas aeruginosa* can be sensed by TLR2, TLR4, TLR5, and NLRC4, which induce an overall inflammatory host response (Travassos et al., 2004; Sutterwala et al., 2007). With the present knowledge, we can conclude that chlamydiae are consistently recognized by surface TLR2 and intracellular TLR2, while TLR4 plays a minor role in recognition. Among the intracellular receptors, NOD1 is a prominent receptor, which along with TLR2 can induce the MAPK and NF-κB activation pathway, leading to inflammatory cytokine expression. Meanwhile, chlamydial nucleic acids or second messengers are sampled by multiple receptors, which have yet to be identified. These receptors use the STING pathway to primarily activate IRF3 and induce IFN-β expression. The NLRP cytosolic sensors are also sampling chlamydial effectors, leading to caspase-1 activation, but it is likely that other inflammasomes are also engaged. Globally, it appears that cooperative recognition by TLR2, NOD, the STING pathway, and NLRP3 occurs following chlamydial infection. All these pathways have protective and/or pathological consequences. For instance, IL-1β, a cytokine resulting from TLR2- and NLRP3-mediated recognition, is necessary for an optimal Th1 response. However, IL-1β can also produce a pathological response (Prantner et al., 2009). Therefore, it is important to regulate these cytokines and the upstream receptors to achieve an optimal protective response with minimal pathology. This is particularly important for vaccine development, since a chlamydial vaccine candidate should not be a TLR2 ligand and should induce optimal levels of cyto-kines like IL-1β, to induce adaptive immunity without causing tissue damage. Chlamydial and host factors that determine the immune protective or immunopathogenic nature of the host response are described in chapter 14, "*Chlamydia* vaccine: progress and challenges."

Overall, chlamydial infection of epithelial cells results in a predominantly proinflammatory response (Derbigny et al., 2005; Buchholz and Stephens, 2006; Burian et al., 2010). Some of these responses can be linked to a TLR2-mediated response based on transcriptosome analysis (Burian et al., 2010). However, at this point it is not possible to equate the overall cytokine response during infection to signaling from all the recognition receptors discussed, due to several gaps in our knowledge. First, the downstream effects of signaling are not additive. Independent studies using knockdown of specific receptors or gene KO cells indicate the use of the receptor but do not tell us the extent of this use. Second, a number of studies described in this chapter used human or mouse cell lines or mouse macrophages but not primary cervical epithelial cells. Although cell lines and mouse macrophages provide clues about the specific use of receptors and signaling pathways, the results have to be confirmed with the cells that are infected in the natural host. Innate immunity acquires its specificity by variable expression of receptors and signaling molecules in different cell types. Therefore, it is important to study the receptors and signaling cascades implicated in chlamydial infection at the natural infection site, for each chlamydial species.

The receptors described in this chapter were all identified and characterized in the last 5 to 8 years, which demonstrates that this field is still very young. We expect additional DNA sensors will be discovered in the coming years and more will be learned about cell type-specific use of these receptors. Further, a number of receptors that function as negative regulators have been discovered recently (reviewed by Flannery and Bowie [2010] and Steevels and Meyaard [2011]), and their role during chlamydial infection has not been studied. Of foremost importance is the identification of specific chlamydial ligands that are recognized by host receptors. Possible approaches for the future include global proteomic and lipidomic analyses and the development of an experimental genetic system to create chlamydial mutants. This work has the potential to open up avenues for taming the host inflammatory response during vaccination, so that it is protective and not deleterious, and to optimize a strong memory response.

ACKNOWLEDGMENTS

Thanks go to Daniel Prantner for his contribution, Laxmi Yeruva for critical reading, and John Gregan (ACHRI) for editorial changes.

REFERENCES

Abdul-Sater, A. A., E. Koo, G. Hacker, and D. M. Ojcius. 2009. Inflammasome-dependent caspase-1 activation in cervical epithelial cells stimulates growth of the intracellular pathogen *Chlamydia trachomatis*. *J. Biol. Chem.* **284:**26789–26796.

Ablasser, A., F. Bauernfeind, G. Hartmann, E. Latz, K. A. Fitzgerald, and V. Hornung. 2009. RIG-I-dependent sensing of poly(dA:dT) through the induction of an RNA polymerase III-transcribed RNA intermediate. *Nat. Immunol.* **10:**1065–1072.

Amer, A., L. Franchi, T. D. Kanneganti, M. Body-Malapel, N. Ozoren, G. Brady, S. Meshinchi, R. Jagirdar, A. Gewirtz, S. Akira, and G. Nunez. 2006. Regulation of *Legionella* phagosome maturation and infection through flagellin and host Ipaf. *J. Biol. Chem.* **281:** 35217–35223.

Bas, S., L. Neff, M. Vuillet, U. Spenato, T. Seya, M. Matsumoto, and C. Gabay. 2008. The proinflammatory cytokine response to *Chlamydia trachomatis* elementary bodies in human macrophages is partly mediated by a lipoprotein, the macrophage infectivity potentiator, through TLR2/TLR1/TLR6 and CD14. *J. Immunol.* **180:**1158–1168.

Boyden, E. D., and W. F. Dietrich. 2006. Nalp1b controls mouse macrophage susceptibility to anthrax lethal toxin. *Nat. Genet.* **38:**240–244.

Buchholz, K. R., and R. S. Stephens. 2006. Activation of the host cell proinflammatory interleukin-8 response by *Chlamydia trachomatis*. *Cell. Microbiol.* **8:**1768–1779.

Buchholz, K. R., and R. S. Stephens. 2007. The extracellular signal-regulated kinase/mitogen-activated protein kinase pathway induces the inflammatory factor interleukin-8 following *Chlamydia trachomatis* infection. *Infect. Immun.* **75:** 5924–5929.

Buchholz, K. R., and R. S. Stephens. 2008. The cytosolic pattern recognition receptor NOD1 induces inflammatory interleukin-8 during *Chlamydia trachomatis* infection. *Infect. Immun.* **76:**3150–3155.

Bulut, Y., E. Faure, L. Thomas, H. Karahashi, K. S. Michelsen, O. Equils, S. G. Morrison, R. P. Morrison, and M. Arditi. 2002. Chlamydial heat shock protein 60 activates macrophages and endothelial cells through Toll-like receptor 4 and MD2 in a MyD88-dependent pathway. *J. Immunol.* **168:**1435–1440.

Burian, K., V. Endresz, J. Deak, Z. Kormanyos, A. Pal, D. Nelson, and D. P. Virok. 2010. Transcriptome analysis indicates an enhanced activation of adaptive and innate immunity by *Chlamydia*-infected murine epithelial cells treated with interferon gamma. *J. Infect. Dis.* **202:**1405–1414.

Buss, C., B. Opitz, A. C. Hocke, J. Lippmann, V. van Laak, S. Hippenstiel, M. Krull, N. Suttorp, and J. Eitel. 2010. Essential role of mitochondrial antiviral signaling, IFN regulatory factor (IRF)3, and IRF7 in *Chlamydophila pneumoniae*-mediated IFN-beta response and control of bacterial replication in human endothelial cells. *J. Immunol.* **184:**3072–3078.

Cao, F., A. Castrillo, P. Tontonoz, F. Re, and G. I. Byrne. 2007. *Chlamydia pneumoniae*-induced macrophage foam cell formation is mediated by Toll-like receptor 2. *Infect. Immun.* **75:**753–759.

Cerretti, D. P., L. T. Hollingsworth, C. J. Kozlosky, M. B. Valentine, D. N. Shapiro, S. W. Morris, and N. Nelson. 1994. Molecular characterization of the gene for human interleukin-1 beta converting enzyme (IL1BC). *Genomics* **20:**468–473.

Chaix, J., M. S. Tessmer, K. Hoebe, N. Fuseri, B. Ryffel, M. Dalod, L. Alexopoulou, B. Beutler, L. Brossay, E. Vivier, and T. Walzer. 2008. Cutting edge: priming of NK cells by IL-18. *J. Immunol.* **181:**1627–1631.

Chamaillard, M., M. Hashimoto, Y. Horie, J. Masumoto, S. Qiu, L. Saab, Y. Ogura, A. Kawasaki, K. Fukase, S. Kusumoto, M. A. Valvano, S. J. Foster, T. W. Mak, G. Nunez, and N. Inohara. 2003. An essential role for NOD1 in host recognition of bacterial peptidoglycan containing diaminopimelic acid. *Nat. Immunol.* **4:**702–707.

Chen, J. C., J. P. Zhang, and R. S. Stephens. 1996. Structural requirements of heparin binding to *Chlamydia trachomatis*. *J. Biol. Chem.* **271:**11134–11140.

Cheng, W., P. Shivshankar, Z. Li, L. Chen, I. T. Yeh, and G. Zhong. 2008. Caspase-1 contributes to *Chlamydia trachomatis*-induced upper urogenital tract inflammatory pathologies without affecting the course of infection. *Infect. Immun.* **76:**515–522.

Chin, A. I., P. W. Dempsey, K. Bruhn, J. F. Miller, Y. Xu, and G. Cheng. 2002. Involvement of receptor-interacting protein 2 in innate and adaptive immune responses. *Nature* **416:**190–194.

Christian, J., J. Vier, S. A. Paschen, and G. Hacker. 2010. Cleavage of the NF-κB family protein p65/RelA by the chlamydial protease-like activity factor (CPAF) impairs proinflammatory signaling in cells infected with chlamydiae. *J. Biol. Chem.* **285:**41320–41327.

Christie, P. J. 2001. Type IV secretion: intercellular transfer of macromolecules by systems ancestrally related to conjugation machines. *Mol. Microbiol.* **40:**294–305.

Crimmins, G. T., A. A. Herskovits, K. Rehder, K. E. Sivick, P. Lauer, T. W. Dubensky, Jr., and D. A. Portnoy. 2008. *Listeria monocytogenes* multidrug resistance transporters activate a cytosolic surveillance pathway of innate immunity. *Proc. Natl. Acad. Sci. USA* **105:**10191–10196.

Daffis, S., M. A. Samuel, B. C. Keller, M. Gale, Jr., and M. S. Diamond. 2007. Cell-specific IRF-3 responses protect against West Nile virus infection by interferon-dependent and -independent mechanisms. *PLoS Pathog.* **3:**e106.

Darville, T., J. M. O'Neill, C. W. Andrews, Jr., U. M. Nagarajan, L. Stahl, and D. M. Ojcius. 2003. Toll-like receptor-2, but not Toll-like receptor-4, is essential for development of oviduct pathology in chlamydial genital tract infection. *J. Immunol.* **171:**6187–6197.

Deng, L., C. Wang, E. Spencer, L. Yang, A. Braun, J. You, C. Slaughter, C. Pickart, and Z. J. Chen. 2000. Activation of the IκB kinase complex by TRAF6 requires a dimeric ubiquitin-conjugating enzyme complex and a unique polyubiquitin chain. *Cell* **103:**351–361.

Derbigny, W. A., S. C. Hong, M. S. Kerr, M. Temkit, and R. M. Johnson. 2007. *Chlamydia muridarum* infection elicits a beta interferon response in murine oviduct epithelial cells dependent on interferon regulatory factor 3 and TRIF. *Infect. Immun.* **75:**1280–1290.

Derbigny, W. A., R. M. Johnson, K. S. Toomey, S. Ofner, and K. Jayarapu. 2011. The *Chlamydia muridarum*-induced IFN-beta response is TLR3-dependent in murine oviduct epithelial cells. *J. Immunol.* **185:**6689–6697.

Derbigny, W. A., M. S. Kerr, and R. M. Johnson. 2005. Pattern recognition molecules activated by *Chlamydia muridarum* infection of cloned murine oviduct epithelial cell lines. *J. Immunol.* **175:**6065–6075.

Dreher, D., M. Kok, C. Obregon, S. G. Kiama, P. Gehr, and L. P. Nicod. 2002. *Salmonella* virulence factor SipB induces activation and release of IL-18 in human dendritic cells. *J. Leukoc. Biol.* **72:**743–751.

Erridge, C., A. Pridmore, A. Eley, J. Stewart, and I. R. Poxton. 2004. Lipopolysaccharides of *Bacteroides fragilis*, *Chlamydia trachomatis* and *Pseudomonas aeruginosa* signal via toll-like receptor 2. *J. Med. Microbiol.* **53:**735–740.

Fan, T., H. Lu, H. Hu, L. Shi, G. A. McClarty, D. M. Nance, A. H. Greenberg, and G. Zhong. 1998. Inhibition of apoptosis in chlamydia-infected cells: blockade of mitochondrial cytochrome c release and caspase activation. *J. Exp. Med.* **187:**487–496.

Fantuzzi, G., D. A. Reed, and C. A. Dinarello. 1999. IL-12-induced IFN-gamma is dependent on caspase-1 processing of the IL-18 precursor. *J. Clin. Investig.* **104:**761–767.

Faustin, B., L. Lartigue, J. M. Bruey, F. Luciano, E. Sergienko, B. Bailly-Maitre, N. Volkmann, D. Hanein, I. Rouiller, and J. C. Reed. 2007. Reconstituted NALP1 inflammasome reveals two-step mechanism of caspase-1 activation. *Mol. Cell* **25:**713–724.

Feldmann, J., A. M. Prieur, P. Quartier, P. Berquin, S. Certain, E. Cortis, D. Teillac-Hamel, A. Fischer, and G. de Saint Basile. 2002. Chronic infantile neurological cutaneous and articular syndrome is caused by mutations in CIAS1, a gene highly expressed in polymorphonuclear cells and chondrocytes. *Am. J. Hum. Genet.* **71:**198–203.

Feldmeyer, L., M. Keller, G. Niklaus, D. Hohl, S. Werner, and H. D. Beer. 2007. The inflammasome mediates UVB-induced activation and secretion of interleukin-1beta by keratinocytes. *Curr. Biol.* **17:**1140–1145.

Fitzgerald, K. A., S. M. McWhirter, K. L. Faia, D. C. Rowe, E. Latz, D. T. Golenbock, A. J. Coyle, S. M. Liao, and T. Maniatis. 2003. IKKepsilon and TBK1 are essential components of the IRF3 signaling pathway. *Nat. Immunol.* **4:** 491–496.

Flannery, S., and A. G. Bowie. 2010. The interleukin-1 receptor-associated kinases: critical regulators of innate immune signalling. *Biochem. Pharmacol.* **80:**1981–1991.

Fox, A., J. C. Rogers, J. Gilbart, S. Morgan, C. H. Davis, S. Knight, and P. B. Wyrick. 1990. Muramic acid is not detectable in *Chlamydia psittaci* or *Chlamydia trachomatis* by gas chromatography-mass spectrometry. *Infect. Immun.* **58:**835–837.

Franchi, L., A. Amer, M. Body-Malapel, T. D. Kanneganti, N. Ozoren, R. Jagirdar, N. Inohara, P. Vandenabeele, J. Bertin, A. Coyle, E. P. Grant, and G. Nunez. 2006. Cytosolic flagellin requires Ipaf for activation of caspase-1 and interleukin 1beta in salmonella-infected macrophages. *Nat. Immunol.* **7:**576–582.

Frendéus, B., C. Wachtler, M. Hedlund, H. Fischer, P. Samuelsson, M. Svensson, and C. Svanborg. 2001. *Escherichia coli* P fimbriae utilize the Toll-like receptor 4 pathway for cell activation. *Mol. Microbiol.* **40:**37–51.

Galle, M., P. Schotte, M. Haegman, A. Wullaert, H. J. Yang, S. Jin, and R. Beyaert. 2008. The *Pseudomonas aeruginosa* Type III secretion system plays a dual role in the regulation of caspase-1 mediated IL-1beta maturation. *J. Cell. Mol. Med.* **12:**1767–1776.

Ghayur, T., S. Banerjee, M. Hugunin, D. Butler, L. Herzog, A. Carter, L. Quintal, L. Sekut, R. Talanian, M. Paskind, W. Wong, R. Kamen, D. Tracey, and H. Allen. 1997. Caspase-1 processes IFN-gamma-inducing factor and regulates LPS-induced IFN-gamma production. *Nature* **386:**619–623.

Girardin, S. E., I. G. Boneca, J. Viala, M. Chamaillard, A. Labigne, G. Thomas, D. J. Philpott, and P. J. Sansonetti. 2003. Nod2 is a general sensor of peptidoglycan through muramyl dipeptide (MDP) detection. *J. Biol. Chem.* **278:**8869–8872.

Greten, F. R., M. C. Arkan, J. Bollrath, L. C. Hsu, J. Goode, C. Miething, S. I. Goktuna, M. Neuenhahn, J. Fierer, S. Paxian, N. Van Rooijen, Y. Xu, T. O'Cain, B. B. Jaffee, D. H. Busch, J. Duyster, R. M. Schmid, L. Eckmann, and M. Karin. 2007. NF-κB is a negative regulator of IL-1beta secretion as revealed by genetic and pharmacological inhibition of IKKbeta. *Cell* **130:**918–931.

Hacker, H., V. Redecke, B. Blagoev, I. Kratchmarova, L. C. Hsu, G. G. Wang, M. P. Kamps, E. Raz, H. Wagner, G. Hacker, M. Mann, and M. Karin. 2006. Specificity in Toll-like receptor signalling through distinct effector functions of TRAF3 and TRAF6. *Nature* **439:**204–207.

Halle, A., V. Hornung, G. C. Petzold, C. R. Stewart, B. G. Monks, T. Reinheckel, K. A. Fitzgerald, E. Latz, K. J. Moore, and D. T. Golenbock. 2008. The NALP3 inflammasome is involved in the innate immune response to amyloid-beta. *Nat. Immunol.* **9:**857–865.

Hasegawa, M., Y. Fujimoto, P. C. Lucas, H. Nakano, K. Fukase, G. Nunez, and N. Inohara. 2008. A critical role of RICK/RIP2 polyubiquitination in Nod-induced NF-kappaB activation. *EMBO J.* **27:**373–383.

He, X., S. Mekasha, N. Mavrogiorgos, K. A. Fitzgerald, E. Lien, and R. R. Ingalls. 2010. Inflammation and fibrosis during *Chlamydia pneumoniae* infection is regulated by IL-1 and the NLRP3/ASC inflammasome. *J. Immunol.* **184:**5743–5754.

He, X., A. Nair, S. Mekasha, J. Alroy, C. M. O'Connell, and R. R. Ingalls. 2011. Enhanced virulence of *Chlamydia muridarum* respiratory infections in the absence of TLR2 activation. *PLoS ONE* **6:**e20846.

Heil, F., P. Ahmad-Nejad, H. Hemmi, H. Hochrein, F. Ampenberger, T. Gellert, H. Dietrich, G. Lipford, K. Takeda, S. Akira, H. Wagner, and S. Bauer. 2003. The Toll-like receptor 7 (TLR7)-specific stimulus loxoribine uncovers a strong relationship within the TLR7, 8 and 9 subfamily. *Eur. J. Immunol.* **33:**2987–2997.

Herbst-Kralovetz, M. M., A. J. Quayle, M. Ficarra, S. Greene, W. A. Rose, R. Chesson, R. A. Spagnuolo, and R. B. Pyles. 2008. Quantification and comparison of Toll-like receptor expression and responsiveness in primary and immortalized human female lower genital tract epithelia. *Am. J. Reprod. Immunol.* **59:**212–224.

Hersh, D., D. M. Monack, M. R. Smith, N. Ghori, S. Falkow, and A. Zychlinsky. 1999. The *Salmonella* invasin SipB induces macrophage apoptosis by binding to caspase-1. *Proc. Natl. Acad. Sci. USA* **96:**2396–2401.

Hilbi, H., J. E. Moss, D. Hersh, Y. Chen, J. Arondel, S. Banerjee, R. A. Flavell, J. Yuan, P. J. Sansonetti, and A. Zychlinsky. 1998. Shigella-induced apoptosis is dependent on caspase-1 which binds to IpaB. *J. Biol. Chem.* **273:**32895–32900.

Hiscott, J., J. Marois, J. Garoufalis, M. D'Addario, A. Roulston, I. Kwan, N. Pepin, J. Lacoste, H. Nguyen, G. Bensi, et al. 1993. Characterization of a functional NF-kappa B site in the human interleukin 1 beta promoter: evidence for a positive autoregulatory loop. *Mol. Cell. Biol.* **13:**6231–6240.

Hoebe, K., X. Du, P. Georgel, E. Janssen, K. Tabeta, S. O. Kim, J. Goode, P. Lin, N. Mann, S. Mudd, K. Crozat, S. Sovath, J. Han, and B. Beutler. 2003. Identification of Lps2 as a key transducer of MyD88-independent TIR signalling. *Nature* **424:**743–748.

Hoffman, H. M., J. L. Mueller, D. H. Broide, A. A. Wanderer, and R. D. Kolodner. 2001. Mutation of a new gene encoding a putative pyrin-like protein causes familial cold autoinflammatory syndrome and Muckle-Wells syndrome. *Nat. Genet.* **29:**301–305.

Hornung, V., A. Ablasser, M. Charrel-Dennis, F. Bauernfeind, G. Horvath, D. R. Caffrey, E. Latz, and K. A. Fitzgerald. 2009. AIM2 recognizes cytosolic dsDNA and forms a caspase-1-activating inflammasome with ASC. *Nature* **458:**514–518.

Hornung, V., J. Ellegast, S. Kim, K. Brzozka, A. Jung, H. Kato, H. Poeck, S. Akira, K. K. Conzelmann, M. Schlee, S. Endres, and G. Hartmann. 2006. 5′-Triphosphate RNA is the ligand for RIG-I. *Science* **314:**994–997.

Hvid, M., A. Baczynska, B. Deleuran, J. Fedder, H. J. Knudsen, G. Christiansen, and S. Birkelund. 2007. Interleukin-1 is the initiator of Fallopian tube destruction during *Chlamydia trachomatis* infection. *Cell. Microbiol.* **9:**2795–2803.

Igietseme, J. U., G. A. Ananaba, J. Bolier, S. Bowers, T. Moore, T. Belay, F. O. Eko, D. Lyn, and C. M. Black. 2000. Suppression of endogenous IL-10 gene expression in dendritic cells enhances antigen presentation for specific Th1 induction: potential for cellular vaccine development. *J. Immunol.* **164:**4212–4219.

Ingalls, R., P. Rice, N. Qureshi, K. Takayama, J. Lin, and D. Golenbock. 1995. The inflammatory cytokine response to *Chlamydia trachomatis* infection is endotoxin mediated. *Infect. Immun.* **63:**3125–3130.

Ishikawa, H., and G. N. Barber. 2008. STING is an endoplasmic reticulum adaptor that facilitates innate immune signalling. *Nature* **455:**674–678.

Ishikawa, H., Z. Ma, and G. N. Barber. 2009. STING regulates intracellular DNA-mediated, type I interferon-dependent innate immunity. *Nature* **461:**788–792.

Jin, M. S., and J.-O. Lee. 2008. Structures of the Toll-like receptor family and its ligand complexes. *Immunity* **29:**182–191.

Johnson, R. M. 2004. Murine oviduct epithelial cell cytokine responses to *Chlamydia muridarum* infection include interleukin-12-p70 secretion. *Infect. Immun.* **72:**3951–3960.

Jorgensen, I., M. M. Bednar, V. Amin, B. K. Davis, J. P. Ting, D. G. McCafferty, and R. H. Valdivia. 2011. The *Chlamydia* protease CPAF regulates host and bacterial proteins to maintain pathogen vacuole integrity and promote virulence. *Cell Host Microbe* **10:**21–32.

Kagami, S., T. Kakinuma, H. Saeki, Y. Tsunemi, H. Fujita, K. Sasaki, K. Nakamura, T. Takekoshi, M. Kishimoto, H. Mitsui, M. Komine, A. Asahina, and K. Tamaki. 2005. Increased serum CCL28 levels in patients with atopic dermatitis, psoriasis vulgaris and bullous pemphigoid. *J. Investig. Dermatol.* **124:**1088–1090.

Kanneganti, T. D., M. Body-Malapel, A. Amer, J. H. Park, J. Whitfield, L. Franchi, Z. F. Taraporewala, D. Miller, J. T. Patton, N. Inohara, and G. Nunez. 2006a. Critical role for Cryopyrin/Nalp3 in activation of caspase-1 in response to viral infection and double-stranded RNA. *J. Biol. Chem.* **281:**36560–36568.

Kanneganti, T. D., N. Ozoren, M. Body-Malapel, A. Amer, J. H. Park, L. Franchi, J. Whitfield, W. Barchet, M. Colonna, P. Vandenabeele, J. Bertin, A. Coyle, E. P. Grant, S. Akira, and G. Nunez. 2006b. Bacterial RNA and small antiviral compounds activate caspase-1 through cryopyrin/Nalp3. *Nature* **440:**233–236.

Kato, H., O. Takeuchi, E. Mikamo-Satoh, R. Hirai, T. Kawai, K. Matsushita, A. Hiiragi, T. S. Dermody, T. Fujita, and S. Akira. 2008. Length-dependent recognition of double-stranded ribonucleic acids by retinoic acid-inducible gene-I and melanoma differentiation-associated gene 5. *J. Exp. Med.* **205:**1601–1610.

Kato, H., O. Takeuchi, S. Sato, M. Yoneyama, M. Yamamoto, K. Matsui, S. Uematsu, A. Jung, T. Kawai, K. J. Ishii, O. Yamaguchi, K. Otsu, T. Tsujimura, C. S. Koh, C. Reis e Sousa, Y. Matsuura, T. Fujita, and S. Akira. 2006. Differential roles of MDA5 and RIG-I helicases in the recognition of RNA viruses. *Nature* **441:**101–105.

Kawai, T., and S. Akira. 2006. TLR signaling. *Cell Death Differ.* **13:**816–825.

Kim, Y. G., J. H. Park, M. H. Shaw, L. Franchi, N. Inohara, and G. Nunez. 2008. The cytosolic sensors Nod1 and Nod2 are critical for bacterial recognition and host defense after exposure to Toll-like receptor ligands. *Immunity* **28:**246–257.

Kuida, K., J. A. Lippke, G. Ku, M. W. Harding, D. J. Livingston, M. S. Su, and R. A. Flavell. 1995. Altered cytokine export and apoptosis in mice deficient in interleukin-1 beta converting enzyme. *Science* **267:**2000–2003.

Lad, S. P., J. Li, J. da Silva Correia, Q. Pan, S. Gadwal, R. J. Ulevitch, and E. Li. 2007. Cleavage of p65/RelA of the NF-κB pathway by *Chlamydia*. *Proc. Natl. Acad. Sci. USA* **104:**2933–2938.

Lamkanfi, M., R. K. Malireddi, and T. D. Kanneganti. 2009. Fungal zymosan and mannan activate the cryopyrin inflammasome. *J. Biol. Chem.* **284:**20574–20581.

Latz, E., A. Schoenemeyer, A. Visintin, K. A. Fitzgerald, B. G. Monks, C. F. Knetter, E. Lien, N. J. Nilsen, T. Espevik, and D. T. Golenbock. 2004. TLR9 signals after translocating from the ER to CpG DNA in the lysosome. *Nat. Immunol.* **5:**190–198.

Lilo, S., Y. Zheng, and J. B. Bliska. 2008. Caspase-1 activation in macrophages infected with *Yersinia pestis* KIM requires the type III secretion system effector YopJ. *Infect. Immun.* **76:**3911–3923.

Lu, H., C. Shen, and R. C. Brunham. 2000. *Chlamydia trachomatis* infection of epithelial cells induces the activation of caspase-1 and release of mature IL-18. *J. Immunol.* **165:**1463–1469.

Mariathasan, S., K. Newton, D. M. Monack, D. Vucic, D. M. French, W. P. Lee, M. Roose-Girma, S. Erickson, and V. M. Dixit. 2004. Differential activation of the inflammasome by caspase-1 adaptors ASC and Ipaf. *Nature* **430:** 213–218.

Mariathasan, S., D. S. Weiss, K. Newton, J. McBride, K. O'Rourke, M. Roose-Girma, W. P. Lee, Y. Weinrauch, D. M. Monack, and V. M. Dixit. 2006. Cryopyrin activates the inflammasome in response to toxins and ATP. *Nature* **440:**228–232.

Martin, M. U., and H. Wesche. 2002. Summary and comparison of the signaling mechanisms of the Toll/interleukin-1 receptor family. *Biochim. Biophys. Acta* **1592:**265–280.

Martinon, F., K. Burns, and J. Tschopp. 2002. The inflammasome: a molecular platform triggering activation of inflammatory caspases and processing of proIL-beta. *Mol. Cell* **10:**417–426.

Martinon, F., V. Petrilli, A. Mayor, A. Tardivel, and J. Tschopp. 2006. Gout-associated uric acid crystals activate the NALP3 inflammasome. *Nature* **440:**237–241.

Martinon, F., and J. Tschopp. 2004. Inflammatory caspases: linking an intracellular innate immune system to autoinflammatory diseases. *Cell* **117:**561–574.

Matsumoto, M., K. Funami, M. Tanabe, H. Oshiumi, M. Shingai, Y. Seto, A. Yamamoto, and T. Seya. 2003. Subcellular localization of Toll-like receptor 3 in human dendritic cells. *J. Immunol.* **171:**3154–3162.

McCoy, A. J., N. E. Adams, A. O. Hudson, C. Gilvarg, T. Leustek, and A. T. Maurelli. 2006. l,l-Diaminopimelate aminotransferase, a trans-kingdom enzyme shared by *Chlamydia* and plants for synthesis of diaminopimelate/lysine. *Proc. Natl. Acad. Sci. USA* **103:**17909–17914.

McCoy, A. J., and A. T. Maurelli. 2005. Characterization of *Chlamydia* MurC-Ddl, a fusion protein exhibiting D-alanyl-D-alanine ligase activity involved in peptidoglycan synthesis and D-cycloserine sensitivity. *Mol. Microbiol.* **57:**41–52.

McWhirter, S. M., R. Barbalat, K. M. Monroe, M. F. Fontana, M. Hyodo, N. T. Joncker, K. J. Ishii, S. Akira, M. Colonna, Z. J. Chen, K. A. Fitzgerald, Y. Hayakawa, and R. E. Vance. 2009. A host type I interferon response is induced by cytosolic sensing of the bacterial second messenger cyclic-di-GMP. *J. Exp. Med.* **206:**1899–1911.

Meixenberger, K., F. Pache, J. Eitel, B. Schmeck, S. Hippenstiel, H. Slevogt, P. N'Guessan, M. Witzenrath, M. G. Netea, T. Chakraborty, N. Suttorp, and B. Opitz. 2010. *Listeria monocytogenes*-infected human peripheral blood mononuclear cells produce IL-1beta, depending on listeriolysin O and NLRP3. *J. Immunol.* **184:**922–930.

Miao, E. A., C. M. Alpuche-Aranda, M. Dors, A. E. Clark, M. W. Bader, S. I. Miller, and A. Aderem. 2006. Cytoplasmic flagellin activates caspase-1 and secretion of interleukin 1beta via Ipaf. *Nat. Immunol.* **7:**569–575.

Miao, E. A., R. K. Ernst, M. Dors, D. P. Mao, and A. Aderem. 2008. *Pseudomonas aeruginosa* activates caspase 1 through Ipaf. *Proc. Natl. Acad. Sci. USA* **105:**2562–2567.

Miao, E. A., D. P. Mao, N. Yudkovsky, R. Bonneau, C. G. Lorang, S. E. Warren, I. A. Leaf, and A. Aderem. 2010. Innate immune detection of the type III secretion apparatus through the NLRC4 inflammasome. *Proc. Natl. Acad. Sci. USA* **107:**3076–3080.

Nagarajan, U. M., D. M. Ojcius, L. Stahl, R. G. Rank, and T. Darville. 2005. *Chlamydia tracho-*

matis induces expression of IFN-gamma-inducible protein 10 and IFN-beta independent of TLR2 and TLR4, but largely dependent on MyD88. *J. Immunol.* **175:**450–460.

Nagarajan, U. M., D. Prantner, J. D. Sikes, C. W. Andrews, Jr., A. M. Goodwin, S. Nagarajan, and T. Darville. 2008. Type I interferon signaling exacerbates *Chlamydia muridarum* genital infection in a murine model. *Infect. Immun.* **76:**4642–4648.

Nagarajan, U. M., J. Sikes, D. Prantner, C. W. Andrews, Jr., L. Frazer, A. Goodwin, J. N. Snowden, and T. Darville. 2011. MyD88 deficiency leads to decreased NK cell gamma interferon production and T cell recruitment during *Chlamydia muridarum* genital tract infection, but a predominant Th1 response and enhanced monocytic inflammation are associated with infection resolution. *Infect. Immun.* **79:**486–498.

Nagarajan, U. M., J. D. Sikes, L. Yeruva, and D. Prantner. 2012. Significant role of IL-1 signaling, but limited role of inflammasome activation, in oviduct pathology during *Chlamydia muridarum* genital infection. *J. Immunol.* **188:**2866-2875.

Naiki, Y., K. S. Michelsen, N. W. Schroder, R. Alsabeh, A. Slepenkin, W. Zhang, S. Chen, B. Wei, Y. Bulut, M. H. Wong, E. M. Peterson, and M. Arditi. 2005. MyD88 is pivotal for the early inflammatory response and subsequent bacterial clearance and survival in a mouse model of *Chlamydia pneumoniae* pneumonia. *J. Biol. Chem.* **280:**29242–29249.

Nedwin, G. E., S. L. Naylor, A. Y. Sakaguchi, D. Smith, J. Jarrett-Nedwin, D. Pennica, D. V. Goeddel, and P. W. Gray. 1985. Human lymphotoxin and tumor necrosis factor genes: structure, homology and chromosomal localization. *Nucleic Acids Res.* **13:**6361–6373.

Netea, M. G., B. J. Kullberg, J. M. Galama, A. F. Stalenhoef, C. A. Dinarello, and J. W. Van der Meer. 2002. Non-LPS components of *Chlamydia pneumoniae* stimulate cytokine production through Toll-like receptor 2-dependent pathways. *Eur. J. Immunol.* **32:**1188–1195.

O'Connell, C. M., Y. M. AbdelRahman, E. Green, H. K. Darville, K. Saira, B. Smith, T. Darville, A. M. Scurlock, C. R. Meyer, and R. J. Belland. 2011. Toll-like receptor 2 activation by *Chlamydia trachomatis* is plasmid dependent, and plasmid-responsive chromosomal loci are coordinately regulated in response to glucose limitation by *C. trachomatis* but not by *C. muridarum*. *Infect. Immun.* **79:**1044–1056.

O'Connell, C. M., R. R. Ingalls, C. W. Andrews, Jr., A. M. Scurlock, and T. Darville. 2007. Plasmid-deficient *Chlamydia muridarum* fail to induce immune pathology and protect against oviduct disease. *J. Immunol.* **179:**4027–4034.

O'Connell, C. M., I. A. Ionova, A. J. Quayle, A. Visintin, and R. R. Ingalls. 2006. Localization of TLR2 and MyD88 to *Chlamydia trachomatis* inclusions. Evidence for signaling by intracellular TLR2 during infection with an obligate intracellular pathogen. *J. Biol. Chem.* **281:**1652–1659.

Ohmori, Y., S. Fukumoto, and T. A. Hamilton. 1995. Two structurally distinct kappa B sequence motifs cooperatively control LPS-induced KC gene transcription in mouse macrophages. *J. Immunol.* **155:**3593–3600.

Ohmori, Y., and T. A. Hamilton. 1995. The interferon-stimulated response element and a kappa B site mediate synergistic induction of murine IP-10 gene transcription by IFN-gamma and TNF-alpha. *J. Immunol.* **154:**5235–5244.

Opitz, B., S. Forster, A. C. Hocke, M. Maass, B. Schmeck, S. Hippenstiel, N. Suttorp, and M. Krull. 2005. Nod1-mediated endothelial cell activation by *Chlamydophila pneumoniae*. *Circ. Res.* **96:**319–326.

Ozoren, N., J. Masumoto, L. Franchi, T. D. Kanneganti, M. Body-Malapel, I. Erturk, R. Jagirdar, L. Zhu, N. Inohara, J. Bertin, A. Coyle, E. P. Grant, and G. Nunez. 2006. Distinct roles of TLR2 and the adaptor ASC in IL-1beta/IL-18 secretion in response to *Listeria monocytogenes*. *J. Immunol.* **176:**4337–4342.

Patin, D., J. Bostock, D. Blanot, D. Mengin-Lecreulx, and I. Chopra. 2009. Functional and biochemical analysis of the *Chlamydia trachomatis* ligase MurE. *J. Bacteriol.* **191:**7430–7435.

Prantner, D., T. Darville, and U. M. Nagarajan. 2010. Stimulator of IFN gene is critical for induction of IFN-beta during *Chlamydia muridarum* infection. *J. Immunol.* **184:**2551–2560.

Prantner, D., T. Darville, J. D. Sikes, C. W. Andrews, Jr., H. Brade, R. G. Rank, and U. M. Nagarajan. 2009. Critical role for interleukin-1beta (IL-1beta) during *Chlamydia muridarum* genital infection and bacterial replication-independent secretion of IL-1beta in mouse macrophages. *Infect. Immun.* **77:**5334–5346.

Prantner, D., and U. M. Nagarajan. 2009. Role for the chlamydial type III secretion apparatus in host cytokine expression. *Infect. Immun.* **77:**76–84.

Prebeck, S., C. Kirschning, S. Durr, C. da Costa, B. Donath, K. Brand, V. Redecke, H. Wagner, and T. Miethke. 2001. Predominant role of toll-like receptor 2 versus 4 in *Chlamydia pneumoniae*-induced activation of dendritic cells. *J. Immunol.* **167:**3316–3323.

Qiu, H., Y. Fan, A. G. Joyee, S. Wang, X. Han, H. Bai, L. Jiao, N. Van Rooijen, and X. Yang. 2008. Type I IFNs enhance susceptibility to *Chlamydia muridarum* lung infection by enhancing apoptosis of local macrophages. *J. Immunol.* **181:**2092–2102.

Rank, R. G., H. M. Lacy, A. Goodwin, J. Sikes, J. Whittimore, P. B. Wyrick, and U. M. Nagarajan. 2010. Host chemokine and cytokine response in the endocervix within the first developmental cycle of *Chlamydia muridarum*. *Infect. Immun.* **78:**536–544.

Rasmussen, S. J., L. Eckmann, A. J. Quayle, L. Shen, Y. X. Zhang, D. J. Anderson, J. Fierer, R. S. Stephens, and M. F. Kagnoff. 1997. Secretion of proinflammatory cytokines by epithelial cells in response to *Chlamydia* infection suggests a central role for epithelial cells in chlamydial pathogenesis. *J. Clin. Investig.* **99:**77–87.

Reiling, N., S. Ehlers, and C. Holscher. 2008. MyDths and un-TOLLed truths: sensor, instructive and effector immunity to tuberculosis. *Immunol. Lett.* **116:**15–23.

Robinson, D., K. Shibuya, A. Mui, F. Zonin, E. Murphy, T. Sana, S. B. Hartley, S. Menon, R. Kastelein, F. Bazan, and A. O'Garra. 1997. IGIF does not drive Th1 development but synergizes with IL-12 for interferon-gamma production and activates IRAK and NFκB. *Immunity* **7:**571–581.

Rodriguez, N., R. Lang, N. Wantia, C. Cirl, T. Ertl, S. Dürr, H. Wagner, and T. Miethke. 2008. Induction of iNOS by *Chlamydophila pneumoniae* requires MyD88-dependent activation of JNK. *J. Leukoc. Biol.* **84:**1585–1593.

Sabbah, A., T. H. Chang, R. Harnack, V. Frohlich, K. Tominaga, P. H. Dube, Y. Xiang, and S. Bose. 2009. Activation of innate immune antiviral responses by Nod2. *Nat. Immunol.* **10:**1073–1080.

Saha, S. K., E. M. Pietras, J. Q. He, J. R. Kang, S. Y. Liu, G. Oganesyan, A. Shahangian, B. Zarnegar, T. L. Shiba, Y. Wang, and G. Cheng. 2006. Regulation of antiviral responses by a direct and specific interaction between TRAF3 and Cardif. *EMBO J.* **25:**3257–3263.

Sasu, S., D. LaVerda, N. Qureshi, D. T. Golenbock, and D. Beasley. 2001. *Chlamydia pneumoniae* and chlamydial heat shock protein 60 stimulate proliferation of human vascular smooth muscle cells via toll-like receptor 4 and p44/p42 mitogen-activated protein kinase activation. *Circ. Res.* **89:**244–250.

Seth, R. B., L. Sun, C. K. Ea, and Z. J. Chen. 2005. Identification and characterization of MAVS, a mitochondrial antiviral signaling protein that activates NF-kappaB and IRF 3. *Cell* **122:**669–682.

Shao, W., G. Yeretssian, K. Doiron, S. N. Hussain, and M. Saleh. 2007. The caspase-1 digestome identifies the glycolysis pathway as a target during infection and septic shock. *J. Biol. Chem.* **282:**36321–36329.

Soulat, D., A. Bauch, S. Stockinger, G. Superti-Furga, and T. Decker. 2006. Cytoplasmic *Listeria monocytogenes* stimulates IFN-beta synthesis without requiring the adapter protein MAVS. *FEBS Lett.* **580:**2341–2346.

Srinivasula, S. M., J.-L. Poyet, M. Razmara, P. Datta, Z. Zhang, and E. S. Alnemri. 2002. The PYRIN-CARD protein ASC is an activating adaptor for Caspase-1. *J. Biol. Chem.* **277:**21119–21122.

Steevels, T. A. M., and L. Meyaard. 2011. Immune inhibitory receptors: essential regulators of phagocyte function. *Eur. J. Immunol.* **41:**575–587.

Stetson, D. B., and R. Medzhitov. 2006. Recognition of cytosolic DNA activates an IRF3-dependent innate immune response. *Immunity* **24:**93–103.

Sutterwala, F. S., L. A. Mijares, L. Li, Y. Ogura, B. I. Kazmierczak, and R. A. Flavell. 2007. Immune recognition of *Pseudomonas aeruginosa* mediated by the IPAF/NLRC4 inflammasome. *J. Exp. Med.* **204:**3235–3245.

Suzuki, T., L. Franchi, C. Toma, H. Ashida, M. Ogawa, Y. Yoshikawa, H. Mimuro, N. Inohara, C. Sasakawa, and G. Nunez. 2007. Differential regulation of caspase-1 activation, pyroptosis, and autophagy via Ipaf and ASC in *Shigella*-infected macrophages. *PLoS Pathog.* **3:**e111.

Tabeta, K., P. Georgel, E. Janssen, X. Du, K. Hoebe, K. Crozat, S. Mudd, L. Shamel, S. Sovath, J. Goode, L. Alexopoulou, R. A. Flavell, and B. Beutler. 2004. Toll-like receptors 9 and 3 as essential components of innate immune defense against mouse cytomegalovirus infection. *Proc. Natl. Acad. Sci. USA* **101:**3516–3521.

Takaoka, A., Z. Wang, M. K. Choi, H. Yanai, H. Negishi, T. Ban, Y. Lu, M. Miyagishi, T. Kodama, K. Honda, Y. Ohba, and T. Taniguchi. 2007. DAI (DLM-1/ZBP1) is a cytosolic DNA sensor and an activator of innate immune response. *Nature* **448:**501–505.

Takeda, K., T. Kaisho, and S. Akira. 2003. Toll-like receptors. *Annu. Rev. Immunol.* **21:**335–376.

Tapping, R. I., S. Akashi, K. Miyake, P. J. Godowski, and P. S. Tobias. 2000. Toll-like receptor 4, but not toll-like receptor 2, is a signaling receptor for *Escherichia* and *Salmonella* lipopolysaccharides. *J. Immunol.* **165:**5780–5787.

Terra, J. K., C. K. Cote, B. France, A. L. Jenkins, J. A. Bozue, S. L. Welkos, S. M. LeVine, and K. A. Bradley. 2010. Cutting edge: resistance to *Bacillus anthracis* infection mediated by a lethal toxin sensitive allele of Nalp1b/Nlrp1b. *J. Immunol.* **184:**17–20.

Thirumalai, K., K. S. Kim, and A. Zychlinsky. 1997. IpaB, a *Shigella flexneri* invasin, colocalizes with interleukin-1 beta-converting enzyme in the cytoplasm of macrophages. *Infect. Immun.* **65:**787–793.

Thornberry, N. A., H. G. Bull, J. R. Calaycay, K. T. Chapman, A. D. Howard, M. J.

Kostura, D. K. Miller, S. M. Molineaux, J. R. Weidner, J. Aunins, et al. 1992. A novel heterodimeric cysteine protease is required for interleukin-1 beta processing in monocytes. *Nature* **356:**768–774.

Ting, J. P. Y., S. B. Willingham, and D. T. Bergstralh. 2008. NLRs at the intersection of cell death and immunity. *Nat. Rev. Immunol.* **8:**372–379.

Toshchakov, V., B. W. Jones, P. Y. Perera, K. Thomas, M. J. Cody, S. Zhang, B. R. Williams, J. Major, T. A. Hamilton, M. J. Fenton, and S. N. Vogel. 2002. TLR4, but not TLR2, mediates IFN-beta-induced STAT1alpha/beta-dependent gene expression in macrophages. *Nat. Immunol.* **3:**392–398.

Travassos, L. H., S. E. Girardin, D. J. Philpott, D. Blanot, M. A. Nahori, C. Werts, and I. G. Boneca. 2004. Toll-like receptor 2-dependent bacterial sensing does not occur via peptidoglycan recognition. *EMBO Rep.* **5:**1000–1006.

Trumstedt, C., E. Eriksson, A. M. Lundberg, T. B. Yang, Z. Q. Yan, H. Wigzell, and M. E. Rottenberg. 2007. Role of IRAK4 and IRF3 in the control of intracellular infection with *Chlamydia pneumoniae*. *J. Leukoc. Biol.* **81:**1591–1598.

Tseng, P. H., A. Matsuzawa, W. Zhang, T. Mino, D. A. Vignali, and M. Karin. 2010. Different modes of ubiquitination of the adaptor TRAF3 selectively activate the expression of type I interferons and proinflammatory cytokines. *Nat. Immunol.* **11:**70–75.

Unterholzner, L., S. E. Keating, M. Baran, K. A. Horan, S. B. Jensen, S. Sharma, C. M. Sirois, T. Jin, E. Latz, T. S. Xiao, K. A. Fitzgerald, S. R. Paludan, and A. G. Bowie. 2010. IFI16 is an innate immune sensor for intracellular DNA. *Nat. Immunol.* **11:**997–1004.

Vance, R. E. 2010. Inflammasome activation: how macrophages watch what they eat. *Cell Host Microbe* **7:**3–5.

Vignola, M. J., D. F. Kashatus, G. A. Taylor, C. M. Counter, and R. H. Valdivia. 2010. cPLA2 regulates the expression of type I interferons and intracellular immunity to *Chlamydia trachomatis*. *J. Biol. Chem.* **285:**21625–21635.

Wathelet, M. G., C. H. Lin, B. S. Parekh, L. V. Ronco, P. M. Howley, and T. Maniatis. 1998. Virus infection induces the assembly of coordinately activated transcription factors on the IFN-beta enhancer in vivo. *Mol. Cell* **1:**507–518.

Welter-Stahl, L., D. M. Ojcius, J. Viala, S. Girardin, W. Liu, C. Delarbre, D. Philpott, K. A. Kelly, and T. Darville. 2006. Stimulation of the cytosolic receptor for peptidoglycan, Nod1, by infection with *Chlamydia trachomatis* or *Chlamydia muridarum*. *Cell. Microbiol.* **8:**1047–1057.

Widmer, U., K. R. Manogue, A. Cerami, and B. Sherry. 1993. Genomic cloning and promoter analysis of macrophage inflammatory protein (MIP)-2, MIP-1 alpha, and MIP-1 beta, members of the chemokine superfamily of proinflammatory cytokines. *J. Immunol.* **150:**4996–5012.

Wolf, K., H. J. Betts, B. Chellas-Gery, S. Hower, C. N. Linton, and K. A. Fields. 2006. Treatment of *Chlamydia trachomatis* with a small molecule inhibitor of the *Yersinia* type III secretion system disrupts progression of the chlamydial developmental cycle. *Mol. Microbiol.* **61:** 1543–1555.

Woodward, J. J., A. T. Iavarone, and D. A. Portnoy. 2010. c-di-AMP secreted by intracellular *Listeria monocytogenes* activates a host type I interferon response. *Science* **328:**1703–1705.

Yamamoto, M., S. Sato, H. Hemmi, K. Hoshino, T. Kaisho, H. Sanjo, O. Takeuchi, M. Sugiyama, M. Okabe, K. Takeda, and S. Akira. 2003. Role of adaptor TRIF in the MyD88-independent toll-like receptor signaling pathway. *Science* **301:**640–643.

Yang, P., H. An, X. Liu, M. Wen, Y. Zheng, Y. Rui, and X. Cao. 2010. The cytosolic nucleic acid sensor LRRFIP1 mediates the production of type I interferon via a beta-catenin-dependent pathway. *Nat. Immunol.* **11:**487–494.

Yao, S. Y., A. Ljunggren-Rose, C. W. Stratton, W. M. Mitchell, and S. Sriram. 2001. Regulation by IFN-beta of inducible nitric oxide synthase and interleukin-12/p40 in murine macrophages cultured in the presence of *Chlamydia pneumoniae* antigens. *J. Interferon Cytokine Res.* **21:** 137–146.

Yaraei, K., L. A. Campbell, X. Zhu, W. C. Liles, C.-c. Kuo, and M. E. Rosenfeld. 2005. Effect of *Chlamydia pneumoniae* on cellular ATP content in mouse macrophages: role of Toll-like receptor 2. *Infect. Immun.* **73:**4323–4326.

Zhong, B., Y. Yang, S. Li, Y. Y. Wang, Y. Li, F. Diao, C. Lei, X. He, L. Zhang, P. Tien, and H. B. Shu. 2008. The adaptor protein MITA links virus-sensing receptors to IRF3 transcription factor activation. *Immunity* **29:**538–550.

CHLAMYDIA IMMUNOPATHOGENESIS

Toni Darville and Catherine M. O'Connell

11

INTRODUCTION

In the realm of infectious diseases, it has often been observed that an overly aggressive inflammatory host response can be more problematic than the infection that initiated it. This is certainly true in the case of infection with *Chlamydia trachomatis*, where the pathology that leads to the serious morbidities of blindness after conjunctival infection and infertility after female genital tract infection is the result of the host inflammatory response. A brief review of clinical data reveals the dual roles of the host response in pathogenesis and protection. The clinical aspects of chlamydial infection and disease are discussed in more detail in chapter 1, "*Chlamydia* infection and epidemiology."

Infection of the ocular conjunctiva with *C. trachomatis* leads to trachoma, which remains the commonest infectious cause of blindness worldwide. Infection of the conjunctiva with *C. trachomatis* is confined to the epithelium. It triggers release of proinflammatory cytokines and influx of inflammatory cells. Clinically, this manifests as papillary and/or follicular inflammation of the tarsal conjunctiva. Eventually the infection resolves, and the clinically visible inflammation gradually subsides. Studies from communities where trachoma is endemic have found lower rates of reinfection and disease at older ages, as well as significant decline in the duration of active disease with age, suggesting that a protective adaptive immune response develops as individuals are repeatedly exposed to infection (Bailey et al., 1999; Grassly et al., 2008). However, because the immune response is relatively ineffective, repeated infection is common in an endemic environment, leading to recurrent inflammation and the development of scar tissue within the conjunctiva over many years (Dawson et al., 1990; West et al., 2001). Over time the scar tissue contracts, causing the eyelid to roll inwards towards the eye (entropion) and the eyelashes to scratch the ocular surface (trichiasis). Constant abrasion of the surface of the eye causes irreversible corneal opacity and blindness. The degree to which conjunctival scarring develops appears to depend on a complex interplay between infectious load, infection frequency, and host-specific immunogenetic factors (Natividad et al., 2005, 2006, 2007). Elegant comparative transcriptome analyses of conjunctival genes from persons with and without trichiasis living in communities of hyperendemicity are

Toni Darville and Catherine M. O'Connell, Children's Hospital of Pittsburgh of UPMC, Rangos Research Center, One Children's Hospital Drive, 4401 Penn Ave., Pittsburgh, PA 15224.

Intracellular Pathogens I: Chlamydiales
Edited by Ming Tan and Patrik M. Bavoil © 2012 ASM Press, Washington, DC
10.1128/9781555817329.ch11

helping to identify host responses involved in this immunofibrogenic disease process as well as those involved in protection (Natividad et al., 2010; Burton et al., 2011).

Although men and women are both susceptible to genital tract infection by *C. trachomatis*, women are uniquely susceptible to reproductive sequelae such as chronic pelvic pain, infertility, and ectopic pregnancy because the bacteria can ascend from the endocervix to the upper genital tract and cause pelvic inflammatory disease (PID). Although *C. trachomatis* is a frequent pathogen associated with PID and is isolated in 10 to 27% of PID patients, rates of progression vary widely between patients (Heinonen and Miettinen, 1994; Bachmann et al., 1999; Haggerty et al., 2003; Wiesenfeld et al., 2005). Among high-risk populations, the progression to symptomatic PID within 2 weeks of untreated infection in asymptomatic *C. trachomatis*-positive women is generally low, at 2 to 5% (Hook et al., 1994; Bachmann et al., 1999; Geisler et al., 2008; Haggerty et al., 2010). Actual rates of progression of chlamydial infection to reproductive sequelae are not well defined. Antibodies to *C. trachomatis* antigens have been found to be more common among infertile women (Haggerty et al., 2010). Clinical data support a role for delayed treatment and ongoing infection in enhancing disease. Hillis et al. reported that delay of treatment for more than 3 days increases the risk of impaired fertility following an episode of PID (Hillis et al., 1993). Further, the association was strongest among women infected with *C. trachomatis*. Among a large cohort of patients with clinically suspected PID, pregnancy rates were significantly lower among women whose antibody titers to chlamydial elementary bodies (EBs) collected at the end of the study were in the highest third (Ness et al., 2008). Even if the rate of progression is low, the high incidence of lower genital tract infection in women 14 to 25 years of age indicates that many are at risk of developing serious reproductive morbidities.

The majority of men and women with *C. trachomatis* genital infections are asymptomatic (Miller et al., 2004). This high rate of asymptomatic infection contributes to the extremely high prevalence of chlamydial infection because infection goes unnoticed, undiagnosed, and therefore untreated. Despite an absence of symptoms, infected women may have histologic evidence of endometrial inflammation by biopsy. In a cross-sectional study of 100 women positive for chlamydial cervical infection but negative for symptoms of acute PID (cervical motion tenderness, lower abdominal pain, or adnexal tenderness), 27 had biopsy evidence of endometritis (Wiesenfeld et al., 2002). This reflects the presence of "subclinical PID" that may promote reproductive sequelae. Evidence supporting the important role of subclinical PID in the development of fallopian tube damage has emerged from several retrospective studies of women with tubal factor infertility (infertility due to damaged fallopian tubes). Serologic evidence of prior chlamydial infections was documented in 23 to 91% of women, the majority of whom had no history of acute PID (Punnonen et al., 1979; Jones et al., 1982; Tjiam et al., 1985). Subclinical upper genital tract inflammation appears sufficient to promote fallopian tube damage similar to that seen in acute PID. Using light and electron microscopy to assess tubal damage in groups of infertile women with and without a history of overt PID, Patton and colleagues observed similar degrees of flattened mucosal folds, extensive deciliation, reduced ciliary beat frequency, and degeneration of secretory epithelial cells in both groups (Patton et al., 1989).

The frequency with which reinfection occurs seems important for both ocular and genital disease development. Repeated conjunctival infection clearly leads to an increased risk of conjunctival scarring and trichiasis and is likely to be a prerequisite for development of blinding trachoma. Repeated chlamydial genital infection leads to PID and other reproductive sequelae. However, it is difficult to determine whether the risk per infection increases with each recurrent episode (Kimani et al., 1996; Hillis et al., 1997).

A role for host factors as key determinants of pathology is clearly evident from studies of communities where ocular *C. trachomatis* infection is hyperendemic, i.e., where every person is exposed and repeatedly infected but only some individuals develop blinding trachoma while others are spared (Gambhir et al., 2007). The largely asymptomatic nature of genital tract infection in men and women reveals the level of success that *Chlamydia* has reached with respect to masquerading as an innocuous commensal. The problem is that despite the lack of overt symptoms, *Chlamydia* may be acting as a covert injurious parasite. In vitro cellular and molecular investigations are rapidly dissecting the "push and shove" that occurs between the host cell and the microorganism as it invades, establishes its territory, and scrounges nutrients and structural materials that enable it to grow, prosper, and divide. The cell rapidly recognizes the presence of an invader and responds, but the response is thwarted and often unsuccessful, leading to a state of chronic parasitism. In vivo immunological studies in animal models and immunoepidemiological studies in humans indicate that resolution of infection can occur with minimal to no disease development provided that the correct responses are induced in the right amount. Recognizing that the immune response to this organism leads to tissue damage, it is important to delineate the specific host responses involved in disease promotion both for rational vaccine design and the discovery of biomarkers to monitor the effectiveness of candidate therapeutics and vaccines. This chapter will review data related to host, bacterial, and environmental factors that affect this complicated multidimensional process and how they relate to tipping the balance towards chronic disease development.

INNATE IMMUNE MECHANISMS CONTRIBUTING TO PATHOLOGY

Host Epithelial Cells and Mucosa

The importance of the epithelial cell as a critical, early component of the host response to *Chlamydia* infection was first identified by Rasmussen et al. (1997), who demonstrated that in vitro infection of cervical and colonic epithelial cells with *C. trachomatis* induced the secretion of an array of cytokines that have chemoattractant and proinflammatory functions. Moreover, in contrast to invasion by other bacteria, which induce a rapid but transient proinflammatory cytokine response upon entry (Eckmann et al., 1993), chlamydial internalization was insufficient to elicit a response. Rather, intracellular replication was required, and the epithelial cytokine response was sustained throughout the chlamydial developmental cycle. Endocervical epithelial cells released interleukin 1α (IL-1α) after chlamydial infection, and specific anti-IL-1α antibodies inhibited the induced proinflammatory cytokine cascade (Rasmussen et al., 1997). IL-1α release may act to amplify the inflammatory response by stimulating additional cytokine production, even by adjacent, uninfected cells. These findings formed the basis for Richard Stephens's "cellular paradigm of chlamydial pathogenesis." Stephens theorized that "the inflammatory processes of chlamydial pathogenesis are elicited by infected host cells and are necessary and sufficient to account for chronic inflammation and the promotion of cellular proliferation, tissue remodeling and scarring—the ultimate cause of disease sequelae" (Stephens, 2003).

The cellular paradigm of pathogenesis invokes the host epithelial cell as the central player in pathogenesis. Epithelial cells possess surface and intracellular innate immune receptors that enable them to recognize conserved chlamydial ligands and initiate inflammation. Thus, the infected epithelial cell serves as a key innate responder cell. Tissue-damaging responses begin soon after the bacterium infects the epithelial cell, and the infected epithelial cells continue to drive inflammation as long as they remain infected. Untreated, chlamydial infections can last from months to years (Molano et al., 2005), resulting in scarring and fibrosis of the conjunctiva in both men and women and of the fallopian tubes in women.

Transcriptional profiling studies of Ethiopians with trachomatous trichiasis revealed enhanced expression of proinflammatory cytokine genes (*IL1*, *CXCL5*, and *S100A7*) and matrix metalloproteases compared to individuals without trichiasis (Burton et al., 2011). Cases showed evidence of ongoing inflammation and tissue remodeling, which were more marked when clinical inflammation was also present. Significantly, *C. trachomatis* was detected in only 1 of 772 subjects, showing that these processes were active in the absence of current *C. trachomatis* infection. There was limited evidence of a Th1 response (*INDO* and *NOS2A*) and no association between a Th2 response and cases. These data confirm an active role for the epithelium in late cicatricial (scarring) stages of trachoma and indicate that even after active infection has resolved, residual chlamydial products may continue to drive scarring disease or that once the inflammatory process is initiated by infection, scarring responses are inevitable in certain individuals.

Chemokines and Cytokines

Recognizing the significance of early inflammatory events in chlamydial pathogenesis, the murine model of genital tract infection was adapted to characterize the early chemokine and cytokine response and correlate it with the chlamydial developmental cycle in vivo (Rank et al., 2010). Direct inoculation into the cervix allowed for induction of a high-level synchronous infection. Observation by transmission electron microscopy revealed that EBs were associated with epithelial cell surfaces at 3 hours postinfection (hpi) but that by 12 hpi EBs had entered host cells, transitioned to reticulate bodies, and begun replicating. Active replication by the infecting chlamydiae was confirmed at 12 hpi via 16s rRNA transcripts, which were not detected earlier. However, host responses could be detected very early in this infection because transcriptional profiling of inflammatory genes at 3 hpi revealed expression of 11 genes encoding the chemokine receptors CCR2 and CCR6, the chemokines CCL3 (MIP-1α), CCL20 (MIP-3α), CCL24, CCL25, and CXCL15, and the cytokines IL-1F8, IL-13, and tumor necrosis factor alpha (TNF-α) in cervical tissues. CCL3 (MIP-1α) and CCL24 (eotaxin-2) are chemotactic for immature dendritic cells, and CXCL15 is chemotactic for polymorphonuclear neutrophils (PMNs). This response amplified and diversified by 12 hpi, with the detection of numerous actively transcribed genes encoding chemokine receptors, chemokines, and cytokines. Importantly, UV-inactivated organisms did not elicit a chemokine or cytokine response at these early times, even though free lipopolysaccharide (LPS) or other proinflammatory bacterial products present in the inoculum should have stimulated host cell or resident professional innate immune cell pathogen recognition receptors. Therefore, these data indicate that infection and replication within the host cell are critical for activation of innate immune receptors, which is similar to what has been reported to occur in vitro (O'Connell et al., 2006). PMNs were easily observed infiltrating the infected epithelium by 24 hpi (Rank et al., 2010), and natural killer cell activity has been noted as early as 12 hours after vaginal inoculation of mice (Tseng and Rank, 1998), confirming that the chemokine gradient develops rapidly after infection. This intracervical infection model utilized a high inoculating dose of chlamydiae (10^7), which may stimulate a more vigorous response than human infection, where it is estimated that 10^2 organisms are an average inoculation dose in ejaculate (Geisler et al., 2001; Rank et al., 2003). However, it highlights the rapidity and complexity of the inflammatory response, demonstrating that multiple inflammatory mediators (among just 80 analyzed) are induced in infected tissue cells during a single chlamydial developmental cycle.

Proinflammatory chemokine and cytokine secretions in response to genital tract infection have also been documented in guinea pig and macaque models (Darville et al., 1995, 1997; Van Voorhis et al., 1996; Belay et al., 2002). Furthermore, infection of human fallopian tube explants also resulted in the production of TNF-α (Ault et al., 1996) and increased expression of adhesion molecules on oviduct endothelial cells (Kelly et al., 2001). This milieu leads to activation and recruitment of innate and, later, adaptive immune cells to resolve the infection, but subsets of these responses may also induce collateral damage to genital tract tissue. Defining and differentiating the specific responses that promote tissue damage or benign resolution of infection are important ongoing research goals.

Interleukin-1

Infection of fallopian tube organ cultures results in release of IL-1 by epithelial cells and cellular damage, independent of inflammatory cell influx. The addition of IL-1 receptor antagonist to the cultures eliminated tissue destruction induced by infection, indicating a direct role for this cytokine in pathogenesis (Hvid et al., 2007). Further documentation of a role for this cytokine in pathogenesis of genital tract disease has been gleaned from recent studies using murine models. First, caspase-1 knockout (KO) mice, which lack the protease required to activate IL-1β, exhibit decreased oviduct pathology but a normal course of infection (Cheng et al., 2008). Secondly, IL-1β KO mice exhibit decreased frequency of hydrosalpinx (cystic dilated oviducts) despite delayed clearance of chlamydiae (Prantner et al., 2009). Interestingly, the major producers of IL-1β in this study were macrophages and neutrophils rather than epithelial cells when assessed via flow cytometry. Evidence for the role of IL-1β in trachoma pathogenesis was revealed by a study that found IL-1β was significantly associated with trachomatous disease and concurrent *C. trachomatis* infection compared with age- and sex-matched controls from a region of hyperendemicity who did not have trachoma (Skwor et al., 2008). Recently, activation of IL-1R expressed on gut epithelium has been demonstrated to trigger epithelial damage via Rho kinase-dependent degradation of tight junctional claudin-4 (Lapointe et al., 2010). Oviduct and conjunctival epithelia both express claudin-4, as well as other claudin proteins. It is possible that IL-1R-mediated degradation of these proteins is a mechanism that contributes to disruption of both of these mucosal epithelial surfaces, promoting disease development. Thus, IL-1 may have a specific role in eliciting tissue damage through the destruction of epithelial tight junction proteins.

Tumor Necrosis Factor Alpha

TNF-α is a proinflammatory cytokine released primarily from monocytes and macrophages upon invasion of the host by a wide variety of pathogens. In the late 1980s, Manor and Sarov reported that human monocyte-derived macrophages inhibited growth of *C. trachomatis* (L2/4434/Bu) in human laryngeal carcinoma cells (HEp-2 cells) and this inhibition was reduced by the addition of anti-TNF-α antibodies (Manor and Sarov, 1988). Further, Shemer-Avni et al. reported a direct inhibitory effect of human recombinant TNF-α on the in vitro growth of *C. trachomatis* in HEp-2 cells (Shemer-Avni et al., 1988). In murine genital tract infection, genetic deficiency for the p55 type I receptor for TNF-α results in a statistically significant delay in clearance of *C. muridarum* and *C. trachomatis* serovar D from the genital tract (Perry et al., 1999). Since a direct inhibitory effect of TNF-α was not demonstrated in vitro, the investigators proposed that the delay in clearance was due to indirect effects of TNF-α on local effectors of host defense.

Recent data by Murthy et al. reveal an important role for TNF-α and $CD8^+$ T cells in induction of pathology in the mouse model of genital tract infection (Murthy et al., 2011). Mice genetically deficient in perforin (perforin KO mice) or TNF-α production (TNF-α KO mice) displayed comparable vaginal chlamydial clearance but significantly reduced oviduct pathology when compared to wild-type C57BL/6 mice. Since both perforin and TNF-α are effector molecules released from $CD8^+$ T cells, mice deficient in $CD8^+$ T cells were infected and were also found to exhibit reduced hydrosalpinx. Repletion of $CD8^+$ KO mice with wild-type or perforin KO, but not TNF-α KO, $CD8^+$ T cells at the time of challenge resulted in hydrosalpinx rates comparable to those observed in wild-type C57BL/6 mice, suggesting that TNF-α production from $CD8^+$ T cells is important for pathogenesis. Additionally, transfer of TNF-α^+ $CD8^+$ T cells into TNF-α KO mice significantly enhanced the incidence of hydrosalpinx and oviduct dilatation compared to TNF-α KO mice, but not to the level of wild-type mice, suggesting that TNF-α production from $CD8^+$ T cells and TNF-α production from non-$CD8^+$ cells cooperate to induce oviduct pathology. These results provide compelling new evidence supporting the contribution of TNF-α production to *Chlamydia*-induced reproductive tract pathology.

Several trachoma-focused studies support a role for TNF-α in conjunctival disease as well. Genetic variation at the TNF locus conferred an increased risk of trichiasis in a large matched-pair case-control study of Gambians (Natividad et al., 2007). The TNF-308A allele correlated with increased TNF production in lymphocyte cultures stimulated with EBs, suggesting that TNF-308A is a marker of a high TNF production phenotype associated with increased risk of disease.

Neutrophils

Murine genital tract studies indicate that the intensity of neutrophil influx into the oviduct correlates directly with development of hydrosalpinx after the infection has resolved (Shah et al., 2005). Prolonged infiltration of neutrophils into the oviduct also correlated with an increased incidence of severe hydrosalpinx in an immunologically normal mouse strain that exhibits increased disease susceptibility (Darville et al., 2001; Ramsey et al., 2005). A common feature in all mouse strains evaluated was the observation of a marked neutrophilic inflammatory response coincident with ascending infection of the genital tract. Activation of phagocyte oxidase in myeloid cells causes release of superoxide molecules that are damaging to cells and tissues. Interestingly, mice with a deletion of a key component of phagocyte oxidase (p47phox$^{-/-}$) have lower rates of hydrosalpinx (Ramsey et al., 2001) following infection, suggesting that activation of neutrophils by chlamydiae contributes directly to oviduct pathology.

The importance of neutrophil activation in the inflammatory process associated with *C. trachomatis* infection of the female reproductive tract was also supported by a study of women at risk for PID. In these women, vaginal levels of neutrophil α-defensins (HNP1-3), which are markers of neutrophil activation, were strongly associated with endometritis, but only in the presence of elevated defensin levels (Wiesenfeld et al., 2002). Although oculogenital serovars are killed by neutrophils, residual chlamydial envelopes containing major outer membrane protein (MOMP) and LPS may continue to stimulate neutrophil chemotaxis. This has been observed in vitro by using human endometrial epithelial cell cultures monitored for more than 1 month after exposure to neutrophils loaded with the antibiotic azithromycin (Wyrick et al., 1999). Thus, antigens may persist for some time after the organisms are killed, inducing continued neutrophil inflammation and ongoing release of tissue-damaging molecules in the host.

Matrix Metalloproteinases

Among the potentially tissue-damaging molecules expressed by neutrophils are matrix metalloproteinases (MMPs), which are involved in the proteolysis and repair of the extracellular matrix. Ramsey et al. have implicated neutrophil production of MMP-9 in the development of scarring and fibrosis of the murine oviduct after chlamydial infection (Ramsey et al., 2005). Studies involving humans also indicate a role for MMPs in production of tissue damage. Fallopian tube epithelial cells infected in vitro with *C. trachomatis* produce MMP-2, and infected oviduct stromal cells produce MMP-9 (Ault et al., 2002).

Expression and activity of MMP-9 are enhanced in the inflamed conjunctiva of trachoma subjects (Burton et al., 2004). Genetic variation within the MMP-9 gene affects in vitro MMP-9 expression levels, enzymatic activity, and susceptibility to various inflammatory and fibrotic conditions (Zhang et al., 1999). Genotyping for polymorphisms in the MMP-9 gene of 651 case-control pairs from villages in the Gambia where trachoma is endemic revealed that a Q279R mutation was associated with lower risk for severe disease sequelae after ocular *C. trachomatis* infection (Natividad et al., 2006). This mutation leads to a nonsynonymous amino acid change within the active site of the enzyme that may reduce MMP-9-induced degradation of structural components of the extracellular matrix of the conjunctiva and associated fibrosis.

MMPs can be released by other immune cells besides neutrophils. Monocyte/macrophage populations staining positive for gelatinase B, which is another MMP, were present in high numbers in conjunctival biopsy specimens from patients with active trachoma compared to healthy controls or patients with vernal keratoconjunctivitis (El-Asrar et al., 2000). Activated macrophages with giant cell morphology clearly stained with gelatinase B-specific monoclonal antibody were observed in trachoma specimens. Further, zymography revealed that gelatinase B levels in trachoma specimens were significantly higher than the levels found in normal conjunctiva, suggesting that this enzyme contributes to induction of conjunctival scarring.

Clinical Implications of the Cellular Paradigm

Recognizing that the infected host epithelial cell initiates the inflammatory response that ultimately leads to disease development, it follows that infection of conjunctival or oviduct epithelia is a prerequisite for pathology to occur in the eye or the fallopian tubes, respectively. This seems fairly obvious in the case of eye disease and leads to the simple conclusion that prevention or rapid clearance of *C. trachomatis* conjunctival infection will prevent blinding trachoma. It also has important implications for preventing the long-term morbidities of chlamydial genital infection, which primarily results from tissue damage of the oviduct rather than the primary site of inoculation, which is the cervix. Early diagnosis and administration of antibiotics before chlamydiae have ascended and infected the upper genital tract could prevent oviduct pathology, which is the rationale for screening and treatment programs. Better yet, induction of an adaptive immune response (during infection of the cervix or by vaccination) that prevents ascension of chlamydiae to the oviduct, upon initial or repeated infection, could also effectively prevent disease. These ideas have implications for the design of control programs for women, including the possibility that induction of sterilizing immunity might not be required to protect women from some of the more severe morbidities caused by chlamydial infection.

ROLE OF PATHOGEN RECOGNITION RECEPTOR SIGNALING IN PATHOGENESIS

TLRs 2 and 4

Researchers have begun to identify the cellular receptors involved in *C. trachomatis*-induced stimulation of cytokine release. Toll-like receptors (TLRs) act as pathogen recognition receptors (PRRs) that enable cells to recognize conserved bacterial, viral, and fungal structural elements. In cell culture, *C. trachomatis* infection of HEK cells transfected with the pathogen molecular pattern receptors TLR2 and TLR4/MD-2 revealed that TLR2 was required for IL-8 secretion while the role of TLR4/MD-2 was minimal. This result was reproduced with infection of immortalized human ectocervical cells (O'Connell et al., 2006). Activation was dependent on live, replicating bacteria because infection with UV-inactivated bacteria or treatment of infected cells with the antibiotic chloramphenicol abrogated the induction of IL-8 secretion. The response was largely dependent on the MyD88 adaptor molecule. Confocal microscopy experiments revealed that both TLR2 and MyD88 colocalize with the intracellular chlamydial inclusion, suggesting that TLR2 is actively engaged in signaling from this intracellular location (O'Connell et al., 2006).

Examination of the outcomes of genital tract infection in wild-type mice and mice genetically deficient in TLR2 or TLR4 confirmed a dominant role for TLR2 in the pathologic response to *C. muridarum* in the genital tract (Darville et al., 2003). TLR4 KO mice responded to infection similarly to wild-type controls and developed similar pathology. In TLR2 KO mice, although the course and extent of lower genital tract infection were equivalent to those of wild-type mice, a significant reduction in oviduct and mesosalpinx pathology was detected. The TLR2 KO mice had significantly lower levels of TNF-α and MIP-2 in genital tract secretions during the first week of infection. Although the cytokine response that remained in the absence of TLR2 signaling was sufficient to recruit effector cells to the genital tract that resolved infection, a reduction in inflammation and pathology resulted from the lack of TLR2 signaling. This role of TLR2 in chlamydial pathology is supported by the observation that primary infection with plasmid-cured *C. muridarum*, which fails to signal via TLR2, does not induce oviduct pathology. Upon infection with plasmid-cured strains, immunologically normal mice develop a Th1 adaptive response and appear to be as resistant to secondary challenge infection as mice infected with wild-type, plasmid-containing *C. muridarum* (O'Connell et al., 2007). Thus, although pathology is abrogated, *Chlamydia*-induced signaling that is independent of TLR2 leads to an effective acquired immune response. It is possible that the downstream immune response elicited by infection with these strains is globally muted in intensity or that the quality of the response is altered and specific mediators of tissue damage are no longer induced. Regardless of the mechanism, these data confirm the ability of the host epithelial cell-initiated response to induce downstream immune effectors that eradicate infection without eliciting chronic upper genital tract pathology.

Examination of human tissues for expression of TLRs has revealed the presence of mRNA for TLRs 1 through 9 in uterine epithelium (Schaefer et al., 2005). TLR2 is highly expressed in fallopian tubes and the cervix (Pioli et al., 2004), whereas TLR4 is weakly expressed in fallopian tubes (Pioli et al., 2004) and transcripts of TLR4 and its coadaptor MD2 are not detected in human cervical epithelial cells (Fichorova et al., 2002). Thus, by virtue of its relative expression, TLR2 may be a primary PRR available in the lower genital tract and oviducts to drive the pathology-inducing inflammatory response to infection. Interestingly, TLR10, which is reportedly expressed by fallopian tube epithelium (Hart et al., 2009), may function as a "decoy" receptor by binding ligand as a heterodimer with TLR2 but ultimately failing to transduce a proinflammatory signal (Guan et al., 2010). In vivo, this interaction could act to dampen the inflammatory response to infection. Mice do not express a functional TLR10 (Hasan et al., 2005), which impedes investigation of the role of this TLR in chlamydial infection. However, it is possible that this deficiency results in heightened sensitivity to chlamydial ligand and contributes to the frequency and severity of pathology observed in the murine genital tract model.

PRR Adaptors MyD88 and TRIF

Recognition of bacterium-associated molecular patterns by each of the cellular PRRs recruits their distinct adaptor proteins, which activate signaling cascades to induce cytokine production. In the absence of MyD88, an adaptor molecule downstream of TLR2, TLR4, and TLRs 7 through 9, mice exhibit prolonged infection and enhanced pathology in the upper genital tract (Chen et al., 2010b; Nagarajan et al., 2011). A blunted Th1 response and ineffective suppression of the Th2 response were observed in MyD88 KO mice. Although multiple proinflammatory chemokines and cytokines were reduced (IL-17, CXCL1, or KC, a neutrophil-attracting chemokine) or absent (TNF-α) in MyD88 KO mice, there was enhanced oviduct disease (Chen et al., 2010b) and marked granulomatous inflammation in the uterine horns (Nagarajan et al., 2011). This was associated with increased CCL2 (monocyte chemotactic protein-1 or MCP-1) levels that likely led to the increased recruitment of macrophages. Further deletion of TLR4-TRIF signaling in MyD88 KO mice, using TLR4/MyD88 double-KO mice, did not further compromise host defense against chlamydiae, suggesting that compensatory mechanisms are TLR independent. Despite some polarization toward a Th2 response, a Th1 response remained predominant in the absence of MyD88 and was sufficient to provide protection against challenge infection equivalent to that observed in wild-type mice (Chen et al., 2010b; Nagarajan et al., 2011). Genetic deficiency of MyD88 removes *Chlamydia*-induced TLR2 and TLR4 signaling via this adaptor, and in TLR4/MyD88 double-KO mice, signaling via TLR2-MyD88, TLR4-MyD88, and TLR4-TRIF is absent. Despite the absence of these signaling pathways, the infection, although prolonged, eventually resolves. Chlamydial activation of alternative PRRs results in a Th1 response that is decreased but intact, and enhanced CCL2 and monocyte/macrophage responses likely contribute to infection resolution.

Chlamydial PRR Ligands

Chlamydial heat shock protein 60 (Hsp60) is recognized by host PRR TLR4 (Bulut et al., 2002), and the intracellular PRR NOD1 is also activated in response to infection, although the corresponding chlamydial ligands have not been identified (Buchholz and Stephens, 2006). Most bacterial TLR2 ligands are glycolipids or lipopeptides, although exceptions have been noted. Signaling via TLR2 may be heavily influenced by the nature of the acyl chains attached to the lipoprotein ligand via interactions with TLR1 or TLR6. The current model suggests that triacylated lipopeptides signal via TLR2/TLR1 heteromers, whereas diacylated lipopeptides signal via TLR2/TLR6 heteromers (Akira and Sato, 2003). Bas et al. have recently demonstrated that the chlamydial surface-expressed protein Mip is a TLR2 ligand when expressed as a recombinant protein in *Escherichia coli* (Bas et al., 2008). The biochemical characterization of chlamydial Mip revealed that the types and distribution of fatty acids on the recombinant Mip expressed in *E. coli* resembled those of the native lipoprotein (Neff et al., 2007), although native Mip appeared to have a slightly higher proportion of amide-linked acyl chains. No difference in transcription or expression of Mip has been detected in plasmid-cured *C. trachomatis* (O'Connell et al., 2011), suggesting that this lipoprotein is not a predominant chlamydial TLR2 ligand. However, alterations of its lipid modification could account for the inability of plasmid-cured *C. trachomatis* to elicit TLR2-dependent signaling. Pgp3, a protein expressed from the chlamydial plasmid, has been shown to exhibit proinflammatory effects (Li et al., 2008). It is possible that multiple TLR2 ligands are expressed by chlamydiae and the degree of expression determines pathogenicity. Further investigation is required to enable identification of the pathogenic TLR2 ligand(s) of *C. trachomatis*.

PRRs, upon ligand binding, activate a variety of inflammatory signaling pathways including NF-κB, NF-IL-6, and mitogen-activated protein (MAP) kinases. Instead of using the NF-κB pathway (Lad et al., 2007a), chlamydiae activate the MAP kinase ERK1/2 pathway (Buchholz and Stephens, 2007), likely because this pathway is needed for nutrient acquisition (Fukuda et al., 2005). Inhibition of this pathway suppresses chlamydial growth and acquisition of triglycerophospholipids, host cPLA2 activity, and *Chlamydia*-induced cytokine production (Su et al., 2004; Fukuda et al., 2005; Buchholz and Stephens, 2007). Activation of the ERK1/2/cPLA2 signaling pathway contributes to inflammation via release of arachidonic acid, which is converted into prostaglandins, including prostaglandin E2 (PGE2), which can induce cytokine production. The MAP kinase ERK1/2 pathway also leads directly to cytokine gene activation. Although these responses should enhance host defense, the chronicity of chlamydial infections in humans suggests they are often ineffective, and ongoing release of these mediators during chronic infection may, in fact, promote tissue pathology. Chlamydial recognition and PRR signaling are reviewed in more detail in chapter 10, "Immune recognition and host cell response during *Chlamydia* infection."

CHLAMYDIA-SPECIFIC VIRULENCE MECHANISMS

It has proven challenging to identify chlamydial products that directly contribute to the development of immune pathology. Chlamydial LPS (Ingalls et al., 1995) and Hsp60 (Ohashi et al., 2000) have both been demonstrated to stimulate the innate response via TLR4. However, chlamydial LPS has low endotoxic activity (Brade et al., 1986; Heine et al., 2003), and the lack of significant TLR4 expression in the cervix and upper reproductive tract (Pioli et al., 2004) suggests that the contribution of these molecules to proinflammatory signaling during female reproductive tract infection may be small. More significant are ligands expressed by the bacteria that stimulate TLR2, but these have yet to be identified, although comparisons between wild-type and plasmid-deficient bacteria may provide a route to their identification. The role of the plasmid in their expression is also unclear. Loss of the conserved plasmid from *C. trachomatis* and *C. muridarum* is pleiotropic, resulting in not only reduced innate inflammatory signaling (O'Connell et al., 2007) but also impaired glycogen accumulation (Matsumoto et al., 1998; O'Connell and Nicks, 2006) and infectivity (O'Connell et al., 2006). Studies using the murine model revealed that infection with plasmid-cured *C. muridarum* CM972 resulted in reduced bacterial burden in the oviduct when compared to the wild-type parent (O'Connell et al., 2007) (Russell et al., 2011) and in significantly reduced lower genital tract shedding when plasmid-deficient *C. trachomatis* was used for infection (Carlson et al., 2008; Kari et al., 2008). Carlson et al. (2008) proposed that the plasmid encodes a transcriptional regulator of chlamydial virulence. A conserved group of chromosomal loci under the control of this putative regulator have been identified (O'Connell et al., 2011), and these genes may encode effectors of infectivity.

Chlamydial proteins are also secreted into the host cell cytoplasm by a number of mechanisms, including a type III secretion system (Fields and Hackstadt, 2000), while others are translocated via the Sec-dependent pathway to the chlamydial exterior (Chen et al., 2010a), raising the possibility that some may have immunomodulatory potential. A striking example is the serine protease, chlamydial protease/proteasome-like activity factor (CPAF), which is secreted relatively late in the chlamydial developmental cycle into the cytoplasm of the host cell and which has a multiplicity of targets. Degradation of host proteins by CPAF is proposed to facilitate immune evasion by impairing major histocompatibility complex antigen expression (Zhong et al., 2001) and to support bacterial replication by delaying apoptosis, promoting inclusion expansion, and replenishing intracellular amino acid pools (reviewed by Zhong [2009]). Similarly, the chlamydial tail-specific protease Tsp (CT441) may impede NF-κB activation of inflammatory signaling by targeting NF-κB p65 for cleavage in infected cells (Lad et al., 2007b), although CPAF itself has recently been implicated in this proteolytic activity (Christian et al., 2010).

CONTRIBUTION OF THE ADAPTIVE RESPONSE TO THE DEVELOPMENT OF CHLAMYDIAL DISEASE

Although it seems counterintuitive, the cellular adaptive response induced after chlamydial infection may contribute to the ultimate development of pathology. It is well documented that trachoma results from an immunopathologic response to repeated or chronic infection of the conjunctival epithelium. Reports of enhanced disease in individuals previously vaccinated against *C. trachomatis* highlight the need for caution when approaching the development of a chlamydial vaccine. Although administration of inactivated EBs as a trachoma vaccine to Punjab Indian children resulted in significant short-term protection from infection and disease and no deleterious effects (Dhir et al., 1967; Werner and Sareen, 1977), enhanced disease was observed in some Taiwanese children previously vaccinated with a low dose (10^7) of inactivated EBs (Woolridge et al., 1967). In a trachoma vaccine trial in monkeys, there was protection from infection and disease when the monkeys were challenged with the same serovar, but higher disease scores were seen when the challenge was with a *heterologous* serovar (Wang et al., 1967). These findings have raised concern that a whole-cell vaccine could induce pathogenic responses and have led to a focus on the development of a subunit vaccine. Vaccine development is discussed in more detail in chapter 14, "*Chlamydia* vaccine: progress and challenges."

Monkey models (Van Voorhis et al., 1997) and guinea pig models (Rank et al., 1995b) of repeated genital tract infection indicate that $CD4^+$ and $CD8^+$ T cells infiltrate more rapidly and in larger numbers than neutrophils during repeat oviduct infections, and this recurrent inflammatory reaction ultimately culminates in fibrosis and scarring. Human epidemiologic studies indicate increased risk of disease with repeated infection (Kimani et al., 1996; Hillis et al., 1997; Bakken et al., 2007). However, this may reflect an ongoing fibrotic/healing process in response to the initial infection rather than the cumulative effect of episodes of tissue destruction triggered with each reoccurrence. The application of more sophisticated immunological methods to murine and monkey models as well as to human clinical samples is beginning to provide a clearer picture of specific components of the adaptive immune response that may contribute to pathology. The continuation of such investigations should eventually provide sufficient detail that a safe and effective vaccine can be designed.

$CD4^+$ Th1 Cells

Studies using the murine model of *C. trachomatis* genital tract infection have established that resolution of genital infection is dependent upon an influx of gamma interferon (IFN-γ)-producing $CD4^+$ Th1 cells (Cain and Rank, 1995; Morrison et al., 1995, 2000; Cotter et al., 1997; Perry et al., 1997). Multiple studies in the murine model reveal a negative correlation between the induction of a strong *Chlamydia*-specific $CD4^+$ Th1 response and induction of oviduct pathology. Examples include the abbreviated infection and decreased pathology observed in IL-10 KO mice (Igietseme et al., 2000), reduced oviduct pathology in murine strains with stronger $CD4^+$ Th1 responses (Darville et al., 1997), and the ability of vaccines that induce vigorous $CD4^+$ Th1 responses to protect mice from development of oviduct pathology upon challenge (Murthy et al., 2007). It is possible that protection from oviduct disease has been observed in these studies because a strong Th1 response reduces and abbreviates chlamydial infection at the level of the oviduct.

A single *C. muridarum* infection of immunologically normal mice results in such severe pathology that it is difficult to quantify enhanced pathology upon challenge. However, primary infection with plasmid-deficient *C. muridarum* protects mice from developing oviduct damage upon challenge with wild-type *C. muridarum* Nigg. Recent examination of site-specific bacterial burden and cellular infiltrates in mice primarily infected with *C. muridarum* Nigg and mice primarily infected with plasmid-deficient *C. muridarum* and then challenged with *C. muridarum* Nigg revealed a 2- to 4-log decrease in oviduct bacterial burden in both groups upon secondary infection. Significantly decreased numbers and percentages of infiltrating neutrophils but significantly increased percentages of $CD4^+$ Th1 cells were detected in the oviducts and cervical tissues of challenged mice (M. Riley and T. Darville, unpublished data). These data indicate that a heightened $CD4^+$ Th1 response, together with an intact antibody response, limits infection of the oviduct and prevents pathology. Thus, an anamnestic $CD4^+$ Th1 response appears to be protective in the mouse model.

Data from humans also indicate a protective role for $CD4^+$ T cells. Among HIV-seropositive women, a $CD4^+$ lymphocyte count of $<400/mm^3$ was determined to be an independent risk factor for *C. trachomatis* PID (Kimani et al., 1996). A study by Agrawal and colleagues examined cervical lymphocyte cytokine responses of 255 *C. trachomatis* antibody-positive women with or without fertility disorders (infertility and multiple spontaneous abortions) as well as of healthy controls negative for *C. trachomatis* serum immunoglobulin M or G (Agrawal et al., 2009). Flow cytometric analysis revealed a significant increase in the mean number of $CD4^+$ T lymphocytes in the cervical mucosa of fertile women compared to those with fertility disorders and to negative controls. The $CD8^+$ T-cell population was only slightly increased in the two *Chlamydia*-infected groups.

It has been speculated that *Chlamydia*-specific antigens induce host $CD4^+$ T-cell responses that are increased with subsequent infections, promoting tissue damage and scarring (Grayston et al., 1985). Chlamydial Hsp60 has been investigated as a potential antigen responsible for induction of delayed type hypersensitivity-induced disease. This molecule attracted attention as a candidate "pathogenic" antigen after studies with immune guinea pigs and monkeys suggested that direct eye inoculation with this "sensitizing" antigen promoted heightened inflammation of the conjunctiva (Watkins et al., 1986; Morrison et al., 1989). However, residual Triton X detergent contaminating the extracts proved to be the inducer of disease. Later studies revealed a protective role for vaccination with Hsp60 (Rank et al., 1995a) in the guinea pig model of trachoma. Although human studies have revealed detection of elevated titers of antibody to Hsp60 in those with more-severe disease (Toye et al., 1993; Peeling et al., 1997), this may simply indicate increased exposure to chlamydiae through chronic or repeated infection. A recent large prospective study of women with PID did not reveal a correlation between increased antibody titers to Hsp60 and worse outcome (Ness et al., 2008).

$CD4^+$ Th17 Cells

The Th17 response has been implicated in the development of infection-induced immunopathology in other models of infection (Koenders et al., 2005; Dubin and Kolls, 2007). Th17 cells produce IL-17 that recruits neutrophils to inflammatory sites and induces neutrophil-promoting chemokines. Cytokines that influence Th1 and Th2 lineage commitment, specifically IFN-γ and IL-4, inhibit the development of Th17 cells (Kolls and Linden, 2004; Cruz et al., 2006). Thus, IFN-γ KO mice have a significantly increased Th17 response that is associated with a marked increase in neutrophilic responses and chronic pathology (Scurlock et al., 2010). A role for IL-17 in neutrophil recruitment was indicated by the detection of reduced infiltration of neutrophils into the genital tract of *C. muridarum*-infected mice genetically deficient in the IL-17 receptor (Scurlock et al., 2010). Despite reduced neutrophil infiltration, compared to infection of wild-type mice, chronic oviduct pathology and the course of infection were not altered. An important role for IL-17 in the induction of Th1 cells had been previously demonstrated during respiratory infection with *C. muridarum* (Bai et al., 2009). In addition, Scurlock and colleagues found that IL-17 receptor KO mice exhibited a reduced *Chlamydia*-specific Th1 response after genital tract infection (Scurlock et al., 2010). The compensatory increase in macrophage influx and TNF-α production observed in IL-17 receptor KO mice may explain the similar course of infection and pathological outcome observed in these mice when compared with wild-type controls (Scurlock et al., 2010). Taken together, these data indicate that although IL-17 contributes to neutrophil influx and enhancement of the Th1 response during genital infection, it is not required for either of these responses. In addition, absence of IL-17-induced responses does not protect from chronic oviduct pathology, likely due to IL-17-independent release of chemokines and cytokines from infected epithelial cells and monocytes/macrophages.

Increased levels of IL-17 and IL-22 (a cytokine produced by Th17 cells) have been detected in the cervical washes of *C. trachomatis*-infected women when compared to uninfected controls (Jha et al., 2011). Several studies have shown that both IL-17 and IL-22 increase protection against some bacteria in experimental models (Higgins et al., 2006; Khader et al., 2007; Scriba et al., 2008; Zheng et al., 2008) and cooperatively enhance expression of antimicrobial peptides associated with host defense (Liang et al., 2006). However, these cytokines can be proinflammatory and tissue damaging depending on the environment (Eyerich et al., 2010). Further experimentation is warranted to determine the role of Th17 cells in human chlamydial infection.

$CD8^+$ T Cells

Past correlative data from monkey and guinea pig models and more recent direct data from the mouse model indicate that adaptive $CD8^+$ T cells may be culprits in disease development. Reinfection studies performed using the macaque salpingeal pocket model suggest rapid lymphocytic influx during a chlamydial infection, with $CD8^+$ T cells making up two-thirds of the infiltrating lymphocytes (Van Voorhis et al., 1997). Tertiary infection of pockets reignites lymphocytic activity with focal areas of epithelial destruction and the formation of extensive fibrosis and lymphoid follicles in the deep stroma (Van Voorhis et al., 1997). Perforin mRNA was also detected, further suggesting a pathogenic role for cytolytic $CD8^+$ T cells. Additional data from this model indicate that cytotoxic $CD8^+$ T cells primed against Hsp60 play a role in tissue damage (Lichtenwalner et al., 2004). Thus, this direct-inoculation, nonhuman primate model displays immune-mediated destruction of genital tract tissue that is enhanced with each repeat infection. Unfortunately, TNF-α, which has been shown to be the mediator of $CD8^+$ T-cell-induced damage in the mouse primary infection model (Murthy et al., 2011), was not measured in the macaque studies. A controlled analysis of production of TNF-α and perforin from $CD8^+$ T cells in tissues from macaques and humans with chlamydial infection is needed. $CD8^+$ T cells may play a dual role, since their cytolytic function may aid in limiting infection (Gervassi et al., 2003; Jayarapu et al., 2010).

B Cells

Histological examination of tissues from women with PID caused by *C. trachomatis* revealed neutrophils in endometrial epithelium and within gland lumens, dense stromal lymphocytic infiltration, stromal plasma cells, and germinal centers containing transformed lymphocytes (Kiviat et al., 1985). Germinal center formation generally depends on cooperative interactions of helper T cells and B cells. Helper T cells express the surface molecule CD40 ligand and secrete cytokines upon activation by antigen-presenting dendritic cells. CD40 ligand expressed by activated helper T cells then binds to CD40 on B cells and initiates B-cell proliferation and differentiation. Within a few days of antigen exposure, a single lymphocyte within a germinal center may give rise to 5,000 progeny. Thus, detection of follicles and germinal centers in infected female genital tract tissues is evidence of local stimulation of adaptive immune responses to chlamydial infection. Whether follicles persist after infection has resolved is unknown. It is possible that such local proliferation of adaptive immune cells enhances resolution of infection and limits pathology. No data indicate a role for B cells in the development of pathology. Indeed, abundant data from the mouse model of genital tract infection reveal that chlamydia-specific B cells and antibody effectively lower the bacterial burden upon challenge, thereby protecting the oviduct from infection and the mouse from disease (Su et al., 1997; Morrison and Morrison, 2005).

GENETIC, PHYSIOLOGIC, AND ENVIRONMENTAL FACTORS INFLUENCING SEVERITY OF DISEASE OUTCOME

Both host and chlamydiae have the potential to drive disease development, but additional cofactors may potentially amplify or diminish these effects. Such elements include aspects of human genetics and local physiologic and/or environmental factors that render the immunologic signaling environment more or less proinflammatory or facilitate chlamydial infection or replication.

Genetic Factors

A number of studies have suggested an association between specific host immune responses and susceptibility to *Chlamydia* infection and/or disease. These host responses may influence the development of an effective adaptive response or possibly the strength of the innate inflammatory response to infection. In a small study of Australian women attending a sexually transmitted diseases clinic, it was determined that low peripheral blood mononuclear cell IFN-γ and high IL-10 responses to Hsp60 were markers for increased risk of *Chlamydia* infection and PID (Debattista et al., 2002). Among women with *Chlamydia*-related tubal infertility, T-cell responses to chlamydial Hsp60 were associated with a specific IL-10 promoter polymorphism (IL-10-1082AA) and with specific HLA class II DQ alleles (HLA-DQA1*0102 and HLA-DQB1*0602) (Kinnunen et al., 2002). Thus, genetic factors that regulate the induction and activation of Th1 cells may be important in directing a protective host response.

In contrast, a study by Barr and coworkers provides genetic support for the role of tissue-infiltrating memory T cells in chlamydial pathogenesis (Barr et al., 2005). A chemokine receptor deletion mutation, CCR5-Δ32, correlated significantly with protection from tubal damage, although the study was limited by a sample size of only 41 *C. trachomatis*-seropositive patients. Chemokine receptor 5 is crucial for T-cell activation and function. The investigators also compared the outcome of chlamydial genital infection in wild-type mice and mice genetically deficient for CCR5. Although the CCR5-deficient mice sustained an infection course of greater intensity and duration, they were protected from infertility compared to immunologically intact mice (Barr et al., 2005). Together, these data reveal the potential double-edged effect of T-cell effectors activated in response to chlamydial infection.

Limited studies investigating human genetic functional polymorphisms related to innate immune molecules have been conducted with Dutch Caucasian women. Investigation of TLR4 (Morré et al., 2003), the LPS-sensing TLR4 coreceptor CD14 (Ouburg et al., 2005), IL-1β, and the IL-1 receptor (Murillo et al., 2003) have so far revealed no association between functional polymorphisms in the genes encoding these molecules and tubal infertility. Karimi et al. examined the role of two TLR2 single nucleotide polymorphisms (SNPs) in 468 Dutch women (Karimi et al., 2009). This study determined that a haplotype formed by −16934 T>A and +2477 G>A SNPs was significantly associated with protection against tubal disease following *C. trachomatis* infection. Further, this haplotype was associated with a decrease in severity of *C. trachomatis* infection, suggesting a possible protective function.

Investigation of 18 tagging SNPs in four TLR genes (TLR1, TLR2, TLR4, and TLR6) and two adaptor molecules (TIRAP and MyD88) was conducted among 205 African Americans with clinically suspected PID from the PID Evaluation and Clinical Health Study (Ness et al., 2008). Analysis revealed that women with PID who carried the TLR4 rs1927911 CC genotype had increased odds of chlamydial infection (case:control frequency, 36:14; odds ratio, 3.7; 95% confidence interval [CI], 1.6 to 8.8; $P = 0.002$), (B. Taylor, T. Darville, R. Ferrell, C. Kammerer, J. Zmuda, R. Ness, and C. L. Haggerty, unpublished data). The TLR1 rs5743618 TT genotype was also associated with *C. trachomatis* (case:control frequency, 84:65; odds ratio, 2.8; 95% CI, 1.3 to 6.2; $P = 0.008$). No significant associations were found between any other TLR or adaptor molecule SNP and chlamydial infection. In addition, there were no significant associations between TLR or adaptor molecule SNPs and endometritis. However, the power of the analysis was limited by the small sample size. Additional studies are needed involving women with chlamydial infection with and without evidence of endometritis.

Wild-type homozygosity for mannose-binding lectin allele A has been associated with protection against tubal occlusion in Hungarian Caucasian women (Sziller et al., 2007). Binding of mannose-binding lectin to sugar groups of the *C. trachomatis* MOMP blocks attachment of organisms to host cells (Swanson et al., 1998). Low levels of this lectin may lead to an increased infectious burden and/or increased ascension to the oviducts triggering enhanced innate or adaptive inflammatory responses that could increase the risk for tubal factor infertility.

Physiologic Factors

Physiologic factors may also influence the severity of pathology after chlamydial infection. Human epidemiologic studies have indicated an increased risk of chlamydial infection in women using oral contraceptives (combined estrogen/progesterone) or depot medroxyprogesterone acetate (Baeten et al., 2001; Morrison et al., 2009), although no enhancement of upper tract infection has been detected (Ness et al., 2001). However, a recent prospective study of 948 Kenyan sex workers revealed that women using depot medroxyprogesterone acetate had a significantly decreased risk of PID (hazard ratio, 0.4; 95% CI, 0.2 to 0.7) (Baeten et al., 2001). These hormonal associations may impact disease development in a number of ways. Possibilities include influencing susceptibility to infection as has been demonstrated in vitro (Sweet et al., 1986; Guseva et al., 2003) or more indirectly by modulating the expression of important innate receptors such as TLR2 (reviewed by Beagley and Gockel [2003]) and inflammatory signaling mechanisms, e.g., nitrous oxide (reviewed by Shao et al. [2010]), leading to an altered immune response.

Coinfection with Other Pathogens

The female lower genital tract is not a sterile site and carries a diverse and dynamic population of commensal organisms (Ravel et al., 2011). The impact this microflora may have on chlamydial infection or on inflammatory responses to chlamydiae has yet to be examined. Furthermore, the contribution of coinfection with other sexually transmitted pathogens is also not well understood. It is not uncommon for women with chlamydial infection to be coinfected with *Neisseria gonorrhoeae*, a pathogen capable of causing PID and serious reproductive sequelae in its own right (Hook and Handsfield, 1999). Interestingly, a mouse model of coinfection using *C. muridarum* revealed that preexisting chlamydial infection appeared to enhance subsequent colonization by *N. gonorrhoeae*, likely as a result of *Chlamydia*-induced immune modulation (Vonck et al., 2011). Unfortunately, the severity of upper tract pathology caused by *C. muridarum* alone prevented evaluation of the impact of coinfection on the development of oviduct disease.

Human studies have revealed that women infected with HIV suffer worse chlamydial disease, which correlates with $CD4^+$ cell counts, further highlighting the importance of this component of the adaptive response in the control and clearance of chlamydial infection (Kimani et al., 1996). However, in vitro coculture studies suggest that the presence of the virus does not affect bacterial replication (Broadbent et al., 2011). Nevertheless, Mitchell et al. have detected higher production of proinflammatory cytokines in the genital tract of HIV-infected women independent of microbiota, including chlamydiae (Mitchell et al., 2011), possibly leading to synergistic effects that could potentiate inflammatory damage from a chlamydial infection. *C. trachomatis* appears to produce aberrant "persistent" developmental forms when cultured in HSV-2-infected cells in vitro (Deka et al., 2006).

The mechanism regulating this process is not completely understood, but interaction of the viral glycoprotein D with the host cell appears sufficient to trigger this event (Vanover et al., 2010). Whether these developmental forms are induced during human infection is unknown.

Environmental Factors

Rapid multiplication of bacteria is essential to establish infection, and this is particularly important for chlamydiae since they are dependent on the host cell for essential nutrients and energy intermediates. One effective host defense mechanism that targets acquisition of critical substrates is IFN-γ because, in addition to its role in immune regulation, it targets chlamydial replication directly via induction of indoleamine 2,3-deoxygenase (IDO) in infected cells (Beatty et al., 1994). This enzyme catalyzes the initial step in the degradation of tryptophan and, by depleting the cytoplasmic pool of this essential amino acid, causes inhibition of chlamydial cell division and the appearance of "persistent" developmental forms (Beatty et al., 1993). Furthermore, the different tissue tropisms displayed by ocular and genital chlamydiae have been correlated with differences in the functionality of the tryptophan biosynthetic pathways within the serovars. The genital tract strains retain the capacity to utilize exogenous indole presumed to be generated by the vaginal microbiota, while ocular strains lack a functional TrpA subunit for tryptophan synthase and cannot (Caldwell et al., 2003). More nuanced still, the low-oxygen environment of the female genital tract also contributes to the suppression of IFN-γ-mediated regulation of IDO via impaired activation of the JAK-Stat pathway (Roth et al., 2010), reducing the effectiveness of the IFN-γ host defense mechanism at this mucosal site.

Chlamydiae must also compete for energy substrates. *C. trachomatis* cannot transport glucose, because it is deficient in the glucose subunits of the phosphotransferase system (McClarty and Stephens, 1999), nor can it synthesize the intermediate glucose-6-phosphate, because it lacks a hexokinase (Vender and Moulder, 1967). Consequently, chlamydiae are dependent on host glycolysis as their sole source for this metabolic intermediate. *C. trachomatis* cultured in cells chemically inhibited for glucose-6-phosphate biosynthesis displayed reduced TLR2-activating potential in a reporter assay system (O'Connell et al., 2011), suggesting that chlamydiae have the capacity to downregulate this virulence-associated phenotype under limiting conditions. Oviduct epithelial cells are the primary source of glucose in follicular fluid, the concentration of which varies over the course of the menstrual cycle (reviewed by Leese et al. [2001]). Consequently, the cytoplasmic availability of glucose-6-phosphate in these cells may vary in response to hormonal signaling and could become limiting for infecting chlamydiae, leading to reduced TLR2 activation, dampened inflammation, and milder disease.

CONCLUSIONS

The cellular paradigm of *Chlamydia* pathogenesis (Stephens, 2003) states that the host response to chlamydiae is initiated and sustained by epithelial cells, which are the primary targets of chlamydial infection. Infected host epithelial cells act as first responders, initiating and propagating immune responses through recognition of various chlamydial ligands via pathogen recognition receptors. They secrete chemokines that recruit inflammatory leukocytes to the site of infection, as well as cytokines that induce and augment the cellular inflammatory response. Unfortunately, this response is frequently ineffective at resolving infection, and ongoing stimulation of the host cells and bystander cells leads to continued release of mediators that induce direct tissue damage. Data implicate IL-1 as a potential key pathologic mediator. The release of proteases, clotting factors, and tissue growth factors from infected host cells and infiltrating inflammatory cells leads to tissue damage and eventual scarring—the hallmark of *Chlamydia*-induced disease. Since reinfection with chlamydiae is a frequent occurrence, repeated inflammatory responses may lead to repeated insult to the tissues and further promote tissue scarring. The cellular paradigm makes no distinction between damage induced by professional innate immune cells (neutrophils and monocytes) and adaptive lymphocyte populations but assumes that both cell populations contribute to pathogenesis. Since the host cell response to bacteria is the inciting inflammatory event, increased and prolonged bacterial burden correlates directly with disease development. Pathogen-specific and environmental factors that promote infection and bacterial survival lead to enhanced disease. Plasmid-encoded factors and type III secretion effectors appear to be key bacterial virulence factors.

Fortunately, immunological studies in mice indicate that select adaptive immune responses can effectively limit infection while producing no or minimal collateral tissue damage. Antibody- and IFN-γ-producing $CD4^+$ T cells appear to be the most effective host defense mechanisms in this regard. Immunoepidemiologic and genetic studies in humans support a primarily protective role for $CD4^+$ Th1 responses. In contrast, studies with both animals and humans reveal that neutrophils and their products are primarily involved in tissue damage. Studies with monkeys and mice indicate that $CD8^+$ T cells may contribute to pathology, while in vitro and murine studies indicate that cytolytic $CD8^+$ T cells inhibit bacterial propagation. The availability of high-throughput genomics and proteomics, together with sophisticated immunological methods, should help us to delineate the specific cells and responses that promote tissue damage and to provide for rational vaccine design.

REFERENCES

Agrawal, T., R. Gupta, R. Dutta, P. Srivastava, A. R. Bhengraj, S. Salhan, and A. Mittal. 2009. Protective or pathogenic immune response to genital chlamydial infection in women—a possible role of cytokine secretion profile of cervical mucosal cells. *Clin. Immunol.* **130:**347–354.

Akira, S., and S. Sato. 2003. Toll-like receptors and their signaling mechanisms. *Scand. J. Infect. Dis.* **35:**555–562.

Ault, K. A., K. A. Kelly, P. E. Ruther, A. A. Izzo, L. S. Izzo, I. M. Sigar, and K. H. Ramsey. 2002. *Chlamydia trachomatis* enhances the expression of matrix metalloproteinases in an in vitro model of the human fallopian tube infection. *Am. J. Obstet. Gynecol.* **187:**1377–1383.

Ault, K. A., O. W. Tawfik, M. M. Smith-King, J. Gunter, and P. F. Terranova. 1996. Tumor necrosis factor-alpha response to infection with *Chlamydia trachomatis* in human fallopian tube organ culture. *Am. J. Obstet. Gynecol.* **175:**1242–1245.

Bachmann, L. H., C. M. Richey, K. Waites, J. R. Schwebke, and E. W. Hook III. 1999. Patterns of *Chlamydia trachomatis* testing and follow-up at a University Hospital Medical Center. *Sex. Transm. Dis.* **26:**496–499.

Baeten, J. M., P. M. Nyange, B. A. Richardson, L. Lavreys, B. Chohan, H. L. Martin, Jr., K. Mandaliya, J. O. Ndinya-Achola, J. J. Bwayo, and J. K. Kreiss. 2001. Hormonal contraception and risk of sexually transmitted disease acquisition: results from a prospective study. *Am. J. Obstet. Gynecol.* **185:**380–385.

Bai, H., J. Cheng, X. Gao, A. G. Joyee, Y. Fan, S. Wang, L. Jiao, Z. Yao, and X. Yang. 2009. IL-17/Th17 promotes type 1 T cell immunity against pulmonary intracellular bacterial infection through modulating dendritic cell function. *J. Immunol.* **183:**5886–5895.

Bailey, R., T. Duong, R. Carpenter, H. Whittle, and D. Mabey. 1999. The duration of human ocular *Chlamydia trachomatis* infection is age dependent. *Epidemiol. Infect.* **123:**479–486.

Bakken, I. J., F. E. Skjeldestad, and S. A. Nordbo. 2007. *Chlamydia trachomatis* infections increase the risk for ectopic pregnancy: a population-based, nested case-control study. *Sex. Transm. Dis.* **34:**166–169.

Barr, E. L., S. Ouburg, J. U. Igietseme, S. A. Morre, E. Okwandu, F. O. Eko, G. Ifere, T. Belay, Q. He, D. Lyn, G. Nwankwo, J. Lillard, C. M. Black, and G. A. Ananaba. 2005. Host inflammatory response and development of complications of *Chlamydia trachomatis* genital infection in CCR5-deficient mice and subfertile women with the CCR5delta32 gene deletion. *J. Microbiol. Immunol. Infect.* **38:**244–254.

Bas, S., L. Neff, M. Vuillet, U. Spenato, T. Seya, M. Matsumoto, and C. Gabay. 2008. The proinflammatory cytokine response to *Chlamydia trachomatis* elementary bodies in human macrophages is partly mediated by a lipoprotein, the macrophage infectivity potentiator, through TLR2/TLR1/TLR6 and CD14. *J. Immunol.* **180:**1158–1168.

Beagley, K. W., and C. M. Gockel. 2003. Regulation of innate and adaptive immunity by the female sex hormones oestradiol and progesterone. *FEMS Immunol. Med. Microbiol.* **38:**13–22.

Beatty, W. L., T. A. Belanger, A. A. Desai, R. P. Morrison, and G. I. Byrne. 1994. Tryptophan depletion as a mechanism of gamma interferon-mediated chlamydial persistence. *Infect. Immun.* **62:**3705–3711.

Beatty, W. L., G. I. Byrne, and R. P. Morrison. 1993. Morphologic and antigenic characterization of interferon gamma-mediated persistent *Chlamydia trachomatis* infection in vitro. *Proc. Natl. Acad. Sci. USA* **90:**3998–4002.

Belay, T., F. O. Eko, G. A. Ananaba, S. Bowers, T. Moore, D. Lyn, and J. U. Igietseme. 2002. Chemokine and chemokine receptor dynamics during genital chlamydial infection. *Infect. Immun.* **70:**844–850.

Brade, L., S. Schramek, U. Schade, and H. Brade. 1986. Chemical, biological, and immunochemical properties of the *Chlamydia psittaci* lipopolysaccharide. *Infect. Immun.* **54:**568–574.

Broadbent, A., P. Horner, G. Wills, A. Ling, R. Carzaniga, and M. McClure. 2011. HIV-1 does not significantly influence *Chlamydia trachomatis* serovar L2 replication in vitro. *Microbes Infect.* **13:**575–584.

Buchholz, K. R., and R. S. Stephens. 2006. Activation of the host cell proinflammatory interleukin-8 response by *Chlamydia trachomatis*. *Cell. Microbiol.* **8:**1768–1779.

Buchholz, K. R., and R. S. Stephens. 2007. The extracellular signal-regulated kinase/mitogen-activated protein kinase pathway induces the inflammatory factor interleukin-8 following *Chlamydia trachomatis* infection. *Infect. Immun.* **75:**5924–5929.

Bulut, Y., E. Faure, L. Thomas, H. Karahashi, K. S. Michelsen, O. Equils, S. G. Morrison, R. P. Morrison, and M. Arditi. 2002. Chlamydial heat shock protein 60 activates macrophages and endothelial cells through Toll-like receptor 4 and MD2 in a MyD88-dependent pathway. *J. Immunol.* **168:**1435–1440.

Burton, M. J., R. L. Bailey, D. Jeffries, D. C. Mabey, and M. J. Holland. 2004. Cytokine and fibrogenic gene expression in the conjunctivas of subjects from a Gambian community where trachoma is endemic. *Infect. Immun.* **72:**7352–7356.

Burton, M. J., S. N. Rajak, J. Bauer, H. A. Weiss, S. B. Tolbert, A. Shoo, E. Habtamu, A. Manjurano, P. M. Emerson, D. C. Mabey, M. J. Holland, and R. L. Bailey. 2011. Conjunctival transcriptome in scarring trachoma. *Infect. Immun.***79:**499–511.

Cain, T. K., and R. G. Rank. 1995. Local Th1-like responses are induced by intravaginal infection of mice with the mouse pneumonitis biovar of *Chlamydia trachomatis*. *Infect. Immun.* **63:**1784–1789.

Caldwell, H. D., H. Wood, D. Crane, R. Bailey, R. B. Jones, D. Mabey, I. Maclean, Z. Mohammed, R. Peeling, C. Roshick, J. Schachter, A. W. Solomon, W. E. Stamm, R. J. Suchland, L. Taylor, S. K. West, T. C. Quinn, R. J. Belland, and G. McClarty. 2003. Polymorphisms in *Chlamydia trachomatis* tryptophan synthase genes differentiate between genital and ocular isolates. *J. Clin. Investig.* **111:**1757–1769.

Carlson, J. H., W. M. Whitmire, D. D. Crane, L. Wicke, K. Virtaneva, D. E. Sturdevant, J. J. Kupko III, S. F. Porcella, N. Martinez-Orengo, R. A. Heinzen, L. Kari, and H. D. Caldwell. 2008. The *Chlamydia trachomatis* plasmid is a transcriptional regulator of chromosomal genes and a virulence factor. *Infect. Immun.* **76:** 2273–2283.

Chen, D., L. Lei, C. Lu, R. Flores, M. P. DeLisa, T. C. Roberts, F. E. Romesberg, and G. Zhong. 2010a. Secretion of the chlamydial virulence factor CPAF requires the Sec-dependent pathway. *Microbiology* **156:**3031–3040.

Chen, L., L. Lei, X. Chang, Z. Li, C. Lu, X. Zhang, Y. Wu, I. T. Yeh, and G. Zhong. 2010b. Mice deficient in MyD88 Develop a Th2-dominant response and severe pathology in the upper genital tract following *Chlamydia muridarum* infection. *J. Immunol.* **184:**2602–2610.

Cheng, W., P. Shivshankar, Z. Li, L. Chen, I. T. Yeh, and G. Zhong. 2008. Caspase-1 contributes to *Chlamydia trachomatis*-induced upper urogenital tract inflammatory pathologies without affecting the course of infection. *Infect. Immun.* **76:**515–522.

Christian, J., J. Vier, S. A. Paschen, and G. Hacker. 2010. Cleavage of the NF-κB family protein p65/RelA by the chlamydial protease-like activity factor (CPAF) impairs proinflammatory signaling in cells infected with chlamydiae. *J. Biol. Chem.* **285:**41320–41327.

Cotter, T. W., K. H. Ramsey, G. S. Miranpuri, C. E. Poulsen, and G. I. Byrne. 1997. Dissemination of *Chlamydia trachomatis* chronic genital tract infection in gamma interferon gene knockout mice. *Infect. Immun.* **65:**2145–2152.

Cruz, A., S. A. Khader, E. Torrado, A. Fraga, J. E. Pearl, J. Pedrosa, A. M. Cooper, and A. G. Castro. 2006. Cutting edge: IFN-gamma regulates the induction and expansion of IL-17-producing CD4 T cells during mycobacterial infection. *J. Immunol.* **177:**1416–1420.

Darville, T., C. W. Andrews, Jr., K. K. Laffoon, W. Shymasani, L. R. Kishen, and R. G. Rank. 1997. Mouse strain-dependent variation in the course and outcome of chlamydial genital tract infection is associated with differences in host response. *Infect. Immun.* **65:**3065–3073.

Darville, T., C. W. Andrews, Jr., J. D. Sikes, P. L. Fraley, and R. G. Rank. 2001. Early local cytokine profiles in strains of mice with different outcomes from chlamydial genital tract infection. *Infect. Immun.* **69:**3556–3561.

Darville, T., K. K. Lafoon, L. R. Kishen, and R. G. Rank. 1995. Tumor necrosis factor-alpha activity in genital tract secretions of guinea pigs infected with chlamydiae. *Infect. Immun.* **63:**4675–4681.

Darville, T., J. M. O'Neill, C. W. Andrews, Jr., U. M. Nagarajan, L. Stahl, and D. M. Ojcius. 2003. Toll-like receptor-2, but not Toll-like receptor-4, is essential for development of oviduct pathology in chlamydial genital tract infection. *J. Immunol.* **171:**6187–6197.

Dawson, C. R., R. Marx, T. Daghfous, R. Juster, and J. Schachter. 1990. What clinical signs are critical in evaluating the intervention in trachoma?, p. 271–278. *In* W. R. Bowie (ed.), *Chlamydial Infections*. Cambridge University Press, Cambridge, United Kingdom.

Debattista, J., P. Timms, and J. Allan. 2002. Reduced levels of gamma-interferon secretion in response to chlamydial 60 kDa heat shock protein amongst women with pelvic inflammatory disease and a history of repeated *Chlamydia trachomatis* infections. *Immunol. Lett.* **81:**205–210.

Deka, S., J. Vanover, S. Dessus-Babus, J. Whittimore, M. K. Howett, P. B. Wyrick, and R. V. Schoborg. 2006. *Chlamydia trachomatis* enters a viable but non-cultivable (persistent) state within herpes simplex virus type 2 (HSV-2) co-infected host cells. *Cell. Microbiol.* **8:**149–162.

Dhir, S. P., L. P. Agarwal, R. Detels, S. P. Wang, and J. T. Grayston. 1967. Field trial of two bivalent trachoma vaccines in children of Punjab Indian villages. *Am. J. Ophthalmol.* **63**(Suppl.)**:**1639–1644.

Dubin, P. J., and J. K. Kolls. 2007. IL-23 mediates inflammatory responses to mucoid *Pseudomonas aeruginosa* lung infection in mice. *Am. J. Physiol. Lung Cell Mol. Physiol.* **292:**L519–L528.

Eckmann, L., M. F. Kagnoff, and J. Fierer. 1993. Epithelial cells secrete the chemokine interleukin-8 in response to bacterial entry. *Infect. Immun.* **61:**4569–4574.

El-Asrar, A. M., K. Geboes, S. A. Al-Kharashi, A. A. Al-Mosallam, L. Missotten, L. Paemen, and G. Opdenakker. 2000. Expression of gelatinase B in trachomatous conjunctivitis. *Br. J. Ophthalmol.* **84:**85–91.

Eyerich, S., K. Eyerich, A. Cavani, and C. Schmidt-Weber. 2010. IL-17 and IL-22: siblings, not twins. *Trends Immunol.* **31:**354–361.

Fichorova, R. N., A. O. Cronin, E. Lien, D. J. Anderson, and R. R. Ingalls. 2002. Response to *Neisseria gonorrhoeae* by cervicovaginal epithelial cells occurs in the absence of toll-like receptor 4-mediated signaling. *J. Immunol.* **168:**2424–2432.

Fields, K. A., and T. Hackstadt. 2000. Evidence for the secretion of *Chlamydia trachomatis* CopN by a type III secretion mechanism. *Mol. Microbiol.* **38:**1048–1060.

Fukuda, E. Y., S. P. Lad, D. P. Mikolon, M. Iacobelli-Martinez, and E. Li. 2005. Activation of lipid metabolism contributes to interleukin-8 production during *Chlamydia trachomatis* infection of cervical epithelial cells. *Infect. Immun.* **73:**4017–4024.

Gambhir, M., M. G. Basanez, F. Turner, J. Kumaresan, and N. C. Grassly. 2007. Trachoma: transmission, infection, and control. *Lancet Infect. Dis.* **7:**420–427.

Geisler, W. M., R. J. Suchland, W. L. Whittington, and W. E. Stamm. 2001. Quantitative culture of *Chlamydia trachomatis*: relationship of inclusion-forming units produced in culture to clinical manifestations and acute inflammation in urogenital disease. *J. Infect. Dis.* **184:**1350–1354.

Geisler, W. M., C. Wang, S. G. Morrison, C. M. Black, C. I. Bandea, and E. W. Hook III. 2008. The natural history of untreated *Chlamydia trachomatis* infection in the interval between screening and returning for treatment. *Sex. Transm. Dis.* **35:**119–123.

Gervassi, A. L., P. Probst, W. E. Stamm, J. Marrazzo, K. H. Grabstein, and M. R. Alderson. 2003. Functional characterization of class Ia- and non-class Ia-restricted *Chlamydia*-reactive $CD8^+$ T cell responses in humans. *J. Immunol.* **171:**4278–4286.

Grassly, N. C., M. E. Ward, S. Ferris, D. C. Mabey, and R. L. Bailey. 2008. The natural history of trachoma infection and disease in a Gambian cohort with frequent follow-up. *PLoS Negl. Trop. Dis.* **2:**e341.

Grayston, J. T., S. P. Wang, L. J. Yeh, and C. C. Kuo. 1985. Importance of reinfection in the pathogenesis of trachoma. *Rev. Infect. Dis.* **7:**717–725.

Guan, Y., D. R. Ranoa, S. Jiang, S. K. Mutha, X. Li, J. Baudry, and R. I. Tapping. 2010. Human TLRs 10 and 1 share common mechanisms of innate immune sensing but not signaling. *J. Immunol.* **184:**5094–5103.

Guseva, N. V., S. T. Knight, J. D. Whittimore, and P. B. Wyrick. 2003. Primary cultures of female swine genital epithelial cells in vitro: a new approach for the study of hormonal modulation of *Chlamydia* infection. *Infect. Immun.* **71:**4700–4710.

Haggerty, C. L., S. L. Gottlieb, B. D. Taylor, N. Low, F. Xu, and R. B. Ness. 2010. Risk of sequelae after *Chlamydia trachomatis* genital infection in women. *J. Infect. Dis.* **201**(Suppl. 2): S134–S155.

Haggerty, C. L., R. B. Ness, A. Amortegui, S. L. Hendrix, S. L. Hillier, R. L. Holley, J. Peipert, H. Randall, S. J. Sondheimer, D. E. Soper, R. L. Sweet, and G. Trucco. 2003. Endometritis does not predict reproductive morbidity after pelvic inflammatory disease. *Am. J. Obstet. Gynecol.* **188:**141–148.

Hart, K. M., A. J. Murphy, K. T. Barrett, C. R. Wira, P. M. Guyre, and P. A. Pioli. 2009. Functional expression of pattern recognition receptors in tissues of the human female reproductive tract. *J. Reprod. Immunol.* **80:**33–40.

Hasan, U., C. Chaffois, C. Gaillard, V. Saulnier, E. Merck, S. Tancredi, C. Guiet, F. Briere, J. Vlach, S. Lebecque, G. Trinchieri, and E. E. Bates. 2005. Human TLR10 is a functional receptor, expressed by B cells and plasmacytoid dendritic cells, which activates gene transcription through MyD88. *J. Immunol.* **174:**2942–2950.

Heine, H., S. Muller-Loennies, L. Brade, B. Lindner, and H. Brade. 2003. Endotoxic activity and chemical structure of lipopolysaccharides from *Chlamydia trachomatis* serotypes E and L2 and *Chlamydophila psittaci* 6BC. *Eur. J. Biochem.* **270:**440–450.

Heinonen, P. K., and A. Miettinen. 1994. Laparoscopic study on the microbiology and severity of acute pelvic inflammatory disease. *Eur. J. Obstet. Gynecol. Reprod. Biol.* **57:**85–89.

Higgins, S. C., A. G. Jarnicki, E. C. Lavelle, and K. H. Mills. 2006. TLR4 mediates vaccine-induced protective cellular immunity to *Bordetella pertussis*: role of IL-17-producing T cells. *J. Immunol.* **177:**7980–7989.

Hillis, S. D., R. Joesoef, P. A. Marchbanks, J. N. Wasserheit, W. Cates, Jr., and L. Westrom. 1993. Delayed care of pelvic inflammatory disease as a risk factor for impaired fertility. *Am. J. Obstet. Gynecol.* **168:**1503–1509.

Hillis, S. D., L. M. Owens, P. A. Marchbanks, L. E. Amsterdam, and W. R. MacKenzie. 1997. Recurrent chlamydial infections increase the risks of hospitalization for ectopic pregnancy and pelvic inflammatory disease. *Am. J. Obstet. Gynecol.* **176:**103–107.

Hook, E. W., and H. H. Handsfield. 1999. Gonococcal infection in the adult, p. 451–466. *In* K. Holmes, P. F. Sparling, P. A. Mardh, S. M. Lemon, W. E. Stamm, and J. N. Wasserheit (ed.), *Sexually Transmitted Diseases*. McGraw-Hill Book Co., New York, NY.

Hook, E. W., III, C. Spitters, C. A. Reichart, T. M. Neumann, and T. C. Quinn. 1994. Use of cell culture and a rapid diagnostic assay for *Chlamydia trachomatis* screening. *JAMA* **272:**867–870.

Hvid, M., A. Baczynska, B. Deleuran, J. Fedder, H. J. Knudsen, G. Christiansen, and S. Birkelund. 2007. Interleukin-1 is the initiator of fallopian tube destruction during *Chlamydia trachomatis* infection. *Cell. Microbiol.* **9:**2795–2803.

Igietseme, J. U., G. A. Ananaba, J. Bolier, S. Bowers, T. Moore, T. Belay, F. O. Eko, D. Lyn, and C. M. Black. 2000. Suppression of endogenous IL-10 gene expression in dendritic cells enhances antigen presentation for specific Th1 induction: potential for cellular vaccine development. *J. Immunol.* **164:**4212–4219.

Ingalls, R. R., P. A. Rice, N. Qureshi, K. Takayama, J. S. Lin, and D. T. Golenbock. 1995. The inflammatory cytokine response to *Chlamydia trachomatis* infection is endotoxin mediated. *Infect. Immun.* **63:**3125–3130.

Jayarapu, K., M. Kerr, S. Ofner, and R. M. Johnson. 2010. Chlamydia-specific CD4 T cell clones control *Chlamydia muridarum* replication in epithelial cells by nitric oxide-dependent and -independent mechanisms. *J. Immunol.* **185:**6911–6920.

Jha, R., P. Srivastava, S. Salhan, A. Finckh, C. Gabay, A. Mittal, and S. Bas. 2011. Spontaneous secretion of interleukin-17 and -22 by human cervical cells in *Chlamydia trachomatis* infection. *Microbes Infect.* **13:**167–178.

Jones, R. B., B. R. Ardery, S. L. Hui, and R. E. Cleary. 1982. Correlation between serum antichlamydial antibodies and tubal factor as a cause of infertility. *Fertil. Steril.* **38:**553–558.

Kari, L., W. M. Whitmire, J. H. Carlson, D. D. Crane, N. Reveneau, D. E. Nelson, D. C. Mabey, R. L. Bailey, M. J. Holland, G. McClarty, and H. D. Caldwell. 2008. Pathogenic diversity among *Chlamydia trachomatis* ocular strains in nonhuman primates is affected by subtle genomic variations. *J. Infect. Dis.* **197:**449–456.

Karimi, O., S. Ouburg, H. J. de Vries, A. S. Pena, J. Pleijster, J. A. Land, and S. A. Morré. 2009. TLR2 haplotypes in the susceptibility to and severity of *Chlamydia trachomatis* infections in Dutch women. *Drugs Today* **45**(Suppl. B):67–74.

Kelly, K. A., S. Natarajan, P. Ruther, A. Wisse, M. H. Chang, and K. A. Ault. 2001. *Chlamydia trachomatis* infection induces mucosal addressin cell adhesion molecule-1 and vascular cell adhesion molecule-1, providing an immunologic link between the fallopian tube and other mucosal tissues. *J. Infect. Dis.* **184:**885–891.

Khader, S. A., G. K. Bell, J. E. Pearl, J. J. Fountain, J. Rangel-Moreno, G. E. Cilley, F. Shen, S. M. Eaton, S. L. Gaffen, S. L. Swain, R. M. Locksley, L. Haynes, T. D. Randall, and A. M. Cooper. 2007. IL-23 and IL-17 in the establishment of protective pulmonary $CD4^+$ T cell responses after vaccination and during *Mycobacterium tuberculosis* challenge. *Nat. Immunol.* **8:**369–377.

Kimani, J., I. W. Maclean, J. J. Bwayo, K. MacDonald, J. Oyugi, G. M. Maitha, R. W. Peeling, M. Cheang, N. J. Nagelkerke, F. A. Plummer, and R. C. Brunham. 1996. Risk factors for *Chlamydia trachomatis* pelvic inflammatory disease among sex workers in Nairobi, Kenya. *J. Infect. Dis.* **173:**1437–1444.

Kinnunen, A. H., H. M. Surcel, M. Lehtinen, J. Karhukorpi, A. Tiitinen, M. Halttunen, A. Bloigu, R. P. Morrison, R. Karttunen, and J. Paavonen. 2002. HLA DQ alleles and interleukin-10 polymorphism associated with *Chlamydia trachomatis*-related tubal factor infertility: a case-control study. *Hum. Reprod.* **17:**2073–2078.

Kiviat, N. B., M. Peterson, E. Kinney-Thomas, M. Tam, W. E. Stamm, and K. K. Holmes. 1985. Cytologic manifestations of cervical and vaginal infections. II. Confirmation of *Chlamydia trachomatis* infection by direct immunofluorescence using monoclonal antibodies. *JAMA* **253:**997–1000.

Koenders, M. I., J. K. Kolls, B. Oppers-Walgreen, B. L. van den, L. A. Joosten, J. R. Schurr, P. Schwarzenberger, W. B. Van Den Berg, and E. Lubberts. 2005. Interleukin-17 receptor deficiency results in impaired synovial expression of interleukin-1 and matrix metalloproteinases 3, 9, and 13 and prevents cartilage destruction during chronic reactivated streptococcal cell wall-induced arthritis. *Arthritis Rheum.* **52:**3239–3247.

Kolls, J. K., and A. Linden. 2004. Interleukin-17 family members and inflammation. *Immunity* **21:**467–476.

Lad, S. P., J. Li, J. da Silva Correia, Q. Pan, S. Gadwal, R. J. Ulevitch, and E. Li. 2007a. Cleavage of p65/RelA of the NF-kappaB pathway by *Chlamydia. Proc. Natl. Acad. Sci. USA* **104:**2933–2938.

Lad, S. P., G. Yang, D. A. Scott, G. Wang, P. Nair, J. Mathison, V. S. Reddy, and E. Li. 2007b. Chlamydial CT441 is a PDZ domain-containing tail-specific protease that interferes with the NF-kappaB pathway of immune response. *J. Bacteriol.* **189:**6619–6625.

Lapointe, T. K., P. M. O'Connor, N. L. Jones, D. Menard, and A. G. Buret. 2010. Interleukin-1 receptor phosphorylation activates Rho kinase to disrupt human gastric tight junctional claudin-4 during *Helicobacter pylori* infection. *Cell. Microbiol.* **12:**692–703.

Leese, H. J., J. I. Tay, J. Reischl, and S. J. Downing. 2001. Formation of fallopian tubal fluid: role of a neglected epithelium. *Reproduction* **121:**339–346.

Li, Z., D. Chen, Y. Zhong, S. Wang, and G. Zhong. 2008. The chlamydial plasmid-encoded protein pgp3 is secreted into the cytosol of *Chlamydia*-infected cells. *Infect. Immun.* **76:**3415–3428.

Liang, S. C., X. Y. Tan, D. P. Luxenberg, R. Karim, K. Dunussi-Joannopoulos, M. Collins, and L. A. Fouser. 2006. Interleukin (IL)-22 and IL-17 are coexpressed by Th17 cells and cooperatively enhance expression of antimicrobial peptides. *J. Exp. Med.* **203:**2271–2279.

Lichtenwalner, A. B., D. L. Patton, W. C. Van Voorhis, Y. T. Sweeney, and C. C. Kuo. 2004. Heat shock protein 60 is the major antigen which stimulates delayed-type hypersensitivity reaction in the macaque model of *Chlamydia trachomatis* salpingitis. *Infect. Immun* **72:**1159–1161.

Manor, E., and I. Sarov. 1988. Inhibition of *Chlamydia trachomatis* replication in HEp-2 cells by human monocyte-derived macrophages. *Infect. Immun.* **56:**3280–3284.

Matsumoto, A., H. Izutsu, N. Miyashita, and M. Ohuchi. 1998. Plaque formation by and plaque cloning of *Chlamydia trachomatis* biovar trachoma. *J. Clin. Microbiol.* **36:**3013–3019.

McClarty, G., and R. S. Stephens. 1999. Chlamydial metabolism as inferred from the complete genome sequence, p. 69–100. *In* R. S. Stephens (ed.), Chlamydia*: Intracellular Biology, Pathogenesis, and Immunity*. American Society for Microbiology, Washington, DC.

Miller, W. C., C. A. Ford, M. Morris, M. S. Handcock, J. L. Schmitz, M. M. Hobbs, M. S. Cohen, K. M. Harris, and J. R. Udry. 2004. Prevalence of chlamydial and gonococcal infections among young adults in the United States. *JAMA* **291:**2229–2236.

Mitchell, C., J. Hitti, K. Paul, K. Agnew, S. E. Cohn, A. E. Luque, and R. Coombs. 2011. Cervicovaginal shedding of HIV type 1 is related to genital tract inflammation independent of changes in vaginal microbiota. *AIDS Res. Hum. Retrovir.* **27:**35–39.

Molano, M., C. J. Meijer, E. Weiderpass, A. Arslan, H. Posso, S. Franceschi, M. Ronderos, N. Munoz, and A. J. van den Brule. 2005. The natural course of *Chlamydia trachomatis* infection in asymptomatic Colombian women: a 5-year follow-up study. *J. Infect. Dis.* **191:**907–916.

Morré, S. A., L. S. Murillo, C. A. Bruggeman, and A. S. Pena. 2003. The role that the functional Asp299Gly polymorphism in the toll-like receptor-4 gene plays in susceptibility to *Chlamydia trachomatis*-associated tubal infertility. *J. Infect. Dis.* **187:**341–342.

Morrison, C. S., A. N. Turner, and L. B. Jones. 2009. Highly effective contraception and acquisition of HIV and other sexually transmitted infections. *Best Pract. Res. Clin. Obstet. Gynaecol.* **23:**263–284.

Morrison, R. P., K. Feilzer, and D. B. Tumas. 1995. Gene knockout mice establish a primary protective role for major histocompatibility complex class II-restricted responses in *Chlamydia trachomatis* genital tract infection. *Infect. Immun.* **63:**4661–4668.

Morrison, R. P., K. Lyng, and H. D. Caldwell. 1989. Chlamydial disease pathogenesis. Ocular hypersensitivity elicited by a genus-specific 57-kD protein. *J. Exp. Med.* **169:**663–675.

Morrison, S. G., and R. P. Morrison. 2005. A predominant role for antibody in acquired immunity to chlamydial genital tract reinfection. *J. Immunol.* **175:**7536–7542.

Morrison, S. G., H. Su, H. D. Caldwell, and R. P. Morrison. 2000. Immunity to murine *Chlamydia trachomatis* genital tract reinfection involves B cells and CD4(+) T cells but not CD8(+) T cells. *Infect. Immun.* **68:**6979–6987.

Murillo, L. S., J. A. Land, J. Pleijster, C. A. Bruggeman, A. S. Pena, and S. A. Morré. 2003. Interleukin-1B (IL-1B) and interleukin-1 receptor antagonist (IL-1RN) gene polymorphisms are not associated with tubal pathology and *Chlamydia trachomatis*-related tubal factor subfertility. *Hum. Reprod.* **18:**2309–2314.

Murthy, A. K., J. P. Chambers, P. A. Meier, G. Zhong, and B. P. Arulanandam. 2007. Intranasal vaccination with a secreted chlamydial protein enhances resolution of genital *Chlamydia muridarum* infection, protects against oviduct pathology, and is highly dependent upon endogenous gamma interferon production. *Infect. Immun.* **75:**666–676.

Murthy, A. K., W. Li, B. K. Chaganty, S. Kamalakaran, M. N. Guentzel, J. Seshu, T. G. Forsthuber, G. Zhong, and B. P. Arulanandam. 2011. TNF-α production from $CD8^+$ T cells mediates oviduct pathological sequelae following primary genital *Chlamydia muridarum* infection. *Infect. Immun.* **79:**2928–2935.

Nagarajan, U. M., J. Sikes, D. Prantner, C. W. Andrews, Jr., L. Frazer, A. Goodwin, J. N. Snowden, and T. Darville. 2011. MyD88 deficiency leads to decreased NK cell gamma interferon production and T cell recruitment during *Chlamydia muridarum* genital tract infection, but a predominant Th1 response and enhanced monocytic inflammation are associated with infection resolution. *Infect. Immun.* **79:**486–498.

Natividad, A., G. Cooke, M. J. Holland, M. J. Burton, H. M. Joof, K. Rockett, D. P. Kwiatkowski, D. C. Mabey, and R. L. Bailey. 2006. A coding polymorphism in matrix metalloproteinase 9 reduces risk of scarring sequelae of ocular *Chlamydia trachomatis* infection. *BMC Med. Genet.* **7:**40.

Natividad, A., T. C. Freeman, D. Jeffries, M. J. Burton, D. C. Mabey, R. L. Bailey, and M. J. Holland. 2010. Human conjunctival transcriptome analysis reveals the prominence of innate defense in *Chlamydia trachomatis* infection. *Infect. Immun.* **78:**4895–4911.

Natividad, A., N. Hanchard, M. J. Holland, O. S. Mahdi, M. Diakite, K. Rockett, O. Jallow, H. M. Joof, D. P. Kwiatkowski, D. C. Mabey, and R. L. Bailey. 2007. Genetic variation at the TNF locus and the risk of severe sequelae of ocular *Chlamydia trachomatis* infection in Gambians. *Genes Immun.* **8:**288–295.

Natividad, A., J. Wilson, O. Koch, M. J. Holland, K. Rockett, N. Faal, O. Jallow, H. M. Joof, M. J. Burton, N. D. Alexander, D. P. Kwiatkowski, D. C. Mabey, and R. L. Bailey. 2005. Risk of trachomatous scarring and trichiasis in Gambians varies with SNP haplotypes at the interferon-gamma and interleukin-10 loci. *Genes Immun.* **6:**332–340.

Neff, L., S. Daher, P. Muzzin, U. Spenato, F. Gulacar, C. Gabay, and S. Bas. 2007. Molecular characterization and subcellular localization of macrophage infectivity potentiator, a *Chlamydia trachomatis* lipoprotein. *J. Bacteriol.* **189:**4739–4748.

Ness, R. B., D. E. Soper, R. L. Holley, J. Peipert, H. Randall, R. L. Sweet, S. J. Sondheimer, S. L. Hendrix, A. Amortegui, G. Trucco, D. C. Bass, and S. F. Kelsey. 2001. Hormonal and barrier contraception and risk of upper genital tract disease in the PID Evaluation and Clinical Health (PEACH) study. *Am. J. Obstet. Gynecol.* **185:**121–127.

Ness, R. B., D. E. Soper, H. E. Richter, H. Randall, J. F. Peipert, D. B. Nelson, D. Schubeck, S. G. McNeeley, W. Trout, D. C. Bass, K. Hutchison, K. Kip, and R. C. Brunham. 2008. *Chlamydia* antibodies, chlamydia heat shock protein, and adverse sequelae after pelvic inflammatory disease: the PID Evaluation and Clinical Health (PEACH) Study. *Sex. Transm. Dis.* **35:**129–135.

O'Connell, C. M., Y. M. Abdelrahman, E. Green, H. K. Darville, K. Saira, B. Smith, T. Darville, A. M. Scurlock, C. R. Meyer, and R. J. Belland. 2011. TLR2 activation by *Chlamydia trachomatis* is plasmid dependent, and plasmid-responsive chromosomal loci are coordinately regulated in response to glucose limitation by *C. trachomatis* but not by *C. muridarum*. *Infect. Immun.* **79:**1044–1056.

O'Connell, C. M., R. R. Ingalls, C. W. Andrews, Jr., A. M. Skurlock, and T. Darville. 2007. Plasmid-deficient *Chlamydia muridarum* fail to induce immune pathology and protect against oviduct disease. *J. Immunol.* **179:**4027–4034.

O'Connell, C. M., I. A. Ionova, A. J. Quayle, A. Visintin, and R. R. Ingalls. 2006. Localization of TLR2 and MyD88 to *Chlamydia trachomatis* inclusions. Evidence for signaling by intracellular TLR2 during infection with an obligate intracellular pathogen. *J. Biol. Chem.* **281:**1652–1659.

O'Connell, C. M., and K. M. Nicks. 2006. A plasmid-cured *Chlamydia muridarum* strain displays altered plaque morphology and reduced infectivity in cell culture. *Microbiology* **152:**1601–1607.

Ohashi, K., V. Burkart, S. Flohe, and H. Kolb. 2000. Cutting edge: heat shock protein 60 is a putative endogenous ligand of the toll-like receptor-4 complex. *J. Immunol.* **164:**558–561.

Ouburg, S., J. Spaargaren, J. E. den Hartog, J. A. Land, J. S. Fennema, J. Pleijster, A. S. Pena, and S. A. Morre. 2005. The CD14 functional gene polymorphism -260 C > T is not involved in either the susceptibility to *Chlamydia trachomatis* infection or the development of tubal pathology. *BMC Infect. Dis.* **5:**114.

Patton, D. L., D. E. Moore, L. R. Spadoni, M. R. Soules, S. A. Halbert, and S. P. Wang. 1989. A comparison of the fallopian tube's response to overt and silent salpingitis. *Obstet. Gynecol.* **73:**622–630.

Peeling, R. W., J. Kimani, F. Plummer, I. Maclean, M. Cheang, J. Bwayo, and R. C. Brunham. 1997. Antibody to chlamydial hsp60 predicts an increased risk for chlamydial pelvic inflammatory disease. *J. Infect. Dis.* **175:**1153–1158.

Perry, L. L., K. Feilzer, and H. D. Caldwell. 1997. Immunity to *Chlamydia trachomatis* is mediated by T helper 1 cells through IFN-gamma-dependent and -independent pathways. *J. Immunol.* **158:**3344–3352.

Perry, L. L., H. Su, K. Feilzer, R. Messer, S. Hughes, W. Whitmire, and H. D. Caldwell. 1999. Differential sensitivity of distinct *Chlamydia trachomatis* isolates to IFN-γ -mediated inhibition. *J. Immunol.* **162:**3541–3548.

Pioli, P. A., E. Amiel, T. M. Schaefer, J. E. Connolly, C. R. Wira, and P. M. Guyre. 2004. Differential expression of Toll-like receptors 2 and 4 in tissues of the human female reproductive tract. *Infect. Immun.* **72:**5799–5806.

Prantner, D., T. Darville, J. D. Sikes, C. W. Andrews, Jr., H. Brade, R. G. Rank, and U. M. Nagarajan. 2009. Critical role for interleukin-1beta (IL-1beta) during *Chlamydia muridarum* genital infection and bacterial replication-independent secretion of IL-1beta in mouse macrophages. *Infect. Immun.* **77:**5334–5346.

Punnonen, R., P. Terho, V. Nikkanen, and O. Meurman. 1979. Chlamydial serology in infertile women by immunofluorescence. *Fertil. Steril.* **31:**656–659.

Ramsey, K. H., I. M. Sigar, S. V. Rana, J. Gupta, S. M. Holland, and G. I. Byrne. 2001. Role for inducible nitric oxide synthase in protection from chronic *Chlamydia trachomatis* urogenital disease in mice and its regulation by oxygen free radicals. *Infect. Immun.* **69:**7374–7379.

Ramsey, K. H., I. M. Sigar, J. H. Schripsema, N. Shaba, and K. P. Cohoon. 2005. Expression of matrix metalloproteinases subsequent to urogenital *Chlamydia muridarum* infection of mice. *Infect. Immun.* **73:**6962–6973.

Rank, R. G., A. K. Bowlin, R. L. Reed, and T. Darville. 2003. Characterization of chlamydial genital infection resulting from sexual transmission from male to female guinea pigs and determination of infectious dose. *Infect. Immun.* **71:**6148–6154.

Rank, R. G., C. Dascher, A. K. Bowlin, and P. M. Bavoil. 1995a. Systemic immunization with Hsp60 alters the development of chlamydial ocular disease. *Investig. Ophthalmol. Vis. Sci.* **36:**1344–1351.

Rank, R. G., H. M. Lacy, A. Goodwin, J. Sikes, J. Whittimore, P. B. Wyrick, and U. M. Nagarajan. 2010. Host chemokine and cytokine response in the endocervix within the first developmental cycle of *Chlamydia muridarum*. *Infect. Immun.* **78:**536–544.

Rank, R. G., M. M. Sanders, and D. L. Patton. 1995b. Increased incidence of oviduct pathology in the guinea pig after repeat vaginal inoculation with the chlamydial agent of guinea pig inclusion conjunctivitis. *J. Sex. Transm. Dis.* **22:**48–54.

Rasmussen, S. J., L. Eckmann, A. J. Quayle, L. Shen, Y. X. Zhang, D. J. Anderson, J. Fierer, R. S. Stephens, and M. F. Kagnoff. 1997. Secretion of proinflammatory cytokines byepithelial cells in response to *Chlamydia* infection suggests a central role for epithelial cells in chlamydial pathogenesis. *J. Clin. Investig.* **99:**77–87.

Ravel, J., P. Gajer, Z. Abdo, G. M. Schneider, S. S. Koenig, S. L. McCulle, S. Karlebach, R. Gorle, J. Russell, C. O. Tacket, R. M. Brotman, C. C. Davis, K. Ault, L. Peralta, and L. J. Forney. 2011. Vaginal microbiome of reproductive-age women. *Proc. Natl. Acad. Sci. USA* **108**(Suppl. 1)**:**4680–4687.

Roth, A., P. Konig, G. van Zandbergen, M. Klinger, T. Hellwig-Burgel, W. Daubener, M. K. Bohlmann, and J. Rupp. 2010. Hypoxia abrogates antichlamydial properties of IFN-γ in human fallopian tube cells in vitro and ex vivo. *Proc. Natl. Acad. Sci. USA* **107:**19502–19507.

Russell, M., T. Darville, K. Chandra-Kuntal, B. Smith, C. W. Andrews, Jr., and C. M. O'Connell. 2011. Infectivity acts as in vivo selection for maintenance of the chlamydial cryptic plasmid. *Infect. Immun.* **79:**98–107.

Schaefer, T. M., J. V. Fahey, J. A. Wright, and C. R. Wira. 2005. Innate immunity in the human female reproductive tract: antiviral response of uterine epithelial cells to the TLR3 agonist poly(I:C). *J. Immunol.* **174:**992–1002.

Scriba, T. J., B. Kalsdorf, D. A. Abrahams, F. Isaacs, J. Hofmeister, G. Black, H. Y. Hassan, R. J. Wilkinson, G. Walzl, S. J. Gelderbloem, H. Mahomed, G. D. Hussey, and W. A. Hanekom. 2008. Distinct, specific IL-17- and IL-22-producing $CD4^+$ T cell subsets contribute to the human anti-mycobacterial immune response. *J. Immunol.* **180:**1962–1970.

Scurlock, A. M., L. C. Frazer, C. W. Andrews, Jr., C. M. O'Connell, I. P. Foote, S. L. Bailey, K. Chandra-Kuntal, J. K. Kolls, and T. Darville. 2010. IL-17 contributes to generation of Th1 immunity and neutrophil recruitment during *Chlamydia muridarum* genital tract infection but is not required for macrophage influx or normal resolution of infection. *Infect. Immun.* **79:**1349–1362.

Shah, A. A., J. H. Schripsema, M. T. Imtiaz, I. M. Sigar, J. Kasimos, P. G. Matos, S. Inouye, and K. H. Ramsey. 2005. Histopathologic changes related to fibrotic oviduct occlusion after genital tract infection of mice with *Chlamydia muridarum. Sex. Transm. Dis.* **32:**49–56.

Shao, R., S. X. Zhang, B. Weijdegard, S. Zou, E. Egecioglu, A. Norstrom, M. Brannstrom, and H. Billig. 2010. Nitric oxide synthases and tubal ectopic pregnancies induced by *Chlamydia* infection: basic and clinical insights. *Mol. Hum. Reprod.* **16:**907–915.

Shemer-Avni, Y., D. Wallach, and I. Sarov. 1988. Inhibition of *Chlamydia trachomatis* growth by recombinant tumor necrosis factor. *Infect. Immun.* **56:**2503–2506.

Skwor, T. A., B. Atik, R. P. Kandel, H. K. Adhikari, B. Sharma, and D. Dean. 2008. Role of secreted conjunctival mucosal cytokine and chemokine proteins in different stages of trachomatous disease. *PLoS Negl. Trop. Dis.* **2:**e264.

Stephens, R. S. 2003. The cellular paradigm of chlamydial pathogenesis. *Trends Microbiol.* **11:**44–51.

Su, H., K. Feilzer, H. D. Caldwell, and R. P. Morrison. 1997. *Chlamydia trachomatis* genital tract infection of antibody-deficient gene knockout mice. *Infect. Immun.* **65:**1993–1999.

Su, H., G. McClarty, F. Dong, G. M. Hatch, Z. K. Pan, and G. Zhong. 2004. Activation of Raf/MEK/ERK/cPLA2 signaling pathway is essential for chlamydial acquisition of host glycerophospholipids. *J. Biol. Chem.* **279:**9409–9416.

Swanson, A. F., R. A. Ezekowitz, A. Lee, and C. C. Kuo. 1998. Human mannose-binding protein inhibits infection of HeLa cells by *Chlamydia trachomatis. Infect. Immun.* **66:**1607–1612.

Sweet, R. L., M. Blankfort-Doyle, M. O. Robbie, and J. Schachter. 1986. The occurrence of chlamydial and gonococcal salpingitis during the menstrual cycle. *JAMA* **255:**2062–2065.

Sziller, I., O. Babula, A. Ujhazy, B. Nagy, P. Hupuczi, Z. Papp, I. M. Linhares, W. J. Ledger, and S. S. Witkin. 2007. *Chlamydia trachomatis* infection, fallopian tube damage and a mannose-binding lectin codon 54 gene polymorphism. *Hum. Reprod.* **22:**1861–1865.

Tjiam, K. H., G. H. Zeilmaker, A. T. Alberda, B. Y. van Heijst, J. C. de Roo, A. A. Polak-Vogelzang, T. van Joost, E. Stolz, and F. Michel. 1985. Prevalence of antibodies to *Chlamydia trachomatis, Neisseria gonorrhoeae,* and *Mycoplasma hominis* in infertile women. *Genitourin. Med.* **61:**175–178.

Toye, B., C. Laferriäre, P. Claman, P. Jessamine, and R. Peeling. 1993. Association between antibody to the chlamydial heat-shock protein and tubal infertility. *J. Infect. Dis.* **168:**1236–1240.

Tseng, C. T., and R. G. Rank. 1998. Role of NK cells in early host response to chlamydial genital infection. *Infect. Immun.* **66:**5867–5875.

Vanover, J., J. Kintner, J. Whittimore, and R. V. Schoborg. 2010. Interaction of herpes simplex virus type 2 (HSV-2) glycoprotein D with the host cell surface is sufficient to induce *Chlamydia trachomatis* persistence. *Microbiology* **156:**1294–1302.

Van Voorhis, W. C., L. K. Barrett, Y. T. Sweeney, C. C. Kuo, and D. L. Patton. 1996. Analysis of lymphocyte phenotype and cytokine activity in the inflammatory infiltrates of the upper genital tract of female macaques infected with *Chlamydia trachomatis. J. Infect. Dis.* **174:**647–650.

Van Voorhis, W. C., L. K. Barrett, Y. T. Sweeney, C. C. Kuo, and D. L. Patton. 1997. Repeated *Chlamydia trachomatis* infection of *Macaca nemestrina* fallopian tubes produces a Th1-like cytokine response associated with fibrosis and scarring. *Infect. Immun.* **65:**2175–2182.

Vender, J., and J. W. Moulder. 1967. Initial step in catabolism of glucose by the meningopneumonitis agent. *J. Bacteriol.* **94:**867–869.

Vonck, R. A., T. Darville, C. M. O'Connell, and A. E. Jerse. 2011. Chlamydial infection increases gonococcal colonization in a novel murine coinfection model. *Infect. Immun.* **79:**1566–1577.

Wang, S. P., J. T. Grayston, and E. R. Alexander. 1967. Trachoma vaccine studies in monkeys. *Am. J. Ophthalmol.* **63(**Suppl.)**:**1615–1630.

Watkins, N. G., W. J. Hadlow, A. B. Moos, and H. D. Caldwell. 1986. Ocular delayed hypersensitivity: a pathogenetic mechanism of chlamydial-conjunctivitis in guinea pigs. *Proc. Natl. Acad. Sci. USA* **83:**7480–7484.

Werner, G. T., and D. K. Sareen. 1977. Trachoma in Punjab: a study of the prevalence and of mass treatment. *Trop. Geogr. Med.* **29:**135–140.

West, S. K., B. Munoz, H. Mkocha, Y. H. Hsieh, and M. C. Lynch. 2001. Progression of active trachoma to scarring in a cohort of Tanzanian children. *Ophthalmic Epidemiol.* **8:**137–144.

Wiesenfeld, H. C., R. P. Heine, M. A. Krohn, S. L. Hillier, A. A. Amortegui, M. Nicolazzo, and R. L. Sweet. 2002. Association between elevated neutrophil defensin levels and endometritis. *J. Infect. Dis.* **186:**792–797.

Wiesenfeld, H. C., R. L. Sweet, R. B. Ness, M. A. Krohn, A. J. Amortegui, and S. L. Hillier. 2005. Comparison of acute and subclinical pelvic inflammatory disease. *Sex. Transm. Dis.* **32:**400–405.

Woolridge, R. L., J. T. Grayston, I. H. Chang, C. Y. Yang, and K. H. Cheng. 1967. Long-term follow-up of the initial (1959–1960) trachoma vaccine field trial on Taiwan. *Am. J. Ophthalmol.* **63**(Suppl.)**:**1650–1655.

Wyrick, P. B., S. T. Knight, T. R. Paul, R. G. Rank, and C. S. Barbier. 1999. Persistent chlamydial envelope antigens in antibiotic-exposed infected cells trigger neutrophil chemotaxis. *J. Infect. Dis.* **179:**954–966.

Zhang, B., S. Ye, S. M. Herrmann, P. Eriksson, M. de Maat, A. Evans, D. Arveiler, G. Luc, F. Cambien, A. Hamsten, H. Watkins, and A. M. Henney. 1999. Functional polymorphism in the regulatory region of gelatinase B gene in relation to severity of coronary atherosclerosis. *Circulation* **99:**1788–1794.

Zheng, Y., P. A. Valdez, D. M. Danilenko, Y. Hu, S. M. Sa, Q. Gong, A. R. Abbas, Z. Modrusan, N. Ghilardi, F. J. de Sauvage, and W. Ouyang. 2008. Interleukin-22 mediates early host defense against attaching and effacing bacterial pathogens. *Nat. Med.* **14:**282–289.

Zhong, G. 2009. Killing me softly: chlamydial use of proteolysis for evading host defenses. *Trends Microbiol.* **17:**467–474.

Zhong, G., P. Fan, H. Ji, F. Dong, and Y. Huang. 2001. Identification of a chlamydial protease-like activity factor responsible for the degradation of host transcription factors. *J. Exp. Med.* **193:**935–942.

CHLAMYDIAL PERSISTENCE REDUX

Gerald I. Byrne and Wandy L. Beatty

12

INTRODUCTION

In the third edition of the Wilson et al. text on bacterial pathogenesis (Wilson et al., 2011) the authors make a point of explaining that a major reason for bacteria to be in the public health spotlight in the 21st century is their capacity to develop resistance to antibiotics, to acquire genes from other bacteria and thereby accrue new virulence traits, or to emerge as opportunistic pathogens due to host abatement to combat innocuous microbes. These arguments are in fact true for many infectious disease pathogens, but they categorically do not apply to chlamydiae. The genus *Chlamydia* is essentially pathogenic by definition (it has no other niche but the eukaryotic host to exploit) and virtually ubiquitous; yet it fits none of the criteria deemed to be essential for success as an infectious disease agent these days.

First, at present antibiotic resistance is not a major issue with chlamydiae. It is true that a tetracycline resistance cassette has been identified in the genome of some isolates of *Chlamydia suis*, a porcine pathogen that is constantly exposed to tetracycline in hog feed; but this finding has not emerged as a problem in hogs, let alone other species. Certainly we should be ever vigilant for emergent drug resistance against both human and animal infections, since the repertoire of effective classes of drugs against chlamydiae is limited. But other than for porcine infections, the reasons that chronic chlamydial infections are difficult to effectively manage are not related to the development of true genetic resistance, but rather to a drug tolerant phenotype. By and large, human isolates have remained fully susceptible to tetracyclines, erythromycin, and azalides.

Second, the *Chlamydiaceae* are not very genetically diverse. All members of the *Chlamydiaceae* have very similar genomes (reviewed in chapter 2, "Deep and wide: comparative genomics of *Chlamydia*"). They code for roughly 1,000 proteins, and most of the same genes are not only present in all strains but also preserved in the same order in the genome. This fact argues strongly that transfer of genes to chlamydiae from other bacteria does not occur with any degree of frequency. Even development of experimental gene transfer systems for chlamydiae has been problematical (reviewed in chapter 15, "Chlamydial genetics: decades of efforts,

Gerald I. Byrne, Department of Microbiology, Immunology, and Biochemistry, University of Tennessee Health Science Center, Memphis, TN 38163. *Wandy L. Beatty*, Department of Molecular Microbiology, Washington University at St. Louis, St. Louis, MO 63110.

Intracellular Pathogens I: Chlamydiales
Edited by Ming Tan and Patrik M. Bavoil © 2012 ASM Press, Washington, DC
10.1128/9781555817329.ch12

very recent successes"). This research specialty field has been fraught with stops and starts, with hopeful signs followed by long periods of silence. A tractable gene transfer system has not been developed to study chlamydial virulence and pathogenesis. Thus, exchange of genes is something chlamydiae neither do on their own nor can be coerced to do experimentally under idealized laboratory conditions.

Finally, the chlamydiae are anything but opportunistic. Some of the earliest recorded accurate descriptions of an infectious disease are a hieroglyphic description of remedies for trachoma from an ancient Egyptian text written some 3,400 years ago (Stern, 1875) and a written comment about eye disease from the Chang dynasty in China from more than 3,000 years ago (Taylor, 2008). *Chlamydia* continues to be a serious reproductive health problem for women globally despite quantum-like improvements in screening and treatment. In the United States, *Chlamydia trachomatis* is the most frequently reported bacterial infectious disease, period; and similar reporting frequencies are seen everywhere around the world. *Chlamydia* spp. continue to cause problems both in domesticated animals and in sylvatic populations (e.g., koalas).

How is it that *Chlamydia*, a bacterial genus that has none of the features associated with textbook traits for success as a pathogen, has been successfully causing disease in eukaryotes for the past 700 million years or so (Horn et al., 2004)? Obviously the chlamydiae have found another way. The relationships that chlamydiae have evolved with their eukaryotic hosts, in terms of both successfully invading epithelial or mononuclear phagocytic cells and dealing with the innate and acquired immune response generated in their presence, are key elements in understanding the success that chlamydiae enjoy. We have made dramatic strides in the postgenomic era in understanding what chlamydiae can and cannot do biochemically and metabolically and how the host responds to the presence of these pathogens. However, a still unresolved question is whether the encounter of chlamydiae with the host involves what has come to be known as persistence, which has been defined as a reversible interruption in the productive intracellular chlamydial growth cycle that is mediated by environmental factors (Fig. 1). It is the purpose of this chapter to define what we do and do not know about chlamydial persistence with the goal of determining the extent of the role that persistence plays in chlamydial disease (if at all) and what we should be doing to better understand this chlamydial attribute.

PERSISTENCE AS A FUNCTION OF MICROBIAL PATHOGENESIS: WHAT DO OTHER MICROBIAL PATHOGENS DO, AND ARE THERE ANALOGIES FOR CHLAMYDIAE?

One of the problems encountered in the study of chlamydiae is that often the basic biology of the pathogen, worked out in cell culture systems, is then applied to speculate about actual disease pathogenesis, without knowing if there are strong in vivo correlates to cell culture observations. In contrast, there are a number of important microbial pathogens that clearly have a persistent, dormant, or latent stage associated with disease. These infections may serve as examples of well-characterized persistence phenotypes associated with disease. For the pathogens described below, persistence was first documented as a feature of the disease, and this was followed by characterization of persistence mechanisms, which is just the opposite of how the problem of persistence has been developed in the case of *Chlamydia*.

Vivax Malaria

Malaria is arguably the most significant infectious disease on the planet. Although four species of *Plasmodium* cause infections in people (*P. falciparum*, *P. vivax*, *P. malariae*, and *P. ovale*), two dominate: *P. falciparum* and *P. vivax*. *P. falciparum*, which causes a form of malaria called malignant tertian malaria, is considered a major killer due to its capacity to modify the surface of infected erythrocytes such that they stick to each other and to the endothelial cells lining vessels and capillaries.

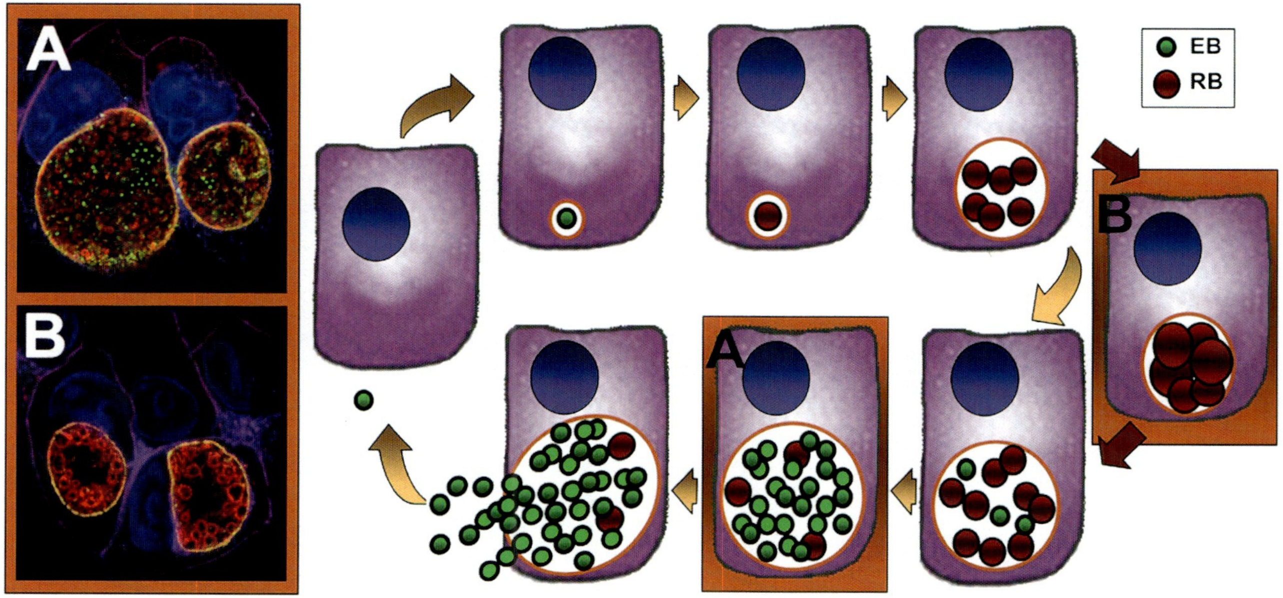

FIGURE 1 Schematic diagram of the chlamydial developmental cycle. Red arrows indicate altered intracellular chlamydial development mediated by environmental factors. (A) Immunofluorescence images and schematic show mature *C. trachomatis* serovar B inclusions with elementary bodies (EBs) in green (anti-OmcB), reticulate bodies (RBs) in red (anti-major outer membrane protein), and the inclusion membrane in orange (anti-incG). (B) Persistence in vitro in response to IFN-γ results in enlarged, aberrant RBs that can be maintained in this state for extended periods of time, with subsequent reversion to normal intracellular development. doi:10.1128/9781555817329.ch12.f1

This sequestration of infected erythrocytes has numerous consequences important for both diagnosis (e.g., only ring stages are found in the circulation) and pathogenesis, as blockage of capillaries causes ischemic consequences. The most serious capillary blockage complication is cerebral malaria, the principal immediate cause of malaria mortality in children. But vivax malaria also can have severe consequences, often disproportional to the level of parasitemia, which is generally less than for *P. falciparum* due to the *P. vivax* preference for young erythrocytes. *P. vivax* and *P. ovale* also have the unusual capacity to cause a clinical relapse months to years after the initial illness has cleared, even though the individual has left the area of endemicity and is not at risk of becoming reinfected. Although all plasmodia are intracellular pathogens of erythrocytes, they must go through an obligatory stage in the liver before merozoites are released into the circulation. This preerythrocytic stage is a one-time-only portion of the infectious cycle for falciparum malaria, but an added developmental form is produced by *P. vivax* and *P. ovale*, called the hypnozoite, that remains within hepatic cells in a nongrowing dormant state. This form of the parasite can reactivate to produce red blood cell-invading merozoites from weeks to months and even years after the initial circulating parasitemia has been cleared or treated. Although the nature of the hypnozoite remains mysterious, recent findings (Wells et al., 2010; Imwong et al., 2007; Chen et al., 2007) demonstrate that these "dormant" forms are clonal and are frequently genetic variants of the original infecting strain. The existence of hypnozoites not only is well founded in the clinical literature but also has practical ramifications for malaria management because additional therapeutics are required to eradicate the liver-bound hypnozoites.

WHAT ABOUT CHLAMYDIAE?

Vivax malaria represents a classic example of persistence of a unique developmental form that can emerge at a later time and cause disease. Are there data to suggest that something similar exists for chlamydiae? It is important to recognize that the first consideration should be whether or not clinical conditions or indications suggest the presence of a dormant persistent form similar to the *P. vivax* hypnozoite. For example, there are a smattering of reports describing individuals with reactivation of ocular disease years after they have left regions of the world where trachoma is endemic (reviewed by Taylor [2008]). One might predict that if a dormant form of *C. trachomatis* could reactivate and cause trachoma, this type of presentation would be frequently reported; but it is not. In sharp contrast, everyone who is infected with vivax malaria runs the risk of reactivated infections. The hypnozoite is part of the *P. vivax* developmental process within the mammalian host; but there is little evidence that there is an analogous dormant but reactivatable stage of chlamydial infection.

Interestingly, there may be a developmental form of *Chlamydia*, at least for *Chlamydia psittaci*, that resembles a hypnozoite. This is the "cryptic" form originally described by Moulder in the late 1970s (Moulder et al., 1980) in a mouse fibroblast cell culture model, which has been completely understudied since. Cryptic chlamydiae emerge after the developmental cycle is complete and the host cell population experiences "wipeout." That is to say, virtually all of the host cells in the cell culture population support a productive infection and are then destroyed, releasing infectious elementary bodies (EBs). However, if one saves the flasks and adds fresh medium that is allowed to incubate, a new population of host cells will emerge in about 2 weeks. Interestingly, these cells are not capable of being reinfected by the same strain that yielded the initial wipeout. This might very well be because the surviving host cells are variants that no longer take up the invading strain; but the curious finding that Moulder noticed was that if the incubation time is extended for another 2 weeks or so without the reinfection step, the cells eventually become infected with chlamydiae again and another wipeout soon follows. The conclusion drawn from these studies was that the surviving host cells harbor cryptic chlamydiae, which initiate a new round of productive infection under favorable environmental conditions. The search for cryptic chlamydiae during the initial stages of host cell regrowth proved futile, and therefore the cryptic form of the pathogen was proposed by inference. These interesting observations have never been properly followed up and are worthy of reexamination.

Are there cryptic chlamydiae that emerge during a natural infection and act in a manner similar to that of hypnozoites? Unfortunately, the appropriate in vivo investigation has not been done. An important lesson that can be drawn from these studies is that it may be dangerous to base any type of disease property exclusively on in vitro findings, but unfortunately this has been done time and time again for *Chlamydia*. The case of cryptic chlamydiae may simply represent a cell culture-based mechanism for survival of the pathogen. If cryptic chlamydiae exist in nature, they might be unique to *C. psittaci* and related species (Moulder never found the equivalent in *C. trachomatis* infections [Lee and Moulder, 1981]). Therefore, reactivation of bird infections, genital infections in sheep, or intestinal infections in swine might be a good method to look for these enigmatic chlamydial developmental forms with unique properties.

Toxoplasmosis

Toxoplasma gondii is an obligate intracellular protozoan parasite that is acquired by eating undercooked meat (mainly pork and lamb, but also beef) containing *Toxoplasma* cysts or by ingesting mature oocysts associated with cat feces, felines being the definitive host for *T. gondii* (Montoya and Liesenfeld, 2004). Congenital transmission places the fetus at risk for serious disease as the parasite can enter the fetal circulation by infection of the placenta when maternal primary disease occurs during gestation. Infection also can occur as a result of organ transplant from an asymptomatic, seropositive donor to a seronegative recipient. This is because once the acute stage of the disease has resolved (often asymptomatic in immunocompetent individuals), the infection goes into a persistent phase characterized by the presence of slow-growing forms, contained within an intracellular cyst-like structure. These developmental forms are called bradyzoites to distinguish them from the fast-growing tachyzoites that characterize acute infections. This latent or persistent stage of *T. gondii* development is a critical component in the pathogenesis and transmission of this parasite. The significance of the bradyzoite form of the parasite became apparent when it was recognized that transmission frequently occurs via ingestion of contaminated meat (reviewed by Sullivan et al. [2009]), but the clinical significance of tissue cysts was not recognized until immunosuppression became more frequent as a result of organ transplantation or cancer therapy. Then, when AIDS emerged as a major new immunosuppressive disease, reactivated toxoplasmosis became one of the most frequently encountered opportunistic infections in AIDS patients with low $CD4^+$ cell counts (Sullivan et al., 2009; Weiss and Kim, 2000).

WHAT ABOUT CHLAMYDIAE?

There are many analogies between *Toxoplasma* and *Chlamydia*, as these two intracellular pathogens represent remarkable examples of convergent evolution of very different life forms. *Toxoplasma* and *Chlamydia* both invade a variety of host cell types, ranging from epithelial cells to mononuclear phagocytes. They both reside within the confines of a cytoplasmic vesicle termed the parasitophorous vacuole in the case of *Toxoplasma* and the inclusion in the case of *Chlamydia*. They both initiate a sequence of events that probably starts during the uptake process to prevent lysosomes from fusing with the pathogen-containing cytoplasmic vesicle, thus sequestering themselves from host cell-specific antimicrobial factors. They both cause extensive remodeling of vesicle membranes and initiate reprogramming events that alter host cell physiology to the benefit of the pathogen.

Given the similarities of their intracellular niches, it may not be too surprising that intracellular growth of both of these pathogens can be curtailed by similar immune mechanisms. Both are affected by intracellular starvation conditions imposed by induction of a tryptophan-decyclizing enzyme IDO (indoleamine 2,3 dioxygenase), which is induced in the presence of the immune-regulated cytokine gamma interferon (IFN-γ) (Byrne et al., 1986; Pfefferkorn et al., 1986). This may not be completely unique to *Chlamydia* and *Toxoplasma* (for example, see Peng and Monack, 2010), yet it represents a nutrient starvation-based mechanism to slow intracellular growth without pathogen eradication. The idea that the induction of IDO by the host cell, through the innate or acquired immune response, and the reversible inhibition of intracellular growth contribute to the development of *Toxoplasma* bradyzoites and development of slow-growing aberrant chlamydial forms is a compelling model for the development of persistence. Unfortunately this model is also overly simplistic. Bradyzoite development in *Toxoplasma* is by no means fully understood but is thought to be a stress response regulated by a multilayered cascade of steps that involves production of an intracellular protective "cyst" coat around the parasitophorous vacuole, a shutdown of general protein synthesis, and interference with replication (Montoya and Liesenfeld, 2004; Sullivan et al., 2009; Weiss and Kim, 2000). It is also known that the likelihood of bradyzoite formation is to some extent strain dependent, with slower-growing, less virulent strains being more adept at shifting from the tachyzoite to the bradyzoite stage (Soete et al., 1994; Howe and Sibley, 1995). Depletion of intracellular tryptophan via IDO may contribute to this morphogenesis, but the reversible shift between a tachyzoite and a bradyzoite is thought to be stochastic, with intact immunity better able to control the tachyzoite stage.

The details of cell culture models of chlamydial persistence have been reviewed extensively (for examples, see Beatty et al., 1994; Hogan et al., 2004; Wyrick, 2010; and Schoborg, 2011), and correlations to in vivo conditions have been developed (Hogan et al., 2004). The overall process involves virtually complete absence of intracellular replication, the appearance of greatly enlarged aberrant reticulate body (RB) forms, and dramatic reprogramming of protein expression patterns. There are some data to support altered transcriptional patterns (Mathews et al., 2001; Gerard et al., 2004; Goellner et al., 2006; Maurer et al., 2007), disconnected transcription and translation (Ouellette et al., 2006), and regulated gene expression via noncoding small RNAs during cell culture-induced chlamydial persistence (AbdelRahman et al., 2011). However, little is known about structural changes to the inclusion membrane, and there are no reports of clinical complications stemming from chlamydial reactivation in immunocompromised individuals. In fact, in most natural animal infections in which persistence has been documented, the persistent state is associated with asymptomatic infection (Pospischil et al., 2009; Reinhold et al., 2010, 2011). A notable exception may be persistence of chlamydiae in ewes subsequent to infectious abortion, where the persistence of the pathogen is thought to contribute to diminished reproductive capacity and genital tract pathology (Papp and Shewen, 1996, 1997). Thus, although stress-induced (through nutrient deprivation, temperature shifts, or changes in host cell physiology as a result of immune-regulated cytokines), slow-growing developmental forms of *Toxoplasma* and *Chlamydia* share common features, the bradyzoite-containing *Toxoplasma* cyst and the aberrant RB-containing inclusion are clearly not equivalent. The *Toxoplasma* cyst is the main reason why *Toxoplasma* infections are incurable. Cysts survive passage through the stomach, are impervious to host immunity and drug treatments, and allow the parasite to persist benignly. Yet bradyzoites remain infectious and are capable of reactivating and being efficiently transmitted. In contrast, aberrant RBs observed in the cell culture model of chlamydial persistence have not been shown to reactivate in the context of chlamydial disease. If persistence were a major feature of chlamydial genital tract disease, associations between reactivated infections and immune suppression should be evident. This is not the case in general, although there is at least one report showing that HIV-infected sex workers with depleted $CD4^+$ T cells have a higher risk of pelvic inflammatory disease (Kimani et al., 1996). It may be that the interaction between chlamydiae and their host is more subtle, but compelling arguments for in vivo persistence in human chlamydial diseases have not been made. In addition, there remains an unexplained gap between in vitro descriptions of persistence and clinical disease syndromes, and more details are needed about the growth state of persistent chlamydiae in vivo and its pathologic consequences on the host.

Syphilis

Treponema pallidum is a human-specific microaerophilic spirochete that causes syphilis. Since the mid-19th century, this sexually transmitted infection has been recognized to progress through three distinct stages, referred to as primary, secondary, and tertiary syphilis (reviewed by Singh and Romanowski [1999]). Syphilis is one of the best-described infectious diseases, with a long history of academic and practical interest, including the infamous self-infection study carried out by the well-intentioned 18th century British physician John Hunter, who set out to demonstrate that the signs and symptoms of what turned out to be a mixed infection by *N. gonorrhoeae* and *T. pallidum* were caused by *N. gonorrhoeae* (Hunter, 1818). This experiment seems to be a little like testing the laws of gravity by jumping off a bridge to assess the effects of acceleration. The foolhardy nature of the investigation was made worse by the fact that Dr. Hunter was wrong. The good doctor suffered not only from gonococcal infection but also from syphilis, an unfortunate segue since syphilis was an incurable disease with life-threatening complications at that time (Singh and Romanowski, 1999; Baughn and Musher, 2005; LaFond and Lukehart, 2006). Other 18th and 19th century investigators completed their studies without delusions of grandeur. Philippe Ricord is credited with the first accurate account of primary, secondary, and tertiary syphilis in 1837, a time that predated popularization of the germ theory of disease by such luminaries as Robert Koch, Louis Pasteur, and Rudolf Virchow (reviewed by Baughn and Musher [2005]).

Syphilis is a multistage disease. The progression of syphilis, from the primary lesion through the tertiary stages of disease, is an extremely well-studied example of the natural history of a chronic infection. The primary lesion appears within 2 or 3 weeks of sexual exposure to an infected individual. It initially appears as a red, painless papule that progresses to an ulcerative chancre, which is loaded with treponemes and often has an exudate and an indurated margin. The primary lesion occurs most frequently in the genital, perianal, anal, or oral area as a reflection of recent sexual activity. At this early stage of the infection the organisms have already disseminated throughout the body via the lymphatics and the bloodstream, although it is not yet clinically apparent. Unless present in an obviously visible location, primary lesions often go unnoticed and will spontaneously heal within a month or two without treatment.

Despite the disappearance of the primary lesion, the infection may not have been cleared, and 25% of patients with untreated primary syphilis will develop secondary syphilis. The lesions of secondary syphilis result from hematogenous dissemination of treponemes from the primary chancre. Generally, secondary lesions appear 2 to 3 months after the initial appearance of the primary lesion and manifest as a variety of skin rashes containing transmissible treponemes. The general sequence is for a macular rash to erupt into papular lesions. Secondary syphilis is basically a systemic disease that also includes a variety of nonspecific complaints that accompany the dermatologic manifestations. Although the presence of treponemes in secondary lesions is plentiful, as demonstrated by dark-field microscopy or nucleic acid detection methods, the lesions spontaneously resolve within 1 or 2 months.

The next stage in the progression of disease is latency, characterized by continuous seroreactivity in the absence of symptoms. Sometimes reactivation of secondary syphilis can occur during this time, usually within 1 year of the initial secondary episode. If latency continues for more than a year, this is referred to as late latent syphilis. Individuals who have late latent syphilis are refractory to reinfection. However, following successful treatment, patients are susceptible to reinfection.

Tertiary syphilis, although mainly of historic importance, represents an excellent example of how progressive inflammation may manifest decades after the initial infection. As early as 2 years after primary infection, granulomatous nodular lesions may develop in the skin, bone, or other tissue. These are referred to as gummas. *T. pallidum* can be identified in these lesions, and the lesions resolve when antibiotics are administered (LaFond and Lukehart, 2006). Cardiovascular and late neurologic complications can develop 20 to 30 years after exposure. Historically, cardiovascular syphilis was responsible for the majority of deaths due to this infection. Late neurologic complications lead to personality disorder (e.g., delusions of grandeur—get it now!), impaired movement (general paresis), and muscle weakness resulting from nerve damage (tabes dorsalis). The presence of the pathogen has only rarely been identified in tertiary syphilis, but PCR methods have been used to demonstrate the presence of the organism in cardiovascular disease (O'Regan et al., 2002).

T. pallidum may be one of only a very few bacterial pathogens that is more difficult to work with than *Chlamydia*. We are more than 100 years removed from the initial specific identification of *T. pallidum* as an important human pathogen, and it still has not been routinely cultivated except by inoculation into experimental animals (reviewed by LaFond and Lukehart [2006]). The organism is virtually impossible to keep alive in vitro. However, in an in vivo infection, it rapidly disseminates from the skin via lymphatic and bloodstream invasion. Within an infected individual, it replicates and survives for lengthy periods of time, initiating an inflammatory cascade that may have dire consequences for the host in the absence of any well-established traditional virulence factors. It is almost entirely unclear how this pathogen has been so successful in the human host, although it is likely that a small proportion of infecting pathogens resist macrophage ingestion and killing. It is this subpopulation that survives subsequent to secondary stages of disease, maintains treponemal latency, and ultimately contributes to the development of tertiary syphilis. There are indications that a family of treponemal outer envelope proteins (Tprs) may undergo antigenic variation via gene conversion mechanisms and that expression of different Tpr variants are important in changes the pathogen undergoes in surviving through to latency and tertiary stages of disease (LaFond and Lukehart, 2006).

WHAT ABOUT CHLAMYDIAE?

Syphilis is the ultimate example of a chronic clinical syndrome (primary, secondary, latency, and tertiary phases) that was well described before anything was known about the properties of the pathogen that might contribute to the disease process. Even today, very little mechanistically is known about how *T. pallidum* orchestrates progression from a primary chancre to tertiary disease. Perhaps the stages of trachoma come closest to mimicking the distinct stages of syphilis, but for trachoma, the pathologic changes that accompany disease progression are very different from those seen with syphilis. The presence of the pathogen is required for each stage of syphilis. For trachoma, it is clear that follicular scarring, a process that occurs relatively early in the pathogenesis of disease, sets the stage for eyelid deformities, trichiasis, and corneal abrasion. Blinding trachoma is a direct function of mechanical malfunctions of the eyelid and eyelashes that may be exacerbated by events unrelated to chlamydial infection per se but are rather a function of other infections or even irritants like sand or grit. Certainly there is nothing described in the pathogenesis of trachoma that would suggest either dissemination or long-term persistence of the pathogen. Trachoma is best thought of as an acute conjunctival infection of the very young, where repeated exposures lead to follicular scarring resulting in mechanical deformities that culminate in abrasions of the cornea and may lead to blindness. If ocular *C. trachomatis* infection does persist, it is rare, because literally millions of exposed individuals leave regions of endemicity for trachoma, yet reports of positive cultures in individuals having lived for years afterwards in regions of nonendemicity are few; and even in these cases aberrant forms (persistence) have not been identified.

It is known that some women who recover from chlamydial genital tract infections may harbor the organism for more than a year (McCormack et al., 1979; Oriel and Ridgway, 1982b). However, in these cases it is always difficult to sort out reinfection from persistent infections; and, as of yet, there is no clear chlamydial persistence phenotype that is associated with chronically infected women. In addition, unlike syphilis, a distinct clinical syndrome that requires long-term persistence from a primary infection for chlamydial genital tract infection has not been described.

Infections caused by *Chlamydia pneumoniae* may have disseminating chronic manifestations (Watson and Alp, 2008), and these conditions (e.g., heart disease and atopic asthma) implicate the presence of persisting forms of the pathogen, especially in either macrophages, endothelial cells, or smooth muscle cells in atherosclerotic plaque of the coronary arteries (Watson and Alp, 2008; Mahoney and Coombes, 2001). *C. pneumoniae* has been identified, although rarely cultivated, in atheromas (Watson and Alp, 2008; Mahoney and Coombes, 2001), although these findings have been disputed (Regan et al., 2002; Ieven and Hoymans, 2005). If these organisms causally contribute to heart disease, then the microbiology of the disease must be very different from that of chronic syphilis since secondary and tertiary stages of syphilis respond to antibiotics, while, at least for treatment of heart disease patients, antibiotics effective against chlamydiae do not prevent secondary events (collated by Song et al. [2008]). A role for *C. pneumoniae* in other chronic diseases, such as nonatopic asthma or neurologic diseases, has not been studied in a prospective manner and remains highly speculative at this time both in terms of a true role for chlamydiae and for the presence of persistent chlamydiae.

On the other hand, veterinary chlamydial infections, including *C. suis* infections of swine, *C. psittaci* infections of birds, or *Chlamydia abortus* and *Chlamydia pecorum* infection of livestock may have true persistent components. For example, the presence of classic aberrant chlamydial developmental forms (Fig. 2) has been documented in the gastrointestinal tract of pigs that have *C. suis* infection (Pospischil et al., 2009), and short-term antibiotic treatment of infected animals is not effective in eliminating subclinical infection (Reinhold et al., 2011). Evidence for the development of antibiotic resistance to tetracyclines via acquisition of resistance gene cassettes has been reported for *C. suis* (Lenart et al., 2001). Is it possible that one of the consequences of the presence of persistent forms is competence for genetic exchange? *C. abortus*, a sexually acquired infection in sheep, is a major cause of infectious abortion. Chronic chlamydial infection of the reproductive tract of ewes that experience pregnancy failure has been reported to limit the breeding life of affected ewes and eventually results in upper genital tract pathology (Papp and Shewen, 1997). It is unfortunate that a detailed characterization of the nature of chlamydial development in chronically infected ewes has not been done. Are these persistent infections, and is there an equivalent in the upper genital tract of women who harbor *C. trachomatis*?

Are there chlamydial attributes that are similar to those perceived to be important in the development of syphilis latency? The chlamydial Pmp family of proteins may share features with the *T. pallidum* Tpr family. It is known that some chlamydial Pmp family members are surface expressed and may be developmentally regulated. Pmp expression in vivo has been

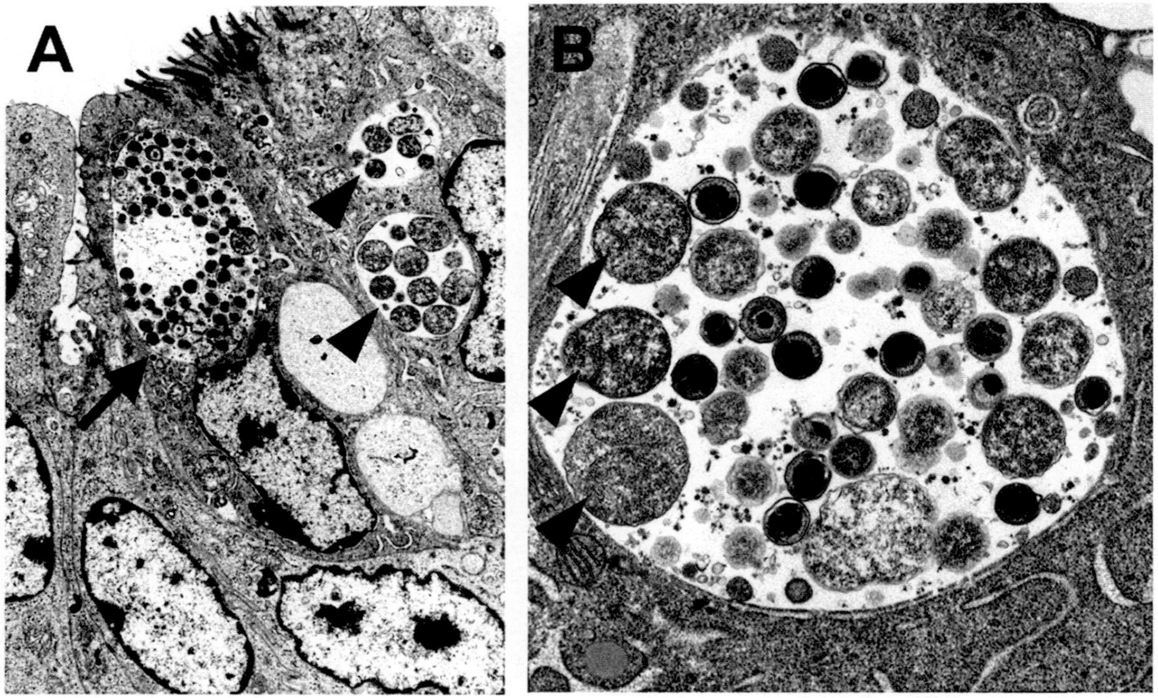

FIGURE 2 Chlamydial inclusions in intestinal enterocytes of naturally infected swine. (A) Lower-magnification electron micrograph showing both a typical inclusion (indicated by arrow) with an abundance of dense EBs and inclusions containing somewhat enlarged RBs (arrowheads). (B) Higher-magnification electron micrograph showing an inclusion containing typical EBs and RBs, in addition to enlarged, aberrant RBs (arrowheads). It is interesting that both normal and abnormal developmental forms are seen, as if persistence is a stochastic event for *C. suis* infecting the pig intestine. Images generously provided by Andreas Pospischil, University of Zurich. doi: 10.1128/9781555817329.ch12.f2

investigated by assessing immune responses to these proteins (Tan et al., 2009). In vivo studies are needed to determine if subsets of chlamydial Pmps are associated with acute versus chronic disease.

Tuberculosis

Tuberculosis is a chronic disease caused by the bacterium *Mycobacterium tuberculosis*. Recognition of asymptomatic carriage punctuated by episodes of active infection is very well described in the literature, with a rich history of both clinical and basic investigations. Control of tuberculosis is a function of the immune response and the development of granulomas, generally in the lung, consisting of a necrotic core surrounded by immune effector cells. Tubercle bacilli are often found in the necrotic core, existing under highly hypoxic conditions (Chao and Rubin, 2010). Reactivation of disease is a result of interference with the immune response. Identification of immune mechanisms that control primary infection and prevent reactivation is an area of intense investigation (Lin and Flynn, 2010). Studies of the natural history of tuberculosis and the cell and molecular biology of the pathogen have provided us not only with a fundamentally precise understanding of disease progression but also with an excellent lexicon of terms that help to precisely define disease status and characterizations of the pathogen in ways that help understand shifts in disease state. For example, according to Chao and Rubin (Chao and Rubin, 2010) latency and reactivation reflect disease states, whereas dormancy and resuscitation reflect pathogen phenotypes. Latency is defined as "a state of asymptomatic infection characterized by low bacterial counts and a lack of clinical signs of disease." Reactivation is defined as "the transition from asymptomatic *M. tuberculosis* infection to visible signs of active disease." Dormancy is defined as "a state of nonreplication that is characterized by long-term viability despite metabolic down-regulation." Resuscitation is defined as "a metabolic transformation from a relatively inert, nonreplicating state into an actively dividing state." According to these definitions, pathogen dormancy correlates with latency and pathogen resuscitation correlates with reactivation.

Tuberculosis researchers have successfully exploited molecular genetic techniques to understand conditions that elicit pathogen dormancy. Knowledge of how *M. tuberculosis* modulates gene expression and metabolism to enter and exit the nonreplicating dormant state has then been applied to develop in vivo models of latency and reactivation. Environmental factors that trigger dormancy include starvation for nutrients and hypoxia (Chao and Rubin, 2010; Shleeva et al., 2004; Wayne and Sohaskey, 2001). When these stress-related conditions are applied to *M. tuberculosis*, striking metabolic reprogramming occurs, including upregulation of stress response genes and downregulation of many central metabolism genes. Microarray studies have provided evidence for a hypoxic regulon called *dos* (an acronym for dormancy survival) (Voskuil et al., 2004). The *dos* regulon includes activation of genes needed for catabolism of novel carbon sources, including fatty acids and cholesterol (Chao and Rubin, 2010). Mutants unable to control the *dos* regulon do not resuscitate from hypoxia normally (Kumar et al., 2007) and have a defect in a two-component signaling system regulator called DosR (also known as DevR). Activity controlled by *dos* eventually matures into a longer-term hypoxia response controlled by a stress-related sigma factor (Rustad et al., 2008). Some of the *dos*-regulated genes are thought to restrain virulence by slowing down bacterial growth. DosR-mediated events also may confer phenotypic resistance to antibiotics via toxin-antitoxin mechanisms that involve rapid mRNA degradation when the levels of the toxin gene product exceed those of the antitoxin (Ramage et al., 2009).

Resuscitation from dormancy involves upregulation of enzymes that alter peptidoglycan structure. It is known that dormancy results in a peptidoglycan structure that is cross-linked differently from that found in vegetative cells.

It is thought that this creates a more stable cell wall for dormant-phase organisms. Switching to the vegetative form of peptidoglycan produces a muropeptide ligand that initiates a signal transduction cascade, which leads to metabolic upregulation and thereby facilitates resuscitation (Chao and Rubin, 2010; Lavollay et al., 2008).

It is very clear that *M. tuberculosis* reactivates from a latent state and that this results in episodes of active disease. The roles of in vitro-defined dormancy and resuscitation have not yet been proven to be the mechanisms that account for latency/reactivation disease states of tuberculosis. However, the investigation of programmed changes in the pathogen's growth status is a logical strategy toward understanding active versus latent tuberculosis, as it involves the pathogen responding to environmental cues that ultimately reflect changes in the immune status of the host.

WHAT ABOUT CHLAMYDIAE?

Clinical data for chlamydial infections do not provide strong support for alternation between active and latent states, except perhaps for *C. psittaci* infections in birds under highly stressed conditions in the abattoir, which may cause reactivation of latent gastrointestinal infection. This potentially interesting example of reactivation has not been systematically studied, and very little is known about either the status of the host or the status of the pathogen under conditions of latency or reactivation for avian infections.

An alternative possibility is that active chlamydial infection may irreversibly transition into latent disease without any possibility of reactivation. Perhaps female genital tract infections best exemplify this scenario. Active chlamydial infection must occur in the lower genital tract. It is there that chlamydiae are required to complete a productive infectious cycle (EB → RB → EB) to be effectively transmitted. In contrast, transmission is never associated with upper genital tract disease, and therefore the necessity of completing the developmental cycle to yield infectious EB populations is not an essential component of chlamydial infection of the upper genital tract. Thus, latency in the form of persistence or development of other metabolically dormant phenotypes would not compromise transmissibility, since the upper genital tract is, in effect, a dead-end locale.

The problem with this scenario is that data have not been obtained to determine if RBs, cryptic chlamydiae, or noninfectious aberrant forms are enriched in upper genital tract disease. But can we take some cues from the work done in tuberculosis research to assess pathogen phenotypes that are programmed according to environmental modulation to gain insights into the pathogenesis of chlamydial genital tract infections in women? It turns out that much like what has been seen in studies on *M. tuberculosis* dormancy in vitro, chlamydiae may have a way of regulating a productive versus a nonproductive growth status. Indeed the development of acquired immunity in the form of Th1-dependent immune-regulated cytokines induces changes in the physiology of the infected host cell known to elicit persistent chlamydial growth in vitro. We come once again to the IFN-γ-mediated induction of IDO, the intracellular degradation of tryptophan, and its effects on chlamydial growth (Beatty et al., 1994). The development of aberrant chlamydial growth under these conditions, all done in cell culture models, provides a potentially physiologically relevant mechanism for the induction of chlamydial latency. The relevance of this model of chlamydial persistence was bolstered by chlamydial proteome studies (Shaw et al., 2000) and genome-sequencing studies (Caldwell et al., 2003). These studies demonstrated that (i) *C. trachomatis* has a partial tryptophan operon that is activated under conditions of tryptophan starvation and (ii) the required substrate for the production of tryptophan, a required amino acid, is indole (Wood et al., 2003). Neither chlamydiae nor the mammalian host can make indole, but intriguingly, members of the vaginal microbiota are very often indole producers. With these findings, in vitro observations made without in vivo correlations begin to make sense.

Now we have a scenario where chlamydiae have evolved a mechanism that utilizes the polymicrobial environment in the lower genital tract to overcome tryptophan starvation induced by the host immune response. This mechanism involves the ability of chlamydiae to sense tryptophan availability through the requirement of this amino acid as a corepressor for the transcriptional regulator TrpR. When tryptophan is readily available, TrpR represses expression of the enzyme tryptophan synthase from the partial tryptophan operon of genital *C. trachomatis* (Carlson et al., 2006; Akers and Tan, 2006). However, when the host cell induces IDO and degrades tryptophan to inhibit chlamydial growth, the absence of tryptophan prevents TrpR from functioning as a repressor and allows tryptophan synthase to be expressed. Indole, provided by the polymicrobial environment of the lower genital tract, can then be used as a substrate for production of tryptophan, allowing normal chlamydial growth and generation of infectious progeny. In the upper genital tract, however, other microbial flora may not be present and chlamydiae may be unable to overcome IDO-induced tryptophan starvation. But this is of little importance to the chlamydiae that reside in this locale, since a productive infection in the upper genital tract is not important for transmission. These differences in the pathogen phenotype at different sites in the female reproductive tract may explain some fundamental features of chlamydial genital disease: productive infection in the lower genital tract leads to spread, while a nonproductive persistent infection may lead to smoldering inflammation, which is a hallmark of chlamydial upper genital tract disease in women.

This is a compelling story that builds on cell culture systems of immune effector functions, the development of chlamydial persistence in vitro, and chlamydial genomic and gene regulation studies. Indeed, the implication that the evolution of this genital tract pathogen involves the need to acquire tryptophan in a way that engages polymicrobial involvement is often overlooked in infectious disease research. But let us step back from this hypothetical example of how chlamydial genital tract disease in women may occur and consider what is actually known. We speculate about the role of immunoregulated cytokines in altering chlamydial growth to induce persistence. All data about this part of the story are from cell culture systems, and in vivo correlates have not been established. We discuss the inducible partial tryptophan operon and indole requirements as if this is what actually is occurring in vivo, but correlates between indole-producing microbiota and chlamydial EB production as defined by infectivity assays have not been obtained. We consider the differences in indole-producing microbiotas between upper and lower tract compartments, but this remains speculative; and we conclude that persistent forms may account for the insidious inflammation associated with upper genital tract disease in women, but we have not identified persistent infections as contributing factors to chlamydial genital tract disease in women. The problem is not that the hypotheses concerning chlamydial persistence and disease status are not attractive but that required in vivo data are lacking. If nonproductive, long-lasting chlamydial phenotypes are residing in cells of the upper genital tract of women with chlamydial disease, then these forms must be identified either in an appropriate animal model or in specimens taken from patients. If conversion from productive to nonproductive, persistent infections is impacted by the genital tract microbiota and its capacity to produce indole, then investigation of the metabolome of the genital microbiota in women and the status of chlamydial infections should be done. These studies are necessary to support or refute the role of persistent infections in chlamydial disease.

Typhoid Fever

Typhoid fever is a systemic infection caused by *Salmonella enterica* serovar Typhi and acquired via the oral route (Monack et al., 2004b). Infection is characterized by inflammation in the small intestine at Peyer's patches. This intestinal compartment, comprised of lymph nodules designed to provide antigen sampling for macrophages, dendritic cells, and T and B lymphocytes, is the site of *Salmonella* serovar Typhi dissemination, where phagocytes are infected within the lamina propria. Infected phagocytes gain access to the lymphatics and bloodstream, resulting in spread of the infection to the liver, spleen, gall bladder, and bone marrow. Some individuals become lifelong carriers, periodically shed *Salmonella* Typhi in their stools, and are therefore important reservoirs of infection (Sinnott and Teall, 1987). The carrier state is characterized by a robust immune response to serovar Typhi and the absence of disease symptoms. Studies of persistent *Salmonella* disease in mice indicate that persistence is a function of the capacity of the pathogen to survive in macrophages (Monack et al., 2004a), and at least for nontyphoidal salmonellae, macrophage persistence contributes to increased disease severity in AIDS patients (Gordon, 2008). It is also known that macrophage persistence requires the SPI1 and SPI2 type three secretion (T3S) systems and effector gene products that help resist the effects of antimicrobial peptides and phagocyte oxidases and other antimicrobial effects (Monack et al., 2004b; Hensel, 2000).

WHAT ABOUT CHLAMYDIAE?

Persistent salmonellae reside in monocytes and macrophages. There are reports of chlamydiae existing as persistent, aberrant RBs in macrophages in the joints of patients with reactive arthritis (Carter and Hudson, 2010), but the characteristics of the pathogen within infected joints have not been established. Reactive arthritis is an illness characterized by joint inflammation occurring shortly after gastroenteritis caused by gram-negative bacteria or genital infections caused by *C. trachomatis* (Townes, 2010).

Members of the genus *Salmonella* are able to effect changes in host cell function after infection by virtue of their T3S system, which is designed to deliver pathogen-produced effectors into membrane structures and cytoplasm of the host cell. *Chlamydia*, like *Salmonella*, possesses a T3S system, and secreted effectors have been shown to modulate host cell function (Hower et al., 2009; Mitel et al., 2010). One important criterion for long-term persistence is the ability to maintain a stable relationship with the infected host cell over a long term. Intracellular chlamydiae are reported to program infected host cell physiology to actively elicit an antiapoptotic state, especially under conditions of nonproductive, persistent growth (Byrne and Ojcius, 2004). This activity may involve the secretion of chlamydial effector molecules into the host cell cytoplasm via T3S. It is not known if this mechanism of host cell reprogramming occurs in vivo or if it in any way contributes to the development or maintenance of chronic chlamydial infections, but it is a feature shared with *Salmonella*, a well-known cause of persistent disease.

WHAT IN VIVO CONDITIONS FALL UNDER THE RUBRIC OF CHLAMYDIAL PERSISTENCE?

Chlamydial infections of animals, including birds, sheep, swine, and bovines are common, often subclinical and chronic, and are reported to have a measureable adverse impact on the health and well-being of economically important livestock (Pospischil et al., 2009; Reinhold et al., 2010, 2011; Papp and Shewen, 1996, 1997). Unfortunately, very little is known about the growth status of the pathogen in these chronic infections. Work on *C. suis* infections in swine has clearly demonstrated the presence of chlamydial inclusions containing morphotypes that have features in common with classical cell culture-based chlamydial persistence (Pospischil et al., 2009). This provides a rationale for study of veterinary chlamydial infections to gain a perspective on how modulation of pathogen function and productive versus persistent infection may relate to the problem of chronic chlamydial disease. Roger Rank and colleagues (see chapter 13, "In vivo chlamydial infection") have shown that *C. muridarum* infection of mice depleted of neutrophils promotes the development of the aberrant chlamydial phenotype and that Nu/

Nu nude mice maintain long-term chronic *C. trachomatis* infections. Study of these models should be exploited more fully.

For human chlamydial infections, postgonococcal urethritis may provide an example of clinical persistence. Postgonococcal urethritis is defined as a persistent or recurrent sexually acquired urethral infection occurring in men who had been successfully treated for gonorrhea with beta-lactam antibiotics (Oriel and Ridgway, 1982a). Most of these patients were found to be positive for *Chlamydia*. This syndrome is now of historical significance because antichlamydial antibiotics are now routinely added to the treatment of gonococcal infections. However, when intracellular chlamydiae are exposed to penicillin or other beta-lactam antibiotics in vitro, they stop dividing and display the large aberrant phenotype associated with cell culture persistence (Matsumoto and Manire, 1970). While aberrant chlamydial forms have not been identified in postgonococcal urethritis, the detection of chlamydiae after treatment with beta-lactam antibiotics suggests that *C. trachomatis* may persist in the human host and reactivate to cause disease. This is not an insignificant observation, although it is curious that the best example of chlamydial persistence in human disease arises from an exogenous source of persistence induction (penicillin) rather than as a result of a pathogen-driven mechanism.

Curiously, there were no reports of the equivalent of post-gonococcal-treatment disease in women when it was established that *C. trachomatis* caused postgonococcal urethritis in the 1970s (Richmond et al., 1972). At that time there were reports to indicate that chlamydial cervicitis could persist for more than a year in untreated women who denied having had intercourse since their initially positive examination (McCormack et al., 1979). Geisler recently commented on the practical and ethical issues associated with human natural history of infection studies (Geisler, 2010). It is difficult to know with certainty the time at which initial exposure to the pathogen occurs or whether reinfections play a role. Most importantly, once the pathogen is identified, treatment, if available, must begin. In the absence of natural history studies, but with reasonable indications that persistence is important in upper genital tract disease in women, it is of critical importance to establish reliable biomarkers to help characterize the presence of chronic chlamydiae with a high degree of specificity and sensitivity.

WHERE DO WE GO FROM HERE?

We began this chapter by suggesting that chlamydiae were different from other pathogens. This chapter has posed the question of how similar or different chlamydiae are from other microbial pathogens that are also different from run-of-the-mill pathogens. All comparative examples chosen represent pathogens well established in causing persistent, chronic, or latent disease (Fig. 3). The pathogens selected for comparison here were chosen because we thought they provided excellent examples of chronic disease that have correlates with chlamydiae as assessed by cell culture models. But this list is by no means complete, and we may not even have chosen the best examples. Why not take the time to make your own list and see how chlamydiae are similar to or different from the pathogens that you select. It is fun—and educational!

Evidence suggests that chlamydiae also are different from the group of persistent pathogens described here, although they share features with many of them. There are very few examples of chlamydial infections that cause reactivation of infection in ways that have been described for vivax malaria, toxoplasmosis, or tuberculosis, except when the conditions for persistence are established exogenously, such as via induction of the persistence phenotype by administration of penicillin, setting the stage for postgonococcal urethritis, but apparently (and curiously) with no recognized equivalent in women. Similarly, chlamydial infections do not appear to progress through defined stages of primary, secondary, and tertiary disease in a manner similar to *T. pallidum* infection and syphilis, where stealth in avoiding immune

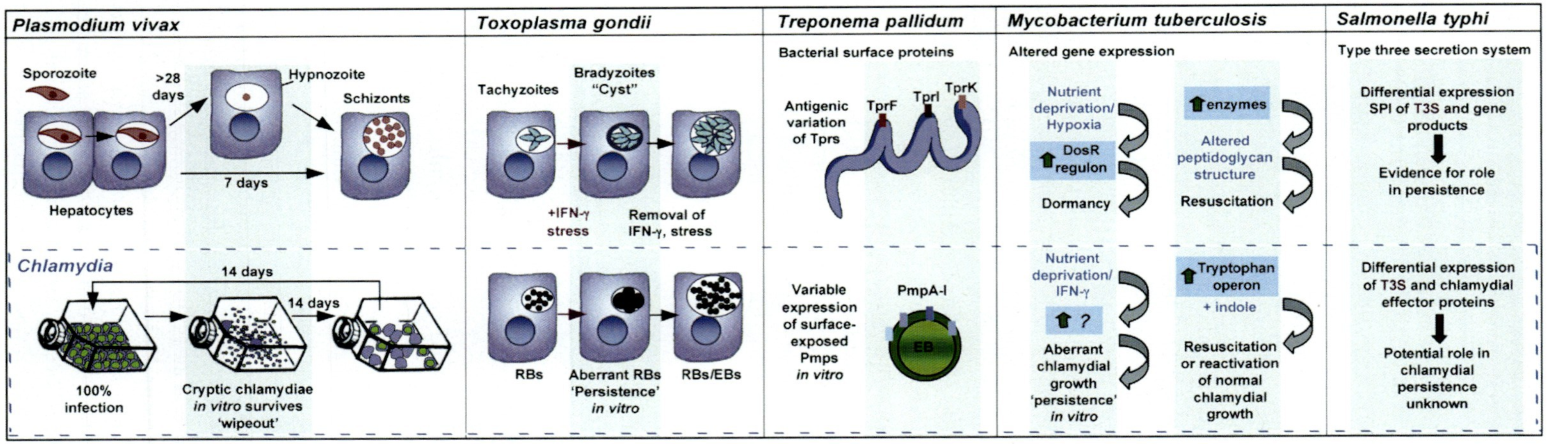

FIGURE 3 Factors contributing to persistence of other microorganisms and potential correlates in *Chlamydia*. Attributes of other microorganisms are shown in upper panels, and potential correlates in *Chlamydia* are shown in the corresponding lower panels. Please see the text for details.
doi:10.1128/9781555817329.ch12.f3

elimination from an extracellular residence must be paramount.

Established examples of chronic chlamydial infections may share similarities with typhoid fever. The carrier state in typhoid fever is basically that of an asymptomatic shedder. This condition seems similar to chlamydial infections reported in swine and other livestock. In these cases, severe acute infections (e.g., infectious abortion) mature into chronic relationships between the microbe and the host with debatable pathogenic potential (Pospischil et al., 2009; Reinhold et al., 2011; Papp and Shewen, 1997). There are a number of available animal models that could be developed to study chronic chlamydial infections using natural infection models. These include studies of *C. suis* in pigs (Pospischil et al., 2009) and *C. abortus* in sheep (Papp and Shewen, 1996, 1997). Development of these models is likely to provide direct information relevant to swine and sheep infectious diseases and may provide insight relevant to human chlamydial infections. At the very least, we will learn how animal-infecting *Chlamydia* strains and diseases of livestock are similar to, or different from, human chlamydial strains and chlamydial diseases of humans.

But is there more to the relationship between chlamydiae and their hosts than currently meets the eye? Chimera is an interesting word. Sometimes it refers to something that is real and very abnormal, but usually it conceptualizes a creation in one's mind, a figment of the imagination, a phantom from lost worlds—not an everyday reality. Antonyms for chimera include on the one hand terms like reality and truth but on the other hand terms like normal, ordinary, and regular. The mythological fire-breathing monster with a lion's head, a goat's body, and a serpent's tail is the most common vision of a chimera; clearly a creature of dreams (or nightmares). But what of a microbial pathogen with the dormancy potential of *P. vivax*, the capacity to survive an immune response like *T. gondii* and *M. tuberculosis*, and a tendency toward asymptomatic carriage like *Salmonella* but with the potential for causing long-term consequences like *T. pallidum*? Is this a chimera or the definition of a highly successful, well-adapted pathogen that has found a novel way to survive within its host? Is this the definition of *Chlamydia*?

A major question continues to be where human *C. trachomatis* infections fit in the spectrum of acute versus chronic infection. Chronic stages of blinding trachoma represent a mechanical disorder where the pathogen is irrelevant once the lid distortion process begins. Reactive arthritis is clearly a chronic disease in which chlamydiae may be involved, and the story of *C. pneumoniae* and heart disease is well known. Urethritis and cervicitis may evolve into more chronic conditions, but natural history studies are difficult to conduct and relevant animal models are not well established vis-à-vis persistence. Upper genital tract disease in women may reflect chronic disease, but sorting through the process in the natural host is not practical or ethical. Animal models may be useful tools to better understand chlamydial persistence in the upper genital tract, but in vivo approaches have provided only limited information, thus far.

Systems biology studies are needed to address key questions that will link details regarding pathogen phenotypes and the development of chronic disease. In developing these types of strategies, it will be important first to carefully select an appropriate, tractable model host that reflects conditions relevant to the experimental questions under investigation. Mouse models probably represent the most valuable available tool because these animals can be genetically modified to suit the type of infection under investigation. The mouse as host may not be the only experimental system likely to yield a body of systems biology data; but it is relevant and highly tractable in that production of designer mouse strains featuring traits specific for best mimicking human disease is feasible, and data sets using mice lend themselves readily to bioinformatic and systems biology approaches. When "omics" tools are applied to the study of chlamydial disease, the problem of persistence and chlamydial disease severity may finally mature to the point where reliable biomarkers are discovered that will enable translation to human infections in ways that will help us to better understand the true role of chronic infections in the repertoire of important human diseases that are caused by chlamydiae.

ACKNOWLEDGMENTS

We thank Patrik Bavoil, Ming Tan, Andreas Pospischil, Roger Rank, and Byron Batteiger for critically reviewing the text. Their suggestions have made the chapter better. Lingering deficiencies are the fault of the authors. This work was supported by PHS grant AI19782 (GIB) and Department of Defense award W81XWH-09-0391 (GIB).

REFERENCES

AbdelRahman, Y. M., L. A. Rose, and R. J. Belland. 2011. Developmental expression on non-coding RNAs in *Chlamydia trachomatis* during normal and persistent growth. *Nucleic Acids Res.* **39:**1843–1854.

Akers, J. C., and M. Tan. 2006. Molecular mechanism of tryptophan-dependent transcriptional regulation in *Chlamydia trachomatis. J. Bacteriol.* **188:**4236–4243.

Baughn, R. E., and D. M. Musher. 2005. Secondary syphylitic lesions. *Clin. Microbiol. Rev.* **18:**205–216.

Beatty, W. L., R. P. Morrison, and G. I. Byrne. 1994. Persistent chlamydiae: from cell culture to a paradigm for chlamydial pathogenesis. *Microbiol. Rev.* **58:**686–699.

Byrne, G. I., L. K. Lehmann, and G. L. Landry. 1986. Induction of tryptophan catabolism is the mechanism for gamma-interferon-mediated inhibition of intracellular *Chlamydia psittaci* replication in T24 cells. *Infect. Immun.* **53:**347–351.

Byrne, G. I., and D. Ojcius. 2004. *Chlamydia* and apoptosis: life and death decisions of an intracellular pathogen. *Nat. Rev. Microbiol.* **2:**802–808.

Caldwell, H. D., H. Wood, D. Crane, R. Bailey, R. B. Jones, D. Mabey, I. MacLean, Z. Mohammed, R. Peeling, C. Roshick, J. Schachter, A. W. Solomon, W. E. Stamm, R. L. Suchland, L. Taylor, S. K. West, T. C. Quinn, R. J. Belland, and G. McClarty. 2003. Polymorphisms in *Chlamydia trachomatis* tryptophan synthase genes differentiate between genital and ocular isolates. *J. Clin. Investig.* **111:**1757–1769.

Carlson. J. H., H. Wood, C. Roshick, H. D. Caldwell, and G. McClarty. 2006. *In vivo* and *in vitro* studies of *C. trachomatis* TrpR:DNA interactions. *Mol. Microbiol.* **59:**1678–1691.

Carter, J. D., and A. P. Hudson. 2010. The evolving story of *Chlamydia*-induced reactive arthritis. *Curr. Opin. Rheumatol.* **22:**424–430.

Chao, M. C., and E. J. Rubin. 2010. Letting sleeping *dos* lie: does dormancy play a role in tuberculosis? *Annu. Rev. Microbiol.* **64:**293–311.

Chen, N., A. Auliff, K. Riekmann, M. Gatton, and Q. Cheng. 2007. Relapses of *Plasmodium vivax* result from clonal hypnozoites activated at predetermined intervals. *J. Infect. Dis.* **195:**934–941.

Geisler, W. M. 2010. Duration of untreated, uncomplicated *Chlamydia trachomatis* genital infection and factors associated with chlamydia resolution: a review of human studies. *J. Infect. Dis.* **201**(S2): S104–S113.

Gerard, H., J. Whittum-Hudson, H. R. Schumacher, and A. Hudson. 2004. Differential expression of three *C. trachomatis* hsp-60-encoding genes in active vs persistent infections. *Microb. Pathog.* **36:**35–39.

Goellner, S., E. Schubert, E. Lieber-Tenorio, H. Hotzel, H. P. Saluz, and K. Sache. 2006. Transcriptional response patterns of *Chlamydophila psittaci* in different in vitro models of persistent infection. *Infect. Immun.* **74:**4801–4808.

Gordon, M. A. 2008. *Salmonella* infections in immunocompromised adults. *J. Infect.* **56:**413–422.

Hensel, M. 2000. *Salmonella* pathogenicity island 2. *Mol. Microbiol.* **36:**1015–1023.

Hogan, R. J., S. A. Mathews, S. Mukhopadhyay, J. T. Summersgill, and P. Timms. 2004. Chlamydial persistence: beyond the basic paradigm. *Infect. Immun.* **72:**1843–1855.

Horn, M., A. Collingro, S. Schmitz-Esser, C. L. Beier, U. Purkhold, B. Fartmann, P. Brandt, G. J. Nyakatura, M. Droege, D. Frishman, T. Rattei, H.-W. Mewes, and M. Wagner. 2004. Illuminating the evolutionary history of *Chlamydia*. *Science* **304:**728–730.

Howe, D. K., and D. L. Sibley. 1995. *Toxoplasma gondii* comprises three clonal lineages: correlation of parasite genotype with human disease. *J. Infect. Dis.* **172:**1561–1566.

Hower, S., K. Wolf, and K. Fields. 2009. Evidence that CT694 is a novel *Chlamydia trachomatis* T3S substrate capable of functioning during invasion or early cycle development. *Mol. Microbiol.* **72:**1423–1437.

Hunter, J. 1818. *Treatise on Venereal Disease*, 2nd ed. Sherwood, Neely, and Jones, London, United Kingdom.

Ieven, M. M., and V. Y. Hoymans. 2005. Involvement of *Chlamydia pneumoniae* in atherosclerosis: more evidence for lack of evidence. *J. Clin. Microbiol.* **43:**19–24.

Imwong, M., G. Snounou, S. Pukrittayakamee, N. Tanomsing, J. R. Kim, A. Nandy, J. P. Guthmann, F. Nosten, J. Carlton, S. Looareesuwan, S. Nair, D. Sudimack, N. P. J.

Day, T. J. C. Anderson, and N. J. White. 2007. Relapses of *Plasmodium vivax* infection usually result from activation of heterologous hypnozoites. *J. Infect. Dis.* **195:**927–933.

Kimani, J., I. W. Maclean, J. J. Bwayo, K. MacDonald, J. Oyugi, G. M. Maitha, R. W. Peeling, M. Cheang, N. J. D. Nagelkerke, F. A. Plummer, and R. C. Brunham. 1996. Risk factors for *Chlamydia trachomatis* pelvic inflammatory disease among sex workers in Nairobi, Kenya. *J. Infect. Dis.* **173:**1437–1444.

Kumar, A., J. C. Toleda, R. P. Patel, J. R. Lancaster, Jr., and A. J. Steyn. 2007. *Mycobacterium tuberculosis* DosS is a redo sensor and DosT is a hypoxia sensor. *Proc. Natl. Acad. Sci. USA* **104:**11568–11574.

LaFond, R. E., and S. A. Lukehart. 2006. Biological basis for syphilis. *Clin. Microbiol. Rev.* **19:**29–49.

Lavollay, M., M. Arthur, M. Fourgeaud, L. Dubost, A. M. N. Veziris, D. Blanot, L. Gutmann, and J.-L. Mainardi. 2008. The peptidoglycan of stationary phase *Mycobacterium tuberculosis* predominantly contains cross-links generated by L,D-transpeptidation. *J. Bacteriol.* **190:**4360–4366.

Lee, C. K., and J. W. Moulder. 1981. Persistent infection of mouse fibroblasts (McCoy cells) with a trachoma strain of *Chlamydia trachomatis*. *Infect. Immun.* **32:**822-829.

Lenart, J., A. A. Anderson, and D. D. Rockey. 2001. Growth and development of tetracycline-resistant *Chlamydia suis*. *Antimicrob. Agents Chemother.* **42:**2198–2203.

Lin, P. L., and J. L. Flynn. 2010. Understanding latent tuberculosis: a moving target. *J. Immunol.* **185:**15–22.

Mahoney, J. B., and B. K. Coombes. 2001. *Chlamydia pneumoniae* and atherosclerosis: does the evidence support a causal or contributory role? *FEMS Microbiol. Lett.* **197:**1–9.

Mathews, S., C. George, C. Flegg, D. Stenzel, and P. Timms. 2001. Differential expression of *ompA, ompB, pyk, nlpD* and Cpn0585 genes between normal and interferon-gamma treated cultures of *Chlamydia pneumoniae*. *Microb. Pathog.* **30:**337–345.

Matsumoto, A., and G. P. Manire. 1970. Electron microscopic observations on the effect of penicillin on the morphology of *Chlamydia psittaci*. *J. Bacteriol.* **101:**278–285.

Maurer, A. P., A. Mehlitz, H. J. Mollenkopf, and T. F. Meyer. 2007. Gene expression profiles of *Chlamydophila pneumoniae* during the developmental cycle and iron-depleted mediated persistence. *PLoS Pathog.* **3:**e83.

McCormack, W. M., S. Alpert, D. E. McComb, R. L. Nichols, Z. Semine, and S. H. Zinner. 1979. Fifteen-month follow-up study of women infected with *C. trachomatis*. *N. Engl. J. Med.* **300:**123–125.

Mitel, S., N. J. Miller, E. R. Fischer, and T. Hackstadt. 2010. Specific chlamydial inclusion membrane proteins associated with active Src family kinases in microdomains that interact with the host microtubule network. *Cell. Microbiol.* **12:**1235–1249.

Monack, D. M., D. M. Bouley, and S. Falkow. 2004a. *Salmonella typhimurium* persists within macrophages in the mesenteric lymph nodes of chronically infected Nramp1+/+ mice and can be reactivated by IFN-γ neutralization. *J. Exp. Med.* **199:**231–241.

Monack, D. M., A. Mueller, and S. Falkow. 2004b. Persistent bacterial infections: the interface of the pathogen and the host immune system. *Nat. Rev. Microbiol.* **2:**747–765.

Montoya, J. G., and O. Liesenfeld. 2004. Toxoplasmosis. *Lancet* **363:**1965–1976.

Moulder, J. W., N. J. Levy, and L. P. Shulman. 1980. Persistent infection of mouse fibroblasts (L cells) with *Chlamydia psittaci*: evidence for a cryptic chlamydial form. *Infect. Immun.* **30:**874–883.

O'Regan, A. W., C. Castro, S. A. Lukehart, J. M. Kasznica, P. A. Rice, and M. F. Joyce-Brady. 2002. Barking up the wrong tree? Use of polymerase chain reaction to diagnose syphylitic aortitis. *Thorax* **57:**917–918.

Oriel, J. D., and G. L. Ridgway. 1982a. Genital infection by *Chlamydia trachomatis*. *Curr. Topics Infect. Dis.* **2:**41–52.

Oriel, J. D., and G. L. Ridgway. 1982b. Genital infection by *Chlamydia trachomatis*. *Curr. Top. Infect. Dis.* **2:**53–67.

Ouellette, S. P., T. P. Hatch, Y. M. AbdelRahman, L. A. Rose, R. J. Belland, and G. I. Byrne. 2006. Global transcriptional up-regulation in the absence of increased translation in *Chlamydia* during IFN-γ-mediated host cell tryptophan starvation. *Mol. Microbiol.* **62:**1387–1401.

Papp, J. R., and P. E. Shewen. 1996. Localization of chronic *Chlamydia psittaci* infection in the reproductive tract of sheep. *J. Infect. Dis.* **174:**1296–1302.

Papp, J. R., and P. E. Shewen. 1997. *Chlamydia psittaci* infection in sheep: a paradigm for human reproductive tract infection. *J. Reprod. Immunol.* **34:**185–202.

Peng, K., and D. M. Monack. 2010. Indoleamine 2,3-dioxygenase 1 is a lung-specific innate immune defense mechanism that inhibits growth of *Francisella tularensis* tryptophan auxotrophs. *Infect. Immun.* **78:**2723–2733.

Pfefferkorn, E. R., M. Eckel, and S. Rebhun. 1986. Interferon-gamma suppresses the growth of *Toxoplasma gondii* in human fibroblasts through starvation for tryptophan. *Mol. Biochem. Parasitol.* **20:**215–224.

Pospischil, A., N. Borel, E. H. Chowdury, and F. Guscetti. 2009. Aberrant chlamydial developmental forms in the gastrointestinal tract of pigs spontaneously and experimentally infected with *Chlamydia suis*. *Vet. Microbiol.* **135:**147–156.

Ramage, H. R., L. E. Connolly, and J. S. Cox. 2009. Comprehensive functional analysis of *Mycobacterium tuberculosis* toxin-antitoxin systems: implications for pathogenesis, stress responses, and evolution. *PLoS Genet.* **5:**e1000767.

Regan, M. J., B. J. Wood, Y. H. Hsieh, M. L. Theodore, T. C. Quinn, D. B. Hellman, R. Green, C. A. Gaydos, and J. H. Stone. 2002. Temporal arteritis and *Chlamydia pneumoniae*. Failure to detect the organism by polymerase chain reaction in ninety cases and ninety controls. *Arthritis Rheum.* **46:**1056–1060.

Reinhold, P., E. Liebler-Tenorio, S. Sattler, and K. Sachse. 2011. Recurrence of *Chlamydia suis* infection in pigs after short-term antimicrobial treatment. *Vet. J.* **187:**405–407.

Reinhold, P., K. Sachse, and B. Kaltenboeck. 2011. *Chlamydiaceae* in cattle: commensals, trigger organisms, or pathogens? *Vet. J.* **189:**257–267. doi:10.1016/j.tvjl.2010.09.003.

Richmond, S. J., A. L. Hilton, and S. K. R. Clarke. 1972. Chlamydial infection. Role of *Chlamydia* sub-group A in non-gonococcal and post-gonococcal urethritis. *Br. J. Vener. Dis.* **48:**437–444.

Rustad, T. R., M. I. Harrell, R. Liao, and D. R. Sherman. 2008. The enduring hypoxic response of *Mycobacterium tuberculosis*. *PLoS One* **3:**e1502.

Schoborg, R. V. 2011. *Chlamydia* persistence—a tool to dissect *Chlamydia*-host cell interactions. *Microbes Infect.* **13:**649–662.

Shaw, A. C., G. Christiansen, P. Roepstorff, and S. Birkelund. 2000. Genetic differences in the *Chlamydia trachomatis* tryptophan synthase a-subunit can explain variations in serovar pathogenesis. *Microbes Infect.* **2:**581–592.

Shleeva, M. O., G. V. Mukamolova, M. Young, H. D. Williams, and A. S. Kaprelyants. 2004. Formation of "non-culturable" cells of *Mycobacterium smegmatis* in stationary phase in response to growth under suboptimal conditions and their Rpf-mediated resuscitation. *Microbiology* **150:**1687–1697.

Singh, A. E., and B. Romanowski. 1999. Syphilis: review with emphasis on clinical, epidemiologic, and some biologic features. *Clin. Microbiol. Rev.* **12:**187–209.

Sinnott, C. R., and A. J. Teall. 1987. Persistent gallbladder carriage of *Salmonella typhi*. *Lancet* **i:**976.

Soete, M., D. Camus, and J. F. Dubremetz. 1994. Experimental induction of bradyzoite-specific antigen expression and cyst formation by the RH strain of *Toxoplasma gondii* in vitro. *Exp. Parasitol.* **78:**361–370.

Song, Z., P. Brassard, and J. M. Brophy. 2008. A meta-analysis of antibiotic use for the secondary prevention of cardiovascular diseases. *Can. J. Cardiol.* **24:**391–395.

Stern, L. C. 1875. *Papyros Ebers: Das hermetische Buch uber die Arzeneimittel der alten Agypter in hieratischer Schrift, herausgegeben mit Inhaltsangabe und Einleitung versehen von Georg Ebers, mit Heiroglyphisch-Lateinischem Gloisser von Ludwig Stern, mit Unterstutzung des Koniglich Sachsischen Cultusministerium*, 2. G Ebers ed., Leipzig, Germany.

Sullivan, W. J., Jr., A. T. Smith, and B. R. Joyce. 2009. Understanding mechanisms of differentiation and a role in the pathogenesis of *Toxoplasma gondii*: a review. *Mem. Inst. Oswaldo Cruz* **104:**155–161.

Tan, C., R.-C. Hsia, H. Shou, C. Haggerty, C. Gaydos, D. Dean, A. Scurlock, D. P. Wilson, and P. M. Bavoil. 2009. *Chlamydia trachomatis*-infected patients display variable antibody profiles against the nine-member polymorphic membrane protein family. *Infect. Immun.* **77:**3218–3226.

Taylor, H. R. 2008. *Trachoma, a Blinding Scourge from the Bronze Age to the Twenty-First Century*. Haddington Press, South Yarra, Australia.

Townes, J. M. 2010. Reactive arthritis after enteric infections in the United States: the problem of definition. *Clin. Infect. Dis.* **50:**247–254.

Voskuil, M. I., K. C. Visconti, and G. K. Schoolnik. 2004. *Mycobacterium tuberculosis* gene expression during adaptation to stationary phase and low-oxygen dormancy. *Tuberculosis* **84:**218–227.

Watson, C., and N. J. Alp. 2008. Role of *Chlamydia pneumoniae* in atherosclerosis. *Clin. Sci.* **114:**509–531.

Wayne, L. G., and C. D. Sohaskey. 2001. Nonreplicating persistence of *Mycobacterium tuberculosis*. *Annu. Rev. Microbiol.* **55:**139–163.

Weiss, L. M., and K. Kim. 2000. The development and biology of bradyzoites of *Toxoplasma gondii*. *Front. Biosci.* **5:**391–405.

Wells, T. N. C., J. N. Burrows, and J. K. Baird. 2010. Targeting the hypnozoite reservoir of *Plasmodium vivax*: the hidden obstacle to malaria elimination. *Trends Parasitol.* **26:**145–151.

Wilson, B. A., A. A. Salyers, D. D. Whitt, and M. E. Winkler. 2011. *Bacterial Pathogenesis: a Molecular Approach*, 3rd ed. ASM Press, Washington, DC.

Wood, H., C. Fehlner-Gardner, J. Berry, E. Fischer, B. Graham, T. Hackstadt, C. Roshick, and G. McClarty. 2003. Regulation of tryptophan synthase gene expression in *Chlamydia trachomatis*. *Mol. Microbiol.* **49:**1347–1359.

Wyrick, P. B. 2010. *Chlamydia* persistence *in vitro*: an overview. *J. Infect. Dis.* **201**(S2)**:**S88–S95.

IN VIVO CHLAMYDIAL INFECTION

Roger G. Rank

13

INTRODUCTION

While it is clearly not a huge revelation, in order to thoroughly understand the pathogenesis of infection and host response to an infectious agent, one must utilize relevant animal models for the disease, i.e., "*in vivo veritas.*" In vitro studies on pathogenesis and host response, by definition, drastically limit the variables that occur in vivo and simplify the study to the point where the results may bear little relevance to actual clinical disease. That is not to say that in vitro studies do not have a role; on the contrary, they can provide valuable insights into molecular processes, particularly at the intracellular level, that are difficult to address when an animal is used. Nevertheless, new technology now gives us the opportunity to answer basic questions regarding cell biology and molecular pathways in the animal.

Unlike some infectious diseases in which there is only a single animal model or none at all, there are several excellent models of chlamydial infection. The question then becomes which is the best model to use. Many investigators fall into the trap of trying to answer every question with a single model. A better approach is to determine exactly what research question(s) one wants to answer, and only then choose the model that is best suited to answer it.

Generally, a chapter on animal models describes the various models and how they represent the human disease. There are ample detailed reviews that describe the available models and how these address the pathogenesis of chlamydial disease and the host response to the organism (Darville and Hiltke, 2010; Miyairi et al., 2010; Morrison and Caldwell, 2002; Patton, 1992; Rank, 1988, 1999, 2007, 2009). Therefore, this chapter will only briefly summarize the various models and then address their potential uses. These include the modeling of aspects of chlamydial disease that have either just come to the forefront because of new technologies or have been avoided because it was felt that either a given model could not address the hypothesis or models were simply overlooked because the question did not fit into the traditional use of a model. Although there is a wide spectrum of diseases caused by *Chlamydia*, the majority of animal model research has centered on models of genital infection; therefore, the focus of this chapter will be on models of genital infections,

Roger G. Rank, Department of Microbiology and Immunology, University of Arkansas for Medical Sciences and Arkansas Children's Hospital Research Institute, Little Rock, AR 72202.

Intracellular Pathogens I: Chlamydiales
Edited by Ming Tan and Patrik M. Bavoil © 2012 ASM Press, Washington, DC
10.1128/9781555817329.ch13

but the other models will be discussed where relevant.

SUMMARY OF ANIMAL MODELS FOR CHLAMYDIAL DISEASE

While various animals have been tried and evaluated over the years as potential models for chlamydial diseases, the only ones that have been extensively used and have provided the vast majority of in vivo data are the mouse, the guinea pig, and the nonhuman primate. This is not to say that the other models have not served a purpose or resulted in valuable data—it is just that these are the most commonly used. Each of these models has advantages and disadvantages. In this section, I will briefly review the major models with emphasis on their applicability to a particular disease state as well as the interpretive problems that may arise because of specific aspects of the models. The animal models and the infection sites for which they have been used are presented in Table 1.

Mouse

The mouse model has many general strengths as well as specific advantages for *Chlamydia* research. Mice are relatively inexpensive, and multiple inbred strains are available, together with an ever-expanding repertoire of knockout and transgenic mice and a vast array of reagents. A major advantage of the mouse is that mice can be infected with four different species of *Chlamydia*: *C. muridarum*, *C. trachomatis*, *C. pneumoniae*, and *C. psittaci*. The mouse continues to be the most used animal model for studies of the host response and pathogenesis of chlamydial infection.

C. muridarum, which causes a natural infection of mice in nature, was originally derived from the respiratory tract of laboratory mice in the 1940s and termed the agent of mouse pneumonitis (Gordon et al., 1938; Nigg, 1942). Although it was first isolated from the lung, its natural site of infection is likely to be the gut, and transmission in nature is via the fecal-oral route (Rank, 2007). There has been some debate over whether the organism can be detected in nature, but recently, Ramsey and colleagues (K. H. Ramsey, personal communication) have titered 256 serum samples from a reference bank for specimens from deer mice (*Peromyscus* spp.) and found that 74% were positive for antibody to *C. muridarum*. In addition, they probed serum samples from a colony of specific-pathogen-free *Peromyscus* and found a 50% reactivity. Specificity for *C. muridarum* was confirmed by immunoblotting and by PCR analysis of tissues for chlamydial 16S rRNA. These data indicate that *C. muridarum* is endemic at least in some natural mouse populations. Because the mouse is the natural host for *C. muridarum*, the relationships and responses that are uncovered with this animal model are more likely to be predictive of the natural relationship between humans and *C. trachomatis*.

The *C. muridarum* genital tract model was first described by Barron and colleagues in 1981 (Barron et al., 1981) and is currently the most commonly used animal model for chlamydial genital infection. It is primarily used for the study of the host response in immunity and disease (Morrison and Caldwell, 2002; Rank, 1999, 2007), and both male (Pal et al., 2004)

TABLE 1 Major animal models

Animal species	*Chlamydia* species	Site of infection
Mouse	*C. muridarum*	Genital (male/female), respiratory, gut
	C. trachomatis	Genital (female), respiratory
	C. pneumoniae	Respiratory
	C. psittaci	Systemic
Guinea pig	*C. caviae*	Genital (male/female), respiratory, conjunctiva, rectum
Nonhuman primate	*C. trachomatis*	Genital (male/female), conjunctiva, rectum

and female mice can be infected via the genital route. The majority of studies has been, and continues to be, with female animals because of ease of infection and because of the emphasis on understanding the disease and the immune response in the reproductive tract as a primary target of morbidity in humans. In general, with some variation according to the mouse strain used, mice resolve the infection in 3 to 4 weeks and are completely immune to reinfection in the short term. However, the longer the time span between resolution of infection and the challenge infection, the more likely it is that the mice will become reinfected. The challenge infection is shorter in duration and has a lower peak number of chlamydiae. The target tissue of the infection is the cervix, but ascending infection to the endometrium and oviducts readily occurs with resultant endometritis, salpingitis, and infertility (de la Maza et al., 1994; Rank, 2007).

C. muridarum can also infect the respiratory tract producing a lethal pneumonitis if sufficient organisms are inoculated. Much of our understanding of the basic roles of antibody- and cell-mediated immunity to chlamydial infection was first published by Dwight Williams in a series of elegant papers using this model (Rank, 1999; Williams et al., 1984, 1986). While there are clear differences between *C. muridarum* infection of the respiratory tract and that of the genital tract, there are also many similarities, including a major role for cell-mediated immunity. The respiratory model is still being used for basic immunologic studies in the lung and has been used as a convenient tool for screening vaccine candidates, since the severity of disease is directly related to the body weight during the course of infection (Brunham and Zhang, 1999; Murdin et al., 2000). However, one must always pursue caution in vaccine studies because an antigen that elicits protective immunity against *C. muridarum* in the mouse may not necessarily be protective against *C. trachomatis* in humans.

Ideally, one should employ an infection model that uses the actual human pathogen. Initially, mice were infected via the respiratory route with *C. trachomatis* serovars L2, B, C, D, and G (Chen and Kuo, 1980; Kuo and Chen, 1980). Female mice can also be infected genitally with *C. trachomatis*, but a high inoculum of IFU (inclusion-forming units) is required, and fewer than 100% of the animals will become infected. The infection is limited in length to 2 to 10 days unless one pretreats the mice with progesterone (Depo-Provera) to produce a state of anestrus, thereby eliminating the variable of the estrous cycle (Tuffrey and Taylor-Robinson, 1981). Also, mice cannot be easily infected when in estrus because the thick cellular material present in the genital tract inhibits attachment of chlamydiae to their target tissue. When mice are pretreated with progesterone, infection with *C. trachomatis* will last 3 to 7 weeks, although the duration is dependent upon the serovar used (Ito et al., 1990). Progesterone pretreatment is commonly used in *C. muridarum* infections as well to eliminate the variable of the estrous cycle, although it is not as essential for establishing the infection as it is for *C. trachomatis*. Even with progesterone treatment, *C. trachomatis* does not elicit severe upper genital tract pathology and produces only minimal, if any, oviductal pathology, which is clearly a disadvantage when studying infertility that results from salpingitis. Another limitation is that the reproductive physiologies of mice and humans are vastly different, and one cannot realistically use the mouse to determine the impact of reproductive hormones on chlamydial genital disease in humans.

The mouse has also been used as a model to study pneumonia elicited by *C. pneumoniae*. Mice can be inoculated intranasally, and a chronic pneumonia very similar to the human infection results, although severity and length of infection are dependent upon the mouse strain used (Kaukoranta-Tolvanen et al., 1993; Yang et al., 1993). The association of *C. pneumoniae* with atherosclerosis prompted Campbell and Kuo to study atherogenesis in apolipoprotein E-deficient mice, which spontaneously develop atherosclerosis. Following intranasal infection with *C. pneumoniae*, chlamydiae were detected in lung, spleen, and more importantly, aorta for 20 weeks following infection and were associated with more-severe arterial disease than is characteristic of human heart disease (Campbell et al., 1998; Moazed et al., 1997).

Finally, the mouse has been used to study systemic infection with *C. psittaci*, and some basic concepts regarding the host immune response, particularly the role of gamma interferon (IFN-γ), were described using this model (Byrne et al., 1988; Byrne and Faubion, 1982; Byrne and Krueger, 1983). Because it can be potentially used as a bioweapon, there has been some resurgence of interest in studying *C. psittaci* in recent years.

Guinea Pig

Like the mouse, the guinea pig is naturally infected with a chlamydial species, *Chlamydia caviae*. Murray first isolated *C. caviae* from a conjunctival infection of young laboratory guinea pigs and termed it the agent of guinea pig inclusion conjunctivitis (Murray, 1964). The organism was later independently isolated by Schachter (Kazdan et al., 1967). While all experimental work primarily utilizes the Murray isolate, the organism is undoubtedly still endemic in guinea pig populations as evidenced by the occasional appearance in the veterinary literature of case reports of chlamydial ocular infections (Schmeer et al., 1985). *C. caviae* was originally used as a model of inclusion conjunctivitis and trachoma and is still an excellent model to study chlamydial ocular disease (Monnickendam et al., 1980b). Mount and Barron first reported that *C. caviae* could be inoculated into the female genital tract to produce a disease analogous to human disease (Mount et al., 1972). They also reported that the infection could be transmitted sexually from males to females and that infection of pregnant females resulted in perinatal transmission to newborns with resultant neonatal conjunctivitis (Mount et al., 1973). The model has been further used to characterize ascending genital tract infection to the endometrium and oviduct, urethral infection in males, the effect of reproductive hormones and the estrous cycle on the infection course, and production of pneumonitis of the newborn (Rank et al., 1982, 1985, 1993; Rank and Sanders, 1992; Wang et al., 2010). Guinea pigs are an excellent model for studying the pathogenesis of chlamydial genital disease in humans because of the similar length of the estrous cycle (17 days) and the very close similarity of the reproductive physiology and histology. The infection course in both the genital tract and conjunctiva is about 3 to 4 weeks followed by short-term complete immunity and long-lasting partial immunity. As in the mouse, the target tissue in the genital tract is the cervix and ascending infection to the endometrium and oviduct occurs in about 80% of the animals. The pathology in all parts of the genital tract is virtually the same as seen in humans.

The major disadvantage of the guinea pig has been the lack of a wide range of immunological reagents, knockout animals, and easily accessible inbred guinea pig strains. Nevertheless, the model has been used for studies of the adaptive immune response as well as immunization studies. The guinea pig is ideal for vaccine studies to show proof of concept in prevention of disease and pathology. However, the infectious species in the guinea pig model suffers from the same disadvantage as *C. muridarum*, in that it is not *C. trachomatis*; so one must be concerned that a protective antigen or vaccine in the guinea pig may not be readily transferable to humans and that a *C. trachomatis* vaccine candidate may not protect in the guinea pig if it does not have genus-wide epitopes.

Nonhuman Primate

Without a doubt, the animal model that most closely resembles human infection is the nonhuman primate, which has the advantage that it can be infected with *C. trachomatis*. Several different species of nonhuman primates have been infected in the female genital tract with *C. trachomatis*, including pig-tailed macaques (*Macaca nemestrina)* (Patton et al., 1983, 1987a), grivet monkeys (*Chlorocebus aethiops*) (Moller and Mardh, 1980b; Ripa et al.,

1979), cynomolgus (*Macaca fasicularis*) (Patton et al., 1987b), rhesus monkeys (*Macaca mulatta*) (Patton et al., 1987b), marmosets (*Callithrix jaccus*) (Johnson et al., 1980), and baboons (*Papio* sp.) (Alexander and Chiang, 1967; Darougar et al., 1977). Additional studies have demonstrated infection of the male urethra in baboons (Digiacomo et al., 1975), chimpanzees (Jacobs et al., 1978), and grivet monkeys (Moller and Mardh, 1980a). In general, infection in the cervix lasts 6 to 15 weeks, which is unlike that in smaller animals and more similar to that in women. Ascending infection and pathology in the upper genital tract can be demonstrated but require multiple inoculations (Moller and Mardh, 1980b; Patton et al., 1987a).

The nonhuman primate has also been used extensively as a model for trachoma. Typical trachoma can be elicited in rhesus monkeys by repeated weekly inoculation with *C. trachomatis* (Taylor et al., 1981). This regimen produces long-lasting disease (>35 weeks) with follicular development and inflammation. Just as in human trachoma, organisms can only be isolated for the first few weeks of infection and are then difficult to detect, even after a fresh challenge. Thus, the nonhuman primate has been very valuable in our understanding of chlamydial disease, but the major disadvantage is still the extreme cost and availability, as well as the inability to utilize larger numbers of animals for statistically significant analysis.

BIOLOGY OF CHLAMYDIAL INFECTION: OVERVIEW

The major application of animal models of chlamydial infection has been directed toward the characterization of the host response and the understanding of the mechanisms of protective immunity, all oriented toward the development of an effective vaccine. Directly related to these studies has been the extensive use of animal models for the testing of potential vaccine candidates. There are many reviews on the immunology of chlamydial disease and the development of a vaccine (Darville, 2006; Kelly, 2006; Morrison and Caldwell, 2002; Ramsey, 2006; Rank, 1988, 1994, 1999, 2007, 2009; Rank and Whittum-Hudson, 2010). In addition, much work has been devoted to the characterization of the pathologic response to infection and the cellular components that elicit this response. These are significant questions with direct translational applications and are an important use of the animal model systems; however, there are many fundamental questions about the biology of chlamydiae and chlamydial infections that either remain unanswered or have only been addressed in vitro and remain unconfirmed in live animals. In this section, various aspects of chlamydial biology will be discussed in the context of animal models.

Use of Animal Models To Study the Chlamydial Developmental Cycle

Perhaps the most intriguing biological characteristic of chlamydiae is their biphasic developmental cycle. It is also one of the most studied aspects of chlamydial biology, and the developmental regulation of chlamydial genes as three major temporal groups is discussed in more detail in chapter 7, "Temporal gene regulation during the chlamydial developmental cycle." The vast majority of our knowledge of the developmental cycle has been derived from experimentation in cell culture. Beginning with the morphologic appearance of the various stages of the cycle, there has been an immense amount of ultrastructural work performed in vitro to characterize the structure of reticulate bodies (RBs) and elementary bodies (EBs), which has been reviewed in detail by Matsumoto (Matsumoto, 1988). In addition, Ward presented an elegant stepwise review of the developmental cycle with ultrastructural images representing each stage in the process (Ward, 1988). Under optimal conditions, it is unlikely that there are significant differences in the morphology of individual EBs, RBs, and inclusions in vitro compared to in vivo. However, the absence of essential molecules (e.g., tryptophan) or the presence of certain compounds (e.g., penicillin and IFN-γ) both in vitro and in vivo can alter the morphology.

Interestingly, the first description of the chlamydial developmental cycle by Bedson and Bland was developed by observations of splenic impression smears from mice infected with *C. psittaci* (then known as the "psittacosis virus") (Bedson and Bland, 1932). The first ultrastructural images from human genital infection were published by Swanson and colleagues and demonstrated classic *C. trachomatis* inclusions with RBs and EBs in cervical epithelial cells from a cervical biopsy specimen and were no different from what was seen in tissue culture (Swanson et al., 1975). Low-magnification pictures showed a strong acute inflammatory response with polymorphonuclear leukocytes (PMNs) often in the vicinity of infected cells. Typical RBs, intermediate bodies (IBs), and EBs were further demonstrated by electron microscopy in intestinal tissue of *Chlamydia*-infected calves (Doughri et al., 1972), cervical, urethral, oviductal, and bladder epithelia from guinea pigs infected with *C. caviae* (Soloff et al., 1982, 1985), conjunctival tissue from nonhuman primates infected with *C. trachomatis* (Patton and Taylor, 1986), and cervical and oviductal tissue from mice infected with *C. muridarum* (Phillips et al., 1984; Phillips and Burillo, 1998).

While the morphology of the inclusions in vivo was basically the same as that observed in vitro, few clues as to how the cycle terminated in vivo were available. It was always assumed that the cells either lysed, extruded their inclusion, or underwent apoptosis as was observed in vitro (Byrne and Ojcius, 2004; Hybiske and Stephens, 2007; Neeper et al., 1990). A study by Doughri (Doughri et al., 1972) examined ileum tissue infected with a strain of *Chlamydia* isolated from the joint of a newborn calf with polyarthritis and suggested mechanisms by which the developmental cycle was initiated and terminated. With regard to entry, they visualized EBs adsorbed to microvilli and EBs being taken into cells by pinocytosis. Based on their observations, Doughri and colleagues further suggested that chlamydiae are released from infected cells by three different mechanisms: (i) the inclusion membrane disappears, and the apical cell membrane ruptures, releasing chlamydiae into the intestinal lumen; (ii) the entire infected cell loses its close apposition to the basal lamina, and PMNs migrating into the space under the cell appear to aid in the detachment of the cell from the epithelium; and (iii) occasionally, chlamydiae are present in a cytoplasmic protuberance that appeared to be pinched off into the lumen.

Soloff and colleagues made similar observations in cervical tissue from guinea pigs infected with *C. caviae* (Soloff et al., 1985). They observed that in some cases, cells with apparently late stage inclusions lost the integrity of the inclusion membrane and the cell detached from the epithelium. Furthermore, they also noted that PMNs entered into enlarged intercellular spaces between the infected cell at the surface and the underlying cells, suggesting that they were engaged in removing the cell from the epithelium. Loss of integrity of the inclusion membrane and detachment from the epithelial surface thus appears to be a common exit strategy of late-stage inclusions in vivo. Using *C. muridarum*, Phillips and colleagues (Phillips et al., 1984) noted that mature inclusions in oviduct cells often had disrupted inclusion membranes with chlamydiae free in the cytoplasm, and they suggested that at this stage, the cells would break apart and release the bacteria into the lumen of the oviduct. Patton and Taylor also showed evidence of this mechanism in the conjunctiva of cynomolgus monkeys infected with *C. trachomatis* (Patton and Taylor, 1986).

These studies have been most informative about the developmental cycle in vivo and, in general, support in vitro observations, especially with regard to the development of the inclusion within the host cell. Nevertheless, in each of these studies, the tissues were collected at a single time after the infection had been established, and as a consequence, multiple stages of the infection were present because chlamydial infections are asynchronous with respect to the developmental cycle. Therefore, in order to develop a synchronous infection to characterize the sequential events in the developmental cycle in vivo, we modified the mouse model to inoculate *C. muridarum* directly into the cervix to infect as many target cells as possible in a short time (Rank et al., 2010). Standard intravaginal inoculation results in the loss of a large amount of the inoculum through vaginal drainage, and not all of the EBs penetrate through the cervical os at the same time, resulting in an asynchronous infection. In the intracervical infection, mice are anesthetized, an incision is made in the lower abdomen, and the genital tract is exposed. The uterine horns are ligated to prevent the inoculum from spreading away from the cervix, and the inoculum is injected into the base of the uterine horn, depositing the organisms directly at the cervix.

When cervical tissue was collected at 18, 24, 30, 36, 42, and 48 hours after inoculation and examined by transmission electron microscopy, virtually all of the inclusions were at the same stage for each time point (Rank et al., 2011). This synchronous infection presented the opportunity to determine cellular and molecular events in relation to the developmental cycle, and since it was the initial developmental cycle of the infection, it was possible to examine sequential events at the target site. We previously measured 16S RNA levels at 3, 12, and 24 hours after intracervical infection and found that it was detectable at 12 hours but not at 3 hours (Rank et al., 2010). The ultrastructural analysis at 18 hours only detected luminal EBs and an occasional internalized IB or RB just inside the cell membrane, suggesting that actual infection of the host cell occurs between 12 and 18 hours (Rank et al., 2011). This would seem slow compared to in vitro studies, but one must remember that cervical mucus likely impedes attachment to target cells in vivo. By 24 hours, many cells were infected with multiple inclusions, each inclusion containing 1 to 5 RBs with evidence that some of the inclusions were beginning to fuse (Fig. 1A). By 30 hours, only single inclusions containing solely RBs were found with cells, indicating that the fusion process occurred between 24 to 30 hours after infection (Fig. 1B). In vitro, fusion occurs between 10 and 12 hours, but if one accounts for a 10- to 14-hour-delayed cell infection in vivo,

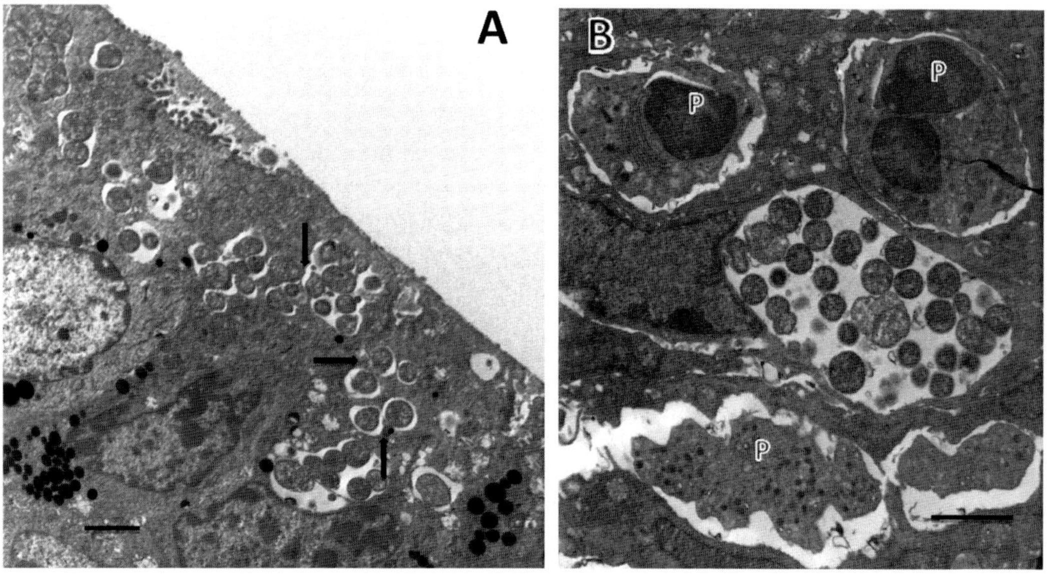

FIGURE 1 Events occurring 24 hours and 30 hours after infection. (A) At 24 hours, multiple early inclusions containing one to a few RBs are seen. Some epithelial cells contain multiple inclusions. RBs are clearly dividing, and the proximity of inclusions suggests that the fusion process is under way (arrows). (B) Inclusion with multiple RBs at 30 hours after infection and PMNs (P) in contact with the infected epithelial host cell. Scale bars for both panels, 2 μm. doi:10.1128/9781555817329.ch13.f1

the time frame is about the same (Hackstadt et al., 1999; Matsumoto et al., 1991). This is the first documentation that fusion of inclusions occurs at an actual mucosal site in vivo. At 36 hours, the inclusions were larger and EBs were now apparent in addition to RBs (Fig. 2A). Between 42 and 48 hours, the initial developmental cycle was clearly terminating. Just as was seen in previous studies, the inclusion membranes in many cells were breaking down so that the chlamydiae were free in the cytoplasm (Doughri et al., 1972; Phillips and Burillo, 1998; Rank et al., 2011; Soloff et al., 1985). The cells containing those late-stage inclusions were often vacuolated, lacking in microvilli, and obviously dying (Fig. 2B). Cells in the process of breaking apart could be seen spilling their chlamydiae into the lumen. It was not uncommon for these cells to be detaching from the epithelium, thus confirming both in vivo and in vitro observations that this is one mechanism by which the developmental cycle terminates and liberates EBs.

It has been well documented in numerous studies that an acute inflammatory response is invariably present in the early stages of chlamydial infection at a mucosal site. Using the intracervical model, we also observed that PMNs began to enter into the submucosa by 12 hours after infection and were present in the epithelium by 24 hours (Rank et al., 2010). Both Soloff and colleagues (Soloff et al., 1985) and Doughri and colleagues (Doughri et al., 1972) had commented upon the apparent role that PMNs play in the termination of the developmental cycle. We also have investigated PMNs in an ultrastructural analysis of the guinea pig conjunctiva, 4 days after infection with *C. caviae* (Rank et al., 2008). An intense acute inflammatory response was noted, but importantly, PMNs were always observed in association with infected cells. It was not unusual to find packets of multiple PMNs in intercellular spaces directly beneath the infected cells as if in the process of actively dislodging the infected cells from the epithelium. One could even find multiple infected cells, still connected by desmosomes with the entire set of infected cells seemingly being lifted off the surface by PMNs. We also observed healthy-looking infected cells in the lumen that contained only early inclusions. We hypothesized that PMNs not only are important for the termination of the cycle but also play a role in the distribution of chlamydiae to other sites in the epithelium. There is no ultrastructural evidence in vivo for direct transmission of EBs to adjacent cells, and it was observed in the guinea pig conjunctiva that large areas of tissue destruction ensued in areas of infection and inflammation. Therefore, it would seem logical that the chlamydiae, in order to continue the infection, need some mechanism to "move" to other epithelial sites with available target cells, since they obviously lack their own motility mechanisms. Thus, even though chlamydiae may be killed by PMNs, the PMNs may actually serve an important function by dislodging infected cells from the epithelium. These cells may then move via peristalsis or fluid dynamics to other sites or may become available for transmission to new hosts upon sexual activity, or by mechanical means as in trachoma.

In fact, this mandatory pursuit of uninfected epithelial cells by chlamydiae in order to begin another round of replication may be a major factor in ascending infection in the female genital tract. That PMNs likely play a crucial role in ascending infection was demonstrated by Ramsey and his colleagues when they observed that mice depleted of, or deficient in, matrix metalloprotease 9, which is produced in abundance by PMNs, did not develop ascending infection with *C. muridarum* even though the cervical infection course was not altered (Imtiaz et al., 2006).

In addition to both of the above mechanisms for the termination of the developmental cycle, we also observed a new mechanism using the intracervical infection model. At 42 to 48 hours after infection, infected cells could be seen containing an intracellular PMN (Fig. 3) (Rank et al., 2011). In some cases, the PMN was abutting the inclusion, but in others, the inclusion membrane was no longer present

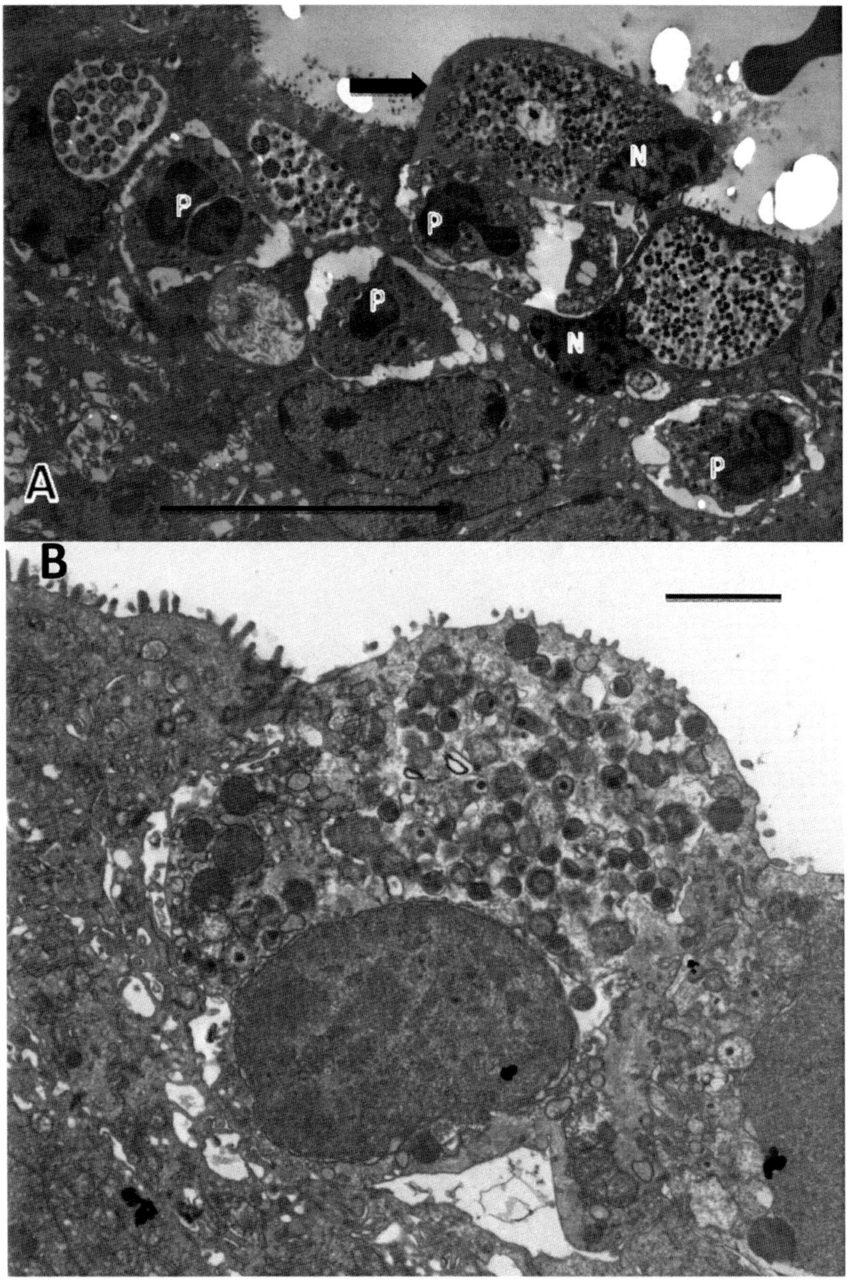

FIGURE 2 Events at 36 hours after infection. (A) Multiple large mature inclusions are seen. In particular, one infected cell (arrow) is being detached from the epithelium with a PMN (P) directly in the space beneath the cell, giving the appearance of "pushing" the cell off of the epithelial surface; the PMN appears to have phagocytized numerous RBs. Nuclei (N) of the infected cells are visible. (B) Terminal infected cell with chlamydiae distributed in the cytosol. The cell appears to be in the process of being dislodged from the epithelial layer. Scale bars: panel A, 10 μm; panel B, 2 μm. doi:10.1128/9781555817329.ch13.f2

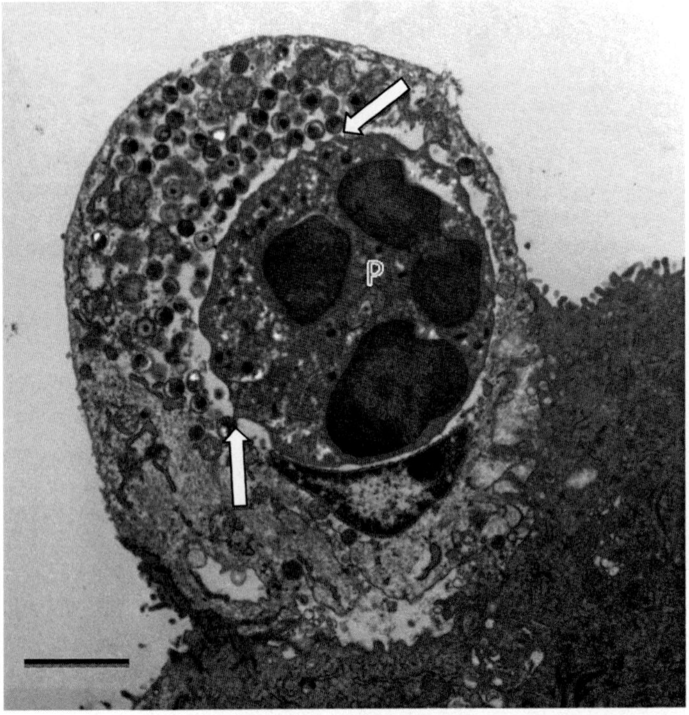

FIGURE 3 PMNs enter into the infected host epithelial cell to make direct contact with chlamydiae. At 42 hours after infection, the inclusion membrane is no longer visible and the intracellular PMN is in direct contact with chlamydiae. Note the projections from the PMNs touching the IBs or EBs (arrows). Scale bar: 2 μm.
doi:10.1128/9781555817329.ch13.f3

and the PMN was in actual contact with the chlamydiae. It was apparent that the PMN had penetrated into the infected cell in order to engage the chlamydiae. This is a unique mechanism that had not been reported for any other intracellular bacterium. That PMNs can actually enter into cells, leaving the cells intact (emperipoiesis), has been demonstrated in other systems as a means for the PMN to leave a venule or capillary and pass into the extravascular space (Feng et al., 1998).

Thus, based on these studies, there is strong evidence that in vivo, the chlamydial developmental cycle terminates by three distinct mechanisms, two of which require the participation of PMNs (Fig. 4). In vitro studies have reported that termination of the cycle occurs by shedding of the infected cells from the monolayer and ultimately destruction of the cell (Neeper et al., 1990), by actual extrusion of the inclusion from the cell (Hybiske and Stephens, 2007), and/or by apoptosis of the infected cell (Byrne and Ojcius, 2004; Ojcius et al., 1998). The in vivo electron microscopy studies support the shedding mechanisms, but there has been no evidence thus far for a mechanism involving apoptosis. Doughri and colleagues reported seeing extrusion of inclusions (Doughri et al., 1972); however, this has not been supported by any other in vivo study to date.

One of the important questions concerning the chlamydial developmental cycle is the determination of the events that cue RBs to begin the process of converting into EBs. There has been an extensive amount of work done on this using strictly in vitro cell culture. Recently, Wilson and Bavoil have carefully studied electron micrographs by Matsumoto and have developed a hypothesis that attachment of RBs to the inclusion membrane by the type III secretion (T3S) injectisomes is critical for RBs to continue replicating (Wilson et al., 2006). The T3S needle complex gives the replicating RB direct communication and access to the host cell cytosol, and when physical

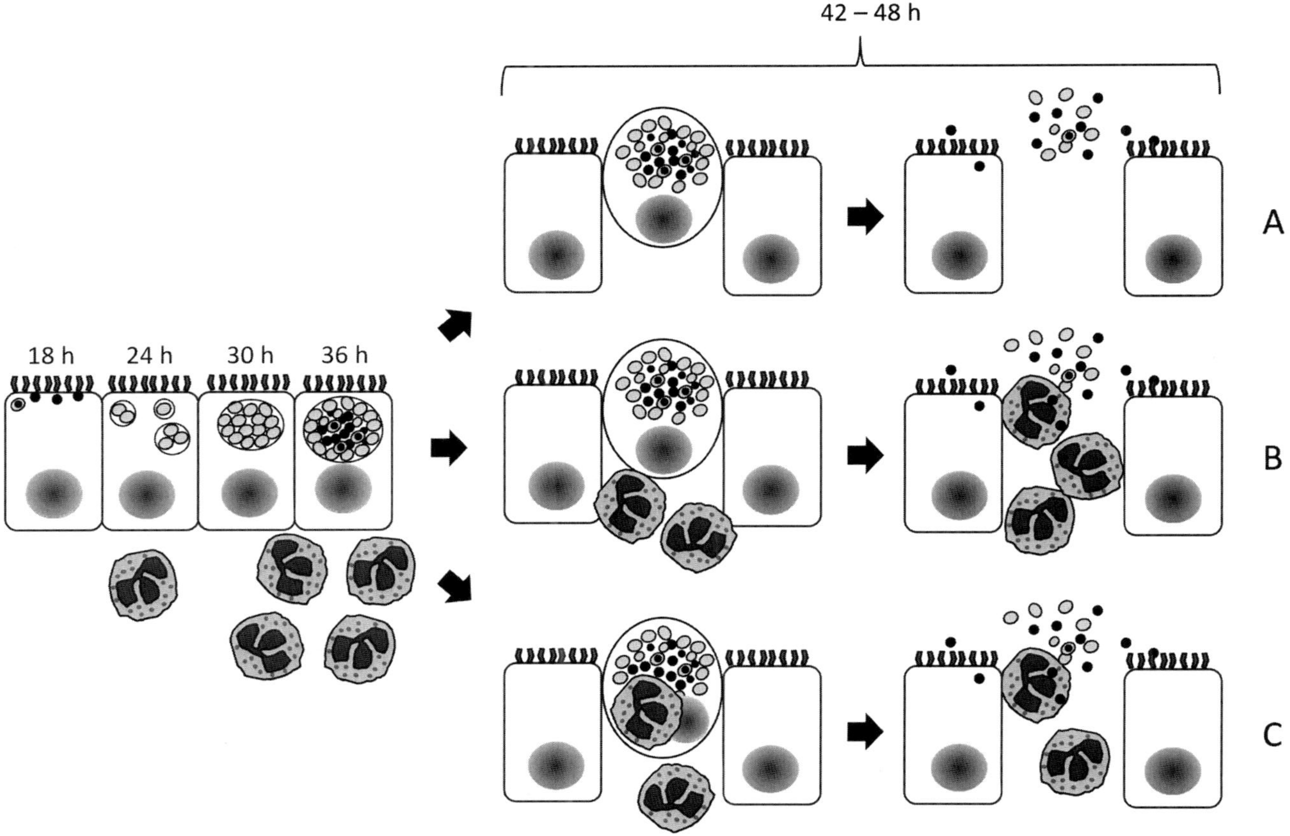

FIGURE 4 Schematic in vivo development of *C. muridarum*. At 18 h, host cells containing internalized EBs are shown, with some in the process of becoming RBs. At 24 h, multiple small inclusions are present. At 30 h, fusion of inclusions has occurred and inclusions contain primarily RBs. At 36 h, inclusions have increased in size with ongoing differentiation of RBs into EBs. At 42 to 48 h, termination of the developmental cycle occurs by three specific mechanisms. (A) The inclusion membrane disintegrates, leaving organisms free in the cytosol. The cell is dislodged from the epithelial layer and breaks apart, allowing EBs to attach to uninfected cells. (B) PMNs attracted to the site effect detachment of the infected cell from the epithelial layer likely through the action of matrix-metalloprotease 9. In some cases, the host cell and inclusion are intact so that the entire cell becomes free in the lumen. Ultimately, the cell dies, releasing EBs at the same or a distant site. (C) The PMN may actually enter the host cell in "pursuit" of chlamydiae. The host cell will die as a result of the enzymes released by the PMN with release of EBs into the lumen. doi:10.1128/9781555817329.ch13.f4

detachment occurs, coupled disruption of T3S activity somehow provides a signal for the RB to begin the transformation into an EB. Using known or estimated physical and biological parameters (e.g., the surface area of an RB hypothesized to be in contact with the inclusion membrane, the inclusion membrane surface area, and the replication rate of chlamydiae), they developed a mathematical model that supports this "contact-dependent" hypothesis. Implicit in this hypothesis is that RBs attached to the inclusion membrane are larger than those that are detached, as these have begun the transition to EBs.

In our intracervical infection model with *C. muridarum*, we routinely observed inclusions at 36 to 42 hours that had a ring of larger RBs attached to the inclusion membrane, while those RBs in the interior of the inclusion were smaller (Rank et al., 2011). Moreover, the vast majority of EBs were in the interior of the inclusion (Fig. 5). To confirm the contact-dependent hypothesis in vivo, we measured the area of all RBs in multiple electron micrograph images with ImageJ software from 24 to 36 hours after infection. Since the infection was synchronous, we were able to determine with confidence the relative age of the inclusions. The median size of the RBs decreased as the inclusions progressed through the developmental cycle. Furthermore, in inclusions at 36 and 42 hours postinfection, the area of membrane-attached RBs was significantly larger than the area of detached RBs. Therefore, these data from in vivo cervical infection support the contact-dependent hypothesis developed with in silico modeling from in vitro observations.

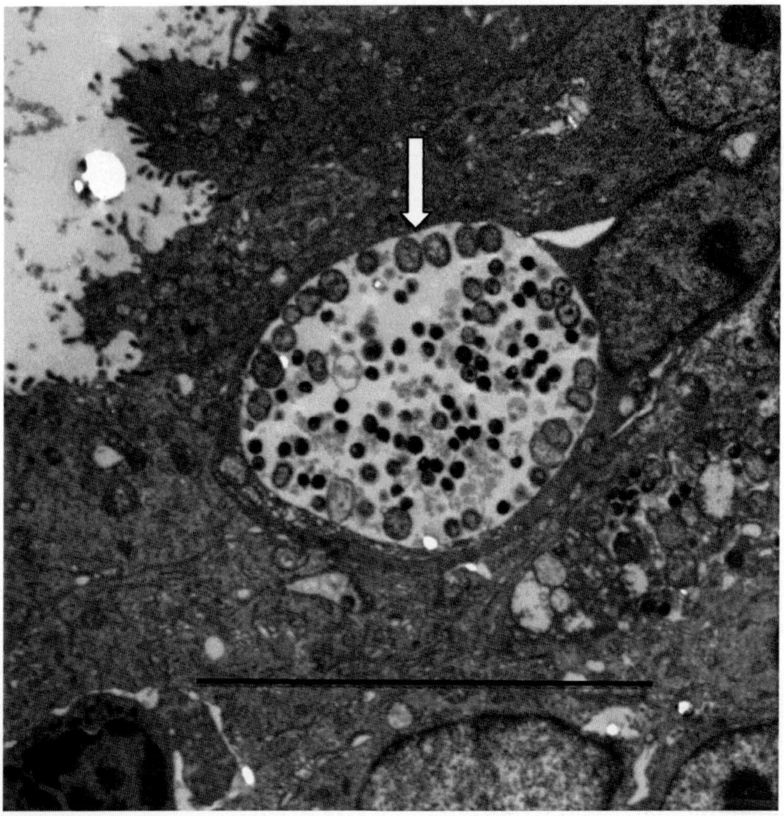

FIGURE 5 Inclusion at 42 hours after infection. Note that the majority of the RBs in this inclusion (arrow) are along the periphery of the inclusion adjacent to the inclusion membrane while the EBs are in the internal area. Scale bar: 10 μm.
doi:10.1128/9781555817329.ch13.f5

The intracervical model has also been used to determine the relationship between the developmental cycle and the initiation of the acute inflammatory response (Rank et al., 2010). There have been many such studies performed in vitro to determine when various chemokines and cytokines are produced in relation to the developmental cycle, but those studies were performed with cultured cells. Even if those cells were derived from the natural site of infection, the fact remains that they are still in tissue culture without all of the components of the microenvironment existing at the target site in the animal. Cytokine/chemokine arrays were performed with RNA derived from the mouse cervix at 3, 12, and 24 hours after intracervical infection with *C. muridarum*, and the expression was compared to that of mice inoculated with UV-inactivated organisms or mice receiving a sham inoculation. There was no significant response in the control animals, but CCR2 and CCR6 as well as tumor necrosis factor alpha (TNF-α) were upregulated as early as 3 hours after infection, a time when chlamydial 16S rRNA could not be detected. However, by 12 hours after infection, 16S rRNA was expressed and PMNs were beginning to enter the tissue site but were not yet in the epithelium. Surprisingly, 41 different chemokine receptors, chemokines, and cytokines were upregulated, indicating that chlamydiae merely in the process of infecting cells are able to activate mechanisms that produce this large array of host response molecules. Chemokines for PMNs, NK cells, activated T cells, and mononuclear cells were all upregulated. By 24 hours, there was a significant influx of PMNs and expression of many proinflammatory cytokines, such as TNF-α, interleukin-1α, interleukin-1β, and IFN-γ, continued to increase.

Because of the synchronicity of the initial infection, this model has the potential to be used for a number of studies that could yield data unavailable from in vitro studies. For instance, as the technology and sensitivity for assessing the transcriptome improves, one could potentially assess both host cell and chlamydial transcriptomes in the same tissue from a single animal and relate the results to the developmental cycle. This would lead directly to other questions such as how different variants of the same chlamydial strain may alter the host response or how the response may differ in different strains of mice. What are the chlamydial and host factors that mediate infection site specificity (e.g., cervix versus endometrium versus oviduct)?

Use of Animal Models To Study Persistence

It is commonly accepted that chlamydiae can persist in the host, whether human or animal, for long periods of time, but the mechanism(s) of persistence is unknown (discussed in more detail in chapter 12, "Chlamydial persistence redux"). Chlamydial persistence has come to be defined in terms of a cell culture phenomenon of large aberrant RBs that do not replicate, with decreased amounts of glycogen in the inclusion and the presence of "miniature" RBs. In addition, expression of Hsp60 appears to be upregulated, while that of major outer membrane protein is downregulated (Beatty et al., 1995). These "persistent" inclusions, which perhaps are better described as "stressed inclusions," have been classically elicited by adding penicillin or IFN-γ to the culture media. However, whether these are truly representative of an in vivo mechanism for persistence is not known. Aberrant RBs have been detected in vivo but only in the very early stage of infection. Pospischil and colleagues recently produced electron micrographs of inclusions in the gut of pigs infected with *Chlamydia suis*, approximately 2 to 7 days after infection (Pospischil et al., 2009). These inclusions had all the characteristics of stressed inclusions and were found commonly in the epithelium, although normal inclusions were also observed in the same tissue. We have also detected stressed inclusions 36 to 42 hours after intracervical infection of mice treated with antibodies to PMNs to deplete those cells (Rank et al., 2011). Both stressed inclusions and normal inclusions could be found in the tissue, even in adjacent cells (Fig. 6). It is not surprising that stressed inclusions were detected soon after infection, since it is very likely that there is a strong IFN-γ response early in the infection (Tseng and Rank, 1998). Nevertheless, the simple presence of stressed inclusions during early infection says nothing about the state of the organism in a true long-term persistent infection.

Perhaps, instead of "persistent" infection, a long-term infection should be simply referred to as a chronic infection. One of the problems in investigating this phenomenon has been the lack of a convenient animal model. Chronic subclinical infections have been documented in psittacine birds and in sheep and cattle, but these are not ideal for experimentation. In all commonly used animal models, the infection eventually resolves, but its length varies depending on the chlamydial species or animal strain. Studies on persistent infections in animal models have been limited. Yang and Kuo observed that after apparent resolution of *C. trachomatis* respiratory infection in mice, the infection could be reactivated by treatment with cortisone (Yang et al., 1983); however, treatment was initiated very shortly after resolution of infection. Similar reactivation by treatment with cortisone was also demonstrated in *C. pneumoniae* infection of mice (Laitinen et al., 1996; Malinverni et al., 1995). Cotter and colleagues (Cotter et al., 1997a) reported that inapparent *C. muridarum* infections in mice

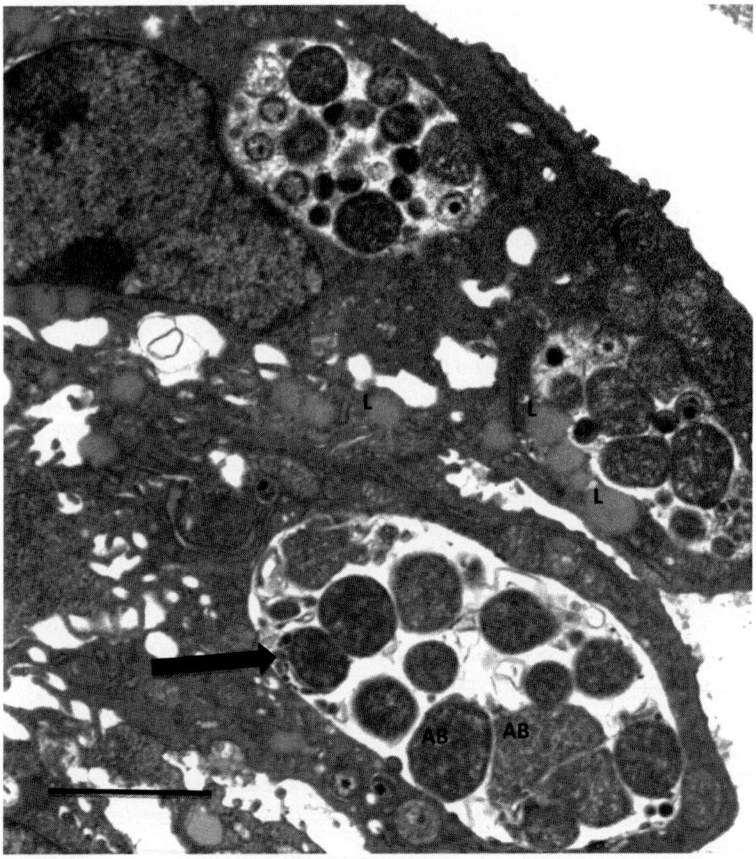

FIGURE 6 Stressed inclusions were observed in the cervix of a mouse depleted of PMNs by antibody treatment. Note the large aberrant RBs (AB) as well as packets of miniature RBs (black arrow). Also, the stressed inclusion is devoid of granular material representing glycogen accumulation. Lipid bodies (L) can be observed associated with the inclusion and apparently entering the inclusion. Scale bar: 2 μm.
doi:10.1128/9781555817329.ch13.f6

could be reactivated up to 9 weeks after infection by immunosuppressive treatment with cyclophosphamide, begun 7 weeks after infection. The number of mice reactivated was low, but the results indicate that an infection could be present in the absence of detectable organisms. Nevertheless, reactivation could not be elicited after 9 weeks, indicating total cure of infection. That cyclophosphamide/cortisone treatment in these studies induced reactivation suggests that the immune response was controlling the infection. Whether the organisms were actively replicating at undetectable levels or were in a bacteriostatic state, such as in stressed inclusions through the action of IFN-γ, is not known.

The obvious problem in defining a true chronic or persistent infection lies, in fact, with the definition that is used. Must stressed forms be found to define an infection as a persistent infection? It is just as likely that the infection is restrained by the immune response at low levels of replication that are below detection. There is ample precedent for such a mechanism in other infectious diseases, and an obvious example is malaria.

In humans, it has always been assumed that persistent infections reside in the genital tract, based on numerous clinical studies showing isolation and reisolation of chlamydiae from this site, often over long periods of time with intervening periods when the organism cannot be isolated. In animal models in which persistent infections have been studied, mice were either infected in the lung with organisms that are not natural parasites of the mouse (*C. trachomatis* or *C. pneumoniae*) or with a natural parasite of the mouse (*C. muridarum*) in an unnatural site, i.e., the genital tract. Thus, there are concerns that we are not studying persistence in the correct anatomical site. The site of infection in nature for virtually all *Chlamydia* spp. infecting mammals and birds is the gastrointestinal tract. Although *C. muridarum* was first isolated from the lungs of mice (Nigg, 1942), efforts to prove horizontal transmission via the respiratory route were negative (Karr, 1943). Karr then added infected lung material to drinking water and added infected mouse carcasses to cages of uninfected mice and observed that chlamydiae could be isolated from the lungs of mice, suggesting oral transmission of infection and aspiration of chlamydiae into the lungs (Karr, 1943). Similarly, Cotter and colleagues housed uninfected mice with mice infected intravaginally with *C. muridarum* in the same cage and found that chlamydiae could be isolated only from the mesenteric lymph nodes of the naïve mice and from no other site, clearly indicating oral transmission (Cotter et al., 1997b). Therefore, transmission in mice in nature occurs either through grooming or more likely through coprophagia (ingesting feces). Infection of the lung probably occurs through aspiration during eating.

If the gastrointestinal site is indeed the site of persistent infection, then one would predict that organisms are present in the gut for a substantial period of time. Infection in the gut would facilitate transmission and maintenance of the organism in the population. Conversely, if the time frame of gut infection were short, the likelihood of the organism remaining in the population would be considerably less. That antibodies to *C. muridarum* can be found in a high percentage of wild field mice suggests that the infection is endemic in this population (K. H. Ramsey, personal communication). In fact, we have detected antibodies to *C. caviae* in 75 of 76 sera from domestic guinea pigs from four different regions in Peru, indicating that *C. caviae* is endemic in these populations as well (R. G. Rank and A. Hung Chaparro, unpublished data). To compare enteric and nonenteric mucosal infections, Igietseme and colleagues infected mice with *C. muridarum* by either the respiratory, genital, or oral routes and attempted to isolate organisms from the various tissues for as long as 260 days (Igietseme et al., 2001). Not surprisingly, respiratory and genital infections elicited a strong inflammatory response and resolved within 25 days; in contrast, viable chlamydiae were isolated from the small intestine for up to 50 days and from the large intestine for least 260 days. Moreover, there was no pathology evident in the intestines of the animals at any time point. Therefore, not only did this study provide solid evidence for long-term infection in the gut, but it also suggested that there was active chlamydial replication because viable organisms could easily be isolated over the entire time span. In fact, chlamydiae would have to be continually dividing since gut epithelium has an extremely rapid turnover rate of approximately 24 to 36 hours.

From the perspective of chlamydiae, one could not design a better site to maintain a long-term infection than the gastrointestinal tract. Not only is there a continual rapid replenishing of fresh host cells, but there are also multiple mechanisms that downregulate the host response in the gut in contrast to other mucosal tissues (Chehade and Mayer, 2005). In addition, the constant flow of fecal matter to the outside allows for transmission of the organism to new hosts. Perry and Hughes suggested that the intestine may be a reservoir for chronic reinfection of other sites (Perry and Hughes, 1999). Along with others, they found that initial infection of genital, respiratory, or oral sites did not stay restricted to those sites. For example, mice infected genitally developed respiratory and oral infections within 10 days of genital infection, and mice infected orally also became infected in the lung and genital tract (Cotter et al., 1997b; Perry and Hughes, 1999). Surprisingly, spread to the intestinal tract still occurred when genitally infected mice were restricted from grooming by attachment of "Elizabethan" collars (Perry and Hughes, 1999). This finding suggests that there was external exchange of contaminated fluids between the genital and gastrointestinal tracts.

These are very significant observations that have often been overlooked, but they indicate that the host response at a given site may be influenced by infection at other mucosal sites. Moreover, assessment of systemic host responses such as serum antibody or peripheral blood T-cell responses is not necessarily predictive of the local response at a given mucosal site. If other mucosal sites are indeed infected, they will also contribute to the systemic response. More importantly, in women, the lower gastrointestinal tract may become infected by contamination from the vagina or via oral or anal sexual activity and may serve as the reservoir for reinfection of the genital tract. Clearly, women develop cystitis (bladder inflammation) far more commonly than men, and the primary culprit in urinary tract infections in women is *Escherichia coli*, a common gut commensal. Thus, we hypothesize that true persistent or chronic *C. trachomatis* infections in humans occur in the gastrointestinal tract with reseeding of the genital tract. This model is supported by the observation that the gut is the natural site of chlamydial infection for virtually all mammals and birds.

Use of Animal Models To Study Determinants of Pathogenicity

It has been well documented that chlamydiae elicit disease through the induction of a host response to the infection and that it is the host response that actually causes tissue damage (Ramsey, 2006; Stephens, 2003) (reviewed in chapter 11, "*Chlamydia* immunopathogenesis"). As described above, the acute inflammatory response is initiated within a few hours of infection and the early stages of any primary chlamydial infection are characterized by an intense infiltration of PMNs. This inflammation results in tissue damage through the interaction of the PMN with the host cell and collateral damage elicited by the release of the multitude of products from the PMN, such as matrix metalloprotease 9 and myeloperoxidase (Imtiaz et al., 2006; Ramsey, 2006). This damage may be reparable at some tissue sites, but in the oviduct/fallopian tube, fibrotic tissue that causes blockage or constriction may be formed, resulting in involuntary sterility. There are also substantial data showing that host pathology is caused by the cell-mediated immune response and that repeated infection elicits a mononuclear cell response of T cells and monocytes, which can result in tissue damage. This mechanism has been demonstrated in animal models of trachoma and genital tract disease

(Grayston et al., 1985; Monnickendam et al., 1980a; Patton et al., 1987a; Rank et al., 1995b; Taylor, 1985; Tuffrey et al., 1990; Wolner-Hanssen et al., 1986). In repeated infections, the number of organisms is often diminished, but antigen alone is sufficient to induce the cell-mediated pathologic response.

Despite a large amount of data on how cellular and molecular host response effectors cause disease, the actual mechanisms by which chlamydiae initiate the host response are just beginning to be understood. Over the years, there has been a quest for "virulence" factors, with Hsp60 and lipopolysaccharide being prime candidates but ultimately not proving to be key factors (Darville et al., 2003; Rank et al., 1995a; Stephens, 2003). It is unlikely that there is a single virulence factor but more likely that multiple host and chlamydial factors contribute to disease. Currently, those known factors that can influence the development of chlamydial disease include genetic variation in the host response or host physiology, the size of the inoculating dose, the growth rate of the organism, and the ability of the organism to activate the host response. Much of the information on these mechanisms has been derived from in vitro studies, but the real impact of these factors can only be defined with studies in animal models.

Genetically determined variations in host response or host physiology have been proposed to be key factors in chlamydial pathogenesis because of correlations between severe disease and the presence or absence of genomic polymorphisms. As an example, Conway and colleagues noted that scarring trachoma was more common in individuals with a specific polymorphism for TNF-α who produce increased levels of TNF-α (Conway et al., 1997). Also, specific strains of mice have more-severe disease and longer infections, clearly indicating a significant host genetic contribution to disease (Darville et al., 1997, 2001; de la Maza et al., 1994; Qiu et al., 2005). While earlier studies using inbred strains of mice restricted the search for genetic differences to the major histocompatibility complex genes, it is now possible to delve into the entire mouse genome. In separate studies, Miyairi and Bernstein-Hanley and their colleagues have employed a forward genetic approach in recombinant inbred mice to identify gene loci associated with resistance to chlamydial infection. Miyairi and colleagues utilized recombinant inbred mouse strains generated from parents susceptible or resistant to *C. psittaci* infection by inbreeding F2 progeny for more than 20 generations by sibling mating (Miyairi et al., 2007). With many such recombinant strains, they were able to identify a 1.5-Mbp region of chromosome 11 containing a cluster of three p47 GTPases that were associated with resistance to infection. Bernstein-Hanley and colleagues utilized a similar approach to identify p47 GTPases that were associated with resistance to systemic *C. muridarum* infection (Bernstein-Hanley et al., 2006a, 2006b). These observations not only indicate that host genetics are critical in determining susceptibility to infection but also demonstrate the powerful capabilities of this recombinant mouse system to identify genes that may be associated with a given trait.

The most obvious chlamydial factor affecting pathogenesis is the number of organisms in the inoculating dose, which to some extent may also depend on the mode of transmission. For instance, in sexual transmission, the inoculum from an infected male is likely greater than the inoculum from an infected female because of the mechanics of sexual intercourse. With trachoma, the inoculum may vary depending on whether there is direct contact with ocular secretions or transmission by a fly serving as a mechanical vector. The effects of dose have been demonstrated in many ways including earlier onset of infection, death of the subject at higher doses with survival at lower doses for respiratory infections with *C. muridarum* (Ramsey et al., 2009), increased peak levels of organisms isolated from mice infected with *C. muridarum* (Maxion et al., 2004), earlier onset with increased peak levels of organisms in guinea pig infections with *C. caviae* (Rank et al., 2003), and simply establishment of infection or not (infectious dose 50 determination).

Interestingly, Maxion showed an inverse relationship between the intravaginal infection dose and the development of oviduct pathology, suggesting that higher doses activated the adaptive immune response more quickly, eliminating the organisms before they could ascend to the upper genital tract (Maxion et al., 2004).

Closely related to the dose is the ability of the bacterium to replicate. While it is not surprising that different chlamydial species or even *C. trachomatis* serovars have different growth rates, it has become apparent that these growth rates can correlate with the type of chlamydial disease. Miyairi and colleagues evaluated ocular and genital strains of *C. trachomatis* for the length of their developmental cycle and growth curve kinetics (Miyairi et al., 2006). Interestingly, they observed that ocular serovars had longer developmental cycles and slower growth curves than the genital serovars, thus showing a correlation between biologic characteristics of *C. trachomatis* serovars and their "pathotype" or type of disease that they cause. Kari and Caldwell took this concept one step further by examining two different serovar A isolates for their plaque size, growth curve kinetics, and IFN-γ sensitivity and related those parameters to their in vivo virulence by inoculation into the conjunctiva of nonhuman primates (Kari et al., 2008). One isolate had smaller plaques, slower growth, and increased sensitivity to IFN-γ than the other, and upon conjunctival infection, produced a lower peak IFU titer with shorter infection course and less pathology than did the other isolate. Thus, it is clear that even within a given serovar there can be significant variation in virulence. A corollary of these observations is that a virulence factor for chlamydiae could be any polymorphism that results in an enhanced growth rate, whether that polymorphism affects metabolism, replication, T3S, or other factors.

We have reported a similar observation, using *C. caviae* in which plaques were isolated from a conjunctival swab of an infected guinea pig (Binet et al., 2010). The plaques varied considerably in size, suggesting that the population of chlamydiae was mixed with regard to their growth rates. Upon subculture of individual plaques onto medium with and without azithromycin, azithromycin-resistant mutants were obtained. The plaques of the mutants were smaller than those produced by the parents, and the growth rate was also significantly slower with a smaller burst size. Upon inoculation of mutants into the guinea pig conjunctiva, significantly less pathology was observed in the animals infected with the mutants, even though the number of IFU produced by each was virtually identical to that of the parent strain. Importantly, the isolation of variable-sized plaques and even a mutant strain from a conjunctival swab indicates that there are multiple variants within a given population. The existence of genomic variants in a chlamydial population is discussed in chapter 2, "Deep and wide: comparative genomics of *Chlamydia*."

Support of the concept that variants within populations exist naturally was presented by Ramsey and colleagues in their examination of potential differences in the Nigg and Weiss strains of *C. muridarum* (Ramsey et al., 2009). While not previously appreciated, a thorough examination of the literature shows that there were two separate isolates of *C. muridarum* (then the agent of mouse pneumonitis) at the University of Chicago in 1938-1940, one by Nigg (Nigg, 1942) and the other by Gordon (Gordon et al., 1938; Rank, 2007). Both strains continued to be passaged with the Gordon or Chicago strain becoming known as the Weiss strain (J. Moulder, personal communication). Ramsey and colleagues demonstrated that the Weiss strain grows faster, has a larger burst size, and is more virulent than the Nigg strain. The history of these isolates suggests that they represent distinct *C. muridarum* variants, since it is highly likely that both isolates came from the same source of mice at the University of Chicago. Plaquing of each strain produced multiple plaques of variable size (Ramsey et al., 2009). It is interesting that the strains have apparently maintained their same relative degree of virulence over the years, even though they were only passaged initially in yolk sacs and have then been maintained in tissue culture without any selection by the host response.

Therefore, in understanding the production of chlamydial disease, it is important to consider that individuals are likely to be infected with a mixed population of chlamydial variants, some growing more efficiently than others, or some having a given polymorphism that is either advantageous or disadvantageous in competitive growth within the population. Thus, the actual virulence of an inoculum may be equivalent to that of the sum of the variants contained within the inoculum. This concept leads to several questions. Do some individuals develop more-severe pathology because they have a dominant variant that is more virulent than the other variants within the population? Is there a shifting of the proportional representation of individual variants within a population as a result of in vivo selective pressures? Do faster-growing variants, or variants that are more "fit," have a competitive advantage within the overall population? Does the ability of one variant to activate a strong host response affect the growth or survival of other variants within the population?

To address some of these questions, we have begun to examine the dynamics of variants within an in vivo population. To do this, we inoculated an azithromycin-resistant mutant and its parent strain in equal numbers into the guinea pig conjunctiva and determined the infection kinetics and gross pathologic response compared to animals inoculated singly with either the mutant or parent strain (Binet et al., 2010). Alone, each had the same infection course, although the mutant elicited considerably less pathology. However, when inoculated together, the peak IFU titer of the parent strain was significantly greater than that of the mutant, and the pathologic response was no different than that of the parent strain alone. These data indicated that the slower-replicating mutant strain could not compete as well with the parent strain when contained in the same inoculum. More importantly, the pathologic response had the phenotype of the parent strain. How the proportion of each of the two variants might change through passage in animals or whether it would stay the same remains to be determined. Nevertheless, this initial study suggests that understanding the dynamics of mixed populations of variants in vivo may be important in understanding the contribution of variants to clinical disease. Such studies can only be performed in vivo where host selective pressures are available.

Use of Animal Models To Study the Impact of Host Physiologic Factors

There is a tendency to think only about effects of the host response or differences among the chlamydiae as the key factors in the control of infection and the production of disease. However, there are a multitude of physiologic factors within the host that can affect the course of infection and resultant disease, and these factors can be evaluated in animal models. The most obvious is the effect of reproductive hormones produced in women. The isolation of chlamydiae from women taking oral contraceptives was described some time ago and confirmed by in vivo experiments in the guinea pig: *C. caviae* model (Kinghorn and Waugh, 1981; Rank et al., 1982; Rank and Barron, 1982; Washington et al., 1985). In addition, Sweet and coworkers reported that the majority of cases of chlamydial or gonococcal salpingitis occur within the first week after menses, suggesting a hormonal association (Sweet et al., 1986). Treatment of guinea pigs with estradiol resulted in a longer infection course with higher IFU titers but, more significantly, profoundly exacerbated upper genital tract pathology. Similar data were obtained even when physiologic levels of estradiol or actual oral contraceptives were used (Barron et al., 1988; Pasley et al., 1985a). Moreover, when animals were infected at different times during the estrous cycle, there was increased oviduct pathology when the infection occurred just before the onset of estrus (Rank et al., 1993). Treatment with heavy doses of progesterone did not alter the infection course (Pasley et al., 1985b), but in a separate study, when guinea pigs were treated with progesterone and infected with lower doses of *C. caviae*, an increased infection course was seen (R. G. Rank, unpublished data). In mice, progesterone is routinely used to facilitate genital infection by forcing a condition of anestrus, and its use is essential for successful infection of mice with *C. trachomatis*.

Another aspect of chlamydial infection that has long been neglected in animal studies is the effect of infection with more than one sexually transmitted agent, which commonly occurs in humans. The problem with such studies is the limited animal models that are available for the various sexually transmitted agents and the manipulations that certain models require to establish infection. Recently, however, Jerse and Darville and their groups have established a mouse model for coinfection with *C. muridarum* and *Neisseria gonorrhoeae* (Vonck et al., 2011). Mice were first infected intravaginally with *C. muridarum* and then injected with estradiol, followed by intravaginal inoculation of *N. gonorrhoeae*. The estradiol treatment reduced the proinflammatory cytokine and chemokine levels in mice infected with chlamydiae, but there was an increase in PMNs in the genital tract in mice coinfected with both organisms compared to mice infected with either organism alone. While the mechanism has not yet been delineated, this model represents a major step forward in the study of the complexities of dual infections.

Coinfection with herpes simplex virus (HSV) can be studied with *C. caviae* in guinea pigs and *C. trachomatis* or *C. muridarum* in mice. In a preliminary study, mice were inoculated with approximately 10^6 IFU of *C. trachomatis* serovar E or *C. muridarum* after standard progesterone treatment. In parallel, a group of similarly treated mice were mock inoculated with equal volumes of phosphate-buffered saline. Animals were swabbed on day 3 postinfection and assessed by PCR for active chlamydial infection. Three days later, each animal was challenged with HSV type 2 and monitored for vaginal viral titer and signs of disease. There was a substantial reduction in HSV type 2 vaginal titer and disease signs in mice that had been infected with *C. trachomatis* or *C. muridarum* (N. Bourne and R. B. Pyles, personal communication). Both of these coinfection models have the potential to provide important information on a common scenario in human sexually transmitted infections.

DEVELOPMENT OF BIOMATHEMATICAL MODELS USING ANIMAL EXPERIMENTATION

Perhaps the newest use of an animal model will ultimately result in the demise of animal models. With the capabilities today to accumulate massive amounts of qualitative and quantitative data from many different components of host and pathogen physiology, it should be possible to construct biomathematical models that can then be used to evaluate hypotheses in silico. Biomathematical models give one the opportunity to do thousands of simulations to test a hypothesis without performing an actual animal infection. Nevertheless, at least initially, it will still be important to validate the models with animal experiments. Recently, Wilson and colleagues constructed a biomathematical model that incorporated dose-response data from *C. caviae* conjunctival infection in guinea pigs, which included IFU levels, conjunctival pathology scores, and in vitro data about the replication rates of chlamydiae within an inclusion (Wilson et al., 2006, 2009). The resulting model was able to utilize IFU scores in the early stage of the infection to predict the level of pathology that would result. The model was validated by predicting pathology levels in animals inoculated with *C. caviae* and its phage, φCPG1, which is lytic for *C. caviae* (Hsia et al., 2000). *Chlamydia* mathematical models are discussed in more detail in chapter 16, "Biomathematical modeling of *Chlamydia* infection and disease."

In another study using data from *C. caviae* genital infection, Mallet and coworkers developed a biomathematical model that described the movement of chlamydiae up the genital tract (Mallet et al., 2009). The model utilized known information about the burst size from the chlamydial developmental cycle, IFU levels from cervicovaginal swabs, and isolation data from cervix, uterus, and oviduct over time. The model indicated that ascending "movement" of chlamydiae from the endocervix to the endometrium was related to the destruction of *Chlamydia*-infected cells and the availability of uninfected cells. The availability of uninfected cells is more limited in chlamydial genital infections, since the organism is not invasive and can only access cells in the most superficial epithelial layer. They also used the model to demonstrate that, not surprisingly, the spread of chlamydiae up the genital tract is modulated by the innate host response and ultimately limited by the onset of the adaptive response. Of importance, the model was able to predict the actual IFU data obtained from various genital tract tissues as the infection progressed.

These two models are relatively straightforward and do not include the myriad of variables that exist in vivo. Nevertheless, as new information about the in vivo roles of specific molecular and cellular components becomes available, it can be added to the models. While it may be difficult to imagine at this time, it is not inconceivable that in the future, it may be possible to construct a set of equations that can accurately define the interaction of host and chlamydiae in vivo and predict the impact of a given intervention (antimicrobial or vaccine) or a specific genetic polymorphism. This will be the true legacy of animal models.

REFERENCES

Alexander, E. R., and W.-T. Chiang. 1967. Infection of pregnant monkeys and their offspring with TRIC agents. *Am. J. Ophthalmol.* **53:**1145–1153.

Barron, A. L., J. N. Pasley, R. G. Rank, H. J. White, and R. E. Mrak. 1988. Chlamydial salpingitis in female guinea pigs receiving oral contraceptives. *Sex. Transm. Dis.* **15:**169–173.

Barron, A. L., H. J. White, R. G. Rank, B. L. Soloff, and E. B. Moses. 1981. A new animal model for the study of *Chlamydia trachomatis* genital infections: infection of mice with the agent of mouse pneumonitis. *J. Infect. Dis.* **143:**63–66.

Beatty, W. L., R. P. Morrison, and G. I. Byrne. 1995. Reactivation of persistent *Chlamydia trachomatis* infection in cell culture. *Infect. Immun.* **63:**199–205.

Bedson, S. P., and J. O. W. Bland. 1932. A morphological study of psittacosis virus, with the description of a developmental cycle. *Br. J. Exp. Pathol.* **13:**461–466.

Bernstein-Hanley, I., Z. R. Balsara, W. Ulmer, J. Coers, M. N. Starnbach, and W. F. Dietrich. 2006a. Genetic analysis of susceptibility to *Chlamydia trachomatis* in mouse. *Genes Immun.* **7:**122–129.

Bernstein-Hanley, I., J. Coers, Z. R. Balsara, G. A. Taylor, M. N. Starnbach, and W. F. Dietrich. 2006b. The p47 GTPases Igtp and Irgb10 map to the *Chlamydia trachomatis* susceptibility locus Ctrq-3 and mediate cellular resistance in mice. *Proc. Natl. Acad. Sci. USA* **103:**14092–14097.

Binet, R., A. K. Bowlin, A. T. Maurelli, and R. G. Rank. 2010. Impact of azithromycin resistant mutations on the virulence and fitness of *Chlamydia caviae* in guinea pigs. *Antimicrob. Agents Chemother.* **54:**1094–1101.

Brunham, R. C., and D. Zhang. 1999. Transgene as vaccine for *Chlamydia*. *Am. Heart J.* **138:** S519–S522.

Byrne, G. I., and C. L. Faubion. 1982. Lymphokine-mediated microbistatic mechanisms restrict *Chlamydia psittaci* growth in macrophages. *J. Immunol.* **128:**469–474.

Byrne, G. I., L. E. Guagliardi, R. E. Huebner, and D. M. Paulnock. 1988. Immunomodulation and *Chlamydia*: immunosuppression and the protective immune response to *C. psittaci* in mice. *Adv. Exp. Med. Biol.* **239:**343–352.

Byrne, G. I., and D. A. Krueger. 1983. Lymphokine-mediated inhibition of *Chlamydia* replication in mouse fibroblasts is neutralized by anti-gamma interferon immunoglobulin. *Infect. Immun.* **42:**1152–1158.

Byrne, G. I., and D. M. Ojcius. 2004. *Chlamydia* and apoptosis: life and death decisions of an intracellular pathogen. *Nat. Rev. Microbiol.* **2:**802–808.

Campbell, L. A., T. C. Moazed, C. C. Kuo, and J. T. Grayston. 1998. Preclinical models for *Chlamydia pneumoniae* and cardiovascular disease: hypercholesterolemic mice. *Clin. Microbiol. Infect.* **4**(Suppl. 4):S23–S32.

Chehade, M., and L. Mayer. 2005. Oral tolerance and its relation to food hypersensitivities. *J Allergy Clin. Immunol.* **115:**3–12.

Chen, W., and C. Kuo. 1980. A mouse model of pneumonitis induced by *Chlamydia trachomatis*: morphologic, microbiologic, and immunologic studies. *Am. J. Pathol.* **100:**365–382.

Conway, D. J., M. J. Holland, R. L. Bailey, A. E. Campbell, O. S. Mahdi, R. Jennings, E. Mbena, and D. C. Mabey. 1997. Scarring trachoma is associated with polymorphism in the tumor necrosis factor alpha (TNF-alpha) gene promoter and with elevated TNF-alpha levels in tear fluid. *Infect. Immun.* **65:**1003–1006.

Cotter, T. W., G. S. Miranpuri, K. H. Ramsey, C. E. Poulsen, and G. I. Byrne. 1997a. Reactivation of chlamydial genital tract infection in mice. *Infect. Immun.* **65:**2067–2073.

Cotter, T. W., K. H. Ramsey, G. S. Miranpuri, C. E. Poulsen, and G. I. Byrne. 1997b. Dissemination of *Chlamydia trachomatis* chronic genital tract infection in gamma interferon gene knockout mice. *Infect. Immun.* **65:**2145–2152.

Darougar, S., M. A. Monnickendam, H. El-Sheikh, J. D. Treharne, R. M. Woodland, and B. R. Jones. 1977. Animal models for the study of chlamydial infections of the eye and genital tract, p. 186–198. *In* D. Hobson and K. K. Holmes (ed.), *Nongonococcal Urethritis and Related Infections.* American Society for Microbiology, Washington, DC.

Darville, T. 2006. Innate immunity, p. 339–364. *In* P. M. Bavoil and P. B. Wyrick (ed.), Chlamydia*: Genomics and Pathogenesis.* Horizon Bioscience, Norfolk, United Kingdom.

Darville, T., C. W. Andrews, Jr., K. K. Laffoon, W. Shymasani, L. R. Kishen, and R. G. Rank. 1997. Mouse strain-dependent variation in the course and outcome of chlamydial genital tract infection is associated with differences in host response. *Infect. Immun.* **65:**3065–3073.

Darville, T., C. W. Andrews, Jr., J. D. Sikes, P. L. Fraley, L. Braswell, and R. G. Rank. 2001. Mouse strain-dependent chemokine regulation of the genital tract T helper cell type 1 immune response. *Infect. Immun.* **69:**7419–7424.

Darville, T., and T. Hiltke. 2010. Pathogenesis of *Chlamydia trachomatis* genital infection: an overview. *J. Infect. Dis.* **201**(Suppl. 2)**:**S114–S125.

Darville, T., J. M. O'Neill, C. W. Andrews, Jr., U. M. Nagarajan, L. Stahl, and D. M. Ojcius. 2003. Toll-Like receptor-2, but not toll-like receptor-4, is essential for development of oviduct pathology in chlamydial genital tract infection. *J. Immunol.* **171:**6187–6197.

de la Maza, L., S. Pal, A. Khamesipour, and E. M. Peterson. 1994. Intravaginal inoculation of mice with the *Chlamydia trachomatis* mouse pneumonitis biovar results in infertility. *Infect. Immun.* **62:**2094–2097.

Digiacomo, R. F., J. L. Gale, S. P. Wang, and M. D. Kiviat. 1975. Chlamydial infection of the male baboon urethra. *Br. J. Vener. Dis.* **51:** 310–313.

Doughri, A. M., J. Storz, and K. P. Altera. 1972. Mode of entry and release of *Chlamydiae* in infections of intestinal epithelial cells. *J. Infect. Dis.* **126:**652–657.

Feng, D., J. A. Nagy, K. Pyne, H. F. Dvorak, and A. M. Dvorak. 1998. Neutrophils emigrate from venules by a transendothelial cell pathway in response to FMLP. *J. Exp. Med.* **187:**903–915.

Gordon, F. B., G. Freeman, and J. M. Clampit. 1938. A pneumonia-producing filtrable agent from stock mice. *Proc. Soc. Exp. Biol. Med.* **39:**450–453.

Grayston, J. T., S. P. Wang, L. J. Yeh, and C. C. Kuo. 1985. Importance of reinfection in the pathogenesis of trachoma. *Rev. Infect. Dis.* **7:** 717–725.

Hackstadt, T., M. A. Scidmore-Carlson, E. I. Shaw, and E. R. Fischer. 1999. The *Chlamydia trachomatis* IncA protein is required for homotypic vesicle fusion. *Cell. Microbiol.* **1:**119–130.

Hsia, R., H. Ohayon, P. Gounon, A. Dautry-Varsat, and P. M. Bavoil. 2000. Phage infection of the obligate intracellular bacterium, *Chlamydia psittaci* strain guinea pig inclusion conjunctivitis. *Microbes Infect.* **2:**761–772.

Hybiske, K., and R. S. Stephens. 2007. Mechanisms of host cell exit by the intracellular bacterium *Chlamydia. Proc. Natl. Acad. Sci. USA* **104:** 11430–11435.

Igietseme, J. U., J. L. Portis, and L. L. Perry. 2001. Inflammation and clearance of *Chlamydia trachomatis* in enteric and nonenteric mucosae. *Infect. Immun.* **69:**1832–1840.

Imtiaz, M. T., J. H. Schripsema, I. M. Sigar, J. N. Kasimos, and K. H. Ramsey. 2006. Inhibition of matrix metalloproteinases protects mice from ascending infection and chronic disease manifestations resulting from urogenital *Chlamydia muridarum* infection. *Infect. Immun.* **74:**5513–5521.

Ito, J. I., Jr., J. M. Lyons, and L. P. Airo-Brown. 1990. Variation in virulence among oculogenital serovars of *Chlamydia trachomatis* in experimental genital tract infection. *Infect. Immun.* **58:** 2021–2023.

Jacobs, N. F., Jr., E. S. Arum, and S. J. Kraus. 1978. Experimental infection of the chimpanzee urethra and pharynx with *Chlamydia trachomatis. Sex. Transm. Dis.* **5:**132–136.

Johnson, A. P., C. M. Hetherington, M. F. Osborn, B. J. Thomas, and D. Taylor-Robinson. 1980. Experimental infection of the marmoset genital tract with *Chlamydia trachomatis. Br. J. Exp. Pathol.* **61:**291–295.

Kari, L., W. M. Whitmire, J. H. Carlson, D. D. Crane, N. Reveneau, D. E. Nelson, D. C. Mabey, R. L. Bailey, M. J. Holland, G. McClarty, and H. D. Caldwell. 2008. Pathogenic diversity among *Chlamydia trachomatis* ocular strains in nonhuman primates is affected by subtle genomic variations. *J. Infect. Dis.* **197:**449–456.

Karr, H. V. 1943. Study of a latent pneumotropic virus of mice. *J. Infect. Dis.* **72:**108–116.

Kaukoranta-Tolvanen, S.-S. E., A. L. Laurila, P. Saikku, M. Leinonen, L. Liesirova, and K. Laitinen. 1993. Experimental infection of *Chlamydia pneumoniae* in mice. *Microb. Pathog.* **15:** 293–302.

Kazdan, J. J., J. Schachter, and M. Okumoto. 1967. Inclusion conjunctivitis in the guinea pig. *Am. J. Opthalmol.* **64:**116–124.

Kelly, K. A. 2006. T lymphocyte trafficking to the female reproductive mucosa, p. 413–434. *In* P. M. Bavoil and P. B. Wyrick (ed.), Chlamydia*: Genomics and Pathogenesis.* Horizon Biosciences, Norfolk, United Kingdom.

Kinghorn, G. R., and M. A. Waugh. 1981. Oral contraceptive use and prevalence of infection with *Chlamydia trachomatis* in women. *Br. J. Vener. Dis.* **57:**187–190.

Kuo, C., and W. J. Chen. 1980. A mouse model of *Chlamydia trachomatis* pneumonitis. *J. Infect. Dis.* **141:**198–202.

Laitinen, K., A. L. Laurila, M. Leinonen, and P. Saikku. 1996. Reactivation of *Chlamydia pneumoniae* infection in mice by cortisone treatment. *Infect. Immun.* **64:**1488–1490.

Malinverni, R., C. Kuo, L. A. Campbell, and J. T. Grayston. 1995. Reactivation of *Chlamydia pneumoniae* lung infection in mice by cortisone. *J. Infect. Dis.* **172:**593–594.

Mallet, D. G., K.-J. Heymer, R. G. Rank, and D. P. Wilson. 2009. Chlamydial infection and spatial ascension of the female genital tract: a novel hybrid cellular automata and continuum mathematical model. *FEMS Immunol. Med. Microbiol.* **57:**173–182.

Matsumoto, A. 1988. Structural characteristics of chlamydial bodies, p. 21–45. *In* A. Barron (ed.), *Microbiology of Chlamydia.* CRC Press, Boca Raton, FL.

Matsumoto, A., H. Bessho, K. Uehira, and T. Suda. 1991. Morphological studies of the association of mitochondria with chlamydial inclusions and the fusion of chlamydial inclusions. *J. Electron. Microsc.* (Tokyo) **40:**356–363.

Maxion, H. K., W. Liu, M. H. Chang, and K. A. Kelly. 2004. The infecting dose of *Chlamydia muridarum* modulates the innate immune response and ascending infection. *Infect. Immun.* **72:**6330–6340.

Miyairi, I., O. S. Mahdi, S. P. Ouellette, R. J. Belland, and G. I. Byrne. 2006. Different growth rates of *Chlamydia trachomatis* biovars reflect pathotype. *J. Infect. Dis.* **194:**350–357.

Miyairi, I., K. H. Ramsey, and D. L. Patton. 2010. Duration of untreated chlamydial genital infection and factors associated with clearance: review of animal studies. *J. Infect. Dis.* **201**(Suppl. 2)**:** S96–S103.

Miyairi, I., V. R. Tatireddigari, O. S. Mahdi, L. A. Rose, R. J. Belland, L. Lu, R. W. Williams, and G. I. Byrne. 2007. The p47 GTPases Iigp2 and Irgb10 regulate innate immunity and inflammation to murine *Chlamydia psittaci* infection. *J. Immunol.* **179:**1814–1824.

Moazed, T. C., C. Kuo, J. T. Grayston, and L. A. Campbell. 1997. Murine models of *Chlamydia pneumoniae* infection and atherosclerosis. *J. Infect. Dis.* **175:**883–890.

Moller, B. R., and P. A. Mardh. 1980a. Experimental epididymitis and urethritis in grivet monkeys provoked by *Chlamydia trachomatis. Fertil. Steril.* **34:**275–279.

Moller, B. R., and P. A. Mardh. 1980b. Experimental salpingitis in grivet monkeys by *Chlamydia trachomatis.* Modes of spread of infection to the Fallopian tubes. *Acta Pathol. Microbiol. Scand.* **88:**107–114.

Monnickendam, M. A., S. Darougar, J. D. Treharne, and A. M. Tilbury. 1980a. Development of chronic conjunctivitis with scarring and pannus, resembling trachoma, in guinea pigs. *Br. J. Ophthalmol.* **64:**284–290.

Monnickendam, M. A., S. Darougar, J. D. Treharne, and A. M. Tilbury. 1980b. Guinea pig inclusion conjunctivitis as a model for the study of trachoma: clinical, microbiological, serological, and cytological studies of primary infection. *Br. J. Ophthalmol.* **64:**279–283.

Morrison, R. P., and H. D. Caldwell. 2002. Immunity to murine chlamydial genital infection. *Infect. Immun.* **70:**2741–2751.

Mount, D. T., P. E. Bigazzi, and A. L. Barron. 1972. Infection of genital tract and transmission of ocular infection to newborns by the agent of guinea pig inclusion conjunctivitis. *Infect. Immun.* **5:**921–926.

Mount, D. T., P. E. Bigazzi, and A. L. Barron. 1973. Experimental genital infection of male guinea pigs with the agent of guinea pig inclusion conjunctivitis and transmission to females. *Infect. Immun.* **8:**925–930.

Murdin, A. D., P. Dunn, R. Sodoyer, J. Wang, J. Caterini, R. C. Brunham, L. Aujame, and R. Oomen. 2000. Use of a mouse lung challenge model to identify antigens protective against *Chlamydia pneumoniae* lung infection. *J. Infect. Dis.* **181**(Suppl. 3):S544–S551.

Murray, E. S. 1964. Guinea pig inclusion conjunctivitis. I. Isolation and identification as a member of the Psittacosis-Lymphogranuloma-Trachoma group. *J. Infect. Dis.* **114:**1–12.

Neeper, I. D., D. L. Patton, and C.-C. Kuo. 1990. Cinematographic observations of growth cycles of *Chlamydia trachomatis* in primary cultures of human amniotic cells. *Infect. Immun.* **58:** 2042–2047.

Nigg, C. 1942. An unidentified virus which produces pneumonia and systemic infection in mice. *Science* **95:**49–50.

Ojcius, D. M., P. Souque, J. L. Perfettini, and A. Dautry-Varsat. 1998. Apoptosis of epithelial cells and macrophages due to infection with the obligate intracellular pathogen *Chlamydia psittaci*. *J. Immunol.* **161:**4220–4226.

Pal, S., E. M. Peterson, and L. M. de la Maza. 2004. New murine model for the study of *Chlamydia trachomatis* genitourinary tract infections in males. *Infect. Immun.* **72:**4210–4216.

Pasley, J. N., R. G. Rank, A. J. Hough, Jr., C. Cohen, and A. L. Barron. 1985a. Effects of various doses of estradiol on chlamydial genital infection in ovariectomized guinea pigs. *Sex. Transm. Dis.* **12:**8–13.

Pasley, J. N., R. G. Rank, A. J. Hough, Jr., C. Cohen, and A. L Barron. 1985b. Absence of progesterone effects on chlamydial genital infection in female guinea pigs. *Sex. Transm. Dis.* **12:**156-158.

Patton, D. L. 1992. Microbiology and pathology of pelvic inflammatory disease, p. 23–33. *In* G. S. Berger and L. V. Westrom (ed.), *Pelvic Inflammatory Disease*. Raven Press, Ltd., New York, NY.

Patton, D. L., S. A. Halbert, C. C. Kuo, S. P. Wang, and K. K. Holmes. 1983. Host response to primary *Chlamydia trachomatis* infection of the fallopian tube in pig-tailed monkeys. *Fertil. Steril.* **40:**829–840.

Patton, D. L., C.-C. Kuo, S.-P. Wang, and S. A. Halbert. 1987a. Distal tubal obstruction induced by repeated *Chlamydia trachomatis* salpingeal infection in pig-tailed macaques. *J. Infect. Dis.* **155:**1292–1299.

Patton, D. L., C.-C. Kuo,S.-P. Wang, R. M. Brenner, M. D. Sternfeld, S. A. Morse, and R. C. Barnes. 1987b. Chlamydial infection of subcutaneous fimbrial transplants in cynomolgus and rhesus monkeys. *J. Infect. Dis.* **155:**229–235.

Patton, D. L., and H. R. Taylor. 1986. The histopathology of experimental trachoma: ultrastructural changes in the conjunctival epithelium. *J. Infect. Dis.* **153:**870–878.

Perry, L. L., and S. Hughes. 1999. Chlamydial colonization of multiple mucosae following infection by any mucosal route. *Infect. Immun.* **67:** 3686–3689.

Phillips, D. M., and C. A. Burillo. 1998. Ultrastructure of the murine cervix following infection with *Chlamydia trachomatis*. *Tissue Cell* **30:**446–452.

Phillips, D. M., C. E. Swenson, and J. Schachter. 1984. Ultrastructure of *Chlamydia trachomatis* infection of the mouse oviduct. *J. Ultrastruct. Res.* **88:**244–256.

Pospischil, A., N. Borel, E. H. Chowdhury, and F. Guscetti. 2009. Aberrant chlamydial developmental forms in the gastrointestinal tract of pigs spontaneously and experimentally infected with *Chlamydia suis*. *Vet. Microbiol.* **135:**147–156.

Qiu, H., S. Wang, J. Yang, Y. Fan, A. G. Joyee, X. Han, L. Jiao, and X. Yang. 2005. Resistance to chlamydial lung infection is dependent on major histocompatibility complex as well as nonmajor histocompatibility complex determinants. *Immunology* **116:**499–506.

Ramsey, K. H. 2006, Alternative mechanisms of pathogenesis, p. 435–473. *In* P. M. Bavoil and P. B. Wyrick (ed.), Chlamydia*: Genomics and Pathogenesis*. Horizon Bioscience, Norfolk, United Kingdom.

Ramsey, K. H., I. M. Sigar, J. H. Schripsema, C. J. Denman, A. K. Bowlin, G. S. A. Myers, and R. G. Rank. 2009. Strain and virulence diversity in the mouse pathogen *Chlamydia muridarum*. *Infect. Immun.* **77:**3284–3293.

Rank, R. G. 1988. Role of the immune response, p. 217–234. *In* A. L. Barron (ed.), *Microbiology of* Chlamydia. CRC Press, Boca Raton, FL.

Rank, R. G. 1994. Animal models for urogenital infections, p. 83–92. *In* V. L. Clark and P. M. Bavoil (ed.), *Bacterial Pathogenesis. Part A. Identification and Regulation of Virulence Factors*, vol. 235. Academic Press, San Diego, CA.

Rank, R. G. 1999. Models of immunity, p. 239–295. *In* R. S. Stephens (ed.), Chlamydia*: Intracellular Biology, Pathogenesis, and Immunity*. American Society for Microbiology,Washington, DC.

Rank, R. G. 2007. Chlamydial diseases, p. 325–348. *In* J. G. Fox et al. (ed.), *The Mouse in Biomedical Research*, 2nd ed. Elsevier, New York, NY.

Rank, R. G. 2009. Chlamydia, p. 845–868. *In* A. Barret and L. Stanberry (ed.), *Vaccines for Biodefense and Emerging and Neglected Diseases*. Elsevier, Oxford, United Kingdom.

Rank, R. G., and A. L. Barron. 1982. Prolonged genital infection by GPIC agent associated with immunosuppression following treatment with estradiol, p. 391–394. *In* P.-A. Mardh et al. (ed.), *Chlamydial Infections*. Elsevier Biomedical Press, New York, NY.

Rank, R. G., A. K. Bowlin, R. L. Reed, and T. Darville. 2003. Characterization of chlamydial genital infection resulting from sexual transmission from male to female guinea pigs and determination of infectious dose. *Infect. Immun.* **71:**6148–6154.

Rank, R. G., C. Dascher, A. K. Bowlin, and P. M. Bavoil. 1995a. Systemic immunization with Hsp60 alters the development of chlamydial ocular disease. *Investig. Ophthalmol. Vis. Sci.* **36:** 1344–1351.

Rank, R. G., A. J. Hough, Jr., R. F. Jacobs, C. Cohen, and A. L. Barron. 1985. Chlamydial pneumonitis induced in newborn guinea pigs. *Infect. Immun.* **48:**153–158.

Rank, R. G., H. M. Lacy, A. Goodwin, J. Sikes, J. Whittimore, P. B. Wyrick, and U. M. Nagarajan. 2010. Host chemokine and cytokine response in the endocervix within the first developmental cycle of *Chlamydia muridarum*. *Infect. Immun.* **78:**536–544.

Rank, R. G., and M. M. Sanders. 1992. Pathogenesis of endometritis and salpingitis in a guinea pig model of chlamydial genital infection. *Am. J. Pathol.* **140:**927–936.

Rank, R. G., M. M. Sanders, and A. T. Kidd. 1993. Influence of the estrous cycle on the development of upper genital tract pathology as a result of chlamydial infection in the guinea pig model of pelvic inflammatory disease. *Am. J. Pathol.* **142:**1291–1296.

Rank, R. G., M. M. Sanders, and D. L. Patton. 1995b. Increased incidence of oviduct pathology in the guinea pig after repeat vaginal inoculation with the chlamydial agent of guinea pig inclusion conjunctivitis. *J. Sex. Transm. Dis.* **22:**48–54.

Rank, R. G., H. J. White, A. J. Hough, J. N. Pasley, and A. L. Barron. 1982. Effect of estradiol on chlamydial genital infection of female guinea pigs. *Infect. Immun.* **38:**699–705.

Rank, R. G., J. Whittimore, A. K. Bowlin, S. Dessus-Babus, and P. B. Wyrick. 2008. Chlamydiae and polymorphonuclear leukocytes: unlikely allies in the spread of chlamydial infection. *FEMS Immunol. Med. Microbiol.* **54:**104–113.

Rank, R. G., J. Whittimore, A. K. Bowlin, and P. B. Wyrick. 2011. The intimate relationship between polymorphonuclear leukocytes and the chlamydial developmental cycle: an in vivo ultrastructural analysis. *Infect. Immun.* **79:**3291–3301.

Rank, R. G., and J. Whittum-Hudson. 2010. Protective immunity to chlamydial genital infection: evidence from animal studies. *J. Infect. Dis.* **201**(Suppl. 2):S168-S177.

Ripa, K. T., B. R. Mller, P. A. Mardh, E. A. Freundt, and F. Melsen. 1979. Experimental acute salpingitis in grivet monkeys provoked by *Chlamydia trachomatis*. *Acta Pathol. Microbiol. Scand. B* **87:**65–70.

Schmeer, N., R. Weiss, M. Reinacher, H. Krauss, and M. Karo. 1985. Verlauf einer Chlamydien-bedingten "Meerschweinschen-Einschlusskoerperchen-Konjunktivitis" in einer Versuchstierhaltung. *Z. Versuchstierkunde* **27:**233–240.

Soloff, B. L., R. G. Rank, and A. L. Barron. 1982. Ultrastructural studies of chlamydial infection in guinea pig urogenital tract. *J. Comp. Pathol.* **92:**547–558.

Soloff, B. L., R. G. Rank, and A. L. Barron. 1985. Electron microscopic observations concerning the in vivo uptake and release of the agent of guinea pig inclusion conjunctivitis (*Chlamydia psittaci*) in guinea pig exocervix. *J. Comp. Pathol.* **95:**335–344.

Stephens, R. S. 2003. The cellular paradigm of chlamydial pathogenesis. *Trends Microbiol.* **11:**44–51.

Swanson, J., D. A. Eschenbach, E. R. Alexander, and K. K. Holmes. 1975. Light and electron microscopic study of *Chlamydia trachomatis* infection of the uterine cervix. *J. Infect. Dis.* **131:** 678–687.

Sweet, R. L., M. Blankfort-Doyle, M. O. Robbie, and J. Schachter. 1986. The occurrence of chlamydial and gonococcal salpingitis during the menstrual cycle. *JAMA* **255:**2062–2065.

Taylor, H. R. 1985. Ocular models of chlamydial infection. *Rev. Infect. Dis.* **7:**737–740.

Taylor, H. R., R. A. Prendergast, C. R. Dawson, J. Schachter, and A. M. Silverstein. 1981. An animal model for cicatrizing trachoma. *Investig. Ophthalmol. Vis. Sci.* **21:**422–433.

Tseng, C. K., and R. G. Rank. 1998. Role of NK cells in the early host response to chlamydial genital infection. *Infect. Immun.* **66:**5867–5875.

Tuffrey, M., F. Alexander, and D. Taylor-Robinson. 1990. Severity of salpingitis in mice after primary and repeated inoculation with a human strain of *Chlamydia trachomatis*. *J. Exp. Pathol.* **71:**403–410.

Tuffrey, M., and D. Taylor-Robinson. 1981. Progesterone as a key factor in the development of a mouse model for genital-tract infection with *Chlamydia trachomatis*. *FEMS Microbiol. Lett.* **12:**111–115.

Vonck, R. A., T. Darville, C. M. O'Connell, and A. E. Jerse. 2011. Chlamydial infection increases gonococcal colonization in a novel murine coinfection model. *Infect. Immun.* **79:**1566–1577.

Wang, Y., U. Nagarajan, L. Hennings, A. K. Bowlin, and R. G. Rank. 2010. Local host response to chlamydial urethral infection in male guinea pigs. *Infect. Immun.* **78:**1670–1681.

Ward, M. E. 1988. The chlamydial developmental cycle, p. 71–96. *In* A. L. Barron (ed.), *Microbiology of* Chlamydia. CRC Press, Inc., Boca Raton, FL.

Washington, A. E., S. Gove, J. Schachter, and R. L. Sweet. 1985. Oral contraceptives, *Chlamydia trachomatis* infection, and pelvic inflammatory disease. A word of caution about protection. *JAMA* **253:**2246–2250.

Williams, D. M., T. Kung, and J. Schachter. 1986. Immunity to the mouse pneumonitis agent (murine *Chlamydia trachomatis*), p. 465–468. *In* D.

Oriel et al. (ed.), *Chlamydial Infections*. Cambridge University Press, Cambridge, United Kingdom.

Williams, D. M., J. Schachter, J. J. Coalson, and B. Grubbs. 1984. Cellular immunity to the mouse pneumonitis agent. *J. Infect. Dis.* **149:**630–639.

Wilson, D. P., A. K. Bowlin, P. M. Bavoil, and R. G. Rank. 2009. Ocular pathology elicited by *Chlamydia* and the predictive value of quantitative modeling. *J. Infect. Dis.* **199:**1780–1789.

Wilson, D. P., P. Timms, D. L. McElwain, and P. M. Bavoil. 2006. Type III secretion, contact-dependent model for the intracellular development of *Chlamydia. Bull. Math. Biol.* **68:**161–178.

Wolner-Hanssen, P., D. L. Patton, W. E. Stamm, and K. K. Holmes. 1986. Severe salpingitis in pig-tailed macaques after repeated cervical infections followed by a single tubal inoculation with *Chlamydia trachomatis*, p. 371–374. *In* D. Oriel et al. (ed.), *Chlamydia infections*. Cambridge University Press, New York, NY.

Yang, Y. S., C. C. Kuo, and W. J. Chen. 1983. Reactivation of *Chlamydia trachomatis* lung infection in mice by cortisone. *Infect. Immun.* **39:**655–658.

Yang, Z.-P., C.-C. Kuo, and J. T. Grayston. 1993. A mouse model of *Chlamydia pneumoniae* strain TWAR pneumonitis. *Infect. Immun.* **61:**2037–2040.

CHLAMYDIA VACCINE: PROGRESS AND CHALLENGES

Ashlesh K. Murthy, Bernard P. Arulanandam, and Guangming Zhong

14

INTRODUCTION

Why is an efficacious *Chlamydia* vaccine still a major goal when many antibiotics are effective as therapies to eliminate chlamydial infections? In part, this is due to challenges in widespread screening of infected, often asymptomatic individuals, concerns about eventual antibiotic resistance, and the paradoxical possibility that antibiotics prevent the development of natural immunity following infection (the "arrested immunity" hypothesis (Brunham and Rekart, 2008). All this points to the likelihood that an effective antichlamydial vaccine would still have a large impact on the global control of chlamydial disease. While licensed *Chlamydia trachomatis* or *Chlamydia pneumoniae* vaccines are not currently available for human use, several successful veterinary *Chlamydia* vaccines are in use. These include live attenuated elementary bodies (EBs) or fixed EBs as the immunogen (reviewed by Longbottom and Livingstone [2006]). Since these vaccines are efficacious only for short durations in protecting against infection, booster doses are administered to animals annually to maintain continuous protective immunity. Such repeated vaccination, however, would not be desirable for humans. Additionally, potential adverse reactions to the vaccine, such as upper genital tract pathology, are not as paramount a concern in animals as they are in humans.

A HISTORICAL PERSPECTIVE ON *CHLAMYDIA* VACCINE DEVELOPMENT

Most early vaccines were directed against pathogens that were often lethal or that caused severe damage to the host during an acute infection. Childhood diseases such as diphtheria, measles, mumps, rubella, and poliomyelitis, which used to cause high levels of morbidity and mortality, have been controlled successfully with live-attenuated, killed, or subunit vaccine preparations.

Before chlamydial genital infections were widely recognized in humans, efforts to vaccinate individuals against trachoma took place in the late 1950s through the early 1970s. Four different groups conducted extensive vaccine trials in areas of trachoma endemicity in India, China, Africa, and the Middle East. In these

Ashlesh K. Murthy, Department of Pathology, Midwestern University, Downers Grove, IL 60515. *Bernard P. Arulanandam*, South Texas Center for Emerging Infectious Diseases, Department of Biology, University of Texas at San Antonio, San Antonio, TX 78249. *Guangming Zhong*, Department of Microbiology and Immunology, University of Texas Health Science Center, San Antonio, TX 78229.

Intracellular Pathogens I: Chlamydiales
Edited by Ming Tan and Patrik M. Bavoil © 2012 ASM Press, Washington, DC
10.1128/9781555817329.ch14

studies, a variety of antigen preparations were used at different doses, with and without adjuvants and with single or multiple serovar combinations. These studies have been reviewed in detail previously (Bietti and Werner, 1967; Rank et al., 2009; Schachter and Dawson, 1978). In general, if protection was observed, it only lasted a short time, 3 months to 2 years, and was evidenced by a decreased incidence in onset of disease. Protection was also found to be serovar specific. Interestingly, Bietti and colleagues demonstrated that if individuals already had active trachoma, vaccination reduced the intensity of the disease in a significant number of people, indicating that the vaccine could have a therapeutic benefit (Bietti and Werner, 1967).

In two separate clinical trials, however, immunization resulted in an increased onset of disease (Nichols et al., 1966; Woolridge et al., 1967), although there was no evidence of the vaccine causing more-severe disease. In a single experimental study, several human volunteers who were immunized and then inoculated in the conjunctiva (Grayston et al., 1961) developed disease while two volunteers inoculated with placebo did not, leading the investigators to suggest that the vaccine may have enhanced the disease process. Moreover, a few monkeys developed more-severe disease following challenge in several experimental studies in which nonhuman primates were immunized with certain vaccines (Collier et al., 1967; Wang and Grayston, 1967). Therefore, despite significant success in multiple clinical trials, the short-term effectiveness of the vaccines and their potential harmful effects (primarily gleaned from experimental studies involving monkeys and never confirmed in human clinical trials) led to the discontinuation of further clinical trials with whole-organism vaccines.

At least two important insights were gained from these trachoma vaccine trials. First, inactivated whole-organism vaccines are efficacious in reducing the incidence of infection, albeit for short periods of time. This finding provided the first evidence that protective immunity to chlamydial infection is theoretically achievable by immunization. Second, the vaccines induced serovar-specific immunity, suggesting that the critical determinants of protective immunity involve antigen(s) that govern serovar specificity.

The negative yet unverified perception that whole-cell vaccines may be deleterious to humans resulted in reduced interest in a whole-cell chlamydial vaccine. However, there is ample precedent for the successful use of subcellular vaccines that are directed against specific components of a microbial pathogen. Infections such as diphtheria and tetanus have been controlled by vaccine preparations containing toxoids (inactivated toxins) from *Corynebacterium diphtheriae* and *Clostridium tetani*, respectively. The capsules of *Streptococcus pneumoniae* and *Haemophilus influenzae* type b have been used to induce protective immunity against pneumococcal and *Haemophilus* infections, respectively. However, these examples of toxoid and capsular vaccines are directed against infections caused by extracellular bacterial pathogens, and antibodies primarily mediate the protective immunity. In contrast, *Chlamydia* is an obligate intracellular bacterial pathogen (Zomorodipour and Andersson, 1999), which grows and replicates in an intracellular inclusion. It is inherently difficult to obtain large amounts of chlamydiae or chlamydial components for a vaccine because one inclusion-forming unit (IFU) of *Chlamydia* will only produce a few hundred progeny (Sabet et al., 1984), unlike free-living bacteria, which can be grown in large cultures, or viruses, which produce millions of progeny per infectious cycle.

A reasonable alternative approach therefore would be to express chlamydial subunits/proteins within genetically tractable hosts such as *Escherichia coli* and purify the recombinant proteins in large quantities. Successful recombinant vaccines have been developed against hepatitis B, and more recently against human papillomavirus (HPV). Genital chlamydial disease and HPV infection are the two most common sexually transmitted infections in the United States, and they have many similarities. The majority of HPV infections are initially asymptomatic but later lead to genital warts in young individuals of both sexes and to the severe sequelae of vulval, vaginal, and cervical cancers in women and anal cancer in both men and women (reviewed by Huh [2009]). Chlamydial genital infections also are initially asymptomatic in the majority of infected individuals and when left untreated, lead to severe long-term consequences in women, including pelvic inflammatory disease, ectopic pregnancy, and infertility (reviewed by Geisler et al. [2008]). The prime goal of Gardasil, a quadrivalent recombinant vaccine against HPV types 6, 11, 16, and 18, is to prevent the severe long-term consequences of an HPV infection, such as genital warts and cancer. Gardasil has been shown to protect against vaginal and vulval cancer in young women and against genital warts in young men (Herbert and Coffin, 2008). The objective of a similarly efficacious *Chlamydia* vaccine should be to prevent the serious sequelae, and not just the acute infection.

In summary, the accumulated body of evidence has provided important guiding principles for *Chlamydia* vaccine development: (i) whole-organism vaccine preparations may be efficacious, as animal studies and the trachoma trials have suggested; (ii) a multisubunit component vaccine should likely include combinations of serovar/biovar-specific protective antigens; and (iii) the prevention of pathological sequelae and the reduction of the incidence of infection should be the driving forces in vaccine development.

GOALS FOR A *CHLAMYDIA* VACCINE

An ideal *Chlamydia* vaccine would elicit total resistance to infection. The available evidence, however, suggests that this may not be a realistic goal, since humans with genital infections develop natural immunity against reinfection that is incomplete and relatively short-lived (reviewed by Geisler et al. [2008]). Similarly, in the *Chlamydia muridarum* mouse model of genital infection, not all mice were completely resistant to reinfection (reviewed by Morrison and Caldwell [2002]). Live chlamydial infections induce the best protective immunity in both humans and mice when compared to immunization with either dead organisms or component antigens (Li et al., 2010; Olsen et al., 2010). Therefore, with a vaccine based on a single chlamydial antigen or a combination of antigens, it is more realistic to expect partial control, at best, of the initial infection. The good news may be that sterilizing immunity, which may not be achievable, may not be necessary to prevent the reproductive sequelae of the infection. Indeed, efforts to immunize with selected chlamydial antigens have shown that it is possible to significantly reduce the severity of reproductive sequelae without achieving total resistance to infection (Murthy et al., 2007).

CORRELATES OF PROTECTIVE IMMUNITY AGAINST CHLAMYDIAL INFECTIONS

The particular site of infection usually determines the type of host immune response(s) provoked by the infection and the type of immunity required to clear the infection. Members of the genus *Chlamydia* are primarily mucosal bacterial pathogens that exhibit a strong tropism for mucosal epithelial cells. Remarkable similarities and some differences exist in immune protection against different species of *Chlamydia* and at different sites of infection. However, the vast majority of studies have examined the role of immune responses against *C. trachomatis* genital infections.

B Cells and Antibody

Antibody responses against a number of chlamydial antigens can be detected during and following the infection in humans and in animal models. Serum antibody directed against defined chlamydial antigens has usually been found to correlate not with protective immunity but with pathogenesis and disease severity.

For example, patients with complications following chlamydial infections have been shown to display relatively high titers of antibodies against specific chlamydial antigens, such as the 60-kDa heat shock protein (Hsp60), compared to patients without complications (Punnonen et al., 1979; Brunham et al., 1992; LaVerda et al., 2000). In contrast, the level of mucosal anti-*Chlamydia* immunoglobulin A (IgA) has been shown to correlate with protective immunity against genital *C. trachomatis* infection and its pathological sequelae in humans (Brunham et al., 1983; Morrison and Morrison, 2005; Cotter et al., 1995). Passively transferred IgA antibodies specific for major outer membrane protein (MOMP) have been shown to protect against infection in the mouse model of genital infection (Pal et al., 1997a). However, gamma interferon (IFN-γ)-deficient mice that produce IgA in amounts comparable to those produced by wild-type mice are compromised in their ability to resolve chlamydial infections (Johansson et al., 1997), suggesting that the role for mucosal IgA may not be as significant as previously thought. IgG, not IgA, has been suggested to be the predominant immunoglobulin at the genital mucosa (Masson et al., 1969; Rank and Batteiger, 1989), and a role for passively transferred serum antichlamydial antibodies in protective immunity against primary infection and reinfection with *Chlamydia caviae* in the guinea pig has been demonstrated (Rank et al., 1979; Rank and Barron, 1983). However, mice deficient in antibody or Fc receptors resolve a primary genital chlamydial infection in a manner comparable to that of wild-type animals, although they display suboptimal resolution of secondary infection (Morrison et al., 2000; Morrison and Morrison, 2001, 2005; Moore et al., 2002, 2003). Collectively, these studies support an important role for antibodies in protective immunity against chlamydial genital infections, albeit by yet-to-be-defined mechanisms. Therefore, the induction of antibody responses should be a goal of vaccine development efforts.

T Cells

Cell-mediated immunity has been shown to be important for clearance of chlamydial infections (reviewed by Morrison and Caldwell [2002], Brunham and Rey-Ladino [2005], and Rockey et al. [2009]). α/β T cells (Morrison and Caldwell, 2002), γ/δ T cells (Williams et al., 1996), and NK cells (Tseng and Rank, 1998) respond to the infection; however, the bulk of experimental evidence suggests an important role only for α/β T cells in clearance of chlamydial infections. A protective role for $CD4^+$ T cells is apparent in HIV-infected individuals who display a higher incidence of chlamydial sexually transmitted disease and worsening of upper reproductive tract pathology associated with the HIV-induced reduction of $CD4^+$ T-cell counts (Kimani et al., 1996). Moreover, T helper 1 (Th1)-type $CD4^+$ T-cell responses strongly correlate with protective immunity, whereas Th2-type responses correlate with disease progression and pathology following ocular chlamydial infections (Mabey et al., 1991; Holland et al., 1996). Studies in mouse models have shown that adoptive transfer of enriched *Chlamydia*-specific $CD4^+$ T cells enhances clearance and that certain clones of antigen-specific $CD4^+$ T cells are highly efficacious in inducing early resolution of infection (Ramsey and Rank, 1990; Igietseme et al., 1993; Su and Caldwell, 1995).

In contrast, the role of $CD8^+$ T cells has been suggested to be secondary to that of $CD4^+$ T cells in clearance of infecting chlamydiae (Morrison and Caldwell, 2002; Brunham and Rey-Ladino, 2005). $CD8^+$ T cells, which are typically associated with defense against intracellular pathogens, infiltrate the sites of chlamydial infection (Kim et al., 1999; Roan and Starnbach, 2006), and antigen-specific $CD8^+$ T cell responses can be detected (Wizel et al., 2002; Starnbach et al., 2003). Additionally, several antigen-specific $CD8^+$ T-cell clones have been shown to induce enhanced clearance of chlamydial genital infection (Igietseme et al., 1994). A role for $CD8^+$ T cells has also been described in pulmonary *C. pneumoniae* infections (Rodriguez et al., 2006).

Irrespective of the T-cell type involved in clearance, the production of IFN-γ by these cells appears to be a key aspect of the host response to infection. This is suggested by multiple studies using the mouse model of genital chlamydial infection (Rank et al., 1992; Igietseme et al., 1994; Cotter et al., 1997; Johansson et al., 1997; Perry et al., 1997, 1999b; Ito and Lyons, 1999). Mice genetically deficient in major histocompatibility complex (MHC) class II or α/β T cells (Morrison et al., 1995), but not MHC class I (Morrison et al., 1995) or perforin (Perry et al., 1999a), are unable to clear the infection. Likewise, mice deficient in IFN-γ or the IFN-γ receptor do not resolve a primary genital infection fully (Morrison and Caldwell, 2002), leading to dissemination of the chlamydiae into other intra-abdominal organs (Cotter et al., 1997). These mice also display suboptimal immunity against reinfection. Overall, $CD4^+$ and $CD8^+$ T-cell clones that are able to enhance clearance of the genital infection appear to be dependent upon their ability to produce high levels of IFN-γ (Igietseme et al., 1993; Igietseme et al., 1994). While IFN-γ-producing Th1-type antigen-specific $CD4^+$ T cells have been described as both essential and sufficient for optimal resolution of chlamydial genital infections (Li et al., 2008a), some recent studies have suggested that multifunctional $CD4^+$ T cells, which produce tumor necrosis factor alpha in addition to IFN-γ, may correlate better with protective immune responses (Yu et al., 2011). Moreover, the production of interleukin 17 (IL-17) has been suggested in some studies to be important for the induction of protective Th1-type $CD4^+$ T cells (Scurlock et al., 2011), although the role of Th17-type $CD4^+$ T cells has yet to be clarified. In summary, available data strongly suggest that the induction of Th1-type $CD4^+$ T cells that produce high levels of IFN-γ should be an important target in the development of a *Chlamydia* vaccine.

PROTECTIVE AND PATHOGENIC RESPONSES IN CHLAMYDIAL INFECTION

The human trachoma trials demonstrated that vaccination with killed whole-cell chlamydiae induced transient protection against infection. Later studies in nonhuman primates suggested that both protective and pathogenic immune responses resulted from vaccination. Although the jury is still out on the significance of these studies, the determinants of immune protection and immunopathogenesis should be clearly delineated in order to maximize protection while minimizing the possibility of deleterious consequences. A number of studies have shown that chlamydial antigens *per se* and the nature of the host immune response to those antigens should be important concerns in vaccine design. Immune responses against chlamydial Hsp60 correlate strongly with pathology (Morrison et al., 1989b; Zhong and Brunham, 1992; Toye et al., 1993). In contrast, a strain of *C. muridarum* lacking the chlamydial virulence plasmid has been shown to cause comparable infection, but less-pathological sequelae than the wild-type strain. In addition, the plasmidless strain elicited robust protective immunity against subsequent challenge with wild-type organisms (O'Connell et al., 2007, 2011). Moreover, a similar plasmidless *C. trachomatis* strain has been suggested to be less virulent than the wild-type strain in humans (O'Connell et al., 2011). These studies illustrate that specific chlamydial virulence factors are critical determinants that can tip the balance between protective and deleterious host immune responses.

Host factors are also involved in inducing the pathology associated with chlamydial disease. Studies with mice have shown that the absence of a variety of host determinants of the innate immune response, such as Toll-like receptor 2 (Darville et al., 2003), CXCR2 (Lee et al., 2010a), polymorphonuclear neutrophils (Lee et al., 2010b), matrix metalloproteinase-9 (Imtiaz et al., 2007), or caspase-1 (Cheng et al., 2008), leads to a reduction in pathological sequelae while making little difference to clearance of infection with a wild-type strain of *C. muridarum*. Two hypotheses have been proposed with regard to the involvement of the host inflammatory response in chlamydial pathogenesis. The "cellular hypothesis" suggests that pathology is mainly a consequence of the proinflammatory cytokine response, which includes tumor necrosis factor alpha, IL-1, and IL-6, from persistently infected cells (Rasmussen et al., 1997; Johnson, 2004). On the other hand, the "immunological hypothesis" suggests that the host adaptive immune response itself causes pathology. Multiple mechanisms of pathogenesis have been implicated under the latter hypothesis: (a) a robust Th1-type $CD4^+$ T-cell response helps to clear the infection but also causes collateral tissue damage (Maxion and Kelly, 2002; Van Voorhis et al., 1997; Holland et al., 1996); (b) a robust Th2-type $CD4^+$ T-cell response downregulates the protective Th1-type response (Holland et al., 1996; Wang et al., 1999; Yang et al., 1999; Igietseme et al., 2000); (c) molecular mimicry between host and chlamydial antigens, such as Hsp60, leads to recognition of these antigens by autoreactive T and B cells (Brunham and Rey-Ladino, 2005); and (d) $CD8^+$ T cells induce pathology, as suggested by correlative evidence from individuals with trachoma (Kimani et al., 1996), rhesus macaques with salpingitis (Kimani et al., 1996; Van Voorhis et al., 1996, 1997), and by definitive evidence from a recent study using the mouse model of genital infection (Murthy et al., 2011). While the cellular and the immunologic hypotheses continue to be debated, it is important to consider that these two hypotheses are not mutually exclusive, since proinflammatory cytokines are produced in significant amounts by infiltrating immune cells.

The balance between protective and deleterious immune responses is thus likely to involve both chlamydial virulence factors and host immune determinants. Additional work needs to be done to finely discriminate between factors that induce protection and those that induce pathology. For example, the observation that plasmid-deficient chlamydiae induce less pathology suggests that proteins expressed from or controlled by genes on the plasmid are determinants of pathogenesis. However, immunization with at least one plasmid-encoded protein, pORF5, has been shown to elicit protection (Li et al., 2008d). Similarly, MOMP has been considered the most promising choice as a protective antigen for over 3 decades, but a recent study has suggested that anti-MOMP antibodies, while inhibiting infectivity *in vitro*, may enhance pathogenesis *in vivo* (Cunningham et al., 2011). The choice of the right antigen(s) to induce protective immune responses, with minimal to no pathogenic outcome, will be a crucial aspect of *Chlamydia* vaccine development. The role of the host immune response in chlamydial pathogenesis is discussed in more detail in chapter 11, "*Chlamydia* immunopathogenesis."

IDENTIFICATION OF CHLAMYDIAL ANTIGENS THAT ELICIT PROTECTIVE IMMUNITY

The identification of protective chlamydial subunit antigens has been pursued for nearly 3 decades. Several approaches have been employed for this purpose (reviewed by Rockey et al. [2009]). Empirical methodologies were used in early studies, and a limited number of antigens were identified. However, when the genome sequence of *Chlamydia* became available, genome-wide screening became possible. The ~1-Mb *C. trachomatis* genome encodes approximately 900 proteins, each of which can be cloned and expressed for analysis of antigen-specific host immune responses elicited during chlamydial infections. Such approaches are targeted towards identifying novel antigens that are immunogenic in *Chlamydia*-infected individuals, followed by identification of protective antigens and elimination of those that are associated with pathogenesis (Wang et al., 2010). A scheme for the identification and evaluation of *Chlamydia* vaccine candidates is depicted in Fig. 1.

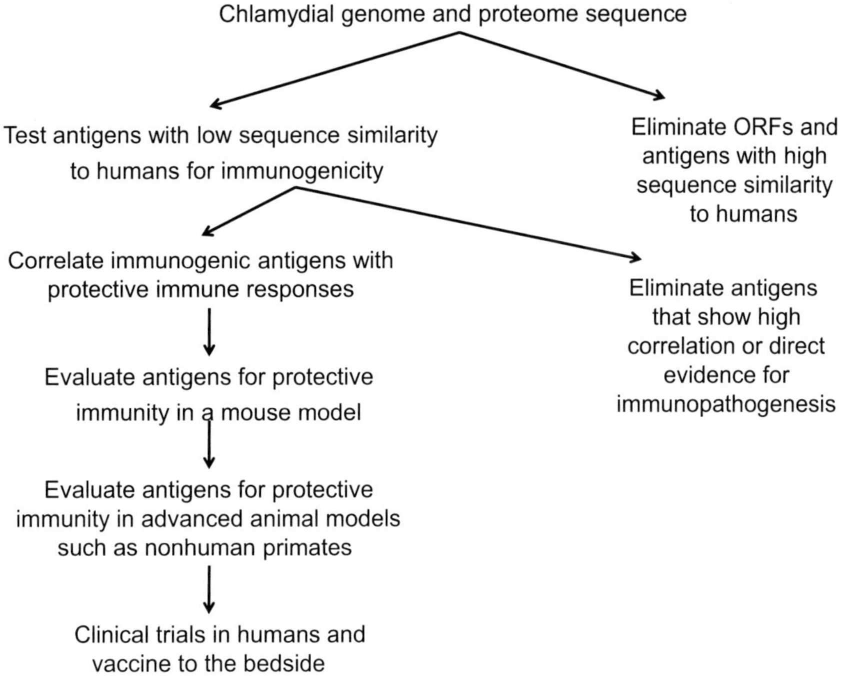

FIGURE 1 Scheme for *Chlamydia* vaccine candidate identification and evaluation. doi:10.1128/9781555817329.ch14.f1

BROAD CLASSES OF CHLAMYDIAL ANTIGENS AND METHODS OF IDENTIFICATION

Irrespective of the methodologies used, chlamydial antigens can be classified into four broad categories, as depicted schematically in Fig. 2, based on location and timing of expression during the developmental cycle, since these properties likely determine the nature of the immune responses that target the respective antigens. The first category includes chlamydial antigens such as MOMP (OmpA) and OmcB (Omp2), which are expressed on the surface of developing forms. These antigens would be optimal targets for antibodies that neutralize infectivity of the pathogen, but can also serve as targets of cell-mediated responses. The other three categories would primarily be targets of cell-mediated immunity. These include antigens expressed in the bacterial cytoplasm such as Hsp60, antigens that localize to the inclusion membrane such as the Inc proteins, and antigens that are secreted into the host cytosol such as chlamydial protease/proteasome-like activity factor (CPAF).

Various methodologies have been used to identify immunogenic antigens, with most initially focused on identifying chlamydial antigens that induce immune responses during infection. The rationale is that these antigens must be exposed to the host immune system, and therefore antigen-specific responses induced by a vaccine have a good likelihood of recognizing these antigens during a subsequent infection. Humoral and cell-mediated immune responses have been evaluated for this purpose with samples from humans and animal models.

Antigen Resolution by 2D-GE and Immunoblotting

In a gel electrophoresis and immunoblotting approach, chlamydial antigens derived from purified organisms or *Chlamydia*-infected cells

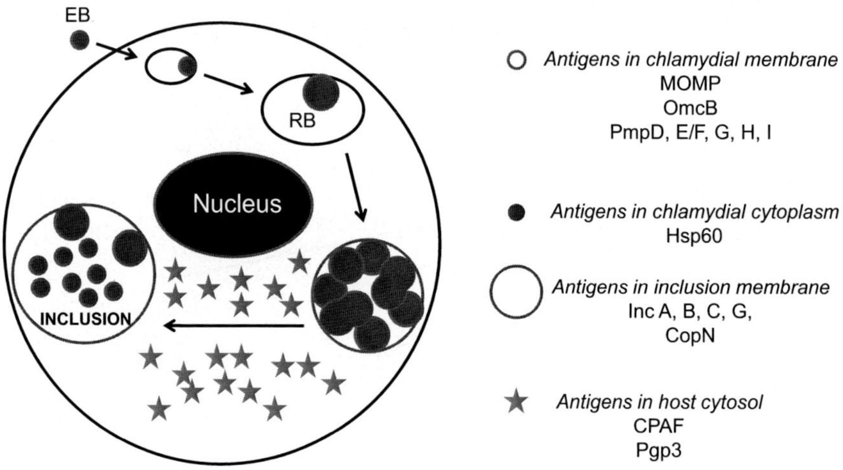

FIGURE 2 Classification scheme showing a *Chlamydia*-infected cell and four classes of chlamydial antigens that are potential vaccine candidates.
doi:10.1128/9781555817329.ch14.f2

are subjected to two-dimensional gel electrophoresis (2D-GE) and then probed with sera from *Chlamydia*-seropositive humans (Caldwell et al., 1975a, 1975b; Sanchez-Campillo et al., 1999). Individual polypeptide spots identified by patient antibodies can be eluted and analyzed by mass spectrometry, followed by molecular identification using published 2D-GE protein maps (Shaw et al., 2002) and the genome sequence. The disadvantages of this approach include the following: (a) proteins are denatured during 2D-GE, and therefore only antibodies against linear epitopes are recognized; the identified linear antigens may not induce protective immunity against proteins such as MOMP that have been shown to require conformation-dependent antibodies for neutralization of infectivity (Wolf et al., 2001); (b) the screening procedure will not recognize protective antigens that are only expressed by reticulate bodies during intracellular growth if purified EBs are used; additionally, targets of cell-mediated responses may be selectively missed, as such screening would eliminate mid-developmental-cycle antigens that are primary targets for antibody-mediated neutralization of infectivity; (c) this approach is not particularly useful for direct identification of T-cell antigens, although development of a robust antibody response would typically require helper T-cell assistance; therefore, this may be only a first step in an indirect approach to identify T-cell antigens; (d) many antigens are undetectable by 2D-GE and therefore would not be identified by this method. Despite these limitations, the approach has been successful in identifying unique chlamydial immunogenic protein antigens (Bunk et al., 2008).

Radioimmunoprecipitation and 2D-GE

Chlamydial proteins are labeled during growth with radioactive amino acids, and individual antigens from lysed chlamydial organisms or *Chlamydia*-infected cells that are recognized by patient sera are first immunoprecipitated and then resolved using 2D-GE, followed by identification as mentioned above (Shaw et al., 2002). Advantages of this approach include the potential for detection of conformational epitopes and the ability to detect intracellular location of specific proteins by cell fractionation. However, this method is not useful when a protein cannot be metabolically labeled or detected over sensitivity thresholds, nor is it useful for direct identification of T-cell antigens.

Genome-Wide Protein Expression for Detecting Both Antibody and T-Cell Responses

Based on the available genome sequence, a global approach can be used to identify immunogenic antigens in humans and in animal models. All chlamydial coding sequences can be expressed, or predictive software can be used to identify subsets of potentially protective antigens. Antibody and T-cell responses to each antigen can then be measured in sera or cellular samples from humans or animals infected with *Chlamydia* (Sharma et al., 2006; Wang et al., 2010). A disadvantage of this approach is that the expression of antigens in a nonnative system may affect the detection of the native conformation.

Immunoproteomics

MHC molecules determine the presentation of specific antigenic epitopes to T cells and the subsequent generation of an immune response. Therefore, peptides presented on MHC molecules may be screened by mass spectrometry, thereby leading to the identification of the specific antigens. Such an approach has been used to identify several antigens presented by either MHC class I or II molecules in mice (Karunakaran et al., 2008). Recently, this approach was used to demonstrate that the superior protective immunity induced by infection with live EBs, compared to immunization with killed EBs, may involve induction of immune responses against different sets of epitopes (Yu et al., 2011). An additional benefit can be built into this approach by the use of transgenic mice expressing commonly occurring human leukocyte antigen (HLA) molecules in place of mouse MHC molecules (Murthy et al., 2006), wherein the identified antigens are likely those presented to T cells in infected human individuals.

Antigen Discovery Using T-Cell Lines

Clones of antigen-specific T cells induced during infection can be isolated from animals, proliferated in vitro, and then evaluated for their ability to adoptively transfer protective immunity to naïve recipient mice (Fling et al., 2001; Starnbach et al., 2003). Subsequently, protective T-cell clones can be matched with target cells expressing clones from a chlamydial genomic DNA expression library to identify the protective antigens. Alternatively, genomic and protein libraries can be screened for MHC binding peptides and then expressed within caged MHC molecules, followed by evaluation of immune responses upon binding of the caged tetramers to T cells from previously *Chlamydia*-infected, haplotype-matched mice (Grotenbreg et al., 2008). Both approaches utilize contemporary technologies for antigen identification, and the T-cell clone approach has the added advantage of allowing direct detection of protective antigens.

Identification of Protective Antigens among Immunogenic Antigens

The approaches described above are useful for identification of highly immunogenic antigens, which may or may not be involved in a protective immune response. Therefore, it is important to further evaluate these immunogenic antigens for induction of protective immunity. Such an approach may involve several steps including negative and positive selection of antigens.

A first screen may be used to eliminate antigens that have a high level of amino acid sequence similarity to published human protein sequences, thereby avoiding the risk of inducing cross-reactive immune responses and consequent autoimmunity. However, it is not clear whether similarity in amino acid sequence is a reliable indicator of a potential autoimmune response or if there is a threshold of defined conservation beyond which such responses are likely to be elicited. Moreover, similarity in linear amino acid sequence may not be of much consequence for cross-reactive responses mediated by antibodies that detect conformational epitopes.

Correlating Immune Response Profiles with Patient Clinical Outcomes

Once immunogenic antigens have been identified, their role in chlamydial infection needs to be determined. One way of doing this is by correlating antibody and cell-mediated responses against individual antigens with observed clinical outcomes in human individuals. The presence of secondary tubal infertility in women serves as an important marker of *Chlamydia*-induced pathological sequelae in women. The rationale for this approach is that high immune reactivity specific for *C. trachomatis*-seropositive women without infertility can identify protective antigens, whereas an elevated response specific to females with infertility could identify pathogenesis-related antigens. The precedent to justify this approach is that high levels of delayed hypersensitivity reactions induced by chlamydial Hsp60 correlate with infertility in women (Peeling et al., 1997; Kinnunen et al., 2002), a feature also supported by animal studies (Morrison et al., 1989a, 1989b; Patton et al., 1994). This approach may prove to be very useful when genome-wide screens are conducted to identify protective and pathogenesis-related chlamydial antigens. However, a drawback is that patients with infertility likely have had several episodes of infection (Westrom et al., 1992), and thus, any immunogenic antigen, irrespective of its role in protection or pathogenesis, could show an association with infertility. Moreover, immunity induced by vaccination may work in ways different from that induced by the infection itself (Igietseme et al., 2009). Therefore, the next step would be to directly evaluate the protective and pathogenic effects of individual antigens in animal models.

Confirmatory Studies To Determine Protective Antigens

There are several animal models to test candidate vaccine antigens including rhesus macaques, grivet monkeys, guinea pigs, and mice (Rank, 1994). Each animal model has distinct advantages and disadvantages; however, mice make a good choice for preliminary studies because it is feasible to use them in large numbers, and a growing number of targeted gene knockout animals and a wide variety of immunological reagents are available. Mice can be immunized via different routes followed by measurement of antibody and cell-mediated immune responses, reduction of vaginal bacterial shedding, and protection against hydrosalpinx formation and infertility (Rank, 1994). The antigen can be administered as a gene carried on a plasmid (Li et al., 2008c) or as a protein, either derived from *Chlamydia* or expressed in a heterologous system such as *Escherichia coli* (Murthy et al., 2007). Th1-type antigen-specific responses can be induced in mice through the use of appropriate adjuvants including, but not limited to, immune stimulating complexes (Igietseme and Murdin, 2000), CpG deoxynucleotides (Berry et al., 2004; Cong et al., 2007), monatanide ISA-720 (Pal et al., 2005), IC-31 (Cheng et al., 2011), or IL-12 (Eko et al., 2003; Murthy et al., 2007) or through the use of *Vibrio cholerae* ghosts expressing the antigen of interest (Eko et al., 2003). Additionally, viral vectors (Penttila et al., 2004), liposomes (Hansen et al., 2008), and vault nanoparticles (Champion et al., 2009) have been used for vaccine delivery in mouse models to stimulate optimal immune responses. The utility of these approaches in humans remains to be established.

Selected candidate antigen(s) can then be tested in other animal models such as guinea pigs and nonhuman primates. However one limitation of animal models in general is that the pathogens and the mechanisms of host-mediated immunity may be different from those that are relevant for human infection. For example, the mouse model of infection typically involves *C. muridarum*, which is different from *C. trachomatis* in several ways, including a proposed different IFN-γ evasion mechanism that may be associated with the differential host tropism of these pathogens (Nelson et al., 2005; McClarty et al., 2007). Moreover, IFN-γ in human cells has been shown to lead to tryptophan depletion, thus starving and eliminating the pathogen, whereas in the mouse, IFN-γ has been proposed to induce GTPases (Nelson et al., 2005). Therefore, extrapolation from one animal model to another requires some caution.

IDENTIFIED CANDIDATE VACCINE ANTIGENS AND IMMUNOGENIC PROTEINS

Several protective and immunogenic chlamydial subunit antigens have been identified, particularly since the availability of the genome sequence made global screening possible. A detailed listing of chlamydial antigens that have been evaluated was described in a recent review (Hafner et al., 2008).

Antigens Expressed on the Surface of Developmental Chlamydial Forms

Given the conventional principle of inducing neutralizing antibodies for vaccine-mediated protection against infection, initial studies focused on evaluating chlamydial antigens expressed on the surface of the EB, including MOMP and the cysteine-rich OmcB. MOMP constitutes ~60% of the total protein mass in the chlamydial outer membrane and displays 84 to 97% amino acid identity among different *C. trachomatis* serovars (Brunham and Peeling, 1994). Anti-MOMP antibodies were shown to induce in vitro neutralization of chlamydial infectivity (Caldwell and Perry, 1982), paving the way for several in vivo studies in mouse models involving full-length *Chlamydia*-derived or recombinant MOMP, MOMP fragments and synthetic peptides, and DNA encoding MOMP, administered with a variety of different adjuvants and by different immunization routes (Morrison and Caldwell, 2002). Only MOMP derived from *Chlamydia* and then refolded to native configuration has been consistently successful in inducing robust protective immunity, including partial resistance to infection, early clearance, and significant reduction in upper genital tract pathology (Pal et al., 2005). The requirement for refolding suggested that MOMP-derived immunity is dependent on conformation and therefore primarily antibody mediated. MOMP-specific antibody also has been used to demonstrate passive transfer of protection in mice (Pal et al., 1997b, 2008).

MOMP includes four variable domains, which are the basis of serovar differentiation (Kim and DeMars, 2001). Therefore, not surprisingly, MOMP-induced resistance to infection, which is presumably due to neutralizing antibody-mediated effects, has been shown to be serovar specific (Byrne et al., 1993; Villeneuve et al., 1994). However, only a few serovars of *C. trachomatis* commonly infect humans, and it may be possible to use a combination of MOMP polypeptides or shared MOMP epitopes from these serovars (Batteiger, 1996) for a human vaccine. MOMP also includes five constant domains that are highly conserved between serovars and contain conserved T-cell epitopes (Kim and DeMars, 2001). Strong T-cell responses are induced following immunization with MOMP that have been shown to contribute to protective immunity (Farris et al., 2010), suggesting that the shared epitopes of MOMP may be able to elicit T-cell-mediated immunity against multiple serovars in a conformation-independent fashion.

MOMP in combination with OmcB expressed as recombinant proteins in *Vibrio cholerae* ghosts has been shown to induce robust immunity including early clearance and reduced pathology against homologous or heterologous chlamydial genital challenge in mice (Eko et al., 2004). OmcB is a surface-exposed adhesin that can bind to glycosaminoglycans (Fadel and Eley, 2007; Ting et al., 1995; Stephens et al., 2001; Moelleken and Hegemann, 2008), and both B- and T-cell epitopes within OmcB have been identified (Gervassi et al., 2004; Frikha-Gargouri et al., 2008). Immunization with an OmcB-encoding plasmid also has been found to significantly reduce *C. pneumoniae* lung burden following challenge (Penttila et al., 2000). However, it has also been reported that T cells recognizing OmcB are found in association with reactive arthritis in patients previously infected with *C. trachomatis* (Goodall et al., 2001). Thus, OmcB, like MOMP and pORF5, may play a dual role in pathogenesis and immune protection. It may be possible to get around this problem by defining epitopes that only induce a protective immune response without any associated pathology.

Additional proteins that are produced in abundant amounts in the chlamydial outer membrane, including PorB (Kawa et al., 2004; MacMillan et al., 2007) and the polymorphic outer membrane proteins (Pmps) PmpD, PmpG, PmpI, and PmpE/F2 (Goodall et al., 2001; Crane et al., 2006; Yu et al., 2006), have been found to be highly immunogenic. PmpD, an autotransporter that can form oligomers during infection (Kiselev et al., 2007; Swanson et al., 2009), was found to contain T-cell epitopes in an expression library screen (Goodall et al., 2001) and B-cell epitopes that induce species-common, panneutralizing antibodies (Crane et al., 2006). The porin PorB also has been found to contain several antigenic clusters that induce neutralizing antibodies (Kawa and Stephens, 2002; Kawa et al., 2004); however, only weak anti-PorB antibody responses are made during a chlamydial infection in mice, in which the humoral response is dominated by anti-MOMP antibodies (Kawa et al., 2004). When administered as a recombinant protein coexpressed with MOMP in *V. cholerae* ghosts, PorB has been shown to induce a degree of protection against genital *C. trachomatis* serovar D challenge (Ifere et al., 2007). A putative outer membrane protein, TC0512, was also identified as a protective immunogen in a *C. muridarum* genomic expression library screen (McNeilly et al., 2007). A YopD homolog expressed on the membrane also has been shown to contain immunogenic T-cell epitopes (Goodall et al., 2001). The chlamydial ADP/ATP translocase was identified in a genomic expression library screen as an immunodominant antigen, and a plasmid carrying DNA for this protein has been shown to induce modest protective immunity against pulmonary *C. pneumoniae* challenge in mice (Murdin et al., 2000). Finally, two nonprotein antigens, the chlamydial lipopolysaccharide and a glycolipid exoantigen, have been shown to be highly immunogenic (Brade et al., 1986; Stuart et al., 1991). It has been reported that the latter can induce protective immunity (Whittum-Hudson et al., 1996, 2001). Further evaluation of these antigens in animal models may help to better characterize their potential as protective vaccine candidates.

Antigens Expressed in the Cytosol of Chlamydial Developmental Forms

Hsp60, a chaperone expressed in the cytosol of *Chlamydia*, has been identified as being highly immunogenic with respect to inducing both $CD4^+$ and $CD8^+$ T-cell responses during infection (Deane et al., 1997; Holland et al., 1997). Immunization with plasmids carrying the chlamydial Hsp60 gene *groEL* also significantly reduced *C. pneumoniae* lung burden (Penttila et al., 2000). However, the association of anti-Hsp60 responses with immunopathogenesis in several studies (Morrison et al., 1989a; Zhong and Brunham, 1992; Toye et al., 1993) has led investigators to question the utility and protective potential of this antigen. A genome expression library screen for immunogenicity followed by DNA immunization has been used to identify four housekeeping genes that induced robust protective immunity against *Chlamydia abortus* lung infection in mice (Stemke-Hale et al., 2005). Interestingly, this study also indicated that some antigens may enhance disease and that nonoverlapping fragments of a single putative gene may induce distinct protective and pathogenic effects, suggesting that subgene level evaluation may be needed to identify protective epitopes within certain antigens.

Antigens That Localize to the Inclusion Membrane

Proteins in the inclusion membrane (Rockey et al., 2009) have been suggested to be important in mediating chlamydial vesicle fusion and other interactions of chlamydiae with host cells (Rockey et al., 2009). Many inclusion proteins are highly immunogenic and thus have potential for vaccine development. Inclusion membrane proteins, such as IncA, IncB, and IncC, are not present in the infectious EB but induce immunodominant antibody responses in *Chlamydia*-infected humans (Rockey et al., 1995; Bannantine and Rockey, 1999; Li et al., 2008b). The C-terminal fragments (Rockey et al., 1997; Hackstadt et al., 1999) and N-terminal fragments of CT529 (Li et al., 2008b) are exposed to the host cytosol and elicit dominant antibody and T-cell responses. N-terminal fragments of CT223 and CT618 also are recognized dominantly by human antibodies (Li et al., 2008b), suggesting that these peptides also may be exposed to the host cytosol. Protective immunity against pulmonary *C. pneumoniae* challenge has been shown to be induced by immunization with DNA encoding T-cell epitopes from inclusion membrane proteins (Pinchuk et al., 2005), and recombinant IncA protein in mice induces protective immunity resembling that mediated by T cells against *C. muridarum* genital challenge (Li et al., 2007). The type III secretion regulator protein, CopN, itself a type III secreted effector of *Chlamydia*, localizes to the inclusion membrane (Fields and Hackstadt, 2000) and is an immunodominant antigen in *Chlamydia*-seropositive women (Sharma et al., 2006). Additionally, intranasal immunization with CopN has been shown to induce protective immunity against *C. pneumoniae* lung infection in mice (Tammiruusu et al., 2007).

Antigens Secreted into the Host Cytosol

Chlamydial antigens secreted into the host cytosol have been identified as immunodominant antigens, with strong potential for induction of protective immunity. The role of CPAF in inducing protective immunity has recently been reviewed in detail (Murthy et al., 2009b). CPAF is secreted into the host cytosol (Zhong et al., 2001). It is produced as a zymogen with dormancy of the proteolytic activity maintained by an internal inhibitory segment (Huang et al., 2008). Upon dimerization and *trans*-autocatalytic cleavage, mature and active CPAF is produced as a serine protease consisting of a homodimer of catalytic domains, each with two distinct subunits (Huang et al., 2008). In the host cytosol, CPAF degrades several host proteins including transcription factors required for MHC gene activation (Zhong et al., 1999, 2000, 2001), proapoptotic proteins (Pirbhai et al., 2006), and cytoskeletal elements (Dong et al., 2004) and thus functions as a virulence factor. However, this protein is one of the most immunodominant antigens in assays utilizing patient sera from *Chlamydia*-infected individuals (Sharma et al., 2004, 2005, 2006). Mice immunized with recombinant CPAF display a high level of protective immunity against genital *C. muridarum* challenge, as indicated by significantly enhanced clearance and minimal oviduct pathology (Murthy et al., 2006, 2007, 2011) and infertility (Murthy et al., 2011). The protective immunity is predominantly mediated by IFN-γ-producing Th1-type antigen-specific $CD4^+$ T cells (Murthy et al., 2007; Murphey et al., 2006; Li et al., 2008a), with a limited contribution of antibody (Li et al., 2008a; Murthy et al., 2009a) and $CD8^+$ T cells (Li et al., 2008a). Heat-denatured CPAF also can induce comparable protective immunity, suggesting that protein conformation may not be a limitation for this antigen (Chaganty et al., 2010). CPAF displays 99% amino acid identity among the different *C. trachomatis* serovars (Dong et al., 2005), which suggests that it may have potential for inducing broadly cross-reactive protective immunity. These findings indicate that CPAF is an attractive candidate for further evaluation in nonhuman primate models.

Most *Chlamydia* spp. carry a 7.3-kb virulence plasmid encoding eight putative open reading frames (ORFs) designated pORF1 to pORF8 (Thomas et al., 1997). pORF5 codes for the plasmid glycoprotein 3 (Pgp3), which is at least partially secreted into the host cytosol (Li et al., 2008c). Pgp3 also is found in the chlamydial envelope as a trimer (Chen et al., 2010), and human patient sera immunodominantly detect the trimeric, but not monomeric, form of Pgp3 (Li et al., 2008c; Chen et al., 2010). The immunogenicity of this protein prompted evaluation of Pgp3 as a vaccine candidate, and plasmid DNA expressing Pgp3 was shown to induce protective immunity against chlamydial challenge (Li et al., 2008c). While an attractive prospect for protective immunity, the association of the chlamydial plasmid with pathogenesis (O'Connell et al., 2007, 2011) underscores the need to further clarify the actual role of Pgp3. Recent studies also have identified two other secreted chlamydial proteins, CT621 (Hobolt-Pedersen et al., 2009) and CT622 (Gong et al., 2011), which localize to the host cytosol and may be useful as vaccine candidates.

Hypothetical Proteins

The availability of the genome sequence and global screening has identified several hypothetical proteins of unknown function that are immunogenic antigens. For example, in a multistep screening approach involving correlation of human immune responses, elimination of antigens with high similarity to human sequences, and evaluation in mouse models, CT875 was identified as an antigen that is capable of inducing robust protective immunity against homologous and heterologous genital chlamydial challenge in mice (Coler et al., 2009). Interestingly, this study also identified CPAF (CT858) as one of the prime candidates to induce protective immunity. Other genome-wide expression library screens have been used to identify immunogenic antigens, including several hypothetical proteins, from *C. muridarum* and *C. trachomatis* (Molina et al., 2010; Cruz-Fisher et al., 2011).

THINKING AHEAD: CLINICAL TRIAL IN HUMANS

As we close the gaps in our understanding of various aspects pertinent to *Chlamydia* vaccine development, the possibility of a human clinical trial appears on the horizon. In preparation, first and foremost, we need to consider safety issues to minimize the chance that a candidate *Chlamydia* vaccine will contribute to pathology. Chlamydial infections are generally nonlethal, making it all the more necessary to produce a vaccine with minimal to no adverse side effects. Towards this end, it will be important to have a good understanding of the protective versus pathogenic potential for each vaccine component. Second, the end point of a clinical trial needs thoughtful evaluation. Based on the results in animal models thus far, the only antigen that reduces the incidence of infection via neutralizing antibodies is native MOMP, and the protection is partial, whereas most recombinant antigens are typically robust T-cell antigens and induce significant reduction of oviduct pathology, but not resistance to initial infection.

A combinatorial approach with multiple subunit antigens may be needed for an effective *Chlamydia* vaccine. Combinatorial vaccines including recombinant OmcB (CT443) and rl16 (CT521) induced no resistance to infection but only enhanced clearance that was dependent upon $CD4^+$ T-cell responses (Olsen et al., 2010). A similar protection phenotype also was found when recombinant MOMP, IncA, and CPAF were used in combination (Li et al., 2007). Since these recombinant antigens have been shown to elicit Th1-type $CD4^+$ T-cell responses that are qualitatively comparable or superior to intranasal infection with live EBs, it may be necessary to enhance their protective immunity by adding native surface proteins that induce antibodies to neutralize chlamydial infectivity (Olsen et al., 2010). However, even a combination of recombinant CPAF and UV-inactivated chlamydial EBs did not induce the level of resistance to infection afforded by intranasal immunization with live EBs, but only induced significant enhancements in clearance of the genital infection and reduction in pathology in mouse models (Li et al., 2010). Instead of assessing the efficacy of a chlamydial vaccine by its ability to reduce the incidence of infection, which is the standard applied to many vaccines, it may be more clinically relevant to aim for reduced incidence of pathological sequelae from a chlamydial infection.

Given that the reproductive period in women is approximately from the ages of 15 to 45 and that multiple chlamydial infection episodes may be required to induce pathology, it may be difficult to gauge within a reasonably short period of time the success of a vaccine candidate in a clinical trial based solely on reproductive pathology. To this end, the identification of early biomarkers to predict impending reproductive damage is important. Also, a significant portion of the population may be already burdened with chlamydial infection and may be on the path towards developing reproductive sequelae. For the benefit of such individuals, a therapeutic antichlamydial vaccine needs to be pursued and, more importantly, the end points of a trial involving these individuals should be explored. Finally, the target population for a vaccine needs to be determined. Based on the similarities between HPV and *Chlamydia*, it would appear that young adults should be vaccinated before the onset of sexual activity in order to realize the full potential of a *Chlamydia* vaccine. In this regard, issues such as mass education and obtaining parental consent need due consideration.

SUMMARY

Several immunogenic antigens have been identified from the ~1,000 chlamydial coding sequences of the chlamydial genome. A variety of approaches and animal models have been used for these screens, each with distinct advantages and disadvantages, leading to confirmatory or conflicting reports about the ability of particular antigens to induce protective immunity. Not surprisingly, only surface-exposed antigens appear to induce antibodies that neutralize initial infectivity, whereas antigens of all categories appear capable of eliciting robust T-cell responses. Some immunodominant antigens are hypothetical proteins of unknown function and location. While the nature of immune responses to antigens can be categorized based on location during the chlamydial developmental cycle, it needs to be noted that a given antigen may elicit responses that are either protective or associated with pathology. A rudimentary understanding of immunopathogenesis, compounded by reliance on correlative studies to identify pathogenesis-related antigens, confounds clear-cut identification of protective antigens. However, when a protective antigen is identified, our current understanding dictates that the goal should be to elicit both antibody and a robust Th1-type $CD4^+$ T-cell response, whereas more evidence is required to support the induction of $CD8^+$ T-cell responses. Collectively, contemporary technologies have enabled considerable progress in *Chlamydia* vaccine development in the recent past, but several issues remain to be addressed before advancing the identified antigens into human clinical trials. It is apparent that a better understanding of chlamydial cell biology and pathogenesis will provide a much-needed boost to *Chlamydia* vaccine development. Additionally, the currently widely used mouse model of *C. muridarum* infection has several distinct differences when compared to *C. trachomatis* infection in humans. Perhaps more intimate collaborative efforts between investigators pursuing basic and clinical research and the use of the nonhuman primate model will provide a more rational basis for antigen choice and the tailored induction of immune responses for development of a safe and efficacious *Chlamydia* vaccine for humans.

ACKNOWLEDGMENT

We thank M. Neal Guentzel, UTSA, for critical review of the manuscript.

REFERENCES

Bannantine, J. P., and D. D. Rockey. 1999. Use of primate model system to identify *Chlamydia trachomatis* protein antigens recognized uniquely in the context of infection. *Microbiology* **145:** 2077–2085.

Batteiger, B. E. 1996. The major outer membrane protein of a single *Chlamydia trachomatis* serovar can possess more than one serovar-specific epitope. *Infect. Immun.* **64:**542–547.

Berry, L. J., D. K. Hickey, K. A. Skelding, S. Bao, A. M. Rendina, P. M. Hansbro, C. M. Gockel, and K. W. Beagley. 2004. Transcutaneous immunization with combined cholera toxin and CpG adjuvant protects against *Chlamydia muridarum* genital tract infection. *Infect. Immun.* **72:**1019–1028.

Bietti, G., and G. H. Werner. 1967. *Trachoma: Prevention and Treatment.* Charles C Thomas, Springfield, IL.

Brade, L., S. Schramek, U. Schade, and H. Brade. 1986. Chemical, biological, and immunochemical properties of the *Chlamydia psittaci* lipopolysaccharide. *Infect. Immun.* **54:**568–574.

Brunham, R. C., C. C. Kuo, L. Cles, and K. K. Holmes. 1983. Correlation of host immune response with quantitative recovery of *Chlamydia trachomatis* from the human endocervix. *Infect. Immun.* **39:**1491–1494.

Brunham, R. C., R. Peeling, I. Maclean, M. L. Kosseim, and M. Paraskevas. 1992. *Chlamydia trachomatis*-associated ectopic pregnancy: serologic and histologic correlates. *J. Infect. Dis.* **165:** 1076–1081.

Brunham, R. C., and R. W. Peeling. 1994. *Chlamydia trachomatis* antigens: role in immunity and pathogenesis. *Infect. Agents Dis.* **3:**218–233.

Brunham, R. C., and M. L. Rekart. 2008. The arrested immunity hypothesis and the epidemiology of chlamydia control. *Sex. Transm. Dis.* **35:**53–54.

Brunham, R. C., and J. Rey-Ladino. 2005. Immunology of *Chlamydia* infection: implications for a *Chlamydia trachomatis* vaccine. *Nat. Rev. Immunol.* **5:**149–161.

Bunk, S., I. Susnea, J. Rupp, J. T. Summersgill, M. Maass, W. Stegmann, A. Schrattenholz, A. Wendel, M. Przybylski, and C. Hermann. 2008. Immunoproteomic identification and serological responses to novel *Chlamydia pneumoniae* antigens that are associated with persistent *C. pneumoniae* infections. *J. Immunol.* **180:**5490–5498.

Byrne, G. I., R. S. Stephens, G. Ada, H. D. Caldwell, H. Su, R. P. Morrison, B. Van der Pol, P. Bavoil, L. Bobo, and S. Everson. 1993. Workshop on *in vitro* neutralization of *Chlamydia trachomatis*: summary of proceedings. *J. Infect. Dis.* **168:**415–420.

Caldwell, H. D., C. C. Kuo, and G. E. Kenny. 1975a. Antigenic analysis of chlamydiae by two-dimensional immunoelectrophoresis. I. Antigenic heterogeneity between *C. trachomatis* and *C. psittaci*. *J. Immunol.* **115:**963–968.

Caldwell, H. D., C. C. Kuo, and G. E. Kenny. 1975b. Antigenic analysis of chlamydiae by two-dimensional immunoelectrophoresis. II. A trachoma-LGV-specific antigen. *J. Immunol.* **115:**969–975.

Caldwell, H. D., and L. J. Perry. 1982. Neutralization of *Chlamydia trachomatis* infectivity with antibodies to the major outer membrane protein. *Infect. Immun.* **38:**745–754.

Chaganty, B. K., A. K. Murthy, S. J. Evani, W. Li, M. N. Guentzel, J. P. Chambers, G. Zhong, and B. P. Arulanandam. 2010. Heat denatured enzymatically inactive recombinant chlamydial protease-like activity factor induces robust protective immunity against genital chlamydial challenge. *Vaccine* **28:**2323–2329.

Champion, C. I., V. A. Kickhoefer, G. Liu, R. J. Moniz, A. S. Freed, L. L. Bergmann, D. Vaccari, S. Raval-Fernandes, A. M. Chan, L. H. Rome, and K. A. Kelly. 2009. A vault nanoparticle vaccine induces protective mucosal immunity. *PLoS One* 4:e5409.

Chen, D., L. Lei, C. Lu, A. Galaleldeen, P. J. Hart, and G. Zhong. 2010. Characterization of Pgp3, a *Chlamydia trachomatis* plasmid-encoded immunodominant antigen. *J. Bacteriol.* **192:** 6017–6024.

Cheng, C., M. I. Cruz-Fisher, D. Tifrea, S. Pal, B. Wizel, and L. M. de la Maza. 2011. Induction of protection in mice against a respiratory challenge by a vaccine formulated with the *Chlamydia* major outer membrane protein adjuvanted with IC31®. *Vaccine* **29:**2437–2443.

Cheng, W., P. Shivshankar, Z. Li, L. Chen, I. T. Yeh, and G. Zhong. 2008. Caspase-1 contributes to *Chlamydia trachomatis*-induced upper urogenital tract inflammatory pathologies without affecting the course of infection. *Infect. Immun.* **76:**515–522.

Coler, R. N., A. Bhatia, J. F. Maisonneuve, P. Probst, B. Barth, P. Ovendale, H. Fang, M. Alderson, Y. Lobet, J. Cohen, P. Mettens, and S. G. Reed. 2009. Identification and characterization of novel recombinant vaccine antigens for immunization against genital *Chlamydia trachomatis*. *FEMS Immunol. Med. Microbiol.* **55:**258–270.

Collier, L. H., W. A. Blyth, N. M. Larin, and J. Treharne. 1967. Immunogenicity of experimental trachoma vaccines in baboons. III. Experiments with inactivated vaccines. *J. Hyg.* (London) **65:**97–107.

Cong, Y., M. Jupelli, M. N. Guentzel, G. Zhong, A. K. Murthy, and B. P. Arulanandam. 2007. Intranasal immunization with chlamydial protease-like activity factor and CpG deoxynucleotides enhances protective immunity against genital *Chlamydia muridarum* infection. *Vaccine* **25:**3773–3780.

Cotter, T. W., Q. Meng, Z. L. Shen, Y. X. Zhang, H. Su, and H. D. Caldwell. 1995. Protective efficacy of major outer membrane protein-specific immunoglobulin A (IgA) and IgG monoclonal antibodies in a murine model of *Chlamydia trachomatis* genital tract infection. *Infect. Immun.* **63:**4704–4714.

Cotter, T. W., K. H. Ramsey, G. S. Miranpuri, C. E. Poulsen, and G. I. Byrne. 1997. Dissemination of *Chlamydia trachomatis* chronic genital tract infection in gamma interferon gene knockout mice. *Infect. Immun.* **65:**2145–2152.

Crane, D. D., J. H. Carlson, E. R. Fischer, P. Bavoil, R. C. Hsia, C. Tan, C. C. Kuo, and H. D. Caldwell. 2006. *Chlamydia trachomatis* polymorphic membrane protein D is a species-common pan-neutralizing antigen. *Proc. Natl. Acad. Sci. USA* **103:**1894–1899.

Cruz-Fisher, M. I., C. Cheng, G. Sun, S. Pal, A. Teng, D. M. Molina, M. A. Kayala, A. Vigil, P. Baldi, P. L. Felgner, X. Liang, and L. M. de la Maza. 2011. Identification of immunodominant antigens by probing a whole *Chlamydia trachomatis* open reading frame proteome microarray using sera from immunized mice. *Infect. Immun.* **79:**246–257.

Cunningham, K. A., A. J. Carey, L. Hafner, P. Timms, and K. W. Beagley. 2011. *Chlamydia muridarum* major outer membrane protein-specific antibodies inhibit *in vitro* infection but enhance pathology *in vivo*. *Am. J. Reprod. Immunol.* **65:** 118–126.

Darville, T., J. M. O'Neill, C. W. Andrews, Jr., U. M. Nagarajan, L. Stahl, and D. M. Ojcius. 2003. Toll-like receptor-2, but not Toll-like receptor-4, is essential for development of oviduct pathology in chlamydial genital tract infection. *J. Immunol.* **171:**6187–6197.

Deane, K. H., R. M. Jecock, J. H. Pearce, and J. S. Gaston. 1997. Identification and characterization of a DR4-restricted T cell epitope within chlamydia heat shock protein 60. *Clin. Exp. Immunol.* **109:**439–445.

Dong, F., H. Su, Y. Huang, Y. Zhong, and G. Zhong. 2004. Cleavage of host keratin 8 by a *Chlamydia*-secreted protease. *Infect. Immun.* **72:**3863–3868.

Dong, F., Y. Zhong, B. Arulanandam, and G. Zhong. 2005. Production of a proteolytically active protein, chlamydial protease/proteasome-like activity factor, by five different *Chlamydia* species. *Infect. Immun.* **73:**1868–1872.

Eko, F. O., Q. He, T. Brown, L. McMillan, G. O. Ifere, G. A. Ananaba, D. Lyn, W. Lubitz, K. L. Kellar, C. M. Black, and J. U. Igietseme. 2004. A novel recombinant multisubunit vaccine against *Chlamydia*. *J. Immunol.* **173:**3375–3382.

Eko, F. O., W. Lubitz, L. McMillan, K. Ramey, T. T. Moore, G. A. Ananaba, D. Lyn, C. M. Black, and J. U. Igietseme. 2003. Recombinant *Vibrio cholerae* ghosts as a delivery vehicle for vaccinating against *Chlamydia trachomatis*. *Vaccine* **21:**1694–1703.

Fadel, S., and A. Eley. 2007. *Chlamydia trachomatis* OmcB protein is a surface-exposed glycosaminoglycan-dependent adhesin. *J. Med. Microbiol.* **56:**15–22.

Farris, C. M., S. G. Morrison, and R. P. Morrison. 2010. CD4+ T cells and antibody are required for optimal major outer membrane protein vaccine-induced immunity to *Chlamydia muridarum* genital infection. *Infect. Immun.* **78:**4374–4383.

Fields, K. A., and T. Hackstadt. 2000. Evidence for the secretion of *Chlamydia trachomatis* CopN by a type III secretion mechanism. *Mol. Microbiol.* **38:**1048–1060.

Fling, S. P., R. A. Sutherland, L. N. Steele, B. Hess, S. E. D'Orazio, J. Maisonneuve, M. F. Lampe, P. Probst, and M. N. Starnbach. 2001. CD8+ T cells recognize an inclusion membrane-associated protein from the vacuolar pathogen *Chlamydia trachomatis*. *Proc. Natl. Acad. Sci. USA* **98:**1160–1165.

Frikha-Gargouri, O., R. Gdoura, A. Znazen, B. Gargouri, J. Gargouri, A. Rebai, and A. Hammami. 2008. Evaluation of an *in silico* predicted specific and immunogenic antigen from the OmcB protein for the serodiagnosis of *Chlamydia trachomatis* infections. *BMC Microbiol.* **8:**217.

Geisler, W. M., C. Wang, S. G. Morrison, C. M. Black, C. I. Bandea, and E. W. Hook III. 2008. The natural history of untreated *Chlamydia trachomatis* infection in the interval between screening and returning for treatment. *Sex. Transm. Dis.* **35:**119–123.

Gervassi, A. L., K. H. Grabstein, P. Probst, B. Hess, M. R. Alderson, and S. P. Fling. 2004. Human CD8+ T cells recognize the 60-kDa cysteine-rich outer membrane protein from *Chlamydia trachomatis*. *J. Immunol.* **173:**6905–6913.

Gong, S., L. Lei, X. Chang, R. Belland, and G. Zhong. 2011. *Chlamydia trachomatis* secretion of hypothetical protein CT622 into host cell cytoplasm via a secretion pathway that can be inhibited by the type III secretion system inhibitor compound 1. *Microbiology* **157:**1134–1144.

Goodall, J. C., G. Yeo, M. Huang, R. Raggiaschi, and J. S. Gaston. 2001. Identification of *Chlamydia trachomatis* antigens recognized by human CD4+ T lymphocytes by screening an expression library. *Eur. J. Immunol.* **31:**1513–1522.

Grayston, J. T., R. L. Woolridge, C. W. Chen, F. A. Assaad, S. Maffei, C. H. Yen, and C. Y. Yang. 1961. Bacterial conjunctivitis caused by an eye ointment base used as a placebo in therapeutic trials. *Am. J. Ophthalmol.* **52:**251–256.

Grotenbreg, G. M., N. R. Roan, E. Guillen, R. Meijers, J. H. Wang, G. W. Bell, M. N. Starnbach, and H. L. Ploegh. 2008. Discovery of CD8+ T cell epitopes in *Chlamydia trachomatis* infection through use of caged class I MHC tetramers. *Proc. Natl. Acad. Sci. USA* **105:**3831–3836.

Hackstadt, T., M. A. Scidmore-Carlson, E. I. Shaw, and E. R. Fischer. 1999. The *Chlamydia trachomatis* IncA protein is required for homotypic vesicle fusion. *Cell. Microbiol.* **1:**119–130.

Hafner, L., K. Beagley, and P. Timms. 2008. *Chlamydia trachomatis* infection: host immune responses and potential vaccines. *Mucosal Immunol.* **1:**116–130.

Hansen, J., K. T. Jensen, F. Follmann, E. M. Agger, M. Theisen, and P. Andersen. 2008. Liposome delivery of *Chlamydia muridarum* major outer membrane protein primes a Th1 response that protects against genital chlamydial infection in a mouse model. *J. Infect. Dis.* **198:**758–767.

Herbert, J., and J. Coffin. 2008. Reducing patient risk for human papillomavirus infection and cervical cancer. *J. Am. Osteopathol. Assoc.* **108:**65–70.

Hobolt-Pedersen, A. S., G. Christiansen, E. Timmerman, K. Gevaert, and S. Birkelund. 2009. Identification of *Chlamydia trachomatis* CT621, a protein delivered through the type III secretion system to the host cell cytoplasm and nucleus. *FEMS Immunol. Med. Microbiol.* **57:**46–58.

Holland, M. J., R. L. Bailey, D. J. Conway, F. Culley, G. Miranpuri, G. I. Byrne, H. C. Whittle, and D. C. Mabey. 1996. T helper type-1 (Th1)/Th2 profiles of peripheral blood mononuclear cells (PBMC); responses to antigens of *Chlamydia trachomatis* in subjects with severe trachomatous scarring. *Clin. Exp. Immunol.* **105:** 429–435.

Holland, M. J., D. J. Conway, T. J. Blanchard, O. M. Mahdi, R. L. Bailey, H. C. Whittle, and D. C. Mabey. 1997. Synthetic peptides based on *Chlamydia trachomatis* antigens identify cytotoxic T lymphocyte responses in subjects from a trachoma-endemic population. *Clin. Exp. Immunol.* **107:**44–49.

Huang, Z., Y. Feng, D. Chen, X. Wu, S. Huang, X. Wang, X. Xiao, W. Li, N. Huang, L. Gu, G. Zhong, and J. Chai. 2008. Structural basis for activation and inhibition of the secreted chlamydia protease CPAF. *Cell Host Microbe* **4:**529–542.

Huh, W. K. 2009. Human papillomavirus infection: a concise review of natural history. *Obstet. Gynecol.* **114:**139–143.

Ifere, G. O., Q. He, J. U. Igietseme, G. A. Ananaba, D. Lyn, W. Lubitz, K. L. Kellar, C. M. Black, and F. O. Eko. 2007. Immunogenicity and protection against genital *Chlamydia* infection and its complications by a multisubunit candidate vaccine. *J. Microbiol. Immunol. Infect.* **40:**188–200.

Igietseme, J. U., G. A. Ananaba, J. Bolier, S. Bowers, T. Moore, T. Belay, F. O. Eko, D. Lyn, and C. M. Black. 2000. Suppression of endogenous IL-10 gene expression in dendritic cells enhances antigen presentation for specific Th1 induction: potential for cellular vaccine development. *J. Immunol.* **164:**4212–4219.

Igietseme, J. U., Q. He, K. Joseph, F. O. Eko, D. Lyn, G. Ananaba, A. Campbell, C. Bandea, and C. M. Black. 2009. Role of T lymphocytes in the pathogenesis of *Chlamydia* disease. *J. Infect. Dis.* **200:**926–934.

Igietseme, J. U., D. M. Magee, D. M. Williams, and R. G. Rank. 1994. Role for CD8$^+$ T cells in antichlamydial immunity defined by *Chlamydia*-specific T-lymphocyte clones. *Infect. Immun.* **62:** 5195–5197.

Igietseme, J. U., and A. Murdin. 2000. Induction of protective immunity against *Chlamydia trachomatis* genital infection by a vaccine based on major outer membrane protein-lipophilic immune response-stimulating complexes. *Infect. Immun.* **68:**6798–6806.

Igietseme, J. U., K. H. Ramsey, D. M. Magee, D. M. Williams, T. J. Kincy, and R. G. Rank. 1993. Resolution of murine chlamydial genital infection by the adoptive transfer of a biovar-specific, Th1 lymphocyte clone. *Reg. Immunol.* **5:**317 321.

Imtiaz, M. T., J. T. Distelhorst, J. H. Schripsema, I. M. Sigar, J. N. Kasimos, S. R. Lacy, and K. H. Ramsey. 2007. A role for matrix metalloproteinase-9 in pathogenesis of urogenital *Chlamydia muridarum* infection in mice. *Microbes Infect.* **9:**1561–1566.

Ito, J. I., and J. M. Lyons. 1999. Role of gamma interferon in controlling murine chlamydial genital tract infection. *Infect. Immun.* **67:**5518–5521.

Johansson, M., K. Schon, M. Ward, and N. Lycke. 1997. Genital tract infection with *Chlamydia trachomatis* fails to induce protective immunity in gamma interferon receptor-deficient mice despite a strong local immunoglobulin A response. *Infect. Immun.* **65:**1032–1044.

Johnson, R. M. 2004. Murine oviduct epithelial cell cytokine responses to *Chlamydia muridarum* infection include interleukin-12-p70 secretion. *Infect. Immun.* **72:**3951–3960.

Karunakaran, K. P., J. Rey-Ladino, N. Stoynov, K. Berg, C. Shen, X. Jiang, B. R. Gabel, H. Yu, L. J. Foster, and R. C. Brunham. 2008. Immunoproteomic discovery of novel T cell antigens from the obligate intracellular pathogen *Chlamydia. J. Immunol.* **180:**2459–2465.

Kawa, D. E., J. Schachter, and R. S. Stephens. 2004. Immune response to the *Chlamydia trachomatis* outer membrane protein PorB. *Vaccine* **22:**4282–4286.

Kawa, D. E., and R. S. Stephens. 2002. Antigenic topology of chlamydial PorB protein and identification of targets for immune neutralization of infectivity. *J. Immunol.* **168:**5184–5191.

Kim, S. K., M. Angevine, K. Demick, L. Ortiz, R. Rudersdorf, D. Watkins, and R. DeMars. 1999. Induction of HLA class I-restricted CD8$^+$ CTLs specific for the major outer membrane protein of *Chlamydia trachomatis* in human genital tract infections. *J. Immunol.* **162:**6855–6866.

Kim, S. K., and R. DeMars. 2001. Epitope clusters in the major outer membrane protein of *Chlamydia trachomatis. Curr. Opin. Immunol.* **13:**429–436.

Kimani, J., I. W. Maclean, J. J. Bwayo, K. MacDonald, J. Oyugi, G. M. Maitha, R. W. Peeling, M. Cheang, N. J. Nagelkerke, F. A. Plummer, and R. C. Brunham. 1996. Risk factors for *Chlamydia trachomatis* pelvic inflammatory disease among sex workers in Nairobi, Kenya. *J. Infect. Dis.* **173:**1437–1444.

Kinnunen, A., P. Molander, R. Morrison, M. Lehtinen, R. Karttunen, A. Tiitinen, J. Paavonen, and H. M. Surcel. 2002. *Chlamydial* heat shock protein 60-specific T cells in inflamed salpingeal tissue. *Fertil. Steril.* **77:**162–166.

Kiselev, A. O., W. E. Stamm, J. R. Yates, and M. F. Lampe. 2007. Expression, processing, and localization of PmpD of *Chlamydia trachomatis* serovar L2 during the chlamydial developmental cycle. *PLoS One* **2:**e568.

LaVerda, D., L. N. Albanese, P. E. Ruther, S. G. Morrison, R. P. Morrison, K. A. Ault, and G. I. Byrne. 2000. Seroreactivity to *Chlamydia trachomatis* Hsp10 correlates with severity of human genital tract disease. *Infect. Immun.* **68:**303–309.

Lee, H. Y., J. H. Schripsema, I. M. Sigar, S. R. Lacy, J. N. Kasimos, C. M. Murray, and K. H. Ramsey. 2010a. A role for CXC chemokine receptor-2 in the pathogenesis of urogenital *Chlamydia muridarum* infection in mice. *FEMS Immunol. Med. Microbiol.* **60:**49–56.

Lee, H. Y., J. H. Schripsema, I. M. Sigar, C. M. Murray, S. R. Lacy, and K. H. Ramsey. 2010b. A link between neutrophils and chronic disease manifestations of *Chlamydia muridarum* urogenital infection of mice. *FEMS Immunol. Med. Microbiol.* **59:**108–116.

Li, W., M. N. Guentzel, J. Seshu, G. Zhong, A. K. Murthy, and B. P. Arulanandam. 2007. Induction of cross-serovar protection against genital chlamydial infection by a targeted multisubunit vaccination approach. *Clin. Vaccine Immunol.* **14:**1537–1544.

Li, W., A. K. Murthy, M. N. Guentzel, J. P. Chambers, T. G. Forsthuber, J. Seshu, G. Zhong, and B. P. Arulanandam. 2010. Immunization with a combination of integral chlamydial antigens and a defined secreted protein induces robust immunity against genital chlamydial challenge. *Infect. Immun.* **78:**3942–3949.

Li, W., A. K. Murthy, M. N. Guentzel, J. Seshu, T. G. Forsthuber, G. Zhong, and B. P. Arulanandam. 2008a. Antigen-specific $CD4^+$ T cells produce sufficient IFN-gamma to mediate robust protective immunity against genital *Chlamydia muridarum* infection. *J. Immunol.* **180:**3375–3382.

Li, Z., C. Chen, D. Chen, Y. Wu, Y. Zhong, and G. Zhong. 2008b. Characterization of fifty putative inclusion membrane proteins encoded in the *Chlamydia trachomatis* genome. *Infect. Immun.* **76:**2746–2757.

Li, Z., D. Chen, Y. Zhong, S. Wang, and G. Zhong. 2008c. The chlamydial plasmid-encoded protein pgp3 is secreted into the cytosol of *Chlamydia*-infected cells. *Infect. Immun.* **76:**3415–3428.

Li, Z., S. Wang, Y. Wu, G. Zhong, and D. Chen. 2008d. Immunization with chlamydial plasmid protein pORF5 DNA vaccine induces protective immunity against genital chlamydial infection in mice. *Sci. China C* **51:**973–980.

Longbottom, D., and M. Livingstone. 2006. Vaccination against chlamydial infections of man and animals. *Vet. J.* **171:**263–275.

Mabey, D. C., M. J. Holland, N. D. Viswalingam, B. T. Goh, S. Estreich, A. Macfarlane, H. M. Dockrell, and J. D. Treharne. 1991. Lymphocyte proliferative responses to chlamydial antigens in human chlamydial eye infections. *Clin. Exp. Immunol.* **86:**37–42.

MacMillan, L., G. O. Ifere, Q. He, J. U. Igietseme, K. L. Kellar, D. M. Okenu, and F. O. Eko. 2007. A recombinant multivalent combination vaccine protects against *Chlamydia* and genital herpes. *FEMS Immunol. Med. Microbiol.* **49:**46–55.

Masson, P. L., J. F. Heremans, and J. Ferin. 1969. Clinical importance of the biochemical changes in the female genital tract. I. Studies on the proteins of cervical mucus. *Int. J. Fertil.* **14:**1–7.

Maxion, H. K., and K. A. Kelly. 2002. Chemokine expression patterns differ within anatomically distinct regions of the genital tract during *Chlamydia trachomatis* infection. *Infect. Immun.* **70:**1538–1546.

McClarty, G., H. D. Caldwell, and D. E. Nelson. 2007. Chlamydial interferon gamma immune evasion influences infection tropism. *Curr. Opin. Microbiol.* **10:**47–51.

McNeilly, C. L., K. W. Beagley, R. J. Moore, V. Haring, P. Timms, and L. M. Hafner. 2007. Expression library immunization confers partial protection against *Chlamydia muridarum* genital infection. *Vaccine* **25:**2643–2655.

Moelleken, K., and J. H. Hegemann. 2008. The *Chlamydia* outer membrane protein OmcB is required for adhesion and exhibits biovar-specific differences in glycosaminoglycan binding. *Mol. Microbiol.* **67:**403–419.

Molina, D. M., S. Pal, M. A. Kayala, A. Teng, P. J. Kim, P. Baldi, P. L. Felgner, X. Liang, and L. M. de la Maza. 2010. Identification of immunodominant antigens of *Chlamydia trachomatis* using proteome microarrays. *Vaccine* **28:** 3014–3024.

Moore, T., G. A. Ananaba, J. Bolier, S. Bowers, T. Belay, F. O. Eko, and J. U. Igietseme. 2002. Fc receptor regulation of protective immunity against *Chlamydia trachomatis*. *Immunology* **105:**213–221.

Moore, T., C. O. Ekworomadu, F. O. Eko, L. MacMillan, K. Ramey, G. A. Ananaba, J. W. Patrickson, P. R. Nagappan, D. Lyn, C. M. Black, and J. U. Igietseme. 2003. Fc receptor-mediated antibody regulation of T cell immunity against intracellular pathogens. *J. Infect. Dis.* **188:**617–624.

Morrison, R. P., R. J. Belland, K. Lyng, and H. D. Caldwell. 1989a. Chlamydial disease pathogenesis. The 57-kD chlamydial hypersensitivity antigen is a stress response protein. *J. Exp. Med.* **170:**1271–1283.

Morrison, R. P., and H. D. Caldwell. 2002. Immunity to murine chlamydial genital infection. *Infect. Immun.* **70:**2741–2751.

Morrison, R. P., K. Feilzer, and D. B. Tumas. 1995. Gene knockout mice establish a primary protective role for major histocompatibility complex class II-restricted responses in *Chlamydia trachomatis* genital tract infection. *Infect. Immun.* **63:**4661–4668.

Morrison, R. P., K. Lyng, and H. D. Caldwell. 1989b. Chlamydial disease pathogenesis. Ocular hypersensitivity elicited by a genus-specific 57-kD protein. *J. Exp. Med.* **169:**663–675.

Morrison, S. G., and R. P. Morrison. 2001. Resolution of secondary *Chlamydia trachomatis* genital tract infection in immune mice with depletion of both $CD4^+$ and $CD8^+$ T cells. *Infect. Immun.* **69:**2643–2649.

Morrison, S. G., and R. P. Morrison. 2005. A predominant role for antibody in acquired immunity to chlamydial genital tract reinfection. *J. Immunol.* **175:**7536–7542.

Morrison, S. G., H. Su, H. D. Caldwell, and R. P. Morrison. 2000. Immunity to murine *Chlamydia trachomatis* genital tract reinfection involves B cells and $CD4(^+)$ T cells but not $CD8(^+)$ T cells. *Infect. Immun.* **68:**6979–6987.

Murdin, A. D., P. Dunn, R. Sodoyer, J. Wang, J. Caterini, R. C. Brunham, L. Aujame, and R. Oomen. 2000. Use of a mouse lung challenge model to identify antigens protective against *Chlamydia pneumoniae* lung infection. *J. Infect. Dis.* **181**(Suppl. 3)**:**S544–S551.

Murphey, C., A. K. Murthy, P. A. Meier, G. M. Neal, G. Zhong, and B. P. Arulanandam. 2006. The protective efficacy of chlamydial protease-like activity factor vaccination is dependent upon $CD4^+$ T cells. *Cell. Immunol.* **242:**110–117.

Murthy, A. K., B. K. Chaganty, W. Li, M. N. Guentzel, J. P. Chambers, J. Seshu, G. Zhong, and B. P. Arulanandam. 2009a. A limited role for antibody in protective immunity induced by rCPAF and CpG vaccination against primary genital *Chlamydia muridarum* challenge. *FEMS Immunol. Med. Microbiol.* **55:**271–279.

Murthy, A. K., J. P. Chambers, P. A. Meier, G. Zhong, and B. P. Arulanandam. 2007. Intranasal vaccination with a secreted chlamydial protein enhances resolution of genital *Chlamydia muridarum* infection, protects against oviduct pathology, and is highly dependent upon endogenous gamma interferon production. *Infect. Immun.* **75:**666–676.

Murthy, A. K., Y. Cong, C. Murphey, M. N. Guentzel, T. G. Forsthuber, G. Zhong, and B. P. Arulanandam. 2006. Chlamydial protease-like activity factor induces protective immunity against genital chlamydial infection in transgenic mice that express the human HLA-DR4 allele. *Infect. Immun.* **74:**6722–6729.

Murthy, A. K., M. N. Guentzel, G. Zhong, and B. P. Arulanandam. 2009b. Chlamydial protease-like activity factor—insights into immunity and vaccine development. *J. Reprod. Immunol.* **83:**179–184.

Murthy, A. K., W. Li, M. N. Guentzel, G. Zhong, and B. P. Arulanandam. 2011. Vaccination with the defined chlamydial secreted protein CPAF induces robust protection against female infertility following repeated genital chlamydial challenge. *Vaccine* **29:**2519–2522.

Nelson, D. E., D. P. Virok, H. Wood, C. Roshick, R. M. Johnson, W. M. Whitmire, D. D. Crane, O. Steele-Mortimer, L. Kari, G. McClarty, and H. D. Caldwell. 2005. Chlamydial IFN-gamma immune evasion is linked to host infection tropism. *Proc. Natl. Acad. Sci. USA* **102:**10658–10663.

Nichols, R. L., S. D. Bell, Jr., E. S. Murray, N. A. Haddad, and A. A. Bobb. 1966. Studies on trachoma. V. Clinical observations in a field trial of bivalent trachoma vaccine at three dosage levels in Saudi Arabia. *Am. J. Trop. Med. Hyg.* **15:**639–647.

O'Connell, C. M., Y. M. Abdelrahman, E. Green, H. K. Darville, K. Saira, B. Smith, T. Darville, A. M. Scurlock, C. R. Meyer, and R. J. Belland. 2011. Toll-like receptor 2 activation by *Chlamydia trachomatis* is plasmid dependent, and plasmid-responsive chromosomal loci are coordinately regulated in response to glucose limitation by *C. trachomatis* but not by *C. muridarum.* *Infect. Immun.* **79:**1044–1056.

O'Connell, C. M., R. R. Ingalls, C. W. Andrews, Jr., A. M. Scurlock, and T. Darville. 2007. Plasmid-deficient *Chlamydia muridarum* fail to induce immune pathology and protect against oviduct disease. *J. Immunol.* **179:**4027–4034.

Olsen, A. W., M. Theisen, D. Christensen, F. Follmann, and P. Andersen. 2010. Protection against *Chlamydia* promoted by a subunit vaccine (CTH1) compared with a primary intranasal infection in a mouse genital challenge model. *PLoS One* **5:**e10768.

Pal, S., J. Bravo, E. M. Peterson, and L. M. de la Maza. 2008. Protection of wild-type and severe combined immunodeficiency mice against an intranasal challenge by passive immunization with monoclonal antibodies to the *Chlamydia trachomatis* mouse pneumonitis major outer membrane protein. *Infect. Immun.* **76:**5581–5587.

Pal, S., E. M. Peterson, and L. M. de la Maza. 2005. Vaccination with the *Chlamydia trachomatis* major outer membrane protein can elicit an immune response as protective as that resulting from inoculation with live bacteria. *Infect. Immun.* **73:**8153–8160.

Pal, S., I. Theodor, E. M. Peterson, and L. M. de la Maza. 1997a. Immunization with an acellular vaccine consisting of the outer membrane complex of *Chlamydia trachomatis* induces protection against a genital challenge. *Infect. Immun.* **65:**3361–3369.

Pal, S., I. Theodor, E. M. Peterson, and L. M. de la Maza. 1997b. Monoclonal immunoglobulin A antibody to the major outer membrane protein of the *Chlamydia trachomatis* mouse pneumonitis biovar protects mice against a chlamydial genital challenge. *Vaccine* **15:**575–582.

Patton, D. L., Y. T. Sweeney, and C. C. Kuo. 1994. Demonstration of delayed hypersensitivity in *Chlamydia trachomatis* salpingitis in monkeys: a pathogenic mechanism of tubal damage. *J. Infect. Dis.* **169:**680–683.

Peeling, R. W., J. Kimani, F. Plummer, I. Maclean, M. Cheang, J. Bwayo, and R. C. Brunham. 1997. Antibody to chlamydial hsp60 predicts an increased risk for chlamydial pelvic inflammatory disease. *J. Infect. Dis.* **175:**1153–1158.

Penttila, T., A. Tammiruusu, P. Liljestrom, M. Sarvas, P. H. Makela, J. M. Vuola, and M. Puolakkainen. 2004. DNA immunization followed by a viral vector booster in a *Chlamydia pneumoniae* mouse model. *Vaccine* **22:**3386–3394.

Penttila, T., J. M. Vuola, V. Puurula, M. Anttila, M. Sarvas, N. Rautonen, P. H. Makela, and M. Puolakkainen. 2000. Immunity to *Chlamydia pneumoniae* induced by vaccination with DNA vectors expressing a cytoplasmic protein (Hsp60) or outer membrane proteins (MOMP and Omp2). *Vaccine* **19:**1256–1265.

Perry, L. L., K. Feilzer, and H. D. Caldwell. 1997. Immunity to *Chlamydia trachomatis* is mediated by T helper 1 cells through IFN-gamma-dependent and -independent pathways. *J. Immunol.* **158:**3344–3352.

Perry, L. L., K. Feilzer, S. Hughes, and H. D. Caldwell. 1999a. Clearance of *Chlamydia trachomatis* from the murine genital mucosa does not require perforin-mediated cytolysis or Fas-mediated apoptosis. *Infect. Immun.* **67:**1379–1385.

Perry, L. L., H. Su, K. Feilzer, R. Messer, S. Hughes, W. Whitmire, and H. D. Caldwell. 1999b. Differential sensitivity of distinct *Chlamydia trachomatis* isolates to IFN-gamma-mediated inhibition. *J. Immunol.* **162:**3541–3548.

Pinchuk, I., B. C. Starcher, B. Livingston, A. Tvninnereim, S. Wu, E. Appella, J. Sidney, A. Sette, and B. Wizel. 2005. A $CD8^+$ T cell heptaepitope minigene vaccine induces protective immunity against *Chlamydia pneumoniae. J. Immunol.* **174:**5729–5739.

Pirbhai, M., F. Dong, Y. Zhong, K. Z. Pan, and G. Zhong. 2006. The secreted protease factor CPAF is responsible for degrading pro-apoptotic BH3-only proteins in *Chlamydia trachomatis*-infected cells. *J. Biol. Chem.* **281:**31495–31501.

Punnonen, R., P. Terho, V. Nikkanen, and O. Meurman. 1979. Chlamydial serology in infertile women by immunofluorescence. *Fertil. Steril.* **31:**656–659.

Ramsey, K. H., and R. G. Rank. 1990. The role of T cell subpopulations in resolution of chlamydial genital infection in mice, p. 241–244. *In* Proceedings of the 7th International Symposium on Human Chlamydial Infection. Cambridge University Press, New York, NY.

Rank, R. G. 1994. Animal models for urogenital infections. *Methods Enzymol.* **235:**83–93.

Rank, R. G., and A. L. Barron. 1983. Humoral immune response in acquired immunity to chlamydial genital infection of female guinea pigs. *Infect. Immun.* **39:**463–465.

Rank, R. G., and B. E. Batteiger. 1989. Protective role of serum antibody in immunity to chlamydial genital infection. *Infect. Immun.* **57:**299–301.

Rank, R. G., A. K. Bowlin, S. Cane, H. Shou, Z. Liu, U. M. Nagarajan, and P. M. Bavoil. 2009. Effect of *Chlamydiaphage* phiCPG1 on the course of conjunctival infection with "*Chlamydia caviae*" in guinea pigs. *Infect. Immun.* **77:**1216–1221.

Rank, R. G., K. H. Ramsey, E. A. Pack, and D. M. Williams. 1992. Effect of gamma interferon on resolution of murine chlamydial genital infection. *Infect. Immun.* **60:**4427–4429.

Rank, R. G., H. J. White, and A. L. Barron. 1979. Humoral immunity in the resolution of genital infection in female guinea pigs infected with the agent of guinea pig inclusion conjunctivitis. *Infect. Immun.* **26:**573–579.

Rasmussen, S. J., L. Eckmann, A. J. Quayle, L. Shen, Y. X. Zhang, D. J. Anderson, J. Fierer, R. S. Stephens, and M. F. Kagnoff. 1997. Secretion of proinflammatory cytokines by epithelial cells in response to *Chlamydia* infection suggests a central role for epithelial cells in chlamydial pathogenesis. *J. Clin. Investig.* **99:**77–87.

Roan, N. R., and M. N. Starnbach. 2006. Antigen-specific CD8+ T cells respond to *Chlamydia trachomatis* in the genital mucosa. *J. Immunol.* **177:** 7974–7979.

Rockey, D. D., D. Grosenbach, D. E. Hruby, M. G. Peacock, R. A. Heinzen, and T. Hackstadt. 1997. *Chlamydia psittaci* IncA is phosphorylated by the host cell and is exposed on the cytoplasmic face of the developing inclusion. *Mol. Microbiol.* **24:**217–228.

Rockey, D. D., R. A. Heinzen, and T. Hackstadt. 1995. Cloning and characterization of a *Chlamydia psittaci* gene coding for a protein localized in the inclusion membrane of infected cells. *Mol. Microbiol.* **15:**617–626.

Rockey, D. D., J. Wang, L. Lei, and G. Zhong. 2009. *Chlamydia* vaccine candidates and tools for chlamydial antigen discovery. *Expert Rev. Vaccines* **8:**1365–1377.

Rodriguez, A., M. Rottenberg, A. Tjarnlund, and C. Fernandez. 2006. Immunoglobulin A and CD8 T-cell mucosal immune defenses protect against intranasal infection with *Chlamydia pneumoniae*. *Scand. J. Immunol.* **63:**177–183.

Sabet, S. F., J. Simmons, and H. D. Caldwell. 1984. Enhancement of *Chlamydia trachomatis* infectious progeny by cultivation of HeLa 229 cells treated with DEAE-dextran and cycloheximide. *J. Clin. Microbiol.* **20:**217–222.

Sanchez-Campillo, M., L. Bini, M. Comanducci, R. Raggiaschi, B. Marzocchi, V. Pallini, and G. Ratti. 1999. Identification of immunoreactive proteins of *Chlamydia trachomatis* by Western blot analysis of a two-dimensional electrophoresis map with patient sera. *Electrophoresis* **20:**2269–2279.

Schachter, J., and C. R. Dawson. 1978. *Human Chlamydial Infections*. PSG Publishing Co. Inc., Littleton, MA.

Scurlock, A. M., L. C. Frazer, C. W. Andrews, Jr., C. M. O'Connell, I. P. Foote, S. L. Bailey, K. Chandra-Kuntal, J. K. Kolls, and T. Darville. 2011. Interleukin-17 contributes to generation of Th1 immunity and neutrophil recruitment during *Chlamydia muridarum* genital tract infection but is not required for macrophage influx or normal resolution of infection. *Infect. Immun.* **79:**1349–1362.

Sharma, J., A. M. Bosnic, J. M. Piper, and G. Zhong. 2004. Human antibody responses to a *Chlamydia*-secreted protease factor. *Infect. Immun.* **72:**7164–7171.

Sharma, J., F. Dong, M. Pirbhai, and G. Zhong. 2005. Inhibition of proteolytic activity of a chlamydial proteasome/protease-like activity factor by antibodies from humans infected with *Chlamydia trachomatis*. *Infect. Immun.* **73:**4414–4419.

Sharma, J., Y. Zhong, F. Dong, J. M. Piper, G. Wang, and G. Zhong. 2006. Profiling of human antibody responses to *Chlamydia trachomatis* urogenital tract infection using microplates arrayed with 156 chlamydial fusion proteins. *Infect. Immun.* **74:**1490–1499.

Shaw, A. C., K. Gevaert, H. Demol, B. Hoorelbeke, J. Vandekerckhove, M. R. Larsen, P. Roepstorff, A. Holm, G. Christiansen, and S. Birkelund. 2002. Comparative proteome analysis of *Chlamydia trachomatis* serovar A, D and L2. *Proteomics* **2:**164–186.

Starnbach, M. N., W. P. Loomis, P. Ovendale, D. Regan, B. Hess, M. R. Alderson, and S. P. Fling. 2003. An inclusion membrane protein from *Chlamydia trachomatis* enters the MHC class I pathway and stimulates a $CD8^+$ T cell response. *J. Immunol.* **171:**4742–4749.

Stemke-Hale, K., B. Kaltenboeck, F. J. DeGraves, K. F. Sykes, J. Huang, C. H. Bu, and S. A. Johnston. 2005. Screening the whole genome of a pathogen *in vivo* for individual protective antigens. *Vaccine* **23:**3016–3025.

Stephens, R. S., K. Koshiyama, E. Lewis, and A. Kubo. 2001. Heparin-binding outer membrane protein of chlamydiae. *Mol. Microbiol.* **40:**691–699.

Stuart, E. S., P. B. Wyrick, J. Choong, S. B. Stoler, and A. B. MacDonald. 1991. Examination of chlamydial glycolipid with monoclonal antibodies: cellular distribution and epitope binding. *Immunology* **74:**740–747.

Su, H., and H. D. Caldwell. 1995. CD4+ T cells play a significant role in adoptive immunity to *Chlamydia trachomatis* infection of the mouse genital tract. *Infect. Immun.* **63:**3302–3308.

Swanson, K. A., L. D. Taylor, S. D. Frank, G. L. Sturdevant, E. R. Fischer, J. H. Carlson, W. M. Whitmire, and H. D. Caldwell. 2009. *Chlamydia trachomatis* polymorphic membrane protein D is an oligomeric autotransporter with a higher-order structure. *Infect. Immun.* **77:**508–516.

Tammiruusu, A., T. Penttila, R. Lahesmaa, M. Sarvas, M. Puolakkainen, and J. M. Vuola. 2007. Intranasal administration of chlamydial outer protein N (CopN) induces protection against pulmonary *Chlamydia pneumoniae* infection in a mouse model. *Vaccine* **25:**283–290.

Thomas, N. S., M. Lusher, C. C. Storey, and I. N. Clarke. 1997. Plasmid diversity in *Chlamydia*. *Microbiology* **143**(Pt. 6)**:**1847–1854.

Ting, L. M., R. C. Hsia, C. G. Haidaris, and P. M. Bavoil. 1995. Interaction of outer envelope proteins of *Chlamydia psittaci* GPIC with the HeLa cell surface. *Infect. Immun.* **63:**3600–3608.

Toye, B., C. Laferriere, P. Claman, P. Jessamine, and R. Peeling. 1993. Association between antibody to the chlamydial heat-shock protein and tubal infertility. *J. Infect. Dis.* **168:**1236–1240.

Tseng, C. T., and R. G. Rank. 1998. Role of NK cells in early host response to chlamydial genital infection. *Infect. Immun.* **66:**5867–5875.

Van Voorhis, W. C., L. K. Barrett, Y. T. Sweeney, C. C. Kuo, and D. L. Patton. 1996. Analysis of lymphocyte phenotype and cytokine activity in the inflammatory infiltrates of the upper genital tract of female macaques infected with *Chlamydia trachomatis*. *J. Infect. Dis.* **174:**647–650.

Van Voorhis, W. C., L. K. Barrett, Y. T. Sweeney, C. C. Kuo, and D. L. Patton. 1997. Repeated *Chlamydia trachomatis* infection of *Macaca nemestrina* fallopian tubes produces a Th1-like cytokine response associated with fibrosis and scarring. *Infect. Immun.* **65:**2175–2182.

Villeneuve, A., L. Brossay, G. Paradis, and J. Hebert. 1994. Determination of neutralizing epitopes in variable domains I and IV of the major outer-membrane protein from *Chlamydia trachomatis* serovar K. *Microbiology* **140**(Pt. 9)**:**2481–2487.

Wang, J., Y. Zhang, C. Lu, L. Lei, P. Yu, and G. Zhong. 2010. A genome-wide profiling of the humoral immune response to *Chlamydia trachomatis* infection reveals vaccine candidate antigens expressed in humans. *J. Immunol.* **185:**1670–1680.

Wang, S., Y. Fan, R. C. Brunham, and X. Yang. 1999. IFN-gamma knockout mice show Th2-associated delayed-type hypersensitivity and the inflammatory cells fail to localize and control chlamydial infection. *Eur. J. Immunol.* **29:** 3782–3792.

Wang, S. P., and J. T. Grayston. 1967. Pannus with experimental trachoma and inclusion conjunctivitis agent infection of Taiwan monkeys. *Am. J. Ophthalmol.* **63**(Suppl.)**:**1133–1145.

Westrom, L., R. Joesoef, G. Reynolds, A. Hagdu, and S. E. Thompson. 1992. Pelvic inflammatory disease and fertility. A cohort study of 1,844 women with laparoscopically verified disease and 657 control women with normal laparoscopic results. *Sex. Transm. Dis.* **19:**185–192.

Whittum-Hudson, J. A., L. L. An, W. M. Saltzman, R. A. Prendergast, and A. B. MacDonald. 1996. Oral immunization with an anti-idiotypic antibody to the exoglycolipid antigen protects against experimental *Chlamydia trachomatis* infection. *Nat. Med.* **2:**1116–1121.

Whittum-Hudson, J. A., D. Rudy, H. Gerard, G. Vora, E. Davis, P. K. Haller, S. M. Prattis, A. P. Hudson, W. M. Saltzman, and E. S. Stuart. 2001. The anti-idiotypic antibody to chlamydial glycolipid exoantigen (GLXA) protects mice against genital infection with a human biovar of *Chlamydia trachomatis*. *Vaccine* **19:**4061–4071.

Williams, D. M., B. G. Grubbs, K. Kelly, E. Pack, and R. G. Rank. 1996. Role of gamma-delta T cells in murine *Chlamydia trachomatis* infection. *Infect. Immun.* **64:**3916–3919.

Wizel, B., B. C. Starcher, B. Samten, Z. Chroneos, P. F. Barnes, J. Dzuris, Y. Higashimoto, E. Appella, and A. Sette. 2002. Multiple *Chlamydia pneumoniae* antigens prime $CD8^+$ Tc1 responses that inhibit intracellular growth of this vacuolar pathogen. *J. Immunol.* **169:**2524–2535.

Wolf, K., E. Fischer, D. Mead, G. Zhong, R. Peeling, B. Whitmire, and H. D. Caldwell. 2001. *Chlamydia pneumoniae* major outer membrane protein is a surface-exposed antigen that elicits antibodies primarily directed against conformation-dependent determinants. *Infect. Immun.* **69:**3082–3091.

Woolridge, R. L., J. T. Grayston, I. H. Chang, C. Y. Yang, and K. H. Cheng. 1967. Long-term follow-up of the initial (1959–1960) trachoma vaccine field trial on Taiwan. *Am. J. Ophthalmol.* **63**(Suppl.)**:**1650–1655.

Yang, X., J. Gartner, L. Zhu, S. Wang, and R. C. Brunham. 1999. IL-10 gene knockout mice show enhanced Th1-like protective immunity and absent granuloma formation following *Chlamydia trachomatis* lung infection. *J. Immunol.* **162:**1010–1017.

Yu, H., K. P. Karunakaran, I. Kelly, C. Shen, X. Jiang, L. J. Foster, and R. C. Brunham. 2011. Immunization with live and dead *Chlamydia muridarum* induces different levels of protective immunity in a murine genital tract model: correlation with MHC class II peptide presentation and multifunctional Th1 cells. *J. Immunol.* **186:**3615–3621.

Yu, H. H., E. G. Di Russo, M. A. Rounds, and M. Tan. 2006. Mutational analysis of the promoter recognized by *Chlamydia* and *Escherichia coli* sigma(28) RNA polymerase. *J. Bacteriol.* **188:**5524–5531.

Zhong, G., and R. C. Brunham. 1992. Antibody responses to the chlamydial heat shock proteins Hsp60 and Hsp70 are H-2 linked. *Infect. Immun.* **60:**3143–3149.

Zhong, G., P. Fan, H. Ji, F. Dong, and Y. Huang. 2001. Identification of a chlamydial protease-like activity factor responsible for the degradation of host transcription factors. *J. Exp. Med.* **193:**935–942.

Zhong, G., T. Fan, and L. Liu. 1999. *Chlamydia* inhibits interferon gamma-inducible major histocompatibility complex class II expression by degradation of upstream stimulatory factor 1. *J. Exp. Med.* **189:**1931–1938.

Zhong, G., L. Liu, T. Fan, P. Fan, and H. Ji. 2000. Degradation of transcription factor RFX5 during the inhibition of both constitutive and interferon gamma-inducible major histocompatibility complex class I expression in chlamydia-infected cells. *J. Exp. Med.* **191:**1525–1534.

Zomorodipour, A., and S. G. Andersson. 1999. Obligate intracellular parasites: *Rickettsia prowazekii* and *Chlamydia trachomatis*. *FEBS Lett.* **452:**11–15.

CHLAMYDIAL GENETICS: DECADES OF EFFORT, VERY RECENT SUCCESSES

Brendan M. Jeffrey, Anthony T. Maurelli, and Daniel D. Rockey

15

INTRODUCTION

Molecular genetic tools are some of the most powerful means for analysis in any system, be it viral/prokaryotic/eukaryotic, big/small, or intracellular/extracellular. *Chlamydia* is one of only a few bacterial pathogens for which a tractable genetic system is lacking or only in a rudimentary stage (others include the spirochete *Treponema pallidum*, which causes syphilis, and members of the genera *Tropheryma*, *Orientia*, and *Lawsonia*) and is arguably the most significant pathogen for which this technology remains unavailable. A consistent theme for these genetically intractable organisms is that they are often obligate intracellular organisms that cannot be propagated axenically or are organisms that cannot be cultured outside host organisms in any way. The absence of an experimental genetic system has severely limited the ability for definitive study of gene and protein function in each of these genera or species.

This has not been for lack of effort; a large set of researchers have worked on this problem. Experts in plasmids of gram-negative bacteria expected that broad-host-range plasmids could be manipulated and transformed into recipient chlamydiae, followed by selection that would allow isolation and propagation of plasmid-positive strains. This approach has been applied with success to genetic modification of *Coxiella burnetii* (Beare et al., 2009; Voth et al., 2011), an obligate intracellular species that has recently been cultured axenically (Omsland et al., 2008, 2009). Plasmid transformation has also been used in other obligate intracellular bacteria, leading to maintenance of plasmids as episomes or to a recombination event targeting the recipient genome (reviewed by Beare et al. [2011]). There have been attempts to introduce vectors containing constitutive chlamydial promoters driving expression of chloramphenicol acetyltransferase into chlamydiae, with a goal of generating chloramphenicol resistance in transformants. These plasmid-based studies have resulted in transient maintenance of exogenous genes (Tam et al., 1994; Binet and Maurelli, 2009). Researchers also attempted to exploit the chlamydial virulence plasmid, a near-ubiquitous, conserved element whose role in infection is currently

Brendan M. Jeffrey and Daniel D. Rockey, Department of Biomedical Sciences and Molecular and Cellular Biology Program, Oregon State University, Corvallis, OR 97331-4804. *Anthony T. Maurelli*, Department of Microbiology and Immunology, F. Edward Hébert School of Medicine, Uniformed Services University, Bethesda, MD 20814-4799.

Intracellular Pathogens I: Chlamydiales
Edited by Ming Tan and Patrik M. Bavoil © 2012 ASM Press, Washington, DC
10.1128/9781555817329.ch15

being elucidated (O'Connell and Nicks, 2006; Chen et al., 2010; Olivares-Zavaleta et al., 2010; Russell et al., 2011). The presence of this plasmid formed the basis for several attempts to develop a chlamydial genetic system, as it logically might be a tool for manipulation and reintroduction into the organism. *Chlamydia* spp. also have bacteriophages (Storey et al., 1989; Hsia et al., 2000; Read et al., 2000; Karunakaran et al., 2002; Rupp et al., 2007), and there are ongoing efforts to exploit these phages as vectors for introduction of genes. Promiscuous transposons have been successfully used in many challenging bacterial systems (Stewart et al., 2004; Beare et al., 2009), and there is precedent for natural introduction of a transposon into a veterinary chlamydial pathogen (Dugan et al., 2004, 2007), which will be discussed below. It is reasonable to expect that these elements should be useful in modification of the chlamydial genome, but they have only recently culminated in an experimental genetic system that can be used to introduce foreign genes into *Chlamydia* in a targeted manner.

Why was it so challenging to develop a routine transformation system for these organisms? The following sections will address a set of challenging aspects of the chlamydial system and point out some avenues that can be exploited, perhaps opening the door to technologies for routine genetic modification of chlamydiae.

CHALLENGES IN TRANSFORMING CHLAMYDIAE

The obligatory intracellular lifestyle and unique biphasic developmental cycle of the *Chlamydiaceae* contribute to the difficulty in genetically manipulating these organisms (the developmental cycle is discussed in more detail in chapter 7, "Temporal gene regulation during the chlamydial developmental cycle"). It is likely that several aspects of this lifestyle have a role in these complications. The rigid, extensively disulfide cross-linked outer membrane complex of the metabolically inactive elementary bodies (EBs) may hinder the introduction of exogenous DNA into this developmental form. Furthermore, the lack of metabolic activity at this stage may preclude interactions with the recipient genome. The challenges may be parallel to those in *Bacillus* and *Clostridium* spp. systems. While the vegetative forms of these bacteria are readily made competent, the spore structures are refractory to acquiring or incorporating introduced DNA (Peter Setlow, personal communication). EBs are similarly resistant to sonic disruption or shear stress (though not similarly resistant to the stresses of heat, desiccation, or time), and this environmental stability may lead to analogous resistance to DNA uptake. Nevertheless, recent efforts have been successful, leading to stable incorporation of foreign DNA into chlamydiae (see below; see also Binet and Maurelli, 2009, and Wang et al., 2011), with the products of the transformation integrated into the chromosome via homologous recombination. There have been other reports of successful electroporation of DNA into EBs (Tam et al., 1994; O'Connell and Maurelli, 1998), but there was no evidence for stable maintenance of the introduced plasmids.

The introduction of DNA into an EB is hypothesized to be a challenging task for several reasons. The cell wall may be physically impermeable to DNA, or the pores introduced by electroporation might be of a size that does not allow DNA molecules to efficiently enter. Another issue is that DNA in both EBs and bacterial spores is bound by small protein molecules—in chlamydiae this involves histone-like proteins (Hackstadt et al., 1991, 1993; Perara et al., 1992; Barry et al., 1993), while in spore-forming bacteria, a set of small acid-soluble proteins is tightly bound to genomic DNA (Setlow, 2007). These DNA-binding proteins may physically wall off the chromosome from introduced donor DNA, and they may render the DNA inaccessible for the chlamydial recombination machinery and any subsequent homologous recombination. It is also conceivable that plasmid DNA without bound histone-like proteins is not recognized by the replication or transcriptional machinery early in development.

After infection and uptake, EBs differentiate into the metabolically active reticulate bodies (RBs), a process concomitant with the modification of the primary endosome into a vacuole tailored for chlamydial development (discussed in more detail in chapter 6, "Initial interactions of chlamydiae with the host cell," and chapter 8, "Cell biology of the chlamydial inclusion"). The RB is an attractive target for introducing exogenous DNA into chlamydiae. RBs have cell envelopes that lack the structural rigidity of EBs, lack peptidoglycan (also missing in EBs), but otherwise are generally similar to typical gram-negative bacteria. RBs divide by binary fission, are metabolically active, and undergo DNA replication. There is also considerable recent evidence that the RB genome undergoes recombination (see below). However, there are unique challenges when considering RBs as targets for DNA exchange. Functional RBs are present only within the inclusion, and there are no reports demonstrating success in using isolated RBs to infect host cells. Therefore, a strategy of transforming purified RBs and then using them to infect cells, similar to what has been accomplished with *Coxiella burnetii* (Voth et al., 2011), is not a viable option at present. It is possible that DNA can be introduced into RBs within cells; however, there are four lipid bilayers (the host plasma membrane, the inclusion membrane, and the RB outer and inner membranes) that exogenous DNA will need to cross to enter the chlamydial cytoplasm. Perhaps a multiple-pulse electroporation strategy might be exploited to overcome these challenges.

Intracellular RBs might be targeted through the use of a second intracellular DNA donor that is genetically tractable, such as *Shigella*, *Salmonella*, or *Legionella*. However, these donor bacteria are likely to be separated from chlamydiae by at least the inclusion membrane, since most intracellular bacterial species do not occupy a common vacuole within host cells (Heinzen et al., 1996; Sauer et al., 2005). Therefore, introduction of DNA into RBs will require the development of technologies to access the lumen of the inclusion, a concept that appears increasingly possible as we learn more about vesicular trafficking in infected host cells (discussed in more detail in chapter 6).

Antibiotic Selection

Antibiotic resistance genes have been commonly used in other bacteria as selectable markers to facilitate the isolation of transformants, and several have been used in chlamydiae (Table 1). However, there are caveats and challenges to the use of antibiotic resistance in this system because of the organism's unusual biology. The antibiotic resistance gene likely must be expressed at all times to provide continuous positive selective pressure, and this is usually accomplished by expressing the

TABLE 1 Antibiotics previously used to generate stable or transient resistance in *Chlamydia* spp.

Compound	Targeted process	Endogenous resistance gene	Reference(s) for use in *Chlamydia* spp.
Chloramphenicol	Protein synthesis	None	Tam et al., 1994
Kasugamycin	Protein synthesis	16S rRNA	Binet and Maurelli, 2009
Nalidixic acid	DNA supercoiling	*gyrA*	Binet and Maurelli, 2009
Ofloxacin	DNA supercoiling	*gyrA*	Demars et al., 2007; Suchland et al., 2009
Rifampin	Transcription	*rpoB*	Demars et al., 2007; Suchland et al., 2009
Spectinomycin	Protein synthesis	16S rRNA	Binet and Maurelli, 2009
Tetracycline	Protein synthesis	*tet*(C)	Lenart et al., 2001; Dugan et al., 2004; Suchland et al., 2009

gene from a constitutive promoter. In EBs, however, there is global silencing of transcription, which is believed to be due to the condensation of chlamydial DNA by histone-like proteins (Barry et al., 1993). Upon entry into the host cell, only a small number of immediate early genes are expressed within 1 hour postinfection (Belland et al., 2003). The majority of chlamydial genes are not transcribed during transition from an EB into a RB and are only expressed in midcycle (Shaw et al., 2002; Belland et al., 2003; Albrecht et al., 2010). Thus, the promoters for the immediate early genes, or perhaps truly constitutive promoters, would appear to be the best choice for expressing the antibiotic marker soon after chlamydial entry into the host cell. This regulation of chlamydial gene expression is discussed in more detail in chapter 7. Failure to provide continuous antibiotic selection during transformation experiments could lead to false positives, especially if the antibiotic only inhibits growth.

It is also critical to consider whether an antibiotic is clinically important. The deliberate transfer of a drug resistance trait to microorganisms that are not known to acquire the trait naturally is prohibited by the NIH's *Guidelines for Research Involving Recombinant DNA Molecules* (available at http://oba.od.nih.gov/rdna/nih_guidelines_oba.html), as acquisition of resistance to the drug could compromise its use to control these disease agents in human or veterinary medicine. The *tet*(C) allele (described below) is an example of an exogenous resistance mechanism that is known to function in *Chlamydia*, but its approval for use as a selectable marker must be formally obtained from the NIH on a laboratory-by-laboratory basis. Clinically important antibiotics that should not be used for selection are shown in Table 2, together with antibiotics that are not suitable because they do not enter or are toxic to mammalian cells.

All other markers used in successful selection for resistance in *Chlamydia* have involved a mutated metabolic target gene for the antibiotic being used (Demars et al., 2007; Demars and Weinfurter, 2008; Binet and Maurelli, 2009; Suchland et al., 2009). If antibiotic selection proves to remain challenging, strategies using non-antibiotic-based selection strategies may be employable (Roshick et al., 2006; Vidal et al., 2008).

When using positive selection to identify transformants, it is also important to generate microbiological clones both for the starting strain and for analyzing the strains that have acquired resistance. Stemming from the recognition that many chlamydial strains are polyclonal, researchers have recently addressed the important concept that isolation of microbiological clones is critical for the evaluation of the biology of individual strains. Different laboratories use either limiting dilution (Suchland and Stamm, 1991) or plaque purification (Gieffers et al., 2002; O'Connell and Nicks, 2006) of individual chlamydial inclusions (i.e., "colonies") when they purify a phenotypically variant strain. It is conceivable that the inability to identify individual transformed chlamydial subpopulations might have previously led to

TABLE 2 Antibiotic resistance markers that cannot be used for selection of transformants in *Chlamydia*

Antibiotics of clinical importance for chlamydial infections[a]	Antibiotics that do not penetrate mammalian cells	Antibiotics that are toxic to eukaryotic cells at concns needed to kill *Chlamydia* spp.[b]
Amoxicillin, doxycycline, azithromycin, erythromycin, sulfisoxazole	Kanamycin, gentamicin, streptomycin	Blasticidin S, hygromycin B, geneticin (G 418), zeocin, phleomycin, nourseothricin

[a]Centers for Disease Control and Prevention, 2010.
[b]A. T. Maurelli, unpublished data.

false-negative results in otherwise successful *Chlamydia* transformation experiments. Indeed the microbiological cloning of chlamydiae as a critical step prior to initiating any attempt at developing a chlamydial genetic system should be considered a priority.

GENOMICS AND OUR UNDERSTANDING OF NATURAL LATERAL GENE TRANSFER IN CHLAMYDIAE

As described in chapter 2, "Deep and wide: comparative genomics of *Chlamydia*," genome-sequencing efforts have proved highly effective in increasing our understanding of the biology of this organism in the absence of a targeted genetic manipulation system. Release of the first genome sequence of the urogenital *Chlamydia trachomatis* strain D/UW3-cx in 1998 (Stephens et al., 1998) provided the initial impetus to addressing many questions about basic chlamydial biology. The availability of this and additional chlamydial genome sequences has allowed researchers to search for hints that chlamydiae might be amenable to DNA uptake. Genome sequences of different *Chlamydia* species are remarkably syntenic (i.e., they share gene order throughout the chromosome), with very limited examples of recently acquired genomic islands. Members of the related *Parachlamydiaceae* encode type IV secretion assemblies on a genomic island (Greub et al., 2004; Horn et al., 2004), which was likely acquired by horizontal transfer, and *Simkania* spp. carry a conjugative plasmid that encodes a type IV secretion system. Different members of this group have a variety of transposon sequences (Greub et al., 2004; Clifton et al., 2005). A single insertion element (IS) is found in isolates of *Chlamydia suis* (Dugan et al., 2007), but the genomes of pathogenic human *Chlamydia* spp. do not contain transposons or recognizable recombination targets (i.e., Chi sites). The relative abundance of candidate gene transfer systems in other members of the *Chlamydiales* suggests that contemporary *Chlamydia* spp. evolved into their genetically intractable niche by reductive evolution from a genetically amenable ancestor (Horn et al., 2004).

When the first *Chlamydia* genome was sequenced, a collection of genes with more similarity to genes from plants than bacteria was noted (Stephens et al., 1998), suggesting possible acquisition via exchange between a chlamydial ancestor and a host cell. Many of these candidate acquired genes are of chloroplast origin, and this association may reflect the ancient relationship between the *Chlamydiaceae* and the chloroplast-cyanobacterium lineage rather than the more recent acquisition of plant-like genes by the *Chlamydiaceae* (Brinkman et al., 2002). This hypothesis is strengthened by a recent report investigating the gene product of *CT390*, which encodes the chlamydial DAP aminotransferase (Brinkman et al., 2002; McCoy et al., 2006). This report provides data demonstrating that this enzyme uniquely fills a metabolic gap in chlamydiae, cyanobacteria, and plants. A recent publication by Horn and colleagues expands on the possible contribution of chlamydial genes to plant genomes, concluding that progenitor chlamydiae likely contributed genetic material to ancestral photosynthetic eukaryotes (Collingro et al., 2011).

Chlamydia spp. encode a complete type III secretion (T3S) machine that is used to secrete effectors outside the bacterium into the inclusion membrane and the cytoplasm of the infected cell (Mabey, 2008). In most systems, the collected T3S genes are found on pathogenicity islands or on plasmids, which were likely acquired via horizontal transmission. In contrast, genes for the chlamydial T3S system are not present on a pathogenicity island but instead are arranged in several loci that are scattered around the chromosome. Unlike other bacteria, there is also no difference in the G+C ratio between T3S genes and the rest of the chlamydial chromosome (Hueck, 1998; Subtil et al., 2000; Kim, 2001). These data suggest that *Chlamydia* did not recently acquire its T3S genes via horizontal transfer from another organism. Overall, the collected genome sequence data suggest that contemporary chlamydial strains are only very rarely recipients of lateral gene transfer events from other species, which might help explain why it has been difficult to introduce DNA into chlamydiae in the laboratory.

EVIDENCE FOR NATURAL LATERAL GENE TRANSFER WITHIN CONTEMPORARY *CHLAMYDIA* SPECIES

While the above discussion stresses that there is little evidence that *Chlamydia* spp. are receptive to acquisition of foreign genes, there is strong evidence for active gene transfer between closely related chlamydial strains and species. The following sections will address the surprising finding that chlamydiae have the tools to naturally acquire and integrate homologous DNA into their genome as long as the DNA is donated by a related chlamydial strain.

Recombination in Clinical Isolates

Studies of genetic diversity within *ompA* provided the first evidence of recombination and horizontal gene transfer in *C. trachomatis* (Lampe et al., 1993; Brunham et al., 1994; Millman et al., 2001). Recombination in *ompA* results in antigenic changes on the surface of the major outer membrane protein, the most abundant surface protein on EBs and RBs and the basis for *C. trachomatis* serotyping (see chapter 4, "The chlamydial cell envelope"). These studies were expanded to other highly variable regions of the genomes such as the plasticity zone and the polymorphic membrane protein (Pmp) gene family (Gomes et al., 2004, 2006, 2007). Gomes and colleagues provided compelling evidence that recombination occurs at a number of hot spots in the *C. trachomatis* genome including the regions around *CT049*, *ompA*, and *pmp* genes as well as the plasticity zone (Gomes et al., 2004, 2006, 2007). The genome sequence of clinical isolates (Jeffrey et al., 2010) revealed evidence of recombination throughout the genome, in addition to sites described by others. An attractive model for these observations is that coinfection of a single human by more than one strain of *Chlamydia* spp. (Suchland et al., 2003) provides the opportunity for different strains to come in contact with one another (Hackstadt et al., 1999; Suchland et al., 2008) and undergo genome shuffling through a currently unknown mechanism. Hot spots identified in previous work (Gomes et al., 2007) seem to favor genetic regions encoding outer membrane proteins, leading to the hypothesis that there is positive selection of recombinants by the host immune system.

A Chlamydial Transposon and Implications for Lateral Gene Transfer

Although it is clear that horizontal gene transfer and recombination occur between chlamydial strains, until recently there was no evidence for *Chlamydia* incorporating foreign DNA into its genome. The genetic characterization of tetracycline resistance in a chlamydial pathogen of swine, *Chlamydia suis,* revealed a recent lateral gene transfer event from a nonchlamydial species (Dugan et al., 2004). Resistant *C. suis* strains were initially identified in the upper Midwest of the United States (Andersen and Rogers, 1998), and similar resistant strains were recently identified in Italy (Di Francesco et al., 2008). Tetracycline-resistant strains carry genomic islands of between 5 and 12 kb that are integrated into their chromosomes. Each "Tet(C) island" encodes the tetracycline efflux pump Tet(C) and contains a set of genes associated with plasmid maintenance and replication (Fig. 1) (Dugan et al., 2004). Most, but not all, of these related Tet(C) islands also carry the bicistronic insertion sequence *IS*cs605, which is derived from the *IS*200 and *IS*1341 insertion sequence families and structurally similar to the *IS*605 family of insertion elements commonly found in *Helicobacter* spp. (Dugan et al., 2004). Functional analysis of *IS*cs605 in an *Escherichia coli* system revealed that the *IS*200 homolog is sufficient to mediate transposition of the element and that integration is targeted nearly universally to the pentameric sequence

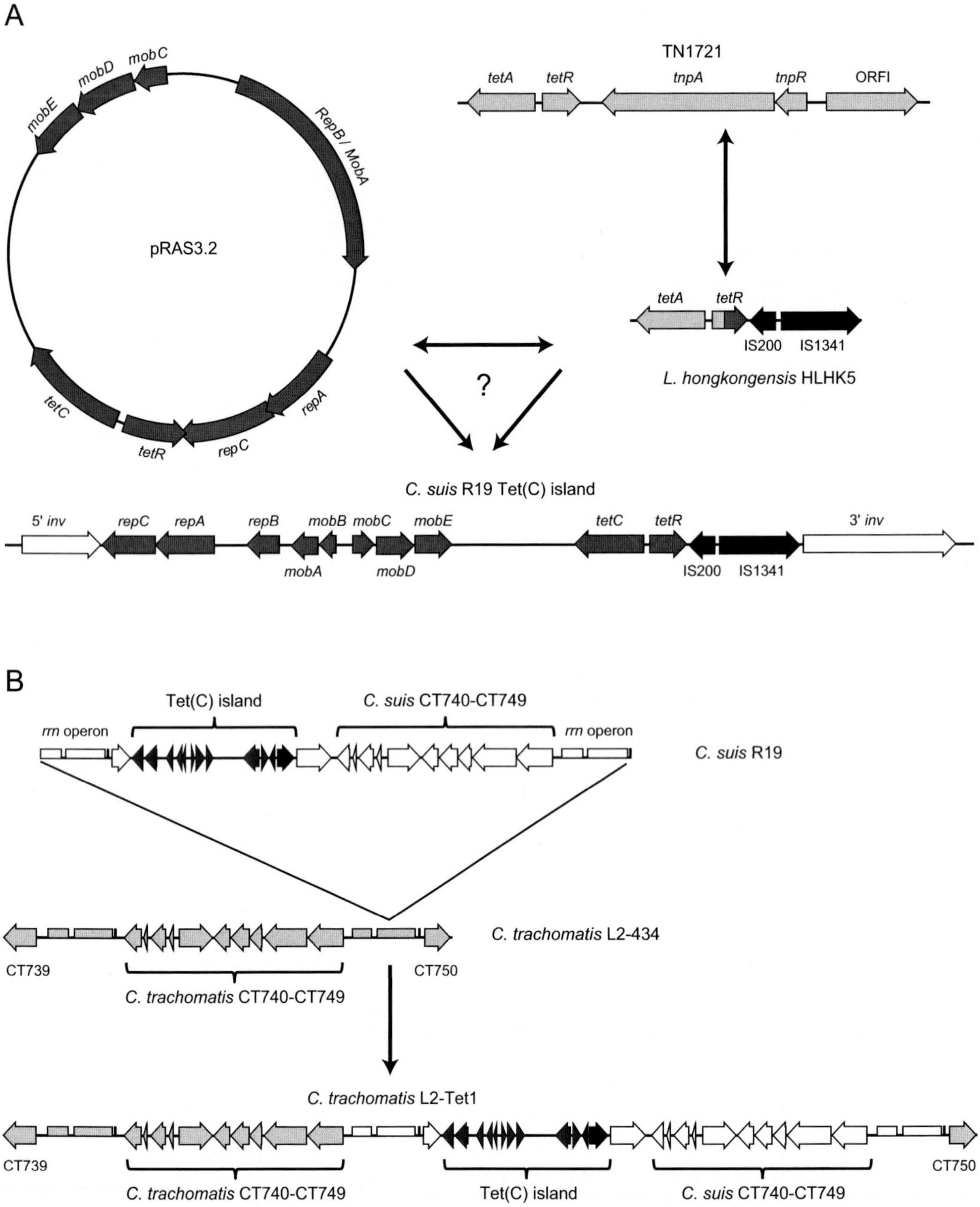

FIGURE 1 Model for the integration of the tetracycline resistance gene into the *inv*-like gene in the *C. suis* genome. (A) The Tet(C) island of *C. suis* contains sequences that are very similar to sequences from other fish bacteria. *tet*R and *IS*cs605, which is composed of *IS*200 and *IS*1341 insertion sequences, are similar to sequences in *Laribacter hongkongensis* HLHK5. Adjacent genes in the Tet(C) island are highly similar to plasmid maintenance genes from pRAS3.2, which is a plasmid from *Aeromonas salmonicida*. (B) The presence of this naturally occurring island in *C. suis* was then exploited as a tool for laboratory-based recombination experiments similar to those conducted by Demars and colleagues (Binet and Maurelli, 2009). rRNA operons in panel B are indicated by half-width open reading frame boxes. doi:10.1128/9781555817329.ch15.f1

TTCAA (Dugan et al., 2007). This pentamer is present at each putative integration site within the Tet(C) islands of the *C. suis* chromosome.

As with the specific origins of most genomic islands, the origin(s) of the Tet(C) island remains unclear. However, the collected data strongly support the hypothesis that lateral gene transfer from a nonchlamydial bacterium followed by recombination between chlamydiae was the mechanism for acquisition of resistance. Within the Tet(C) islands, the sequences adjacent to *IS*cs605 show a high degree of sequence identity with pRAS3.2, a plasmid carried by the fish pathogen *Aeromonas salmonicida* (L'Abee-Lund and Sorum, 2002; Dugan et al., 2004). The very few differences between the shared sequence of pRAS3.2 and the Tet(C) islands include nucleotide changes in possible control elements for expression of *tet*(C) and within the apparent plasmid origin of replication (Dugan et al., 2004). Insertion of these islands into the *C. suis* genome occurred at an identical position in the *inv*-like gene in all of the Tet-resistant *C. suis* strains analyzed (Dugan et al., 2004).

While the plasmid pRAS3.2 shares considerable sequence identity with the Tet(C) islands, it does not carry the *IS*cs605 insertion sequence found in the chlamydial resistance elements. Recent sequence analysis from another fish pathogen, *Laribacter hongkongensis,* demonstrated that this organism carries a genomic island that shares very high identity with *IS*cs605 and *tet*R in the *C. suis* Tet(C) islands (Lau et al., 2008). This discovery suggested that IS elements found in *L. hongkongensis* contributed genes to the *A. salmonicida* pRAS3.2 plasmid or a very similar genetic element. The linkage of pRAS3.2 sequence with *IS*cs605 then facilitated a possible integration event into the *C. suis* genome at the *inv*-like gene, either directly or following passage in an intermediary bacterial host (Fig. 1A).

The acquisition of these Tet(C) islands remains the only example from nature of horizontal acquisition of foreign DNA by a chlamydial strain. How *C. suis* acquired the various forms of the tetracycline resistance islands discussed by Dugan and colleagues remains unclear, but a model can be proposed. The long-term prophylactic use of tetracycline in swine facilities would provide an environment with strong selective pressure for *C. suis* strains harboring the Tet islands. Conversations with Art Andersen of the National Animal Disease Center in Ames, IA, also indicated that pigs on farms in the upper Midwest were often fed waste from fish processing plants, perhaps providing a means for contact between *C. suis* and organisms harboring the resistance plasmids. The resistant strains identified in Italy share *tet*(C) with the U.S. strains, but data are insufficient to assess whether they were derived from similar or distinct integration events.

These findings demonstrate that *Chlamydia* can acquire foreign DNA by horizontal transfer. It appears to be a rare event, likely driven by the long-term use of low levels of antibiotics as growth promoters. There is no evidence that a similar natural process has led to genomic island integration in *Chlamydia* spp. pathogenic for humans. It is possible that the tools used by *C. suis* to integrate this island may be used by investigators to introduce genes of interest into other *Chlamydia* spp., although efforts to use this have not been successful to date (D. D. Rockey, unpublished data).

Outside the acquisition of tetracycline resistance by *C. suis*, *Chlamydia* spp. appear to be driving in the slow lane of evolutionary progress, working through evolution via single base changes or insertion/deletions, incorporating and exploiting several species-wide gene conversion events and deleting a few genes, with significant consequences (Caldwell et al., 2003; Gomes et al., 2006; Nunes et al., 2008). If this interpretation is correct, it might explain why the organism has been so hard to transform: chlamydiae have not retained an evolutionary strategy that involves routine acquisition and sampling of foreign DNA.

EXPERIMENTAL METHODS FOR GENETIC MODIFICATION OF CHLAMYDIAE

The following sections will review exciting new approaches that are being used to genetically manipulate the chlamydial genome and will discuss how these efforts have helped us understand different aspects of chlamydial biology.

Mutant Chlamydial Strains Generated with Chemical Mutagenesis

Although efforts to conduct site-specific gene disruption are in their infancy, early work by Rodolakis and colleagues (Rodolakis and Bernard, 1984) demonstrated that chemical mutagenesis could be used to generate mutant chlamydial strains. This group generated a live attenuated vaccine strain of the virulent strain of *C. abortus* AB7 (AC1) using nitrosoguanidine. The temperature-sensitive mutant strain is effective as a live vaccine and remains in use as a commercial vaccine for the prevention of ovine abortion (sold as CEVAC Chlamydophila® or Intervet Enzovax). A comparative genomics study of the parent strain and the mutant revealed several loci that could play a role in the observed attenuation in vivo (Burall et al., 2009).

Use of Chemical Mutagenesis To Generate Site-Specific, Isogenic Mutant Strains

Even though there remain technical challenges to the introduction of DNA into chlamydiae, there are recent examples of progress in modifying the chlamydial genome. Kari and colleagues applied a mutagenesis and selection strategy called TILLING (targeted induced local lesions in genomes), which was first developed in plant systems (Stemple, 2004). In a proof-of-principle study performed with *C. trachomatis* serovar D, they generated 24 mutants with single nucleotide polymorphisms (SNPs) in *trpBA* (Fig. 2) (Kari et al., 2011). They targeted the *trpBA* operon, because it encodes an enzyme, tryptophan synthase, which is not necessary for chlamydial growth in media containing tryptophan (Fehlner-Gardiner et al., 2002; Caldwell et al., 2003). EBs were first chemically mutagenized at a low concentration of the mutagen to minimize the chance of multiple mutations in a single organism. The mutagenized chlamydiae were then grown in 96-well trays, with each well containing host cells infected with a mixed pool of about 10 mutagenized organisms, which was expanded through three rounds of lysis and reinfection. Mutants were identified with a heteroduplex DNA screen that involved PCR amplification of the *trpBA* operon, cleavage by CEL1 endonuclease (which cleaves both DNA strands at mismatches or indels), and identification of positive pools by the presence of cleavage products on gel electrophoresis. The PCR amplicons from positive pools were then sequenced to identify SNPs in either *trpB* or *trpA*. A strain with a nonsense mutation in *trpB* was cloned by plaque purification, and the genome was sequenced to verify that there were no additional mutations at other sites. At the end of the procedure, Kari and colleagues identified an isogenic *trpB* null mutant that was produced in a targeted manner without a requirement for selection, which represents a giant leap in the genetic manipulation of chlamydiae.

The Kari approach has the potential to be used at the levels of both individual genes and global analysis of the chlamydial genome. In principle, it can be used to generate a targeted mutation in any nonessential chlamydial gene. In addition, a large pool of heavily mutagenized chlamydiae can be coupled with next-generation deep-sequencing methods to identify genes that are essential for chlamydial growth in cell culture (Kari et al., 2011). For example, temperature-sensitive mutants can be isolated from a library, mutagenized, cultured at lower temperatures, and screened for the inability to grow at 37°C. Although the described method is time-consuming and labor-intensive, the technique is workable and the results will be definitive and reproducible. The effort required is similar to that of generating a hybridoma-producing monoclonal antibody, and the cost/benefit analysis should not hinder individuals from applying these technologies. This protocol will be a valuable tool for examining host-microbe interactions following creation of relevant knockouts at loci of interest to researchers. While it has limits, this is

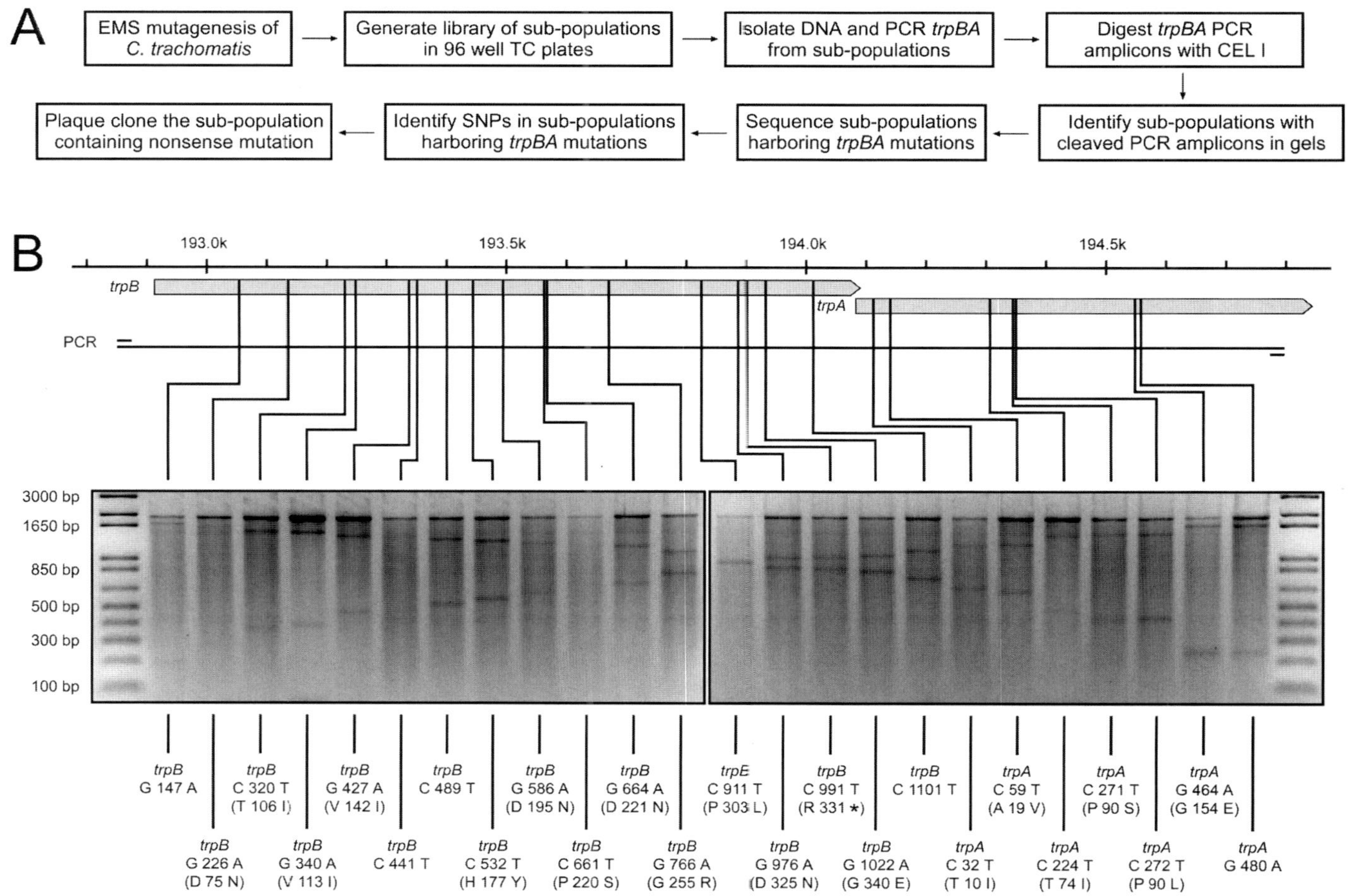

FIGURE 2 The reverse-genetic approach used to generate targeted mutations in the *trpBA* operon. (A) Flowchart for generating and screening libraries. (B) Summary of the *trpBA* mutations identified in the analysis of the library of subpopulations. CEL I digestion of the 24 subpopulations harboring *trpBA* mutations are shown on the gels in the order of their genomic locations. The *trpB* and *trpA* open reading frames are illustrated by the horizontal arrows. The complete sets of SNPs identified by capillary sequencing are indicated below each sample. Locations of SNPs are indicated in the *trpBA* operon above the gel image. Genomic scale and the region corresponding to the PCR amplicon are also shown. The nonsense mutation in *trpB* at position 991 (R331★) truncates the open reading frame by 186 bp. Used by permission of *Proceedings of the National Academy of Sciences USA* and the author. doi:10.1128/9781555817329.ch15.f2

truly the first example of a workable system for generating isogenic mutants in chlamydiae. Certainly the procedure has applications in other challenging systems as well.

Recombinants Created in the Laboratory

Pioneering work by Demars and colleagues demonstrated the remarkable fact that *C. trachomatis* genomes do recombine and recombine regularly and extensively, both within the species and, given the opportunity, with other related *Chlamydia* spp. (Demars et al., 2007). These investigators coinfected HeLa cells with different combinations of ofloxacin-, lincomycin-, rifampin-, and trimethoprim-resistant strains of *C. trachomatis* serovar L1 and selected for doubly resistant mutant strains. The choice of antibiotics was driven by the demonstrated ability of chlamydiae to evolve antibiotic resistance in vitro through single mutations in metabolic genes or rRNA (i.e., *gyrA*, *rpoB*, and 16S rRNA), which allowed the routine generation of parental strains for the crosses. The ease with which mutations could be produced made it important to demonstrate that any doubly resistant progeny were not false positives due to the generation of spontaneous mutants during selection.

Demars and colleagues demonstrated that doubly resistant strains (i.e., progeny of recombination) occurred at frequencies that were significantly higher than the rate to spontaneously generate the resistance mutation. A follow-up study by these investigators determined the genome sequence of the progeny strains and found that surprisingly large regions of the genome (~441 kb to ~671 kb) were derived from each parent (Demars and Weinfurter, 2008). These results are consistent with the homologous recombination of large DNA segments among strains during coinfection of host cells.

More recently, one of our laboratories performed a coinfection with a *C. suis* tetracycline-resistant strain and *C. trachomatis* serovar L2 and successfully created an L2 recombinant strain carrying the tetracycline resistance island (Suchland et al., 2009). Sequence analysis of this recombinant strain revealed a putative homologous recombination event in which donor DNA from *C. suis* had been inserted into one of the *C. trachomatis* ribosomal operons, resulting in a recombinant strain with three ribosomal operons (Suchland et al., 2009). This recombinant strain represents the first documented insertion of a large segment of foreign DNA into the *C. trachomatis* genome under experimental conditions.

These studies provide much-needed evidence that members of the *Chlamydiaceae* can take up foreign DNA from their environment, through an unknown mechanism, and that this DNA can be incorporated into the genome. Genome sequence analysis strongly supports a model where introduced DNA shared between chlamydial strains or species is integrated into the genome via homologous recombination. Genetic exchange has been documented among *C. trachomatis* strains, and also between *C. trachomatis*, *C. suis*, and *Chlamydia muridarum*. Recombination between fusogenic IncA-positive and nonfusogenic IncA-negative strains has been observed, suggesting that fusion of inclusions or cohabitation of a common inclusion is not required for recombination (Suchland et al., 2009). It is noteworthy that we have no evidence for activity of the *C. suis IS*cs605 transposase in any chlamydial recombinant: no target sequence has been identified in over 100 in vitro-generated recombination sites (B. M. Jeffrey and D. D. Rockey, unpublished data).

The generation of doubly resistant recombinant chlamydial strains from parents with differing phenotypic traits can be used as a way to examine the correlation between genotype and phenotype in this system. Recent work in one of our laboratories has taken advantage of this approach to generate recombinant strains from parents that have different phenotypes in vitro. Jeffrey and colleagues first identified parental strains that had differences in inclusion fusion, attachment efficiency, and secondary inclusion formation phenotype (Jeffrey and Rockey, unpublished). These characterized strains were then used to generate recombinants, and phenotypic differences among progeny were determined. The genomes of selected recombinant strains were then sequenced by next-generation sequencing technologies, and a bioinformatics approach was used to correlate phenotypic difference with genotype. Different sets of genetic loci that were associated either with attachment efficiency of the LGV strains or with the formation of secondary inclusions were discovered in these studies.

A parallel approach is being undertaken (Nguyen and Valdivia, 2012). These investigators are using a chemical mutagenesis strategy to introduce lesions in the chlamydial genome, followed by phenotypic characterization of resulting mutants (e.g., slower growth rate or changes in glycogen accumulation). Lateral gene transfer is then used to confirm the allele responsible for the phenotype, by examining recombinant strains for linkage between transferred phenotype and specific genetic lesions.

Introduction of Plasmid DNA into Chlamydiae

The role of recombination in lateral gene transfer is also evident in the first example of stable introduction of exogenous DNA into chlamydiae in a laboratory setting. Binet and Maurelli targeted the single rRNA operon in *Chlamydia psittaci* 6BC by using electroporation to introduce a shuttle plasmid containing a mutated 16S rRNA allele. This allele had been engineered by sequential selection of *C. psittaci* for resistance to the antibiotics kasugamycin and spectinomycin (Binet and Maurelli, 2009). The doubly resistant strain contained two nucleotide substitutions separated by 398 bp within the 16S rRNA sequence (Fig. 3). The use of a multiple selection strategy was important, as doubly resistant random mutant strains were virtually undetectable in untransformed, wild-type *C. psittaci*. This strongly reduced the possibility that false positives were generated in the subsequent transformation experiments.

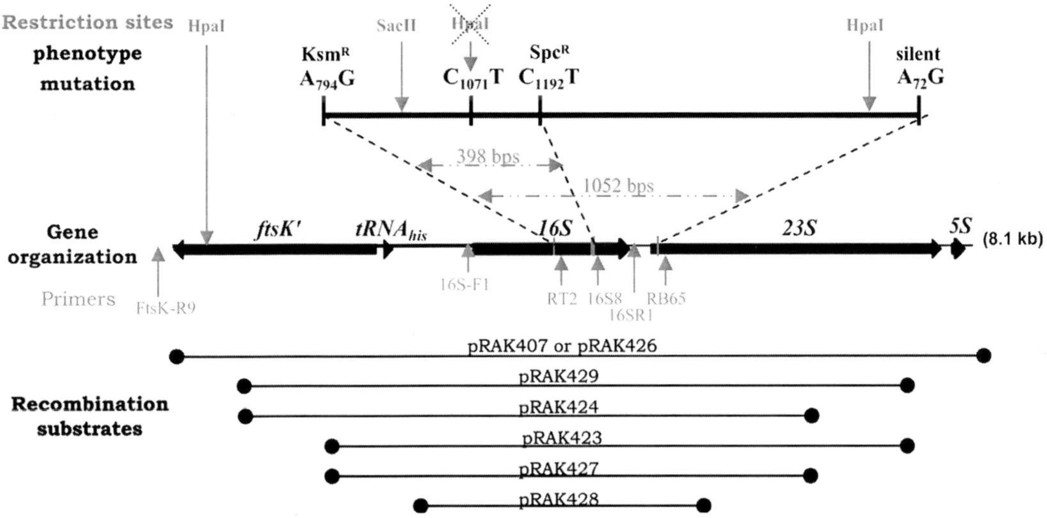

FIGURE 3 Gene organization and plasmid constructs for the transformation vectors used by Binet and Maurelli. Shown are the regions of the ribosomal operons that were present in each of the transformation vectors used in the experiments and the base pair changes in the 16S and 23S rRNA genes that were used as markers in candidate transformants. Kasugamycin (Ksm) resistance was generated by a base pair change at position 794, and spectinomycin (Spc) resistance was associated with a change at position 1192. Two silent mutations were also present in some of the transformation vectors used by these authors. Note that one of the silent mutations (position 1071) allowed the screening of candidate transformants by the removal of an HpaI site in PCR products. The primers used to screen for positives are also indicated. Used by permission of *Proceedings of the National Academy of Sciences USA*. doi:10.1128/9781555817329.ch15.f3

Following this selection process, a set of pUC-based plasmids that contained different lengths of 16S rRNA sequence together with flanking regions from the doubly resistant *C. psittaci* strain were created. The largest of these also contained a silent mutation in a fragment of the 23S rRNA gene included in the construct. A final manipulation introduced an additional silent mutation in the sequence, which eliminated a restriction site and was useful in subsequent analyses of transformed strains. At the end of the process, a set of plasmids that contained sequences with up to four mutations spread out over 1,052 bp of the ribosomal operon was generated.

Electroporation conditions were optimized, leading to conditions that maximized the amount of DNase-resistant vector against a background of minimal killing of chlamydiae. Transformation and selection followed, with a combination of kasugamycin and spectinomycin added at 18 h postinfection with transformed EBs. A total of 329 chlamydial transformants were plaque isolated and characterized by PCR. Sequence analysis and subsequent Southern blotting were conducted for a subset of these clones. The data supported the conclusion that the resistant clones had acquired appropriate base changes from the transforming plasmid, including both the mutations required for resistance and the additional silent mutations that were introduced into the plasmid. All recombinants were stably resistant via integration of target sequences into the chromosome of the

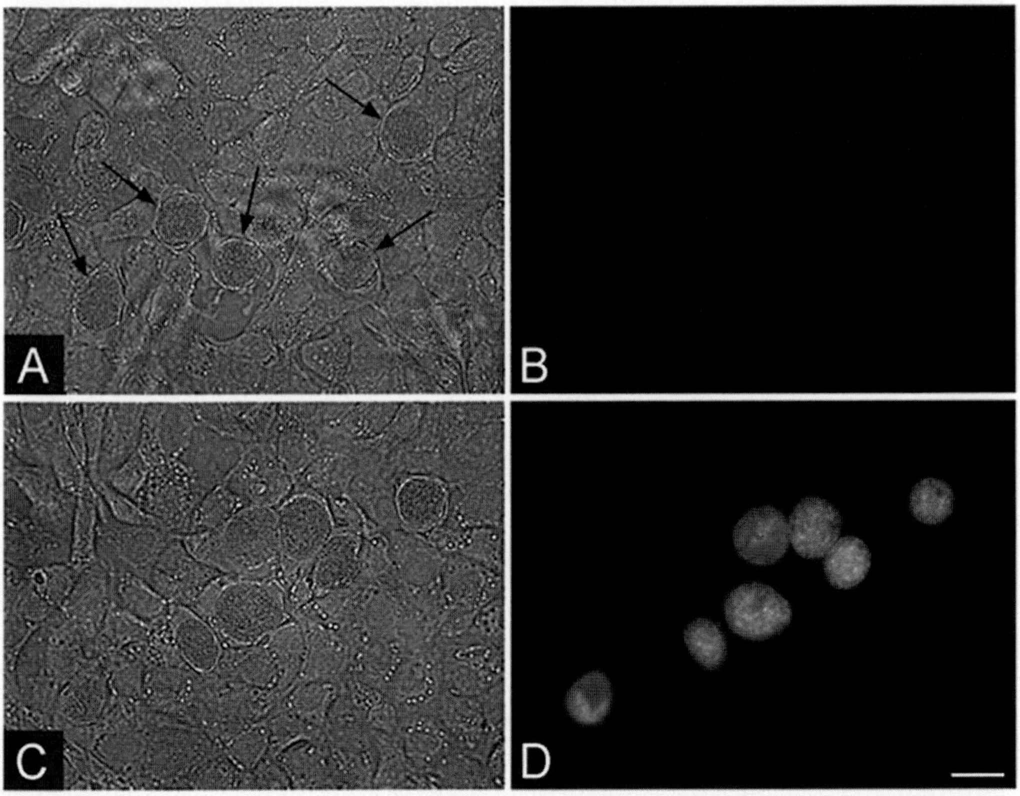

FIGURE 4 Expression of the green fluorescent protein in cells transformed with a plasmid carrying *gfp*. Untransformed *C. trachomatis* L2/434/Bu (control) and *C. trachomatis* L2/434/Bu transformed by pGFP::SW2 were grown on coverslips for two days prior to paraformaldehyde fixation and visualization by fluorescence microscopy. Panel A shows untransformed *C. trachomatis* L2/434/Bu under white light (arrows indicate inclusions). Panel B is the same image, viewed in a fluorescence channel for visualization of GFP. Panel C shows *C. trachomatis* L2/434/Bu transformed with plasmid pGFP::SW2 under white light, and panel D is the same field viewed by fluorescence. The scale bar in panel D represents 20 μm.

recipient bacteria, via homologous recombination. The Southern blots also demonstrated that the pUC sequences were not integrated into the genome, nor were they maintained as episomes in the target chlamydiae.

Toward the end of production of this chapter, Ian Clarke and Simona Kahane and collaborators published a report describing the first stable transformation of chlamydiae by an exogenous plasmid, leading to replication of antibiotic-resistant and green fluorescent protein-expressing *C. trachomatis* (Wang et al., 2011). These researchers constructed hybrid shuttle plasmids containing the chlamydial plasmid from *C. trachomatis* serovar L2 and a pair of different plasmid constructs from Gram-negative bacteria. The first plasmid, pBR325::L2, was created by fusing the chlamydial plasmid with pBR325, which contains a penicillin resistance gene. A construct was generated by ligating pBR325 cleaved just upstream of the tetracycline resistance gene within *orf1* of the chlamydial plasmid. A second construct, pGFP::SW2, contained a high-copy origin of replication (pUC-based), the beta-lactamase gene, and the green fluorescent protein gene under the control of a *Neisseria meningitidis* promoter, ligated within *orf1* of a variant of the chlamydial plasmid. In each case, the constructs were created in *E. coli* and the resulting plasmids were mixed with *C. trachomatis* serovar L2 EBs in a calcium chloride buffer prior to addition of host cells and incubation. Transformed chlamydiae were cultured for several cycles, selecting for resistance to penicillin after the first passage. These authors reported stable transformation to beta-lactam resistance, documented by growth and by Southern blotting for plasmid sequences, and the remarkable result of stable expression of the green fluorescent protein as visualized by green fluorescent inclusions (Fig. 4). The constructs were confirmed by subsequent sequence analysis of recovered plasmids in *E. coli*. The investigators also reported the reacquisition of glycogen accumulation in inclusions of *C. trachomatis* that had previously been cured of its plasmid. This finding adds classical genetic strength to the established premise that the plasmid is directly associated with glycogen accumulation by these organisms.

This is clearly an outstanding result, built upon work by this group and others over several decades. There is much room for discussion of why this work was successful when other similar projects have been less so. Notwithstanding, these early results may finally open the door to the application of standard genetic analyses in the study of chlamydial biology. It is anticipated that many investigators will work to repeat these studies and expand on them in the future.

CONCLUSIONS AND PERSPECTIVES

The described studies demonstrate that chlamydiae can share DNA, can be transformed, and can incorporate introduced DNA into the genome via homologous recombination. The challenge ahead is to develop experimental methods for genetically manipulating chlamydiae so that the functions of individual genes of interest can be elucidated. The procedure developed by Kari and colleagues will be useful for such experiments, and it certainly has tremendous potential in the study of chlamydial biology. The approach is, however, limited by its labor-intensive nature and its current restriction to the inactivation of only nonessential genes. The work of Binet and Maurelli (Binet and Maurelli, 2009) demonstrates that chlamydial EBs can acquire DNA and this DNA can be integrated into the chromosome. The work of Dugan and colleagues demonstrated that chlamydiae can acquire exogenous DNA naturally (Dugan et al., 2007), but efforts designed to exploit this in the laboratory have not been successful. The compelling and rather surprising evidence for abundant and frequent homologous recombination between strains in vivo gives hope to individuals working in this area, but there remains no understanding of how DNA is shared between chlamydiae. Genomics has been a key approach in developing an understanding of the processes involved and will continue to have a significant role. And, of course, the recent work of Wang and colleagues (Wang et al., 2011) has finally opened the door to apparently routine genetic modification of chlamydiae, which will clearly change the way researchers conduct experiments and do research in this system. Thus, decades of hard work have led to recent successes, and we may now have a light to the future for the study of these challenging and important pathogens.

ACKNOWLEDGMENTS

The Rockey laboratory is supported by PHS awards AI069214 and AI086469. The Maurelli laboratory is supported by AI044033 and AI08044. Figures 2 and 3 are used with permission of the *Proceedings of the National Academy of Sciences* and the authors. Laszlo Kari is acknowledged for help in the production of Fig. 2.

REFERENCES

Albrecht, M., C. M. Sharma, R. Reinhardt, J. Vogel, and T. Rudel. 2010. Deep sequencing-based discovery of the *Chlamydia trachomatis* transcriptome. *Nucleic Acids Res.* **38:**868–877.

Andersen, A. A., and D. G. Rogers. 1998. Resistance to tetracycline and sulfadiazine in swine *C. trachomatis* isolates, p. 313–316. *In* R. S. Stephens, G. I. Byrne, I. N. Clarke, et al. (ed.), *Chlamydial Infections: Proceedings of the 9th International Symposium on Human Chlamydial Infection*, Napa, California, June 1998. International Chlamydia Symposium, San Francisco, CA.

Barry, C. E., III, T. J. Brickman, and T. Hackstadt. 1993. Hc1-mediated effects on DNA structure: a potential regulator of chlamydial development. *Mol. Microbiol.* **9:**273–283.

Beare, P. A., D. Howe, D. C. Cockrell, A. Omsland, B. Hansen, and R. A. Heinzen. 2009. Characterization of a *Coxiella burnetii ftsZ* mutant generated by Himar1 transposon mutagenesis. *J. Bacteriol.* **191:**1369–1381.

Beare, P. A., K. M. Sandoz, A. Omsland, D. D. Rockey, and R. A. Heinzen. 2011. Advances in genetic manipulation of obligate intracellular bacterial pathogens. *Front. Microbiol.* **2:**97.

Belland, R. J., D. E. Nelson, D. Virok, D. D. Crane, D. Hogan, D. Sturdevant, W. L. Beatty, and H. D. Caldwell. 2003. Transcriptome analysis of chlamydial growth during IFN-gamma-mediated persistence and reactivation. *Proc. Natl. Acad. Sci. USA* **100:**15971–15976.

Binet, R., and A. T. Maurelli. 2009. Transformation and isolation of allelic exchange mutants of *Chlamydia psittaci* using recombinant DNA introduced by electroporation. *Proc. Natl. Acad. Sci. USA* **106:**292–297.

Brinkman, F. S., J. L. Blanchard, A. Cherkasov, Y. Av-Gay, R. C. Brunham, R. C. Fernandez, B. B. Finlay, S. P. Otto, B. F. Ouellette, P. J. Keeling, A. M. Rose, R. E. Hancock, S. J. Jones, and H. Greberg. 2002. Evidence that plant-like genes in *Chlamydia* species reflect an ancestral relationship between *Chlamydiaceae*, cyanobacteria, and the chloroplast. *Genome Res.* **12:**1159–1167.

Brunham, R., C. Yang, I. Maclean, J. Kimani, G. Maitha, and F. Plummer. 1994. *Chlamydia trachomatis* from individuals in a sexually transmitted disease core group exhibit frequent sequence variation in the major outer membrane protein (*omp1*) gene. *J. Clin. Investig.* **94:**458–463.

Burall, L. S., A. Rodolakis, A. Rekiki, G. S. Myers, and P. M. Bavoil. 2009. Genomic analysis of an attenuated *Chlamydia abortus* live vaccine strain reveals defects in central metabolism and surface proteins. *Infect. Immun.* **77:**4161–4167.

Caldwell, H. D., H. Wood, D. Crane, R. Bailey, R. B. Jones, D. Mabey, I. Maclean, Z. Mohammed, R. Peeling, C. Roshick, J. Schachter, A. W. Solomon, W. E. Stamm, R. J. Suchland, L. Taylor, S. K. West, T. C. Quinn, R. J. Belland, and G. McClarty. 2003. Polymorphisms in *Chlamydia trachomatis* tryptophan synthase genes differentiate between genital and ocular isolates. *J. Clin. Investig.* **111:**1757–1769.

Chen, D., L. Lei, C. Lu, A. Galaleldeen, P. J. Hart, and G. Zhong. 2010. Characterization of *pgp3*, a *Chlamydia trachomatis* plasmid-encoded immunodominant antigen. *J. Bacteriol.* **192:**6017–6024.

Clifton, D. R., C. A. Dooley, S. S. Grieshaber, R. A. Carabeo, K. A. Fields, and T. Hackstadt. 2005. Tyrosine phosphorylation of the chlamydial effector protein Tarp is species specific and not required for recruitment of actin. *Infect. Immun.* **73:**3860–3868.

Collingro, A., P. Tischler, T. Weinmaier, T. Penz, E. Heinz, R. C. Brunham, T. D. Read, P. M. Bavoil, K. Sachse, S. Kahane, M. G. Friedman, T. Rattei, G. S. Myers, and M. Horn. 2011. Unity in variety—the pan-genome of the chlamydiae. *Mol. Biol. Evol.* **28:**3253–3270. doi:10.1093/molbev/msr161.

Demars, R., and J. Weinfurter. 2008. Interstrain gene transfer in *Chlamydia trachomatis* in vitro: mechanism and significance. *J. Bacteriol.* **190:**1605–1614.

Demars, R., J. Weinfurter, E. Guex, J. Lin, and Y. Potucek. 2007. Lateral gene transfer in vitro in the intracellular pathogen *Chlamydia trachomatis*. *J. Bacteriol.* **189:**991–1003.

Di Francesco, A., M. Donati, M. Rossi, S. Pignanelli, A. Shurdhi, R. Baldelli, and R. Cevenini. 2008. Tetracycline-resistant *Chlamydia suis* isolates in Italy. *Vet. Rec.* **163:**251–252.

Dugan, J., A. A. Andersen, and D. D. Rockey. 2007. Functional characterization of IScs605, an insertion element carried by tetracycline-resistant *Chlamydia suis*. *Microbiology* **153:**71–79.

Dugan, J., D. D. Rockey, L. Jones, and A. A. Andersen. 2004. Tetracycline resistance in *Chlamydia suis* mediated by genomic islands inserted into the chlamydial *inv*-like gene. *Antimicrob. Agents Chemother.* **48:**3989–3995.

Fehlner-Gardiner, C., C. Roshick, J. H. Carlson, S. Hughes, R. J. Belland, H. D. Caldwell, and G. McClarty. 2002. Molecular basis defining human *Chlamydia trachomatis* tissue tropism. A possible role for tryptophan synthase. *J. Biol. Chem.* **277:**26893–26903.

Gieffers, J., R. J. Belland, W. Whitmire, S. Ouellette, D. Crane, M. Maass, G. I. Byrne, and H. D. Caldwell. 2002. Isolation of *Chlamydia pneumoniae* clonal variants by a focus-forming assay. *Infect. Immun.* **70:**5827–5834.

Gomes, J. P., W. J. Bruno, M. J. Borrego, and D. Dean. 2004. Recombination in the genome of *Chlamydia trachomatis* involving the polymorphic membrane protein C gene relative to *ompA* and evidence for horizontal gene transfer. *J. Bacteriol.* **186:**4295–4306.

Gomes, J. P., W. J. Bruno, A. Nunes, N. Santos, C. Florindo, M. J. Borrego, and D. Dean. 2007. Evolution of *Chlamydia trachomatis* diversity occurs by widespread interstrain recombination involving hotspots. *Genome Res.* **17:**50–60.

Gomes, J. P., A. Nunes, W. J. Bruno, M. J. Borrego, C. Florindo, and D. Dean. 2006. Polymorphisms in the nine polymorphic membrane proteins of *Chlamydia trachomatis* across all serovars: evidence for serovar Da recombination and correlation with tissue tropism. *J. Bacteriol.* **188:**275–286.

Greub, G., F. Collyn, L. Guy, and C. A. Roten. 2004. A genomic island present along the bacterial chromosome of the *Parachlamydiaceae* UWE25, an obligate amoebal endosymbiont, encodes a potentially functional F-like conjugative DNA transfer system. *BMC Microbiol.* **4:**48.

Hackstadt, T., W. Baehr, and Y. Ying. 1991. *Chlamydia trachomatis* developmentally regulated protein is homologous to eukaryotic histone H1. *Proc. Natl. Acad. Sci. USA* **88:**3937–3941.

Hackstadt, T., T. J. Brickman, C. E. Barry III, and J. Sager. 1993. Diversity in the *Chlamydia trachomatis* histone homologue Hc2. *Gene* **132:**137–141.

Hackstadt, T., M. A. Scidmore-Carlson, E. I. Shaw, and E. R. Fischer. 1999. The *Chlamydia trachomatis incA* protein is required for homotypic vesicle fusion. *Cell. Microbiol.* **1:**119–130.

Heinzen, R. A., M. A. Scidmore, D. D. Rockey, and T. Hackstadt. 1996. Differential interaction with endocytic and exocytic pathways distinguish parasitophorous vacuoles of *Coxiella burnetii* and *Chlamydia trachomatis*. *Infect. Immun.* **64:**796–809.

Horn, M., A. Collingro, S. Schmitz-Esser, C. L. Beier, U. Purkhold, B. Fartmann, P. Brandt, G. J. Nyakatura, M. Droege, D. Frishman, T. Rattei, H.-W. Mewes, and M. Wagner. 2004. Illuminating the evolutionary history of chlamydiae. *Science* **304:**728–730.

Hsia, R. C., L. M. Ting, and P. M. Bavoil. 2000. Microvirus of *Chlamydia psittaci* strain guinea pig inclusion conjunctivitis: isolation and molecular characterization. *Microbiology* **146:**1651–1660.

Hueck, C. J. 1998. Type III protein secretion systems in bacterial pathogens of animals and plants. *Microbiol. Mol. Biol. Rev.* **62:**379–433.

Jeffrey, B. M., R. J. Suchland, K. L. Quinn, J. R. Davidson, W. E. Stamm, and D. D. Rockey. 2010. Genome sequencing of recent clinical *Chlamydia trachomatis* strains identifies loci associated with tissue tropism and regions of apparent recombination. *Infect. Immun.* **78:**2544–2553.

Kari, L., M. M. Goheen, L. B. Randall, L. D. Taylor, J. H. Carlson, W. M. Whitmire, D. Virok, K. Rajaram, V. Endresz, G. McClarty, D. E. Nelson, and H. D. Caldwell. 2011. Generation of targeted *Chlamydia trachomatis* null mutants. *Proc. Natl. Acad. Sci. USA* **108:**7189–7193.

Karunakaran, K. P., J. F. Blanchard, A. Raudonikiene, C. Shen, A. D. Murdin, and R. C. Brunham. 2002. Molecular detection and seroepidemiology of the *Chlamydia pneumoniae* bacteriophage (PhiCpn1). *J. Clin. Microbiol.* **40:**4010–4014.

Kim, J. F. 2001. Revisiting the chlamydial type III protein secretion system: clues to the origin of type III protein secretion. *Trends Genet.* **17:**65–69.

L'Abee-Lund, T. M., and H. Sorum. 2002. A global non-conjugative Tet C plasmid, pRAS3, from *Aeromonas salmonicida*. *Plasmid* **47:**172–181.

Lampe, M. F., R. J. Suchland, and W. E. Stamm. 1993. Nucleotide sequence of the variable domains within the major outer membrane protein gene from serovariants of *Chlamydia trachomatis*. *Infect. Immun.* **61:**213–219.

Lau, S. K., G. K. Wong, M. W. Li, P. C. Woo, and K. Y. Yuen. 2008. Distribution and molecular characterization of tetracycline resistance in *Laribacter hongkongensis*. *J. Antimicrob. Chemother.* **61:**488–497.

Lenart, J., A. A. Andersen, and D. D. Rockey. 2001. Growth and development of tetracycline-resistant *Chlamydia suis*. *Antimicrob. Agents Chemother.* **45:**2198–2203.

Mabey, D. 2008. Trachoma: recent developments. *Adv. Exp. Med. Biol.* **609:**98–107.

McCoy, A. J., N. E. Adams, A. O. Hudson, C. Gilvarg, T. Leustek, and A. T. Maurelli. 2006. L,L-Diaminopimelate aminotransferase, a trans-kingdom enzyme shared by *Chlamydia* and plants for synthesis of diaminopimelate/lysine. *Proc. Natl. Acad. Sci. USA* **103:**17909–17914.

Millman, K. L., S. Tavare, and D. Dean. 2001. Recombination in the *ompA* gene but not the *omcB* gene of *Chlamydia* contributes to serovar-specific differences in tissue tropism, immune surveillance, and persistence of the organism. *J. Bacteriol.* **183:**5997–6008.

Nguyen, B. D., and R. H. Valdivia. 2012. Virulence determinants in the obligate intracellular pathogen *Chlamydia trachomatis* revealed by forward genetic approaches. *Proc. Natl. Acad. Sci. USA* **109**:1263–1268.

Nunes, A., P. J. Nogueira, M. J. Borrego, and J. P. Gomes. 2008. *Chlamydia trachomatis* diversity viewed as a tissue-specific coevolutionary arms race. *Genome Biol.* **9:**R153.

O'Connell, C. M., and K. M. Nicks. 2006. A plasmid-cured *Chlamydia muridarum* strain displays altered plaque morphology and reduced infectivity in cell culture. *Microbiology* **152:**1601–1607.

O'Connell, C. M. C., and A. T. Maurelli. 1998. Introduction of foreign DNA into *Chlamydia* and stable expression of chloramphenicol resistance, p. 519–522. *In* R. S. Stephens, G. I. Byrne, I. N. Clarke, and G. Christiansen (ed.), *Chlamydial Infections: Proceedings of the Ninth International Symposium on Human Chlamydial Infection*. International Chlamydia Symposium, San Francisco, CA.

Olivares-Zavaleta, N., W. Whitmire, D. Gardner, and H. D. Caldwell. 2010. Immunization with the attenuated plasmidless *Chlamydia trachomatis* l2(25667r) strain provides partial protection in a murine model of female genitourinary tract infection. *Vaccine* **28:**1454–1462.

Omsland, A., D. C. Cockrell, E. R. Fischer, and R. A. Heinzen. 2008. Sustained axenic metabolic activity by the obligate intracellular bacterium *Coxiella burnetii*. *J. Bacteriol.* **190:**3203–3212.

Omsland, A., D. C. Cockrell, D. Howe, E. R. Fischer, K. Virtaneva, D. E. Sturdevant, S. F. Porcella, and R. A. Heinzen. 2009. Host cell-free growth of the Q fever bacterium *Coxiella burnetii*. *Proc. Natl. Acad. Sci. USA* **106:**4430–4434.

Perara, E., D. Ganem, and J. N. Engel. 1992. A developmentally regulated chlamydial gene with apparent homology to eukaryotic histone H1. *Proc. Natl. Acad. Sci. USA* **89:**2125–2129.

Read, T. D., R. C. Brunham, C. Shen, S. R. Gill, J. F. Heidelberg, O. White, E. K. Hickey, J. Peterson, T. Utterback, K. Berry, S. Bass, K. Linher, J. Weidman, H. Khouri, B. Craven, C. Bowman, R. Dodson, M. Gwinn, W. Nelson, R. DeBoy, J. Kolonay, G. McClarty, S. L. Salzberg, J. Eisen, and C. M. Fraser. 2000. Genome sequences of *Chlamydia trachomatis* MoPn and *Chlamydia pneumoniae* AR39. *Nucleic Acids Res.* **28:**1397–1406.

Rodolakis, A., and F. Bernard. 1984. Vaccination with temperature-sensitive mutant of *Chlamydia psittaci* against enzootic abortion of ewes. *Vet. Rec.* **114:**193–194.

Roshick, C., H. Wood, H. D. Caldwell, and G. McClarty. 2006. Comparison of gamma interferon-mediated antichlamydial defense mechanisms in human and mouse cells. *Infect. Immun.* **74:**225–238.

Rupp, J., W. Solbach, and J. Gieffers. 2007. Prevalence, genetic conservation and transmissibility of the *Chlamydia pneumoniae* bacteriophage (PhiCpn1). *FEMS Microbiol. Lett.* **273:**45–49.

Russell, M., T. Darville, K. Chandra-Kuntal, B. Smith, C. W. Andrews, Jr., and C. M. O'Connell. 2011. Infectivity acts as in vivo selection for maintenance of the chlamydial cryptic plasmid. *Infect. Immun.* **79:**98–107.

Sauer, J. D., J. G. Shannon, D. Howe, S. F. Hayes, M. S. Swanson, and R. A. Heinzen. 2005. Specificity of *Legionella pneumophila* and *Coxiella burnetii* vacuoles and versatility of *Legionella pneumophila* revealed by coinfection. *Infect. Immun.* **73:**4494–4504.

Setlow, P. 2007. I will survive: DNA protection in bacterial spores. *Trends Microbiol.* **15:**172–180.

Shaw, A. C., K. Gevaert, H. Demol, B. Hoorelbeke, J. Vandekerckhove, M. R. Larsen, P. Roepstorff, A. Holm, G. Christiansen, and S. Birkelund. 2002. Comparative proteome analysis of *Chlamydia trachomatis* serovar A, D and L2. *Proteomics* **2:**164–186.

Stemple, D. L. 2004. TILLING—a high-throughput harvest for functional genomics. *Nat. Rev. Genet.* **5:**145–150.

Stephens, R. S., S. Kalman, C. Lammel, J. Fan, R. Marathe, L. Aravind, W. Mitchell, L. Olinger, R. L. Tatusov, Q. Zhao, E. V. Koonin, and R. W. Davis. 1998. Genome sequence of an obligate intracellular pathogen of humans: *Chlamydia trachomatis*. *Science* **282:**754–759.

Stewart, P. E., J. Hoff, E. Fischer, J. G. Krum, and P. A. Rosa. 2004. Genome-wide transposon mutagenesis of *Borrelia burgdorferi* for identification of phenotypic mutants. *Appl. Environ. Microbiol.* **70:**5973–5979.

Storey, C. C., M. Lusher, and S. J. Richmond. 1989. Analysis of the complete nucleotide sequence of chp1, a phage which infects avian *Chlamydia psittaci*. *J. Gen. Virol.* **70**(Pt. 12)**:**3381–3390.

Subtil, A., A. Blocker, and A. Dautry-Varsat. 2000. Type III secretion system in chlamydia species: identified members and candidates. *Microbes Infect.* **2:**367–369.

Suchland, R. J., L. O. Eckert, S. E. Hawes, and W. E. Stamm. 2003. Longitudinal assessment of infecting serovars of *Chlamydia trachomatis* in Seattle public health clinics: 1988–1996. *Sex. Transm. Dis.* **30:**357–361.

Suchland, R. J., B. M. Jeffrey, M. Xia, A. Bhatia, H. G. Chu, D. D. Rockey, and W. E. Stamm. 2008. Identification of concomitant infection with *Chlamydia trachomatis incA*-negative mutant and wild-type strains by genomic, transcriptional, and biological characterizations. *Infect. Immun.* **76:**5438–5446.

Suchland, R. J., K. M. Sandoz, B. M. Jeffrey, W. E. Stamm, and D. D. Rockey. 2009. Horizontal transfer of tetracycline resistance among *Chlamydia* spp. in vitro. *Antimicrob. Agents Chemother.* **53:**4604–4611.

Suchland, R. J., and W. E. Stamm. 1991. Simplified microtiter cell culture method for rapid immunotyping of *Chlamydia trachomatis. J. Clin. Microbiol.* **29:**1333–1338.

Tam, J. E., C. H. Davis, and P. B. Wyrick. 1994. Expression of recombinant DNA introduced into *Chlamydia trachomatis* by electroporation. *Can. J. Microbiol.* **40:**583–591.

Vidal, L., J. Pinsach, G. Striedner, G. Caminal, and P. Ferrer. 2008. Development of an antibiotic-free plasmid selection system based on glycine auxotrophy for recombinant protein overproduction in *Escherichia coli. J. Biotechnol.* **134:**127–136.

Voth, D. E., P. A. Beare, D. Howe, U. M. Sharma, G. Samoilis, D. C. Cockrell, A. Omsland, and R. A. Heinzen. 2011. The *Coxiella burnetii* cryptic plasmid is enriched in genes encoding type IV secretion system substrates. *J. Bacteriol.* **193:**1493–1503.

Wang, Y., S. Kahane, L. T. Cutcliffe, R. J. Skilton, P. R. Lambden, and I. N. Clarke. 2011. Development of a transformation system for *Chlamydia trachomatis*: restoration of glycogen biosynthesis by acquisition of a plasmid shuttle vector. *PLoS Pathogens* **7:**e1002258.

BIOMATHEMATICAL MODELING OF *CHLAMYDIA* INFECTION AND DISEASE

Andrew P. Craig, Patrik M. Bavoil, Roger G. Rank, and David P. Wilson

16

INTRODUCTION

Biomathematical modeling of infection and disease adds a new dimension of scientific enquiry that has been demonstrated to yield insights into the dynamics, processes, and factors associated with outcomes of biological importance for a variety of systems. There have been an increasing number of experimental studies performed that elicit empirical data on the intracellular development of *Chlamydia*, immune response to associated infection, and consequential pathology of disease. Many of these experiments have been performed in animal models of chlamydial infection or in vitro. Analysis and modeling of the data from these studies provide the opportunity to investigate the mechanisms and implications of chlamydial development, infection, and disease.

Biomathematical modeling of *Chlamydia* is still in its infancy, but foundations have been laid over the last decade. For the biologist, mathematical models can be useful because they can be used to test hypotheses and conduct in silico experiments—which are usually much faster and considerably less expensive to perform than "real" experiments. The results of these in silico experiments can be used to inform further biological research. For the biomathematician, *Chlamydia* research can be an exciting area. There are many areas of *Chlamydia* research in which biomathematical modeling has the potential to make an important contribution. These include exploring hypotheses around the intracellular development of chlamydiae, understanding infection dynamics and interaction with the immune response, gaining insight into the emergence of pathology, and informing the development of vaccines and the mitigation of serious consequences of infection.

In this chapter we begin by reviewing the mathematical models of *Chlamydia* that have been published. We then outline a system of dynamical equations that can be used as a governing framework for examining many aspects of *Chlamydia* infection and pathology, including their calibration to experimental data and insights gained. We then discuss in greater detail some of the more advanced and influential models. We also consider the likely future of biomathematical modeling of *Chlamydia*.

Andrew P. Craig, The Kirby Institute, University of New South Wales, Sydney, NSW 2010, Australia. *Patrik M. Bavoil*, Dept. of Microbial Pathogenesis, University of Maryland, Baltimore, MD 21201. *Roger G. Rank*, Chlamydia Research Group, Arkansas Children's Hospital Research Institute, Little Rock, AR 72202. *David P. Wilson*, The Kirby Institute, University of New South Wales, Sydney, NSW 2010, Australia.

Intracellular Pathogens I: Chlamydiales
Edited by Ming Tan and Patrik M. Bavoil © 2012 ASM Press, Washington, DC
10.1128/9781555817329.ch16

REVIEW OF MATHEMATICAL MODELING OF *CHLAMYDIA*

The field of chlamydial biomathematical modeling has been developing over the past decade and has started to make contributions to the understanding of chlamydial development, infection, and disease. Aspects of relevant biomathematical papers published to date are summarized in Table 1.

The standard initial approach to exploration of the dynamics of infectious agents involves a system of ordinary differential equations (ODEs) that define the numbers of healthy uninfected target cells, infected cells, and infective bodies (Nowak and May, 2000). These equations form the core of some of the published mathematical models of infection dynamics of *Chlamydia*. For example, previously used ODEs have been adapted to *Chlamydia*, and additional features such as a Th1 immune response have been added to the mathematical model of the developmental cycle. Such a model has been used to demonstrate how the rate of emergence of aberrant reticulate bodies (aRBs) is related to the production of overall infectious chlamydiae (Wilson et al., 2003). The model also predicts that the longer the development cycle, the greater the ease with which the cellular immune response can control infections.

Chlamydia infection dynamics have been modeled in space as well as time, by combining partial differential equations (PDEs) of the infection dynamics with a cellular automata grid in order to represent space (Mallet et al., 2009). From this, it was determined that the rate of passive movement of free elementary bodies (EBs) is likely to be the most important factor in deciding whether or not the infection would spread to the upper genital tract. This model was extended by using PDEs in time and two spatial dimensions (M. B. Oskouei, D. G. Mallet, A. Amirshahi, and G. J. Pettet, presented at the Proceedings of the World Congress on Engineering and Computer Science 2010, San Francisco, CA, USA, 20 to 22 October 2010).

Chlamydial infection is important because it can lead to severe pathology. Using experimental data on *Chlamydia caviae* infection in guinea pigs, modeling has been used to predict pathological outcomes based on the time course of inclusion-forming unit (IFU) production, which is a measure of how many infectious progeny are generated (Wilson et al., 2009a). An optimal threshold of IFU-days (i.e., the cumulative area under the curve of a plot of IFU against days) beyond which clear pathology emerged was found. A model function for the IFU time course was defined and matched to the data, and this was used to estimate changes in pathology that would result from different IFU time courses. It was estimated that a 2-$\log_{10}$ reduction in the peak IFU load reduces the chance of observing pathology from 81% to 32% and reduces the chance of observing serious pathology from 33% to 2%.

Epidemiological Modeling

While it is not the main subject of this chapter, epidemiological modeling also has relevance to *Chlamydia* research because of the insights it can provide about vaccine development and immunity. Furthermore, modeling of the infection in an individual host can inform population level models. For example, a population level model that includes an infection time course of chlamydial IFU production showed that the IFU level was the key determinant of the probability of sexual transmission between individuals (Gray et al., 2009). The introduction of prophylactic and therapeutic chlamydial vaccines into the population was then simulated, and the expected population level impact was measured. The model predicts that a vaccine that raises the initial IFU load threshold at which an individual becomes susceptible to infection would be the most effective vaccine.

An unexpected interaction of immunology with a population level *Chlamydia* treatment program was seen in British Columbia, Canada. Based on a rebound in the number of diagnoses of *Chlamydia trachomatis* genital infection after the introduction of scaled up testing and treatment programs, Brunham and

TABLE 1 Summary of the published mathematical models of *Chlamydia* in terms of common features

Model	Reference	ODE based	Explicitly models EB-RB-EB change	Incorporates aRBs	Models pathology	Considers location of infection	Incorporates gene expression
Th1 immune response model	Wilson et al., 2003	•	•	•			
Fab/host cell receptor model	Wilson and McElwain, 2004	•				•	
Gene expression generation time model	Wilson et al., 2004	•	•				•
Chlamydia viral dynamics ODE model	Wilson, 2004	•					
Chlamydia intracellular model	Wilson, 2004	•	•	•			
Contact-dependent T3S model	Wilson et al., 2006	•	•			•	
Delay model	Burns et al., 2007	•	•				
Contact-dependent T3S model extension	Hoare et al., 2008	•	•	•		•	
Cellular automata model	Mallet et al., 2009	•	•			•	
Ocular pathology model	Wilson et al., 2009a				•		
Spatial PDE model	Oskouei et al., presented	•				•	
Viral dynamics ODEs with intermediate stages	Sharomi and Gumel, 2010	•	•				
CD4+ T cell and antibody response model	Vickers et al., 2009	•					

colleagues proposed the arrested immunity hypothesis: that early treatment prevents the building of immunity to *Chlamydia* and thus renders the treated individual more susceptible to reinfection (Brunham and Rekart, 2008). This hypothesis was supported by earlier population level mathematical modeling (Brunham et al., 2005). While the model structure has been debated by the biomathematical modeling community, the underlying concept of arrested immunity was well elucidated through the use of this methodology, and future mathematical modeling in tandem with laboratory research should shed further light on this phenomenon. The clinical and public health aspects of chlamydial infection are discussed in more detail in chapter 1, "*Chlamydia* infection and epidemiology."

MODELING *CHLAMYDIA* WITH VIRAL DYNAMICS EQUATIONS

Chlamydia infection can be considered as having three key components: (i) the bacterium, (ii) the host target cell, and (iii) the host immune response to the bacterium. In order to understand the dynamics of the host-pathogen interaction in *Chlamydia*-infected individuals, it is valuable to develop a quantitative understanding of the various components and their interactions. For example, how effectively and by what mechanism does the cell-mediated immune response control infection? How much of the innate immune response is required to control the initial growth, and how does it act? Establishing a formal system of equations has the benefit of transparency in assumptions, clearly defined factors that can be investigated without confounders, and powerful mathematical tools that can process the equations to gain insights. Mathematical analyses generally take two forms: (i) numerical solution and plotting of the variables in the system; and (ii) formal mathematical (algebraic or calculus-based) manipulation of the equations to elucidate relationships between the parameters and properties of the system.

ODEs have been widely used to describe the dynamics of viral infections, such as HIV. *Chlamydia* is a bacterium, but as an obligate intracellular parasite it has many features in common with viruses: replication takes place only in a host cell, and each round of intracellular infection involves the invasion of a new cell, a lag time for the production of infectious progeny, and then release of the next generation of chlamydiae that can infect another cell in the same individual or transmit the infection to a new individual. Thus, equations that have been used to model viral infections can be readily adapted for use in *Chlamydia* research.

Basic Equations of *Chlamydia* Dynamics

In many ecological systems, there are intriguing predator-prey relationships whereby both predator and prey evolve and their population levels are dictated by the survival of the other species: if there are large numbers of predators, then the number of prey will decrease, which will not be sufficient to sustain the large numbers of predators, so there will be a decrease in their numbers, which subsequently allows the prey population to increase. Predator-prey relationships were described in a quantitative framework with systems of differential equations (Lotka, 1925; Volterra, 1926). These Lotka-Volterra equations have been used extensively in mathematical biology. The host-pathogen interaction is another example of a system that can be considered in a predator-prey framework; e.g., the presence of a pathogen stimulates the generation of an immune response, which controls the pathogen, leading to decreased stimuli for maintaining the same level of immune activation. This simple concept was extended in the application of mathematical modeling to viral infections (Nowak and May, 2000), in what essentially involved adaptations and extensions of the simple predator-prey model. The host-pathogen theoretical framework commences with consideration of three key interacting populations of cells: uninfected target cells, infected cells, and infective bodies.

To describe the basic model mathematically, we define the number of uninfected target cells in a closed region to be denoted by $E(t)$, the number of infected cells to be $I(t)$, and the number of infective bodies (chlamydial EBs measured as IFUs) to be $C(t)$. We denote the rate of production of new uninfected target cells as π, and their natural rate of death in the absence of infection is denoted by δ. Target cells become infected at a rate dependent on the level of chlamydial IFUs, C, and the infectivity (κ) of the infective bodies for the target cells; then, the number of target cells to become infected per unit time can be represented by the mathematical form of κEC. Infected cells are defined to burst (releasing new infective bodies) at a rate of γI and to be removed by the immune system at a rate of ωI. Each bursting infected cell releases N new infective bodies, so new infective bodies are produced at a rate that is dependent on the number of infected cells: $N\gamma I$. These infective bodies are defined to be cleared by the innate immune system or through natural death at a per-particle rate of μ; as chlamydiae must survive away from the site of infection in order to be able to infect other individuals, we suspect that the rate of loss of EBs due to the innate immune system is much higher than the rate of loss of EBs due to natural death. According to these definitions and assumed interactions, the three populations can be described in a system of mathematical equations, specifically, three ODEs (one for each cell population):

$$\overbrace{\frac{dC}{dt}}^{\text{rate of change in number of new infective bodies}} = \overbrace{N\gamma I}^{\text{new infective bodies released}} - \overbrace{\mu C}^{\text{infective bodies die}} - \overbrace{\kappa CE}^{\text{infection}}$$

$$\overbrace{\frac{dE}{dt}}^{\text{rate of change in number of uninfected target cells}} = \overbrace{\pi}^{\text{new healthy cells}} - \overbrace{\delta E}^{\text{natural death}} - \overbrace{\kappa CE}^{\text{infection}}$$

$$\overbrace{\frac{dI}{dt}}^{\text{rate of change in number of infected target cells}} = \overbrace{\kappa CE}^{\text{infection}} - \overbrace{\gamma I}^{\text{infected cells burst}} - \overbrace{\omega I}^{\text{innate immune response}}$$

This system of equations is similar to the basic mathematical model of viral dynamics and is the foundation for understanding the dynamics of *Chlamydia* infections.

DEVELOPING THE EQUATIONS FURTHER FOR *CHLAMYDIA*

Understanding the dynamics of chlamydial infections must start with observation and the measurement of time courses of infections in vivo (see Fig. 1 for an observed IFU time course). If a biomathematical model is to add any value to understanding of the infection dynamics, then it must start by reproducing the observed data. In Fig. 2, we show model-based curves that were generated using the basic equations of *Chlamydia* dynamics, fit to the experimental data shown in Fig. 1; it can be seen that the mathematical model closely matches the data at days 3 and 6, but where the data show a resolving infection the model predicts a nonresolving infection. There is a good reason that the basic equations do not capture this feature of the time course, namely, that it does not adequately represent the adaptive immune response, which is known to start to exert an effect at around day 6 or 7 (the IFU time course predicted by the model resembles IFU time courses observed in RAG$^{-/-}$ mice (R. G. Rank and U. M. Nagarajan, unpublished data), which have no adaptive immune response [Fig. 3]).

Therefore, it is important to extend the basic equations of *Chlamydia* dynamics to incorporate the adaptive immune response. Mathematically, we denote the rates at which the adaptive

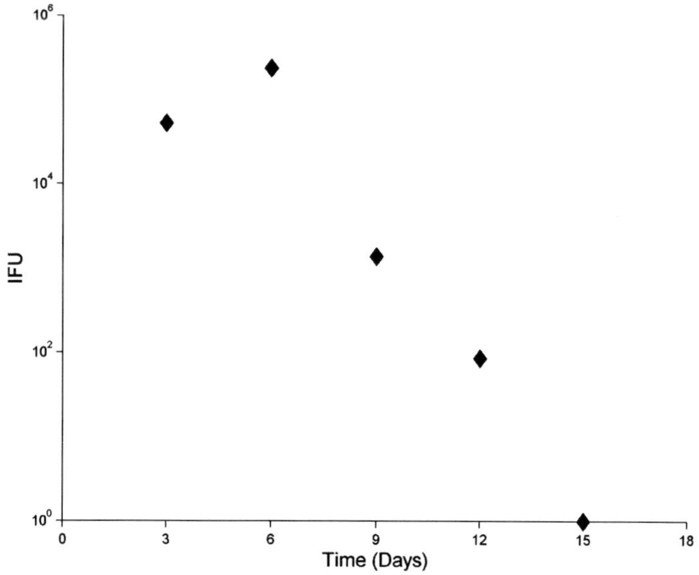

FIGURE 1 Experimental IFU measurements. Mean IFU measurements from swabs taken from guinea pigs inoculated in the eye with 10^3 IFUs of *C. caviae* (Rank and Maurelli, unpublished).
doi:10.1128/9781555817329.ch16.f1

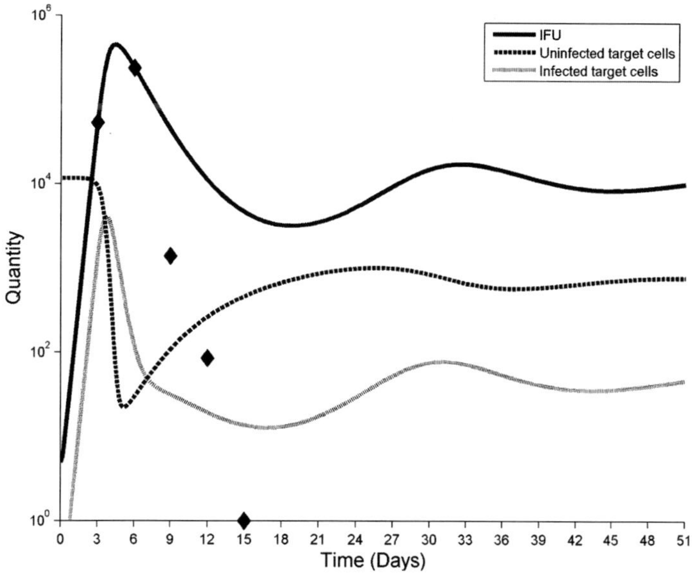

FIGURE 2 Comparison of data and basic model. The data from Fig. 1 are contrasted with curves produced via the basic equations of *Chlamydia* dynamics (without adaptive immune response). Target cell quantities are the numbers of target cells in the area corresponding to the swab from which IFU load was measured. Parameters used were as follows: $\pi = 100$, $\delta = 0.0086$, $\kappa = 1.26 \times 10^{-5}$, $\gamma = 0.5$, $N = 300$, $\omega = 1.549$, $\mu = 0.637$, and $C = 5$ at time $t = 0$.
doi:10.1128/9781555817329.ch16.f2

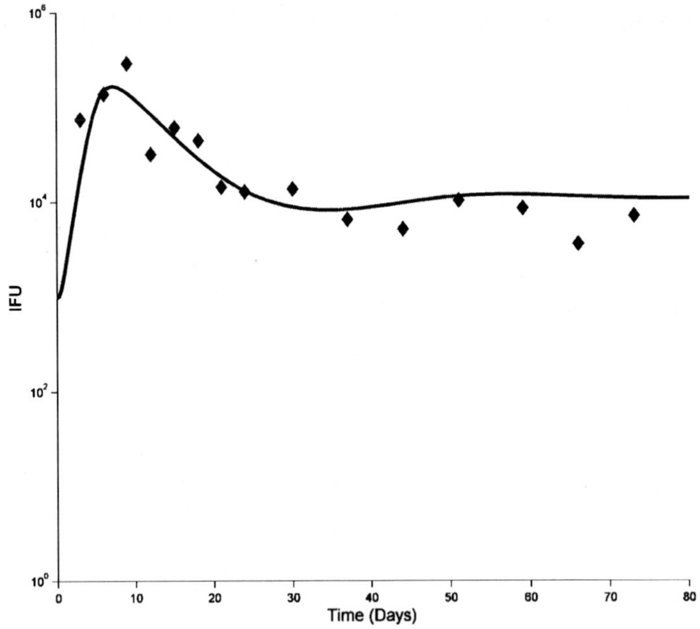

FIGURE 3 Data and model without adaptive immune response. The curve was produced using the basic equations of *Chlamydia* dynamics contrasted with geometric means of IFU measurements from swabs taken from RAG$^{-/-}$ mice inoculated intravaginally with 10^6 IFUs of *Chlamydia muridarum*. RAG$^{-/-}$ mice have no adaptive immune response. Model parameters are as follows: $\pi = 12.1$, $\delta = 0.0086$, $\kappa = 1.28 \times 10^{-5}$, $\gamma = 0.5$, $N = 300$, $\omega = 0.155$, $\mu = 0.235$, and $C = 1{,}000$ at time $t = 0$.
doi:10.1128/9781555817329.ch16.f3

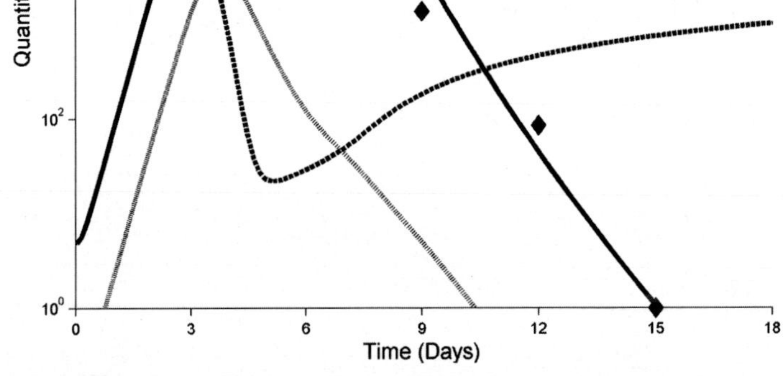

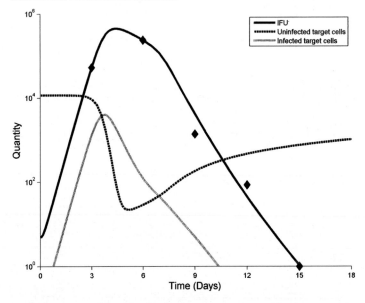

FIGURE 4 Model with adaptive immune response. Shown are curves produced by the basic model of *Chlamydia* dynamics when it included terms for the adaptive immune response. Target cell quantities are the numbers of target cells in the area corresponding to the swab from which IFU load was measured. Parameters used were $\upsilon = 1.2$ and $\rho = 1$, with the other parameters the same as those used in Fig. 2.
doi:10.1128/9781555817329.ch16.f4

immune response kills infected cells and IFUs as ρaI and υaC, respectively, where $a(t)$ varies over time. The equations of *Chlamydia* dynamics then become:

$$\overbrace{\frac{dC}{dt}}^{\text{rate of change in number of new infective bodies}} = \overbrace{N\gamma I}^{\text{new EBs released}} - \overbrace{\mu C}^{\text{EBs die}} - \overbrace{\kappa CE}^{\text{infection}} - \overbrace{\upsilon aC}^{\text{adaptive immune}}$$

$$\overbrace{\frac{dI}{dt}}^{\text{rate of change in number of infected target cells}} = \overbrace{\kappa CE}^{\text{infection}} - \overbrace{\gamma I}^{\text{infected cells burst}} - \overbrace{\omega I}^{\text{innate immune}} - \overbrace{\rho aI}^{\text{adaptive immune}}$$

For approximately the first 6 days, $a(t)$ should be close to zero and then subsequently increase until it is close to 1 by day 8, indicating the maximal immune response; this could be represented by a logistic function such as the following:

$$a(t) = \frac{1}{1 + e^{-4(t-7)}}$$

With this addition of including an adaptive immune response, the model of *Chlamydia* dynamics now reproduces the original data well (Fig. 4).

PARAMETER CHOICES AND ANALYSIS OF THE MODEL

All models require estimates of parameter values (e.g., the natural half-life of an IFU and the duration of the developmental cycle). The purpose of some models is to estimate these very parameters, whereas other models require knowledge or assumptions about parameter values in order to make inferences. *Chlamydia* is studied in a number of animal models, including mice, guinea pigs, and nonhuman primates. In each of these animal models, infection with different strains of *Chlamydia* can and has been investigated, and there are different sites at which infection can occur (conjunctiva, cervix, etc.). Different combinations of animal, infection site, and chlamydial strain produce different—sometimes very different—kinetics. For this reason, parameter choices must be application specific. Model calibration can also be a time-consuming process. Future models should build on the work conducted previously by others. Some of the key parameter estimates used in previous modeling studies are presented in Table 2.

Some parameter values can be assigned based on experimental measurements, such as N, the number of new EBs released by an infected cell through lysis or exocytosis. Properties represented by other parameters are more difficult to observe directly; for example, κ represents the average infectivity rate of each EB (the probability of infecting a target cell per day). To estimate these parameters, mathematical optimization techniques can be used to determine the values that allow the model to predict time courses which most closely match the observed data (as was done to produce the time courses shown in Fig. 2 through 4).

Once the parameters of a model have been established, the model can be used to investigate "what if" scenarios; that is, what does the model predict when specific parameters are altered? In the following section, the parameters that were used to model the time course of an ocular chlamydial infection of guinea pigs (Fig. 4) will be varied or modified in order to gain insights into the kinetics of this chlamydial infection.

RELATIVE CONTRIBUTIONS OF EARLY IFU DYNAMICS

During the early stages of chlamydial infection, prior to a noticeable adaptive immune

TABLE 2 Nonexhaustive lists of key parameters used in published models

Model	Publication	Strain	Burst size	Burst rate	Epithelial cell regeneration	Avg life span of healthy epithelial cells	Doubling time of intracellular RBs
Th1 immune response model	Wilson et al., 2003	*C. trachomatis*	200				
Fab/host cell receptor model	Wilson and McElwain, 2004	*C. trachomatis*	200	0.6/day	40/mm^3/day	0.5 days	
Chlamydia viral dynamics ODE model	Wilson, 2004	General	200–500	0.33–0.6/day	40/mm^3/day	0.5 days	
Contact-dependent T3S model	Wilson et al., 2006	General	100–600				
Contact-dependent T3S model extension	Hoare et al., 2008	General			36–62.4/day		1.5–2.6 h
Cellular automata model	Mallet et al., 2009	*C. caviae*	400–800	Mean: 1/day			1–1.5 h
Spatial PDE model	Oskouei et al., presented	*C. trachomatis*	350	0.45/day		3.3 days	
Viral dynamics ODEs with intermediate stages	Sharomi et al., 2010	*C. trachomatis*	200–500	1/day	24/day	0.4 days	
CD4+ T cell and antibody response model	Vickers et al., 2009	General			1.54/day	0.01/day	

response, chlamydial EBs (IFUs) may enter target cells or die (naturally or because of the innate immune response). Which of these factors dominates the dynamics of IFU numbers? The calibrated mathematical model of *Chlamydia* dynamics was used to determine the values of the rate of IFU infection of target cells ($-\kappa CE$) and the rate of IFU death ($-\mu C$) at various times along the course of the infection. It was determined that the rate of IFU death contributes between 81 and 100% of the combined effect of the two terms. That is, the rate of (natural and innate-immune-induced) death of IFU is always at least 4 times as large as the rate of infectivity in terms of its effect on the IFU dynamics in ocular infection of guinea pigs with *C. caviae*. This result suggests that many more EBs are killed by the innate immune system or natural death than successfully infect cells. These inferences are made solely on the basis of overall IFU levels. Nonetheless, these qualitative conclusions about the relative contributions of factors to early dynamics will hold true even if the quantitative estimates can be refined by further analysis.

VARYING INFECTIVITY

The survival of *Chlamydia* over multiple developmental cycles critically depends on its infectiousness or its ability to enter a target cell. Therefore, we investigated the importance of the infectivity on the infection time course (by varying the parameter κ) using the guinea pig model of conjunctival infection by *C. caviae*. In Fig. 5 we show the expected IFU time courses predicted by the model when the average chlamydial infectivity is doubled or halved. We observe that the peak IFU loads are similar despite different initial growth rates. Doubling the infectivity rate results in a peak IFU load that occurs around 1.5 days earlier but that is only 7% higher. Halving the infectivity rate results in a peak that is around 2 days later and is 23% lower. According to the model, peak IFU levels occur prior to the onset of the adaptive immune response, possibly due to the exhaustion of healthy target cells. The greater reduction in peak IFU associated with a lower infectivity rate is likely to be due to the inability of chlamydiae to replicate fast enough, before the adaptive immune response starts to clear the infection. Simulations involving removal of the adaptive immune response from the model in the situation of reduced (halved) infectivity rate resulted in the peak IFU level still occurring during day 6 but being 10% higher. It seems that the availability of healthy target cells still has a key part in determining the dynamics of the infection.

A modeling study found a predictive relationship between the emergence of pathology and the area under the IFU curve (Wilson et al., 2009a). The total area for the normal curve is 1.20×10^6 IFU-days; when the infectivity is doubled, it is 1.30×10^6; and when the infectivity is halved, it is 7.41×10^5. It was found that a cumulative IFU-days threshold of $\sim 4 \times 10^5$ was associated with the emergence of pathology for guinea pigs infected in the conjunctiva with *C. caviae*; we can apply this result to the model to estimate when pathology would be observed for our different infectivity values. For the baseline scenario, the threshold is passed during day 4. When the infectivity rate is doubled, the threshold is passed during day 3, and when the infectivity rate is halved, the threshold is passed during day 6. Thus, we can estimate how much earlier pathology might occur for an infection with a particularly virulent strain of *Chlamydia*, or how much later it might occur if an intervention reduces infectivity.

The model can also be used to make inferences about how the relationships between the numbers of healthy cells and infected cells vary over the course of the infection (Fig. 6). For both the baseline scenario and the scenario with increased average infectivity, the model simulations suggest that there is an exhaustion of healthy target cells, with the percentage of uninfected target cells decreasing as low as 3% for the baseline case and as low as 1% of initial levels when the average infectivity rate is increased twofold. However, when the average infectivity rate is halved, the percentage of uninfected target cells reaches a minimum of 31%.

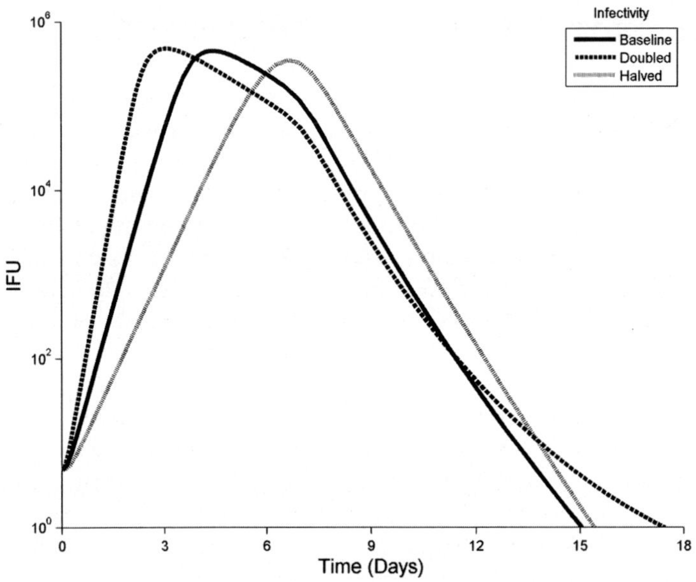

FIGURE 5 Predicted effect of varying infectivity. IFU time courses were produced by varying the infectivity parameter, κ. The effect on the relative timing of the peaks is greater than the effect on the peaks' magnitudes. The baseline value of κ is 1.26×10^{-5}. All other parameters are the same as those used in Fig. 4. doi:10.1128/9781555817329.ch16.f5

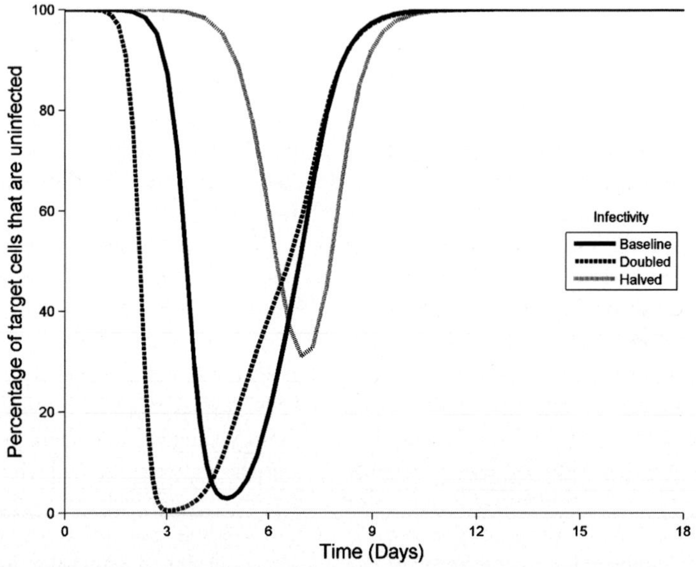

FIGURE 6 Target cell percentages when infectivity is varied. The percentages of target cells that are infected during the infections are shown in Fig. 5. doi:10.1128/9781555817329.ch16.f6

This observation that the availability of target cells may affect the course of chlamydial infection in vivo has not been previously noted but has important implications for understanding the pathogenesis of chlamydial disease. For example, if chlamydiae exhaust the number of target cells in a given site such as the cervix, there would be a selective advantage if the infection could spread to new sites, such as the endometrium and fallopian tubes. A potential mechanism in which the infection can be spread by *Chlamydia*-infected cells that are shed from the epithelium by neutrophils is described in chapter 13, "In vivo chlamydial infection."

VARYING THE INNATE IMMUNE RESPONSE

Very relevant to vaccine development is the question of how boosting the immune system can affect infection dynamics. How would the IFU time course differ if the rate of onset or magnitude of the innate immune response was increased? The model can be used to explore the effects of the innate immune response by adjusting the parameters μ and ω (μ also includes natural EB death, but we assume innate-immune-induced death dominates over the natural life span of EBs). In Fig. 7 we present the time courses produced by doubling or removing the innate immune response. The model predicts that doubling the innate immune response would lead to an estimated reduction in the peak IFU level of approximately 65% (although the peak is actually predicted to occur later, due to the uninfected target cells being exhausted more slowly). Removing the innate immune response increases the peak IFU load five to sixfold, and the infection no longer resolves. Thus, the innate immune response is clearly vital for the resolution of ocular *C. caviae* infection in guinea pigs. Furthermore, according to the predictive pathology model, if the adaptive immune response is doubled in magnitude, then it is possible that the overall

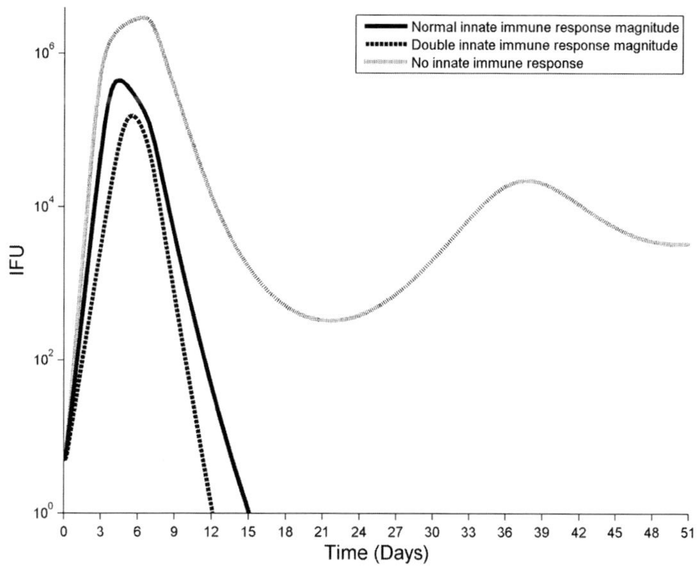

FIGURE 7 Predicted effect of varying innate immunity. Shown are curves produced by the model when the innate immune response is varied. For the normal innate immune response, ω = 1.549 and μ = 0.637; for the doubled innate immune response magnitude, $\omega = 2 \times 1.549$ and $\mu = 2 \times 0.637$; and the curve for no innate immune response is produced with ω = 0 and μ = 0 (other parameters are the same as those used in Fig. 4). doi:10.1128/9781555817329.ch16.f7

area under the IFU curve would not reach the identified threshold associated with the emergence of pathology for guinea pigs infected with *C. caviae* (Wilson et al., 2009a).

VARYING THE ADAPTIVE IMMUNE RESPONSE

Although the innate response is important, it is the adaptive immune response that is vital to the resolution of chlamydial infection (Morrison et al., 2000). However, it may act too slowly to prevent some pathology in the baseline situation. The model simulations suggest that by doubling the magnitude of the adaptive immune response (represented by parameters ρ and υ), there would be no discernable difference in the IFU time course until day 6, but then there would be a more rapid resolution of infection; the infection resolves on day 11 instead of day 15 (Fig. 8).

To quantify the respective rates of IFU decline, we can estimate the IFU half-lives in the presence of the immune response by calculating the points of steepest descent on the IFU curves predicted by the model. For the normal IFU time course, the greatest half-life is 9.7 hours, and when the magnitude of the adaptive immune response is doubled, this changes to 5.8 hours.

We can also model the expected effect of a vaccine that causes the adaptive immune response to activate on day 1, instead of around day 6. (We accomplish this by changing $a(t)$ in the model.) As one would expect, making this change results in the model predicting lower IFU loads and proportions of target cells infected over the course of the infection (results not shown).

The potential effect of such a vaccine was investigated experimentally by infecting with *C. caviae* guinea pigs that had already been immunized with UV-inactivated chlamydiae (Rank et al., 1990). The percentage of sampled target cells infected was observed to be lower in these guinea pigs than in unimmunized, control guinea pigs, and this is

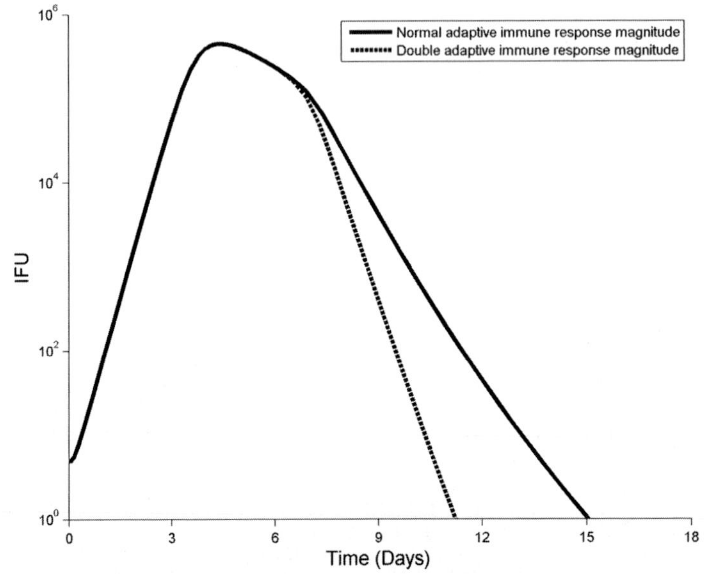

FIGURE 8 Predicted effect of varying adaptive immunity. The IFU curve from Fig. 4 compared to the IFU time course when the magnitude of the adaptive immune response is doubled (i.e., by setting υ = 2.4 and ρ = 2); other parameters for both curves are the same as those used in Fig. 4. doi:10.1128/9781555817329.ch16.f8

attributed to the faster activation of the active immune response in the immunized guinea pigs. The degree of difference in peak percentages of target cells infected in the immunized guinea pigs and in the control guinea pigs was similar to the difference predicted by the mathematical model.

Further mathematical modeling of this sort could contribute to research into vaccine design by estimating the benefits of different immunization strategies, allowing future experiments to target those strategies predicted to be the most effective.

The Basic Reproduction Number and Other Properties of the System

An important concept in mathematical modeling of diseases is the basic reproduction number. The basic reproduction number is defined to be the average number of new cases (infected cells) an existing case (infected cell) will generate when a pathogen is first introduced into a naïve population. If the basic reproduction number is less than unity, then the infection will die out and not be sustained. However, if the basic reproduction number is greater than unity, then the infection will be sustained and continue to grow. In the context of the system of equations presented above, the basic reproduction number (denoted by R_0) is the average number of new infected cells generated as a result of a single infected cell when all of the other cells are uninfected. Ignoring the effect of the adaptive immune system, which is close to zero initially, the value of R_0 is given by the following mathematical expression:

$$R_0 = \frac{N}{\left(1+\frac{\omega}{\gamma}\right)\left(1+\frac{\mu}{\kappa E_0}\right)}$$

where $E_0 = \pi/\delta$ is the number of healthy target cells at time $t = 0$. The basic reproduction number depends on all parameter rates in the system (other than the adaptive immune response parameters ρ and υ), and it will increase with infectivity rates (κ), target cell regeneration (π), and rate of production of new EBs (N and γ) and will decrease with the rate of removal of infected cells (ω), uninfected target cells (δ), and EBs (μ). If $R_0 < 1$, each infected cell will lead to less than one new infected cell on average and the infection will die out. If $R_0 < 1$, then each infected cell will produce more than one new infected cell on average and the infection will be viable.

As the adaptive immune response increases, the reproduction number tends towards:

$$R_a = \frac{N}{\left(1+\frac{\omega+\rho}{\gamma}\right)\left(1+\frac{\mu+\upsilon}{\kappa E_0}\right)}$$

As with R_0, if R_a is <1, each EB will infect less than one healthy cell and the infection will resolve. In the absence of an adaptive response, the populations of cells would converge to the following equilibrium levels:

$$\bar{C} = \frac{\pi\left[(N-1)\gamma-\omega\right]}{\mu(\omega+\gamma)} - \frac{\delta}{\kappa}$$

$$\bar{E} = \frac{\mu(\omega+\gamma)}{\kappa\left[(N-1)\gamma-\omega\right]}$$

$$\bar{I} = \frac{\pi}{\omega+\gamma} - \frac{\delta\mu}{\kappa\left[(N-1)\gamma-\omega\right]}$$

IFU time courses for various R_0 and R_a values are shown in Fig. 9. The concept of the basic reproduction number is an important one. Some cases of *Chlamydia* do indeed resolve naturally, while others do not, so we do see in practice a range of R_0 and R_a values around the critical threshold level. The basic reproduction numbers are particularly relevant to explore thresholds of immune responses required to control infections. If either of R_0 and R_a is reduced below 1, then the infection will resolve. Both R_0 and R_a are determined by several parameters, and each of these can

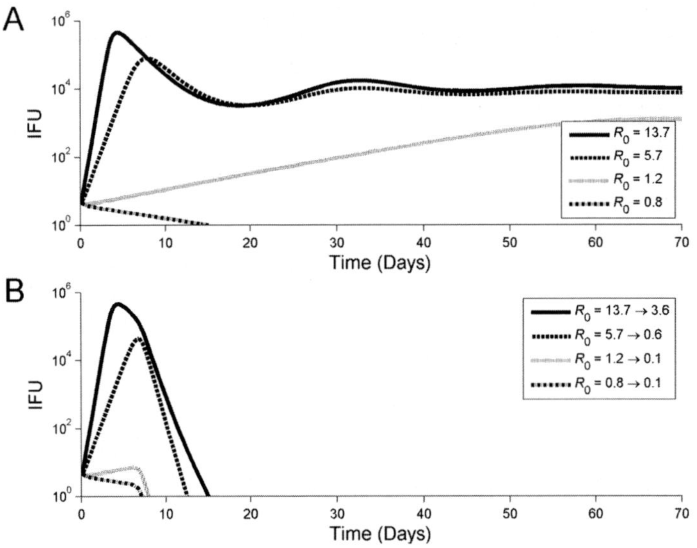

FIGURE 9 Time courses for different R_0 values. (A) Theoretical IFU time courses associated with different R_0 values when the adaptive immune response is not considered. Note that time course with $R_0 < 1$ resolves, while the time courses with $R_0 > 1$ do not resolve. (B) IFU time courses when the adaptive immune response is included; note how the reproduction number changes with all other factors remaining unchanged. There are initially 5 IFUs for all time courses. For the time courses with $R_0 = 13.7$, $\pi = 100$, $\delta = 0.0086$, $\kappa = 1.26 \times 10^{-5}$, $\gamma = 0.5$, $N = 300$, $\omega = 1.549$, and $\mu = 0.637$ (for panel B, $\upsilon = 1.2$ and $\rho = 1$). For the time courses with $R_0 = 5.7$, $\pi = 80$, $\delta = 0.03$, $\kappa = 2 \times 10^{-5}$, $\gamma = 0.5$, $N = 300$, $\omega = 1.549$, and $\mu = 0.637$ (for panel B, $\upsilon = 2.4$ and $\rho = 2$). For the time courses with $R_0 = 1.2$, $\pi = 80$, $\delta = 0.03$, $\kappa = 1.26 \times 10^{-5}$, $\gamma = 0.5$, $N = 100$, $\omega = 1.549$, and $\mu = 0.637$ (for panel B, $\upsilon = 2.4$ and $\rho = 2$). For the time courses with $R_0 = 0.8$, $\pi = 100$, $\delta = 0.0086$, $\kappa = 1.26 \times 10^{-6}$, $\gamma = 0.5$, $N = 150$, $\omega = 1.549$, and $\mu = 0.637$ (for panel B, $\upsilon = 2.4$ and $\rho = 2$). doi:10.1128/9781555817329.ch16.f9

be changed to lower the reproduction numbers. Determining how much of an effect each component of the system would be required to have on respective parameters can be determined in order to decrease R_0 and R_a below unity. As informed by the availability of data, these equations can also be extended to include many additional complexities and variables of different types of cells, cytokines, and chemokines.

Competing *Chlamydia* Infections

In any real-world infection there are likely to be populations of genetic variants. One can construct a biomathematical model of two strains of *Chlamydia* competing in the same infection by keeping separate counts of the numbers of free EBs and infected cells for each strain. This can then be extrapolated to more than two strains. For two different strains, A and B, we let $C_A(t)$ and $C_B(t)$ be their respective numbers of free EBs and $I_A(t)$ and $I_B(t)$ be the respective numbers of infected cells. Each strain has its own version of each parameter. A model to represent the competing strains is as follows:

$$\frac{dC_A}{dt} = \overbrace{N_A\gamma_A I_A}^{\text{new EBs released}} \overbrace{-\mu_A C_A}^{\text{EBs die}} \overbrace{-\kappa_A C_A E}^{\text{infection}} - \overbrace{\upsilon_A a C_A}^{\text{adaptive immune}}$$

$$\frac{dC_B}{dt} = \overbrace{N_B\gamma_B I_B}^{\text{new EBs released}} \overbrace{-\mu_B C_B}^{\text{EBs die}} \overbrace{-\kappa_B C_B E}^{\text{infection}} - \overbrace{\upsilon_B a C_B}^{\text{adaptive immune}}$$

$$\frac{dE}{dt} = \overbrace{\pi}^{\text{new healthy cells}} - \overbrace{\delta E}^{\text{natural death}} - \overbrace{\kappa_A C_A E}^{\text{infection by A}} - \overbrace{\kappa_B C_B E}^{\text{infection by B}}$$

$$\frac{dI_A}{dt} = \overbrace{\kappa_A C_A E}^{\text{infection}} - \overbrace{\gamma_A I_A}^{\text{infected cells burst}} - \overbrace{\omega_A I_A}^{\text{innate immune}} - \overbrace{\rho_A a I_A}^{\text{adaptive immune}}$$

$$\frac{dI_B}{dt} = \overbrace{\kappa_B C_B E}^{\text{infection}} - \overbrace{\gamma_B I_B}^{\text{infected cells burst}} - \overbrace{\omega_B I_B}^{\text{innate immune}} - \overbrace{\rho_B a I_B}^{\text{adaptive immune}}$$

Using these equations as a starting point, we investigated experimental data from guinea pigs with ocular infections caused by wild-type (A) and azithromycin-resistant mutant strains (B) of *C. caviae* (R. G. Rank and A. T. Maurelli, unpublished data).

In the experiment, one group of guinea pigs was inoculated with only the wild-type strain, one group with only the mutant, and one group was coinoculated with both strains. In order to reflect the different levels of immune response in each group, an additional constant parameter, ϕ, the measured immune response, was incorporated into the equations for infected cells:

$$\frac{dI_A}{dt} = \overbrace{\kappa_A C_A E}^{\text{infection}} - \overbrace{\gamma_A I_A}^{\text{infected cells burst}} - \overbrace{\phi\,\omega_A I_A}^{\text{innate immune}} - \overbrace{\phi\,\rho_A a I_A}^{\text{adaptive immune}}$$

$$\frac{dI_B}{dt} = \overbrace{\kappa_B C_B E}^{\text{infection}} - \overbrace{\gamma_B I_B}^{\text{infected cells burst}} - \overbrace{\phi\,\omega_B I_B}^{\text{innate immune}} - \overbrace{\phi\,\rho_B a I_B}^{\text{adaptive immune}}$$

The value of ϕ for each group was set to the interleukin-8 measurement, an important chemokine for neutrophil recruitment, averaged over all measurements at all time points for all animals in that group. We also assume that $\omega_A/\rho_A = \omega_B/\rho_B$ and the average number of EBs released when an infected cell bursts was $N_A = 300$ for the wild-type strain, and we set N_B so that the ratio of N_A to N_B was the same as the ratio of the EB generation rate (fold increase) of the wild-type compared to the mutant found in an in vitro experiment (Binet et al., 2010). Similarly, we assumed that $\gamma_A = 0.5$ and set γ_B so that the ratio of γ_A to γ_B was the same as the ratio of the eclipse times noted in vitro (Binet et al., 2010). We also assumed that $\kappa_A \geq \kappa_B$, and that $\mu_A = \mu_B$. The model was optimized to the data to find values for the parameters; the curves produced by these parameters can be seen in Fig. 10.

Having fit the equations to the data, the model system can be analyzed more thoroughly. First, we can quantify the impact of the immune response on the two strains: it is approximately 10-fold greater for the wild-type than for the mutant strain. Also, despite the similar peak IFU loads, the model predicts that the population of infected target cells reaches a higher peak in the mutant-only infection than it does in the wild-type-only infection. This may be due to the more aggressive action of the immune system on the cells infected with wild-type chlamydiae, resulting in infected cells being cleared before they can accumulate. However, it should be noted that in both monoinoculations the number of uninfected target cells drops by around 98%, resulting in a shortage of target cells.

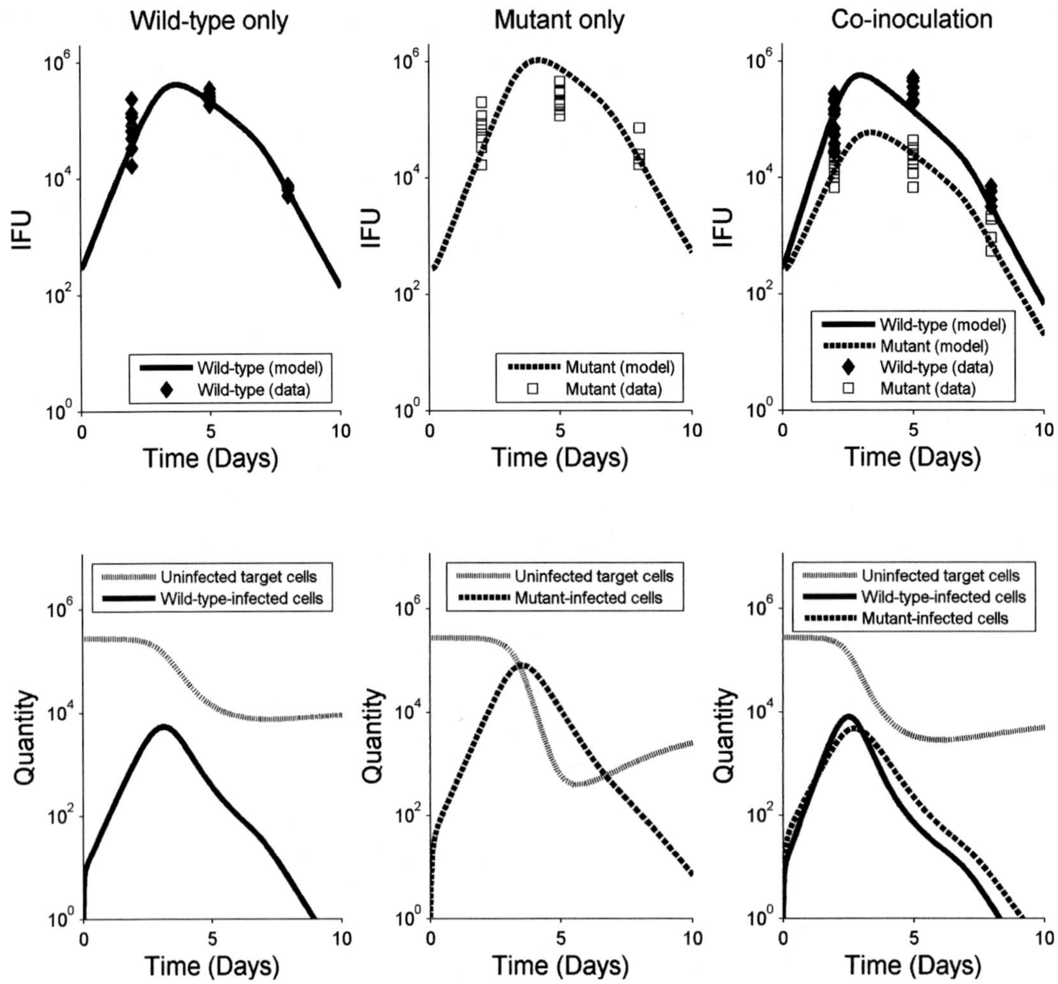

FIGURE 10 Two chlamydial strains in competition. Each graph in the upper row compares IFU time course for model simulation against *C. caviae* ocular infection of guinea pigs for mono- or coinfection of wild-type (SP_6) and mutant azithromycin-resistant (SP_6AZ_2) strains. In the lower graphs, the predicted numbers of target cells are shown. Peak numbers of infected cells are highest in the mutant-only infection, but in all cases the number of target cells drops by at least 97%. The model was first optimized to the data for days 2 and 5 without considering the adaptive immune response, and then those terms were introduced and the model was further optimized to fit the data for day 8. The number of IFUs at time $t = 0$ was 300 in each of the wild-type-only and mutant-only infections, and there were 300 IFUs of each strain for the coinfection. For the wild-type-only infection, $\phi = 296.2$; for the mutant-only infection, $\phi = 158.4$; for the coinfection, $\phi = 242.5$. Other parameters used were $\pi = 681$, $\delta = 0.0025$, $\kappa_A = \kappa_B = 3.61 \times 10^{-6}$, $\gamma_A = 0.5$, $\gamma_B = 0.432$, $N_A = 300$, $N_B = 50.2$, $\omega_A = 0.101$, $\omega_B = 0.0101$, $\mu = 0.96$, $\upsilon = 1.06$, $\rho_A = 0.0058$, and $\rho_B = 5.75 \times 10^{-4}$. doi:10.1128/9781555817329.ch16.f10

The model predicts a higher peak of target cells infected by the wild-type strain in a coinfection than in an infection with the wild-type strain by itself. This is probably due to the lower measured immune response in the coinfection group resulting in a lower clearance rate of infected cells than in the wild-type monoinoculation group. In contrast, the population of target cells infected by the mutant strain in a coinfection has a lower peak than it does in a monoinfection due to the competition for uninfected target cells.

It is apparent in clinical infections that not every individual develops the same degree of pathology. While there are likely multiple reasons for this, including the genetic makeup of the host, it is also possible that competition among variants in the infecting chlamydial population may in part affect the degree of pathology. For example, if the dominant variant has a faster growth rate and elicits a more vigorous inflammatory response, there is likely to be more-severe disease. However, a dominant variant that has a fast growth rate but lacks the ability to elicit a strong inflammatory response may result in less disease. Moreover, an important corollary to this is that competition among variants may result in shifts in the representative populations within the same host or through transmission to new hosts.

OTHER MATHEMATICAL MODELS OF *CHLAMYDIA*

Theoretical frameworks for modeling the infection dynamics of *Chlamydia* have been defined above to describe the relationships between the number of free extracellular EBs, the number of infected cells, and the number of uninfected cells. In this section, other models that use different paradigms to model particular aspects of the kinetics of *Chlamydia* infection are described in further detail.

T3S-Mediated Contact-Dependent Hypothesis Model

A major contribution of biomathematical modeling to *Chlamydia* biology was driven by a biological hypothesis that was not immediately amenable to experimental testing: the type III secretion (T3S)-mediated contact-dependent hypothesis of chlamydial development, which proposes that the transition from EB to RB and back to EB is mediated by T3S via contact of chlamydiae with the inclusion membrane (Wilson et al., 2006). Hackstadt and colleagues first observed RBs in contact with the inclusion membrane that were continuing to divide while a pool of EBs was present in the interior of the inclusion (Hackstadt et al., 1997). This and similar observations led to the development of the T3S-mediated contact-dependent hypothesis of chlamydial development, which suggests that EB-RB-EB transitions are regulated by T3S injectisomes that have been hypothesized to correspond to surface projections on the surface of the chlamydiae (Bavoil and Hsia, 1998).(See Fig. 11 for a visual summary of the hypothesized process.) The T3S system, which is a protein secretion system that *Chlamydia* shares with several other pathogenic gram-negative bacteria, is discussed in more detail in chapter 9, "Protein secretion and *Chlamydia* pathogenesis." The T3S injectisomes allow chlamydiae to inject effector proteins into the host cell cytosol from within the safe confines of the inclusion, modulating the cell's regular operations in ways that support the intracellular infection. The T3S-mediated contact-dependent hypothesis suggests that for *Chlamydia* these surface projections have evolved to play an additional role in determining EB/RB differentiation. Matsumoto observed that the number of projections decreases as the RBs develop (Matsumoto, 1982), and it is suggested that the strength of RB adhesion to the inclusion membrane decreases as the number of projections decreases (Wilson et al., 2006). According to the contact-dependent hypothesis, when a threshold in T3S activity is reached for a given RB (marked by less contact with the inclusion membrane), it becomes untethered from the inclusion membrane, and T3S-mediated translocation of virulence effectors to their cytosolic targets is disrupted, which is the trigger (or is closely associated with the trigger) for differentiation into an infectious EB.

There have been numerous visual observations to provide support for this hypothesis. However, as the hypothesis cannot be directly and easily tested through experimentation, there was motivation to explore the plausibility of the hypothesis through other means. The motivation to construct a mathematical model was to determine whether the observed

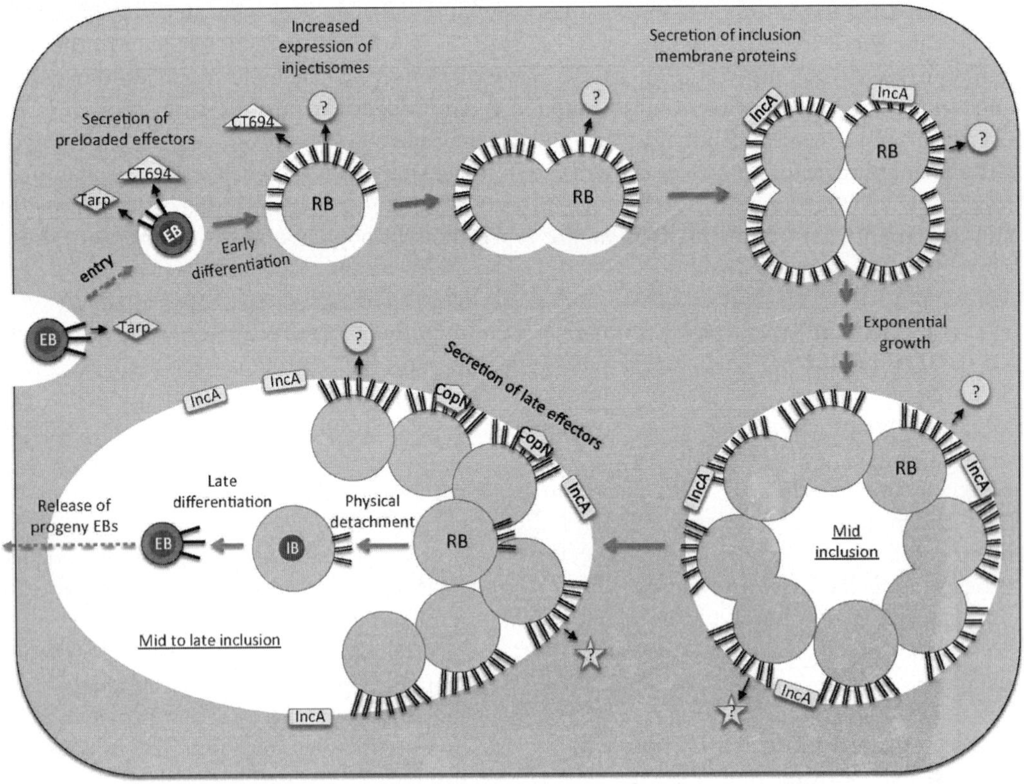

FIGURE 11 The contact-dependent T3S-mediated hypothesis of chlamydial development. The developmental cycle is represented graphically in an infected host cell. Following EB adherence to and entry into a susceptible cell, early differentiation takes place, yielding the first RB. In the contact-dependent hypothesis, at this developmental stage, there is maximum contact between the RB surface and the inclusion membrane surface because the inclusion is small, as well as biosynthesis of additional surface projections (injectisomes). As the developmental cycle progresses, the surface contact between an individual RB and the inclusion membrane is reduced with each round of RB division, and the number of surface projections per RB gradually decreases. By the midstage of the developmental cycle, the RBs are arrayed along, and still in contact with, the inner circumference of the inclusion membrane. With transition to the late stage of the cycle, the surface contact between individual RBs and the inclusion membrane may be reduced beyond a threshold, at which time the RB is no longer able to "stick" and becomes untethered. The contact-dependent hypothesis predicts that physical detachment of a RB in a normally growing inclusion is the trigger, or is closely associated with the trigger, for the late differentiation of the noninfectious RB into the infectious EB. A direct implication of the hypothesis is that the key step in RB-to-EB differentiation is the disruption of the T3S injectisome, whereby effectors are no longer translocated to their cytosolic targets, may accumulate in the chlamydial cytoplasm, or may be secreted into the inclusion lumen. Several T3S effectors (Tarp, CT694, IncA, and CopN) are represented in the figure according to their developmental expression and proposed subcellular location ("?" indicates other effectors). These are described in additional detail in chapter 6, "Initial interactions of chlamydiae with the host cell," and chapter 9, "Protein secretion and *Chlamydia* pathogenesis. (Adapted from Wilson et al., 2006.)
doi:10.1128/9781555817329.ch16.f11

behavior and data can plausibly be produced *as a consequence* of the hypothesis. A mathematical model of the T3S-mediated contact-dependent hypothesis was developed and used to demonstrate that the hypothesis not only is plausible under realistic parameters but also has implications for our understanding of chlamydial development (Wilson et al., 2006).

The mathematical model describes the number of RBs in contact with the inclusion membrane at time *t* postinfection, *R*(*t*); the total number of RBs that have detached from the inclusion membrane in each cell, *I*(*t*); and the total number of EBs in each cell, *E*(*t*). The model applied the following key assumptions of the hypothesis:

1. RBs divide as long as they are in contact with the inclusion membrane.
2. There is a limit to how large an inclusion can grow.
3. Once there is no more space on the inclusion membrane, new RBs will be forced into the lumen of the inclusion.
4. An RB will also lose contact with the inclusion membrane once the number of projections in contact with the membrane goes below a threshold level.
5. RBs out of contact with the inclusion membrane will differentiate into EBs.

The model expressed these "rules" in mathematical terms.

First let us consider the rate of change in the number of RBs, *R*(*t*). RBs on the inclusion membrane divide, but there is a limit to how many can fit on the inside surface of the inclusion membrane (denoted as R_{max}^N). The model assumes that if there is no crowding and plenty of space for growth, there is uninhibited exponential replication, at a per-RB rate, α_0. However, as the number of RBs approaches the maximum number that can physically be in contact with the inclusion membrane, the rate of growth decreases to zero. This is incorporated in the RB growth term as $\alpha_0 R\left(1 - R/R_{max}^N\right)$; when R is low, $\left(1 - R/R_{max}^N\right)$ is close to 1, so the number of RBs will grow exponentially (at rate α_0). As R approaches R_{max}^N, $\left(1 - R/R_{max}^N\right)$ approaches 0, so fewer RBs that are produced can be in contact with the inclusion membrane; instead, they will move to the lumen of the inclusion. This source of detached RBs initiates "intermediate bodies" in the model.

Detachment from the inclusion membrane as the average number of projections decreases, adjusting for volume constraints within the lumen, is modeled by this term:

$$-\frac{\alpha_1 p_{50}^n}{p_{50}^n + p^n(t)} R\left[1 - \frac{V}{V_{max}^N}\right]$$

Here, $p_{50}^n / \left[p_{50}^n + p^n(t)\right]$ represents the average strength of RB membrane attachment due to the average number of projections per chlamydial body, *p*(*t*); the number of projections per RB decreases over time. The parameters, p_{50}^n and $p^n(t)$ are chosen such that the majority of triggering for detachment occurs between 16 and 24 hours postinfection (hpi). The model also makes allowances for the volume of space available in the lumen, *V*, that can be occupied by RBs and EBs, with $V = V_I I + V_E E$, where V_I and V_E are the amount of space in the lumen that each detached RB and EB, respectively, makes unavailable. The inclusion grows to accommodate more chlamydiae as they enter the lumen, but it cannot grow past V_{max}^N; as V approaches V_{max}^N, $\left[1 - V/V_{max}^N\right]$ approaches zero, and so fewer RBs can detach from the inclusion membrane.

Thus, the equation governing the number of RBs, *R*(*t*), is given by:

$$\frac{dR}{dt} = \overbrace{\alpha_0 R\left(1 - \frac{R}{R_{max}^N}\right)}^{\text{RBs on inclusion membrane divide}} \overbrace{- \frac{\alpha_1 p_{50}^n}{p_{50}^n + p^n(t)} R\left[1 - \frac{V}{V_{max}^n}\right]}^{\text{RBs detach from membrane}}$$

The model denotes the number of RBs that have detached from the lumen as *I*(*t*). All RBs lost in the equation for dR/dt are added as a source of detached RBs in the dI/dt equation. Detached RBs are assumed under the hypothesis not to divide, but rather to differentiate into EBs, at a per-chlamydia rate of α_2. The change in the number of detached RBs over time is then given by:

$$\frac{dI}{dt} = N\left(\overbrace{\frac{\alpha_0 R}{R_{max}^N}}^{\text{RBs born into lumen}} + \overbrace{\frac{\alpha_1 p_{50}^n}{p_{50}^n + p^n(t)}}^{\text{RBs detach from membrane}}\right) \times R\left[1 - \frac{V}{V_{max}^N}\right] - \overbrace{\alpha_2 I}^{\text{loose RBs change to EBs}}$$

where N is the number of inclusions in the cell; this is necessary because $R(t)$ is the number of RBs attached to the membrane of each (of possibly multiple) *inclusion(s)*, but $I(t)$ is the total number of detached RBs in each *cell*. Finally, the equation governing the number of EBs in each cell is simply:

$$\frac{dE}{dt} = \overbrace{\alpha_2 I}^{\text{detached RB-to-EB differentiation}}$$

These ODEs represent the conversion into mathematics of the five model rules of the contact-dependent T3S hypothesis. Once appropriate estimates for parameters (such as α_0, p_{50}^n, R_{max}^N, and V_{max}^N) are known, it becomes possible to solve the ODEs numerically and determine whether the levels of RBs and EBs over time match those that are actually observed. If they do, then the T3S-mediated contact-dependent hypothesis remains a plausible hypothesis for the intracellular dynamics of *Chlamydia*.

After identification of appropriate parameter estimates, analysis of the model demonstrated that under realistic conditions the hypothesis was plausible for explaining the crucial trigger of intracellular development (Wilson et al., 2006). Further analysis found that aRBs formed in all simulations when the RB radius exceeded 1.3 μm and that the RB radius that produces the greatest EB yield is approximately 0.575 μm (Hoare et al., 2008).

Further evidence for the T3S-mediated contact-dependent hypothesis was obtained by recording video of *C. trachomatis*-infected McCoy cells and tracking the temporospatial trajectories of the chlamydiae (Wilson et al., 2009b). The magnitude and speed of RB wobbling steadily increased over time during the early and middle stages of development (8 to 20 hpi) and then dramatically between 20 and 24 hpi. The distances traveled by individual chlamydiae were also observed to be greater from 20 to 24 hpi. These findings are consistent with the T3S-mediated contact-dependent hypothesis: we would expect to see the RBs moving more as the amount of contact they have with the inclusion membrane (via the surface projections) decreases, and still more again once they detach completely.

Additionally, only slight wobbling was observed in oversized aRBs produced by exposing chlamydial cultures to penicillin. This suggests that these aRBs remained tethered to the inclusion membrane; their subsequent failure to enter late differentiation was predicted by analysis of the mathematical model proposed by Hoare et al. (2008), further lending weight to the T3S-mediated contact-dependent hypothesis.

The T3S-mediated contact-dependent model illustrates the value of mathematical modeling in biology. While the ideal way to test a hypothesis is usually through experimentation, some hypotheses are difficult to test directly. By making a mathematical model of a hypothesis, we can first examine whether the dynamics predicted by the hypothesis are consistent with the observed behavior and then use the model to make predictions that can be tested experimentally.

MODELING PATHOLOGY

Chlamydial infection and the host response are responsible for eliciting the pathology that is characteristic of chlamydial disease. The effects of a *Chlamydia* infection can be severe, including trachoma-induced blindness in the case of an ocular infection and infertility in the case of a genital tract infection, as is discussed in more detail in chapter 1. A major goal in *Chlamydia* research is to understand what causes the pathology that can arise during a *Chlamydia* infection in order to determine how it can be reduced or prevented. Given that severe pathology occurs in some cases but not all, we would also like to be able to understand the triggers for stages of pathology and predict the level of pathology that will arise based on the parameters that are observed during the early stages of infection.

In this section we discuss a model that does not attempt to follow the underlying behavior of the system but that instead simply allows us to make predictions about one variable (pathology) based on measurements of another (IFU count). Examination of the time courses of IFU yield and pathology from experiments in guinea pigs suggested that higher initial inoculations of *C. caviae* were associated with more-severe ocular pathology (Wilson et al., 2009a). It was hypothesized that the total amount of IFUs at the site of infection since the infection started has a relationship to the development of pathology; that is, different levels of pathology may arise above certain threshold levels of cumulative infection, which takes into account both the magnitude and the duration of infection. This cumulative level of infection can be represented as the area under a graph of IFU count plotted against time postinfection and is measured in IFU-days. The model was constructed in a number of stages:

1. Identify a mathematical function that appropriately represents the IFU time course.
2. Match the function to the experimental data to find reasonable ranges for its parameters.
3. Use the parameters estimated to predict IFU time courses when factors, such as the growth rate/yield of chlamydiae, are affected.
4. Compare the IFU-days for these predicted time courses with the IFU-days for experimental animals to find the expected pathology.

The following function was chosen for the IFU time course generating function:

$$c(t) = \frac{kc_0 \exp\left[\left(\frac{\ln 2}{t_d} - \frac{\ln 2}{t_{1/2}}\right)t\right]}{k - c_0 + c_0 \exp\left[\frac{\ln 2}{t_d}t\right]},$$

where t is the time after infection, t_d is the doubling time for IFU, $t_{1/2}$ is the half-life after peak infection, and c_0 and k are fitting parameters, i.e., c_0 is the initial inoculation and k the theoretical maximal level.

Given that the modeling in this chapter has mostly dealt with coupled systems of ODEs, it might be asked why that approach was eschewed here. The reason lies in the purposes of the different models. The ODE models above seek to *explain* observed behavior by testing whether they can be accounted for by simple rules. The host-pathogen dynamics ODEs explain the exponential growth/peak/equilibrium pattern of infective bodies seen in many infections in terms of the relationships between pathogens, healthy cells, and infected cells; the model of the T3S-mediated contact-dependent hypothesis sought to explain the asynchronous reproduction of intracellular RBs and EBs in terms of their contact with the inclusion membrane. However, the purpose of this study was not to examine the mechanism by which pathology emerges but to understand the quantitative association between the observed IFU load and pathology to make predictions about as-yet-unseen IFU time courses.

A bootstrapping technique was used to fit the IFU time course, $c(t)$, to IFU time courses observed in ocular infections of guinea pigs with *C. caviae* to find ranges for the parameters. In order to determine the threshold infection level at which pathology arose, IFU-days were calculated and compared to the guinea pigs' observed pathology. Biological variation means that there will not be one exact threshold for all individuals and infections, so receiver operating characteristic (ROC) classification was used to provide a best estimate of the threshold. Put simply, the ROC classification method compared all potential threshold levels to the pathology data and determined the percentage of animals correctly classified at each potential threshold level. The threshold level that resulted in the most animals being correctly classified was taken to be the best estimate for the threshold level. A maximum threshold was found at ~4 × 10^5 IFU-days (as cumulative area under the IFU versus time curve). Below ~4 × 10^5 IFU-days, animals were likely to be free of serious pathology, but above ~4 × 10^5 IFU-days, animals were likely to exhibit serious pathology. Thresholds were calculated for other degrees of pathology.

The model was then used to estimate the effects on pathology if the IFU load is reduced. For each simulated IFU time course, the area under the curve was calculated and then compared to the experimental data. It was predicted that reducing the peak IFU level by 50% reduces the chance of seeing pathology from 81% to 72%, and serious pathological findings (one in which exudates are present) were reduced from 33% to 28% (Wilson et al., 2009a). In order to reduce the area under the IFU curve to less than ~4 × 10^5 (the estimated threshold for pathology), it was calculated that the peak IFU load would need to be reduced by at least 88%.To test the predictions of the model, guinea pigs were infected with a mix of *C. caviae* and chlamydiaphage PhiCPG1 to reduce the level of chlamydial infection and alter the IFU time course (Rank et al., 2009). The chlamydiaphage coinfection reduced the level of pathology, and the IFU and pathology levels matched the model predictions with high accuracy (Wilson et al., 2009a).

This model is a proof of concept that shows pathology can be accurately estimated from other measures of a *Chlamydia* infection, such as the IFU time course. There are many further scientific directions to be considered for this biomathematical modeling approach. There is strong evidence that much of the pathology resulting from *Chlamydia* infection is caused by the immune inflammatory response rather than the chlamydiae themselves. The pathology of *C. caviae* infection in guinea pigs with depleted neutrophil levels was noticeably reduced compared to what was observed in control guinea pigs (Lacy et al., 2011). To factor in the effect of neutrophils and other immune cells, new variables (M_1, . . ., M_n) corresponding to levels of different white blood cells could be introduced. Then, the level of pathology might be given by some function f taking as its arguments the variables M_1, . . ., M_n. Various measures of pathology could be used including scores of observed pathology for ocular infections and measurements of swelling for urogenital infection. If validated by experimental data, such a model would enable us to determine the relative contribution of each feature (e.g., cell or effector molecule) in the system to the overall pathology. Knocking out, or reducing, some of these features can be tested and aligned both in silico and in vivo. The model could also be used to predict the reduction in pathology resulting from hypothetical interventions that target particular aspects of the inflammatory immune response.

This proof of concept suggests that similar predictive models of pathology could be developed for other aspects of chlamydial disease. Many women who become infertile as a result of *Chlamydia* infections have never had symptoms, and so it is difficult to predict this complication. A model similar to this guinea pig ocular pathology model, but modified for humans, might allow a medical practitioner to predict the severity of pathology and choose the appropriate intervention.

SPATIAL MODEL

The viral dynamics ODE models described above operate only in the dimension of time. Location also has relevance in *Chlamydia* infections, particularly in infections in the female reproductive tract: the degree and consequences of pathology vary between different areas. Several models of *Chlamydia* infection incorporating space have been constructed, the most advanced one to date being that presented by Oskouei et al. (Oskouei et al., presented). This model takes as its conceptual basis the viral dynamics ODEs and changes them to PDEs to incorporate two spatial dimensions, x and y, to represent a simplified genital tract. This model also incorporates elements of the immune system. Components of the immune system, such as neutrophils, are guided to infected cells by cytokines and chemokines; $K(t)$ represents their concentration. $H(t)$ is the level of immune response that clears free EBs, and $M(t)$ is the level of immune response that clears infected cells.

The PDEs governing the number of infected cells, uninfected cells, and free EBs are:

$$\frac{\partial C}{\partial t} = \overbrace{D_C \nabla^2 C}^{\text{diffusion}} + \overbrace{P\kappa I}^{\text{new EBs released}} - \overbrace{g(t)CE}^{\text{new infections}} - \overbrace{h(t)CH}^{\text{immune response}}$$

$$\frac{\partial E}{\partial t} = \overbrace{D_E \nabla^2 E}^{\text{diffusion}} + \overbrace{E\left(E_{max} - E\right)}^{\text{new cells created}} - \overbrace{g(t)CE}^{\text{new infections}} - \overbrace{\mu_1 E}^{\text{natural death}}$$

$$\frac{\partial I}{\partial t} = \overbrace{D_I \nabla^2 I}^{\text{diffusion}} + \overbrace{g(t)CE}^{\text{new infections}} - \overbrace{\kappa I}^{\text{cells burst/die}} - \overbrace{s(t)IM}^{\text{immune response}}$$

Here, $g(t)$, $h(t)$, and $s(t)$ represent differences in speeds at different times of the estrous cycle, and $g(t)$ is the rate of EB internalization. Cytokines are described by:

$$\frac{\partial K}{\partial t} = \overbrace{D_K \nabla^2 K}^{\text{diffusion}} + \overbrace{\beta_1 CH}^{\text{response to } C} + \overbrace{\beta_2 IM}^{\text{response to } I},$$

and H and M by:

$$\frac{\partial H}{\partial t} = \overbrace{D_H \nabla^2 H}^{\text{diffusion}} + \overbrace{\gamma_1 CH}^{\text{response to C}} - \overbrace{\mu_2 H}^{\text{death}} - \overbrace{\nabla\left(\chi_H \left(H \nabla K\right)\right)}^{\text{move toward chemokines/cytokines}}$$

$$\frac{\partial M}{\partial t} = \overbrace{D_M \nabla^2 M}^{\text{diffusion}} + \overbrace{\gamma_2 IM}^{\text{response to } I} - \overbrace{\mu_3 M}^{\text{death}} - \overbrace{\nabla\left(\chi_M \left(M \nabla K\right)\right)}^{\text{move toward chemokines/cytokines}}$$

Clearly, this model system is mathematically more complex than other models developed to date. The nature of the problem, tracking chlamydial infection over time and space, requires more complex theoretical frameworks. The advantage of this approach is that it takes into account the observation that the severity and consequences of pathology and even the development of infections vary depending on location. A disadvantage is that it is substantially more difficult (or currently impossible) to obtain accurate data that would be necessary to fully inform all features of the model (e.g., the extent of IFU and immune levels at finely specified spatial coordinates and over time). However, broad outcome variables can be obtained from the model simulations and matched with experimental data. The model outcomes were consistent with experimental observations (Oskouei et al., presented). The dynamics of spatial ascension of chlamydial infection from the lower to the upper genital tract are not straightforward, with lower inoculating doses

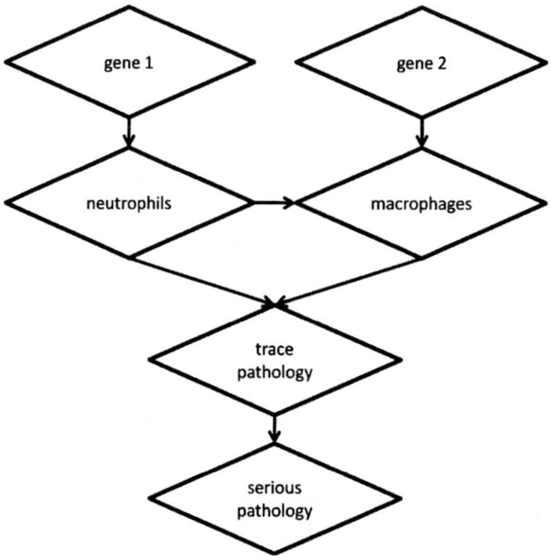

FIGURE 12 A simple Bayesian network. An example of a Bayesian network, representing conditional dependencies of a set of random variables relating chlamydial genes to pathology. The nodes are the random variables and the lines denote the dependencies and their directions.
doi:10.1128/9781555817329.ch16.f12

leading to greater numbers of viable chlamydiae in the oviducts and more-severe oviduct sequelae (Maxion et al., 2004)—and whenever dynamics are not straightforward, there is usually the potential for mathematical modeling to make a contribution. This model is a useful foundational model that could be developed further in the future, including accounting for different severities of pathology.

THE FUTURE OF *CHLAMYDIA* MODELING

Modern technologies have enabled the generation of data about chlamydial infection that were not previously available. For example, there is a large amount of data about the expression of different genes. Real-time PCR has been used to measure 16S rRNA gene transcript levels over time as a marker of changes in the relative replication growth rate of *C. trachomatis* L2 (Wilson et al., 2004). The customized model designed for these data described the intracellular growth of RBs and differentiation into EBs, by use of ODEs, and factors for translating model variables into gene transcript measures. Fitting the model to the data led to estimations of the mean RB doubling time of *C. trachomatis* L2 to be 2.59 hours (95% confidence interval, 2.56 to 2.63). This model has the potential to be used as a basis for applying microarray or RNA-seq technology to identify the gene transcripts that are associated with replicating versus late-differentiating RBs.

To process very large amounts of data, statistical models and other bioinformatics methods are necessary. Bayesian networks in conjunction with quantitative trait loci (QTL) mapping is one of the methodologies that holds some promise for *Chlamydia* research. Briefly, QTL mapping involves finding statistical relationships between genes and phenotypes. The amount of data generated can be very large, with many genes and candidate modulatory relations identified. Bayesian networks relate variables to each other via their conditional dependencies (see Fig. 12 for an example). Applying Bayesian network learning methods to QTL maps can create a much simpler picture of the relationships between the genes and their modulatory relations (Ziebarth et al., 2010). In the case of *Chlamydia* research, a Bayesian network created from QTL mapping can be tested against experimental data just like other types of models. This is an area of ongoing research.

There are several areas in which mathematical modeling has considerable potential to contribute to understanding and control of *Chlamydia* development, dynamics, pathology, and control.

There is evidence that repeated *Chlamydia* infections lead to more serious pathological outcomes (Rank et al., 1995; Hillis et al., 1997; Van Voorhis et al., 1997). *Chlamydia* can also cause a persistent infection in which an infected individual appears to be cleared of *Chlamydia* but then has a recurrence soon after that is not due to reinfection. Persistent/recurrent infections are described in detail in chapter 12, "Chlamydial persistence redux." The mechanism(s) of persistent infection is (are) not clear, but possibilities include the following: (i) the infection survives treatment and remains in a dormant, undetectable state that later reawakens; (ii) the infection is cleared from one part of the body but not from another site, which serves as a reservoir that can later reinfect the initial site; and (iii) the infection survives at a low (undetectable) level, flaring up again later.

The propensity for these putative mechanisms may depend on features of the chlamydiae or the host or interactions between the two. Biomathematical modeling may be able to shed light on how and why persistence occurs by modeling physical processes, as has been done for the T3S-mediated contact-dependent model (Hoare et al., 2008). Here it was shown that an observed phenomenon (large aRBs failing to reproduce or develop) is predicted as the result of a hypothesis (that RB division and conversion into EBs are mediated by contact with the inclusion membrane). An extension of the spatial model may predict the second hypothesis above, or an application of the viral dynamics equations may predict the third hypothesis. Conversely, it may be that an extension of one of those models *disproves* a hypothesis: for example, if the low equilibrium state required for the third hypothesis is predicted by a mathematical model to be so low that it would be unstable or otherwise untenable, then we can reject that hypothesis.

MICROBIOLOGY AND BIOMATHEMATICAL MODELING

The field of microbiology has not traditionally had a strong relationship with mathematics, but biomathematical modeling has considerable potential to add to *Chlamydia* research. *Chlamydia* involves systems with large numbers of agents: in infection kinetics investigations, these agents may be cells and invading bacteria; in intracellular kinetics investigations, the agents may be proteins such as translocated T3S effector proteins. As with all complex systems, changes in behavior at the level of the agent may result in counterintuitive effects on the overall system. Predicting emergent behavior is one of the strengths of mathematics; indeed, it may be very difficult to account for observed system level behavior *without* using mathematics.

Biomathematical modeling has been applied to understanding *Chlamydia* development, infection, and pathology over the last decade; some approaches have been too theoretical; others have been more applied and have involved engagement with the biological community. Biomathematical modeling, by virtue of its fundamental tool, mathematics, is inclined to be largely theoretical. Mathematicians can be accustomed to investigating complex theorems and addressing mathematical curiosities in systems of equations. As such, mathematical modelers can be in danger of focusing on the esoteric rhetoric of mathematical details and miss the great potential for gaining valuable biological insight. A philosophy associated with applied mathematics should not be focused on just the mathematics but rather on its application. Simple models that directly answer questions of relevance are always better than complex models that are abstract, unrealistic, or irrelevant. In order for a model to be truly applied, it must focus on a biological curiosity rather than a mathematical curiosity, and the mathematical modeler must engage with biologists. A massive amount of data has been generated for different biological models of *Chlamydia* infection; therefore, the biomathematical modeler of *Chlamydia* systems does not need to start an investigation with mathematics but with an interrogation of the empirical data. Relatively simple mathematical frameworks can then be developed for exploring mechanisms and possible implications of the data. Foundational models can then be extended both mathematically and in the laboratory. In so doing, biomathematical modeling has great potential as a powerful approach to complement empirical scientific investigation of chlamydial development, infection, immunity, and pathology.

ACKNOWLEDGMENTS

This research was supported by grants from the U.S. National Institutes of Health (NIAID U19 AI084044) to P.B., R.R., and D.W.; by the Arkansas Biosciences Institute to R.R.; and by the Australian Research Council (FT0991990) to D.W. We also thank Daniel Phillips for his contribution to Fig. 11.

REFERENCES

Bavoil, P. M., and R. Hsia. 1998. Type III secretion in *Chlamydia*: a case of déjà vu? *Mol. Microbiol.* **28:**860–862.

Binet, R., A. K. Bowlin, A. T. Maurelli, and R. G. Rank. 2010. Impact of azithromycin resistance mutations on the virulence and fitness of *Chlamydia caviae* in guinea pigs. *Antimicrob. Agents Chemother.* **54:**1094–1101.

Brunham, R. C., B. Pourbohloul, S. Mak, R. White, and M. L. Rekart. 2005. The unexpected impact of a *Chlamydia trachomatis* infection control program on susceptibility to reinfection. *J. Infect. Dis.* **192:**836–844.

Brunham, R. C., and M. L. Rekart. 2008. The arrested immunity hypothesis and the epidemiology of *Chlamydia* control. *Sex. Transm. Dis.* **35:**53–54.

Burns, J. A., E. M. Cliff, and S. E. Doughty. 2007. Sensitivity analysis and parameter estimation for a model of *Chlamydia trachomatis* infection. *J. Inverse Ill-Posed Problems* **15:**243–256.

Gray, R. T., K. W. Beagley, P. Timms, and D. P. Wilson. 2009. Modeling the impact of potential vaccines on epidemics of sexually transmitted *Chlamydia trachomatis* infection. *J. Infect. Dis.* **199:**1680–1688.

Hackstadt, T., E. Fischer, M. Scidmore, D. Rockey, and R. Heinzen. 1997. Origins and functions of the chlamydial inclusion. *Trends Microbiol.* **5:**288–293.

Hillis, S. D., L. M. Owens, P. A. Marchbanks, L. F. Amsterdam, and W. R. MacKenzie. 1997. Recurrent chlamydial infections increase the risks of hospitalization for ectopic pregnancy and pelvic inflammatory disease. *Am. J. Obstet. Gynecol.* **176:**103–107.

Hoare, A., P. Timms, P. Bavoil, and D. Wilson. 2008. Spatial constraints within the chlamydial host cell inclusion predict interrupted development and persistence. *BMC Microbiol.* **8:**5.

Lacy, H. M., A. K. Bowlin, L. Hennings, A. M. Scurlock, U. M. Nagarajan, and R. G. Rank. 2011. Essential role for neutrophils in pathogenesis and adaptive immunity in *Chlamydia caviae* ocular infections. *Infect. Immun.* **79:**1889–1897.

Lotka, A. J. 1925. *Elements of Physical Biology.* Williams and Wilkins, Baltimore, MD. [Reprint; Dover, New York, NY, 1956.]

Mallet, D. G., K.-J. Heymer, R. G. Rank, and D. P. Wilson. 2009. Chlamydial infection and spatial ascension of the female genital tract: a novel hybrid cellular automata and continuum mathematical model. *FEMS Immunol. Med. Microbiol.* **57:**173–182.

Matsumoto, A. 1982. Electron microscopic observations of surface projections on *Chlamydia psittaci* reticulate bodies. *J. Bacteriol.* **150:**358–364.

Maxion, H. K., W. Liu, M.-H. Chang, and K. A. Kelly. 2004. The infecting dose of *Chlamydia muridarum* modulates the innate immune response and ascending infection. *Infect. Immun.* **72:**6330–6340.

Morrison, S. G., H. Su, H. D. Caldwell, and R. P. Morrison. 2000. Immunity to murine *Chlamydia trachomatis* genital tract reinfection involves B cells and CD4+ T cells but not CD8+ T cells. *Infect. Immun.* **68:**6979–6987.

Nowak, M. A., and R. M. May. 2000. *Virus Dynamics: Mathematical Principles of Immunology and Virology*, 1st ed. Oxford University Press, New York, NY.

Oskouei, M. B., D. G. Mallet, A. Amirshahi, and G. J. Pettet. 2010. Mathematical modelling of the interaction of *Chlamydia trachomatis* with the immune system. Presented at the Proceedings of the World Congress on Engineering and Computer Science 2010, San Francisco, CA, USA, 20 to 22 October 2010.

Rank, R. G., B. E. Batteiger, and L. S. Soderberg. 1990. Immunization against chlamydial genital infection in guinea pigs with UV-inactivated and viable chlamydiae administered by different routes. *Infect. Immun.* **58:**2599–2605.

Rank, R. G., A. K. Bowlin, S. Cane, H. Shou, Z. Liu, U. M. Nagarajan, and P. M. Bavoil. 2009. Effect of chlamydiaphage PhiCPG1 on the course of conjunctival infection with *Chlamydia caviae* in guinea pigs. *Infect. Immun.* **77:**1216–1221.

Rank, R. G., M. M. Sanders, and D. L. Patton. 1995. Increased incidence of oviduct pathology in the guinea pig after repeat vaginal inoculation with the chlamydial agent of guinea pig inclusion conjunctivitis. *Sex. Transm. Dis.* **22:**48–54.

Sharomi, O., and A. B. Gumel. 2010. Mathematical study of in-host dynamics of *Chlamydia trachomatis*. *IMA J. Appl. Math.* 1–31. doi: 10.1093/imamat/hxq057.

Van Voorhis, W., L. Barrett, Y. Sweeney, C. Kuo, and D. Patton. 1997. Repeated *Chlamydia trachomatis* infection of *Macaca nemestrina* fallopian tubes produces a Th1-like cytokine response associated with fibrosis and scarring. *Infect. Immun.* **65:**2175–2182.

Vickers, D. M., Q. Zhang, and N. D. Osgood. 2009. Immunobiological outcomes of repeated chlamydial infection from two models of within-host population dynamics. *PLoS ONE* **4**(9)**:**e6886. doi:10.1371/journal.pone.0006886.

Volterra, V. 1926. Fluctuations in the abundance of a species considered mathematically. *Nature* **118:**558–560.

Wilson, D. 2004. Mathematical modelling of *Chlamydia*. *ANZIAM J.* **45:**C201–C214.

Wilson, D., S. Mathews, C. Wan, A. Pettitt, and D. McElwain. 2004. Use of a quantitative gene expression assay based on micro-array techniques and a mathematical model for the investigation of chlamydial generation time. *Bull. Math. Biol.* **66:**523–537.

Wilson, D., P. Timms, D. McElwain, and P. Bavoil. 2006. Type III secretion, contact-dependent model for the intracellular development of *Chlamydia*. *Bull. Math. Biol.* **68:**161–178.

Wilson, D. P., A. K. Bowlin, P. M. Bavoil, and R. G. Rank. 2009a. Ocular pathologic response elicited by *Chlamydia* organisms and the predictive value of quantitative modeling. *J. Infect. Dis.* **199:**1780–1789.

Wilson, D. P., and D. L. S. McElwain. 2004. A model of neutralization of *Chlamydia trachomatis* based on antibody and host cell aggregation on the elementary body surface. *J. Theor. Biol.* **226:**321–330.

Wilson, D. P., P. Timms, and D. L. S. McElwain. 2003. A mathematical model for the investigation of the Th1 immune response to *Chlamydia trachomatis*. *Math. Biosci.* **182:**27–44.

Wilson, D. P., J. A. Whittum-Hudson, P. Timms, and P. M. Bavoil. 2009b. Kinematics of intracellular chlamydiae provide evidence for contact-dependent development. *J. Bacteriol.* **191:**5734–5742.

Ziebarth, J. D., L. Bao, I. Miyairi, and Y. Cui. 2010. Linking genotype to phenotype with Bayesian network modeling of *Chlamydia* infection. *BMC Bioinformatics* **11**(Suppl 4)**:**P19. http://www.biomedcentral.com/1471-2105/11/S4/P19.

INDEX

C

E

F

G

H

U

V

W

X

Y

Z